AF360715

Isolation, Structure Elucidation and Biological Activity of Natural Products

Isolation, Structure Elucidation and Biological Activity of Natural Products

Editor

Jacqueline Aparecida Takahashi

Basel • Beijing • Wuhan • Barcelona • Belgrade • Novi Sad • Cluj • Manchester

Editor
Jacqueline Aparecida Takahashi
Department of Chemistry
Universidade Federal de Minas Gerais
Belo Horizonte
Brazil

Editorial Office
MDPI
St. Alban-Anlage 66
4052 Basel, Switzerland

This is a reprint of articles from the Special Issue published online in the open access journal *Molecules* (ISSN 1420-3049) (available at: www.mdpi.com/journal/molecules/special_issues/Np_isolation).

For citation purposes, cite each article independently as indicated on the article page online and as indicated below:

Lastname, A.A.; Lastname, B.B. Article Title. *Journal Name* **Year**, *Volume Number*, Page Range.

ISBN 978-3-0365-8973-2 (Hbk)
ISBN 978-3-0365-8972-5 (PDF)
doi.org/10.3390/books978-3-0365-8972-5

Contents

About the Editor

Jacqueline Aparecida Takahashi

Jacqueline Aparecida Takahashi has held the position of Professor of Organic Chemistry since 1996 at the Universidade Federal de Minas Gerais (UFMG, Brazil). UFMG is one of the major public universities mantained by the Brazilian Government and has over 50,000 students distributed in 91 undegraduate courses and 90 Graduate programs. Prof. Takahashi has a Bachelor's Degree in Industrial Pharmacy, and a PhD in Science/Organic Chemistry, both from UFMG, and worked for two years at the University of Sussex (Brighton, UK) under the supervision of Prof. James Hanson during her PhD. She spent one year as a Research Scholar at the University of Arizona (Tucson, US) in the group of Prof. Leslie Gunatilaka. Her main research interests over the last 30 years have mainly been the use of fungi to produce biologically active natural products and the microbial functionalization of terpenes (biotransformations), as well as plants of the Brazilian biodiversity. Lately she has also dedicated time to study Brazilian and International Laws on the use and preservation of Brazilian and world biodiversity, including aspects related to climate changes, traditional knowledge and regulatory issues.

Preface

The Special Issue "Isolation, Structure Elucidation and Biological Activity of Natural Products" was proposed to bring together work and reviews on natural products of plant and microbial origin, with special interest in the biological activity of these products. With the various current problems related to the loss of biodiversity associated with climate change and the exploitation of natural areas for economic activities, it is urgent to show the role biodiversity plays for the scientific community and society. Based on this context, the contributions published in this Issue show several works refarding the possible industrial applications of plants and microorganisms, while the review articles show the vastness of existing natural products, many of which deserve further studies for technological development. This Issue is addressed to natural product chemistry researchers, including young researchers enthusiastic about the potential of biodiversity for diverse applications. Likewise, authorities interested in innovation in the area of natural products, companies, and industries will be able to benefit from the articles published in this Issue. I would like to thank the authors of the published articles, as well as the editorial assistance and MDPI's invitation for me to act as guest editor of this Issue.

Jacqueline Aparecida Takahashi
Editor

Editorial

Special Issue—"Isolation, Structure Elucidation and Biological Activity of Natural Products"

Jacqueline Aparecida Takahashi

Exact Sciences Institute, Universidade Federal de Minas Gerais (UFMG), Av. Antonio Carlos, 6627, Belo Horizonte 31270-901, Brazil; takahashi.ufmg@gmail.com

Citation: Takahashi, J.A. Special Issue—"Isolation, Structure Elucidation and Biological Activity of Natural Products". *Molecules* **2023**, *28*, 5392. https://doi.org/10.3390/molecules28145392

Received: 10 July 2023
Accepted: 12 July 2023
Published: 14 July 2023

This Special Issue of *Molecules* gathers fourteen research studies and three review papers covering developments in the scope of the isolation, structure elucidation and biological activity of natural products. Plants are undoubtedly the most well-known sources of bioactive natural products, since plants have been used medicinally by traditional communities for centuries, defining their specific benefits to human health. In this Special Issue, the plants addressed are *Bauhinia forficata, Placolobium vietnamense, Tamarix chinensis, Peperomia obtusifolia, Cymbidium ensifolium, Dendrobium delacourii, Turnera subulata, Eruca sativa, Miconia chamissois* and *Persea americana*, with the aim of achieving the identification of their chemical compositions, the prospection of biological activities, the standardization of extracts and the use of agro-industrial residues. Some studies involved hyphenated techniques such as UPLC-IT-MS^n [1], UHPLC-ESI-QTOF [2], UHPLC/UV/MS/MS [3], PSMS [4] and UPLC–MS/MS [5], which are very useful tools in research on natural products.

B. forficata is a tropical species popularly used in Brazil to treat type II diabetes, rheumatism, local pain, uric acid and uterine problems. Jung et al (2022) carried out a comparative study between locally collected leaves of this species and commercial samples [6]. Using LC-HRMS, a very useful hyphenated analytical technique, the authors were able to identify the presence of flavonoids, phenolic acids and other phenolic volatile compounds, including flavonoid O-glycosides in the plant, and these metabolites were related to the pharmacological actions reported for *B. forficata*. The plant inhibited the α-amylase enzyme, and the results converged to reinforce the biological and pharmacological potential of *B. forficata* as a hypoglycemiant agent [6]. Studies of this type are of great value, especially considering that *B. forficata* is an easily accessible plant for the population and is already commercialized.

P. vietnamense, known in Vietnam as "Rang Rang", is also distributed throughout the world's tropical regions and is used as a folk medicine to treat snakebites, debility and to increase strength after childbirth. Stems of *P. vietnamense* were studied by Do, Huynh, and Sichaem (2022), who successfully isolated eight natural constituents of this species, including a new isoflavone derivative and three new benzyl derivatives, together with four known compounds of the pyranoisoflavone type [7]. The authors conducted a cytotoxic evaluation of the effects of the aforementioned natural products on a human hepatocellular carcinoma (Hep G2) cell line, determined the cytotoxic effects and measured the production of nitric oxide (NO) by RAW 264.7 cells. Among the compounds tested, placovinone A exhibited the most significant cytotoxicity toward the Hep G2 cell line and also strongly inhibited LPS-induced NO production [7]. The authors concluded that placovinone A is a promising lead for discovering potential anticancer and anti-inflammatory agents.

The work of Jiao et al. (2022), conducted with the Chinese plant *T. chinensis*, which is used to treat rheumatoid arthritis, measles and measles complicated with pneumonia, is an excellent example of the success of activity-guided extraction, isolation and purification targeting a specific class of compounds [1]. In this case, the authors were focused on the bioactive polysaccharides present in *T. chinensis*. Flavonoids, triterpenoids, organic acids and volatile oils with anti-inflammatory, bacteriostatic, antioxidant and hepatoprotective effects have already been described from this plant. An activity-guided approach, followed by careful spectroscopic study (HPGPC-ELSD, UPLC-IT-MS^n and NMR analysis), led to the isolation and identification of two novel natural flavonoid-substituted polysaccharides. Both polysaccharides were substituted by quercetin and exhibited anticomplement activities in vitro. Since the authors were focusing on inflammatory responses in viral pneumonia, antioxidant activities were also determined for the polysaccharides. The authors indicated that molecules containing both flavonoids and polysaccharides have advantageous drug delivery. Structure–activity studies showed that multiple monosaccharides also contribute to the anticomplement activity of *T. chinensis* flavonoid-substituted polysaccharides [1].

P. obtusifolia is an ornamental plant species that is widely distributed in tropical areas. Plants from the *Peperomia* genus have been traditionally used to treat asthma, gastric ulcers, bacterial infection, pain and inflammation. Ware et al. (2022) isolated two new phenolic compounds from *P. obtusifolia* (peperomic ester and peperoside), along with other five known metabolites that were screened for their anthelmintic (*Caenorhabditis elegans*), antifungal (*Botrytis cinerea*, *Septoria tritici* and *Phytophthora infestans*) and antibacterial (*Bacillus subtilis*, *Aliivibrio fisheri*) activities, as well as for their cytotoxicity against human prostate (PC-3) and colorectal (HT-29) cancer cell lines [8]. One of the metabolites tested, peperobtusin A, strongly affected the viability and growth of PC-3 cells in MTT and a CV fast-screening assay and was moderately active against the HT-29 cell line [8]. The results are an indicative that ornamental plants can also be sources of bioactive metabolites.

The roots of *Cymbidium ensifolium*, another ornamental plant known as "nang kham" or "chulan" in Thailand, are used in traditional Thai medicine to alleviate liver dysfunction and nephropathy. The aerial parts of *C. ensifolium* were studied by Jimoh et al (2022), who isolated three novel dihydrophenanthrene derivatives (cymensifins A, B and C), along with two known compounds, cypripedin and gigantol [9]. The chemical structures of the dihydrophenanthrene derivatives were elucidated, mainly via HMBC and NOESY correlations. Their activity against human lung cancer H460, breast cancer MCF7 and colon cancer CaCo$_2$ cells were evaluated in cell viability assays and for apoptosis/necrosis. The most promising anticancer compound was cymensifin A, which was found to be active against different cancer cells with higher safety profiles compared with the control, cisplatin [9].

Another orchid, *Dendrobium delacourii*, named "Ueang Dok Ma Kham" in Thailand, was studied by Thant et al. (2022) [10]. They isolated 11 compounds and identified them as hircinol, ephemeranthoquinone, densifloral B, moscatin, 4,9-dimethoxy-2,5-phenanthrenediol, gigantol, batatasin III, lusianthridin, 4,4',7,7'-tetrahydroxy-2,2'-dimethoxy-9,9',10,10'-tetrahydro-1,1'-biphenanthrene, phoyunnanin E and phoyunnanin C. Dimeric phenanthrene derivatives presented stronger α-glucosidase inhibitory activity than the monomers. A kinetic study indicated the active metabolites to be non-competitive enzyme inhibitors. The authors argued the benefits of non-competitive inhibitors over competitive inhibitors. Regarding anti-adipogenic action, ephemeranthoquinone and densifloral B showed the highest levels of activity; the latter restrained adipocyte differentiation in 3T3-L1 cells in a dose-dependent manner [10]. The authors suggested that densifloral B might inhibit adipocyte differentiation via the suppression of the Akt-mediating GSK3β and AMPK–ACC signals. These results are of great interest since diabetes and obesity are major problems worldwide.

Extracts from flowers of the tropical plant *Turnera subulata* were studied by Luz et al. (2022) [5]. Their in vitro immunomodulatory effects in RAW 264.7 macrophages stimulated via LPS were evaluated in the search for anti-inflammatory drugs with low side effects. Vitexin-2-*O*-rhamnoside was identified in the extracts via UPLC–MS/MS, as were methoxy-isoflavones, pheophorbides, octadecatrienoic, stearidonic, ferulic acids and some amino acids. The results demonstrate the immunomodulatory effects of aqueous and hydroalcoholic extracts of *T. subulata* flowers and leaves through the inhibition of inflammatory TNF-α and IL-1β cytokine secretion [5]. Increases in anti-inflammatory IL-10 cytokine levels also supported the activity and corroborate the ethnopharmacological use of plants from the *Turnera* genus in folk medicine as anti-inflammatory remedies.

In addition to essentially medicinal and ornamental plants, one of the works published in this Special Issue focused on a plant used mainly as food, *Eruca sativa* (rocket). Based on works using animal models that associated rocket with biological activities, such as antihypertensive, nephroprotective and antidiabetic activities, Teixeira et al. (2022) hypothesized that this plant could have anti-hyperuricemic effects, combatting a causal factor of hypertension and diabetes [2]. Nine compounds were detected via UHPLC-ESI-QTOF: kaempferol-3-O-β-glucoside, kaempferol-3,4′-di-O-β-glucoside, kaempferol-3-O-(2-sinapoyl-β-glucoside)-4′-O-glucoside, glucosativin glucosinolate, glucoraphanin glucosinolate, leucine, tryptophan, angustione and erucamide. Kaempferol-3,4′-di-O-β-glucoside was elegantly characterized via NMR and quantified in the extract. the anti-hyperuricemic activities of the extracts were mainly related to uricosuric action [2]. Although other mechanisms are yet to be elucidated, the authors suggested that the results indicate the potential use of *E. sativa* in the treatment of hyperuricemia and its comorbidities.

Ferreira et al. (2022) raised an important issue in the study of medicinal plants, which is the standardization of extracts and the variations relating to seasonal variations. The authors focused on *Miconia chamissois*, known as "Folha de Bolo", "Sabiazeira", or "Pixirica" in Brazil, a plant associated with antimicrobial and antioxidant activities, the in vitro inhibition of the enzymes tyrosinase and alpha-amylase, the inhibition of MMP-2 and MMP-9 and cytotoxicity against human cervical cancer cell lines [3]. A standardized extract was formulated, and some major constituents, such as apigenin C-glycosides (vitexin/isovitexin), luteolin C-glycosides (orientin/isoorientin), miconioside B, matteucinol-7-O-β-apiofuranosyl (1 → 6)-β-glucopyranoside and farrerol, were identified (UHPLC-MS/MS). Ferreira et al. evaluated samples collected in the autumn, winter and spring, seasons with different environmental factors such as temperature, ultraviolet radiation, rain index and soil nutrients. No significant correlations between the meteorological data and the biological potential were observed, demonstrating that the species studied was well adapted to the different environmental conditions targeted in the study.

An important aspect linked to the use of plants, medicinal or not, was contextualized in the work of Silva et al. (2022), who studied the peels and seeds of avocado (*Persea americana*), which represent 30% of the fruit and are usually discarded as waste, generating environmental problems and the loss of bioactive compounds that remain in the biomass after processing [4]. The authors showed that avocado residues retain several nutrients such as minerals (Ca, Mg, Mn and Zn) and high amounts of essential fatty acids such as linoleic, palmitic and oleic acids. The chemical profile obtained via paper spray mass spectrometry (PSMS) showed fifty-five metabolites, including phenolic compounds, hydroxycinnamic acids, flavonoids and alkaloids, in the residues. The ethanol extract of the peels was the best acetylcholinesterase inhibitor, with no significant difference ($p > 0.05$) compared to the control, eserine. The seed extracts exhibited an in vivo neuroprotective effect against rotenone-induced damage in an in vivo *Drosophylla melanogaster* model [4]. This work contributed some interesting points to the scope of research on natural products, such as a preoccupation with sustainability and a circular economy, the choice of environmentally favorable solvents for extraction and the role of complementarity in vivo assays. Although toxicity, bioavailability and suitable formulations should be further investigated before using avocado residues in pharmacotherapy, this work showed the potential use of avocado residues as a bioresource in the development of low-cost drugs and functional foods with neuroprotective effects.

Bacteria and fungi are also important producers of bioactive metabolites. The purification of a catechol-type siderophore from *Streptomyces tricolor* was studied with the aim of inducing recovery from iron-deficiency-induced anemia in in vivo rat model (Barakat et al., 2022) [11]. Siderophores are low-molecular-weight natural compounds secreted by microorganisms that act as iron chelators. Due to this characteristic, siderophores have useful therapeutic applications, especially in iron-overload diseases (hemosiderosis, β-thalassemia, hemochromatosis and accidental iron poisoning). Iron chelators are also useful in cancer therapy because cancer cells have higher requirements of iron when compared to healthy cells.

Barakat et al. (2022) focused on applications related to the damages caused by iron-deficiency-induced anemia, showing that sidereophores improved weight gain and were effective during recovery from anemia [11]. The results led the authors to propose some hypotheses and created opportunities to delineate the detailed mechanism of the changes observed and elucidate iron pathways.

Lacticaseibacillus rhamnosus XN2, a bacteriocin-producing strain isolated from the naturally fermented yak yoghurt produced in Xining and Qinghai Provinces in China, was the target of the work of Wei et al. (2022) [12]. The source of the bacteria studied, yogurt from yak milk farmed in the Himalayan region, exemplifies the vastness of the area of natural products and the local value of the studies, as well as the global importance of the results in expanding knowledge in the area of biodiversity and ecosystem services. Bacteriocins from lactic acid bacteria are peptides secreted by some lactic acid bacteria with natural antimicrobial activities against other microorganisms, including food spoilage and pathogens. *L. rhamnosus* XN2 demonstrated antibacterial activity against *Bacillus subtilis*, *B. cereus*, *Micrococcus luteus*, *Brochothrix thermosphacta*, *Clostridium butyricum*, *S. aureus*, *Listeria innocua*, *L. monocytogenes* and *Escherichia coli*. Semi-purified, cell-free supernatants of *L. rhamnosus* XN2 showed bactericidal activity, probably due to the disruption of the sensitive bacteria membrane, as suggested by a flow cytometry analysis. The production of α-haemolysin and biofilm formation were observed for sub-lethal concentrations of the semi-purified material. Bacteriocin was further purified via reversed-phase high-performance liquid chromatography (RP-HPLC), and its amino acid sequence was determined to be Met-Lue-Lys-Lys-Phe-Ser-Thr-Ala-Tyr-Val [12]. The authors concluded that *L. rhamnosus* XN2 and its bacteriocin showed antagonistic activity at both cellular and quorum-sensing levels.

In relation to fungi, their secondary metabolites usually have diverse applications in the food, cosmetic, beverage and textile industries, and their biological activities enable their use in the development of new drugs. Some fungal pigments, such as azaphilones and isolated β-carotene, have already found commercial and industrial applications, and most of the fungi explored for the production of pigments are from a few genera such as *Monascus*, *Talaromyces*, *Aspergillus*, *Penicillium* and *Fusarium*. Lagashetti et al. (2022) focused on a less-common fungal species, isolated from the infected leaves of *Maytenus rothiana* and identified via morphological and molecular methods as *Gonatophragmium triuniae* [13]. Its growth under different conditions revealed the conditions suitable for pigment production, and biological screening demonstrated the antibacterial and antioxidant activities of *G. triuniae* extracts. The major orange-colored pigment produced by the species was identified as 1,2-dimethoxy-3 H-phenoxazin-3-one. The authors note that pigments and other bioactive secondary metabolites of *G. triuniae* have potential applications in the textile and pharmaceutical industries.

Fungi of the genus *Penicillium* are widely studied as sources of bioactive compounds, but research on these fungi and their metabolites is far from complete, as shown in the work of Cadelis et al. (2022) [14]. Studying *P. bissettii* and *P. glabrum*, this group isolated five known polyketide metabolites, penicillic acid, citromycetin, penialdin A, penialdin F and myxotrichin B. During the derivatization of penicillic acid, a novel dihydro derivative was produced, providing evidence for the existence of an open-chained γ-keto acid tautomer in the starting material. Penicillic acid and penialdin F were found to inhibit the growth of methicillin-resistant *S. aureus*, which is important in view of the clinical problems associated with resistant microbial strains, which mainly involve hospitalized patients. Two other metabolites, penialdin F and citromycetin, were active against *Mycobacterium abscessus* and *M. marinum* [14].

This Special Issue also received three interesting review contributions. The first of these addressed recently discovered secondary metabolites from *Streptomyces* species and was prepared by Lacey and Rutledge [15]. Their review, with a particular focus on the year 2020, presented 74 novel secondary metabolites from *Streptomyces* species, with a wide range of chemical scaffold variability, including the cyclic peptides ulleungamide, viennamycins A and B and pentaminomycins C–E, metabolites with complex chemical structures. Linear peptides such as spongiicolazolicins A and B were isolated from marine species, and linear polyketides (e.g., adipostatins E–J, trichostatic acid B, trichostatin A and chresdihydrochalcone), terpenoids (e.g., napyradiomycins and flaviogeranins), polyaromatics (e.g., baikalomycins A–C and gardenomycins A and B), macrocycles (e.g., conglobatins and somamycins) and furans (e.g., furamycins) are among the compounds discussed in the review [16]. The authors noted that the *Streptomyces* species reported in their review were isolated from a wide range of environments, possess a diversity of novel chemical structures and represent a thriving and multifaceted area of drug discovery research.

The second review contribution focused on ^{13}C-NMR data from 504 pentacyclic triterpenoids isolated from plants of the Celastraceae family, covering the period of 2001 to 2021 [15]. This class of secondary metabolites is of the utmost importance as they are reported to possess varied biological potential as antiviral, antimicrobial, analgesic, anti-inflammatory and cytotoxic agents against various tumor cell lines. The review covered the pentacyclic triterpenoids of friedelane, quinonemethide, aromatic, dimer, lupane, oleanane and ursane, among other classes. The data reported by Camargo et al. (2022) highlighted the amazing structural diversity of pentacyclic triterpenes and the complexity of some representatives, such as the dimeric molecules [15]. The ^{13}C-NMR data presented in this review are an enormous contribution to the structural elucidation of new compounds of this class of terpenes.

Last, but not least, Amen and colleagues from Prof. Shimizu's group reviewed the sources, bioactivities, biosynthesis and spectroscopic features of naturally occurring chromone glycosides. The compounds addressed in the review were described from plants (angiosperms) of thirty-three families, three families of ferns, four species of lichens, three species of fungi and three families of actinobacteria [17]. O-glycosides or C-glycosides were analyzed separately; phenyl and isoprenyl chromone glycosides and phenyl ethyl chromone glycosides were then analyzed, followed by the class of chromone glycosides with additional heterocyclic moieties. Hybrids of chromones with other classes of secondary metabolites, hybrids of furano-chromones with cycloartane triterpenes and hybrids of chromones with secoiridoids were presented in sequence, and finally, representants of the groups of chromone alkaloids and aminoglycosides, were discussed. In the second part, the spectroscopic features of the chromones were carefully described, including UV, IR, ^{1}H and ^{13}C-NMR data of the 192 chromones [17]. This is indeed a great collaboration for the prompt identification of chromone metabolites and an incentive to conduct deeper studies with this class of metabolites.

In view of the papers that comprise this Special Issue, we believe that our objectives have been achieved. We are thankful for all the contributions received, and hope that the papers will be of interest to all readers of *Molecules*. Finally, natural products still have much to contribute to humanity, and we wish great results and good discoveries to all authors and readers in their upcoming research.

Funding: We acknowledge the grants from CNPq 31150/2022 and FAPEMIG 00255-18.

Conflicts of Interest: The author declares no conflict of interest.

References

1. Jiao, Y.; Yang, Y.; Zhou, L.; Chen, D.; Lu, Y. Two Natural Flavonoid Substituted Polysaccharides from *Tamarix chinensis*: Structural Characterization and Anticomplement Activities. *Molecules* **2022**, *27*, 4532. [CrossRef] [PubMed]
2. Ferreira, J.d.F.; López, M.H.M.; Gomes, J.V.D.; Martins, D.H.N.; Fagg, C.W.; Magalhães, P.O.; Davies, N.W.; Silveira, D.; Fonseca-Bazzo, Y.M. Seasonal Chemical Evaluation of *Miconia chamissois* Naudin from Brazilian Savanna. *Molecules* **2022**, *27*, 1120. [CrossRef] [PubMed]
3. Teixeira, A.F.; de Souza, J.; Dophine, D.D.; de Souza Filho, J.D.; Saúde-Guimarães, D.A. Chemical Analysis of *Eruca sativa* Ethanolic Extract and Its Effects on Hyperuricaemia. *Molecules* **2022**, *27*, 1506. [CrossRef] [PubMed]
4. da Silva, G.G.; Pimenta, L.P.S.; Melo, J.O.F.; Mendonça, H.d.O.P.; Augusti, R.; Takahashi, J.A. Phytochemicals of avocado residues as potential acetylcholinesterase inhibitors, antioxidants, and neuroprotective agents. *Molecules* **2022**, *27*, 1892. [CrossRef] [PubMed]

5. da Luz, J.R.D.; Barbosa, E.A.; Nascimento, T.E.S.D.; de Rezende, A.A.; Ururahy, M.A.G.; Brito, A.d.S.; Araujo-Silva, G.; López, J.A.; Almeida, M.d.G. Chemical characterization of flowers and leaf extracts obtained from *Turnera subulata* and their immunomodulatory effect on LPS-activated RAW 264.7 macrophages. *Molecules* **2022**, *27*, 1084. [CrossRef] [PubMed]

6. Jung, E.P.; de Freitas, B.P.; Kunigami, C.N.; Moreira, D.D.L.; de Figueiredo, N.G.; Ribeiro, L.D.O.; Moreira, R.F.A. *Bauhinia forficata* Link infusions: Chemical and bioactivity of volatile and non-volatile fractions. *Molecules* **2022**, *27*, 5415. [CrossRef] [PubMed]

7. Do, L.T.; Huynh, T.T.; Sichaem, J. New Benzil and Isoflavone Derivatives with Cytotoxic and NO Production Inhibitory Activities from Placolobium vietnamens E. *Molecules* **2022**, *27*, 4624. [CrossRef] [PubMed]

8. Ware, I.; Franke, K.; Hussain, H.; Morgan, I.; Rennert, R.; Wessjohann, L.A. Bioactive phenolic compounds from *Peperomia obtusifolia*. *Molecules* **2022**, *27*, 4363. [CrossRef] [PubMed]

9. Jimoh, T.O.; Costa, B.C.; Chansriniyom, C.; Chaotham, C.; Chanvorachote, P.; Rojsitthisak, P.; Likhitwitayawuid, K.; Sritularak, B. Three new dihydrophenanthrene derivatives from *Cymbidium ensifolium* and their cytotoxicity against cancer cells. *Molecules* **2022**, *27*, 2222. [CrossRef] [PubMed]

10. Thant, M.T.; Khine, H.E.E.; Nealiga, J.Q.L.; Chatsumpun, N.; Chaotham, C.; Sritularak, B.; Likhitwitayawuid, K. α-Glucosidase inhibitory activity and anti-adipogenic effect of compounds from Dendrobium delacourii. *Molecules* **2022**, *27*, 1156. [CrossRef] [PubMed]

11. Barakat, H.; Qureshi, K.A.; Alsohim, A.S.; Rehan, M. The Purified Siderophore from *Streptomyces tricolor* HM10 Accelerates Recovery from Iron-Deficiency-Induced Anemia in Rats. *Molecules* **2022**, *27*, 4010. [CrossRef] [PubMed]

12. Wei, Y.; Wang, J.; Liu, Z.; Pei, J.; Brennan, C.; Abd El-Aty, A.M. Isolation and characterization of bacteriocin-producing *Lacticaseibacillus rhamnosus* XN2 from yak yoghurt and its bacteriocin. *Molecules* **2022**, *27*, 2066. [CrossRef] [PubMed]

13. Lagashetti, A.C.; Singh, S.K.; Dufossé, L.; Srivastava, P.; Singh, P.N. Antioxidant, antibacterial and dyeing potential of crude pigment extract of *Gonatophragmium triuniae* and its chemical characterization. *Molecules* **2022**, *27*, 393. [CrossRef] [PubMed]

14. Cadelis, M.M.; Nipper, N.S.; Grey, A.; Geese, S.; van de Pas, S.J.; Weir, B.S.; Copp, B.R.; Wiles, S. Antimicrobial Polyketide Metabolites from *Penicillium bissettii* and *P. glabrum*. *Molecules* **2021**, *27*, 240. [CrossRef] [PubMed]

15. Lacey, H.J.; Rutledge, P.J. Recently discovered secondary metabolites from *Streptomyces* species. *Molecules* **2022**, *27*, 887. [CrossRef] [PubMed]

16. Camargo, K.C.; de Aguilar, M.G.; Moraes, A.R.A.; de Castro, R.G.; Szczerbowski, D.; Miguel, E.L.M.; Oliveira, L.R.; Sousa, G.F.; Vidal, D.M.; Duarte, L.P. Pentacyclic Triterpenoids Isolated from Celastraceae: A Focus in the ^{13}C-NMR Data. *Molecules* **2022**, *27*, 959. [CrossRef] [PubMed]

17. Amen, Y.; Elsbaey, M.; Othman, A.; Sallam, M.; Shimizu, K. Naturally occurring chromone glycosides: Sources, bioactivities, and spectroscopic features. *Molecules* **2021**, *26*, 7646. [CrossRef] [PubMed]

Article

Bauhinia forficata Link Infusions: Chemical and Bioactivity of Volatile and Non-Volatile Fractions

Eliane Przytyk Jung [1], Beatriz Pereira de Freitas [2], Claudete Norie Kunigami [1], Davyson de Lima Moreira [3,*], Natália Guimarães de Figueiredo [4], Leilson de Oliveira Ribeiro [1,*] and Ricardo Felipe Alves Moreira [5]

1. Laboratory of Organic and Inorganic Chemical Analysis, National Institute of Technology, Rio de Janeiro 20081-312, Brazil
2. Faculty of Chemical Engineering, Federal University of Fluminense, Niterói 24210-240, Brazil
3. Laboratory of Natural Products, Rio de Janeiro Botanical Garden Research Institute, Rio de Janeiro 22460-030, Brazil
4. Laboratory of Tobacco and Derivatives, National Institute of Technology, Rio de Janeiro 20081-312, Brazil
5. Food and Nutrition Graduate Program, Federal University of Rio de Janeiro State (UNIRIO), Rio de Janeiro 22290-250, Brazil
* Correspondence: davysonmoreira@jbrj.gov.br (D.d.L.M.); leilson.oliveira@int.gov.br (L.d.O.R.)

Abstract: This study aimed to evaluate *Bauhinia forficata* infusions prepared using samples available in Rio de Janeiro, Brazil. As such, infusions at 5% (w/v) of different brands and batches commercialized in the city (CS1, CS2, CS3, and CS4) and samples of plant material botanically identified (BS) were evaluated to determine their total phenolic and flavonoid contents (TPC and TFC), antioxidant capacity (ABTS$^{\bullet+}$, DPPH$^{\bullet}$, and FRAP assays), phytochemical profile, volatile compounds, and inhibitory effects against the α-amylase enzyme. The results showed that infusions prepared using BS samples had lower TPC, TFC and antioxidant potential than the commercial samples ($p < 0.05$). The batch averages presented high standard deviations mainly for the commercial samples, corroborating sample heterogeneity. Sample volatile fractions were mainly composed of terpenes (40 compounds identified). In the non-volatile fraction, 20 compounds were identified, with emphasis on the CS3 sample, which comprised most of the compounds, mainly flavonoid derivatives. PCA analysis demonstrated more chemical diversity in non-volatile than volatile compounds. The samples also inhibited the α-amylase enzyme (IC$_{50}$ value: 0.235–0.801 mg RE/mL). Despite the differences observed in this work, *B. forficata* is recognized as a source of bioactive compounds that can increase the intake of antioxidant compounds by the population.

Keywords: "pata-de-vaca"; phytochemical profile; bioactive compounds; antioxidant capacity; α-amylase inhibition; SPME technique

Citation: Jung, E.P.; de Freitas, B.P.; Kunigami, C.N.; Moreira, D.d.L.; de Figueiredo, N.G.; Ribeiro, L.d.O.; Moreira, R.F.A. *Bauhinia forficata* Link Infusions: Chemical and Bioactivity of Volatile and Non-Volatile Fractions. *Molecules* **2022**, *27*, 5415. https://doi.org/10.3390/molecules27175415

Academic Editor: Jacqueline Aparecida Takahashi

Received: 20 July 2022
Accepted: 20 August 2022
Published: 24 August 2022

Publisher's Note: MDPI stays neutral with regard to jurisdictional claims in published maps and institutional affiliations.

1. Introduction

Bauhinia is a genus comprising over 300 species widely distributed in tropical and subtropical forests. In Brazil, 64 species belonging to the Fabaceae family were identified and are commonly known as "pata-de-vaca" due to the shape of their leaves. Most species are of Asian origin; however, *Bauhinia longifolia* (Bong.) Steud. and *Bauhinia forficata* Link are native species from Brazil [1,2].

B. forficata is widely used in Brazilian folk medicine due to its beneficial effects on different diseases and human disorders such as rheumatism, local pain, uric acid, and uterine problems [3], but it is primarily used to treat type II diabetes [4]. The beneficial effects are associated with various biocompounds present in *B. forficata*, such as flavonoids, alkaloids, and terpenes/terpenoids [2,5]. The flavonoid compounds are highlighted since they are the major class in *B. forficata* extracts. Farag et al. [6] registered the presence of quercetin and kaempferol derivatives in different species of the *Bauhinia* genus, including *B. forficata*.

In Brazil, *B. forficata* is mainly commercialized dried and used to prepare infusions. Thus, under Brazilian law, the *Bauhinia* tea is associated to food products, so it is not mandatory to indicate the content of bioactive or toxic compounds, as in a limited manner in herbal products [7]. *B. forficata* infusions were used in different in vivo studies, such as that reported by Salgueiro et al. [8], who evaluated the effects of infusions on oxidative stress, liver damage, and glycemia in mice. Nevertheless, data on the content of bioactive compounds, antioxidant capacity, and volatile compounds, among other parameters of this plant, to compare botanically identified and commercialized samples and their infusions are scarce in the literature. Since it is well known that various factors such as climate, processing, and storage conditions may influence the content of bioactive compounds and the volatile fraction of medicinal plants [9,10], there is a clear need for further studies.

Despite that, to date, there are no data available on the volatile composition of *B. forficata* infusions. This fraction cannot be underestimated since *B. forficata* is prepared by infusion or decoction and, therefore, some of the volatile content may disperse in the beverage (hydrolate) and contribute to its beneficial actions besides the aroma. This approach has already been evaluated for other medicinal plants, and the migration of terpenoid and other compounds classes present in the essential oil of the plant for infusion was observed [11,12].

For such an evaluation, headspace solid-phase microextraction coupled to gas chromatography–mass spectrometry (HS-SPME/GC–MS) has been reported as a fast, sensitive, and solvent-free technique for analyzing the extraction and isolation of volatile and semi-volatile compounds, and it has been widely used since its invention in 1989 [13,14]. Furthermore, interference from the infusion matrix may be drastically reduced while the headspace analytes are trapped in the fiber [15]. Thus, this technique has been successfully applied to analyze volatile compounds in infusions and teas [16,17].

In this sense, this work aimed to perform a comprehensive chemical characterization of the volatile and non-volatile fractions of botanically identified and commercial samples of *B. forficata* used to prepare infusions at 5%. The antioxidant capacity measured by ABTS$^{\bullet+}$, DPPH$^{\bullet}$ and FRAP assays and inhibitory activity of α-amylase of the samples were also determined.

2. Results

2.1. Bioactive Compounds and Antioxidant Capacity of B. forficata Infusions

The TPC, TFC and antioxidant capacity of the *B. forficata* infusions are summarized in Table 1. It should be pointed out that the results presented in this study for TPC and TFC are expressed as rutin equivalents (RE) since this compound belongs to the flavonoids class, which is the major class in this species [6]. The values of TPC varied from 1923 to 6355 mg RE/100 g. Compared to the literature, the highest value found in this study, which was for the dry basis (7222 mg RE/100 g), is superior to that reported by Port's et al. [18], who evaluated different infusions of herbs from the Brazilian Amazonian region. Even though these authors did not evaluate *B. forficata*. However, their approach was the closest to this study, reporting results of the chemical evaluation for a *B. ungulata* infusion at 2% (g/mL) (2367 mg GAE/100 g dry basis). By calculation, at 5%, 5918 mg GAE/100 g dry basis would be found. Comparisons with data from the literature are difficult since few studies used the same species, and even when the species were the same, the results were expressed using different chemical standards, as in the example above. Additionally, it is easier to find data on *B. forficata* extracted with organic solvent than with hot water (infusion). Thus, our discussion will be focused on the differences observed among the brands and respective batches evaluated herein.

Table 1. Total phenolic content (TPC), total flavonoid content (TFC) and antioxidant capacity of *B. forficata* infusions.

Samples	Assays				
	TPC [1]	TFC [1]	DPPH• [2]	ABTS•+ [2]	FRAP [3]
BSB1	2126 ± 15 [g,h]	648 ± 19 [e]	20 ± 2 [e,f]	27 ± 4 [f]	89 ± 3 [h]
BSB2	2126 ± 29 [g,h]	630 ± 9 [e]	19 ± 0 [f]	30 ± 2 [e,f]	85 ± 6 [h]
BSB3	2772 ± 49 [e]	832 ± 11 [d,e]	21 ± 1 [e,f]	30 ± 2 [e,f]	136 ± 3 [g]
Overall average	2342 ± 324 [B]	703 ± 97 [B]	20 ± 1 [C]	29 ± 3 [B]	103 ± 25 [B]
CS1B1	2364 ± 164 [f,g]	1026 ± 4 [d]	34 ± 1 [d,e,f]	41 ± 0 [d,e,f]	127 ± 2 [g]
CS1B2	2733 ± 55 [e,f]	1042 ± 24 [d]	39 ± 2 [d,e]	46 ± 2 [d]	133 ± 4 [g]
Overall average	2549 ± 230 [B]	1034 ± 18 [B]	36 ± 3 [B,C]	43 ± 3 [B]	130 ± 4 [B]
CS2B1	4740 ± 69 [c]	3122 ± 114 [b]	108 ± 1 [c]	99 ± 1 [c]	242 ± 5 [e]
CS2B2	2245 ± 79 [g,h]	944 ± 40 [d]	37 ± 3 [d,e,f]	45 ± 2 [d,e]	120 ± 2 [g]
CS2B3	3203 ± 215 [d]	626 ± 30 [e]	45 ± 2 [d]	39 ± 1 [d,e,f]	176 ± 1 [f]
OverallAverage	3396 ± 1097 [B]	1564 ± 1178 [B]	63 ± 34 [B,C]	61 ± 29 [B]	179 ± 53 [B]
CS3B1	4681 ± 251 [c]	2006 ± 64 [d]	114 ± 2 [c]	109 ± 10 [c]	330 ± 11 [d]
CS3B2	5448 ± 144 [b]	2422 ± 147 [c]	173 ± 3 [b]	135 ± 12 [b]	385 ± 11 [c]
CS3B3	4833 ± 166 [c]	3700 ± 161 [a]	206 ± 2 [a]	204 ± 7 [a]	571 ± 4 [b]
Overall average	4987 ± 389 [A]	2710 ± 773 [A]	164 ± 42 [A]	149 ± 44 [A]	429 ± 109 [A]
CS4B1	2169 ± 89 [g,h]	1026 ± 18 [d]	45 ± 2 [d]	47 ± 3 [d]	129 ± 1 [g]
CS4B2	6355 ± 137 [a]	2628 ± 90 [c]	185 ± 8 [b]	149 ± 6 [b]	644 ± 19 [a]
CS4B3	1923 ± 4 [h]	482 ± 15 [f]	25 ± 1 [e,f]	27 ± 1 [f]	86 ± 5 [h]
Overallaverage	3483 ± 2158 [A,B]	1378 ± 967 [B]	85 ± 76 [B]	74 ± 57 [B]	286 ± 269 [A,B]

Abbreviations in the "Samples" column represent the different batches of each one of the brands evaluated. Different lowercase letters in the same column indicate that the results are statistically different ($p < 0.05$). Different uppercase letters in the same column indicate a statistically significant difference among groups (BS, CS1, CS2, CS3 and CS4) ($p < 0.05$). [1] Results expressed as mg RE/ 100 g. [2] Results expressed as µmol Trolox/g. [3] Results expressed as µmol Fe^{2+}/g. BSB1 = botanical sample batch 1; BSB2 = botanical sample batch 2; BSB3 = botanical sample batch 3; CS1B1 = commercial sample 1 batch 1; CS1B2=commercial sample 1 batch 2; CS2B1 = commercial sample 2 batch 1; CS2B2 = commercial sample 2 batch 2; CS2B3 = commercial sample 2 batch 3; CS3B1 = commercial sample 3 batch 1; CS3B2 = commercial sample 3 batch 2; CS3B3 = commercial sample 3 batch 3; CS4B1 = commercial sample 4 batch 1; CS4B2 = commercial sample 4 batch 2; CS4B3 = commercial sample 4 batch 3. Results as the mean ± standard deviation (triplicate).

The values for TPC, TFC, and antioxidant capacity measured by DPPH•, ABTS•+, and FRAP assays varied from 1923 to 6355 mg RE/100 g, 482 to 3700 mg RE/100 g, 19 to 206 µmol Trolox/g, 27 to 204 µmol Trolox/g, and 85 to 644 µmol Fe^{2+}/g, respectively. This corroborates that variations among samples and batches were high (Table 1). Among batches of the commercial samples, the highest values for TPC, TFC and antioxidant capacity (CS4B2 and CS3B3) were observed. These were higher than values reported to botanically identified sample (BS), which may be explained by differences in cultivation practices and the way the plants were processed. For example, the drying time may increase the degradation of plant bioactive compounds, whereas soil characteristics and precipitation conditions may affect the biosynthesis of secondary metabolites [9,10].

CS4B2 presented the highest TPC and FRAP values. For the TFC and DPPH• and ABTS•+ assays, CS3B3 presented the highest values (Table 1). The literature points to a direct relationship between TPC and antioxidant capacity; however, in this study, the sample that presented the highest TPC did not show the highest values for antioxidant capacity measured by all assays employed. This corroborates that the phytochemical composition of plant extracts may interact differently with radical species, which helps explain the results found.

High standard deviations were observed in CS2 and CS4 samples. The variation coefficient for the TFC reached 75% in CS2, for example, confirming the heterogeneity among the sample batches. The low standard deviation of the BS may be associated mainly with the standardization of the processing, which was followed from the harvest of leaves to drying. In addition, the harvest was from the same tree, although it took place in different seasons. This may also justify the low standard deviation of CS1 and CS3. Furthermore, conditions such as storage time, temperature, and kind of package have influence on the stability of bioactive compounds.

Since the samples showed heterogeneous batches according to the statistical analysis for this set of experiments, two groups were observed from their averages: one composed of the BS, CS1, CS2, and CS4 groups, for which no statistically significant differences were observed for the TPC, TFC, DPPH$^\bullet$, ABTS$^{\bullet+}$, and FRAP assays ($p > 0.05$), and the other represented by CS3 alone. These data provide important information about the production chain of *B. forficata*, rendering evident the need to standardize the steps that involve from harvest to distribution to deliver to consumers a product that guarantees its bioactive properties. *B. forficata* is widely used in Brazilian folk medicine due to its beneficial effects for treating rheumatism, local pain, uric acid, uterine problems [3], and, especially, type II diabetes [4]. This is possible due to the phytochemical profile of *B. forficata*, which is mainly composed of flavonoids, recognized for their antioxidant capacity [19].

2.2. LC-HRMS Analysis

A total of 20 phenolic compounds (Table 2), among flavonoids, phenolic acids, and other phenolic compounds, were tentatively identified in the samples. The majority are kaempferol and quercetin derivatives. The samples comprised flavonoid *O*-glycosides, thus in accordance with previously reported results, which prove its pharmacological action [20,21]. Additionally, polar compounds were identified in the samples in accordance with the polarity of the infusions.

Table 2. Tentatively identified compounds of *B. forficata* infusions.

	Compounds	*m/z* [M–H]$^-$ exp.	MS^2	Molecular Formula [M–H]$^-$	BS	CS1	CS2	CS3	CS4
1	Caffeoyl tartarate	311.0401	179; 135	$C_{13}H_{11}O_9$	+				
2	*Epi*-Catechin	289.0718	245; 203	$C_{15}H_{13}O_6$				+	
3	Galloyl hexose	331.0670	169; 125	$C_{13}H_{15}O_{10}$	+	+	+		+
4	Hydroxibenzoic acid	137.0244	-	C_7H5O_3	+	+	+		+
5	Dihydroxibenzoic acid hexoside	315.0719	108; 152	$C_{13}H_{15}O_9$			+	+	
6	3-Caffeoyl quinic acid	353.0875	191	$C_{16}H_{17}O_9$	+	+		+	+
7	Kaempferol 3-*O*-rhamnosyl-rutinoside	739.2136	284	$C_{33}H_{39}O_{19}$				+	
8	Rutin	609.1468	300	$C_{27}H_{29}O_{16}$	+	+	+	+	+
9	Myricitrin	463.0880	316	$C_{21}H_{29}O_{12}$			+		
10	Quercetin 3-*O*-glucopyranoside (Isoquercetin)	463.0917	301; 300	$C_{21}H_{29}O_{12}$	+	+	+	+	+
11	Quercetin-*O*-pentoside (Quercetin-*O*-arabinoside)	433.0780	300; 301	$C_{20}H_{17}O_{11}$	+	+	+	+	+
12	Quercetin 3-*O*-rhamnoside	447.0933	284; 285	$C_{21}H_{29}O_{11}$	+	+	+	+	+
13	Kaempferol 3-*O*-glucoside	447.0975	-	$C_{21}H_{29}O_{11}$	+	+	+	+	+
14	Kaempferol 3-*O*-rutinoside	593.1533	327; 284; 285	$C_{27}H_{29}O_{15}$	+	+		+	+
15	Isorhamnetin	315.0502	300	$C_{16}H_{11}O_7$	+	+	+	+	+
16	Isorhamnetin 3-*O*-rutinoside	623.1638	300; 315	$C_{28}H_{31}O_{16}$	+				
17	Quercetin 3-*O*-rhamnosyl-rutinoside	755.2087	300; 489	$C_{33}H_{39}O_{20}$	+			+	
18	Isorhamnetin 3-*O*-rhamnosyl-rutinoside	769.2201	605; 315	$C_{34}H_{41}O_{20}$	+			+	
19	Kaempferol 3-*O*-dirhamnoside	577.1595	431, 285, 284	$C_{27}H_{29}O_{14}$				+	
20	Kaempferol-*O*-pentoside	417.0833	285, 284, 255, 227	$C_{20}H_{17}O_{10}$				+	

BS: botanic sample; CS1: commercial sample 1; CS2: commercial sample 2; CS3: commercial sample 3; CS4: commercial sample 4. *m/z*—mass to charge ratio; MS^2—fragments of the second stage of mass spectrometry.

Most phenolic compounds identified in this study were free phenolic compounds, esterified with sugars or other compounds with low molecular masses, such as quercetin 3-*O*-rhamnoside, Kaempferol 3-*O*-glucoside, and Isorhamnetin.

Rutin, Isoquercetin, Quercetin-*O*-pentoside, Quercetin 3-*O*-rhamnoside, Kaempferol 3-*O*-glucoside, Kaempferol 3-*O*-rutinoside and Isorhamnetin were the compounds detected in all samples. CS3 is the infusion with the greatest number of compounds that vary according to the batch. In CS1, CS2, and CS4, the same 11 flavonoids were identified with differences in the relative abundance of the ions. Compound 3 showed a precursor ion [M–H]$^-$ at 331.0670 *m/z* and a typical loss of a hexose in MS^2 resulting in a [M–H]$^-$

m/z 169 fragment. It was assigned as galloyl hexose. Compound 4 was assigned as hydrox-ibenzoic acid based on precursor ion [M–H]⁻ at 137.0244 *m/z* and a very low error between experimental and theoretical mass of 0.1 ppm [22]. Compound 6 showed a precursor ion [M–H]⁻ at 353.0875 *m/z* and the quinic acid fragment in MS² at *m/z* 191, been identified as 3-Caffeoyl quinic acid. Compound 10 was assigned as Quercetin 3-*O*-glucopyranoside by comparison with literature records (1 ppm error ([M–H]⁻ *m/z* 463.0878) [23]. Compound 19 was assigned as Isorhamnetin 3-*O*-rhamnosyl-rutinoside based on precursor ion [M–H]⁻ *m/z* 769.2190 and based in the loss of Isorhamnetin fragment at *m/z* 315. Kaempferol fragment ion at *m/z* 284 was used to identify compound 20 as Kaempferol 3-*O*-dirhamnoside along with the precursor ion [M–H]⁻ *m/z* 577.1595 [6]. Identification of the other listed compounds by fragmentation data and exact mass were previously described by the authors [24,25].

The UPLC-ESI-Q-TOF MS/MS chromatographic technique was an efficient tool to characterize and identify the phenolic compounds in *B. forficata* infusions. It is important to highlight that the advantage of this technique is that, although it is not quantitative, one may relatively quantify the compounds, even the isomeric forms (e.g., Catechin, *Epi*-catechin, and Quercetin-*O*-pentoside), and, in case of a lack of standards, the compound assignments may be made by comparison of UV spectra and MS data (accurate mass and fragmentation) with previous literature reports [6,22,23].

A PCA analysis of the non-volatile chemical composition showed three distinct groups: I—CS3B2, CS3B3 and CS4B2; II—CS1B1, CS1B2, CS2B1, CS2B2, CS4B1, CS4B3 and BSB1; III—CS3B1, BSB2 (Figure 1A). These results demonstrate great chemical variability between the different samples, although flavonoids Rutin, Isoquercetin, Quercetin-*O*-pentoside, Quercetin 3-*O*-rhamnoside, Kaempferol 3-*O*-glucoside, Kaempferol 3-*O*-rutinoside and Isorhamnetin were detected in all samples.

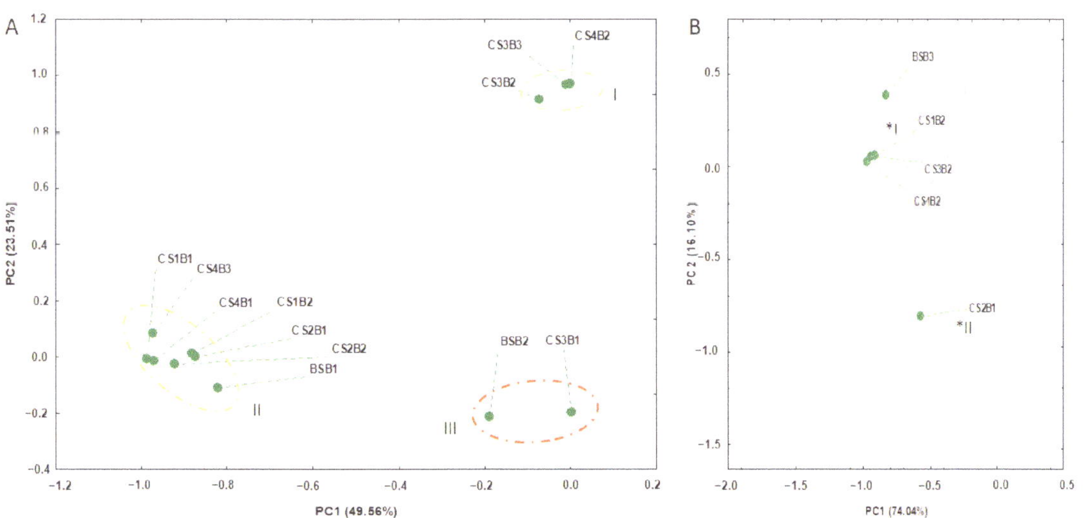

Figure 1. Principal component analysis of (**A**) non-volatile compounds and (**B**) volatile compounds. BS: botanic sample; CS1: commercial sample 1; CS2: commercial sample 2; CS3: commercial sample 3; CS4: commercial sample 4. B is relative to the batch. *I—BSB3 (rich in Caryophyllene oxide), CS1B2, CS3B2 and CS4B2 (rich in Spathulenol); *II—CS2B1 (rich in 2-Propyl-1-heptanol).

2.3. HS-SPME/CG–MS

The identification and relative concentrations of the volatile compounds in the five herbal infusions of *B. forficata* are shown in Table 3, in order of retention time (Rt), and increasing Linear Retention Index (LRI). Forty volatile compounds were tentatively identified, of which only seven were detected in all samples: 2-Propyl-heptanol (7.69–19.42%), Geranyl acetone (4.38–7.31%), Dodecanol (3.11–11.37%), β-Ionone (0.71–5.84%), Spathulenol (11.78–30.87%), Caryophyllene oxide (2.76–17.46%), and Benzoic acid 2-ethylhexyl ester (1.12–20.14%). The volatile compounds included terpenoids, represented by C_{13}-norisoprenoids, sesquiterpenes, and monoterpenes, as well as hydrocarbons, alcohols, esters, aldehydes, ketones, and acids. Among all chemical groups found in the volatiles of the *B. forficata* infusions, sesquiterpenes (hydrocarbon and oxygenated) were present in a higher number (17) and represented most of the composition of the BS (63%), CS1 (61%), CS3 (50%), and CS4 (53%). Esters (31%) and alcohols (29%) accounted for most of the composition of CS2.

Table 3. Tentatively identified compounds of *B. forficata* infusions with their respective relative percentage (%).

Rt (min)	LRI [a]	Compound	Chemical Class	BSB3	CS1B2	CS2B1	CS3B2	CS4B2
14.00	1185	1-Decanal	A	0.10 ± 0.04	-	0.57 ± 0.22	-	-
14.30	1193	2-Propyl-1-heptanol	AL	3.35 ± 0.35	7.69 ± 0.62	19.42 ± 2.52	9.66 ± 4.93	8.96 ± 3.07
16.40	1195	Estragole	PP	-	0.30 ± 0.00	-	-	0.44 ± 0.21
18.44	1357	Eugenol	PP	0.24 ± 0.00	-	-	-	-
20.10	1428	β-Caryophyllene	S	0.85 ± 0.10	-	-	-	-
20.30	1429	α-Ionone	N	3.59 ± 0.47	1.55 ± 0.08	1.64 ± 0.04	-	-
20.90	1448	Geranyl acetone	N	6.88 ± 1.08	7.31 ± 0.00	5.18 ± 0.76	5.02 ± 1.08	4.38 ± 1.25
20.92	1452	α-Humulene	S	1.22 ± 0.45	-	-	-	-
21.00	1461	Alloaromadendrene	S	0.70 ± 0.03	-	-	-	-
21.20	1472	*p*-Benzoquinone	K	-	0.66 ± 0.04	1.50 ± 0.22	0.99 ± 0.03	-
21.40	1480	Dodecanol	AL	4.00 ± 3.75	3.11 ± 0.51	7.14 ± 1.39	3.94 ± 1.27	8.37 ± 0.01
21.70	1485	Deydro-β-ionone	N	-	-	5.30 ± 0.50	-	1.17 ± 0.36
21.80	1486	β-Ionone	N	4.24 ± 0.05	3.08 ± 0.11	0.71 ± 0.23	2.54 ± 0.38	5.84 ± 1.19
21.99	1499	Germacrene D	S	-	0.99 ± 0.02	-	-	-
22.70	1530	δ-Cadinene	S	2.03 ± 0.24	2.72 ± 0.22	-	2.25 ± 0.28	-
22.80	1538	Dihydroactinidiolide	OM	0.70 ± 0.10	-	-	-	-
22.90	1545	Eudesma-3,7(11-diene)	S	0.38 ± 0.07	-	-	-	-
23.20	1554	Nerolidol oxygenated	S	-	-	-	3.55 ± 0.76	-
23.20	1554	Nerolidol oxygenated	S	-	-	-	3.55 ± 0.76	-
24.00	1582	Spathulenol	OS	11.78 ± 1.02	30.87 ± 0.15	8.53 ± 2.34	13.98 ± 1.39	25.86 ± 1.76
24.10	1585	Caryophyllene oxide	OS	15.80 ± 0.42	14.32 ± 0.66	2.76 ± 2.58	17.46 ± 1.48	14.11 ± 0.28
24.40	1598	Ledol	OS	4.05 ± 0.21	-	-	-	-
24.50	1603	Globulol	OS	1.47 ± 0.04	-	-	-	-
24.70	1607	Humulene epoxide II	OS	14.15 ± 0.78	-	1.58 ± 0.14	5.71 ± 0.14	7.08 ± 006
25.20	1631	1,7,7-Trimethyl-2-vinylbicyclo [2.2.1]hept-2-ene (Vinylbornene)	-	-	5.21 ± 0.27	-	-	-
25.40	1634	Longipinocarveol	OS	1.68 ± 0.03	-	-	2.57 ± 0.26	-
25.50	1647	τ-Muurolol	OS	1.75 ± 0.50	-	-	-	-
25.70	1659	α-Cadinol	OS	5.04 ± 0.50	11.99 ± 0.39	-	4.36 ± 0.01	6.32 ± 0.44
27.70	1745	Octanal 2-phenylmethylene	A	-	-	-	0.85 ± 0.27	0.31 ± 0.16
27.90	1768	Tetradecanoic acid	CA	-	0.36 ± 0.25	1.87 ± 0.93	1.58 ± 1.17	0.54 ± 0.63
28.30	1785	Anthracene	H	-	-	-	-	0.68 ± 0.05
28.70	1800	Octadecane	H	-	-	1.04 ± 0.38	-	-
29.60	1850	4,8,12-Tetradecatrienal-5,9,13-trimethyl	A	-	-	1.91 ± 0.59	-	1.08 ± 0.05
30.40	1880	1-Hexadecanol	AL	-	0.63 ± 0.01	3.25 ± 1.85	1.49 ± 1.57	1.87 ± 1.02
34.20	1881	Cyclohexadecane	H	-	1.39 ± 0.00	-	0.99 ± 0.54	-
34.50	1900	Nonadecane	H	-	-	0.81 ± 0.24	-	-
34.80	1909	Methyl hexadecanoate	E	-	-	1.89 ± 0.47	-	-
35.00	1922	Dibutyl phtalate	E	-	-	9.24 ± 3.89	-	-
35.90	2108	Bisphenol A	PH	-	0.10 ± 0.06	3.82 ± 0.48	2.16 ± 0.30	-
39.60	2360	2-Methyltricosane	H	-	-	1.23 ± 1.15	-	-

[a] Linear Retention Index (LRI) calculated for all components using a homologous series of *n*-alkanes analyzed under the same conditions as the samples; (-) not detected. A—aldehyde, AL—alcohol, PP—phenylpropanoid, S—sesquiterpene, N—norisoprenoid, K—ketone, OM—oxygenated monoterpene, OS—oxygenated sesquiterpene, CA—carboxylic acid, HC—hydrocarbon, E—ester, and PH—phenol. BSB3 = botanically identified sample, batch 3; CS1B2 = commercial sample brand 1, batch 2; CS2B1 = commercial sample brand 2, batch 1; CS3B2 = commercial sample brand 3, batch 2; CS4B2 = commercial sample brand 4, batch 2. Relative percentage as the mean ± standard deviation (duplicate).

There are no data in the literature on the volatile composition of *B. forficata* infusions or any species of the *Bauhinia* genus. However, there are two studies that identified constituents of essential oils of this species and demonstrated that they are essentially composed of sesquiterpenoids. Duarte-Almeida et al. [26] and Sartorilli and Correa [27] evaluated the composition of essential oils in *B. forficata* and reported that the content of sesquiterpenoids was 87% and 96%, respectively. Our results and those from essential oils [26,27] are a great evidence that a mostly sesquiterpenic volatile fraction composition may be characteristic of this species.

It is well established that many sesquiterpenes and their alcohol, aldehyde, and ketone derivatives are biologically active or precursors of metabolites with biological functions, while others have desirable fragrance and flavoring properties [28]. Spathulenol (8.53–25.86%) and Caryophyllene oxide (2.76–17.46%) were two of the major compounds in all samples. Both compounds are known to possess several biological activities. Nascimento et al. [29] demonstrated antioxidant, anti-inflammatory, antiproliferative, and antimycobacterial activities of spathulenol, and a moldy and herbaceous odor is attributed to this compound [30]. In turn, Caryophyllene oxide has a floral and woody odor [31,32], and biological activities such as anticholinesterase, analgesic, anti-inflammatory, and antifungal activities were also reported [33,34]. Regarding the class of norisoprenoids (C_{13}), they were detected in all samples at concentrations ranging from 7.56% to 14.71%, highlighting Geranyl acetone and β-Ionone. It is reported that they present a significant aromatic impact in fruits such as grapes, apples, lychee, and mango [35,36], with a floral odor being attributed to them [37].

Attention is drawn to the identification of Bisphenol A (BPA) and Dibutyl phthalate (DBP) in some samples evaluated here, especially CS2, which showed important concentrations of these contaminants in its volatile fraction (9.24% and 3.82%, respectively). As any agricultural product, these herbs may be subjected to chemical contaminations due to agricultural practices, especially in stages when a plastic material is used as packaging or support or due to soil treatment, cultivation in contaminated soil, and other factors [38,39]. Furthermore, the migration of these plasticizers that constitute the packaging cannot be ruled out since it is known that this is the main source of exposure to this type of contaminant [39]. Di Bella et al. [39] and Lo Turco et al. [40] evaluated the BPA contamination of spices and herbs from different origins and found it to be present in several samples. Despite concluding that the ingestion of these contaminants does not imply a risk to human health, one cannot disregard their existence, and mechanisms to mitigate them must be evaluated, such as proposing other packaging materials free from them.

In general, the observed differences among the volatile fraction patterns of the infusions were lower than those observed for non-volatile (Figure 1B). Only CS2B2 formed another group by PCA analysis (Figure 1B). Indeed, different origins of the samples with their unique ecological settings as well as features intrinsic to the medicinal herbs may explain this difference [12]. Moreover, Arsenijević et al. [12] stressed that compounds present in the volatile fraction of infusions play an important role in the antioxidant capacity of these products, thus rendering this evaluation relevant, although it was still not possible to measure it in this work. Once again, we highlight that the results obtained herein are the first step towards revealing the beneficial health effects of *B. forficata* infusions through chemical diversity after evaluating their non-volatile and volatile fractions.

2.4. Assay for α-Amylase Inhibition

In this set of experiments the effect of *B. forficata* infusions that presented better results for TPC, TFC and antioxidant capacity was investigated. The results revealed that all infusions inhibited the α-amylase activity. Based on the IC_{50} values, which represent the concentration required to inhibit 50% of the enzyme activity, the CS2B1 sample was the one that showed the greatest potential for enzyme inhibition, as it showed the lowest IC_{50} value (0.235 mg RE/mL). The IC_{50} values were 0.235 mg RE/mL, 0.245 mg RE/mL, 0.287 mg RE/mL, 0.489 mg RE/mL, and 0.801 mg RE/mL for CS2B1, CS4B2, CS1B2, BSB3, and CS3B2, respectively. Even though CS4B2 presented the highest TPC, this sample exhibited a higher IC_{50} value. It is suggested that the inhibition of α-amylase activity may be due to other phytochemicals also present in the infusions such as terpenoids, which were detected in the samples by HS-SPME/CG–MS. However, it is well known that phenolic compounds, mainly flavonoids, are excellent inhibitors of digestive enzymes. Flavonoids and their derivatives have the ability to reduce the potency of α-amylase and α-glucosidase by either interacting with or inhibiting specific positions of the enzyme [41]. However, other classes of compounds should not be neglected as published by Papoutsis et al. [42], which reported in their review the positive effects of terpenoids, carotenoids, among others compounds on inhibition of α-amylase activity. It is important to note that these compounds should be bioavailable after digestion to act on digestive enzymes. Thus, future studies on this subject should be addressed.

Acarbose is widely used in medicine as an inhibitor of digestive enzymes related to the breakout of polysaccharides. As these enzymes are inhibited, there is a reduction in glucose absorption and, consequently, a decrease in the postprandial blood glucose level elevation, which helps reduce the risk of Diabetes mellitus, for example [42]. Its IC_{50} value was found to be 0.034 mg/mL. Thus, a lower concentration of this substance is required to inhibit 50% of the α-amylase activity when compared to *B. forficata* infusions. However, it should be noted that this medicinal plant is widely used in folk medicine as an adjuvant in treating hyperglycemia by the population, especially those in vulnerable conditions [43].

It is important to demonstrate that infusions prepared from commercially available herbs showed an important inhibitory action on the enzyme despite being less potent than acarbose. Furthermore, cytotoxicity was not observed when different fractions from *B. forficata* were evaluated by Franco et al. [44]. These facts reinforce the biological and pharmacological potential of *B. forficata* as hypoglycemiant agent, which has an important role in Brazilian folk medicine, primarily because it is abundant and easily accessible.

3. Material and Methods

3.1. Plant Material

B. forficata leaves were collected in Petropolis, Rio de Janeiro, Brazil (22°30′04.63″ S, 43°07″58.20″ W, altitude: 958 m) in different seasons (winter, spring, and summer-2018/2019). Voucher specimens were deposited at the Herbarium of the Department of Botany of the Federal University of Rio de Janeiro, under registration number RFA 40.615. The samples were dried in an oven with forced air circulation at 45 °C, then disintegrated in a domestic blender to obtain a powered material, which was used to prepare the infusions. These samples were named BSB1 (winter), BSB2 (spring), and BSB3 (summer).

Four commercial samples purchased from local markets in the city of Rio de Janeiro were also evaluated. Two batches of commercial sample 1 (CS1) and three batches of the other samples (CS2, CS3, and CS4) were acquired, resulting in samples CS1B1, CS1B2, CS2B1, CS2B2, CS2B3, CS3B1, CS3B2, CS3B3, CS4B1, CS4B2, and CS4B3, which were used to prepare the infusions.

3.2. Preparing the Infusions

The infusions were prepared by adding 50 mL of boiling water to 2.5 g of the samples (5% *w/v*). After that, they were allowed rest at room temperature for 20 min. The extracts were filtered and transferred to a volumetric flask, in which the volume was quenched with distilled water until reaching 50 mL [45].

3.3. Analysis

3.3.1. Total Phenolic Content (TPC)

The TPC analysis was performed using the Folin-Ciocalteu reagent (Imbralab, Ribeirão Preto, Brazil), following the method described by Singleton and Rossi [46]. For the reactions, 250 µL of the filtered and appropriately diluted extract was mixed with 1250 µL of 10% Folin-Ciocalteu reagent and 1000 µL of a 7.5% (*w/v*) sodium carbonate solution. Thereafter, the samples were heated at 50 °C for 15 min and cooled at room temperature. The absorbance was measured at 760 nm. A calibration curve was constructed using the rutin (Sigma-Aldrich, St Louis, MO, USA) standard with concentrations ranging from 16 mg/L to 166 mg/L (linear regression: y = 0.0034x−0.0128; R^2 = 0.9988). The TPC is expressed as milligrams of rutin equivalent per 100 g (mg RE/100 g).

3.3.2. Total Flavonoid Content (TFC)

The TFC was determined based on the method described by Zhishen et al. [47] with minor modifications. Here, 0.5 mL of extract was mixed with 3.2 mL of ultrapure water and 150 μL of $NaNO_2$ (5%, *w/v*). After homogenization, the mixture was left to rest for 5 min. Thereafter, 150 μL of $AlCl_3$ (10%, *w/v*) was added to the mixture, and 1 mL of NaOH (1 M) was added after 1 min. The absorbance was recorded at 510 nm with a spectrophotometer (Metash, Shanghai, China) using ultrapure water as a blank. The TFC was calculated using the calibration curve of rutin (Sigma-Aldrich, St. Louis, MO, USA) standard, with the concentration ranging from 99 mg/L to 595 mg/L (linear regression: y = 0.001x + 0.013; R^2 = 0.9974). The results are expressed as mg RE/100 g.

3.3.3. ABTS$^{\bullet+}$ Assay

The antioxidant capacity was determined by the reduction of radical monocation, 2,2′-azinobis-(3-ethylbenzothiazoline-6-sulfonic acid) (ABTS$^{\bullet+}$), according to the procedure described by Gião et al. [48]. The radical was obtained after the addition of 7 mmol/L of ABTS (2,2′-azinobis-(3-ethylbenzothiazoline-6-sulfonic acid) diammonium salt (Sigma-Aldrich, Saint Louis, MO, USA) to 2.45 mmol/L of a potassium persulfate solution (1:1 (*v/v*)). The mixture was left to react in the dark for 16 h. To obtain an absorbance of 0.700 ± 0.020 at 734 nm, the ABTS$^{\bullet+}$ solution was diluted using ultrapure water. For the reactions, 30 μL of each filtered and diluted extract was mixed with 3000 μL of the ABTS$^{\bullet+}$ solution. After 6 min, the absorbance was measured at 734 nm with a spectrophotometer (Metash, Shanghai, China) using ultrapure water as a blank. The ABTS$^{\bullet+}$ antiradical activity was calculated using Trolox solutions (Sigma-Aldrich, Buchs, Switzerland) with different concentrations ranging from 240 to 2000 μmol (linear regression: y = 0.0003x + 0.0094; R^2 = 0.9989). The results are expressed as μmol of Trolox equivalents per gram (μmol TE/g).

3.3.4. DPPH$^{\bullet}$ Assay

The 2,2′-diphenyl-β-picrylhydrazyl radical (DPPH$^{\bullet}$) (Sigma-Aldrich, Steinheim, Germany) scavenging activity of the extracts was determined according to the method described by Hidalgo et al. [49]. For the reactions, 100 μL of each diluted extract was added to 2900 μL of a DPPH$^{\bullet}$ solution (6×10^{-5} M in methanol and diluted to an absorbance of 0.700 at 517 nm). The resulting solutions were allowed to stand for 30 min in the dark at room temperature. Then, the absorbance was measured at 517 nm with a spectrophotometer (Metash, Shanghai, China) using methanol as a blank. The DPPH$^{\bullet}$ scavenging activity was calculated using Trolox solutions (Sigma-Aldrich, Buchs, Switzerland) with different concentrations ranging from 80 to 680 μmol (linear regression: y = 0.0008x + 0.017; R^2 = 0.9962). The results are expressed as μmol TE/g.

3.3.5. FRAP Assay

The ferric reducing/ antioxidant power (FRAP) assay was performed according to the procedure reported by Benzie and Strain [50] with minor modifications. The stock solutions included 300 mM of an acetate buffer (pH 3.6), 10 mM of 2,4,6-tri(2-pyridyl)-s-triazine (Sigma-Aldrich, Buchs, Switzerland) in 40 mM of HCl, and 20 mM of $FeCl_3 \cdot 6H_2O$. The working solution was prepared by mixing 25 mL of the acetate buffer, 2.5 mL of the TPTZ solution, and 2.5 mL of $FeCl_3 \cdot 6H_2O$. Thereafter, 100 μL of each extract was reacted with 3000 μL of the working solution at 37 °C for 30 min, and the absorbance was measured at 593 nm. The FRAP activity was calculated using $FeSO_4 \cdot 7H_2O$ solutions with different concentrations ranging from 150 to 1200 μmol of Fe^{2+} (linear regression: y = 0.0008x + 0.0042; R^2 = 0.9992). The results are expressed as μmol of Fe^{2+} per gram (μmol Fe^{2+}/g).

3.3.6. LC-HRMS Analysis

The sample extract was dissolved in an aqueous solution containing formic acid (0.1%, *v/v*) and subjected to an ultra-performance liquid chromatography-quadrupole/time-of-flight mass spectrometry (UPLCqTOF/MS; maXis Impact, Bruker Daltonics, Billerica, MA, USA) analysis. The separation was performed using a Hypersil C18 column (3 μm particle size, 2.1 mm × 150 mm). The column temperature was maintained at 40 °C. Subsequently, an aliquot of 20 μL was injected into the UPLC-ESI-qTOF system with a flow rate of 0.27 mL/min. The linear gradient elution of A (0.1% formic acid in water) and B (acetonitrile) was applied by employing the following method: 5% of B at the beginning; 5% to 9% of B for 5 min, 9% to 16% of B for 10 min, 16% to 36% of B for 18 min, 36% to 95% of B for 1 min, 95% of B for 12 min, 95% to 5% of B for 1 min, and 5% of B for 13 min. Data Analysis 4.2 software (Bruker Daltonics, Billerica, MA, USA) was used to interpret the data. The MS data were acquired in the negative mode using an electrospray ionization (ESI) source. The data were scanned for each test sample at a mass-to-charge ratio (m/z) from 50 to 1200. Highly pure nitrogen was used as the nebulizing gas and ultrahigh purity helium as the collision gas, and the capillary voltage was set at 5000 V. The ESI parameters included dry gas at 200 °C at a flow rate of 8 L/min and a nebulizer pressure of two bar [25].

3.3.7. HS-SPME/CG–MS

The infusions that presented better results for TPC, TFC and antioxidant capacity were subjected to an analysis of the volatile fraction by Headspace Solid-Phase microextraction followed by gas chromatography–mass spectrometry (HS-SPME/GC–MS).

The headspace volatiles analysis using SPME described by Wang et al. [51] was adopted with minor modifications. Volumes of 10 mL of freshly prepared infusions were placed into 20 mL clear glass vials and immediately capped and placed on a temperature-controlled water bath at 60 °C for 60 min with a SPME fiber coated with 100 μm of PDMS (100% polydimethylsiloxane; Supelco®, Bellefonte, PA, USA) pre-conditioned at 250 °C for 60 min and inserted into the headspace above the liquid surface. A system blank with an empty vial was run as a control assay. SPME fibers were desorbed at 250 °C for 5 min in the injection port of the chromatographic system described below.

The GC–MS analysis of the volatile fractions was carried out using an Agilent 6890N gas chromatograph (Agilent Technologies, Palo Alto, CA, USA) with an HP-5MS 5% phenylmethylsiloxane capillary column (30 m × 0.25 mm, 0.25 μm film thickness; Restek, Bellefonte, PA, USA) equipped with an Agilent 5975 mass selective detector in the electron impact mode (ionization energy: 70 eV) operating according to the following conditions. The oven temperature was initially maintained at 60 °C for one 1 min, then raised at the rate of 8 °C/min to 300 °C, staying at this temperature for 15 min. The injector and detector temperatures were set at 250 °C and 260 °C, respectively. The samples were injected in the splitless mode. A normalization technique was used to obtain quantitative data. Linear retention indices (LRI) were calculated for all components using a homologous series of n-alkanes (C7–C30, Sigma-Aldrich, Laramie, WY, USA) analyzed under the same conditions as the samples. The identification of the volatile fraction components was based on LRI relative to n-alkanes and computer matching with the Wiley275.L and Wiley7n.L libraries and comparisons of the fragmentation patterns of the mass spectra with published data [52].

3.3.8. Assay for α-Amylase Inhibition

The infusions that presented better results for TPC, TFC and antioxidant capacity were subjected to the inhibition assay for α-amylase, performed as reported by Meng et al. [53] with minor modifications. Briefly, 100 μL of extract was mixed with an α-amylase solution (100 μL, 1.0 U/mL) (Sigma-Aldrich, St. Louis, MO, USA) in a phosphate buffer (pH 6.9) and 250 μL of a 1% starch solution. The incubation was carried out for 5 min at 37 °C. The enzyme reaction was stopped by adding dinitrosalicylic acid reagent (250 μL) (Sigma-Aldrich, Steinheim, Germany), and incubation was carried out for 15 min in boiling water. For the dilution, 2 mL of distilled water was added to the final reaction mixture. The absorbance was measured at 540 nm. The inhibitory effect was calculated according to Equation (1), where $Abs_{control-1}$ results from the reaction without adding the enzyme, which was replaced by the buffer solution, while the mixture of the enzyme and starch solution without extract was $Abs_{control-2}$. The results were expressed as IC_{50} (mg RE/mL). Acarbose (Supelco, Laramie, WY, USA) was used as a positive control to compare the inhibitory effects.

$$\text{Inhibition percentage (\%)} = [1 - (Abs_{sample} - Abs_{control-1})/Abs_{control-2}] \times 100 \qquad (1)$$

3.4. Statistical Analysis

The data were statistically analyzed using Statistica software version 13 (Dell Inc., Tulsa, OK, USA), performing an analysis of variance (ANOVA) and Tukey's test to verify the differences among averages, considering the 95% confidence level. Experiments were performed in duplicate/triplicate, and the results are presented as the average $\pm$ standard deviation. Additionally, the principal component analysis (PCA) were used to assess the variance in the non-volatile and volatile samples. Results were processed using STATISTICA software version 10 (StartSoft Inc., Tulsa, OK, USA).

4. Conclusions

It is concluded that the samples presented different TPC, TFC and antioxidant potentials. The commercial CS4B2 and CS3B3 samples showed higher values for bioactive compounds and antioxidant capacity than botanically identified samples. However, both were mostly composed of flavonoid derivatives. PCA analysis demonstrated more chemical diversity in non-volatile than volatile compounds. This analysis may justify the differences observed in the results of the performed assays. To the best of our knowledge, this is the first time that volatile fraction obtained from *B. forficata* infusions has been carried out. It is very clear that it is an important fraction with regard to the aroma besides possible contribution to the biological properties. An inhibitory effect of all *B. forficata* infusions on the α-amylase enzyme was observed. Despite the differences reported in this work, *B. forficata* presents itself as a source of bioactive compounds that may increase the intake of antioxidant compounds by the population.

Author Contributions: Conceptualization, E.P.J., L.d.O.R. and R.F.A.M.; methodology, C.N.K., L.d.O.R. and E.P.J.; formal analysis, B.P.d.F., L.d.O.R. and N.G.d.F.; investigation, C.N.K., B.P.d.F., E.P.J., D.d.L.M. and L.d.O.R.; resources, L.d.O.R. and R.F.A.M.; data curation, C.N.K., L.d.O.R., D.d.L.M. and R.F.A.M.; writing—original draft preparation, E.P.J., L.d.O.R., N.G.d.F. and B.P.d.F.; writing—review and editing, D.d.L.M., L.d.O.R., N.G.d.F. and R.F.A.M.; supervision, L.d.O.R. and R.F.A.M. All authors have read and agreed to the published version of the manuscript.

Funding: This research received no external funding.

Institutional Review Board Statement: Not applicable.

Informed Consent Statement: Not applicable.

Data Availability Statement: Data is contained within this article.

Acknowledgments: The authors acknowledge the support from the National Institute of Technology (INT), Federal University of Rio de Janeiro State (UNIRIO), Conselho Nacional de Desenvolvimento Científico e Tecnológico (CNPq), Fundação de Amparo à Pesquisa no Estado do Rio de Janeiro (FAPERJ) and Coordenação de Aperfeiçoamento de Pessoal de Nível Superior (CAPES).

Conflicts of Interest: The authors declare no conflict of interest.

Sample Availability: Samples of the compounds are available from the authors.

References

1. Vaz, A.M.S.F.; Tozzi, A.M.G.A. Sinopse de *Bauhinia sect. Pauletia* (Cav.) DC. (Leguminosae: Caesalpinioideae: Cercideae) no Brasil. *Braz. J. Bot.* **2005**, *28*, 477–491. [CrossRef]
2. López, R.E.S.; Santos, B.C. *Bauhinia forficata* Link (Fabaceae). *Ver. Fitos.* **2015**, *9*, 217–232.

3.   Cechinel-Zanchett, C.C.; de Andrade, S.F.; Cechinel-Filho, V. Ethnopharmacological, Phytochemical, Pharmacological and Toxicological Aspects of *Bauhinia Forficata*: A Mini-Review Covering the Last Five Years. *Nat. Prod. Commun.* **2018**, *13*, 1934578X1801300732. [CrossRef]

4.   Tonelli, C.A.; de Oliveira, S.Q.; da Silva Vieira, A.A.; Biavatti, M.W.; Ritter, C.; Reginatto, F.H.; de Campos, A.M.; Dal-Pizzol, F. Clinical Efficacy of Capsules Containing Standardized Extract of *Bauhinia Forficata* Link (Pata-de-Vaca) as Adjuvant Treatment in Type 2 Diabetes Patients: A Randomized, Double Blind Clinical Trial. *J. Ethnopharmacol.* **2022**, *282*, 114616. [CrossRef]

5.   Maffioletti, N.S.; Rossato, E.A.; Dal-B'o, S.; Amaral, P.A.; Zanette, V.C. *Bauhinia forficata* Link (Fabaceae) no combate ao diabetes mellitus: Aspectos taxonômicos, agroecológicos, etnobotânicos e terapêuticos. *Ver. Tecnol. Ambiente.* **2012**, *18*, 1–18.

6.   Farag, M.A.; Sakna, S.T.; El-fiky, N.M.; Shabana, M.M.; Wessjohann, L.A. Phytochemical, Antioxidant and Antidiabetic Evaluation of Eight *Bauhinia* L. Species from Egypt Using UHPLC–PDA–qTOF-MS and Chemometrics. *Phytochemistry* **2015**, *119*, 41–50. [CrossRef]

7.   Ministério da Saúde; Agência Nacional de Vigilância Sanitária. Resolução da Diretoria Colegiada n. 267, de 22 de setembro de 2005. In *Approves the Technical Regulation of Plant Species for the Preparation of Teas Diário Oficial*; União da República Federativa do Brasil: Brasília, Brasil, 2005.

8.   Salgueiro, A.C.F.; Folmer, V.; da Silva, M.P.; Mendez, A.S.L.; Zemolin, A.P.P.; Posser, T.; Franco, J.L.; Puntel, R.L.; Puntel, G.O. Effects of *Bauhinia Forficata* Tea on Oxidative Stress and Liver Damage in Diabetic Mice. *Oxid. Med. Cell. Longev.* **2016**, *2016*, 8902954. [CrossRef]

9.   Sotiropoulou, N.S.D.; Flampouri, E.; Skotti, E.; Pappas, C.; Kintzios, S.; Tarantilis, P.A. Bioactivity and Toxicity Evaluation of Infusions from Selected Greek Herbs. *Food Biosci.* **2020**, *35*, 100598. [CrossRef]

10.   Lamien-Meda, A.; Nell, M.; Lohwasser, U.; Börner, A.; Franz, C.; Novak, J. Investigation of Antioxidant and Rosmarinic Acid Variation in the Sage Collection of the Genebank in Gatersleben. *J. Agric. Food Chem.* **2010**, *58*, 3813–3819. [CrossRef] [PubMed]

11.   Tschiggerl, C.; Bucar, F. Investigation of the Volatile Fraction of Rosemary Infusion Extracts. *Sci. Pharm.* **2010**, *78*, 483–492. [CrossRef] [PubMed]

12.   Arsenijević, J.; Drobac, M.; Šoštarić, I.; Ražić, S.; Milenković, M.; Couladis, M.; Maksimović, Z. Bioactivity of Herbal Tea of Hungarian Thyme Based on the Composition of Volatiles and Polyphenolics. *Ind. Crops Prod.* **2016**, *89*, 14–20. [CrossRef]

13.   Ma, C.; Li, J.; Chen, W.; Wang, W.; Qi, D.; Pang, S.; Miao, A. Study of the Aroma Formation and Transformation during the Manufacturing Process of Oolong Tea by Solid-Phase Micro-Extraction and Gas Chromatography–mass Spectrometry Combined with Chemometrics. *Food Res. Int.* **2018**, *108*, 413–422. [CrossRef] [PubMed]

14.   Lin, S.-Y.; Lo, L.-C.; Chen, I.-Z.; Chen, P.-A. Effect of Shaking Process on Correlations between Catechins and Volatiles in Oolong Tea. *J. Food Drug Anal.* **2016**, *24*, 500–507. [CrossRef] [PubMed]

15.   Du, L.; Li, J.; Li, W.; Li, Y.; Li, T.; Xiao, D. Characterization of Volatile Compounds of Pu-Erh Tea Using Solid-Phase Microextraction and Simultaneous Distillation–extraction Coupled with Gas Chromatography–mass Spectrometry. *Food Res. Int.* **2014**, *57*, 61–70. [CrossRef]

16.   Lv, H.-P.; Zhong, Q.-S.; Lin, Z.; Wang, L.; Tan, J.-F.; Guo, L. Aroma Characterisation of Pu-Erh Tea Using Headspace-Solid Phase Microextraction Combined with GC/MS and GC–olfactometry. *Food Chem.* **2012**, *130*, 1074–1081. [CrossRef]

17.   Augusto, F.; Luiz Pires Valente, A. Applications of Solid-Phase Microextraction to Chemical Analysis of Live Biological Samples. *TrAC Trends Anal. Chem.* **2002**, *21*, 428–438. [CrossRef]

18.   Port's, P.S.; Chisté, R.C.; Godoy, H.T.; Prado, M.A. The Phenolic Compounds and the Antioxidant Potential of Infusion of Herbs from the Brazilian Amazonian Region. *Food Res. Int.* **2013**, *53*, 875–881. [CrossRef]

19.   Ferreres, F.; Gil-Izquierdo, A.; Vinholes, J.; Silva, S.T.; Valentão, P.; Andrade, P.B. *Bauhinia Forficata* Link Authenticity Using Flavonoids Profile: Relation with Their Biological Properties. *Food Chem.* **2012**, *134*, 894–904. [CrossRef]

20.   Hwang, D.; Kang, M.; Kang, C.; Kim, G. Kaempferol-3-O-β-rutinoside Suppresses the Inflammatory Responses in Lipopolysaccharide-stimulated RAW264.7 Cells via the NF-κB and MAPK Pathways. *Int. J. Mol. Med.* **2019**, *44*, 2321–2328. [CrossRef]

21.   Jiang, H.; Yamashita, Y.; Nakamura, A.; Croft, K.; Ashida, H. Quercetin and Its Metabolite Isorhamnetin Promote Glucose Uptake through Different Signalling Pathways in Myotubes. *Sci. Rep.* **2019**, *9*, 2690. [CrossRef]

22.   Aquino, A.J.; Alves, T.d.C.; Oliveira, R.V.; Ferreira, A.G.; Cass, Q.B. Chemical Secondary Metabolite Profiling of *Bauhinia longifolia* Ethanolic Leaves Extracts. *Ind. Crops Prod.* **2019**, *132*, 59–68. [CrossRef]

23.   Engels, C.; Gräter, D.; Esquivel, P.; Jiménez, V.M.; Gänzle, M.G.; Schieber, A. Characterization of Phenolic Compounds in Jocote (*Spondias Purpurea* L.) Peels by Ultra High-Performance Liquid Chromatography/electrospray Ionization Mass Spectrometry. *Food Res. Int.* **2012**, *46*, 557–562. [CrossRef]

24.   de Oliveira Ribeiro, L.; Conrado Thomaz, G.F.; de Brito, M.; de Figueiredo, N.; Przytyk Jung, E.; Norie Kunigami, C. Siriguela Peels Provide Antioxidant Compounds-Rich Extract by Solid–liquid Extraction. *J. Food Process. Preserv.* **2020**, *44*, e14719. [CrossRef]

25.   Jung, E.P.; Conrado Thomaz, G.F.; de Brito, M.O.; de Figueiredo, N.G.; Kunigami, C.N.; de Oliveira Ribeiro, L.; Alves Moreira, R.F. Thermal-Assisted Recovery of Antioxidant Compounds from *Bauhinia Forficata* Leaves: Effect of Operational Conditions. *J. Appl. Res. Med. Aromat. Plants* **2021**, *22*, 100303. [CrossRef]

26.   Duarte-Almeida, J.M.; Negri, G.; Salatino, A. Volatile Oils in Leaves of Bauhinia (Fabaceae Caesalpinioideae). *Biochem. Syst. Ecol.* **2004**, *32*, 747–753. [CrossRef]

27.   Sartorilli, P.; Correa, D.S. Constituents of Essential Oil from *Bauhinia Forficata* Link. *J. Essent. Oil Res.* **2007**, *19*, 468–469. [CrossRef]

28. Butnariu, M. Plants as Source of Essential Oils and Perfumery Applications. In *Bioprospecting of Plant Biodiversity for Industrial Molecules*; John Wiley & Sons Ltd.: Hoboken, NJ, USA, 2021; pp. 261–292.

29. Nascimento, K.F.; Moreira, F.M.F.; Alencar Santos, J.; Kassuya, C.A.L.; Croda, J.H.R.; Cardoso, C.A.L.; do Carmo Vieira, M.; Góis Ruiz, A.L.T.; Ann Foglio, M.; de Carvalho, J.E.; et al. Antioxidant, Anti-Inflammatory, Antiproliferative and Antimycobacterial Activities of the Essential Oil of *Psidium Guineense* Sw. and Spathulenol. *J. Ethnopharmacol.* **2018**, *210*, 351–358. [CrossRef]

30. Zellner, B.D.; Amorim, A.C.L.; Miranda, A.L.P.; Alves, R.J.V.; Barbosa, J.P.; Costa, G.L.; Rezende, C.M. Screening of the odour-activity and bioactivity of the essential oils of leaves and flowers of *Hyptis Passerina* Mart. from the Brazilian Cerrado. *J. Braz. Chem. Soc.* **2009**, *20*, 322–332. [CrossRef]

31. Eyres, G.; Dufour, J.-P. 22-Hop Essential Oil: Analysis, Chemical Composition and Odor Characteristics. In *Beer in Health and Disease Prevention*; Preedy, V.R., Ed.; Academic Press: San Diego, CA, USA, 2009; pp. 239–254. ISBN 978-0-12-373891-2.

32. He, Z.; Fan, W.; Xu, Y.; He, S.; Liu, X. Aroma Profile of *Folium Isatidis* Leaf as a Raw Material of Making Bingqu Chixiang Aroma- and Flavor-Type Baijiu. In *Sex, Smoke, and Spirits: The Role of Chemistry*; ACS Symposium Series; American Chemical Society: Washington, DC, USA, 2019; Volume 1321, pp. 16–263. ISBN 9780841234673.

33. Chavan, M.J.; Wakte, P.S.; Shinde, D.B. Analgesic and Anti-Inflammatory Activity of Caryophyllene Oxide from *Annona Squamosa* L. Bark. *Phytomedicine Int. J. Phyther. Phytopharm.* **2010**, *17*, 149–151. [CrossRef]

34. Yang, D.; Michel, L.; Chaumont, J.-P.; Millet-Clerc, J. Use of Caryophyllene Oxide as an Antifungal Agent in an in Vitro Experimental Model of Onychomycosis. *Mycopathologia* **2000**, *148*, 79–82. [CrossRef]

35. Kotseridis, Y.; Baumes, R. Identification of Impact Odorants in Bordeaux Red Grape Juice, in the Commercial Yeast Used for Its Fermentation, and in the Produced Wine. *J. Agric. Food Chem.* **2000**, *48*, 400–406. [CrossRef] [PubMed]

36. Ong, P.K.C.; Acree, T.E. Gas Chromatography/Olfactory Analysis of Lychee (*Litchi Chinesis* Sonn.). *J. Agric. Food Chem.* **1998**, *46*, 2282–2286. [CrossRef]

37. Mahattanatawee, K.; Rouseff, R.; Valim, M.F.; Naim, M. Identification and Aroma Impact of Norisoprenoids in Orange Juice. *J. Agric. Food Chem.* **2005**, *53*, 393–397. [CrossRef]

38. Rozentale, I.; Yan Lun, A.; Zacs, D.; Bartkevics, V. The Occurrence of Polycyclic Aromatic Hydrocarbons in Dried Herbs and Spices. *Food Control* **2018**, *83*, 45–53. [CrossRef]

39. Di Bella, G.; Ben Mansour, H.; Ben Tekaya, A.; Beltifa, A.; Potortì, A.G.; Saiya, E.; Bartolomeo, G.; Dugo, G.; Lo Turco, V. Plasticizers and BPA Residues in Tunisian and Italian Culinary Herbs and Spices. *J. Food Sci.* **2018**, *83*, 1769–1774. [CrossRef]

40. Lo Turco, V.; Potortì, A.G.; Ben Mansour, H.; Dugo, G.; Di Bella, G. Plasticizers and BPA in Spices and Aromatic Herbs of Mediterranean Areas. *Nat. Prod. Res.* **2020**, *34*, 87–92. [CrossRef]

41. Rohn, S.; Rawel, H.M.; Kroll, J. Inhibitory Effects of Plant Phenols on the Activity of Selected Enzymes. *J. Agric. Food Chem.* **2002**, *50*, 3566–3571. [CrossRef] [PubMed]

42. Papoutsis, K.; Zhang, J.; Bowyer, M.C.; Brunton, N.; Gibney, E.R.; Lyng, J. Fruit, Vegetables, and Mushrooms for the Preparation of Extracts with α-Amylase and α-Glucosidase Inhibition Properties: A Review. *Food Chem.* **2021**, *338*, 128119. [CrossRef] [PubMed]

43. Marmitt, D.J.; Bitencourt, S.; Silva, A.d.C.e.; Rempel, C.; Goettert, M.I. The Healing Properties of Medicinal Plants Used in the Brazilian Public Health System: A Systematic Review. *J. Wound Care* **2018**, *27*, S4–S13. [CrossRef]

44. Franco, R.R.; Mota Alves, V.H.; Ribeiro Zabisky, L.F.; Justino, A.B.; Martins, M.M.; Saraiva, A.L.; Goulart, L.R.; Espindola, F.S. Antidiabetic Potential of *Bauhinia Forficata* Link Leaves: A Non-Cytotoxic Source of Lipase and Glycoside Hydrolases Inhibitors and Molecules with Antioxidant and Antiglycation Properties. *Biomed. Pharmacother.* **2020**, *123*, 109798. [CrossRef]

45. Gastaldi, B.; Marino, G.; Assef, Y.; Silva Sofrás, F.M.; Catalán, C.A.N.; González, S.B. Nutraceutical Properties of Herbal Infusions from Six Native Plants of Argentine Patagonia. *Plant Foods Hum. Nutr.* **2018**, *73*, 180–188. [CrossRef] [PubMed]

46. Singleton, V.L.; Rossi, J.A. Colorimetry of Total Phenolics with Phosphomolybdic-Phosphotungstic Acid Reagents. *Am. J. Enol. Vitic.* **1965**, *16*, 144.

47. Zhishen, J.; Mengcheng, T.; Jianming, W. The Determination of Flavonoid Contents in Mulberry and Their Scavenging Effects on Superoxide Radicals. *Food Chem.* **1999**, *64*, 555–559. [CrossRef]

48. Gião, M.S.; González-Sanjosé, M.L.; Rivero-Pérez, M.D.; Pereira, C.I.; Pintado, M.E.; Malcata, F.X. Infusions of Portuguese Medicinal Plants: Dependence of Final Antioxidant Capacity and Phenol Content on Extraction Features. *J. Sci. Food Agric.* **2007**, *87*, 2638–2647. [CrossRef] [PubMed]

49. Hidalgo, M.; Sánchez-moreno, C.; Pascual-teresa, S. De Flavonoid–Flavonoid Interaction and Its Effect on Their Antioxidant Activity. *Food Chem.* **2010**, *121*, 691–696. [CrossRef]

50. Benzie, I.F.F.; Strain, J.J. The Ferric Reducing Ability of Plasma (FRAP) as a Measure of Antioxidant Power: The FRAP Assay. *Anal. Biochem.* **1996**, *76*, 70–76. [CrossRef] [PubMed]

51. Wang, C.; Zhang, W.; Li, H.; Mao, J.; Guo, C.; Ding, R.; Wang, Y.; Fang, L.; Chen, Z.; Yang, G. Analysis of Volatile Compounds in Pears by HS-SPME-GC×GC-TOFMS. *Molecules* **2019**, *24*, 1795. [CrossRef] [PubMed]

52. Adams, R.P. *Identification of Essential Oil Components by Gas Chrochromatography/Quadrupole Mass Spectroscopy*; Allured Publishing Corporation: Carol Stream, IL, USA, 2001.

53. Meng, Y.; Su, A.; Yuan, S.; Zhao, H.; Tan, S.; Hu, C.; Deng, H.; Guo, Y. Evaluation of Total Flavonoids, Myricetin, and Quercetin from *Hovenia Dulcis* Thunb. As Inhibitors of α-Amylase and α-Glucosidase. *Plant Foods Hum. Nutr.* **2016**, *71*, 444–449. [CrossRef]

molecules

Article

New Benzil and Isoflavone Derivatives with Cytotoxic and NO Production Inhibitory Activities from *Placolobium vietnamense*

Lien T. M. Do [1], Tuyet T. N. Huynh [1,2] and Jirapast Sichaem [3,*]

[1]   Institute of Environment-Energy Technology, Sai Gon University, Ho Chi Minh City 748355, Vietnam; liendo.ieet@sgu.edu.vn (L.T.M.D.); lamhuynh1579@gmail.com (T.T.N.H.)
[2]   Lu Gia Secondary School, Distr. 11, Ho Chi Minh City 70000, Vietnam
[3]   Research Unit in Natural Products Chemistry and Bioactivities, Faculty of Science and Technology, Thammasat University Lampang Campus, Lampang 52190, Thailand
*   Correspondence: jirapast@tu.ac.th; Tel.: +66-54237986

Abstract: The phytochemical investigation of *Placolobium vietnamense* stems led to the isolation of a new isoflavone derivative (**1**) and three new benzil derivatives (**2–4**), together with four known pyranoisoflavones (**5–8**). The structures of all isolated compounds were determined on the basis of extensive spectroscopic analyses, including NMR and HRMS spectral data, as well as comparison of their spectroscopic data with those reported in the literature. The cytotoxicity of all isolated compounds was assessed against the human liver hepatocellular carcinoma (Hep G2) cell line, and compound **1** displayed the most significant cytotoxicity with an IC_{50} value of 8.0 µM. Furthermore, all isolated compounds were also tested for their inhibitory activity against NO production in RAW 264.7 macrophages. Of these, compound **1** exhibited the strongest inhibitory efficacy against the LPS-induced NO production with the IC_{50} value of 13.7 µM.

Keywords: *Placolobium vietnamense*; placovinones A–D; benzil and isoflavone derivatives; cytotoxicity; NO production inhibition

Citation: Do, L.T.M.; Huynh, T.T.N.; Sichaem, J. New Benzil and Isoflavone Derivatives with Cytotoxic and NO Production Inhibitory Activities from *Placolobium vietnamense*. *Molecules* **2022**, *27*, 4624. https://doi.org/10.3390/molecules27144624

Academic Editor: Jacqueline Aparecida Takahashi

Received: 6 June 2022
Accepted: 12 July 2022
Published: 20 July 2022

Publisher's Note: MDPI stays neutral with regard to jurisdictional claims in published maps and institutional affiliations.

1. Introduction

Placolobium is a genus of plants in the family Fabaceae, which contains three accepted species. These are distributed throughout the world's tropical regions, some extending into temperate zones, especially in East Asia [1]. *Placolobium vietnamense* N.D.Khoi & Yakovlev is an indigenous plant species, known in Vietnam as 'Rang Rang'. It is a perennial tree with a straight, cylindrical trunk, and brown bark. The fruit is a small pod with a single seed. This plant is used as a folk remedy for snakebites, debility, and to increase strength after childbirth [1]. There has only been one investigation into the chemical constituents of *P. vietnamense* [1]. Previously, our group reported the isolation and structure elucidation of six isoflavonoids, including afrormosin, cladrastin, 8-*O*-methylretusin, millesianin C, barbigerone, and durallone from the EtOAc stem extract of this plant, together with their cytotoxicity. Encouraged by structurally diverse bioactive compounds from *Placolobium* species [2], the aim of this investigation is to revisit *P. vietnamense* in order to search for new bioactive compounds. We report herein the isolation and characterization of benzil and isoflavone derivatives from the stems of *P. vietnamense*. All isolated compounds were assessed for their cytotoxicity against human liver hepatocellular carcinoma (Hep G2) cell line, which is one of the most fatal cancers and has spread to the liver from other organs. Additionally, the inhibitory activity toward NO production in RAW 264.7 macrophages of all isolated compounds was also evaluated.

2. Results and Discussion

2.1. Structural Elucidation of the Isolated Compounds

Chromatographic separation of benzil and isoflavone derivatives from *P. vietnamense* stems allowed for the isolation of eight compounds, including a new isoflavone derivative, placovinone A (**1**), and three new benzil derivatives, placovinones B-D (**2–4**), along with four known pyranoisoflavones (**5–8**) (Figure 1). The structures of all isolated compounds were elucidated based on NMR and HRMS spectral data, as well as a comprehensive comparison of their spectroscopic and physical data with values from the published literature. The known isolated pyranoisoflavones were characterized as ichthynone (**5**) [3], durmillone (**6**) [4], calopogoniumisoflavone B (**7**) [5], and 4′,5′-dimethoxy-6,6-dimethylpyranoisoflavone (**8**) [6].

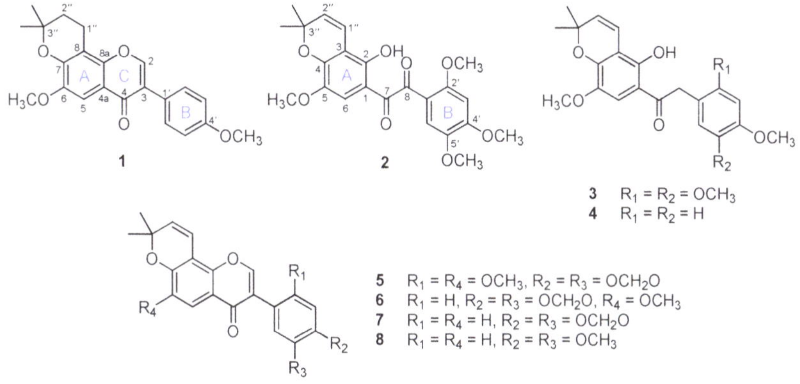

Figure 1. Chemical structures of **1–8**.

Compound **1** was isolated as a colorless gum. The HRESIMS revealed a protonated molecular ion peak at *m/z* 367.1549 [M + H]$^+$ (calcd for $C_{22}H_{23}O_5$ 367.1545) corresponding to the formula $C_{22}H_{22}O_5$. The ^{1}H NMR signal at δ_H 8.42 (s, H-2) and ^{13}C NMR signal at δ_C 152.7 (C-2) were characteristic of the isoflavone skeleton [7]. The existence of AA′BB′ spin-system indicated *para*-substituted B-ring. The presence of a 2,2-dimethyldihydropyrano [8] and two methoxy substituents was identified from the ^{1}H and ^{13}C NMR spectral data (Table 1). A singlet resonance at δ_H 7.32 was assigned to the aromatic proton H-5 on the basis of the long-range coupling to C-4 (δ_C 174.4), C-7 (δ_C 148.3), and C-8a (δ_C 109.9), observed in the HMBC spectrum (Figure 2). The methoxy group δ_H 3.84 (s) was assigned as 6-OCH$_3$ according to the HMBC correlation between 6-OCH$_3$ and C-6 (δ_C 147.1). The long-range correlations observed in the HMBC spectrum of H-1″ (δ_H 2.87, t, *J* = 6.5 Hz) to C-7 and C-8a were key correlations that revealed the position of 2,2-dimethyldihydropyrano moiety was fused to C-7 and C-8, with the anticipated oxygenation at C-7 being supported by the HMBC correlation from H-5 to C-7. Its 1D and 2D NMR spectral data were similar to those of 6-methoxycalopogonlum isoflavone A [9], except for the replacement of a double bond at C-1″ and C-2″ of the 2,2-dimethylpyrano substituent in 6-methoxycalopogonlum isoflavone A by a C-C single bond in **1**. Based on the above spectral evidence, the structure of **1** was established and trivially named as placovinone A.

Table 1. ^{1}H (500 MHz) and ^{13}C (125 MHz) NMR spectroscopic data of **1** recorded in DMSO-d_6 (δ in ppm).

Position	δ_H (*J* in Hz)	δ_C	Position	δ_H (*J* in Hz)	δ_C
2	8.42, s	152.7	3′	6.99, d (8.3)	113.6
3		122.8	4′		158.9
4		174.4	5′	6.99, d (8.3)	113.6
4a		115.9	6′	7.52, d (8.3)	130.0
5	7.32, s	101.8	1″	2.87, t (6.5)	16.4
6		147.1	2″	1.87, t (6.5)	30.6
7		148.3	3″		75.9
8		109.9	4″	1.35, s	26.3
8a		149.5	5″	1.35, s	26.3
1′		124.4	6-OCH$_3$	3.84, s	55.1
2′	7.52, d (8.3)	130.0	4′-OCH$_3$	3.79, s	55.5

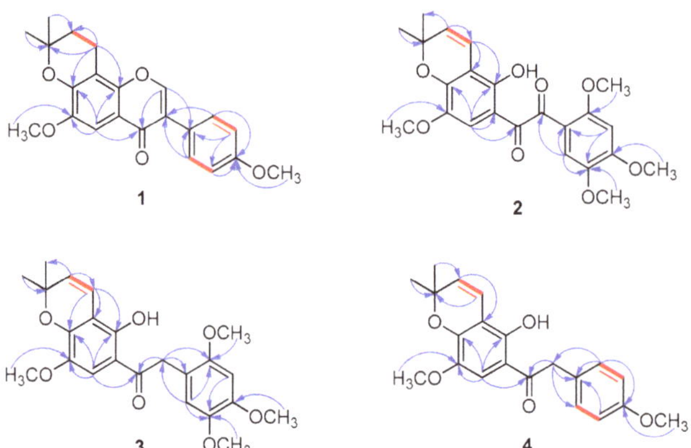

Figure 2. Key COSY (red bold line) and HMBC (blue arrow) correlations of **1–4**.

Compound **2** was obtained as a white amorphous powder. Its molecular formula was determined to be C$_{23}$H$_{24}$O$_8$ based on a protonated molecular ion peak at *m/z* 429.1566 (calcd for C$_{23}$H$_{25}$O$_8$ 429.1549). The signal of a hydroxyl group at δ_H 10.11 (s, 2-OH) in the ^{1}H NMR spectrum, together with those of two carbonyl groups at δ_C 190.7 (C-7) and 191.4 (C-8) in the ^{13}C NMR spectrum, indicated that **2** was a derivative of 1,2-diphenyl-1,2-ethanedione [10]. The ^{1}H and ^{13}C NMR spectral data (Table 2) further revealed the presence of a 2,2-dimethylpyrano fragment and four methoxy substituents. In the ^{1}H NMR spectrum, two singlet protons at δ_H 6.76 and 7.41, were assigned to the two *para*-positioned aromatic protons H-3′ and H-6′ of the B-ring [11], indicating the B-ring of **2** with 2′,4′,5′-trimethoxy substituent. This was also supported by the strong correlations in the HMBC spectrum (Figure 2). The singlet of the aromatic proton at δ_H 7.24 was identified as H-6 on the basis of the HMBC correlations from H-6 to C-1 (δ_C 112.5), C-2 (δ_C 149.7), C-4 (δ_C 148.2), and C-7 (δ_C 190.7). Consequently, the remaining methoxy group (δ_H 3.84, s) was located at C-5, confirmed by the key HMBC correlation between 5-OCH$_3$ and C-5 (δ_C 142.5). Hence the location of the 2,2-dimethylpyrano moiety was found to be at C-3 (δ_C 108.9) and C-4, with the anticipated oxygenation at C-4 being confirmed by the HMBC correlation from H-6 to C-4. A careful comparison of the ^{1}H and ^{13}C NMR spectral data (Table 2) of **2** with dielsianone [12] identified similar signals, distinguished by the presence of two methoxy groups at C-2′ and C-5′. The existence of these two methoxy substituents was confirmed by the HMBC correlations from 2′-OCH$_3$ (δ_H 3.33, s) and 5′-OCH$_3$ (δ_H 3.89, s) to C-2′ (δ_C 156.8) and C-5′ (δ_C 155.6), respectively (Figure 2). From the aforementioned results, the structure of **2** was identified and named as placovinone B.

Table 2. ^{1}H (600 MHz) and ^{13}C (125 MHz) NMR spectroscopic data of **2–4** recorded in DMSO-d_6 (δ in ppm).

Position	2		3		4	
	δ_H (*J* in Hz)	δ_C	δ_H (*J* in Hz)	δ_C	δ_H (*J* in Hz)	δ_C
1		112.5		110.8		110.8
2		149.7		153.6		154.1
3		108.9		109.3		109.5
4		148.2		149.1		149.5
5		142.5		140.8		141.1
6	7.24, s	108.7	7.43, s	113.0	7.41, s	113.7
7		190.7		202.9		203.3
8		191.4	4.20, s	38.8	4.26, s	43.6
1′		114.8		114.2		126.9
2′		156.8		151.3	7.21, d (8.7)	130.7
3′	6.76, s	98.0	6.70, s	98.4	6.88, d (8.7)	114.1
4′		143.6		142.5		158.2
5′		155.6		140.8	6.88, d (8.7)	114.1
6′	7.41, s	110.0	6.83, s	115.7	7.21, d (8.7)	130.7
1″	6.55, d (9.9)	116.0	6.58, d (9.9)	115.1	6.56, d (9.6)	113.7
2″	5.63, d (9.9)	130.0	5.75, d (9.9)	129.1	6.74, d (9.6)	129.3
3″		76.8		77.8		78.0
4″	0.92, s	26.3	1.14, s	27.4	1.39, s	27.9
5″	0.92, s	26.3	1.14, s	27.4	1.39, s	27.9
5-OCH$_3$	3.84, s	56.1	3.76, s	56.1	3.76, s	56.5
2′-OCH$_3$	3.33, s	56.7	3.73, s	56.2		
4′-OCH$_3$	3.55, s	55.9	3.67, s	56.3	3.47, s	55.2
5′-OCH$_3$	3.89, s	56.1	3.79, s	55.8		
2-OH	10.11, s		12.76, s		12.76, s	

Compound **3** was isolated as a white amorphous powder. Its molecular formula, $C_{23}H_{26}O_7$, was determined from its protonated molecular ion peak at *m/z* 415.1759 $[M + H]^+$ (calcd for $C_{23}H_{27}O_7$ 415.1757). This was further confirmed by the ^{13}C NMR spectral data, which disclosed one methylene, two methyl, two olefinic, three aromatic methine, four methoxy, and ten quaternary carbons. The spectroscopic ^{1}H and ^{13}C NMR patterns of **3** (Table 2) were very similar to those of **2**, with the only difference being that the keto carbonyl group at C-8 in **2** (δ_C 191.4) was replaced by a methylene substituent in **3**. This deduction was supported by the HMBC correlations from H-8 (δ_H 4.20, s) to C-7 (δ_C 202.9) and C-1′ (δ_C 114.2). Based on the above spectral evidence, compound **3** was identified and named placovinone C.

Compound **4** was obtained as a white amorphous powder. The molecular formula $C_{21}H_{22}O_5$ was obtained from its HRESIMS, which showed a protonated molecular ion peak at *m/z* 355.1553 $[M + H]^+$ (calcd for $C_{21}H_{23}O_5$ 355.1545). ^{13}C NMR and HSQC spectra of **4** indicated 21 signals, including one carbonyl, one methylene, two methyl, two methoxy, seven methine, and eight quaternary carbons. Two signals at δ_H 7.21 (d, *J* = 8.7 Hz, H-2′, 6′) and 6.88 (d, *J* = 8.7 Hz, H-3′, 5′) appearing as an AA′BB′ type confirmed the presence of a simple *para*-substituted B-ring, with a methoxy group (δ_H 3.47, s) being positioned at C-4′ (δ_C 158.2). The careful comparison of the ^{1}H and ^{13}C NMR spectral data (Table 2) of **4** was shown to be similar to those of **3**, differing only in the absence of two methoxy groups at C-2′ (δ_C 130.7) and C-5′ (δ_C 114.1) on the B-ring of **4**, which was supported by the COSY and HMBC correlations (Figure 2). On the basis of these spectral data, the structure of **4** was unambiguously established and named as placovinone D.

2.2. Cytotoxicity

The cytotoxicity of each isolated compound against Hep G2 cell line was assessed [13–15] and the IC_{50} values are listed in Table 3. Compounds **1–8** exhibited different degrees of cytotoxicity toward Hep G2 cell line. Among them, compound **1** exhibited the most significant cytotoxicity against Hep G2 cell line with an IC_{50} value of 8.0 μM. Compounds **2–4** and **8** showed moderate cytotoxicity with the IC_{50} values of 19.8, 22.9, 23.4, and 35.6 μM, respectively, while compounds **5–7** exhibited weak cytotoxicity with the IC_{50} values of 99.1, 71.6, and 66.6 μM, respectively. Based on the above cytotoxic results, the presence of the 2,2-dimethyldihydropyrano ring in the case of **1** might be responsible for enhancing the activity.

Table 3. Cytotoxicity against Hep G2 cells and inhibition of NO production in macrophage RAW 264.7 cells of **1–8**.

Compound	Cytotoxicity (IC_{50}, μM) [a]	NO Production (IC_{50}, μM) [a]
1	8.0 ± 0.2	13.7 ± 0.5
2	19.8 ± 1.5	31.0 ± 0.3
3	22.9 ± 0.5	47.4 ± 0.3
4	23.4 ± 0.5	15.5 ± 0.4
5	99.1 ± 0.9	>100
6	71.6 ± 0.6	>100
7	66.6 ± 0.5	>100
8	35.6 ± 0.3	54.7 ± 0.2
Ellipticine [b]	0.43 ± 0.03	
Celastrol [b]		1.00 ± 0.10

[a] IC_{50} values were expressed as the mean values of three experiments ± SD. [b] Positive control.

2.3. Inhibition of Nitric Oxide Production

To determine the inhibitory effects of the isolated compounds on NO production (Table 3), LPS-stimulated RAW 264.7 cells were treated with various concentrations of tested compounds [16]. Additionally, the viability of RAW 264.7 cells using an MTT assay to avoid the cytotoxic effects of the isolated compounds was evaluated. Among eight isolated compounds, compounds **1** and **4** highly inhibited NO production in RAW 264.7 cells with the IC_{50} values of 13.7 and 15.5 μM, respectively, whereas compounds **2**, **3**, and **8** moderately inhibited NO production with the IC_{50} values of 31.0, 47.4, and 54.7 μM, respectively. Compounds **1** and **4** demonstrated cytotoxicity toward RAW 264.7 cells with the IC_{50} values of 79.2 and 42.6 μM, respectively, while most of the other compounds showed no obvious cytotoxicity (IC_{50} >100 μM). These results demonstrate that the presence of the *para*-substituted B-ring of **1** and **4** might be responsible for inhibiting NO production.

3. Materials and Methods

3.1. General Experimental Procedures

The NMR spectra were recorded on Bruker AvanceNEO 600 MHz and Bruker Avance III™ HD 500 MHz NMR spectrometers in DMSO-d_6 (Merck, Darmstadt, Germany). Optical rotations were measured on a A.KRÜSS Optronic P8000 polarimeter (KRÜSS, Hamburg, Germany). The IR data were obtained with a Jasco 6600 FT-IR spectrometer using an ATR technique (Jasco, Japan). The HRESIMS spectral data were generated with a X500$_R$ QTOF model mass spectrometer (Sciex, Framingham, MA, USA) and Dionex Ultimate 3000 HPLC system hyphenated with a QExactive Hybrid Quadrupole Orbitrap MS (Thermo Fisher Scientific, Waltham, MA, USA). Silica gel 70–230 mesh (Merck) and Sephadex LH-20 gel (GE Healthcare Bio-Sciences AB, Uppsala, Sweden) were used for column chromatography.

3.2. Plant Material

The stems of *P. vietnamense* were collected in Dak Nong province, Vietnam, in February 2017. The plant material was identified by botanist Vo Van Chi (former lecturer at the University of Medicine and Pharmacy, Ho Chi Minh City, Vietnam). A voucher specimen (No. SGU-A001) has been deposited in the Herbarium of the Laboratory of Chemistry-Biology-Environment, Sai Gon University, Ho Chi Minh City, Vietnam.

3.3. Extraction and Isolation

The air-dried *P. vietnamense* stems (23 kg) were powdered prior to being extracted with 95% EtOH (45 L × 5) at room temperature. The filtered solution was concentrated in vacuo to afford EtOH crude extract (1200 g). This crude extract was suspended in water and partitioned with *n*-hexane and then EtOAc to yield *n*-hexane (271.2 g) and EtOAc (301.3 g) extracts, respectively. The *n*-hexane extract was subjected to silica gel column chromatography (CC) and eluted with *n*-hexane–EtOAc (9:1–0:10, *v/v*) and then EtOAc–MeOH (10:0–0:10, *v/v*). Based on their TLC behavior, the eluted fractions were grouped into fractions HEX.1–HEX.7. Fraction HEX.4 (34.5 g) was subjected to further silica gel CC and eluted with *n*-hexane–EtOAc (8:2, *v/v*) to give subfractions HEX.4.1–HEX.4.8. Subfraction HEX.4.1 (3.0 g) was subjected to silica gel CC and eluted with *n*-hexane–EtOAc (85:15, *v/v*) to yield **3** (7.0 mg), **5** (8.0 mg), and **6** (9.7 mg). Subfraction HEX.4.2 (0.9 g) was further purified using silica gel CC and eluted with *n*-hexane–EtOAc (8:2, *v/v*) to yield **2** (6.5 mg), **7** (6.4 mg), and **8** (11.4 mg). Subfraction HEX.4.3 (1.1 g) was selected for further purification using Sephadex LH-20 gel CC and eluted with MeOH to afford **1** (5.8 mg) and **4** (6.4 mg).

Placovinone A (**1**). Colorless gum. UV (CH$_3$OH) λ_{max} (log ε) 210 (4.49), 231 (4.25), 278 (4.81), 334 (3.47) nm; IR (ATR) ν_{max} 2975, 1718, 1619, 1457, 1343, 1279, 1203, 1150, 1013, 757 cm^{-1}; HRESIMS *m/z* 367.1549 [M + H]$^+$ (calcd for C$_{22}$H$_{23}$O$_5$ 367.1545); ^{1}H NMR (DMSO-d_6, 500 MHz) and ^{13}C NMR (DMSO-d_6, 125 MHz) see Table 1.

Placovinone B (**2**). White amorphous powder. UV (CH$_3$OH) λ_{max} (log ε) 250 (4.39), 270 (4.72), 296 (4.30), 337 (3.18) nm; IR (ATR) ν_{max} 3392, 2977, 2904, 1713, 1635, 1451, 1372, 1288, 1246, 900 cm^{-1}; HRESIMS *m/z* 429.1566 [M + H]$^+$ (calcd for C$_{23}$H$_{25}$O$_8$ 429.1549); ^{1}H NMR (DMSO-d_6, 600 MHz) and ^{13}C NMR (DMSO-d_6, 125 MHz) see Table 2.

Placovinone C (**3**). White amorphous powder. UV (CH$_3$OH) λ_{max} (log ε) 205 (4.07), 272 (4.87), 339 (2.98) nm; IR (ATR) ν_{max} 3394, 2977, 2889, 1710, 1642, 1447, 1333, 1289, 1216, 763 cm^{-1}; HRESIMS *m/z* 415.1759 [M + H]$^+$ (calcd for C$_{23}$H$_{27}$O$_7$ 415.1757); ^{1}H NMR (DMSO-d_6, 600 MHz) and ^{13}C NMR (DMSO-d_6, 125 MHz) see Table 2.

Placovinone D (**4**). White amorphous powder. UV (CH$_3$OH) λ_{max} (log ε) 205 (4.11), 270 (4.87), 333 (3.06) nm; IR (ATR) ν_{max} 3395, 2977, 2896, 1712, 1643, 1448, 1339, 1287, 1218, 763 cm^{-1}; HRESIMS *m/z* 355.1553 [M + H]$^+$ (calcd for C$_{21}$H$_{23}$O$_5$ 355.1545); ^{1}H NMR (DMSO-d_6, 600 MHz) and ^{13}C NMR (DMSO-d_6, 125 MHz) see Table 2.

3.4. Cytotoxicity Assay

According to a previous procedure [17], the cytotoxic evaluation of **1–8** against the growth of human hepatocellular carcinoma (Hep G2) cell line was carried out. The positive control was ellipticine, a powerful anticancer medication with various modes of action. The cancer cells were grown in Dulbecco's Modified Essential Medium (DMEM) at 37 °C in a 5 % CO$_2$ environment with 10% fetal bovine serum (FBS), 1% penicillin and streptomycin, and 1% L-glutamine. The investigated compounds were added at concentrations ranging from 0.5 to 128 µg/mL by dissolving in DMSO (20 mg/mL), and the incubation was carried out once more for 72 h under the same conditions. Following the procedure, an MTT solution (10 µL, 5 mg/mL) was added to each well. The percentage of cell viability vs. sample concentration was plotted using SigmaPlot 10 (Systat Software Inc., San Jose, CA, USA) to calculate the IC$_{50}$ values.

3.5. Inhibition of Nitric Oxide Production Assay
3.5.1. Cell Culture

RAW 264.7 cells were stocked in Dulbecco's Modified Essential and grown at the condition of 37 °C in DMEM supplemented with 10% heat-inactivated FBS, streptomycin sulfate (100 µg/mL), and penicillin (100 units/mL) in a humidified environment of 5% CO$_2$. The RAW 264.7 cells were pre-incubated every two days.

3.5.2. Cell Viability Assay on RAW 264.7 Cells

The cell viability assay was used to determine the cytotoxic effect of the isolated compounds on RAW 264.7 cells. At a density of 1×10^5 cells per well, RAW 264.7 cells were seeded on a 96-well plate and allowed to adhere for 4 h. Then, the cells were treated with 0.5% DMSO, celastrol, and isolated compounds at the indicated concentrations. Celastrol was used as a positive control [16]. After incubating 24 h, the viable cells were measured with a colorimetric assay based on the mitochondria's ability in viable cells to reduce MTT [18]. The viability cells were treated with vehicle only and were defined as 100% viable. $[OD_{570}$ (treated cell culture) $\times 100]/OD_{570}$ was the formula used to determine the percentage of macrophage surviving cells after treatment (vehicle control).

3.5.3. Measurement of Nitric Oxide (NO) Production

The RAW 264.7 cells were stimulated with or without 1 µg/mL of LPS (lipopolysaccharide), which was purchased from Sigma Chemical Co. (St. Louis, MO, USA), for 24 h with or without 0.5% DMSO, celastrol, and isolated compounds at the indicated concentrations. The culture supernatant (100 µL) was then reacted with 100 µL of Griess reagent [16]. After the Griess assay, the remaining cells were used to screen for their viability using colorimetric assay-MTT (Sigma Chemical Co., St. Louis, MO, USA).

4. Conclusions

In conclusion, we have conducted the successful isolation of eight compounds, including a new isoflavone derivative (**1**) and three new benzil derivatives (**2–4**), together with four known pyranoisoflavones (**5–8**) from *P. vietnamense* stems. To the best of our knowledge, compounds **1–8** were isolated for the first time from the genus *Placolobium*. The biological evaluations showed that **1** exhibited the most significant cytotoxicity toward Hep G2 cell line and the strongest inhibitory activity against the LPS-induced NO production. According to these investigation results, the structure of **1** is a promising candidate and could be used as a template for discovering potential anticancer and anti-inflammatory agents.

Supplementary Materials: The following supporting information can be downloaded at: https://www.mdpi.com/article/10.3390/molecules27144624/s1, Figures S1–S23: HRESIMS, 1D, and 2D NMR spectra of **1–4**.

Author Contributions: Conceptualization, J.S. and L.T.M.D.; methodology, L.T.M.D. and T.T.N.H.; formal analysis, J.S.; data curation, J.S. and L.T.M.D.; writing—original draft preparation, J.S. and L.T.M.D.; funding acquisition, J.S. All authors have read and agreed to the published version of the manuscript.

Funding: This work was supported by Thammasat University Research Unit in Natural Products Chemistry and Bioactivities.

Institutional Review Board Statement: Not applicable.

Informed Consent Statement: Not applicable.

Data Availability Statement: The supporting information can be found in the Supplementary Materials.

Conflicts of Interest: The authors declare no conflict of interest.

Sample Availability: Samples of the compounds are not available from the authors.

References

1. Iakovlev, G.P. Notes on the Systematics of Sweetia Spreng., Machaerium Pers., *Angylocalyx* Taub., *Fedorovia* Yakov., *Placolobium* Miq., and *Ormosia* Jacks. (Fabaceae). *Nov. Sist. Nizshikh Rastenii* **1973**, *10*, 190–196.
2. Lien, D.T.M.; Tuyet, H.T.N.; Quyen, T.H.N.; Truong, T.L.; Hang, N.T.T.; Thao, N.T.N.; Dung, N.T.M. Isoflavones from *Placolobium vietnamense*, an Indigenous Plant of Vietnam. *Vietnam J. Chem.* **2021**, *59*, 7–11.
3. Ingham, J.L.; Tahara, S.; Shibaki, S.; Mizutani, J. Isoflavonoids from the Root Bark of *Piscidia erythrina* and a Note on the Structure of Piscidone. *Z. Nat. C* **1989**, *44*, 905–913. [CrossRef]

4. Ollis, W.D.; Rhodes, C.A.; Sutherland, I.O. The Extractives of *Millettia dura* (Dunn): The Constitutions of Durlettone, Durmillone, Milldurone, Millettone and Millettosin. *Tetrahedron* **1967**, *23*, 4741–4760. [CrossRef]
5. Schuda, P.F.; Price, W.A. Total Synthesis of Isoflavones: Jamaicin, Calopogonium Isoflavone-B, Pseudobaptigenin, and Maxima Substance-B. Friedel-Crafts Acylation Reactions with Acid-Sensitive Substrates. *J. Org. Chem.* **1987**, *52*, 1972–1979. [CrossRef]
6. Ye, H.; Chen, L.; Li, Y.; Peng, A.; Fu, A.; Song, H.; Tang, M.; Luo, H.; Luo, Y.; Xu, Y. Preparative Isolation and Purification of Three Rotenoids and One Isoflavone from the Seeds of *Millettia pachycarpa* Benth by High-Speed Counter-Current Chromatography. *J. Chromatogr. A* **2008**, *1178*, 101–107. [CrossRef] [PubMed]
7. Ito, C.; Itoigawa, M.; Kojima, N.; Tokuda, H.; Hirata, T.; Nishino, H.; Furukawa, H. Chemical Constituents of *Millettia taiwaniana*: Structure Elucidation of Five New Isoflavonoids and Their Cancer Chemopreventive Activity. *J. Nat. Prod.* **2004**, *67*, 1125–1130. [CrossRef] [PubMed]
8. Ngadjui, B.T.; Ngameni, B.; Dongo, E.; Kouam, S.F.; Abegaz, B.M. Prenylated and Geranylated Chalcones and Flavones from the Aerial Parts of *Dorstenia ciliata*. *Bull. Chem. Soc. Ethiop.* **2002**, *16*, 157–164.
9. Yenesew, A.; Midiwo, J.O.; Waterman, P.G. 6-Methoxycalpogonium Isoflavone A: A New Isoflavone from the Seed Pods of *Millettia dura*. *J. Nat. Prod.* **1997**, *60*, 806–807. [CrossRef]
10. Magalhães, A.F.; Tozzi, A.M.G.; Magalhães, E.G.; Souza-Neta, L.C. New Prenylated Metabolites of *Deguelia longeracemosa* and Evaluation of Their Antimicrobial Potential. *Planta Med.* **2006**, *72*, 358–363. [CrossRef] [PubMed]
11. Na, Z.; Fan, Q.-F.; Song, Q.-S.; Hu, H.-B. Three New Flavonoids from *Millettia pachyloba*. *Phytochem. Lett.* **2017**, *19*, 215–219. [CrossRef]
12. Gong, T.; Wang, D.-X.; Chen, R.-Y.; Liu, P.; Yu, D.-Q. Novel Benzil and Isoflavone Derivatives from *Millettia dielsiana*. *Planta Med.* **2009**, *75*, 236–242. [CrossRef] [PubMed]
13. Borradaile, N.M.; Carroll, K.K.; Kurowska, E.M. Regulation of HepG2 Cell Apolipoprotein B Metabolism by the Citrus Flavanones Hesperetin and Naringenin. *Lipids* **1999**, *34*, 591–598. [CrossRef] [PubMed]
14. Zhang, X.; Khalidi, O.; Kim, S.Y.; Wang, R.; Schultz, V.; Cress, B.F.; Gross, R.A.; Koffas, M.A.; Linhardt, R.J. Synthesis and Biological Evaluation of 5,7-Dihydroxyflavanone Derivatives as Antimicrobial Agents. *Bioorg. Med. Chem. Lett.* **2016**, *26*, 3089–3092. [CrossRef] [PubMed]
15. Petlevski, R.; Knežević, T.; Ris, M.; Filipašić, A. Protective Effect of Flavanone-Naringenin on High Glucose-Induced Hepatotoxicity to Hep G2 Cells. *Farm. Glas.* **2017**, *73*, 85–97.
16. Cao, T.Q.; Han, J.H.; Lee, H.-S.; Ha, M.T.; Woo, M.H.; Min, B.S. Anti-Inflammatory and Immunosuppressive Effects of *Panax notoginseng*. *Nat. Prod. Sci.* **2019**, *25*, 317–325. [CrossRef]
17. Nguyen, T.T.; Nguyen, H.L.; Pham, T.N.; Nguyen, P.K.; Huynh, T.T.; Sichaem, J.; Do, L.T. Bougainvinones N–P, Three New Flavonoids from *Bougainvillea spectabilis*. *Fitoterapia* **2021**, *149*, 104832. [CrossRef] [PubMed]
18. Cao, T.Q.; Tran, M.H.; Kim, J.A.; Tran, P.T.; Lee, J.-H.; Woo, M.H.; Lee, H.-K.; Min, B.S. Inhibitory Effects of Compounds from *Styrax obassia* on NO Production. *Bioorg. Med. Chem. Lett.* **2015**, *25*, 5087–5091. [CrossRef] [PubMed]

 *molecules* 

Article

Two Natural Flavonoid Substituted Polysaccharides from *Tamarix chinensis*: Structural Characterization and Anticomplement Activities

Yukun Jiao, Yiting Yang, Lishuang Zhou, Daofeng Chen * and Yan Lu *

School of Pharmacy, Institutes of Integrative Medicine, Fudan University, Shanghai 201203, China;
17111030034@fudan.edu.cn (Y.J.); 17211030024@fudan.edu.cn (Y.Y.); 20111030064@fudan.edu.cn (L.Z.)
* Correspondence: dfchen@shmu.edu.cn (D.C.); luyan@fudan.edu.cn (Y.L.)

Abstract: Two novel natural flavonoid substituted polysaccharides (MBAP-1 and MBAP-2) were obtained from *Tamarix chinensis* Lour. and characterized by HPGPC, methylation, ultra-high-performance liquid chromatography-ion trap tandem mass spectrometry (UPLC-IT-MS^n), and NMR analysis. The results showed that MBAP-1 was a homogenous heteropolysaccharide with a backbone of 4)-β-D-Glc*p*-(1→ and →3,4,6)-β-D-Glc*p*-(1→. MBAP-2 was also a homogenous polysaccharide which possessed a backbone of →3)-α-D-Glc*p*-(1→, →4)-β-D-Glc*p*-(1→ and →3,4)-β-D-Glc*p*-2-OMe-(1→. Both the two polysaccharides were substituted by quercetin and exhibited anticomplement activities in vitro. However, MBAP-1 (CH$_{50}$: 0.075 ± 0.004 mg/mL) was more potent than MBAP-2 (CH$_{50}$: 0.249 ± 0.006 mg/mL) and its reduced product, MBAP-1R (CH$_{50}$: 0.207 ± 0.008 mg/mL), indicating that multiple monosaccharides and uronic acids might contribute to the anticomplement activity of the flavonoid substituted polysaccharides of *T. chinensis*. Furthermore, the antioxidant activity of MBAP-1 was also more potent than that of MBAP-2. In conclusion, these two flavonoid substituted polysaccharides from *T. chinensis* were found to be potential oxidant and complement inhibitors.

Keywords: *Tamarix chinensis* Lour.; flavonoid substituted polysaccharides; structural characterization; anticomplement activity; quercetin

Citation: Jiao, Y.; Yang, Y.; Zhou, L.;
Chen, D.; Lu, Y. Two Natural
Flavonoid Substituted
Polysaccharides from *Tamarix
chinensis*: Structural Characterization
and Anticomplement Activities.
Molecules **2022**, *27*, 4532. https://
doi.org/10.3390/molecules27144532

Academic Editor: Jacqueline
Aparecida Takahashi

Received: 2 July 2022
Accepted: 12 July 2022
Published: 15 July 2022

Publisher's Note: MDPI stays neutral
with regard to jurisdictional claims in
published maps and institutional affiliations.

1. Introduction

The complement system, as the first defense line of human immune system, plays an irreplaceable role in numerous diseases [1]. Our previous study showed that the overactivation of complement system is involved with H1N1 induced pneumonia in mice [2]. Moreover, the significant elevation of peripheral complement has been recognized as a hallmark of respiratory distress syndrome associated with severe sepsis, cytokine storm, and multiple organ failure [3]. The autopsy results of COVID-19 patients also suggested that hyper complementation in plasma resulted in alveolar capillary wall damage with increased vascular permeability and further enhanced release of inflammatory mediators, and then intensified tissue damage [4]. In addition, two complement inhibitors, eculizumab and compstatin, have been suggested as potential treatments for COVID-19 [5]. In our previous research, anticomplement polysaccharides and flavonoids from several medicinal plants could significantly alleviate lung injury and increase the survival rate of H1N1 induced mice, and were non-toxic [2,6,7]. Hence, the medicinal plants provided a new resource of complement inhibitors for the treatment of viral pneumonia.

Tamarix chinensis Lour. (Tamaricaceae) has been widely used to treat rheumatoid arthritis, measles (Chinese Pharmacopoeia, 2020), and measles complicated with pneumonia [8]. Its flavonoids, triterpenoids, organic acids, and volatile oils showed anti-inflammatory, bacteriostatic, antioxidant, and hepatoprotective effects [9]. However, there were no reports about *T. chinensis* polysaccharides. In our preliminary studies, the crude polysaccharides of *T. chinensis*, MBAP90, showed significant anticomplement activity with CH$_{50}$ value of

0.186 ± 0.003 mg/mL. Interestingly, MBAP90 exhibited the characteristic color reaction of flavonoids even after deproteinization and dialysis (cutting off M_W: 5000 Da). The ^{1}H-NMR signals at δ 6.5–8.0 also indicated that MBAP90 might contain flavonoid substituted polysaccharides.

In recent years, numerous methods have been applied to graft flavonoids and polysaccharides [10]. Some synthesized flavonoid-polysaccharide conjugates possessed unique advantages compared with flavonoids or polysaccharides. For example, quercetin-grafted carboxymethyl chitosan was amphiphilic with a low critical micelle concentration and good stability [11]. Quercetin-grafted hyaluronic was designed as tumor cell-targeted pro-drug for its significant intestinal permeability, oral bioactivity and antitumor efficacy [12]. However, there were no reports of natural flavonoid substituted polysaccharides.

To explore anticomplement polysaccharides in *T. chininsis* and their anticomplement activities, MBAP90 was further purified by DEAE-cellulose and Sepharyl S-200, which led to the isolation of two novel homogenous polysaccharides, MBAP-1 and MBAP-2. This paper describes their structural characterization and anticomplement activities. As oxidative stress is vital for inflammatory responses in viral pneumonia, antioxidant activities were also determined herein as well [13].

2. Results

2.1. Isolation and Purification of MBAP-1 and MBAP-2

MBAP-1 and MBAP-2 were isolated from the most potent anticomplement fraction (Fr. 2) of MBAP90, and the yields were 0.14% and 0.61%, respectively. The detailed elution curves are shown in Figure 1A.

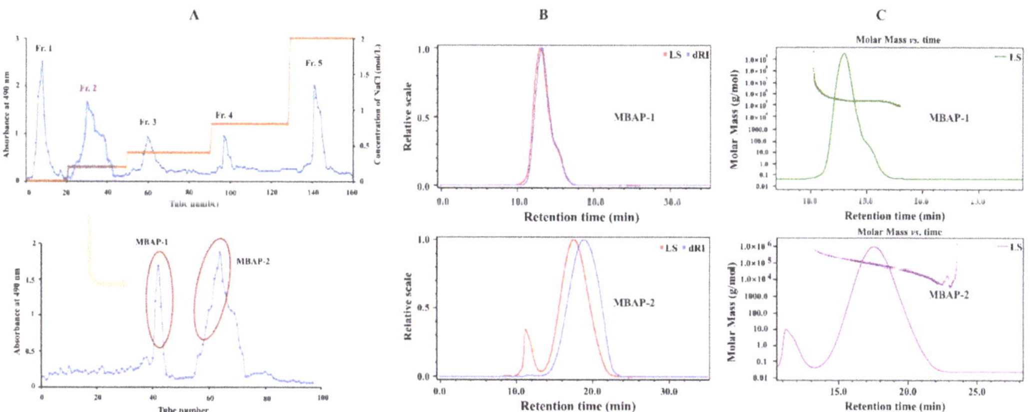

Figure 1. The eluted profiles and HPSEC-MALLS-RI results. (**A**) The eluted profile of MBAP90 on DEAE-52 column and Fr. 2 on Sepharyl S-200 column. (**B**) Superimposed spectra detected using RI and LC at angle of 90° on HPSEC-MALLS-RI. (**C**) Molar mass distribution detected by HPSEC-MALLS-RI.

2.2. Homogeneity and Molecular Weight Assessment

The homogeneity was evaluated by HPGPC-ELSD. As shown in Figure S1 (Supplementary Materials), MBAP-1 and MBAP-2 both showed one single narrow peak, indicating that they were both homogenous. As displayed in Figure 1B,C, the two polysaccharides were both further confirmed to be homogenous using HPSEC-MALLS-RI. Moreover, the results suggested that the relative molecular weights of MBAP-1 and MBAP-2 were 269.3 and 46.5 kDa, respectively. The detailed parameters of MBAP-1 and MBAP-2 were summarized in Table 1.

Table 1. The molecular parameters of MBAP-1 and MBAP-2 determined by SEC-MALLS-RI.

Molecular Characteristics	Parameter	Detection Results	
		MBAP-1	**MBAP-2**
Polydispersity	Mw/Mn	1.01	1.07
	Mz/Mn	1.04	1.05
	Mw	2.693×10^5	4.650×10^4
Molar mass moments (g/mol)	Mn	2.537×10^5	4.611×10^4
	Mz	2.501×10^5	4.520×10^4
	Mp	2.51×10^5	3.91×10^4
Rms radius moments (nm)	Rz	1.5 nm	1.5 nm

2.3. Monosaccharide Composition and Absolute Configuration Analysis

The monosaccharide composition results of MBAP-1 and MBAP-2 are presented in Figure 2A. Obviously, MBAP-1 was composed of glucose, galactose, arabinose, glucuronic acid, and galacturonic acid with the molar ratio of 54.54:4.21:18.18:4.87:4.21. MBAP-2 was mainly consisted of glucose. However, an unknown peak was presented at 24.67 min, which was further identified by GC-MS. The unknown monosaccharide was attributed as 2-*O*-methyl glucose by ion fragments of its alditol acetate (Figure S2). Thus, MBAP-2 was mainly composed of glucose and 2-*O*-methyl glucose with a molar ratio of 88.41:11.59. The *w/w* (%) ratio of each monosaccharide in two polysaccharides is presented in Table 2.

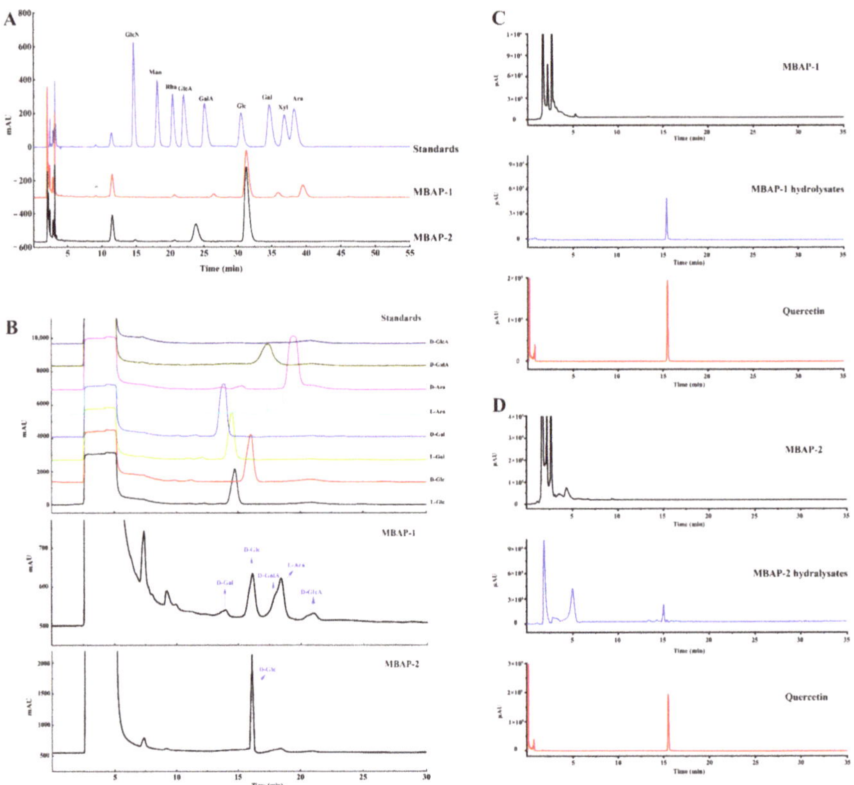

Figure 2. The chromatograms of monosaccharide composition (**A**), monosaccharide absolute configuration analysis (**B**) and UPLC-MS identification results of substituted flavonoids (**C,D**) of MBAP-1 and MBAP-2.

Table 2. The primary chemical characteristics of MBAP-1 and MBAP-2.

Sample	Yield (%)	Total Sugar Content (%)	Uronic Acid (%)	Protein (%)	Flavonoids (%)	Monosaccharide Composition (*w/w*)				
						Glc	GlcA	GalA	Gal	Ara
MBAP-1	0.14	86.06 ± 2.76	9.64 ± 0.56	1.90 ± 0.15	12.03 ± 1.20	65.17 (54.54 [#])	6.27 (4.87 [#])	5.42 (4.21 [#])	5.03 (4.21 [#])	18.10 (18.18 [#])
MBAP-2	0.16	82.03 ± 1.77	-	2.06 ± 0.13	15.96 ± 1.36	88.18 (88.41 [#])	-	-	-	-

[#] The molar ratio of each monosaccharide.

The monosaccharide absolute configurations were also analyzed. As shown in Figure 2B, the monosaccharides of MBAP-1 included D-glucose, D-galactose, L-arabinose, and D-galacturonic acid compared with the standard monosaccharides. Similarly, MBAP-2 consisted of D-glucose and 2-*O*-methyl-D-glucose.

2.4. FT-IR Spectroscopy Assessments

The FT-IR results of MBAP-1 and MBAP-2 are presented in Figure S3. The intense and broad peaks at 3267 and 3307 cm^{-1} indicated the stretching vibrations of hydroxyl groups of MBAP-1 and MBAP-2, respectively. The absorptions at 2924 and 2938 cm^{-1} were assigned to C-H stretching vibration, and the peak at 1593 cm^{-1} in MBAP-1 was due to the asymmetric C=O stretching vibration. The peaks at 1378 and 1405 cm^{-1} and shoulder absorptions were assigned to C-H bending vibration [14]. The typical absorptions of the pyranose ring of polysaccharide presented at 615 or 765 and 632 cm^{-1} [15]. The sharp peaks at 1031, 1078 and 1106 cm^{-1} were stretching vibrations of C-O-C [14].

2.5. Methylation Analysis

Methylation could provide the fundamental glycosidic linkages [16]. The detailed methylation results of MBAP-1, MBAP-1R and MBAP-2 are summarized in Table 3. The uronic acidic residues were confirmed by the comparison of MBAP-1 and MBAP-1R. MBAP-1 was composed of seven kinds of partially methylated alditol acetates (PMAAs): 2-*O*-methyl-D-glucitol, 4,6-di-*O*-methyl-D-glucitol, 2,3-di-*O*-methyl-L-arabinitol, 2,3,5-tri-*O*-methyl-L-arabinitol, 2,6-di-*O*-methyl-D-glucitol, 2,3,6-tri-*O*-methyl-D-galactitol and 2,3,6-tri-*O*-methyl-D-glucitol. While two new PMAAs of 2,3,4,6-tetra-*O*-methyl-D-glucitol and 2,3,4,6-tetra-*O*-methyl-D-galactitol were presented in MBAP-1R, indicating they might be uronic acidic residues. MBAP-2 was composed of 2,4-di-*O*-methyl-D-glucitol, 2,3,4-tri-*O*-methyl-D-glucitol, 2,6-di-*O*-methyl-D-glucitol, 2,3,6-tri-*O*-methyl-D-glucitol, and 2,3,4,6-tetra-*O*-methyl-D-glucitol. More detailed information concerning the residues was confirmed by NMR analysis.

Table 3. The PMAAs results of MBAP-1, MBAP-1R and MBAP-2.

PMAAs	Linkages	Molar Ratio			Major Mass Fragments (*m/z*)
		MBAP-1	MBAP-1R	MBAP-2	
1,3,4,5,6-Penta-*O*-acetyl-1-deuterio-2-*O*-methyl-D-glucitol	1,3,4,6-linked-Glc*p*	1.00	1.00		43, 59, 87, 118, 139, 333
1,2,3,5-Tetra-*O*-acetyl-1-deuterio-4,6-di-*O*-methyl-D-glucitol	1,2,3-linked-Glc*p*	1.03	0.95		43, 59, 71, 87, 101, 129, 161, 202, 262
1,4,5-Tri-*O*-acetyl-1-deuterio-2,3-di-*O*-methyl-L-arabinitol	1,5-linked-Ara*f*	1.87	1.95		43, 59, 87, 102, 118, 129, 189
1,4-Di-*O*-acetyl-1-deuterio-2,3,5-tri-*O*-methyl-L-arabinitol	1-linked-Ara*f*	2.21	2.24		43, 59, 71, 87, 102, 118, 129, 161, 162
1,5-Di-*O*-acetyl-1-deuterio-2,3,4,6-tetra-*O*-methyl-D-galactitol	1-linked-Gal*p*	n.d.	1.12		43, 71, 87, 102, 118, 129, 145, 161, 162, 205
1,3,4,5-Tetra-*O*-acetyl-1-deuterio-2,6-di-*O*-methyl-D-glucitol	1,3,4-linked-Glc*p*	1.98	2.01	0.89	43, 59, 87, 118, 129, 160, 185, 305
1,4,5-Tri-*O*-acetyl-1-deuterio-2,3,6-tri-*O*-methyl-D-galactitol	1,4-linked-Gal*p*	1.03	0.99		43, 59, 71, 87, 102, 118, 129, 162, 233
1,4,5-Tri-*O*-acetyl-1-deuterio-2,3,6-tri-*O*-methyl-D-glucitol	1,4-linked-Glc*p*	8.01	7.89	0.35	43, 59, 71, 87, 102, 118, 129, 162, 233
1,5-Di-*O*-acetyl-1-deuterio-2,3,4,6-tetra-*O*-methyl-D-glucitol	1-linked-Glc*p*	n.d.	1.10	1.20	43, 71, 87, 102, 118, 129, 145, 161, 162, 205
1,3,5,6-Tetra-*O*-acetyl-1-deuterio-2,4-di-*O*-methyl-D-glucitol	1,3,6-linked-Glc*p*			0.38	43, 59, 87, 118, 129, 139, 160, 189, 234, 305
1,5,6-Tri-*O*-acetyl-1-deuterio-2,3,4-tri-*O*-methyl-D-glucitol	1,6-linked-Glc*p*			1.01	43, 59, 87, 99, 101, 118, 129, 162, 189, 233

Note: n.d. means not detected in GC-MS.

2.6. Identification of Substituted Flavonoid with UPLC-IT-MS^{n} and NMR Analysis

The unpredicted signals between δ 6.5–7.5 in ^{1}H-NMR spectra (Figures S6A and S7A) indicated that non-polysaccharide constituents might be existed in MBAP-1 and MBAP-2. Under consideration of their perfect homogeneity, it was speculated that the non-polysaccharide components were linked with the polysaccharide chains. To identify the non-polysaccharide part, UPLC-IT-MS^n was applied for further identification. As shown in Figure 2C, compared with MBAP-1, an obvious peak was presented in its full hydrolysates. The ions of m/z 301.05 [M − H]$^-$ and m/z 303.07 [M + H]$^+$ suggested its relative molecular weight might be 302.04 Da. And typical fragmental ions of m/z 179.00 and 151.00 in negative mode indicated it might be quercetin [17]. Compared with the chromatogram and fragmental ions of quercetin CRS, quercetin was confirmed in MBAP-1. The flavonoid in MBAP-2 was similarly identified to be quercetin as well (Figure 2D). Meanwhile, the contents of quercetin in MBAP-1 and MBAP-2 were 12.03% and 15.96%.

Furthermore, the existence of quercetin in MBAP-1 and MBAP-2 was confirmed by the NMR data. As we all know, the signals of quercetin commonly presented at δ 6.0–8.0 in ^{1}H-NMR spectrum and δ 90–180 in ^{13}C-NMR spectrum [18], most of which were in lower field regions and could be easily distinguished from polysaccharides' signals. The ^{13}C-NMR signals at 177.32 in MBAP-1 (Figure S6B) and 175.57 in MBAP-2 (Figure S7B) were conclusively assigned to carbonyl carbon at C-4 of quercetin. As shown in the grey area of Figures 3 and 4, the signals in lower field mainly belonged to quercetin in the polysaccharides. However, compared with the reported data [18], signals of quercetin herein were slightly shifted, which might be induced by different NMR solvents and spatial effects from polysaccharides. The effect of different solvents on chemical shifts of quercetin was further confirmed by Figure S8. As expected, most of the NMR shifts of quercetin in the mixture of d$_6$-DMSO and D$_2$O (1:1) were higher than the corresponding signals in pure d$_6$-DMSO.

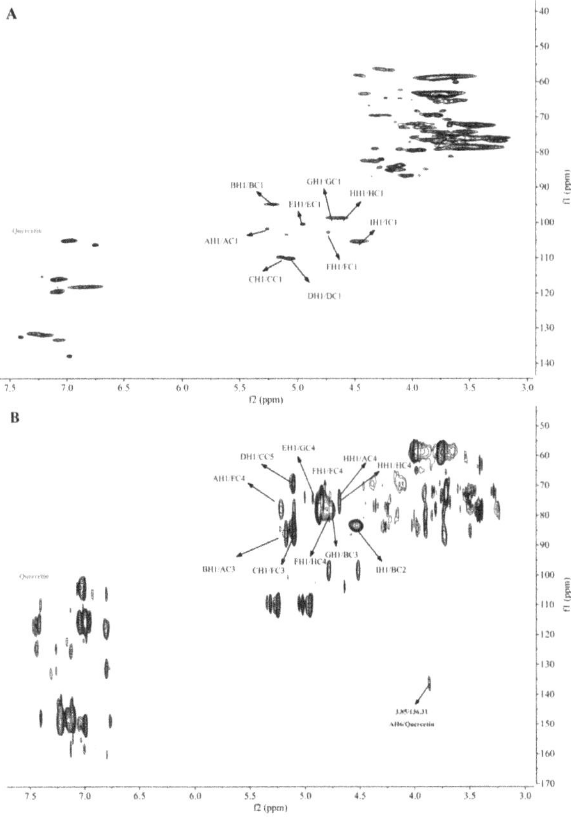

Figure 3. HSQC (**A**) spectrum and HMBC (**B**) spectrum of MBAP-1.

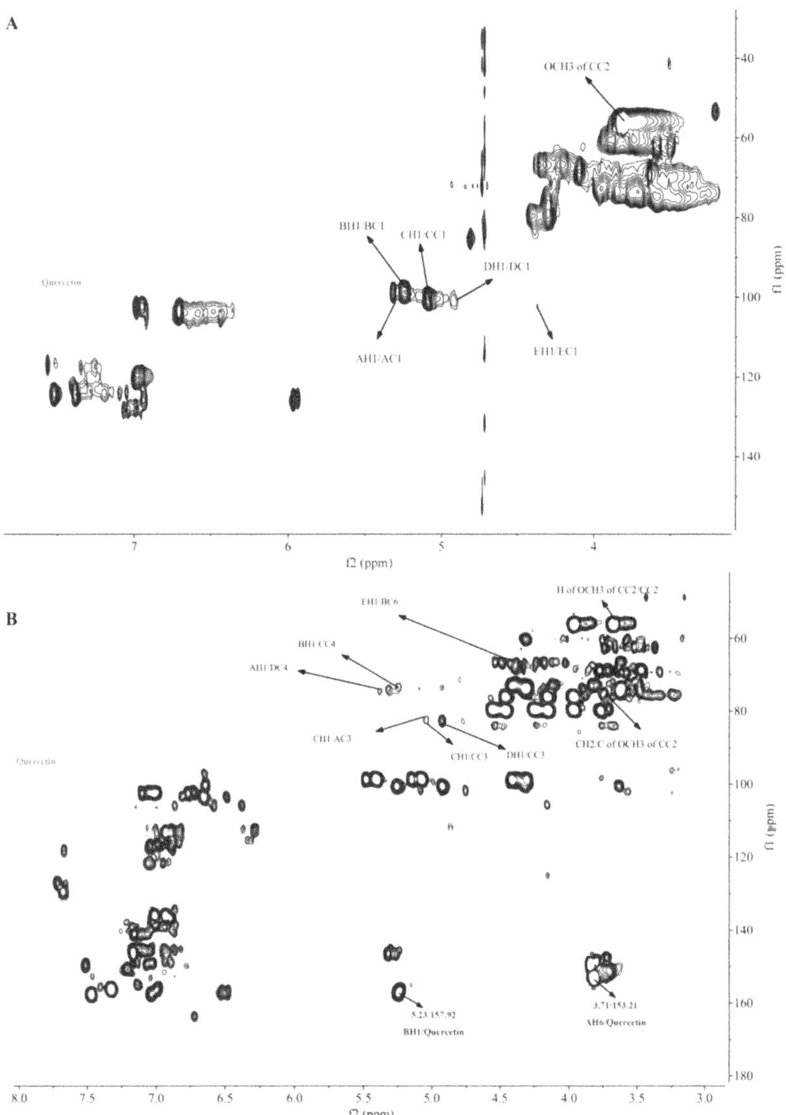

Figure 4. HSQC (**A**) spectrum and HMBC (**B**) spectrum of MBAP-2.

2.7. NMR Analysis of Glycosidic Residues

NMR is very useful in the characterization of anomeric configurations (α- or β-), patterns and sequences of glycosidic linkages. As shown in Figures 3A and S6B, the signals at δ 5.13/109.84 and 5.07/110.23 in HSQC spectrum of MBAP-1 were obviously observed, indicating they might be the anomeric signals of arabinose residues because the anomeric carbon signals of arabinose residues were generally at δ 105–110. GC-MS results indicated MBAP-1 consisted of →5)-α-L-Ara$f$-(1→ and α-L-Ara$f$-(1→, and the anomeric carbon signal of α-L-Ara$f$-(1→ was commonly in lower field than that of →5)-α-L-Ara$f$-(1→ [19]. Thus, the signals at δ 5.13/109.84 and 5.07/110.23 (Figure 3A) were assigned to →5)-α-L-Ara$f$-(1→ and α-L-Ara$f$-(1→, and named as residues C and D, respectively. In addition, the cross signals at δ 5.07/84.16 (DH1/DC2), 4.12/110.23 (DH2/DC1), 4.12/76.51 (DH2/DC3) and

3.88/84.16 (DH3/DC2) in HMBC spectrum and δ 4.12/84.16 (DH2/DC2) and 3.88/76.51 (DH3/DC3) in HSQC spectrum suggested the chemical shifts of H2/C2 and H2/C3 of α-L-Ara*f*-(1→ was δ 4.12/84.16 and 3.88/76.51, respectively. With the same approach, the chemical shifts of H4/C4 and H5/C5 were assigned as δ 4.06/86.96 and 3.72/63.41, respectively. Consequently, the whole chemical shifts of α-L-Ara*f*-(1→ were assigned and further compared with reported literatures [20,21]. For the overlap of anomeric carbon peaks at 90–105 ppm in ^{13}C-NMR spectrum, the anomeric signals of other residues were analyzed with HSQC spectrum (Figure 3A). The intense signal at δ 4.62/98.70 was attributed to →4)-β-D-Glc*p*-(1→ for its abundance in MBAP-1 and noted as residue H. According to the ^{1}H-NMR, ^{13}C-NMR, HSQC, HMBC spectra and reported data [22], δ 4.62 (H1) was correlated to δ 98.70 (HC1) in HSQC and 72.40 (HC2) and 74.20 (HC3) in HMBC spectrum, and δ 3.92 (HH4) was correlated to δ 74.20 (HC3), 71.68 (HC5), and 58.92 (HC6) in HMBC spectrum. Thus, the whole chemical shifts of →4)-β-D-Glc*p*-(1→ were assigned and compared with published data [22]. Accordingly, based on the methylation results and NMR data, the anomeric signals at δ 5.26/101.76, 5.21/94.86, 4.95/100.47, 4.73/102.72, 4.70/98.66, and 4.48/105.32 in HSQC spectrum were assigned to →3,4,6)-α-D-Glc*p*-(1→, →2,3)-α-D-Glc*p*-(1→, β-D-Gal*p*A-(1→, →3,4)-β-D-Glc*p*-(1→, →4)-β-D-Gal*p*-(1→ and β-D-Glc*p*A-(1→, and denoted as A, B, E, F, G, and I, respectively. Then, the complete chemical shifts of glycosidic residues of MBAP-1 were confirmed by HSQC and HMBC spectra, as summarized in Table 4, and were also consistent with reported literature [23–28].

Table 4. ^{1}H-NMR (600 MHz) and ^{13}C-NMR (150 MHz) chemical shifts of MBAP-1 and MBAP-2.

Name	Code	Residues	Chemical Shifts (ppm)					
			H1/C1	H2/C2	H3/C3	H4/C4	H5/C5	H6/C6
MBAP-1	A	→3,4,6)-α-D-Glc*p*-(1→	5.26/101.76	3.36/72.47	4.01/86.21	3.50/77.49	3.98/71.22	3.85/69.46
	B	→2,3)-α-D-Glc*p*-(1→	5.21/94.86	4.33/84.89	4.00/80.62	3.88/72.97	3.92/73.13	3.61/61.54
	C	→5)-α-L-Ara*f*-(1→	5.13/109.84	4.11/79.08	3.99/79.21	4.11/83.30	3.78/68.53	
	D	α-L-Ara*f*-(1→	5.07/110.23	4.12/84.16	3.88/76.51	4.06/86.96	3.72/63.41	
	E	β-D-Gal*p*A-(1→	4.95/100.47	3.77/69.59	3.47/72.12	3.53/73.99	3.73/68.31	-/175.87
	F	→3,4)-β-D-Glc*p*-(1→	4.73/102.72	3.65/76.30	4.06/86.66	3.77/76.40	3.69/78.60	3.92/63.41
	G	→4)-β-D-Gal*p*-(1→	4.70/98.66	3.79/74.14	3.97/75.00	3.65/76.18	4.32/78.82	3.80/59.78
	H	→4)-β-D-Glc*p*-(1→	4.62/98.70	3.30/72.40	3.55/74.20	3.92/78.55	3.63/71.68	3.73/58.92
	I	β-D-Glc*p*A-(1→	4.48/105.32	3.50/72.33	3.60/76.54	3.62/75.58	3.59/77.17	-/178.66
MBAP-2	A	→3,6)-α-D-Glc*p*-(1→	5.32/98.91	4.40/80.62	4.31/81.47	3.76/70.88	3.84/73.76	3.71/69.56
	B	→6)-α-D-Glc*p*-(1→	5.23/98.68	4.25/79.52	3.93/73.76	3.88/70.28	3.90/74.01	3.62/68.56
	C	→3,4)-α-D-2-OCH$_3$-Glc*p*-(1→	5.08/100.60	3.67/75.51	4.39/82.41	3.76/74.51	3.78/74.77	3.53/60.32
	D	→4)-β-D-Glc*p*-(1→	4.92/100.26	3.62/70.51	3.48/72.21	3.64/74.82	3.35/72.40	3.74/60.23
	E	β-D-Glc*p*-(1→	4.37/102.53	3.75/73.10	3.56/73.52	3.40/70.22	3.46/75.83	3.68/60.48

Note: The ^{1}H and ^{13}C chemical shifts of -OMe group were 3.71 and 55.20 ppm.

The monosaccharide composition result showed that MBAP-2 was a polysaccharide mainly composed of glucose. Based on the literature [27–31] and the results of GC-MS, the signals at δ 5.32/98.91, 5.23/98.68, 5.08/100.60, 4.92/100.26 and 4.37/102.53 (Figure 4A) (labeled as residue A, B, C, D, and E) were attributed to →3,6)-α-D-Glc*p*-(1→, →6)-α-D-Glc*p*-(1→, →3,4)-α-D-Glc*p*-(1→, →4)-β-D-Glc*p*-(1→ and β-D-Glc*p*-(1→, respectively. The chemical shifts of glycosidic residues of MBAP-2 were confirmed by HSQC and HMBC spectra and similarly assigned with the method above (Table 4). As shown in Figures 4 and S7, the signal at δ 3.71/55.20 in HSQC spectrum and δ 3.71/75.51 in HMBC spectrum demonstrated the presence of OCH$_3$ [32], which was linked to C-2 of residue C in MBAP-2.

2.8. NMR Analysis of Linkages and Sequences

The sequences and linkage sites among glycosidic residues were confirmed by HMBC analysis. As displayed in Figure 3B, the fiercely cross signal at δ 4.62/78.55 (HH1/HC4) indicated O-1 of residue H was linked to C-4 of residue H in MBAP-1. And the GC-MS results suggested that the residue H accounted for 41%, indicating it was an important part of backbone chain. In addition, the strong signals at δ 4.73/78.55 (FH1/HC4) and 4.62/77.49

(HH1/AC4) revealed that O-1 of residue F was linked to C-4 of residue H and O-1 of residue H was linked to C-4 of residue A. Moreover, the cross peak at δ 5.26/76.40 (AH1/FC4) suggested that O-1 of residue A was linked to C-4 of residue F, and it also indicated that the backbone chains of MBAP-1 was residue A, repeating residue H and residue F in sequence. Accordingly, the signals at δ 4.73/76.40 (FH1/FC4), 5.13/86.66 (CH1/FC3), 5.07/68.53 (DH1/CC5), 5.21/86.21 (BH1/AC3), 4.48/84.89 (IH1/BC2), 4.70/80.62 (GH1/BC3) and 4.95/76.18 (EH1/GC4) suggested that O-1 of residue F was linked to C-4 of residue F; O-1 of residue C was linked to C-3 of residue F; O-1 of residue D was linked to C-5 of residue C; O-1 of residue B was linked to C-3 of residue A; O-1 of residue I was linked to C-2 of residue B; O-1 of residue G was linked to C-3 of residue B; O-1 of residue E was linked to C-4 of residue G. As shown in Figure 3B, the cross peak at δ 3.85/136.31 was obviously observed, indicating that O-6 of residue A was linked to quercetin. Consequently, the proposed primary structure of MBAP-1 is shown in Figure 5.

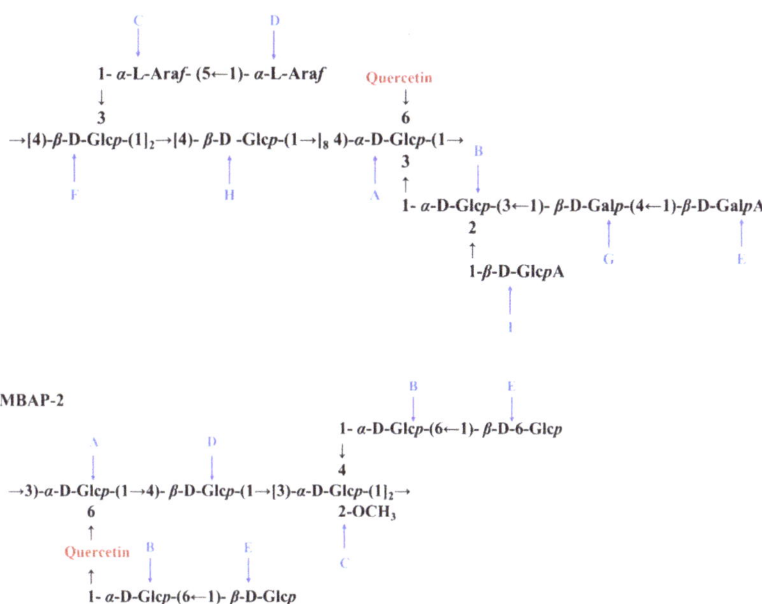

Figure 5. The putative structures of the repeating units of MBAP-1 and MBAP-2. The residues were labeled as A, B, C, D, E, F, G, H and I.

Similarly, the cross peaks at δ 5.32/74.82 (AH1/DC4), 5.23/74.51 (BH1/CC4), 5.08/81.47 (CH1/AC3), 4.92/74.51 (DH1/CC3), and 4.37/68.56 (FH1/BC6) in the HBMC spectrum of MBAP-2 (Figure 4B) indicated the linkage between O-1 of residue A and C-4 of residue D, between O-1 of residue B and C-4 of residue C, between O-1 of residue C and C-3 of residue A, between O-1 of residue D and C-3 of residue C, between O-1 of residue E and C-6 of residue B. The cross signal at 5.08/82.47 (CH1/CC3) and molar ratio result in GC-MS indicated O-1 of residue C was linked at C-3 of residue C and repeated three times. As previous literature reported, the chemical shifts of C-6 and C-8 were lower than C-2′, C-5′ and C-6′ in quercetin [18]. Several cross signals were observed at quercetin signal area in the HSQC spectrum (Figure 4A) and the signals at δ 6.36/102.50 and 6.70/105.36 might belonged to ring A of quercetin for their relevant lower chemical shifts. Moreover, the correlation signals at δ 6.36/157.92 and 6.86/153.21 in the HMBC spectrum (Figure 4B) indicating the signals at δ 157.92 and 153.21 might belong to ring A of quercetin. Thus, the cross signals at δ 5.23/157.92 and 3.71/153.21 in HMBC spectrum indicating that O-1 of residue B and O-6 of residue A were linked to quercetin. Above all, the possible repeating units of MBAP-2 were displayed in Figure 5.

2.9. Morphological Analysis

SEM has become more commonly applied to directly observe the microstructures and aggregation properties of polysaccharides. The morphological pictures of MBAP-1 and MBAP-2 were obtained and shown in Figure 6. MBAP-1 macroaggregated state presented as irregularly curled sheets with smooth surface and spatially as holes of varying sizes. The stacks between sheets were irregular and varied in size, which was consistent with the primary structure of MBAP-1. MBAP-2 appeared as sheets of varying size, with a flocculent surface, and more regularly arranged and loose accumulations at the magnification of 3000×, which was consistent with fewer branches and shorter chains in its primary structure.

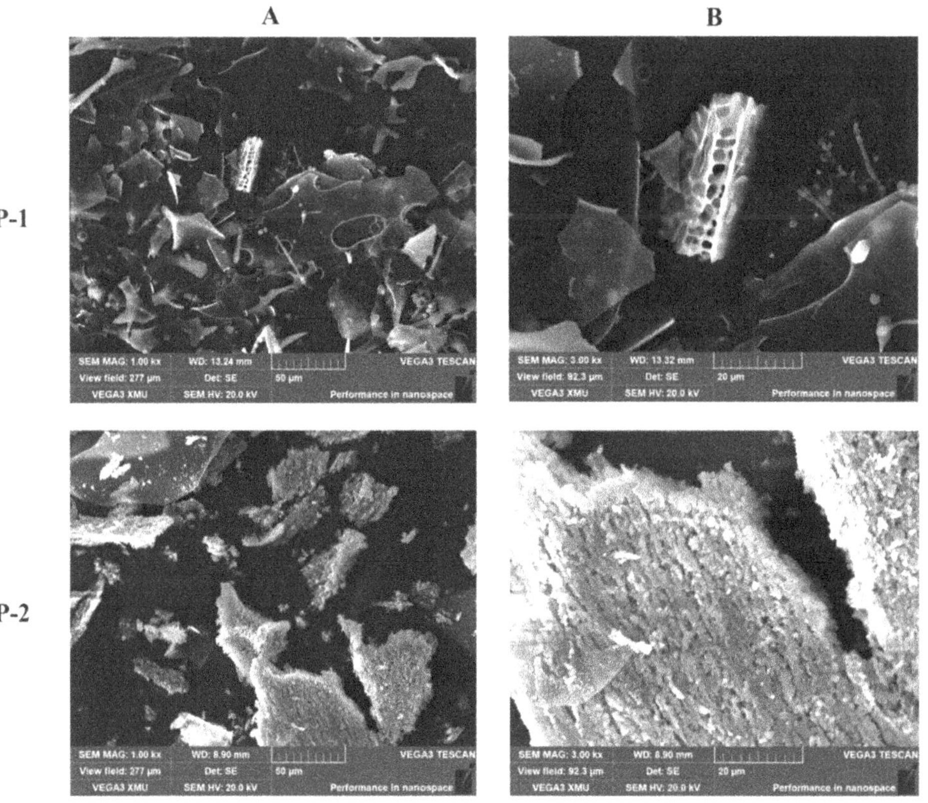

Figure 6. SEM results of MBAP-1 and MBAP-2 at different magnifications ((**A**): 1000×, (**B**): 3000×).

2.10. AFM Analysis

AFM is a powerful technique to observe the direct macromolecular morphology of polysaccharides dissolved in water without interacted effects, which could help us to better understand the natural parameters of polysaccharides [33]. In certain conditions, scanning tunneling microscope could even characterize the structures of polysaccharides [34]. As shown in Figure S9, they were aggregated loosely as a cloudy mass, with pores in the middle. Compared with MBAP-2, MBAP-1 showed looser and more porous, which was consistent with its larger molecular weight and more branches, which was also consistent with their SEM images and primary structures.

2.11. Anticomplement Activity

The anticomplement activities of MBAP-1 and MBAP-2, as well as MBAP-1R and quercetin were evaluated in vitro by hemolytic assay [15]. As shown in Figure 7, MBAP-1 and quercetin presented excellent hemolytic inhibition through classic pathway with CH_{50} values of 0.075 ± 0.004 and 0.085 ± 0.005 mg/mL, respectively, which were comparable to that of heparin (CH_{50}: 0.079 ± 0.003 mg/mL). MBAP-2 (CH_{50}: 0.249 ± 0.006 mg/mL) and MBAP-1R (CH_{50}: 0.207 ± 0.008 mg/mL) showed moderate anticomplement activities. This indicated that different polysaccharide components might result in their different anticomplement activities, even though they were both quercetin substituted. Moreover, the activity of MBAP-1 was weakened when its uronic acids were reduced.

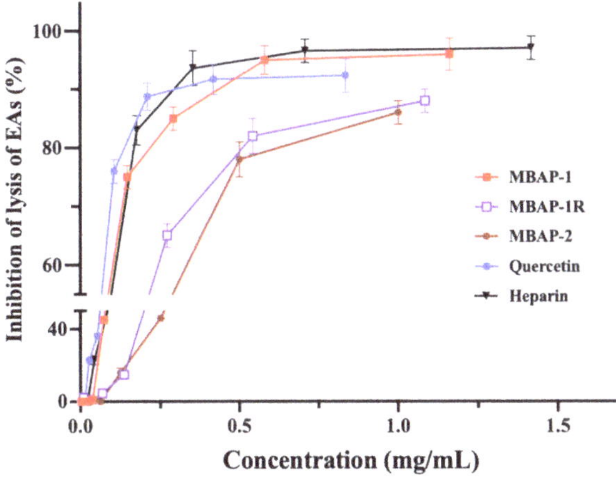

Figure 7. Anticomplement activities of MBAP-1, MBAP-2, MBAP-1R and quercetin.

2.12. Antioxidant Activity

The antioxidant activities of MBAP-1 and MBAP-2 were evaluated in vitro by the ABTS and FARP methods [35]. As displayed in Figure 8A, FRAP values of MBAP-1 were from 0.0145 ± 0.002 mmol/L to 4.483 ± 0.348 mmol/L, while that of MBAP-2 were from 0.124 ± 0.012 mmol/L to 3.455 ± 0.230 mmol/L. As shown in Figure 8B, both MBAP-1 and MBAP-2 exhibited dose-dependent antioxidant activity in the concentration range of 0.31–10.00 mg/mL. The IC_{50} values of MBAP-1 and MBAP-2 were calculated to be 0.935 ± 0.050 mg/mL and 3.286 ± 0.030 mg/mL, respectively. The above results suggested that both MBAP-1 and MBAP-2 exhibited antioxidant activity, and MBAP-1 with multiple monosaccharide composition showed slightly stronger activity.

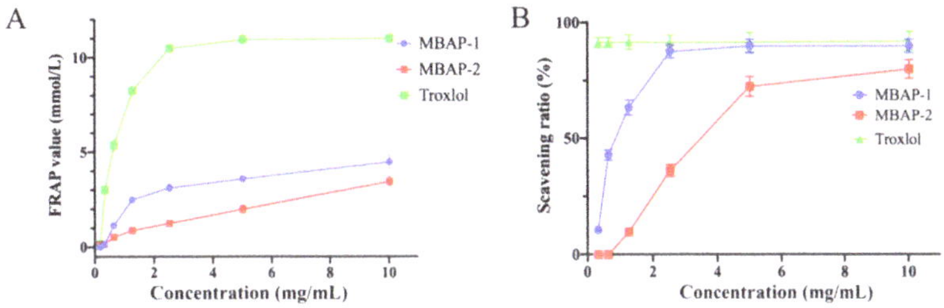

Figure 8. Antioxidant activities of MBAP-1 and MBAP-2. (**A**) The FRAP results of MBAP-1 and MBAP-2. (**B**) The ABTS results of MBAP-1 and MBAP-2.

3. Discussion

Polysaccharides are macromolecules and present various beneficial bioactivities which are directly influenced by their relative molecular weights, uronic acids, and monosaccharide composition [36]. Our previous works showed that polysaccharides with richer monosaccharide composition, higher uronic acid content, or more branches could exhibit stronger anticomplement activity [16,37]. As expected, MBAP-1, with lager molecular weight (263.9 kDa) and multiple monosaccharides, exhibited much more potent anticomplement activity than MBAP-2, which was with smaller molecular weight and had one monosaccharide. Compared with MBAP-1, its reduced product (MBAP-1R) showed obviously weaker activity, revealing that uronic acids played an indispensable role in anticomplement activity. In general, glucan has no anticomplement activity [38]. The moderate activity of MBAP-2 should be owing to its quercetin component, which exhibited potent anticomplement activity. Similarly, it had been demonstrated that the strong antioxidant activity of polysaccharides was related to the composition of galactose and uronic acid β-glycosidic linkages [39], which was consistent with the stronger antioxidant activity of MBAP-1.

In recent years, flavonoid-polysaccharide conjugates have attracted more and more attention for their improved activities or expanded properties in drug delivery attributed to the synergistic effects of the two different components. It was reported that catechin-grafted chitosan showed improved adsorption, controlled release abilities on dye, and increased antioxidant activity compared to chitosan [40]. Quercetin-grafted hyaluronic exhibited significant and sustained pH dependent drug release behaviors and higher selectivity to CD4 cell than free quercetin solution [12]. Moreover, flavonoid grafted onto pectin significantly improved the antioxidant activity of natural pectin [41]. MBAP-1 and MBAP-2 were both quercetin substituted polysaccharides and possessed potent anticomplement and antioxidant activities. They are supposed to have beneficial effects on viral pneumonia based on the potential synergistic mechanism of the active flavonoid and polysaccharide components. Their in vivo effects and mechanisms on H1N1-induced acute lung injury in mice will be investigated and reported in the near future.

4. Materials and Methods

4.1. Materials

The leaves and twigs of *Tamarix chinensis* Lour. were purchased from Bozhou Chinese herb market (Anhui, China) and authenticated by Professor Yan Lu of Fudan University, Shanghai, China. A voucher specimen (TCL20180506) has been deposited at Department of Natural Medicine, School of Pharmacy, Fudan University, Shanghai, China. Information concerning chemical regents and bio-materials used for structural characterization, anticomplement and antioxidant activities is shown in Supplementary Method S1.

4.2. Activity-Guided Extraction, Isolation and Purification

The leaves and twigs of *T. chinensis* (10 kg) were defatted with 10 vols of 95% ethanol (100 L) three times (2 h each time) and extracted with 10 vols of water (100 L) three times (2 h each time). The water extracts were concentrated to 15 L and followed by graded ethanol precipitation at final concentrations of 80% and 90%. The extracts were further mixed with 20% trichloroacetic acid (v/v, 1:1) to deproteinize. After dialyzing with water for 72 h, crude polysaccharides, MBAP80 (239 g) and MBAP90 (120 g) were obtained. Because of the stronger anticomplement activity (CH_{50} of MBAP80: 0.889 ± 0.007 mg/mL, CH_{50} of MBAP90: 0.296 ± 0.006 mg/mL), MBAP90 (100 g) was further purified by chromatography on DEAE-cellulose, eluted with water and successive concentrations of NaCl (0.2, 0.4, 0.8 and 2.0 M). Moreover, the fraction eluted with 0.2 M NaCl (Fr. 2) showed the most potent anticomplement activity (CH_{50}: 0.102 ± 0.006 mg/mL). It was further purified by Sepharyl S-200 to obtain two homogeneous polysaccharides, MBAP-1 (205 mg) and MBAP-2 (920 mg), respectively. The contents of total carbohydrate, uronic acid and protein of MBAP-1 and MBAP-2 were measured according to reported methods [6] and replicated three times.

4.3. Homogeneity and Molecular Weight Determination

The homogeneity and molecular weights of MBAP-1 and MBAP-2 were analyzed by HPGPC-ELSD and HPSEC-MALLS-RI. More details are shown in Supplementary Method S2.

4.4. Monosaccharide Composition and Absolute Configuration Analysis

The monosaccharide composition of MBAP-1 and MBAP-2 was determined using HPLC on an Agilent 1260 HPLC system (Agilent Technologies, Santa Clara, CA, USA) by the method reported before [42]. Briefly, the samples were fully hydrolyzed with TFA (2.5 M) and derivatized with PMP. The PMP-derivates were separated on an Agilent Eclipse Plus C18 column (Agilent Technologies, Santa Clara, CA, USA) (4.6 × 250 mm, 5 μm) at a flow rate of 1.0 mL/min with the mobile phase of a mixture of phosphate buffered saline (0.1 M, pH 6.7) and acetonitrile (83:17), and detected under the wavelength of 245 nm.

To identify the unknown monosaccharide in MBAP-2, MBAP-2 was hydrolyzed with TFA, then converted into alditol acetates according to the methods described in Section 4.5 and then analyzed by GC-MS with a HP-5MS column (0.25 mm × 30 m, 0.25 μm). More details are shown in Supplementary Method S3.

The monosaccharide configuration of MBAP-1 and MBAP-2 was determined according to the literature [43]. Briefly, the sample was hydrolyzed, and reacted with L-cysteine methyl ester and pyridine, follow by O-tolyisothiocyanate. The reaction products were analyzed on an Agilent 1260 HPLC system equipped with the Agilent Eclipse Plus C18 column (4.6 × 250 mm, 5 μm). It was eluted with the mixture of 0.1% acetic acid in water-acetonitrile (25:75) at a flow rate of 0.8 mL/min and detected under the wavelength of 250 nm.

4.5. Reduction, Methylation and GC-MS Analysis

The methylation of MBAP-1 and MBAP-2 was performed with modified Hakomori method according to the literature [18]. Furthermore, in order to confirm the residues of uronic acid in MBAP-1, MABP-1 and its reduced product (MBAP-1R) were both determined. Detailed information is described in Supplementary Method S4.

4.6. Qualitative and Quantitative Analysis of Flavonoid Component

In this part, ultra-high-performance liquid chromatography-ion trap tandem mass spectrometry (UPLC-IT-MS^n) was used to recognize the possible flavonoid component in the two polysaccharides. MBAP-1, MBAP-2 and their full hydrolysates were determined on a Dionex Ultimate 3000 UPLC system (Thermo Fisher Scientific, Waltham, MA, USA) equipped with a LTQ Velos Pro mass spectrometer (Thermo Fisher Scientific, Waltham, MA, USA) using a YMC-Triart C18 column (YMC Europe GmbH, Dinslaken, Germany) (150 mm × 2.1 mm, 1.9 μm). The column temperature and detection wavelength were set at 25 °C and 254 nm, respectively. The mobile phase consisted of water containing 0.1% formic acid (A) and acetonitrile containing 0.1% formic acid (B). An optimized elution gradient was set at follows: 0–2 min, 5% B; 2–35 min, 5–100% B. The injection volume and flow rate were 10 μL and 0.3 mL/min, respectively.

MS data was acquired in both positive and negative ion modes, and the device parameters were set as: auxiliary gas (N$_2$) flow, 10 arb; sheath gas (N$_2$) flow, 40 arb; collision gas (He); source voltage, 3.5 kV; capillary temperature, 350 °C; capillary voltage, 35 V. A data-dependent scan mode was selected for MS^n analysis. The flavonoid components were recognized by the positive and negative MS fragments and further confirmed by comparing with the reference substance.

The quantitative analysis was performed with colorimetry method [44] using quercetin as reference and replicated three times. Briefly, a series of concentrations (0, 0.016, 0.032, 0.048, 0.064 and 0.080 mg/mL) of quercetin and the full hydrolysates were reacted with AlCl$_3$ (0.1%, 2 mL) in ethanol solution for 10 min, and then detected at 254 nm by a

Multiskan Go microplate reader (Themo Fisher Scientific, Waltham, MA, USA). The content of quercetin was calculated based on the standard curve.

4.7. NMR and FT-IR Spectroscopy Analysis

The samples of MBAP-1 and MBAP-2 detected for NMR and FT-IR were prepared according to the method reported before [16].

Briefly, MBAP-1 and MBAP-2 were mixed with KBr, pressed into pellets, and then measured with the FT-IR spectrometer (PerkinElmer, Waltham, MA, USA) from 4000 to $400 \ cm^{-1}$ at a resolution of $2 \ cm^{-1}$ to obtain their FT-IR spectra.

For the NMR experiments, MBAP-1 and MBAP-2 (each 50 mg) were dissolved in D_2O. Then, the ^{1}H-NMR, ^{13}C-NMR, HSQC and HMBC spectra were recorded at 24 °C with an NMR spectrometer (600 M, AVANCE III HD, Bruker, Brno, Switzerland). The detailed setting parameters are shown in the Supplementary Table S1.

4.8. Microstructure and Atomic Force Microscopy (AFM) Analysis

The micromorphology of MBAP-1 and MBAP-2 were examined with an environmental scanning electron microscope (SEM) (VEGA3 XMU, Tescan, Hitacchi, Czech Republic). Briefly, the dried powder was fixed on a silicon wafer, sprayed with gold powder via a sputter-coater under 10 kV acceleration voltage, and SEM images at different magnifications were obtained.

Furthermore, MBAP-1 and MBAP-2 were dissolved in deionized water and diluted to 1.0 ng/mL. Then, 5 µL of the solution was deposited onto a freshly cleaved mica surface and dried under infrared light. The samples were determined by Environment Control Scanning Probe Microscope (Nanonavi E-Sweep, Hitachi High-Technologies, Hitachi, Marunouchi, Japan) on tapping-mode.

4.9. Anticomplement Activity Evaluation

The anticomplement activity through the classical pathway (CP) was determined by hemolytic assay with our laboratory's protocol [16]. Guinea pig sera (GPS) were used as complement source and heparin was used as positive control in this experiment. Hence, 2% SRBCs (sensitized sheep red blood cells) were prepared by washing with BBS three times and then diluted with BBS. The GPS (1:80) was chosen to give sub-maximal lysis in the absence of complement inhibitors. Various dilutions of the tested samples (200 µL) were mixed with 200 µL of GPS, 100 µL of EAs, and 100 µL anti-sheep erythrocyte antibody (1:1000 diluted). Then, the mixture was incubated at 37 °C for 30 min. After incubation, the mixtures were centrifuged immediately (3500 rpm, 4 °C) for 10 min. The optical density of the supernatant (200 µL) was measured at 405 nm by a Multiskan Go microplate reader (Themo Fisher Scientific Inc., Waltham, MA, USA). Inhibition rate of lysis induced by the test samples were calculated by the formula: Hemolytic inhibition rate (%) = $[1 - (A_{test} - A_{sample \ control})/(A_{100\%lysis} - A_{vehicle \ control}] \times 100\%$. The CH_{50} value was the sample concentration resulting in 50% hemolytic inhibition.

4.10. Antioxidant Activity Evaluation

The antioxidant activities were evaluated by ABTS and FARP analysis with Beyotime Elisa kits (Shanghai, China). The experiments were performed based on Elisa kit instructions.

5. Conclusions

Two novel natural flavonoid substituted polysaccharides were isolated and purified from *Tamarix chinensis* Lour. and they both exhibited potent anticomplement and antioxidant activities. Their primary structures were elucidated by monosaccharide composition, IR, D/L configuration, methylation, NMR, and UPLC-MS/MS analysis. In addition, the structure–activity relationship analysis demonstrated that uronic acids, monosaccharide composition, and quercetin were important for their anticomplement activities. In consideration of the advantages of flavonoid substituted polysaccharides in drug delivery and bioactivity, the quercetin substituted polysaccharides of *T. chinensis* might be better complement and oxidation inhibitors.

Supplementary Materials: The following supporting information can be downloaded at: https://www.mdpi.com/article/10.3390/molecules27144532/s1. Methods S1: Materials; Methods S2: Homogeneity and molecular weight determination; Methods S3: Monosaccharide analysis of MBAP-2; Methods S4: Reduction, methylation and GC-MS analysis; Methods S5: NMR spectroscopy analysis; Table S1: The detailed setting parameters in NMR experiments; Figure S1: The HPGPC-ELSD results of MBAP-1 and MBAP-2; Figure S2: The GC-MS spectrogram of monosaccharide of MBAP-2; Figure S3: FT-IR spectroscopy analysis of MBAP-1 and MBAP-2; Figure S4: The mass spectra of PMAAs of MBAP-1; Figure S5: The mass spectra of PMAAs of MBAP-2; Figure S6: The ^{1}H- and ^{13}C-NMR spectra of MBAP-1; Figure S7: The ^{1}H- and ^{13}C-NMR spectra of MBAP-2; Figure S8: The ^{1}H-NMR and ^{13}C-NMR spectra of quercetin in d_6-DMSO and d_6-DMSO: D_2O (1:1).; Figure S9: The AFM results of MBAP-1 and MBAP-2. References [45,46] are cited in the Supplementary Materials.

Author Contributions: Y.J. performed in the initial task of obtainment and bioactivity test of polysaccharides and data analysis and wrote the original draft. Y.Y. and L.Z. performed in the task of data analysis. D.C. and Y.L. acquired the funding and supervised the research work. All authors have read and agreed to the published version of the manuscript.

Funding: This work was supported by grants from the National Natural Science Foundation of China (82173940), the Science and Technology of Tibet Autonomous Region, China (XZ202001YD0003C, XZ201801-GA-13) and the Development Project of Shanghai Peak Disciplines-Integrative Medicine (20180101).

Data Availability Statement: Data will be provided upon request.

Conflicts of Interest: The authors declare no conflict of interest.

Sample Availability: Samples of the MBAP-1 and MBAP-2 are available from the authors.

References

1. Zoltán, P.; Michael, K.; Ashley, F.A. Complement analysis in the era of targeted therapeutics. *Mol. Immunol.* **2018**, *102*, 84–88.
2. Zhu, H.Y.; Lu, X.X.; Ling, L.J.; Li, H.; Ou, Y.Y.; Shi, X.L.; Lu, Y.; Zhang, Y.Y.; Chen, D.F. *Houttuynia cordata* polysaccharides ameliorate pneumonia severity and intestinal injury in mice with influenza virus infection. *J. Ethnopharmacol.* **2018**, *218*, 90–99. [CrossRef] [PubMed]
3. Risitano, A.M.; Mastellos, D.C.; Huber-Lang, M.; Yancopoulou, D.; Lambris, J.D. Complement as a target in COVID-19? *Nat. Rev. Immunol.* **2020**, *20*, 343–344. [CrossRef] [PubMed]
4. Polycarpou, A.; Howard, M.; Farrar, C.A.; Greenlaw, R.; Fanelli, G.; Wallis, R.; Klavinskis, L.S.; Sacks, S. Rationale for targeting complement in COVID-19. *EMBO Mol. Med.* **2020**, *12*, e12642. [CrossRef] [PubMed]
5. Mastaglio, S.; Ruggeri, A.; Risitano, A.M.; Angelillo, P.; Yancopoulou, D.; Mastellos, D.C.; Markus, H.L.; Piemontese, S.; Assanelli, A.; Garlanda, C.; et al. The first case of COVID-19 treated with the complement C3 inhibitor AMY-101. *Clin. Immunol.* **2020**, *215*, 108450. [CrossRef] [PubMed]
6. Fu, Z.L.; Xia, L.; De, J.; Zhu, M.X.; Li, H.; Lu, Y.; Chen, D.F. Beneficial effects on H1N1-induced acute lung injury and structure characterization of anti-complementary acidic polysaccharides from *Juniperus pingii* var. *wilsonii*. *Int. J. Biol. Macro.* **2019**, *129*, 246–253. [CrossRef] [PubMed]
7. Ling, L.J.; Lu, Y.; Zhang, Y.Y.; Zhu, H.Y.; Tu, P.; Li, H.; Chen, D.F. Flavonoids from *Houttuynia cordata* attenuate H1N1-induced acute lung injury in mice via inhibition of influenza virus and Toll-like receptor signaling. *Phytomedicine* **2020**, *67*, 153150. [CrossRef]
8. Zhang, Z.L. Clinical summary of 41 cases of measles complicated with pneumonia *Pract. Clin. J. Integr. Tradit. Chin. West Med.* **1992**, *2*, 41–42.
9. Wang, Z.D.; Zhang, Y.; Dai, Y.D.; Ren, K.; Han, C.; Wang, H.X.; Yi, S.Q. *Tamarix chinensis* Lour. inhibits chronic ethanol-induced liver injury in mice. *World J. Gastroenterol.* **2020**, *26*, 1286–1297. [CrossRef]
10. Liu, J.; Wang, X.; Yong, H.; Kan, J.; Jin, C. Recent advances in flavonoid-grafted polysaccharides: Synthesis, structural characterization, bioactivities and potential applications. *Int. J. Biol. Macro.* **2018**, *116*, 1011–1025. [CrossRef]
11. Wang, X.Y.; Chen, Y.H.; Dahmani, F.Z.; Yin, L.F.; Zhou, J.P.; Yao, J. Amphiphilic carboxymethyl chitosan-quercetin conjugate with P-gp inhibitory properties for oral delivery of paclitaxel. *Biomaterials* **2014**, *35*, 7654–7665. [CrossRef] [PubMed]
12. Pang, X.; Lu, Z.; Du, H.J.; Yang, X.Y.; Zhai, G.X. Hyaluronic acid-quercetin conjugate micelles: Synthesis, characterization, in vitro and in vivo evaluation. *Colloids Surf. B.* **2014**, *123*, 778–786. [CrossRef] [PubMed]

13. Izadi, M.; Cegolon, L.; Javanbakht, M.; Sarafzadeh, A.; Abolghasemi, H.; Alishiri, G.; Zhao, S.; Einollahi, B.; Kashaki, M.; Jonaidi-Jafari, N.; et al. Ozone therapy for the treatment of COVID-19 pneumonia: A scoping review. *Int. Immunopharmacol.* **2021**, *92*, 107307. [CrossRef] [PubMed]

14. Gao, X.; Qu, H.; Shan, S.; Song, C.; Baranenko, D.; Li, Y.Z.; Lu, W.H. A novel polysaccharide isolated from *Ulva Pertusa*: Structure and physicochemical property. *Carbohyd. Polym.* **2020**, *223*, 115849. [CrossRef]

15. Zhang, S.J.; Zhang, Q.; An, L.J.; Zhang, J.J.; Li, Z.G.; Zhang, J.; Li, Y.H.; Tuerhong, M.; Ohizumi, Y.; Jin, J.; et al. A fructan from *Anemarrhena asphodeloides* Bunge showing neuroprotective and immunoregulatory effects. *Carbohyd. Polym.* **2020**, *229*, 115477. [CrossRef]

16. Xia, L.; Deji; Zhu, M.X.; Chen, D.F.; Lu, Y. *Juniperus pingii* var. *wilsonii* acidic polysaccharide: Extraction, characterization and anticomplement activity. *Carbohyd. Polym.* **2020**, *231*, 115728.

17. Martinez-Busi, M.; Arredondo, F.; González, D.; Echeverry, C.; Vega-Teijido, M.A.; Carvalho, D.; Rodríguez-Haralambides, A.; Rivera, F.; Dajas, D.; Abin-Carriquiry, J.A. Purification, structural elucidation, antioxidant capacity and neuroprotective potential of the main polyphenolic compounds contained in *Achyrocline satureioides* (Lam) D.C. (Compositae). *Bioorgan. Med. Chem.* **2019**, *27*, 2579–2591. [CrossRef]

18. Zahmanov, G.; Alipieva, K.; Denev, P.; Todorov, D.; Hinkov, A.; Shishkov, S.; Simova, S. Flavonoid glycosides profiling in dwarf elder fruits (*Sambucus ebulus* L.) and evaluation of their antioxidant and anti-herpes simplex activities. *Ind. Crop. Prod.* **2015**, *63*, 58–64. [CrossRef]

19. Lian, Y.; Zhu, M.; Chen, J.; Yang, B.; Lv, Q.; Wang, L.; Guo, S.C.; Tan, X.B.; Li, C.; Bu, W.Q.; et al. Characterization of a novel polysaccharide from *Moutan Cortex* and its ameliorative effect on AGEs-induced diabetic nephropathy. *Int. J. Biol. Macro.* **2021**, *176*, 589–600. [CrossRef]

20. Huang, F.; Liu, H.; Zhang, R.; Dong, L.; Liu, L.; Ma, Y.; Jia, X.C.; Wang, G.J.; Zhang, M.W. Physicochemical properties and prebiotic activities of polysaccharides from longan pulp based on different extraction techniques. *Carbohyd. Polym.* **2019**, *206*, 344–351. [CrossRef]

21. Chang, X.; Shen, C.Y.; Jiang, J.G. Structural characterization of novel arabinoxylan and galactoarabinan from citron with potential antitumor and immunostimulatory activities. *Carbohyd. Polym.* **2021**, *269*, 118331. [CrossRef] [PubMed]

22. Ye, G.; Li, J.; Zhang, J.; Liu, H.; Ye, Q.; Wang, Z. Structural characterization and antitumor activity of a polysaccharide from *Dendrobium wardianum*. *Carbohyd. Polym.* **2021**, *269*, 118253. [CrossRef] [PubMed]

23. Guo, R.; Xu, Z.; Wu, S.; Li, X.; Li, J.; Hu, H.; Wu, Y.; Ai, L.Z. Molecular properties and structural characterization of an alkaline extractable arabinoxylan from hull-less barley bran. *Carbohyd. Polym.* **2019**, *218*, 250–260. [CrossRef]

24. Zhang, H.; Zou, P.; Zhao, H.; Qiu, J.; Regenstein, J.M.; Yang, X. Isolation, purification, structure and antioxidant activity of polysaccharide from pinecones of *Pinus koraiensis*. *Carbohyd. Polym.* **2021**, *251*, 117078. [CrossRef]

25. Dong, Z.; Zhang, M.; Li, H.; Zhan, Q.; Lai, F.; Wu, H. Structural characterization and immunomodulatory activity of a novel polysaccharide from *Pueraria lobata* (Willd.) Ohwi root. *Int. J. Biol. Macro.* **2020**, *154*, 1556–1564. [CrossRef] [PubMed]

26. Wang, Y.P.; Guo, M.R. Purification and structural characterization of polysaccharides isolated from *Auricularia cornea* var. Li. *Carbohyd. Polym.* **2020**, *230*, 115680. [CrossRef] [PubMed]

27. Gao, X.; Qi, J.; Ho, C.T.; Li, B.; Mu, J.; Zhang, Y.; Hu, H.P.; Mo, W.P.; Chen, Z.Z.; Xie, Y.Z. Structural characterization and immunomodulatory activity of a water-soluble polysaccharide from *Ganoderma leucocontextum* fruiting bodies. *Carbohyd. Polym.* **2020**, *249*, 116874. [CrossRef]

28. Zhang, Y.; Pan, X.; Ran, S.; Wang, K. Purification, structural elucidation and anti-inflammatory activity in vitro of polysaccharides from *Smilax china* L. *Int. J. Biol. Macro.* **2019**, *139*, 233–243. [CrossRef]

29. Xu, S.; Xu, X.; Zhang, L. Branching structure and chain conformation of water-soluble glucan extracted from *Auricularia auricula-judae*. *J. Agr. Food. Chem.* **2012**, *60*, 3498–3506. [CrossRef]

30. Gao, Y.; Guo, Q.; Zhang, K.; Wang, N.; Li, C.; Li, Z.; Zhang, A.L.; Wang, C.L. Polysaccharide from *Pleurotus nebrodensis*: Physicochemical, structural characterization and in vitro fermentation characteristics. *Int. J. Biol. Macro.* **2020**, *165*, 1960–1969. [CrossRef]

31. Wang, X.; Zhang, M.; Zhang, D.; Wang, S.; Yan, C. An *O*-acetyl-glucomannan from the rhizomes of *Curculigo orchioides*: Structural characterization and anti-osteoporosis activity in vitro. *Carbohyd. Polym.* **2017**, *174*, 48–56. [CrossRef] [PubMed]

32. Wang, K.L.; Wang, B.; Hu, R.; Zhao, X.; Li, H.; Zhou, G.; Song, L.L.; Wu, A.M. Characterization of hemicelluloses in *Phyllostachys edulis* (moso bamboo) culm during xylogenesis. *Carbohyd. Polym.* **2019**, *221*, 127–136. [CrossRef]

33. Marszalek, P.E.; Dufrene, Y.F. Stretching single polysaccharides and proteins using atomic force microscopy. *Chem. Soc. Rev.* **2012**, *41*, 3523–3534. [CrossRef] [PubMed]

34. Wu, X.; Delbianco, M.; Anggara, K.; Michnowicz, T.; Pardo-Vargas, A.; Bharate, P.; Sen, S.; Pristl, M.; Rauschenbach, S.; Schlickum, U.; et al. Imaging single glycans. *Nature* **2020**, *582*, 375–378. [CrossRef] [PubMed]

35. Wolosiak, R.; Drużyńska, B.; Derewiaka, D.; Piecyk, M.; Majewska, E.; Ciecierska, M.; Worobiej, E.; Pakosz, P. Verification of the conditions for determination of antioxidant activity by ABTS and DPPH assays—A practical approach. *Molecules* **2021**, *27*, 50. [CrossRef] [PubMed]

36. Qu, J.L.; Huang, P.; Zhang, L.; Qiu, Y.; Qi, H.; Leng, A.J.; Shang, D. Hepatoprotective effect of plant polysaccharides from natural resources: A review of the mechanisms and structure-activity relationship. *Int. J. Biol. Macro.* **2020**, *161*, 24–34. [CrossRef]

37. Lu, Y.; Zhang, J.J.; Huo, J.Y.; Chen, D.F. Structural characterization and anti-complementary activities of two polysaccharides from *Houttuynia cordata*. *Planta Med.* **2019**, *85*, 108–1106. [CrossRef]
38. Wang, H.W.; Li, N.; Zhu, C.; Shi, S.S.; Jin, H.; Wang, S.C. Anti-complementary activity of two homogeneous polysaccharides from *Eclipta prostrata*. *Biochem. Bioph. Res. Commun.* **2017**, *393*, 887–893. [CrossRef]
39. Zeng, F.K.; Chen, W.B.; He, P.; Zhan, Q.P.; Wang, Q.; Wu, H.; Zhang, M.M. Structural characterization of polysaccharides with potential antioxidant and immunomodulatory activities from Chinese water chestnut peels. *Carbohyd. Polym.* **2020**, *246*, 116551. [CrossRef]
40. Zhu, W.L.; Zhang, Z.J. Preparation and characterization of catechin-grafted chitosan with antioxidant and antidiabetic potential. *Int. J. Biol. Macro.* **2014**, *70*, 150–155. [CrossRef]
41. Ahn, S.; Halake, K.; Lee, J. Antioxidant and ion-induced gelation functions of pectins enabled by polyphenol conjugation. *Int. J. Biol. Macro.* **2017**, *101*, 776–782. [CrossRef] [PubMed]
42. Jiao, Y.K.; Hua, D.H.; Huang, D.; Zhang, Q.; Yan, C.Y. Characterization of a new heteropolysaccharide from green guava and its application as an α-glucosidase inhibitor for the treatment of type II diabetes. *Food. Funct.* **2018**, *9*, 3997–4007. [CrossRef] [PubMed]
43. Chen, X.X.; Li, T.Y.; Qing, D.G.; Chen, J.; Zhang, Q.; Yan, C.Y. Structural characterization and osteogenic bioactivities of a novel *Humulus lupulus* polysaccharide. *Food. Funct.* **2020**, *11*, 1165–1175. [CrossRef] [PubMed]
44. Kostalova, D.; Tekel'ová, D.; Czigle, S.; Tóth, J.; Lalkovi, S. The content of flavonoids in selected drug formulations containing *Ginkgo biloba*. *Farmaceutický. Obzor.* **2005**, *74*, 119–123.
45. Chen, X.; Cao, D.; Zhou, L.; Jin, H.; Dong, J.; Yao, J.; Ding, K. Structure of a polysaccharide from *Gastrodia elata* Bl., and oligosaccharides prepared thereof with anti-pancreatic cancer cell growth activities. *Carbohyd. Polym.* **2020**, *86*, 1300–1305. [CrossRef]
46. Huo, J.; Lu, Y.; Jiao, Y.; Chen, D. Structural characterization and anticomplement activity of an acidic polysaccharide from *Hedyotis diffusa*. *Int. J. Biol. Macromol.* **2020**, *155*, 1553–1560. [CrossRef]

Article

Bioactive Phenolic Compounds from *Peperomia obtusifolia*

Ismail Ware [1,2], Katrin Franke [1,*], Hidayat Hussain [1], Ibrahim Morgan [1], Robert Rennert [1] and Ludger A. Wessjohann [1,*]

[1] Department of Bioorganic Chemistry, Leibniz Institute of Plant Biochemistry, D-06120 Halle, Germany; ismailbin.ware@ipb-halle.de (I.W.); hidayat.hussain@ipb-halle.de (H.H.); ibrahim.morgan@ipb-halle.de (I.M.); robert.rennert@ipb-halle.de (R.R.)

[2] Institute of Bioproduct Development, Universiti Teknologi Malaysia, Johor Bahru 81310, Malaysia

* Correspondence: katrin.franke@ipb-halle.de (K.F.); ludger.wessjohann@ipb-halle.de (L.A.W.); Tel.: +49-345-5582-1380 (K.F.); +49-345-5582-1300 (L.A.W.)

Abstract: *Peperomia obtusifolia* (L.) A. Dietr., native to Middle America, is an ornamental plant also traditionally used for its mild antimicrobial properties. Chemical investigation on the leaves of *P. obtusifolia* resulted in the isolation of two previously undescribed compounds, named peperomic ester (**1**) and peperoside (**2**), together with five known compounds, viz. *N*-[2-(3,4-dihydroxyphenyl)ethyl]-3,4-dihydroxybenzamide (**3**), becatamide (**4**), peperobtusin A (**5**), peperomin B (**6**), and arabinothalictoside (**7**). The structures of these compounds were elucidated by 1D and 2D NMR techniques and HREIMS analyses. Compounds **1–7** were evaluated for their anthelmintic (against *Caenorhabditis elegans*), antifungal (against *Botrytis cinerea, Septoria tritici* and *Phytophthora infestans*), antibacterial (against *Bacillus subtilis* and *Aliivibrio fischeri*), and antiproliferative (against PC-3 and HT-29 human cancer cell lines) activities. The known peperobtusin A (**5**) was the most active compound against the PC-3 cancer cell line with IC_{50} values of 25.6 µM and 36.0 µM in MTT and CV assays, respectively. This compound also induced 90% inhibition of bacterial growth of the Gram-positive *B. subtilis* at a concentration of 100 µM. In addition, compound **3** showed anti-oomycotic activity against *P. infestans* with an inhibition value of 56% by using a concentration of 125 µM. However, no anthelmintic activity was observed.

Keywords: Piperaceae; *Peperomia obtusifolia*; isolation; cytotoxicity; anticancer; antibacterial

Citation: Ware, I.; Franke, K.; Hussain, H.; Morgan, I.; Rennert, R.; Wessjohann, L.A. Bioactive Phenolic Compounds from *Peperomia obtusifolia. Molecules* **2022**, *27*, 4363. https://doi.org/10.3390/molecules27144363

Academic Editor: Jacqueline Aparecida Takahashi

Received: 8 June 2022
Accepted: 5 July 2022
Published: 7 July 2022

Publisher's Note: MDPI stays neutral with regard to jurisdictional claims in published maps and institutional affiliations.

1. Introduction

Peperomia is one of the two major genera of the Piperaceae family, with approximately 1700 species common in almost all tropical areas [1]. Plants of the genus *Peperomia* possess numerous uses as traditional medicine to treat asthma, gastric ulcers, bacterial infection, pain, and inflammation [2]. *Peperomia* species produce several classes of secondary metabolites such as flavonoids, secolignans, prenylated phenolic substances, chromenes, polyketides [3], lignans, tetrahydrofurans, terpenoids, meroterpenoids, and amides [1,4,5]. Some compounds from *P. pellucida, P. doclouxii, P. dindygulensis*, and *P. blanda* were reported to exhibit cytotoxic effects [2,6–8].

Peperomia obtusifolia (L.) A.Dietr. is widely distributed from Mexico to the northern part of South America. It is widely known as *baby rubber* or the *paragua* plant. In spite of its dominating decorative use, some communities in Central America use this plant to treat insect and snake bites and also as a skin cleanser [9]. Previous studies on *P. obtusifolia* resulted in the isolation of chromanes, flavonoids, lignans, and other secondary metabolites [10–14]. Several biological activities were described for compounds isolated from this species, such as trypanocidal [11], anti-inflammatory [15], and antibacterial activities [14]. In the present study, we report the isolation, structure elucidation, and biological effects of two previously undescribed compounds (**1** and **2**), along with five known compounds (**3–7**) from the methanolic leaf extract of *P. obtusifolia*.

2. Results and Discussion

The dried leaves of *P. obtusifolia* were extracted with 80% aqueous methanol. The crude extract was suspended in water and successively fractioned by liquid−liquid partition between water and *n*-hexane, and water and ethyl acetate. The constituents were purified by preparative HPLC and repeated column chromatography on silica gel, Sephadex LH20, and Diaion HP20, which yielded two previously undescribed compounds (**1** and **2**), as well as two amide derivatives (**3** and **4**), a chromane (**5**), a secolignan (**6**), and an unusual phenolic nitroalkyl diglycoside (**7**) (Figure 1). The chemical structures of **1** and **2** were elucidated by detailed spectroscopic techniques.

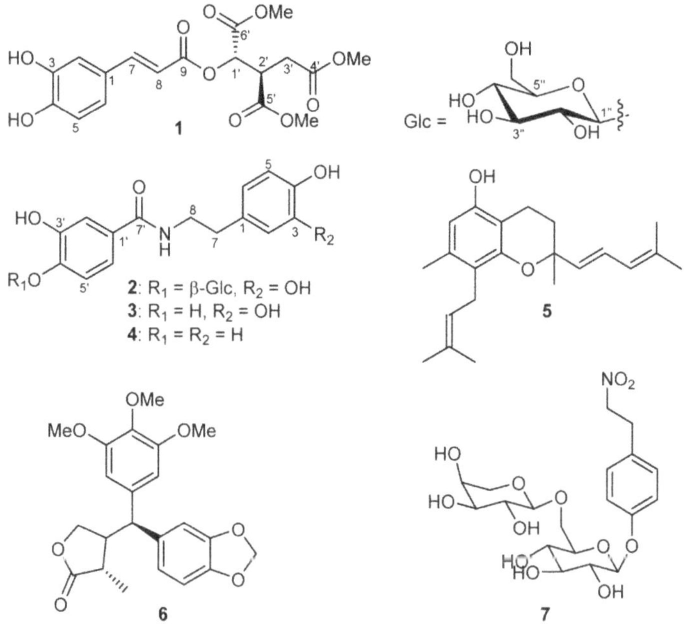

Figure 1. Chemical structures of compounds **1**–**7**.

Compound **1** was obtained as a brownish liquid from the aqueous fraction. The HR-ESI-MS measurements of the deprotonated molecular ion at m/z 395.0975 [M − H]$^-$ indicated the molecular formula $C_{18}H_{20}O_{10}$. The IR spectrum revealed the presence of hydroxyl (3402 cm^{-1}) and carbonyl groups (1715 cm^{-1}). Inspection of the ^{1}H NMR data (Table 1) exhibited characteristic signals of an ABX aromatic system at δ 7.06 (1H, d, J = 2.1 Hz, H-2), 6.79 (1H, d, J = 8.2 Hz, H-5), and 6.97 (1H, d, J = 8.2, 2.1 Hz, H-6) and of one *E*-configured double bond [7.59 (1H, d, J = 15.8 Hz, H-7) at 6.29 (1H, d, J = 15.8 Hz, H-8)]. In addition, ^{1}H NMR data also showed the presence of two saturated methine groups [δ 5.46 (1H, d, J = 3.9 Hz, H-1′) at 3.62 (1H, ddd, J = 9.0, 5.5, 3.9 Hz, H-2′), and one methylene group [δ 2.85 (1H, dd, J = 17.2, 9.0 Hz, H-3′a) at 2.66 (1H, d, J = 17.2, 5.5 Hz, H-3′b)] along with three methoxy groups at δ 3.68, 3.73, and 3.77. The ^{13}C data (Table 1) exhibited one sp^2 quaternary carbon [δ$_C$ 127.4 (C-1)], two oxygenated sp^2 tertiary carbons [δ$_C$ 146.8 (C-3), 150.0 (C-4)], and four carbonyl groups [δ$_C$ 167.5 (C-9), 173.2 (C-4′), 170.2 (C-5′), 172.1 (C-6′)]. The scaffold of compound **1** was established by an HMBC experiment, showing long-range correlations from H-7 to C-1, C-2, C-6, C-8, and C-9, indicating that the aromatic ring and the carbonyl group C-9 were linked to C-7 in the form of a caffeic acid moiety. In addition, COSY correlations of H-1′/H-2′, H-2′/H-1′/H-3′a/H-3′b, H-3′a/H-2′, and H-3b/H-2′, together with HMBC correlations of the three methoxy groups to their respective carbonyls (C-4′, C-5′, and C-6′) and HMBC correlations of H-1′ and C-2′, C-3′, C-5′, C-6′ and of H-2′ and C-1′,

C-3′, C-4′, C-5′, and C-6′ suggested the presence of an isocitrate core. The C-1′ hydroxyl of the isocitrate unit is attached to the caffeic acid moiety at C-9 indicated by the long-range correlation from H-1′ (δ_H 5.46) to C-9 (δ_C 167.5) (Figure 2). The methylated isocitrate moiety of **1** is similar to those found in cryptoporic acids isolated from *Cryptoporus sinensis* [16] and *Cryptoporus volvatus* [17,18], which are also conjugates of isocitrate. Since compound **1** was also detected in an ethanolic crude extract (Figure S9), the methylation is not a processing artifact. The absolute configuration of C-1′ and C-2′ of the isocitrate group was suggested to be (1′*S*,2′*R*) by comparison of its optical rotation sign (negative optical rotation value) with published compounds having a (1′*R*,2′*S*) absolute configuration and opposite-in-sign specific rotation (positive optical rotation value) [16,19]. Thus, the methylated isocitrate moiety is in accordance with an L-threo-isocitric acid core ((1*S*,2*R*)-1-hydroxypropane-1,2,3-tricarboxylic acid). Based on the above-mentioned evidence, the structure of **1** was elucidated as trimethyl (1*S*,2*R*)-1-(((*E*)-3-(3,4-dihydroxyphenyl)acryloyl)oxy)propane-1,2,3-tricarboxylate and given the trivial name peperomic ester.

Table 1. ^{1}H- and ^{13}C-NMR data of **1** (in CD$_3$OD; δ in ppm, *J* in Hz).

No.	δ (H) [a]	δ (C) [b]	HMBC
1	-	127.4	
2	7.06 (d, *J* = 2.1)	115.2	
3	-	146.8	
4	-	150.0	
5	6.79 (d, *J* = 8.2)	116.5	
6	6.97 (dd, *J* = 8.2, 2.1)	123.3	
7	7.59 (d, *J* = 15.8)	148.6	C1, C2, C6, C8, C9
8	6.29 (d, *J* = 15.8)	113.4	C1, C7, C9
9	-	167.5	
1′	5.46 (d, *J* = 3.9)	72.6	C2′, C3′, C5′, C6′, C9′
2′	3.62 (ddd, *J* = 9.0, 5.5, 3.9)	44.2	C1′, C3′, C4′, C5′, C6′
3′a	2.85 (d, *J* = 17.2, 9.0)	32.8	C2′, C1′, C4′, C6′
3′b	2.66 (d, *J* = 17.2, 5.5)	32.8	C2′, C1′, C4′, C6′
4′	-	173.2	
5′	-	170.2	
6′	-	172.1	
OMe	3.73	52.9	C6′
OMe	3.77	53.1	C5′
OMe	3.68	53.4	C4′

[a] Recorded at 400 MHz. [b] Recorded at 100 MHz.

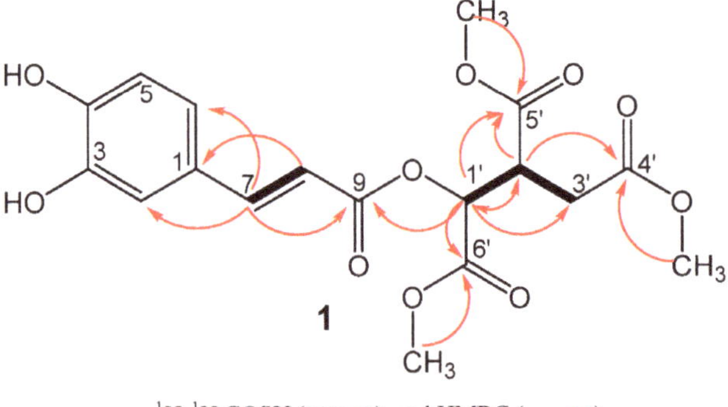

^{1}H-^{1}H COSY (━━), and HMBC (⟶)

Figure 2. Key ^{1}H-^{1}H COSY and HMBC correlation of **1**.

Compound **2** (Figure 3) was obtained as a white amorphous powder from the ethyl acetate fraction. The HR-ESI-MS measurements of the deprotonated molecular ion at *m/z* 450.1379 [M − H]$^-$ indicated the molecular formula C$_{21}$H$_{25}$NO$_{10}$. The IR spectrum displayed absorption for hydroxyl (3250 cm^{-1}) and carbonyl groups (1607 cm^{-1}). The NMR data of compound **2** (Table 2, Figure 3) show the presence of an *N*-[2-(3,4-dihydroxyphenyl)ethyl]-3,4-dihydroxybenzamide moiety [20]. The structure of the aglycon part is the same as in compound **3**.

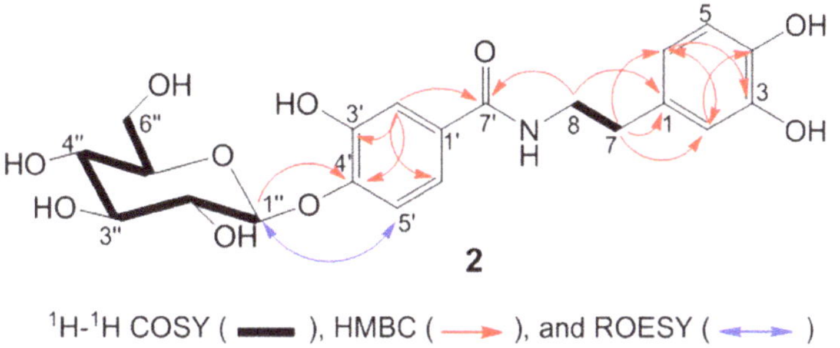

Figure 3. Key ^{1}H-^{1}H COSY, HMBC, and ROESY correlation of **2**.

Table 2. ^{1}H- and ^{13}C-NMR data of **2** (in CD$_3$OD; δ in ppm, *J* in Hz).

No.	δ (H) [a]	δ (C) [b]	HMBC
1	-	132.2	
2	6.70 (d, *J* = 2.1)	117.0	C1, C3, C4, C6, C7
3	-	144.7	
4	-	146.2	
5	6.69 (d, *J* = 8.0)	116.5	C1, C3, C4, C6, C7
6	6.59 (dd, *J* = 8.0, 2.1)	121.1	C2, C3, C7
7	2.75 (t, *J* = 7.6)	35.9	C1, C2, C6, C8
8a	3.57 (dt, *J* = 13.2, 7.6)	42.8	C1, C7, C7′
8b	3.50 (dt, *J* = 13.2, 7.6)	42.8	C1, C7, C7′
1′	-	126.3	
2′	7.25 (d, *J* = 3.1)	120.7	C1′, C3′, C4′, C5′, C6′, C7′
3′	-	154.3	
4′	-	150.1	
5′	7.24 (d, *J* = 8.8)	117.3	C1′, C3′, C4′, C6′
6′	6.86 (dd, *J* = 8.8, 3.1)	120.3	C1′, C3′, C4′, C5′
7′	-	167.8	
1″	4.79 (d, *J* = 7.5)	105.1	C4′, C3″, C5″
2″	3.38 (m)	74.9	
3″	3.41 (m)	78.1	
4″	3.38 (m)	78.5	
5″	3.39 (m)	71.3	
6a″	3.70 (dd, *J* = 12.0, 5.0)	62.5	
6b″	3.89 (d, *J* = 12.0)	62.5	

[a] Recorded at 400 MHz. [b] Recorded at 100 MHz.

In addition, NMR data indicate the presence of an anomeric carbon [(δ$_H$ 4.79 (1H, d, *J* = 7.5 Hz); δ$_C$ 105.1] together with four methine carbon signals between 71.3 and 78.5 ppm, which is typical for glucose. The coupling constant value of the sugar anomeric proton (*J* = 7.5 Hz) revealed the *β*-configuration of the glucosyl moiety. The linkage of C-1″ from the sugar moiety to 4′OH of the 3,4-dihydroxybenzoate moiety was established based on the HMBC long-range correlation from H-1″ (δ$_H$ 4.79) to C-4′ (δ$_C$ 150.1) (Figure 3). This conclusion is supported by the ROESY correlation of H-1″ (δ$_H$ 4.79) with H-5′ (δ$_H$ 7.24). From the above spectroscopic data, the structure of compound **2** was established as *N*-[2-(3,4-dihydroxyphenyl)ethyl]-3,4-dihydroxybenzamide 4′-*O*-*β*-D-glucoside, given the trivial name peperoside.

The phytochemical investigation of the *P. obtusifolia* (L.) A.Dietr. leaves allowed the isolation of five other known compounds, *N*-[2-(3,4-dihydroxyphenyl)ethyl]-3,4-dihydroxybenzamide

(**3**) [20], becatamide (also known as houttuynamide A, **4**) [21], peperobtusin A (**5**) [10], peperomin B (**6**) [22], and arabinothalictoside (**7**) [23]. These compounds were identified by comparing their MS and NMR spectral data with published data, which agree with the data reported for the respective compounds. Among the known constituents, compound **5** was already reported from *P. obtusifolia*, while compounds **3**, **4**, and **6** were isolated from other *Peperomia* species such as *P. duclouxii* [7], *P. tetraphylla* [24], and *P. japonica* [22], respectively. Compound **7** is an unusual nitro natural product which was isolated for the first time from the Japanese plant *Sagittaria trifolia* [25] and later reported in *Aristolochia fordiana* [26], which belong to Piperales and *Annona squamosa* [23].

Peperomia species constituents have previously been reported to exhibit a variety of biological activities, either as extracts or pure compounds [2,27–29]. Thus, the compounds isolated from this species were evaluated for their anthelmintic, antifungal, antibacterial, and anticancer properties. These biological examinations were conducted by using established non-pathogenic model organisms and human cancer cell lines (all BSL-1) in fast-screening assays (Tables 3 and 4).

Table 3. Anthelmintic (*Caenorhabditis elegans*), antifungal (*Botrytis cinerea*, *Septoria tritici* and *Phytophthora infestans*), and antibacterial (*Bacillus subtilis*, *Aliivibrio fisheri*) activities of isolated compounds (**1–7**) from *P. obtusifolia*.

	Anthelmintic Assay	Antifungal Assays			Antibacterial Assays	
	Mortality [%]	Growth Inhibition [%] [a]			Growth Inhibition [%] [a]	
	C. elegans	*B. cinerea*	*S. tritici*	*P. infestans*	*B. subtilis*	*A. fischeri*
Compound	200 µM	125 µM	125 µM	125 µM	100 µM [b]	100 µM [b]
1	1.9 ± 0.8	−13.6 ± 17.3	−12.5 ± 17.9	10.1 ± 2.3	8.0 ± 19.0	−5.6 ± 2.60
2	0.0 ± 0.0	−27.6 ± 7.6	21.8 ± 8.9	−51.5 ± 16.0	7.0 ± 15.0	−3.1 ± 1.80
3	0.0 ± 0.0	−18.5 ± 3.1	−40.0 ± 35.2	56.2 ± 10.3	7.0 ± 2.8	−211.6 ± 27.6
4	2.9 ± 1.4	−32.3 ± 28.7	-9.8 ± 8.3	−66.4 ± 30.4	−21.0 ± 28.0	−35.9 ± 19.4
5	4.8 ± 1.4	27.0 ± 25.8	46.9 ± 8.3	−6.0 ± 64.4	90.0 ± 2.0	−32.7 ± 12.7
6	2.0 ± 1.7	51.4 ± 22.3	−12.9 ± 2.3	−9.4 ± 27.4	−5.0 ± 51.0	30.4 ± 8.4
7	1.7 ± 2.4	2.8 ± 19.5	−7.2 ± 1.9	17.1 ± 12.1	5.0 ± 17.0	1.6 ± 3.0
Pos. control	10 µg/mL ivermectin 98.7 ± 1.9	125 µM epoxiconazole 92.0 ± 1.4		125 µM terbinafine 96.8 ± 1.2	100 µM chloramphenicol 99.0 ± 0	100.0 ± 0

[a] Negative values indicate an increase in fungal or bacterial growth in comparison to the negative control (0% inhibition); [b] Growth inhibition rates below 50% indicate IC_{50} values > 100 µM.

Table 4. Antiproliferative and cytotoxic activities of isolated compounds (**1–7**) from *P. obtusifolia* against human prostate (PC-3) and colorectal (HT-29) cancer cell lines.

	Metabolic cell Viability (MTT)/Cytotoxicity (CV) Assays			
	Cell Viability/Survivals [%]			
	PC-3 (MTT Assay)	PC-3 (CV Assay)	HT-29 (MTT Assay)	HT-29 (CV Assay)
Compound	10 µM	10 µM	10 µM	10 µM
1	101.4 ± 5.8	97.5 ± 4.1	100.3 ± 4.4	91.0 ± 5.1
2	84.1 ± 4.9	84.0 ± 6.9	74.9 ± 6.5	78.2 ± 4.4
3	90.6 ± 3.3	98.3 ± 1.3	82.6 ± 4.8	100.3 ± 3.7
4	114.4 ± 1.2	92.9 ± 5.0	111.9 ± 1.8	89.0 ± 3.0
5	2.4 ± 6.3	−3.4 ± 24.7	59.7 ± 20.5	82.5 ± 4.3
6	120.9 ± 4.3	93.7 ± 2.6	117.1 ± 3.0	88.2 ± 3.9
7	113.7 ± 4.0	112.2 ± 4.7	115.3 ± 3.9	107.5 ± 2.9

Digitonin (125 µM) was used as a positive control, compromising the cells to the point of 0% cell viability/cancer cell survival after 48 h.

The anthelmintic activity was evaluated against *Caenorhabditis elegans*, however, the compounds did not significantly kill the nematodes even when applying a high concentration of 200 µM. Moreover, most of the compounds tested for antifungal effects against the phytopathogens *B. cinerea, S. tritici,* and *P. infestans* were considered as inactive since the inhibition of fungal growth at a concentration of 125 µM was less than 50%. An exception was *N*-[2-(3,4-dihydroxyphenyl)ethyl]-3,4-dihydroxybenzamide (**3**), which displayed interesting anti-oomycotic activity against *P. infestans* with an inhibition value of 56% at the same concentration. For the antibacterial assays, the compounds were tested in concentrations of 1 and 100 µM. None of the compounds inhibited the Gram-negative bacterium *A. fischeri* even at the highest test concentration of 100 µM. However, against the Gram-positive *B. subtilis,* peperobtusin A (**5**) induced a 90% inhibition of bacterial growth when treating at a concentration of 100 µM, prompting further investigations against human pathogens. Prenylated benzopyrans seem to be the active principle of *Peperomia* against Gram-positive bacteria, including resistant *Staphylococcus aureus* and *Enterococcus faecalis* strains [14]. Since all other isolated compounds induced less than 50% inhibition of bacterial growth when applying a concentration of 100 µM (Table 3), their half-maximal inhibitory concentration (IC_{50}) values must be estimated to be greater than 100 µM.

Furthermore, the compounds **1–7** were screened for their potential antiproliferative or cytotoxic activity against two human cancer lines, namely prostate adenocarcinoma cells (PC-3) and colorectal adenocarcinoma cells (HT-29). The compounds' effect on the metabolic cancer cell viability was determined by conducting an MTT assay, general cytotoxic effects were determined by using a CV assay, both after 48 h cancer cell treatment with the compounds under investigation. A very potent permeabilizer of cell membranes, digitonin (125 µM), was used as a positive control, compromising the cells to the point of 0% of cell viability after 48 h. The compounds were tested at two different concentrations, namely 10 nM and 10 µM.

As shown in Table 4, most of the compounds exhibited no significant antiproliferative and cytotoxic effects on PC-3 and HT-29 cancer cells, with the exception of peperobtusin A (**5**), which strongly inhibited the viability and growth of PC-3 cells in an MTT and CV fast-screening assay after treating the cells with 10 µM of the compound for 48 h. Under the same fast-screening assay conditions, a moderate activity of compound **5** against the HT-29 colorectal cancer cell line was also observed.

Based on its antiproliferative and cytotoxic effects, as determined with a 10 µM concentration in the initial fast-screening assay, compound **5** was subjected to further evaluations of its IC_{50} values in both cancer cell lines. The cells were treated with increasing concentrations of up to 100 µM of compound **5** for 48 h, followed by an MTT and CV assay read-out, data analyses, and calculation of IC_{50} values. The IC_{50} values of compound **5** determined by MTT assay, indicating the metabolic viability of the cancer cells, were 25.6 ± 2.8 µM and 45.5 ± 8.8 µM in PC-3 and HT-29 cells, respectively (see Figure 4A). The CV assay, reflecting general cytotoxic and antiproliferative impacts, showed similar IC_{50} values of 36.0 ± 4.1 µM and 58.1 ± 11.2 µM for PC-3 and HT-29 cells, respectively (Figure 4B). However, given the result of the preliminary fast-screening (fixed concentration 10 µM) of compound **5**, which estimated an IC_{50} value of less than 10 µM, the lower efficacy at the above actual calculated IC_{50} values based on the dose-response curves was unexpected, especially in the case of PC-3 cells. However, the IC_{50} values were reproduced in three different assay formats, each with 2-3 biological replicates, namely MTT, CV, and resazurin assays (data not shown), which revealed an IC_{50} range of 25–45 µM. This implies that fast-screening with a single concentration and a single biological replicate, although with technical quadruplicates, is only a very rough, initial estimation of the compounds' activity.

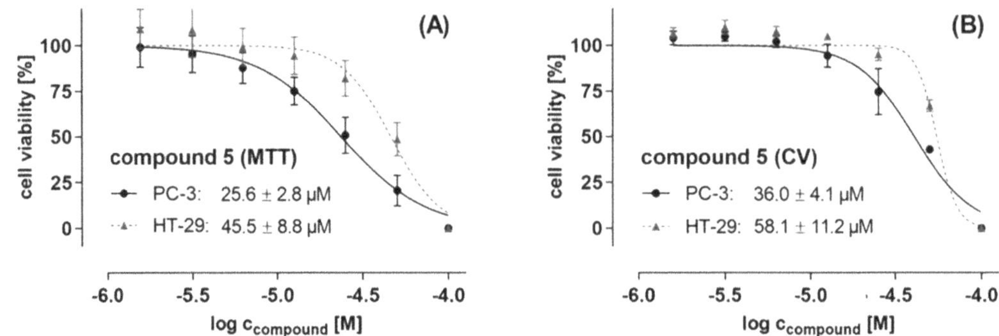

Figure 4. Effect of peperobtusin A (**5**) on (**A**) the metabolic cell viability of prostate PC-3 cancer cells and colon HT-29 cancer cells, as determined by MTT assay after 48 hours of cell treatment. (**B**) General cytotoxic and antiproliferative effect of **5** on both cancer cell lines under comparable treatment conditions, as determined by using crystal violet (CV) assay. Data represent biological duplicates, each comprising technical quadruplicates. IC_{50} curves were analyzed and drawn using SigmaPlot and GraphPad Prism software, respectively. IC_{50} values are given as mean $\pm$ SD.

Peperobtusin A (**5**) is a prenylated benzopyran with a chiral center at C-2. It was obtained with an optical activity of $[\alpha]_D^{20}$ 0 (c 0.28, $CHCl_3$), implying a racemic mixture of the molecule. This result corroborated previously published data [11,30]. The compound was isolated before from the same species [10,11,14] and other *Peperomia* species, such as *P. tetraphylla* [31] and *P. clusiifolia* [32]. Although there are a few existing studies addressing the biological activities of peperobtusin A (**5**), this study is the first report related to the antiproliferative and cytotoxic effect against PC-3 (prostate) and HT-29 (colon) cancer cell lines. Previous investigations indicated the antitrypanosomal and antitumor potential of peperobtusin A. Regarding the antitrypanosomal activity, the compound with an IC_{50} value of 3.1 µM was almost three times more active than the positive control benznidazole [11]. Furthermore, Da Silva Mota and collaborators also evaluated its cytotoxicity in mammalian cells (murine peritoneal macrophages) and found the compound to not be toxic at the concentration level of its trypanocidal activity [11]. A recent study by Shi et al [31] revealed that peperobtusin A (**5**) had antiproliferation effects against the human lymphoma cell line U937 with an IC_{50} value of 63.26 µM at 24 h, however, exerted only minor effects on other human cancer cell lines such as human melanoma (A375), human bladder carcinoma (T24), and human breast epithelial cells (HBL-100). According to Shi et al. [31], the intracellular ROS formation and the activation of caspases and P38 MAPK played significant roles in the apoptosis induced by peperobtusin A. Therefore, in future investigations, the mode of action of peperobtusin A (**5**) in PC3 and HT29 cell lines should be addressed in detail and expanded; toxicological investigations should be performed to validate the safety of this compound before developing it as trypanocide.

3. Materials and Methods

3.1. General Methods

The following instruments were used to obtain physical and spectroscopic data: column chromatography was performed on silica gel (400–630 mesh, Merck, Darmstadt, Germany), Sephadex LH-20 (Fluka, Steinheim, Germany), and Diaion HP20 (Supelco, Bellefonte, PA, USA). Fractions and substances were monitored by TLC. TLC was conducted on precoated Kieselgel 60 F 254 plates (Merck, Darmstadt, Germany) and the spots were detected either by examining the plates under a UV lamp at 254 and 366 nm or by treating the plates with vanillin or natural product reagents. UV spectra were recorded on a Jasco V-770 UV-Vis/NIR spectrophotometer (Jasco, Pfungstadt, Germany), meanwhile specific rotation was measured with a Jasco P-2000 digital polarimeter (Jasco, Pfungstadt,

Germany). The IR (ATR) spectra were carried out in MeOH using a Thermo Nicolet 5700 FT-IR spectrometer (Thermo Nicolet Analytical Instruments, Madison, WI, USA).

NMR spectra were obtained with an Agilent DD2 400 system at +25 °C (Varian, Palo Alto, CA, USA) using a 5 mm inverse detection cryoprobe. The compounds were dissolved in CD_3OD (99.8% D) or $CDCl_3$ (99.8% D) and the spectra were recorded at 399.915 MHz (^1H) and 100.569 MHz (^{13}C). One-dimensional (^1H and ^{13}C) and two-dimensional ($^1H,^{13}C$ HSQC, $^1H,^{13}C$ HMBC, $^1H,^1H$ COSY, and $^1H,^1H$ ROESY) spectra were measured using standard CHEMPACK 8.1 pulse sequences implemented in the Varian VNMRJ 4.2 spectrometer software (Varian, Palo Alto, CA, USA). ^1H chemical shifts are referenced to internal TMS (^1H δ = 0 ppm), while ^{13}C chemical shifts are referenced to CD_3OD (^{13}C δ = 49 ppm) or $CDCl_3$ (^{13}C δ = 77.0 ppm).

The semi-preparative HPLC was performed on a Shimadzu prominence system (Kyoto, Japan) equipped with LabSolutions software, LC-20AT pump, SPD-M20A diode array detector, SIL-20A auto sampler, and FRC-10A fraction collector unit. Chromatographic separation was carried out using a YMC Pack C18 column (5 μm, 120 Å, 150 mm × 10 mm I.D, YMC, USA) using H_2O (A) and CH_3CN (B) as eluents at a flow rate of 2.2 mL min^{-1}.

The high-resolution mass spectra in both positive and negative ion modes were acquired using either an Orbitrap Elite Mass spectrometer or API 3200 Triple Quadrupole System. The Orbitrap Elite Mass spectrometer (Thermofisher Scientific, Bremen, Germany) was equipped with an HESI electrospray ion source (spray voltage 4.0 kV, capillary temperature 275 °C, source heater temperature 80 °C, FTMS resolution 100.000), whereas the API 3200 Triple Quadrupole System (Sciex, Framingham, MA, USA) was equipped with a turbo ion spray source, which performs ionization with an ion spray voltage on 70 eV. During the measurement, the mass/charge range from 50 to 2000 was scanned.

3.2. Plant Material

The aerial part of *P. obtusifolia* (L.) A. Dietr. was collected in April 2018 at the Leibniz Institute of Plant Biochemistry (IPB), Halle (Saale), Germany from an ornamental plant grown for 20 years on an east oriented window sill. A voucher specimen (ISW010) was deposited in the IPB collection.

3.3. Extraction and Isolation

The fresh leaves (267.3 g) of *P. obtusifolia* were dried using a freeze drier and ground to powder form (29.7 g). The powdered material was extracted five times (1.0 L/each) with 80% aqueous methanol at room temperature, yielding 6.4 g of residue after evaporating the solvent in vacuo. The residue was suspended in H_2O (250 mL) and then successively partitioned with *n*-hexane and ethyl acetate, resulting in the two respective fractions (1.4 g *n*-hexane, 0.8 g EtOAc), as well as an aqueous fraction (3.6 g) (See Figure S20 for the isolation scheme).

Part of the EtOAc fraction (0.74 g) was chromatographed on a silica gel column and eluted with a stepwise gradient of CH_2Cl_2: MeOH (1:0 to 0:1) to give seven fractions (E1-E7). Fraction E1 (32.5 mg) underwent column chromatographic separation over a Sephadex LH-20 column (CH_2Cl_2/MeOH 1:1), affording four subfractions (A-D). Further separation of subfraction C (4.3 mg) over silica gel (CH_2Cl_2:MeOH 95:5–90:10) yielded becatamide (**4**, 1.4 mg, R_f = 0.25 in 5% MeOH in CH_2Cl_2) [21]. Fraction E2 (190.5 mg) was chromatographed over a Sephadex-LH20 column (100% MeOH) to afford an amorphous powder that was identified as *N*-[2-(3,4-dihydroxyphenyl)ethyl]-3,4-dihydroxybenz amide (**3**, 142.5 mg, R_f = 0.37 in 10% MeOH in CH_2Cl_2) [20]. Fraction E4 (106.3 mg) was purified over semi-preparative HPLC in a gradient system [H_2O (A), CH_3CN (B); 0 min: 5% B > 0–17 min: 85.5% B > 17–20 min: 100% B, UV at 330 nm, 9.8 min], affording compound **2** (14.5 mg).

The *n*-hexane fraction (1.4 g) was chromatographed over a silica gel column and eluted with a stepwise gradient of *n*-hexane, ethyl acetate, and methanol to give six fractions (H1-H6). Fraction H1 (197.4 mg) was further chromatographed over a Sephadex-LH 20 column (100% CH_2Cl_2) to afford four subfractions (A-D). Purification of subfraction C

(29.1 mg) by silica gel CC (*n*-hex:CH$_2$Cl$_2$ 7:3) yielded peperobtusin A (**5**, 8.1 mg, R$_f$ = 0.76 in 5% CH$_3$OH in CH$_2$Cl$_2$) [10]. Fraction H2 (279.6 mg) was chromatographed over a Sephadex LH-20 column (CH$_2$Cl$_2$:MeOH 1:1) to give five subfractions (A-E). Subfraction C (14.7 mg) was purified by silica gel CC (*n*-hex: CH$_2$Cl$_2$ 3:7–0:1) to afford peperomin B (**6**, 3.1 mg, R$_f$ = 0.04 in CH$_2$Cl$_2$) [22].

The aqueous fraction (2.3 g) was fractionated over a Diaion-HP20 column eluted with an increasing solvent ratio of water and methanol to give five fractions (A1–A5). Fraction A3 (404.7 mg) was chromatographed over a Sepahdex-LH20 column (100% MeOH) to afford seven subfractions (A-G). Subfraction B (55.2 mg) was further separated by silica gel CC (CH$_2$Cl$_2$:MeOH 7:3) to afford arabinothalictoside (**7**, 14.6 mg, R$_f$ = 0.57 in 30% MeOH in CH$_2$Cl$_2$) [23]. Subfraction D (69.4 mg) was further chromatographed over a silica gel column (CH$_2$Cl$_2$:MeOH 8:2–0:1), yielding compound **1** (7.4 mg, R$_f$ = 0.88 in 30% MeOH in CH$_2$Cl$_2$).

Peperomic ester (trimethyl (1*S*,2*R*)-1-(((*E*)-3-(3,4-dihydroxyphenyl)acryloyl)oxy)- propane-1,2,3-tricarboxylate, **1**): brownish liquid; IR (ATR) v_{max}: 3402, 1715 cm^{-1}. UV λ_{max} (MeOH) (log e) 218 (4.15), 333 nm (4.21): ^{1}H-NMR (400 MHz, CD$_3$OD): Table 1; ^{13}C-NMR (100 MHz, CD$_3$OD): Table 1; negative-ion HR-ESI-MS *m/z* 395.0975 [M − H]$^-$ (calcd. for C$_{18}$H$_{19}$O$_{10}$: 395.0984).

Peperoside (*N*-[2-(3,4-dihydroxyphenyl)ethyl]-3,4-dihydroxybenzamide 4′-*O*-β-D-glucoside **2**): white amorphous powder, IR (ATR) v_{max}: 3250, 2931, 2360, and 1607 cm^{-1}. UV λ_{max} (log e) (MeOH) 290 (3.14), 317 nm (3.32). ^{1}H-NMR (400 MHz, CD$_3$OD): Table 2; ^{13}C-NMR (100 MHz, CD$_3$OD): Table 2; negative-ion HR-ESI-MS *m/z* 450.1379 [M − H]$^-$ (calcd. for C$_{21}$H$_{24}$NO$_{10}$: 450.1406).

3.4. Biological activities

3.4.1. Anthelmintic Activity

The anthelmintic bioassay was conducted by applying the method developed by Thomsen et al. using the model organism *C. elegans*, which was previously demonstrated to correlate with anthelmintic activity against parasitic trematodes [33]. The *C. elegans* Bristol N2 wild-type strain was obtained from the Caenorhabditis Genetic Center (CGC), Minnesota University, Minneapolis, USA. The nematodes were grown on petri plates of NGM (Nematode Growth Media) using uracil auxotroph *E. coli* strain OP50 as the food source. The solvent DMSO (2%) and the anthelmintic drug ivermectin (10 µg/mL, nearly 100% dead worms after 30 min incubation) were used as negative and positive controls in all assays, respectively. All assays were carried out in triplicate.

3.4.2. Antifungal Activity

The antifungal activity was examined against the phytopathogenic ascomycetes *B. cinerea* Pars and *S. triciti* Desm. and the oomycete *P. infestans* (Mont.) de Bary in 96-well microtiter plate assays according to protocols from the Fungicide Resistance Action Committee (FRAC) with minor modifications as described by Otto et al. [34]. Briefly, the isolated compounds were tested at the highest concentration of 125 µM, while the solvent DMSO was used as a negative control (max. concentration 2.5%). The commercially used fungicides, epoxiconazole and terbinafine (Sigma-Aldrich, Darmstadt, Germany), served as reference compounds. The pathogen growth was evaluated seven days after inoculation by measurement of the optical density (OD) at λ 405 nm with a TecanGENios Pro microplate reader (five measurements per well using multiple reads in 3 × 3 square). Each experiment was carried out in triplicate.

3.4.3. Antibacterial Activity

The isolated compounds (1 and 100 µM) were evaluated against the Gram-negative *A. fischerii* (DSM507) and the Gram-positive *B. subtilis* 168 (DSM 10), as described by dos Santos et al. [35]. The tests were conducted in 96-well plates based on the bioluminescence (*A. fisheri*) or absorption (*B. subtilis*) read-out. Chloramphenicol (100 µM) was used as a positive control to induce complete inhibition of bacterial growth. The results (mean $\pm$ standard deviation value, $n = 6$) are given in relation to the negative control (bacterial growth, 1% DMSO without test compound) as relative values (percent inhibition). Negative values indicate an increase in bacterial growth.

3.4.4. Cytotoxic Activity

The cytotoxicity and impact on the metabolic cell viability of isolated compounds at 10 nM and 10 µM was evaluated against PC-3 (human prostate adenocarcinoma) and HT-29 (human colorectal adenocarcinoma) cancer cell lines. Both cell lines were purchased from ATCC (Manassas, VA, USA). The cell culture medium RPMI 1640, the supplements FCS and L-glutamine, as well as PBS and trypsin/EDTA were purchased from Capricorn Scientific GmbH (Ebsdorfergrund, Germany). Culture flasks, multi-well plates, and further cell culture plastics were from Greiner Bio-One GmbH (Frickenhausen, Germany) and TPP (Trasadingen, Switzerland), respectively. Resazurin used for the cell viability assays was purchased from Sigma-Aldrich GmbH (Taufkirchen, Germany). PC-3 and HT-29 cells were cultured in RPMI 1640 media supplemented with 10% heat-inactivated FCS, 2 mM L-glutamine, and 1% penicillin/streptomycin in a humidified atmosphere with 5% CO_2 at 37 °C. Routinely, cells were cultured in T-75 flasks until reaching subconfluency (~80%); subsequently, cells were harvested by washing with PBS and detaching by using trypsin/EDTA (0.05% in PBS) prior to cell passaging and seeding for subculturing and assays in 96-well plates, respectively [36].

The cell handling and assay techniques were in accordance with the method described by Khan et al. [36]. In brief, antiproliferative and cytotoxic effects of the compounds were investigated by performing colorimetric MTT (3-(4,5-dimethylthiazol-2-yl)-2,5-diphenyltetrazolium bromide) and CV (crystal violet)-based cell viability assays (Sigma-Aldrich, Taufkirchen, Germany), respectively. For this purpose, cells were seeded in low densities in 96-well plates using the aforementioned cell culture medium. The cells were allowed to adhere for 24 h, followed by the 48 h compound treatment. Based on 20 mM DMSO stock solutions, the compounds were serially diluted in standard growth media to reach final concentrations of 100, 50, 25, 12.5, 6.25, 3.125, and 1.56 µM for cell treatment. For control measures, cells were treated in parallel with 125 µM digitonin (positive control, for data normalization set to 0% cell viability). Each data point was determined in technical quadruplicates and two independent biological replicates. As soon as the 48 h incubation was finished, cell viability was measured.

For the MTT assay, cells were washed once with PBS, followed by incubation with MTT working solution (0.5 mg/mL MTT in culture medium) for 1 h under standard growth conditions. After discarding the MTT solution, DMSO was added in order to dissolve the formed formazan, followed by measuring formazan absorbance at 570 nm, and additionally, at the reference/background wavelength of 670 nm by using a SpectraMax M5 multi-well plate reader (Molecular Devices, San Jose, CA, USA).

For the CV assay, cells were washed once with PBS and fixed with 4% paraformaldehyde (PFA) for 20 min at room temperature (RT). After discarding the PFA solution, the cells were left to dry for 10 min and then stained with 1% crystal violet solution for 15 min at RT. The cells were washed with water and dried overnight at RT. Afterwards, acetic acid (33% in ultrapure water) was added to the stained cells and absorbance was measured at 570 nm and 670 nm (reference wavelength) using a SpectraMax M5 multi-well plate reader (Molecular Devices, San Jose, CA, USA). For data analyses, GraphPad Prism version 8.0.2, SigmaPlot 14.0 and Microsoft Excel 2013 were used.

The results are shown as a percentage of the control values obtained from untreated cultures, i.e. cell viability in percent.

4. Conclusions

The present investigation of a methanol extract from the leaves of *P. obtusifolia* led to the isolation of two new compounds named peperomic ester (**1**) and peperoside (**2**), along with five known compounds. The isolated compounds were evaluated for their anthelmintic, antifungal, antibacterial, and anticancer activities. Peperoside (**2**) and its related compounds (**3, 4**) have the advantage of low activity against all organisms/cells tested. Since similar phenethylamides show taste and neuroactive properties [37,38]—which were not part of this study—their use in such tests, which usually include human panels or animals, appears acceptable.

The known peperobtusin A (**5**) was the most active compound against *B. subtilis* and showed an antiproliferative and cytotoxic effect in the lower micromolar concentration range towards human PC-3 (prostate) and HT-29 (colon) cancer cell lines. The observed antibacterial activity supports the traditional use of *P. obtusifolia* as a skin cleanser. These findings suggest that *P. obtusifolia* might have some general toxicity and thus should be used with care in applications where cytotoxicity is not desired. However, prenylated benzopyran derivatives structurally related to **5** were shown to exhibit selective activity against breast cancer cells but did not influence normal breast cells [39]. The impact of this compound class on the mitochondrial respiratory chain [39,40] and their function as PPAR agonists may play a role in compound selectivity [41]. Thus, more extensive and detailed studies on the anticancer activity of natural prenylated benzopyrans should be undertaken in order to provide an understanding of structural requirements (SAR) for such cytotoxic (or other) activity.

Supplementary Materials: The following supporting information can be downloaded at: https://www.mdpi.com/article/10.3390/molecules27144363/s1, Figure S1: ^{1}H NMR spectrum of compound **1** (400 MHz, MeOH-d_4), Figure S2: ^{13}C NMR spectrum of compound **1** (100 MHz, MeOH-d_4), Figure S3: HSQC spectrum of compound **1** (400 MHz, MeOH-d_4), Figure S4: HMBC spectrum of compound **1** (400 MHz, MeOH-d_4), Figure S5: ^{1}H-^{1}H COSY spectrum of compound **1** (400 MHz, MeOH-d_4), Figure S6: UV spectrum of compound **1** in MeOH, Figure S7: IR spectrum of compound **1** in MeOH, Figure S8: ESI-HRMS spectrum of compound **1** in negative ion mode, Figure S9: HRMS spectrum from ethanol extract of *P. obtusifolia* with selected molecular ion of *m/z* 395.0979 [M − H]$^-$, Figure S10: ^{1}H NMR spectrum of compound **2** (400 MHz, MeOH-d_4), Figure S11: ^{13}C NMR spectrum of compound **2** (100 MHz, MeOH-d_4), Figure S12: DEPT135 spectrum of compound **2** (100 MHz, MeOH-d_4), Figure S13: HSQC spectrum of compound **2** (400 MHz, MeOH-d_4), Figure S14: HMBC spectrum of compound **2** (400 MHz, MeOH-d_4), Figure S15: ^{1}H-^{1}H COSY spectrum of compound **2** (400 MHz, MeOH-d_4), Figure S16: 1D-ROESY spectrum of compound **2** (400 MHz, MeOH-d_4), Figure S17: UV spectrum of compound **2** in MeOH, Figure S18: IR spectrum of compound **2** in MeOH, Figure S19: ESI-HRMS spectrum of compound **2** in negative ion mode, Figure S20. Isolation scheme for compounds **1–7**, Table S1: Polarimeter data of compound **1**.

Author Contributions: Conceptualization, L.A.W.; methodology, I.W. and K.F.; formal analysis, I.W., K.F., H.H., I.M. and R.R.; investigation, I.W.; resources, L.A.W.; data curation, I.W. and K.F.; writing—original draft preparation, I.W., K.F. and R.R.; writing—review and editing, I.W., R.R., K.F. and L.A.W.; supervision, K.F. and L.A.W.; project administration, K.F. and L.A.W.; funding acquisition, I.W. and L.A.W. All authors have read and agreed to the published version of the manuscript.

Funding: This study was supported by the German Academic Exchange Service (DAAD) as a grant scholarship part of the Ph.D. thesis of I.W. Funding program/-ID: Research Grants—Doctoral Programs in Germany, 2017/18 (57299294), ST34. The funder had no role in the study design, decision to publish, or manuscript preparation.

Institutional Review Board Statement: Not applicable.

Informed Consent Statement: Not applicable.

Data Availability Statement: The data presented in this study are available on request from the corresponding author.

Acknowledgments: The authors are indebted to Anke Dettmer, Mthandazo Dube, and Martina Lerbs (all IPB Halle) for their performance of antibacterial, antifungal, anthelmintic, and anticancer bioassays, respectively. Thanks to Andrea Porzel and Gudrun Hahn (IPB Halle) for NMR measurements and Mandy Dorn for HRMS measurements.

Conflicts of Interest: The authors declare that they have no known competing financial interest or personal relationships that could have appeared to influence the work reported in this paper.

References

1. Felippe, L.G.; Baldoqui, D.C.; Kato, M.J.; Bolzani, V.d.S.; Guimarães, E.F.; Cicarelli, R.M.B.; Furlan, M. Trypanocidal tetrahydrofuran lignans from *Peperomia blanda*. *Phytochemistry* **2008**, *69*, 445–450. [CrossRef] [PubMed]
2. Al-Madhagi, W.M.; Mohd Hashim, N.; Awad Ali, N.A.; Alhadi, A.A.; Abdul Halim, S.N.; Othman, R. Chemical profiling and biological activity of *Peperomia blanda* (Jacq.) Kunth. *PeerJ* **2018**, *6*, e4839. [CrossRef] [PubMed]
3. Malquichagua Salazar, K.J.; Delgado Paredes, G.E.; Lluncor, L.R.; Young, M.C.M.; Kato, M.J. Chromenes of polyketide origin from *Peperomia villipetiola*. *Phytochemistry* **2005**, *66*, 573–579. [CrossRef] [PubMed]
4. Seeram, N.P.; Lewis, A.W.; Jacobs, H.; Nair, M.G.; McLean, S.; Reynolds, W.F. Proctoriones A−C: 2-Acylcyclohexane-1,3-dione derivatives from *Peperomia proctorii*. *J. Nat. Prod.* **2000**, *63*, 399–402. [CrossRef] [PubMed]
5. Wu, J.; Li, N.; Hasegawa, T.; Sakai, J.; Kakuta, S.; Tang, W.; Oka, S.; Kiuchi, M.; Ogura, H.; Kataoka, T.; et al. Bioactive tetrahydrofuran lignans from *Peperomia dindygulensis*. *J. Nat. Prod.* **2005**, *68*, 1656–1660. [CrossRef] [PubMed]
6. Xu, S.; Li, N.; Ning, M.M.; Zhou, C.H.; Yang, Q.R.; Wang, M.W. Bioactive compounds from *Peperomia pellucida*. *J. Nat. Prod.* **2006**, *69*, 247–250. [CrossRef] [PubMed]
7. Li, N.; Wu, J.; Hasegawa, T.; Sakai, J.; Bai, L.; Wang, L.; Kakuta, S.; Furuya, Y.; Ogura, H.; Kataoka, T.; et al. Bioactive polyketides from *Peperomia duclouxii*. *J. Nat. Prod.* **2007**, *70*, 998–1001. [CrossRef] [PubMed]
8. Wang, X.; Cheng, Y.; Wu, H.; Li, N.; Liu, R.; Yang, X.; Qiu, Y.Y.; Wen, H.; Liang, J. The natural secolignan peperomin E induces apoptosis of human gastric carcinoma cells via the mitochondrial and PI3K/Akt signaling pathways in vitro and in vivo. *Phytomedicine* **2016**, *23*, 818–827. [CrossRef] [PubMed]
9. Batista, A.N.L.; Santos-Pinto, J.; Batista, J.M., Jr.; Souza-Moreira, T.M.; Santoni, M.M.; Zanelli, C.F.; Kato, M.J.; López, S.N.; Palma, M.S.; Furlan, M. The combined use of proteomics and transcriptomics reveals a complex secondary metabolite network in *Peperomia obtusifolia*. *J. Nat. Prod.* **2017**, *80*, 1275–1286. [CrossRef]
10. Tanaka, T.; Asai, F.; Iinuma, M. Phenolic compounds from *Peperomia obtusifolia*. *Phytochemistry* **1998**, *49*, 229–232. [CrossRef]
11. da Silva Mota, J.; Leite, A.C.; Batista Junior, J.M.; Noelí López, S.; Luz Ambrósio, D.; Duó Passerini, G.; Kato, M.J.; da Silva Bolzani, V.; Barretto Cicarelli, R.M.; Furlan, M. In vitro trypanocidal activity of phenolic derivatives from *Peperomia obtusifolia*. *Planta Med.* **2009**, *75*, 620–623. [CrossRef]
12. Mota, J.d.S.; Leite, A.C.; Kato, M.J.; Young, M.C.M.; Bolzani, V.d.S.; Furlan, M. Isoswertisin flavones and other constituents from *Peperomia obtusifolia*. *Nat. Prod. Res.* **2011**, *25*, 1–7. [CrossRef] [PubMed]
13. Batista, J.M.; Batista, A.N.L.; Kato, M.J.; Bolzani, V.S.; López, S.N.; Nafie, L.A.; Furlan, M. Further monoterpene chromane esters from *Peperomia obtusifolia*: VCD determination of the absolute configuration of a new diastereomeric mixture. *Tetrahedron Lett.* **2012**, *53*, 6051–6054. [CrossRef]
14. Ruiz Mostacero, N.; Castelli, M.V.; Cutró, A.C.; Hollmann, A.; Batista, J.M., Jr.; Furlan, M.; Valles, J.; Fulgueira, C.L.; López, S.N. Antibacterial activity of prenylated benzopyrans from *Peperomia obtusifolia* (Piperaceae). *Nat. Prod. Res.* **2021**, *35*, 1706–1710. [CrossRef] [PubMed]
15. Tamayose, C.I.; Romoff, P.; Toyama, D.O.; Gaeta, H.H.; Costa, C.R.C.; Belchor, M.N.; Ortolan, B.D.; Velozo, L.S.M.; Kaplan, M.A.C.; Ferreira, M.J.P.; et al. Non-clinical studies for evaluation of 8-c-rhamnosyl apigenin purified from *Peperomia obtusifolia* against acute edema. *Int. J. Mol. Sci.* **2017**, *18*, 1972. [CrossRef]
16. Wu, W.; Zhao, F.; Ding, R.; Bao, L.; Gao, H.; Lu, J.C.; Yao, X.S.; Zhang, X.Q.; Liu, H.W. Four new cryptoporic acid derivatives from the fruiting bodies of *Cryptoporus sinensis*, and their inhibitory effects on nitric oxide production. *Chem. Biodivers.* **2011**, *8*, 1529–1538. [CrossRef]
17. Zhou, L.Y.; Yu, X.H.; Lu, B.; Hua, Y. Bioassay-guided isolation of cytotoxic isocryptoporic acids from *Cryptoporus volvatus*. *Molecules* **2016**, *21*, 1692. [CrossRef] [PubMed]
18. Pham, H.T.; Lee, K.H.; Jeong, E.; Woo, S.; Yu, J.; Kim, W.Y.; Lim, Y.W.; Kim, K.H.; Kang, K.B. Species prioritization based on spectral dissimilarity: A case study of polyporoid fungal species. *J. Nat. Prod.* **2021**, *84*, 298–309. [CrossRef] [PubMed]
19. Chepkirui, C.; Yuyama, K.T.; Wanga, L.A.; Decock, C.; Matasyoh, J.C.; Abraham, W.-R.; Stadler, M. Microporenic acids A–G, biofilm inhibitors, and antimicrobial agents from the Basidiomycete *Microporus* species. *J. Nat. Prod.* **2018**, *81*, 778–784. [CrossRef]
20. Liao, Y.; Ye, Y.; Li, S.; Zhuang, Y.; Chen, L.; Chen, J.; Cui, Z.; Huo, L.; Liu, S.; Song, G. Synthesis and SARs of dopamine derivatives as potential inhibitors of influenza virus PAN endonuclease. *Eur. J. Med. Chem.* **2020**, *189*, 112048. [CrossRef]
21. Park, J.B. Protective effects of veskamide, enferamide, becatamide, and oretamide on H_2O_2-induced apoptosis of PC-12 cells. *Phytomedicine* **2011**, *18*, 843–847. [CrossRef] [PubMed]
22. Chen, C.-M.; Jan, F.Y.; Chen, M.-T.; Lee, T.-J. Peperomins A, B, and C, novel secolignans from *Peperomia japonica*. *Heterocycles* **1989**, *29*, 411–414. [CrossRef]
23. Li, H.X.; Yang, S.Y.; Kim, Y.S.; Jang, H.D.; Kim, Y.H.; Li, W. Nitro derivatives and other compounds from sugar apple (*Annona squamosa* L.) leaves exhibit soluble epoxide hydrolase inhibitory activity. *Med. Chem. Res.* **2019**, *28*, 1939–1944. [CrossRef]
24. Li, Y.Z.; Tong, A.P.; Huang, J. Two new norlignans and a new lignanamide from *Peperomia tetraphylla*. *Chem. Biodivers.* **2012**, *9*, 769–776. [CrossRef]

25. Yoshikawa, M.; Yamaguchi, S.; Murakami, T.; Matsuda, H.; Yamahara, J.; Murakami, N. Absolute stereostructures of trifoliones A, B, C, and D, new biologically active diterpenes from the tuber of *Sagittaria trifolia* L. *Chem. Pharm. Bull.* **1993**, *41*, 1677–1679. [CrossRef] [PubMed]

26. Zhou, Z.; Luo, J.; Pan, K.; Kong, L. A new alkaloid glycoside from the rhizomes of *Aristolochia fordiana*. *Nat. Prod. Res.* **2014**, *28*, 1065–1069. [CrossRef]

27. Alves, N.S.F.; Setzer, W.N.; da Silva, J.K.R. The chemistry and biological activities of *Peperomia pellucida* (Piperaceae): A critical review. *J. Ethnopharmacol.* **2019**, *232*, 90–102. [CrossRef]

28. Ng, Z.X.; Than, M.J.Y.; Yong, P.H. *Peperomia pellucida* (L.) Kunth herbal tea: Effect of fermentation and drying methods on the consumer acceptance, antioxidant and anti-inflammatory activities. *Food Chem.* **2021**, *344*, 128738. [CrossRef]

29. Gutierrez, Y.V.; Yamaguchi, L.F.; de Moraes, M.M.; Jeffrey, C.S.; Kato, M.J. Natural products from *Peperomia*: Occurrence, biogenesis and bioactivity. *Phytochem Rev.* **2016**, *15*, 1009–1033. [CrossRef]

30. Batista, J.M., Jr.; Lopez, S.N.; Mota Jda, S.; Vanzolini, K.L.; Cass, Q.B.; Rinaldo, D.; Vilegas, W.; Bolzani Vda, S.; Kato, M.J.; Furlan, M. Resolution and absolute configuration assignment of a natural racemic chromane from *Peperomia obtusifolia* (Piperaceae). *Chirality* **2009**, *21*, 799–801. [CrossRef]

31. Shi, L.; Qin, H.; Jin, X.; Yang, X.; Lu, X.; Wang, H.; Wang, R.; Yu, D.; Feng, B. The natural phenolic peperobtusin A induces apoptosis of lymphoma U937 cells via the Caspase dependent and p38 MAPK signaling pathways. *Biomed. Pharmacother.* **2018**, *102*, 772–781. [CrossRef] [PubMed]

32. Seeram, N.P.; Jacobs, H.; McLean, S.; Reynolds, W.F. A prenylated benzopyran derivative from *Peperomia clusiifolia*. *Phytochemistry* **1998**, *49*, 1389–1391. [CrossRef]

33. Thomsen, H.; Reider, K.; Franke, K.; Wessjohann, L.A.; Keiser, J.; Dagne, E.; Arnold, N. Characterization of constituents and anthelmintic properties of *Hagenia abyssinica*. *Sci. Pharm.* **2012**, *80*, 433–446. [CrossRef] [PubMed]

34. Otto, A.; Porzel, A.; Schmidt, J.; Brandt, W.; Wessjohann, L.; Arnold, N. Structure and absolute configuration of pseudohygrophorones A^{12} and B^{12}, alkyl cyclohexenone derivatives from *Hygrophorus abieticola* (Basidiomycetes). *J. Nat. Prod.* **2016**, *79*, 74–80. [CrossRef] [PubMed]

35. dos Santos, C.H.C.; de Carvalho, M.G.; Franke, K.; Wessjohann, L. Dammarane-type triterpenoids from the stem of *Ziziphus glaziovii* Warm. (Rhamnaceae). *Phytochemistry* **2019**, *162*, 250–259. [CrossRef] [PubMed]

36. Khan, M.F.; Nasr, F.A.; Noman, O.M.; Alyhya, N.A.; Ali, I.; Saoud, M.; Rennert, R.; Dube, M.; Hussain, W.; Green, I.R.; et al. Cichorins D-F: Three new compounds from *Cichorium intybus* and their biological effects. *Molecules* **2020**, *25*, 4160. [CrossRef]

37. Krammer, G.; Ley, J.P.; Geißler, K.; Geißler, T.; Gomoll, F.; Welters, P.; Jach, G.; Wessjohann, L. Method for Biotechnological Production of Methylized Cinnamic Acids and Cinnamic Acid Esters, Methylized Phenethylamines and the Coupling Products Thereof, Particularly of Cinnamic Acid Amides. U.S. Patent US10227620B2, 12 March 2019.

38. Greger, H, Alkamides: A critical reconsideration of a multifunctional class of unsaturated fatty acid amides. *Phytochem. Rev.* **2016**, *15*, 729–770. [CrossRef]

39. Taha, H.; Looi, C.Y.; Arya, A.; Wong, W.F.; Yap, L.F.; Hasanpourghadi, M.; Mohd, M.A.; Paterson, I.C.; Ali, H.A. (6E,10E) Isopolycerasoidol and (6E,10E) isopolycerasoidol methyl ester, prenylated benzopyran derivatives from *Pseuduvaria monticola* induce mitochondrial-mediated apoptosis in human breast adenocarcinoma cells. *PLoS ONE* **2015**, *10*, e0126126. [CrossRef] [PubMed]

40. Zafra-Polo, M.C.; González, M.C.; Tormo, J.R.; Estornell, E.; Cortes, D. Polyalthidin: New prenylated benzopyran inhibitor of the mammalian mitochondrial respiratory chain. *J. Nat. Prod.* **1996**, *10*, 913–916. [CrossRef] [PubMed]

41. Corés Martinez, D.M.; Sanz Ferrando, M.J.; Bermejo del Castillo, A.; Piqueras Ruiz, L.; Collado Sánchez, A.; Gomes Marques, P.; Vila Dasí, L. Prenylated Benzopyranes as PPAR. Agonists Patent WO2019129909, 4 July 2019.

 *molecules*

Article

Three New Dihydrophenanthrene Derivatives from *Cymbidium ensifolium* and Their Cytotoxicity against Cancer Cells

Tajudeen O. Jimoh [1,2], Bruno Cesar Costa [1,3], Chaisak Chansriniyom [4,5], Chatcham Chaotham [3,6], Pithi Chanvorachote [6,7], Pornchai Rojsitthisak [8,9], Kittisak Likhitwitayawuid [4] and Boonchoo Sritularak [4,9,*]

[1] Pharmaceutical Sciences and Technology Program, Faculty of Pharmaceutical Sciences, Chulalongkorn University, Bangkok 10330, Thailand; jimmmypeace@gmail.com (T.O.J.); brunocesarsoares@outlook.com.br (B.C.C.)

[2] Department of Biochemistry, Faculty of Health Sciences, Islamic University in Uganda, Kampala P.O. Box 7689, Uganda

[3] Department of Biochemistry and Microbiology, Faculty of Pharmaceutical Sciences, Chulalongkorn University, Bangkok 10330, Thailand; cchoatham@gmail.com

[4] Department of Pharmacognosy and Pharmaceutical Botany, Faculty of Pharmaceutical Sciences, Chulalongkorn University, Bangkok 10330, Thailand; chaisak.ch@chula.ac.th (C.C.); kittisak.l@chula.ac.th (K.L.)

[5] Natural Products and Nanoparticles Research Unit, Chulalongkorn University, Bangkok 10330, Thailand

[6] Center of Excellence in Cancer Cell and Molecular Biology, Faculty of Pharmaceutical Sciences, Chulalongkorn University, Bangkok 10330, Thailand; pithi.c@chula.ac.th

[7] Department of Pharmacology and Physiology, Faculty of Pharmaceutical Sciences, Chulalongkorn University, Bangkok 10330, Thailand

[8] Department of Food and Pharmaceutical Chemistry, Faculty of Pharmaceutical Sciences, Chulalongkorn University, Bangkok 10330, Thailand; pornchai.r@chula.ac.th

[9] Center of Excellence in Natural Products for Ageing and Chronic Diseases, Faculty of Pharmaceutical Sciences, Chulalongkorn University, Bangkok 10330, Thailand

* Correspondence: boonchoo.sr@chula.ac.th

Citation: Jimoh, T.O.; Costa, B.C.; Chansriniyom, C.; Chaotham, C.; Chanvorachote, P.; Rojsitthisak, P.; Likhitwitayawuid, K.; Sritularak, B. Three New Dihydrophenanthrene Derivatives from *Cymbidium ensifolium* and Their Cytotoxicity against Cancer Cells. *Molecules* **2022**, *27*, 2222. https://doi.org/10.3390/molecules27072222

Academic Editor: Jacqueline Aparecida Takahashi

Received: 7 March 2022
Accepted: 23 March 2022
Published: 29 March 2022

Publisher's Note: MDPI stays neutral with regard to jurisdictional claims in published maps and institutional affiliations.

Abstract: From the aerial parts of *Cymbidium ensifolium*, three new dihydrophenanthrene derivatives, namely, cymensifins A, B, and C (**1–3**) were isolated, together with two known compounds, cypripedin (**4**) and gigantol (**5**). Their structures were elucidated by analysis of their spectroscopic data. The anticancer potential against various types of human cancer cells, including lung, breast, and colon cancers as well as toxicity to normal dermal papilla cells were assessed via cell viability and nuclear staining assays. Despite lower cytotoxicity in lung cancer H460 cells, the higher % apoptosis and lower % cell viability were presented in breast cancer MCF7 and colon cancer CaCo$_2$ cells treated with 50 µM cymensifin A (**1**) for 24 h compared with the treatment of 50 µM cisplatin, an available chemotherapeutic drug. Intriguingly, the half-maximum inhibitory concentration (IC$_{50}$) of cymensifin A in dermal papilla cells at >200 µM suggested its selective anticancer activity. The obtained information supports the further development of a dihydrophenanthrene derivative from *C. ensifolium* as an effective chemotherapy with a high safety profile for the treatment of various cancers.

Keywords: *Cymbidium ensifolium*; Orchidaceae; dihydrophenanthrene; dihydrophenanthrenequinone; anticancer

1. Introduction

Cymbidium spp. (Orchidaceae) comprises about fifty species widely dispersed across tropical and temperate Asia, especially in Thailand, China, and Nepal. Their beautiful flowers and structural appearance have made them more known for ornamental purposes rather than medicinal potentials. However, various parts of *Cymbidium* spp. are used in folkloric medicine to treat/manage many diseases. The leaves, roots, and pseudobulb of some *Cymbidium* spp. are employed in Nepali folkloric medicine for the treatment of boils, fever, paralysis, chronic illnesses, otitis, and as a tonic on rift bones [1]. Some species in this genus have been used as traditional Thai medicine. For example, the leaves of *C. aloifolium* and *C. findlaysonianum* have been used to treat otitis media [2]. *Cymbidium* plants are rich sources of phenanthrenes and bibenzyls, some of which exhibit several pharmacological activities such as cytotoxic, antimicrobial, free radical scavenging, and anti-inflammatory activities [3–5].

Cymbidium ensifolium (L.) Sw., known as "nang kham" or "chulan" in Thai [6], is an elegantly shaped orchid with beautiful aromatic flowers. Its flowers are white blended with yellow and red longitudinal lines along with the sepals and petals (Figure 1). It is distributed in north and northeastern Thailand [7]. The roots of this plant have been used as traditional Thai medicine to alleviate liver dysfunction and nephropathy [2,8]. In our continuing studies on bioactive compounds from orchids [9,10], we have explored the chemical constituents from the aerial parts of *C. ensifolium* with no previous reports on the chemical investigation. In the present study, a methanolic extract prepared from the aerial parts of this plant exhibited a significant cytotoxic effect against various cancer cells. This prompted us to investigate the plant to identify the cytotoxic compounds.

Figure 1. *Cymbidium ensifolium* (L.) Sw.

2. Results and Discussion

2.1. Structure Determination

The phytochemical study of a methanolic extract from the leaves of *Cymbidium ensifolium* resulted in the purification of three novel dihydrophenanthrene derivatives (**1–3**) (Supplementary Materials), along with two known compounds cypripedin (**4**) [11] and gigantol (**5**) [12] (Figure 2). The structures of the new compounds were elucidated through extensive spectroscopic data.

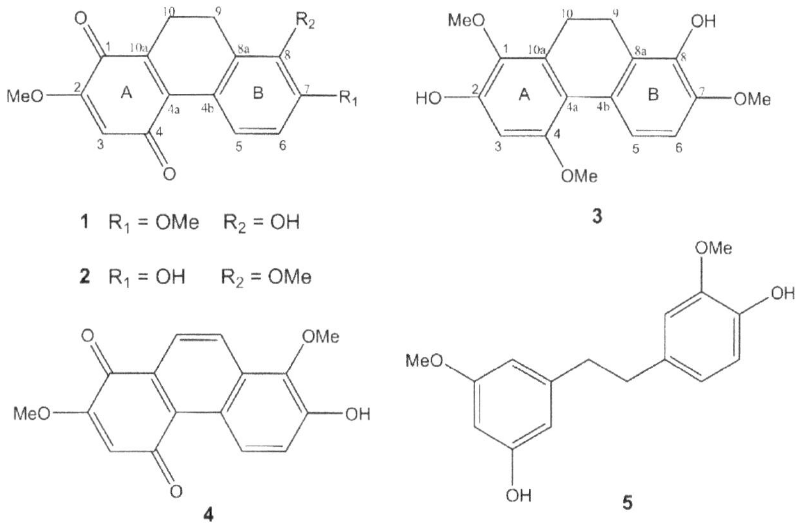

Figure 2. Chemical structures of compounds **1–5** isolated from *Cymbidium ensifolium.*

Compound **1** was purified as a red amorphous solid. The molecular formula $C_{16}H_{14}O_5$ was analyzed from its $[M-H]^-$ at 285.0770 (calcd. for $C_{16}H_{13}O_5$, 285.0763) in the HR–ESI–MS. The IR spectrum displayed absorption bands for hydroxyl (3432 cm^{-1}), aromatic (2924, 1638 cm^{-1}), and ketone (1733 cm^{-1}) groups. The UV absorptions at 255, 330, and 491 nm were indicative of a dihydrophenanthrenequinone nucleus [13]. The chemical structure was supported by the presence of two carbonyl carbons at δ 181.9 (C-1) and 188.0 (C-4). A 9,10-dihydro partial structure of **1** was confirmed by the signals of two methylene carbons at δ 20.0 (C-10) and 20.1 (C-9), which showed HSQC correlations to the protons at δ 2.59 (2H, dd, J = 8.4, 7.2 Hz, H$_2$-10) and 2.80 (2H, m, H$_2$-9), respectively. The ^{1}H NMR spectrum of **1** also exhibited a singlet olefinic proton signal at δ 5.99 (1H, s, H-3), two doublet proton signals at δ 6.90 (1H, d, J = 8.8 Hz, H-6) and 7.63 (1H, d, J = 8.8 Hz, H-5), a singlet hydroxyl signal at δ 7.71 (HO-8) and signals for two methoxyl groups at δ 3.91 (3H, s, MeO-7) and 3.86 (3H, s, MeO-2) (Table 1). The quinone structure of ring A was deduced from the HMBC correlations of C-1 (δ 181.9) with H$_2$-10 and H-3 (Figure 3). The first methoxyl group (δ 3.86) was located on ring A at C-2 based on its NOESY interaction with H-3. For ring B, the doublet proton signal at δ 7.63 was assigned as H-5 according to the HMBC correlations of C-4a (δ 137.3) with H-3 and H-5. The analysis of the ^{1}H-^{1}H COSY spectrum was supported by an *ortho*-coupled correlation between H-5 and H-6. The presence of a hydroxyl group at C-8 (δ 143.4) on ring B was supported by the HMBC correlation of C-8a (δ 125.6) with H-5, H$_2$-10, and HO-8 and the NOESY crosspeak between HO-8 and H$_2$-9. The second methoxyl group (δ 3.91) was substituted at C-7 (δ 150.0) according to its NOESY correlation with H-6 (Figure 3). Based on the above spectral data, **1** was characterized as 8-hydroxy-2,7-dimethoxy-9,10-dihydrophenanthrene-1,4-dione and named cymensifin A.

Compound **2**, a red amorphous solid, showed a similar molecular formula with **1** as $C_{16}H_{14}O_5$ based on its $[M-H]^-$ at 285.0764 (calcd. for $C_{16}H_{13}O_5$, 285.0763) in the HR–ESI–MS. The UV absorptions and IR bands of **2** are similar with **1**, suggestive of a dihydrophenanthrenequinone skeleton. This was confirmed based on two carbonyl carbons at δ 181.8 (C-1) and 188.0 (C-4), and two methylene carbons at δ 20.5 (C-10) and 20.8 (C-9). The ^{1}H NMR of **2** was similar to that of **1** by the presence of a singlet proton signal at δ 5.99 (1H, s, H-3), two aromatic doublet proton signals at δ 6.83 (1H, d, J = 8.8 Hz, H-6) and 7.75 (1H, d, J = 8.8 Hz, H-5), a singlet hydroxyl signal at δ 8.59 (HO-7), two methylene proton signal at δ 2.59 (2H, dd, J = 8.4, 7.6 Hz, H$_2$-10) and 2.78 (2H, m, H$_2$-9), and signals for two methoxyls at δ 3.76 (3H, s, MeO-8) and 3.86 (3H, s, MeO-2) (Table 1). On ring

A of **2**, a methoxyl group (δ 3.86) was placed at C-2 according to its NOESY interaction with H-3. On ring B, the position of the methoxy and hydroxyl groups of **2** was alternated when compared with **1**. The second methyl group (δ 3.76) of **2** was located at C-8, as evidenced by the HMBC correlations of C-8 (δ 145.3) with MeO-8, H-6, H$_2$-9, and HO-7. The HMBC correlations of C-7 (δ 153.1) with H-5 and HO-7 and the NOESY correlation between H-6 and HO-7 support the placement of the hydroxyl group at C-7 (Figure 3). Based on the above spectral evidence, 2 was identified as 7-hydroxy-2,8-dimethoxy-9,10-dihydrophenanthrene-1,4-dione and named cymensifin B.

Table 1. ^{1}H (400 MHz) and ^{13}C-NMR (100 MHz) spectral data of **1–3** in acetone-d_6.

	1		2		3	
Position	δ$_H$ (Multiplicity, *J* in Hz)	δ$_C$	δ$_H$ (Multiplicity, *J* in Hz)	δ$_C$	δ$_H$ (Multiplicity, *J* in Hz)	δ$_C$
1	-	181.9	-	181.8	-	139.3
2	-	159.5	-	159.5	-	149.8
3	5.99 (s)	108.6	5.99 (s)	108.5	6.52 (s)	99.9
4	-	188.0	-	188.0	-	154.9
4a	-	137.3	-	137.2	-	116.7
4b	-	123.8	-	123.0	-	127.6
5	7.63 (d, *J* = 8.8 Hz)	123.8	7.75 (d, *J* = 8.8 Hz)	128.4	7.70 (d, *J* = 8.8 Hz)	120.4
6	6.90 (d, *J* = 8.8 Hz)	109.3	6.83 (d, *J* = 8.8 Hz)	114.9	6.78 (d, *J* = 8.8 Hz)	109.0
7	-	150.0	-	153.1	-	146.3
8	-	143.4	-	145.3	-	142.9
8a	-	125.6	-	134.0	-	124.8
9	2.80 (m)	20.1	2.78 (m)	20.8	2.71 (br s)	23.1
10	2.59 (dd, *J* = 8.4, 7.2 Hz)	20.0	2.59 (dd, *J* = 8.4, 7.6 Hz)	20.5	2.71 (br s)	21.8
10a	-	137.9	-	137.4	-	133.6
MeO-1	-	-	-	-	3.69 (s)	61.1
MeO-2	3.86 (s)	56.6	3.86 (s)	56.6	-	-
MeO-4	-	-	-	-	3.79 (s)	56.1
MeO-7	3.91 (s)	56.3	-	-	3.85 (s)	56.3
MeO-8	-	-	3.76 (s)	61.0	-	-
HO-2	-	-	-	-	8.16 (s)	-
HO-7	-	-	8.59 (s)	-	-	-
HO-8	7.71 (s)	-	-	-	7.36 (s)	-

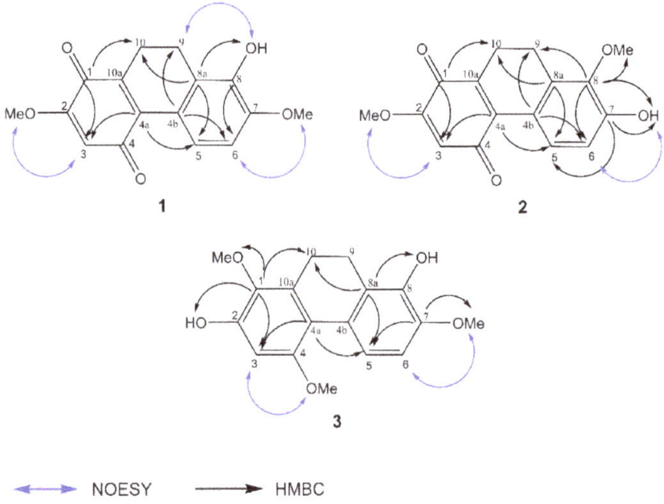

Figure 3. Selected HMBC and NOESY correlations of **1**, **2**, and **3**.

Compound **3**, as a brown amorphous solid, gave a $[M-H]^-$ at 301.1074 (calcd. for $C_{17}H_{17}O_5$, 301.1076) in the HR –ESI–MS, suggesting a molecular formula $C_{17}H_{18}O_5$. The IR spectrum demonstrated the exhibition of hydroxyl (3367 cm^{-1}), aromatic ring (2924, 1606 cm^{-1}), and methylene (1461 cm^{-1}) functionalities. The UV absorptions at 279 and 312 nm and the ^{13}C NMR signals of two methylene carbons at δ 21.8 (C-10) and 23.1 (C-9), which exhibited HSQC correlation with the ^{1}H NMR signal at δ 2.71 (4H, br s, H$_2$-9, H$_2$-10), indicate a dihydrophenanthrene structure [14]. The ^{1}H NMR displayed three aromatic proton signals at δ 6.52-7.70, resonances for three methoxy groups at δ 3.69 (3H, s, MeO-1), 3.79 (3H, s, MeO-4), and 3.85 (3H, s, MeO-7), and signals for two hydroxyl groups at δ 7.36 (1H, s, HO-8) and 8.16 (1H, s, HO-2) (Table 1). For ring A, the assignments of H-3 (δ 6.52, 1H, s), MeO-1, and HO-2 are in accord with the HMBC correlations of C-1 (δ 139.3) with H-3, H$_2$-10, MeO-1, and HO-2 (Figure 3). A NOESY interaction of the second methoxy group at δ 3.79 with H-3 placed this methoxy group at C-4. Compound **3** exhibited structural similarity in ring B to that of **1** by the presence of two doublet proton signals of H-5 and H-6 at δ 7.70 (1H, d, *J* = 8.8 Hz) and 6.78 (1H, d, *J* = 8.8 Hz), respectively. A methoxy group at C-7 was supported by its NOESY interaction with H-6. A hydroxyl group at δ 7.36 was located at C-8 as evidenced by the HMBC correlations of C-8a (δ 124.8) with H-5, H$_2$-10, and HO-8 (Figure 3). Compound **3** was assigned as 2,8-dihydroxy-1,4,7-trimethoxydihydrophananthrene and given the trivial name cymensifin C.

2.2. Cytotoxic Effects against Various Cancer Cells

Because cytotoxic effects of cypripedin (**4**) and gigantol (**5**) were previously demonstrated in lung cancer cells [11,15], compounds **1–3** were evaluated for their anticancer potential against various human cancers, including lung, breast, and colon cancer cells. Cisplatin, a recommended chemotherapeutic agent [16], which at 50 µM caused the 50% reduction of viability in lung cancer cells, was selected for a positive control. The preliminary investigation via MTT assay showed that a culture with methanolic extract from *C. ensifolium* at 50 µg/mL for 24 h significantly diminished viability in lung cancer H460 and breast cancer MCF7 cells, but not in colon cancer CaCo$_2$ cells when compared with control cells which were treated with the solvent vehicle 0.5% DMSO (Table 2). It should be noted that treatment with a lower concentration (10 µg/mL) of the methanolic extract did not obviously decrease cell viability in all cancer cells (data not shown). Among the three new dihydrophenanthrene derivatives, 24 h treatment with compound **1** (50 µM) showed the highest anticancer effect against H460, MCF7, and CaCo$_2$ cells, as evidenced with the lowest % cell viability assessed via an MTT assay. When compared with cisplatin (Sigma-Aldrich Chemical, St. Louis, MO, USA), the higher anticancer potency of compound **1** was indicated with the lower % viability in MCF7 and CaCo$_2$ cells. Additionally, Figure 4a–c illustrate the dose-response relationship of three new dihydrophenanthrene derivatives in H460, MCF7, and CaCo$_2$ cells, respectively. In all cancer cells, compound **1** and cisplatin demonstrated anticancer activity in a concentration dependence. When compared with cisplatin treatment at the same concentration, the significantly lower % cell viability was presented in CaCo$_2$ cells cultured with 50–200 µM of compound **1**.

Table 2. Cell viability percentage in various cancer and normal cells cultured with 50 µM of new dihydrophenanthrene derivatives (**1–3**) from *Cymbidium ensifolium* for 24 h.

Cell Type	% Cell Viability				
	Extract (50 µg/mL)	1 (50 µM)	2 (50 µM)	3 (50 µM)	Cisplatin (50 µM)
H460	43.49 ± 6.42 *	57.65 ± 1.39 *,#	79.48 ± 1.11 #	99.12 ± 2.91 #	46.02 ± 2.07 *
MCF7	52.76 ± 2.09 *	61.51 ± 5.46 *	100.45 ± 9.85 #	96.45 ± 5.11 #	74.05 ± 5.69 *
CaCo$_2$	95.37 ± 2.21	56.92 ± 6.35 *,#	95.26 ± 5.05	90.96 ± 1.44	90.20 ± 4.27
DPCs	73.41 ± 4.72 *	75.66 ± 1.28 *	84.39 ± 5.78 *,#	92.68 ± 2.22 #	66.18 ± 6.44 *

* $p < 0.05$ compared with control cells treated with 0.5% DMSO as the solvent vehicle, # $p < 0.05$ compared with the cisplatin treatment group.

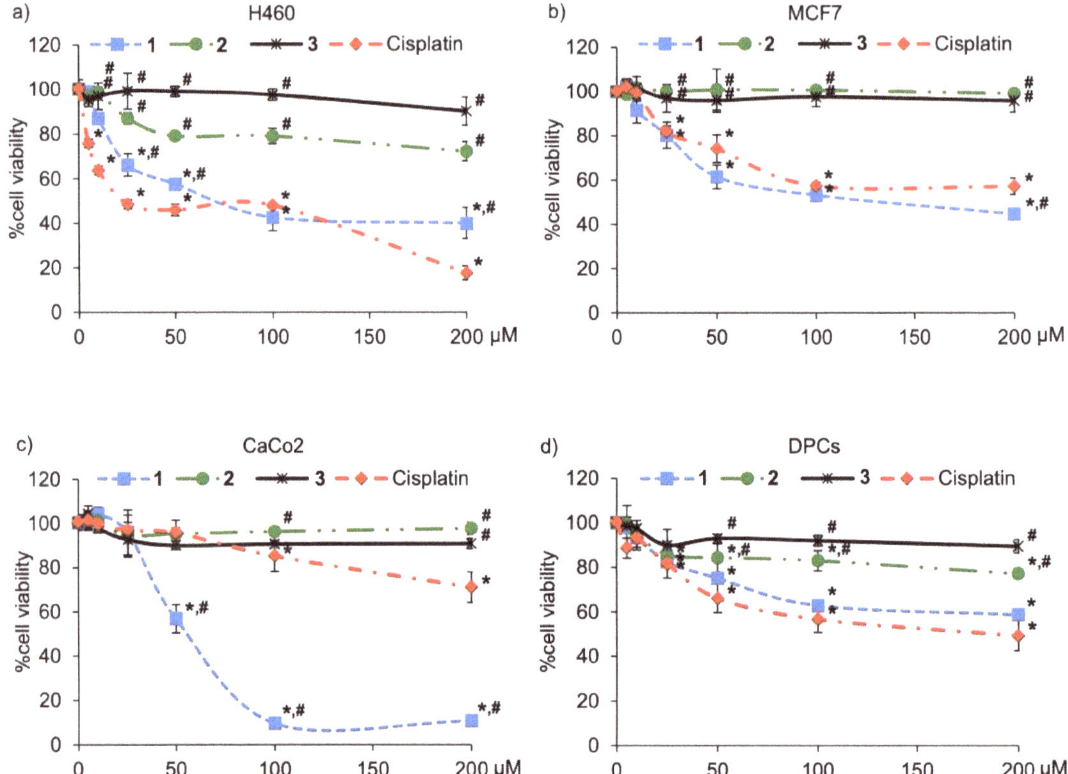

Figure 4. The relationship between cytotoxicity and the concentration of dihydrophenanthrene derivatives (**1–3**) from *Cymbidium ensifolium* in (**a**) lung cancer H460, (**b**) breast cancer MCF7, (**c**) colon cancer $CaCo_2$, and (**d**) dermal papilla cells (DPCs). Cisplatin, a recommended anticancer drug, was used as a positive control. The data were presented as means ± standard error of the mean (SEM) from three independent experiments. * $p < 0.05$ compared with control cells treated with 0.5% DMSO as the solvent vehicle, # $p < 0.05$ compared with treatment of cisplatin at the same concentration.

All data were presented as means ± standard error of the mean (SEM) from three independent experiments.

These were corresponded with apoptosis cell death detected via costaining with Hoechst 33342 and propidium iodide. Figure 5a–c respectively demonstrates apoptosis cells presenting with bright-blue fluorescence of condensed DNA or fragmented nuclei stained with Hoechst 33342 in H460, MCF7, and $CaCo_2$ cells, which were cultured with either compound **1** (50 μM) or cisplatin (50 μM) for 24 h. Notably, there were no necrosis stained with the red fluorescence of propidium iodide in all treated cells. When compared with cisplatin treatment, the culture with compound **1** significantly increased % apoptosis in breast cancer MCF7 and colon cancer $CaCo_2$ cells, although there was lower % apoptosis in lung cancer H460 cells (Figure 6a–c).

It is the fact that the purity of an isolated natural compound critically influences its bioactivity [17]. Composed of known anticancer compounds, cypripedin (**4**) and gigantol (**5**), as well as compound **1**, a new dihydrophenanthrenequinone with a higher potency anticancer effect, the methanolic extract from *C. ensifolium* exhibited potent cytotoxicity against lung cancer and breast cancer cells (Table 2 and Figure 5a,b). Despite no toxicity presented in *C. ensifolium* extract-treated $CaCo_2$ cells (Table 2 and Figure 5c), the dramatic reduction of % cell viability (Figure 4c) and accumulated apoptosis indicated in $CaCo_2$

cells (Figures 5c and 6c) incubated with compound **1** suggested specific cytotoxicity of compound **1** against colon cancer cells.

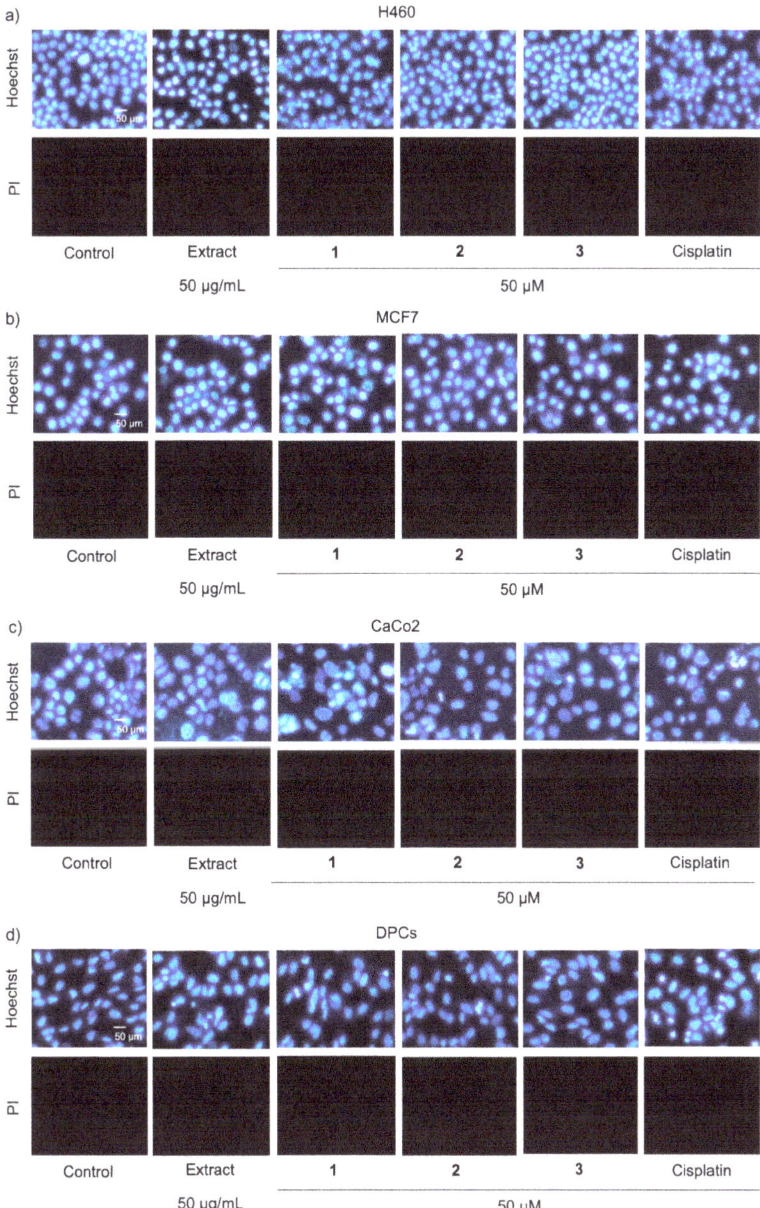

Figure 5. A nuclear staining assay demonstrated apoptosis presented with bright-blue fluorescence of Hoechst 33342 in (**a**) lung cancer H460, (**b**) breast cancer MCF7, (**c**) colon cancer $CaCo_2$, and (**d**) dermal papilla cells (DPCs) cultured with methanolic extract and isolated compounds **1–3** from *Cymbidium ensifolium* for 24 h. Cisplatin, a recommended anticancer drug, was used as a positive control. Notably, there were no detected necrosis cells stained with red fluorescence of propidium iodide (PI) in all treated cells.

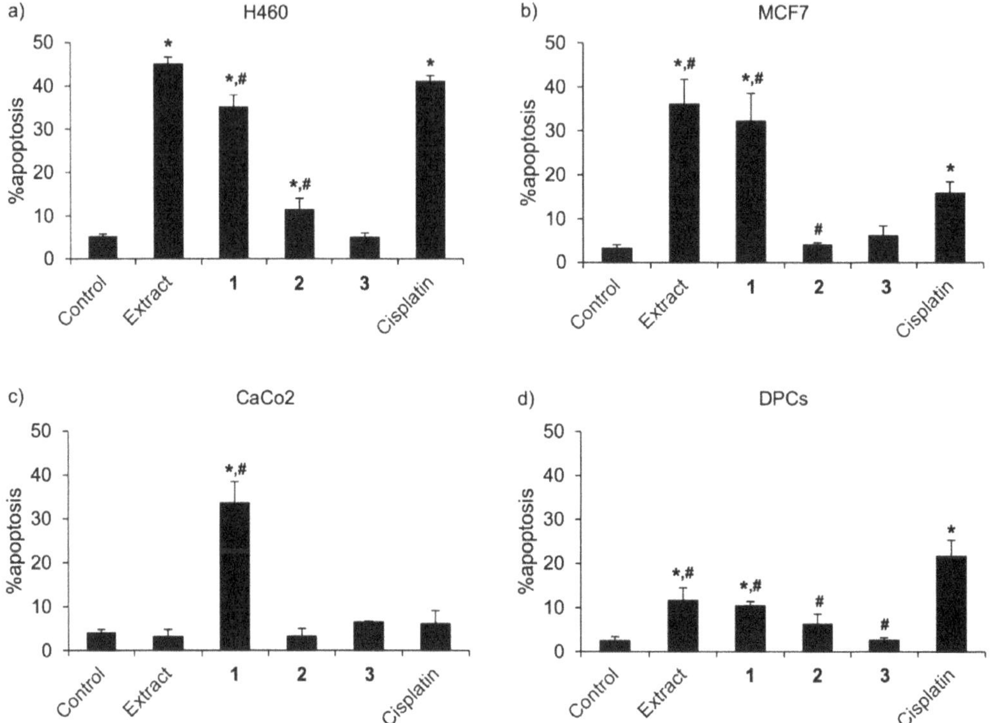

Figure 6. The percent (%) apoptosis calculated from nuclear staining assay in (**a**) lung cancer H460, (**b**) breast cancer MCF7, (**c**) colon cancer $CaCo_2$, and (**d**) dermal papilla cells (DPCs) cultured with methanolic extract and isolated compounds **1–3** from *Cymbidium ensifolium* for 24 h. Cisplatin, a recommended anticancer drug, was used as a positive control. The data were presented as means ± standard error of the mean (SEM) from three independent experiments. * $p < 0.05$ compared with control cells treated with 0.5% DMSO as a solvent vehicle, # $p < 0.05$ compared with the cisplatin treatment group.

As the most affected normal cells, human dermal papilla cells (DPCs) are chosen for evaluating the safety profile of potential anticancer compounds [18,19]. Intriguingly, DPCs cultured with 50 μM of three new dihydrophenanthrene derivatives (**1–3**) exhibited higher % cell viability (Table 2 and Figure 4d) and less apoptosis cells (Figures 5d and 6d) compared with the treatment of cisplatin. Due to the highest anticancer potential among three new dihydrophenanthrene derivatives, the half-maximum inhibitory concentration (IC_{50}) and selectivity index (S.I.) of compound **1** were calculated. Table 3 indicates that when compared with cisplatin, compound **1** possessed lower IC_{50} values in MCF7 (93.04 ± 0.86 μM) and $CaCo_2$ cells (55.14 ± 3.08 μM), but it accounted for higher values of IC_{50} in H460 (66.71 ± 6.62 μM) and DPCs cells (>200 μM). It has been reported that the selectivity index of an anticancer agent, which selectively causes toxicity to cancer cells, should be more than 1 [20]. The higher selectivity index of compound **1** was indicated in MCF7 (>2.15) and $CaCo_2$ cells (>3.62) compared with cisplatin treatment (MCF7: <0.59, $CaCo_2$: <0.59). Although the selectivity index of compound **1** (>3.00) was lower than cisplatin (5.48 ± 0.25) in lung cancer cells, both values are more 1. These data clearly demonstrate the selective anticancer activity of compound **1** against various cancer cells.

Taken together, these results strongly suggest the potent anticancer effect against various types of cancer with the high safety profile of compound **1** isolated from the *C. ensifolium* extract. Nevertheless, the underlying mechanisms involved in the apoptosis inducing effect of compound **1** should be further investigated to clarify the therapeutic target.

Table 3. The half-maximum inhibitory concentration (IC_{50}) and selectivity index (S.I.) of compound **1** from *Cymbidium ensifolium* and cisplatin.

Cell Type	1		Cisplatin	
	IC_{50} (μM)	S.I.	IC_{50} (μM)	S.I.
H460	66.71 ± 6.62	>3.00	21.57 ± 4.13	5.48 ± 0.25
MCF7	93.04 ± 0.86	>2.15	>200	<0.59
CaCo$_2$	55.14 ± 3.08	>3.62	>200	<0.59
DPCs	>200	N/A	117.63 ± 17.19	N/A

N/A: not applicable. All data were presented as means ± standard error of the mean (SEM) from three independent experiments.

3. Materials and Methods

3.1. General Experimental Procedures

UV spectra were measured on a Milton Roy Spectronic 300 Array spectrophotometer (Rochester, Monroe, NY, USA). IR spectra were recorded on a Perkin-Elmer FT-IR 1760X spectrophotometer (Boston, MA, USA). Mass spectra were obtained on a Bruker micro TOF mass spectrometer (ESI-MS) (Billerica, MA, USA). NMR spectra were recorded on a Bruker Avance Neo 400 MHz NMR spectrometer (Billerica, MA, USA). Vacuum liquid column chromatography (VLC) and column chromatography (CC) were performed on silica gel 60 (Merck, Kieselgel 60, 70–320 μm) and silica gel 60 (Merck, Kieselgel 60, 230–400 μm) (Darmstadt, Germany).

3.2. Plant Material

Cymbidium ensifolium was purchased from Chatuchak market, Bangkok, in September 2020. Plant identification was performed by Mr. Yanyong Punpreuk, Department of Agriculture, Bangkok, Thailand. A voucher specimen (BS-CE-092563) has been deposited at the Department of Pharmacognosy and Pharmaceutical Botany, Faculty of Pharmaceutical Sciences, Chulalongkorn University.

3.3. Extraction and Isolation

The air-dried aerial parts of *Cymbidium ensifolium* (798 g) were cut, pulverized, and extracted with MeOH (3 × 6 L) at room temperature. The resultant organic solvent was evaporated under reduced pressure to give a dried mass (97.68 g). The methanolic extract was fractionated by vacuum liquid chromatography on silica gel (hexane–ethyl acetate, gradient) to give five fractions (A–E). Fraction C (23 g) was separated by column chromatography (CC, silica gel hexane–ethyl acetate, gradient) to yield seven fractions (C1–C7). Fraction C3 (46.8 mg) was further separated by CC (silica gel, hexane–dichloromethane, gradient) and then purified on silica gel (hexane–dichloromethane–methanol, gradient) to afford **1** (3 mg). Fractions C5 (32.6 mg) and C7 (25.1 mg) were separated on a hexane–dichloromethane gradient and thereafter, on silica gel (hexane–dichloromethane–methanol gradient) to yield cypripedin (**4**) (3.2 mg) and **2** (4 mg), respectively. Fraction D (16 g) was fractionated by CC (silica gel, hexane–ethyl acetate, gradient) to yield 12 fractions (D1–D12). Fraction D1 (76.3 mg) was further separated on silica gel (hexane–dichloromethane–methanol, gradient) to furnish **3** (1.8 mg). Gigantol (**5**) (2.1 mg) was obtained from fraction D7 (8.0 mg) after purification on Sephadex LH-20 (acetone).

Cymensifin A (**1**): Red amorphous solid; UV (MeOH) λ_{max} (log ε): 255 (4.86), 330 (2.55), 491 (1.28) nm; IR (film) ν_{max}: 3432, 2924, 2854, 1733, 1638, 1461, 1271 cm^{-1}; HR–ESI–MS: $[M-H]^-$ at m/z 285.0770 (calcd. for 285.0763, $C_{16}H_{13}O_5$).

Cymensifin B (**2**): Red amorphous solid; UV (MeOH) λ_{max} (log ε): 254 (4.32), 333 (2.13), 493 (1.01) nm; IR (film) ν_{max}: 3370, 2924, 2854, 1732, 1642, 1460, 1246 cm^{-1}; HR–ESI–MS: $[M-H]^-$ at m/z 285.0764 (calcd. for 285.0763, $C_{16}H_{13}O_5$).

Cymensifin C (**3**): Brown amorphous solid; UV (MeOH) λ_{max} (log ε): 279 (3.74), 312 (2.48) nm; IR (film) ν_{max}: 3367, 2924, 2854, 1606, 1510, 1461, 1246 cm^{-1}; HR–ESI–MS: $[M-H]^-$ at m/z 301.1074 (calcd. for 301.1076, $C_{17}H_{17}O_5$).

3.4. Cell Culture

Human lung cancer H460, breast cancer MCF7, and colon cancer $CaCo_2$ cells were obtained from The American Type Culture Collection (ATCC, Manassas, VA, USA). Lung cancer H460 cells were cultured in the Roswell Park Memorial Institute (RPMI; Gibco, Gaithersburg, MA, USA) medium whereas human MCF7 and $CaCo_2$ cells were maintained in Dulbecco's modified eagle medium (DMEM; Gibco, Gaithersburg, MA, USA). Human dermal papilla cells (DPCs) purchased from Applied Biological Materials Inc. (Richmond, BC, Canada) were cultured in a Prigrow III medium (Applied Biological Materials Inc., Richmond, BC, Canada). All culture mediums were supplemented with 2 mM L-glutamine, 10% FBS (fetal bovine serum) and 100 units/mL penicillin/streptomycin. The cells were maintained in at 37 °C with 5% CO_2 until 70–80% confluence before using in further experiments.

3.5. Cell Viability Assay

After 24 h of indicated treatment, cells seeded at a density of 1×10^4 cells/well in a 96-well plate was further incubated with 0.4 mg/mL of MTT (3-(4,5-dimethylthiazol-2-yl)-2,5-diphenyltetrazolium bromide; Sigma-Aldrich Chemical, St. Louis, MO, USA) for 3 h at 37 °C and kept from light. Then, the formed formazan crystal was solubilized in a DMSO prior measurement of optical density (OD) at 570 nm by using a microplate reader (Anthros, Durham, NC, USA). The OD ratio of treated to control cells, which were incubated with 0.5% DMSO as a solvent vehicle, was calculated and presented as percent cell viability [19,21].

Furthermore, the half-maximum inhibitory concentration (IC_{50}) was calculated and used for determination of selective index in each cancer cell. The selectivity index was represented from the ratio between the IC_{50} value in dermal papilla cells and the IC_{50} value in cancer cells [20].

3.6. Detection of Mode of Cell Death

Apoptosis and necrosis cell death were evaluated by nuclear staining assay. The treated cells were costained with 0.02 µg/mL Hoechst 33342 (Sigma-Aldrich Chemical, St. Louis, MO, USA) and 0.01 µg/mL propidium iodide (Sigma-Aldrich Chemical, St. Louis, MO, USA) at 37 °C for 30 min. The mode of cell death was visualized under a fluorescence microscope (Olympus IX51 with DP70, Olympus, Tokyo, Japan). Apoptosis cells were characterized with a bright-blue fluorescence of Hoechst 33342 which stained fragmented DNA and condensed nuclei. Meanwhile, necrosis cells were distinguished by red fluorescence of propidium iodide. The percentage of apoptosis was represented from the ratio between the number of apoptosis cells to the total cell number [19,22].

3.7. Statistical Analysis

The statistical analysis was performed via SPSS version 22 (IBM Corp., Armonk, NY, USA) with one-way analysis of variance (ANOVA) followed by Tukey's post hoc test. Any p-value under 0.05 was considered as a statistical significance.

4. Conclusions

In this study, three novel dihydrophenanthrenes derivatives, cymensifins A–C (**1–3**), together with two known compounds, cypripedin (**4**) and gigantol (**5**), were isolated from the aerial parts of *C. ensifolium*. The structures of the new compounds were elucidated by spectroscopic analysis. The three new compounds from this plant were evaluated for their cytotoxicity on human lung cancer H460, breast cancer MCF7, and colon cancer $CaCo_2$ cells. Cymensifin A (**1**) showed a promising anticancer effect against various cancer cells with higher safety profiles compared with cisplatin, an available chemotherapy. To the best of our knowledge, this research is the first record on the chemical constituents and biological activity of this plant.

Supplementary Materials: The following supporting information can be downloaded at: https://www.mdpi.com/article/10.3390/molecules27072222/s1, Figure S1: ^{1}H NMR spectrum of **1** (400 MHz) in acetone-d_6, Figure S2: ^{13}C NMR spectrum of **1** (100 MHz) in acetone-d_6, Figure S3: HSQC spectrum of **1** in acetone-d_6, Figure S4: HMBC spectrum of **1** in acetone-d_6, Figure S5: COSY spectrum of **1** in acetone-d_6, Figure S6: NOESY spectrum of **1** in acetone-d_6, Figure S7: HR–ESI–MS spectrum of **1**, Figure S8: ^{1}H NMR spectrum of **2** (400 MHz) in acetone-d_6, Figure S9: ^{13}C NMR spectrum of **2** (100 MHz) in acetone-d_6, Figure S10: HSQC spectrum of **2** in acetone-d_6, Figure S11: HMBC spectrum of **2** in acetone-d_6, Figure S12: COSY spectrum of **2** in acetone-d_6, Figure S13: NOESY spectrum of **2** in acetone-d_6, Figure S14: HR–ESI–MS spectrum of **2**, Figure S15: ^{1}H NMR spectrum of **3** (400 MHz) in acetone-d_6, Figure S16: ^{13}C NMR spectrum of **3** (100 MHz) in acetone-d_6, Figure S17: HSQC spectrum of **3** in acetone-d_6, Figure S18: HMBC spectrum of **3** in acetone-d_6, Figure S19: NOESY spectrum of **3** in acetone-d_6, Figure S20: HR-ESI-MS spectrum of **3**.

Author Contributions: B.S. conceived, designed, and supervised the research project as well as prepared and edited the manuscript; T.O.J. performed the experiments and prepared the manuscript; B.C.C. assisted in cytotoxic activity; C.C. (Chaisak Chansriniyom) performed the NMR operation; C.C. (Chatchai Chaotham) supervised cytotoxic activity and prepared the manuscript; P.C. provided cancer cells. P.R. and K.L. provided comments and suggestions on the preparation of the manuscript. All authors have read and agreed to the published version of the manuscript.

Funding: This research is supported by the 90th Anniversary of the Chulalongkorn University Scholarship under the Ratchadapisek Somphot Endowment Fund.

Institutional Review Board Statement: Not applicable.

Informed Consent Statement: Not applicable.

Data Availability Statement: All data presented in this study are available in the article.

Acknowledgments: T.O.J. is grateful to the Graduate School, Chulalongkorn University for a CUASEAN and NON-ASEAN Ph.D. scholarship. We thank Yanyong Punpreuk for the plant identification. The Research Instrument Center of the Faculty of Pharmaceutical Sciences is acknowledged for providing research facilities.

Conflicts of Interest: The authors declare no conflict of interest.

Sample Availability: Samples of the compounds are not available from authors.

References

1. Shah, S.; Thapa, B.B.; Chand, K.; Pradhan, S.; Singh, A.; Varma, A.; Thakuri, L.S.; Joshi, P.; Pant, B. *Piriformospora indica* promotes the growth of the in-vitro-raised *Cymbidium aloifolium* plantlet and their acclimatization. *Plant Signal. Behav.* **2019**, *14*, e1596716. [CrossRef] [PubMed]
2. Chuakul, W. Ethnomedical uses of Thai Orchidaceous plants. *Mahidol J. Pharm. Sci.* **2002**, *29*, 41–45.
3. Yoshikawa, K.; Baba, C.; Iseki, K.; Ito, T.; Asakawa, Y.; Kawano, S.; Hashimoto, T. Phenanthrene and phenylpropanoid constituents from the roots of *Cymbidium* Great Flower "Marylaurencin" and their antimicrobial activity. *J. Nat. Med.* **2014**, *68*, 743–747. [CrossRef] [PubMed]
4. Lertnitikul, N.; Pattamadilok, C.; Chansriniyom, C.; Suttisri, R. A new dihydrophenanthrene from *Cymbidium finlaysonianum* and structure revision of cymbinodin-A. *J. Asian Nat. Prod. Res.* **2020**, *22*, 83–90. [CrossRef] [PubMed]
5. Lv, S.; Fu, Y.; Chen, J.; Jiao, Y.; Chen, S. Six phenanthrenes from the roots of *Cymbidium faberi* Rolfe. and their biological activities. *Nat. Prod. Res.* **2020**. [CrossRef]
6. Smitinand, T. *Thai Plant. Names (Botanical Names-Vernacular Names), Revised ed.*; The Forest Herbarium, Royal Forest Department: Bangkok, Thailand, 2001; p. 163.
7. Siripiyasing, P.; Kaenratana, K.; Mokkamul, P.; Tanee, T.; Sudmoon, R.; Chaveerach, A. DNA barcoding of the *Cymbidium* species (Orchidaceae) in Thailand. *Afr. J. Agric. Res.* **2012**, *7*, 393–404.
8. Chiakul, W.; Boonpleng, A. Survey on medicinal plants in Ubon Ratchathani province (Thailand). *Thai J. Phytopharm.* **2004**, *11*, 33–54.
9. San, H.T.; Chatsumpun, N.; Juengwatanatrakul, T.; Pornputtapong, N.; Likhitwitayawuid, K.; Sritularak, B. Four novel phenanthrene derivatives with α-glucosidase inhibitory activity from *Gastrochilus bellinus*. *Molecules* **2021**, *26*, 418. [CrossRef]
10. Thant, M.T.; Khine, H.E.E.; Nealiga, J.Q.L.; Chatsumpun, N.; Chaotham, C.; Sritularak, B.; Likhitwitayawuid, K. α-Glucosidase inhibitory activity and anti-adipogenic effect of compounds from *Dendrobium delacourii*. *Molecules* **2022**, *27*, 1156.

11. Wattanathamsan, O.; Treesuwan, S.; Sritularak, B.; Pongrakhananon, V. Cypripedin, a phenanthrenequinone from *Dendrobium densiflorum*, sensitizes non-small cell lung cancer H460 cells to cisplatin-mediated apoptosis. *J. Nat. Med.* **2018**, *72*, 503–513. [CrossRef]
12. Chen, Y.; Xu, J.; Yu, H.; Chen, Q.; Zhang, Y.; Wang, L.; Liu, Y.; Wang, J. Cytotoxic phenolics from *Bulbophyllum odoratissimum*. *Food Chem.* **2008**, *107*, 169–173. [CrossRef]
13. Sritularak, B.; Anuwat, M.; Likhitwitayawuid, K. A new phenanthrenequinone from *Dendrobium draconis*. *J. Asian Nat. Prod. Res.* **2011**, *13*, 251–255. [CrossRef] [PubMed]
14. Lin, Y.; Wang, F.; Yang, L.J.; Chun, Z.; Bao, J.K.; Zhang, G.L. Anti-inflammatory phenanthrene derivatives from *Dendrobium denneanum*. *Phytochemistry* **2013**, *95*, 242–251. [CrossRef] [PubMed]
15. Charoenrungruang, S.; Chanvorachote, P.; Sritularak, B.; Pongrakhananon, V. Gigantol, a bibenzyl from Dendrobium draconis, inhibits the migratory behavior of non-small cell lung cancer cells. *J. Nat. Prod.* **2014**, *77*, 1359–1366. [CrossRef] [PubMed]
16. Ghosh, S. Cisplatin: The first metal based anticancer drug. *Bioorg. Chem.* **2019**, *88*, 102925. [CrossRef] [PubMed]
17. Jaki, B.U.; Franzblau, S.G.; Chadwick, L.R.; Lankin, D.C.; Zhang, F.; Wang, Y.; Pauli, G.F. Purity-activity relationships of natural products: The case of anti-TB active ursolic acid. *J. Nat. Prod.* **2008**, *71*, 1742–1748. [CrossRef]
18. Botchkarev, V.A. Molecular mechanisms of chemotherapy-induced hair loss. *J. Investig. Dermatol. Symp. Proc.* **2003**, *8*, 72–75. [CrossRef]
19. Khine, H.E.E.; Ecoy, G.A.U.; Roytrakul, S.; Phaonakrop, N.; Pornputtapong, N.; Prompetchara, E.; Chanvorachote, P.; Chaotham, C. Chemosensitizing activity of peptide from *Lentinus squarrosulus* (Mont.) on cisplatin-induced apoptosis in human lung cancer cells. *Sci. Rep.* **2021**, *11*, 4060. [CrossRef]
20. Lica, J.J.; Wieczór, M.; Grabe, G.J.; Heldt, M.; Jancz, M.; Misiak, M.; Gucwa, K.; Brankiewicz, W.; Maciejewska, N.; Stupak, A.; et al. Effective drug concentration and selectivity depends on fraction of primitive cells. *Int. J. Mol. Sci.* **2021**, *22*, 4931. [CrossRef]
21. Marks, D.C.; Belov, L.; Davey, M.W.; Davey, R.A.; Kidman, A.D. The MTT cell viability assay for cytotoxicity testing in multidrug-resistant human leukemic cells. *Leuk. Res.* **1992**, *16*, 1165–1173. [CrossRef]
22. Kabakov, A.E.; Gabai, V.L. Cell death and survival assays. *Methods Mol. Biol.* **2018**, *1709*, 107–127. [PubMed]

Article

α-Glucosidase Inhibitory Activity and Anti-Adipogenic Effect of Compounds from *Dendrobium delacourii*

May Thazin Thant [1,2], Hnin Ei Ei Khine [3], Justin Quiel Lasam Nealiga [3], Nutputsorn Chatsumpun [4], Chatchai Chaotham [3,5,*], Boonchoo Sritularak [1,6,*] and Kittisak Likhitwitayawuid [1]

[1] Department of Pharmacognosy and Pharmaceutical Botany, Faculty of Pharmaceutical Sciences, Chulalongkorn University, Bangkok 10330, Thailand; drmaythazinthant@gmail.com (M.T.T.); kittisak.l@chula.ac.th (K.L.)
[2] Department of Pharmacognosy, University of Pharmacy, Yangon 11031, Myanmar
[3] Department of Biochemistry and Microbiology, Faculty of Pharmaceutical Sciences, Chulalongkorn University, Bangkok 10330, Thailand; hnineieikhine12@gmail.com (H.E.E.K.); jquieln@gmail.com (J.Q.L.N.)
[4] Department of Pharmacognosy, Faculty of Pharmacy, Mahidol University, Bangkok 10400, Thailand; nutputsorn.cha@mahidol.ac.th
[5] Preclinical Toxicity and Efficacy Assessment of Medicines and Chemicals Research Unit, Faculty of Pharmaceutical Sciences, Chulalongkorn University, Bangkok 10330, Thailand
[6] Natural Products for Ageing and Chronic Diseases Research Unit, Faculty of Pharmaceutical Sciences, Chulalongkorn University, Bangkok 10330, Thailand
* Correspondence: cchoatham@gmail.com (C.C.); boonchoo.sr@chula.ac.th (B.S.)

Abstract: Chemical investigation of *Dendrobium delacourii* revealed 11 phenolic compounds, and the structures of these compounds were determined by analysis of their NMR and HR-ESI-MS data. All compounds were investigated for their α-glucosidase inhibitory activity and anti-adipogenic properties. Phoyunnanin E (**10**) and phoyunnanin C (**11**) showed the most potent α-glucosidase inhibition by comparing with acarbose, which was used as a positive control. Kinetic study revealed the non-competitive inhibitors against the enzyme. For anti-adipogenic activity, densifloral B (**3**) showed the strongest inhibition when compared with oxyresveratrol (positive control). In addition, densifloral B might be responsible for the inhibition of adipocyte differentiation via downregulating the expression of peroxisome proliferator-activated receptor gamma (PPARγ) and CCAAT enhancer-binding protein alpha (C/EBPα), which are major transcription factors in adipogenesis.

Keywords: *Dendrobium delacourii*; Orchidaceae; α-glucosidase; anti-adipogenic; densifloral B; phoyunnanin E; phoyunnanin C

Citation: Thant, M.T.; Khine, H.E.E.; Nealiga, J.Q.L.; Chatsumpun, N.; Chaotham, C.; Sritularak, B.; Likhitwitayawuid, K. α-Glucosidase Inhibitory Activity and Anti-Adipogenic Effect of Compounds from *Dendrobium delacourii*. *Molecules* **2022**, *27*, 1156. https://doi.org/10.3390/molecules27041156

Academic Editor: Jacqueline Aparecida Takahashi

Received: 18 January 2022
Accepted: 8 February 2022
Published: 9 February 2022

Publisher's Note: MDPI stays neutral with regard to jurisdictional claims in published maps and institutional affiliations.

1. Introduction

Diabetes mellitus (DM) is a chronic metabolic disorder characterized by a high level of blood glucose resulting from a relative or absolute deficiency of insulin action. Type II is the most common type of diabetes, caused by β-cell dysfunction and insulin resistance [1]. The inhibition of α-glucosidase is effective for the treatment of type II diabetes [2]. α-Glucosidase is a membrane-bound enzyme produced from the epithelial cells of the small intestine. This enzyme is capable of converting starch and disaccharides into monosaccharides (glucose). Thus, glucose absorption can be reduced by inhibition of this enzyme, and postprandial blood glucose levels can also be decreased [3,4].

The relation between diabetes and obesity is well established in both traditional and modern therapy. The World Health Organization (WHO) estimates that 44% of diabetes cases are associated with overweightness and obesity [5]. Obesity is a risk factor of type 2 diabetes, coronary heart disease, and hypertension and is becoming a major health problem [6]. As a complex multifactorial chronic disease, obesity is characterized by an excessive adipocyte tissue mass. Adipogenesis is the process of cell differentiation during which fibroblast-like preadipocytes develop into mature adipocytes. Recently, the inhibition of adipogenesis has been proposed as a promising anti-obesity approach [7]. Moreover, anti-obesity is able considerably to reduce the prevalence rate of type 2 diabetes [8].

Currently, drugs from natural sources that specifically inhibit peroxisome proliferator-activated receptor gamma (PPARγ) and CCAAT enhancer-binding protein alpha (C/EBPα) expression are being targeted for the treatment of obesity [9]. PPARγ and C/EBPα increase adipocyte differentiation by activating the gene transcription for generating the adipocyte phenotype [10]. The anti-adipogenic activities of several natural compounds such as catechin [11] and procyanidin [12] via the downregulation of C/EBPα and PPARγ have been reported. Moreover, AMP-activated protein kinase (AMPK) is a common regulator that is involved in various lipid metabolisms [13], and the regulation of the AMPK signaling pathway is essential for anti-obesity [14]. The activation of AMPK can downregulate adipogenic key transcription factors such as PPARγ and C/EBPα, resulting in the suppression of adipocyte differentiation [15]. Acetyl-CoA carboxylase (ACC) is also responsible for the reduction in fatty acid synthesis [16]; in other words, fat accumulation is inhibited by the activation of ACC [17,18]. Adipogenesis can be inhibited by increasing the phosphorylation of both AMPK and ACC [19]. Protein kinase B (Akt) and glycogen synthase kinase-3 beta (GSK3β) are important protein kinases in adipogenesis. The activation of Akt is associated with phosphorylated GSK3β triggering adipogenesis via upregulating C/EBPα and PPARγ [20,21].

The drugs currently available for the treatment of diabetes and obesity are accompanied by severe adverse effects such as insomnia, headache, hypoglycemia, weight gain, constipation, and renal damage [22,23]. There are increasing reports of natural drugs from plant sources, due to their lesser side-effects. *Dendrobium* species are widely used in traditional medicine for the treatment of various diseases, as they possess a variety of pharmacological properties, such as being antidiabetic, antioxidant, anti-inflammatory, antimicrobial, immunomodulatory, and anticancer [24]. In China, *Dendrobium* has been used as traditional medicine for thousands of years as source of tonic, astringent, analgesic, antipyretic, and anti-inflammatory action [25,26]. About 41 species of the *Dendrobium* genus have been recorded in traditional Chinese medicine (TCM); of these, 30 species are collectively known under the Chinese name 'shihu' and are used for nourishing the stomach and increasing bodily fluid production [27]. Some species of this genus have been used in Thai traditional medicine, including *D. cumulatum*, *D. draconis*, *D. indivisum*, *D. trigonopus*, and *D. leonis* [28].

Dendrobium delacourii, named 'Ueang Dok Ma Kham' in Thai, belongs to the family Orchidaceae (Figure 1a). The common name is Delacour's dendrobium, and the plant is native to Thailand, Vietnam, Laos, and Myanmar. There is no previous study on the phytochemical constituents and biological activities of this plant. Our screening of the methanolic extract of *D. delacourii* exhibited α-glucosidase inhibition (80% inhibition at 100 µg/mL). It also showed an inhibitory effect on adipocyte differentiation in 3T3-L1 cells (51% inhibition at 5 µg/mL). In this study, we report the phytochemical constituents of *D. delacourii*, along with their inhibitory effects on α-glucosidase and adipocyte differentiation.

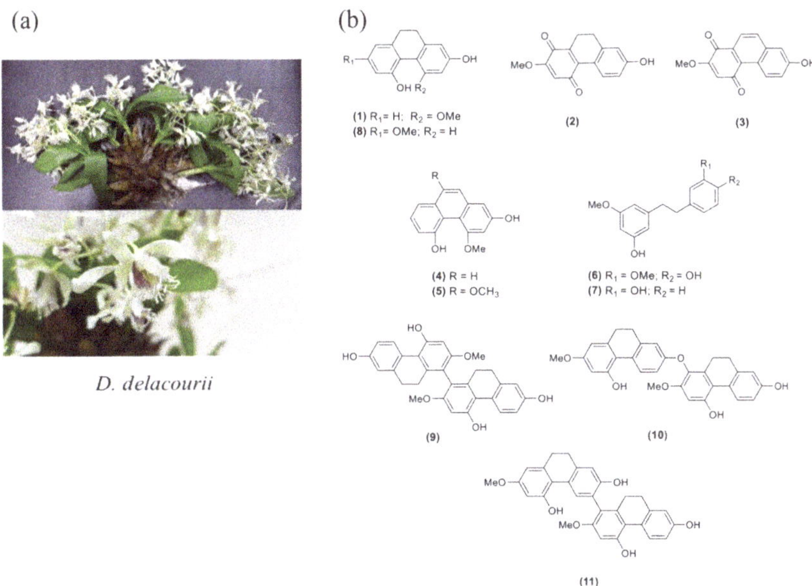

Figure 1. (**a**) *Dendrobium delacourii*. (**b**) The structures of isolated compounds (**1–11**).

2. Results and Discussion

2.1. Structure Determination

A MeOH extract of *D. delacourii* was separated by solvent partition to give ethyl acetate, butanol, and aqueous extracts. These extracts were evaluated for their inhibition of the α-glucosidase enzyme and adipocyte differentiation of 3T3-L1 cells. Only the EtOAc extract showed potent inhibition of α-glucosidase (85.5% inhibition at 100 µg/mL) and an anti-adipogenic effect (49% inhibition at 5 µg/mL). Therefore, the EtOAc extract was selected for further phytochemical investigation. Chromatographic separation of the EtOAc extract resulted in the isolation of 11 compounds. The structures of isolated compounds were characterized through analysis of their spectroscopic data and in comparison with previous reported values and were identified as hircinol (**1**) [29], ephemeranthoquinone (**2**) [30], densifloral B (**3**) [31], moscatin (**4**) [32], 4,9-dimethoxy-2,5-phenanthrenediol (**5**) [33], gigantol (**6**) [34], batatasin III (**7**) [35], lusianthridin (**8**) [36], 4,4′,7,7′-tetrahydroxy-2,2′-dimethoxy-9,9′,10,10′-tetrahydro-1,1′-biphenanthrene (**9**) [36], phoyunnanin E (**10**) [36], and phoyunnanin C (**11**) [37] (Figure 1b).

2.2. α-Glucosidase Inhibitory Activity

All the isolated compounds (**1–11**) were evaluated for their α-glucosidase inhibitory activity. In this study, each compound was initially tested at 100 µg/mL. An IC_{50} was determined if the compound showed more than 50% inhibition of the enzyme. The results are summarized in Table 1. Moscatin (**4**), gigantol (**6**), and lusianthridin (**8**) showed moderate α-glucosidase inhibition, having IC_{50} values of 390.1 ± 9.8 µM, 191.3 ± 6.8 µM, and 195.4 ± 9.6 µM, respectively. In addition, 4,4′,7,7′-tetrahydroxy-2,2′-dimethoxy-9,9′,10,10′-tetrahydro-1,1′-biphenanthrene (**9**), phoyunnanin E (**10**), and phoyunnanin C (**11**) showed stronger α-glucosidase inhibitory activities, with IC_{50} values of 18.4 ± 3.4 µM, 8.9 ± 0.8 µM, and 12.6 ± 0.9 µM, respectively, as compared with the positive control acarbose (IC_{50} 514.4 ± 9.2 µM). It can be observed herein that the dimeric phenanthrene derivatives **9**, **10**, and **11** were more potent than the monomers **4** and **8**.

Table 1. α-Glucosidase inhibitory activity of compounds **1–11** from the EtOAc extract.

Compounds	IC$_{50}$ (μM)
Hircinol (**1**)	NA
Ephemeranthoquinone (**2**)	NA
Densifloral B (**3**)	NA
Moscatin (**4**)	390.1 ± 9.8
4,9-Dimethoxy-2,5-phenanthrenediol (**5**)	NA
Gigantol (**6**)	191.3 ± 6.8
Batatasin III (**7**)	NA
Lusianthridin (**8**)	195.4 ± 9.6
4,4′,7,7′-Tetrahydroxy-2,2′-dimethoxy-9,9′,10,10′-tetrahydro-1,1′-biphenanthrene (**9**)	18.4 ± 3.4
Phoyunnanin E (**10**)	8.9 ± 0.8
Phoyunnanin C (**11**)	12.6 ± 0.9
Acarbose	514.4 ± 9.2

NA = no inhibitory activity.

For further investigation of the mechanism of enzyme inhibition, a kinetic study was carried out on the most potent compounds, phoyunnanin E (**10**) and phoyunnanin C (**11**). The experiment was performed by using Lineweaver–Burk plots of the reciprocal of velocity (1/V) against the reciprocal of substrate concentration (1/[S]) (Figure 2). The substrate *p*-nitrophenol-α-D-glucopyranoside concentration was varied from 0.25 to 2.0 mM in the absence or presence of compound **10** at 12 μM and 22 μM and compound **11** at 12 μM and 24 μM. As summarized in Table 2, the different concentrations of **10** and **11** reduced the V_{max} but did not affect the K_m value, indicating that **10** and **11** are non-competitive types of enzyme inhibitors. On the other hand, the drug acarbose showed an intersection of the lines on the *y*-axis, indicating a competitive type of inhibition. A secondary plot of each compound was then constructed to evaluate the inhibition constant (K_i). We found that the K_i value of acarbose 190.57 μM was obtained, and both **10** (K_i 5.89 μM) and **11** (K_i 5.97 μM) showed much greater affinity to the enzyme than acarbose. Compounds **10** and **11** as non-competitive inhibitors have some benefit over competitive inhibitors according to their binding to the allosteric site of the enzyme; therefore, they do not depend on the substrate concentration [38]. Furthermore, non-competitive inhibitors demand lower concentrations than competitive inhibitors to generate the same result [39].

Table 2. Kinetic parameters of α-glucosidase inhibition in the presence of phoyunnanin E (**10**) and phoyunnanin C (**11**).

Inhibitors	Dose (μM)	V_{max} ΔOD$_/$min	K_m (mM)	K_i (μM)
None	-	0.10	1.22	
10	22	0.024	1.22	5.89
	12	0.049	1.21	
11	24	0.023	1.21	5.97
	12	0.049	1.21	
Acarbose	930	0.11	6.47	190.57
	465	0.10	4.17	

V_{max}, maximum rate of velocity; K_m, Michaelis constant; K_i, inhibitor constant.

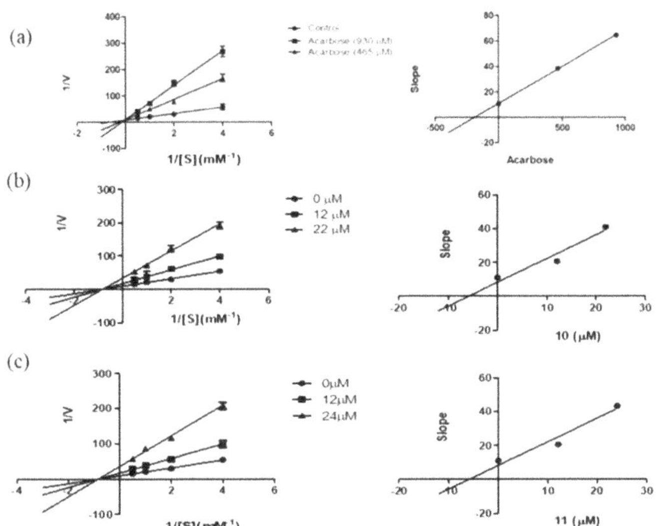

Figure 2. Lineweaver–Burk plots of (**a**) acarbose, (**b**) phoyunnanin E (**10**), and (**c**) phoyunnanin C (**11**). The secondary plot of each compound is on the right.

2.3. Anti-Adipogenic Activity

2.3.1. Ethyl Acetate Extracts from *D. delacourii* Attenuate Lipid Accumulation in Differentiated Adipocytes

To investigate their anti-adipogenic effect, the cytotoxic profile of *D. delacourii* extracts in preadipocytes was determined using MTT (methyl-thiazolyl-diphenyl-tetrazolium bromide) and nuclear staining assays. After culturing with 5 μg/mL of the methanolic, ethyl acetate, or butanolic extracts for 48 h, there were no significant alterations in the viability percentage observed via MTT assay in mouse embryonic preadipocyte 3T3-L1 cells (Figure 3a), compared with the untreated control. It is worth noting that treatment with all extracts at 10–20 μg/mL reduced viability in 3T3-L1 cells to lower than 90% (data not shown). Figure 3b depicts neither apoptosis nor necrosis, which were respectively observed as bright blue fluorescence of Hoechst33342 and propidium iodide red fluorescence in all treated 3T3-L1 cells. Thus, the extracts at 5 μg/mL, which were considered as non-toxic concentration, were chosen for the investigation of anti-adipogenic activity. Notably, treatment with oxyresveratrol (positive control), an anti-adipogenic natural compound, at 20 μM for 48 h also caused no change in cell viability percentage and cell death in preadipocyte 3T3-L1 cells.

For determination of their anti-adipogenic effect, preadipocyte 3T3-L1 cells were incubated with the differentiation medium with or without *D. delacourii* extracts, as mentioned in Materials and Methods. After the differentiation period was complete, the accumulated intracellular lipid droplets were analyzed by oil red O staining. As shown in Figure 3c, the lower level of oil red O percentage was indicated in 3T3-L1 cells cultured either with 5 μg/mL methanolic extract or 5 μg/mL ethyl acetate extract, compared with the control group. Meanwhile the significant decrease in oil red O staining percentage was not demonstrated in the treatment of the butanolic extract. A comparable inhibition of adipocyte differentiation between the methanolic extract, the ethyl acetate extract, and oxyresveratrol (20 μM) was evidenced with not only the reduction in the oil red O percentage but also the diminution of cellular lipid droplets presented in differentiated 3T3-L1 cells stained with oil red O (Figure 3d). Taken together, the ethyl acetate extract that indicated anti-adipogenic potential was selected for further investigation.

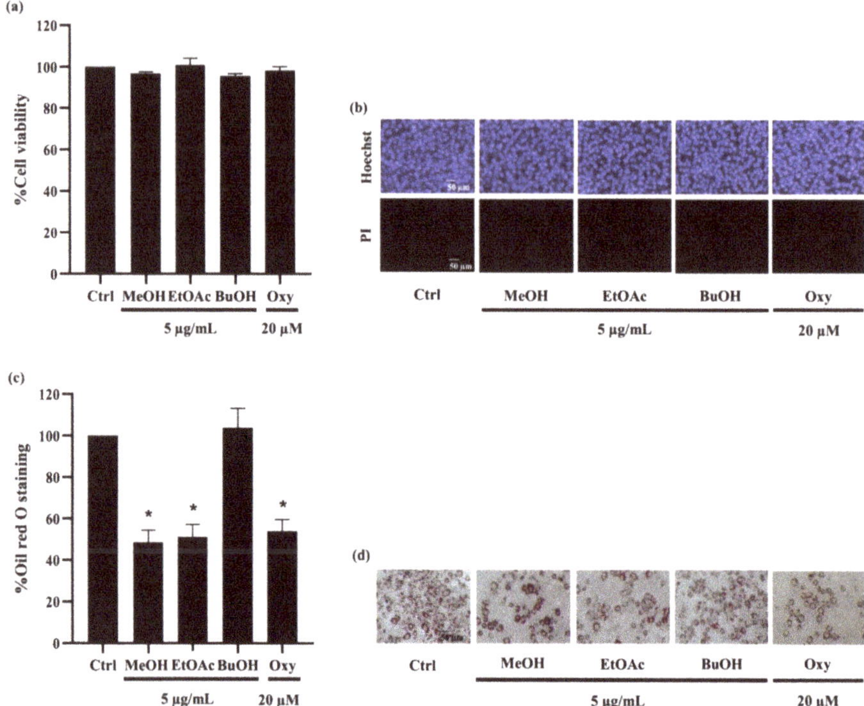

Figure 3. Effect of extracts from *Dendrobium delacourii* on cell viability and intracellular lipid accumulation during adipocyte differentiation. (**a**) MTT viability assay revealed no alteration of cell viability percentage in preadipocyte 3T3-L1 cells cultured with 5 µg/mL of methanolic extract (MeOH), ethyl acetate extract (EtOAc), or butanolic extract (BuOH) from *D. delacourii* for 48 h. (**b**) Treatment with all *D. delacourii* extracts for 48 h did not cause apoptosis or necrosis cell death detected via costaining of Hoechst33342/propidium iodide (PI) in 3T3-L1 cells. The suppression of lipid accumulation during adipogenesis in preadipocyte 3T3-L1 cells incubated with 5 µg/mL MeOH, 5 µg/mL EtOAc or 20 µM oxyresveratrol (Oxy) as a positive control was evidenced with (**c**) lower oil red O staining percentage and (**d**) lower amount of lipid droplets containing cells stained by oil red O compared with untreated control (Ctrl) group. Data are presented as means ± SD from three independent experiments. * $p < 0.05$ versus non-treated control cells.

2.3.2. Screening for Anti-Adipogenic Activity of Compounds (**1–11**)

According to previous report about anti-adipogenic activity of phenolic compound, batatasin I, at 20 µM [40], the suppressive effect of compounds (**1–11**) isolated from ethyl acetate extract of *D. delacourii* at the same concentration (20 µM) was preliminarily demonstrated by the remarkable reduction in both oil red O staining percentage (Figure 4a) and oil red O staining cells (Figure 4b) in differentiated 3T3-L1 cells. Although ephemeranthoquinone (**2**) and densifloral B (**3**) showed the highest anti-adipogenic activity among various isolates as well as positive control (oxyresveratrol), with approximately 64.7% and 65.2% reduction in oil red O staining, respectively, compound **2** at 20 µM significantly decreased viability in 3T3-L1 cells (data not shown). Therefore, compound **3**, which caused no cytotoxicity in 3T3-L1 cells, was selected to further investigate the related anti-adipogenic mechanisms.

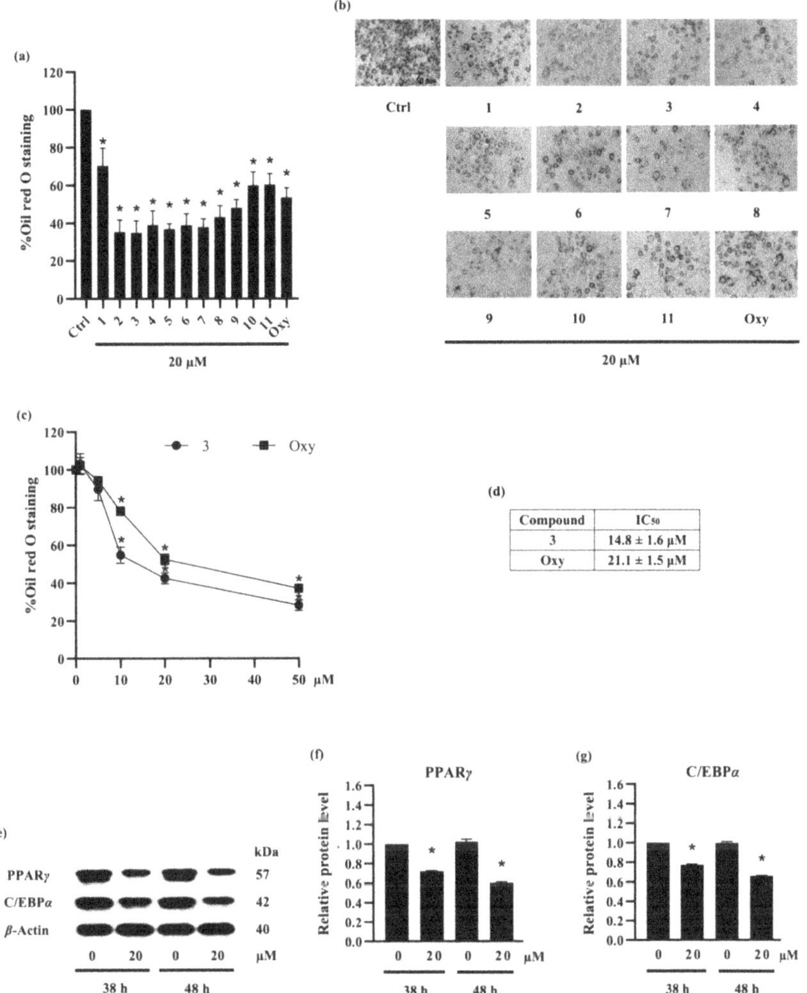

Figure 4. The effect of compounds (**1–11**) from *Dendrobium delacourii* on lipid accumulation in differentiated adipocyte was indicated with (**a**) the percentage of oil red O staining and (**b**) the accumulation of lipid droplets detected by oil red O staining. In comparison with a positive control, oxyresveratrol (Oxy), (**c**) dose–response relationship and (**d**) half-maximum inhibitory concentration (IC_{50}) demonstrated the more potent anti-adipogenic activity of compound **3** in 3T3-L1 cells. (**e**) Immunoblot analysis revealed the decreased protein levels of key lipid transcriptional regulators in 3T3-L1 cells cultured with 20 μM compound **3** for both early (38 h) and late (48 h) time points. The significant reduction in (**f**) PPARγ and (**g**) C/EBPα presented early in preadipocyte 3T3-L1 cells after the incubation with 20 μM compound **3** for 38 h in comparison with the no treatment control (Ctrl) at the same time point. β-actin was used as an internal control. Data are presented as means ± SD from three independent experiments. * $p < 0.05$ versus non-treated control cells at the same time point.

To compare the potency with another natural compound, preadipocyte 3T3-L1 cells were cultured with differentiation medium containing either various concentrations (0–50 μM) of compound **3** or oxyresveratrol. Figure 4c presents the relationship between concentration and oil red O staining percentage in response to compound **3** and oxyresveratrol treatment. The results indicate that compound **3** and oxyresveratrol clearly restrained the adipocyte differentiation in 3T3-L1 cells in a dose-dependent manner. In addition, the effect of a 50% inhibitory concentration (IC_{50}) of compound **3** and oxyresveratrol on adipocyte differentiation was about 14.8 ± 1.6 µM and 21.1 ± 1.5 µM, respectively (Figure 4d). According to IC_{50} data, compound **3** showed higher potency in anti-adipogenesis compared with oxyresveratrol.

2.3.3. Densifloral B (3) Suppresses Adipocyte Differentiation-Related Proteins

Adipogenesis involves a network of transcription factors that contribute to adipocyte differentiation and lipid accumulation [41]. Although the adipogenic gene CCAAT-enhancer-binding protein beta (C/EBPβ) is expressed soon after exposure to the adipogenic inducers, the upregulation of PPARγ and C/EBPα is acquired after 36–48 h of the induction [42]. In this regard, the translational level of PPARγ and C/EBPα was examined to elucidate the anti-adipogenic mechanism in preadipocyte 3T3-L1 cells cultured with 20 µM densifloral B (**3**) at both early (38 h) and late (48 h) stage. Western blot analysis indicated the prolong expression of PPARγ and C/EBPα, the adipogenic transcription factors, in 3T3-L1 cells from early until late stage of adipogenic process. Moreover, the expression level of PPARγ and C/EBPα was restrained in densifloral B-treated 3T3-L1 cells compared with untreated control cells (Figure 4e). Interestingly, the diminution in PPARγ and C/EBPα relative protein levels, which was promptly detected at 38 h, was sustained until 48 h of adipocyte differentiation in the presence of 20 µM densifloral B (Figure 4f,g, respectively).

Recent reports indicate that the downregulation of PPARγ and C/EBPα are correlated with the activation of AMPKα/β and ACC [43]. As presented in Figure 5a, treatment with 20 µM densifloral B (**3**) upregulated the phosphorylated form of AMPKα (p-AMPKα), AMPKβ1 (p-AMPKβ1), and ACC (p-ACC) in 3T3-L1 cells at both 38 h and 48 h of the differentiation period. The increased expression level of p-ACC/ACC (Figure 5b), p-AMPKα/AMPKα (Figure 5c), and p-AMPKβ1/AMPKβ1/2 (Figure 5f) was correlated with suppression of PPARγ and C/EBPα protein expression (Figure 4e) as well as lipid accumulation in differentiated 3T3-L1 cells (Figure 4a). The results presented in this study correspond with the evidence of the stimulation of AMPK and ACC protein being associated with the inhibition of 3T3-L1 adipocyte differentiation [44]. Interestingly, the densifloral B (**3**) also suppressed the activated Akt (p-Akt), an upstream regulator of the AMPK–ACC signal [45] to reduce the adipocyte formation (Figure 5d).

Due to the expression of adipogenic transcription factors also modulated via Akt/GSK3β [46–48], the alteration of the p-GSK3β/GSK3β level was additionally examined in densifloral B-treated 3T3-L1 cells. The phosphorylation by p-Akt results in the inactivation of the GSK3β degradation complex following the initiation of targeted gene expression [49]. Surprisingly, the presence of 20 µM densifloral B (**3**) in differentiation medium significantly lessened p-GSK3β/GSK3β expression level in preadipocyte 3T3-L1 cells at 38–48 h of differentiation time (Figure 5e). Taken together, the present results suggest that densifloral B (**3**) isolated from *D. delacourii* might inhibit adipocyte differentiation via suppression of the Akt-mediating GSK3β and AMPK–ACC signals (Figure 6).

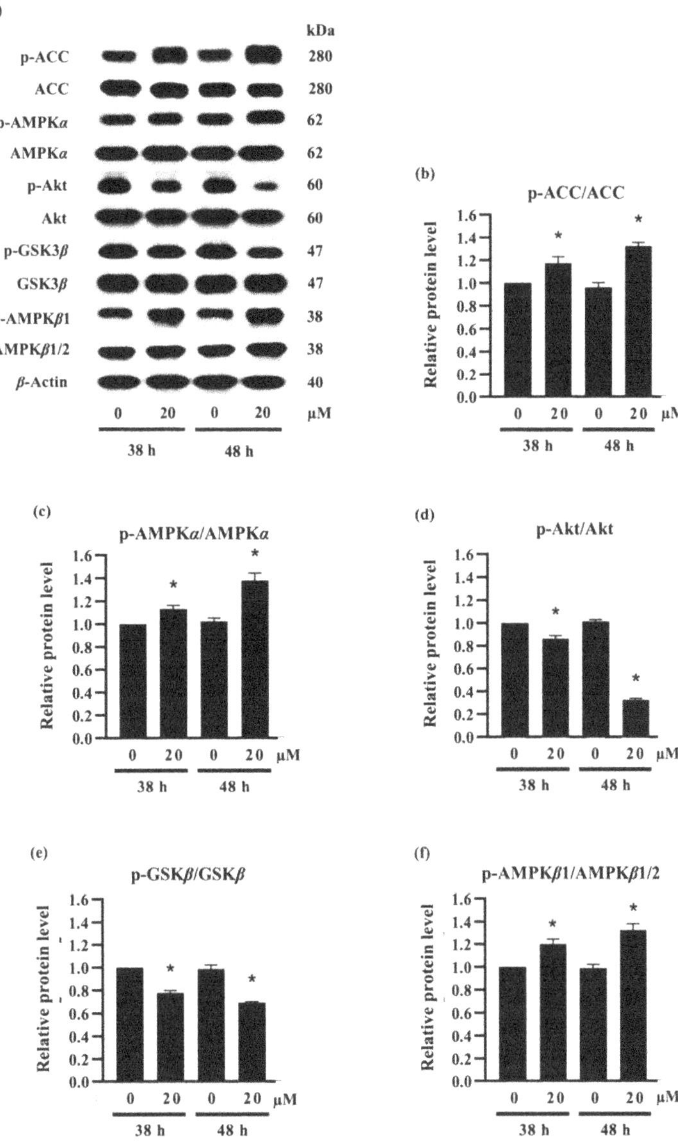

Figure 5. Effect of densifloral B (**3**) on adipocyte differentiation via Akt-related pathways (**a**) The alteration of adipogenesis-related proteins was determined in preadipocyte 3T3-L1 cells after culture with differentiation medium containing 0–20 µM densifloral B for 38–48 h via Western blotting. The upregulated levels of (**b**) p-ACC/ACC, (**c**) p-AMPKα/AMPKα, and (**f**) p-AMPKβ1/AMPKβ1/2, as well as the reduction in (**d**) pAkt/Akt and (**e**) p-GSK3β/GSK3β, were obviously presented in densifloral B-treated 3T3-L1 cells. β-actin served as an internal control. Data are presented as means ± SD from three independent experiments. * $p < 0.05$ versus non-treated control cells at the same time point.

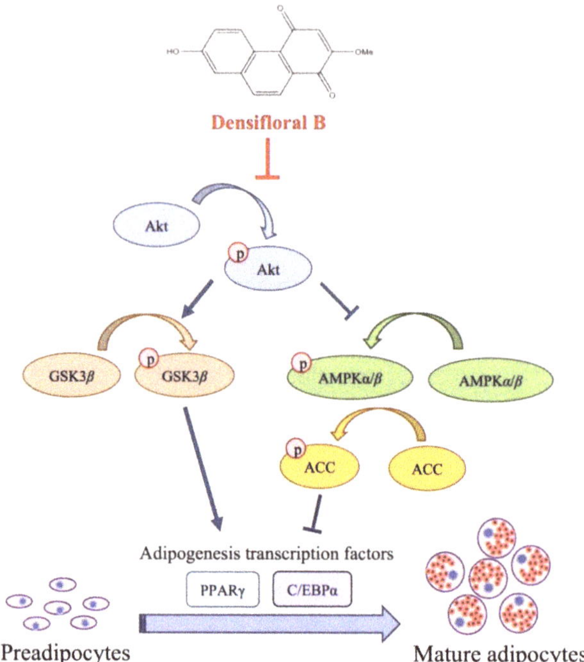

Figure 6. Proposed mechanism of densifloral B (**3**) on inhibition of adipogenesis via modulating Akt-related pathways. Activated Akt (p-Akt) mediates adipocyte differentiation through upregulation of p-GSK3β and suppression of AMPK–ACC signal that both trigger the expression of adipogenic transcription factors PPARγ and C/EBPα. Therefore, the downregulation of p-Akt moderated by densifloral B (**3**), as a consequence of the restraint of p-GSK3β and stimulated AMPK–ACC cascades, effectively inhibits the differentiation of preadipocytes into mature adipocytes.

3. Materials and Methods

3.1. General Experimental Procedures

Mass spectra were recorded on a Bruker micro TOF mass spectrometer (ESI-MS). NMR spectra were recorded on a Bruker Avance DPX-300FT-NMR spectrometer or a Bruker Avance III HD 500 NMR spectrometer. Vacuum-liquid column chromatography (VLC) and column chromatography (CC) were performed on silica gel 60 (Merck, Kieselgel 60, 70–320 mesh), silica gel 60 (Merck, Kieselgel 60, 230–400 mesh) (Darmstadt, Germany), and Sephadex LH-20 (25–100 μm, Pharmacia Fine Chemical Co. Ltd.) (Piscataway, NJ, USA). Acarbose was purchased from Fluka Chemical (Buchs, Switzerland). Mouse embryonic preadipocyte 3T3-L1 cells were acquired from the American Type Culture Collection (ATCC, Manassas, VA, USA). Dulbecco's modified Eagle medium (DMEM), fetal bovine serum (FBS), penicillin/streptomycin, and L-glutamine were obtained from Gibco (Gaithersburg, MA, USA). Yeast alpha-glucosidase enzyme, *p*-nitrophenol-α-D-glucopyranoside, isobutyl-methylxanthine, dexamethasone, MTT (methyl-thiazolyl-diphenyl-tetrazolium bromide), Hoechst33342, propidium iodide, dimethylsulfoxide (DMSO), and oil red O were purchased from Sigma-Aldrich (St. Louis, MO, USA). Insulin was obtained from Himedia (Mumbai, India). The radio-immunoprecipitation assay (RIPA) buffer, chemiluminescent substrates, and the bicinchoninic acid (BCA) protein assay kit were from Thermo Scientific (Rockford, IL, USA), and the nitrocellulose membranes were from Bio-Rad Laboratories (Hercules, CA, USA). All primary and secondary antibodies were purchased from Cell Signaling Technology (Danvers, MA, USA). Oxyresveratrol was provided by Prof. Kittisak Likhitwitayawuid.

3.2. Plant Material

The whole plant of *Dendrobium delacourii* was purchased from a Chatuchak market in May 2018. Plant identification was performed by Dr. Boonchoo Sritularak. A voucher specimen (BS-Ddela-052561) was deposited at the Department of Pharmacognosy and Pharmaceutical Botany, Faculty of Pharmaceutical Sciences, Chulalongkorn University.

3.3. Extraction and Isolation

The dried powder of whole plant *D. delacourii* (3.5 kg) was macerated with MeOH (4 × 15 L), and a methanolic extract (300.6 g) was obtained. This extract was dissolved in water and then partitioned with ethyl acetate (EtOAc) and butanol to give an EtOAc extract (159.7 g), a butanol extract (98.2 g), and an aqueous extract (42.1 g), respectively, after evaporation of the solvent. The EtOAc extract exhibited 85.5% inhibition of α-glucosidase enzyme at 100 μg/mL and also showed 49.0% inhibition of adipocyte differentiation of 3T3-L1 cells at 5 μg/mL, whereas the other extracts were devoid of both activities. Therefore, the EtOAc extract was subjected to further investigation.

The EtOAc extract was separated by vacuum liquid chromatography (silica gel, acetone–hexane, gradient) to give five fractions (A–E). Fraction D (54.2 g) was fractionated on a silica gel column (acetone–hexane, gradient) to give four fractions (DA–DD). Fraction DB (5.4 g) was separated by Sephadex LH-20 (methanol) to yield six fractions (DBA–DBF). Fraction DBB (612.0 mg) was subjected to column chromatography (CC) (silica gel, acetone–hexane, gradient), and then the pure compounds hircinol (**1**) (11.4 mg) and ephemeranthoquinone (**2**) (6.4 mg) were obtained. Densifloral B (**3**) (7.9 mg), moscatin (**4**) (17.1 mg), and 4,9-dimethoxy-2,5-pheneanthrenediol (**5**) (3.6 mg) were obtained from fractions DBD, DBE, and DBF after purification on a silica gel column (acetone–hexane, gradient). Fraction DC (6.1 g) was separated by Sephadex LH-20 (methanol) to yield four fractions (DCA–DCD). The separation of fraction DCA (1.2 g) by CC (silica gel, acetone–hexane, gradient) resulted in gigantol (**6**) (166.5 mg). Fraction DCB (170.9 mg) was separated on a silica gel column (EtOAc-CH_2Cl_2, gradient) to yield batatasin III (**7**). Fraction DCD (428.0 mg) was isolated by a silica gel column (acetone–hexane, gradient) to give lusianthridin (**8**). Fraction DD (6.7 g) was separated by Sephadex LH-20 (methanol) to yield three fractions (DDA-DDC). Quantities of 4,4′,7,7′-tetrahydroxy-2,2′-dimethoxy-9,9′,10,10′-tetrahydro-1,1′-biphenanthrene (**9**) (4.7 mg) and phoyunnanin E (**10**) (7.0 mg) were obtained from fraction DDB, and phoyunnanin C (**11**) (6.8 mg) was obtained from fraction DDC after separation in a silica gel column (methanol–CH_2Cl_2, gradient).

Hircinol (**1**): yellow amorphous solid; HR-ESI-MS: *m/z* 243.1059 [M + H]$^+$ calcd. for $C_{15}H_{15}O_3$, 243.1021, suggesting $C_{15}H_{14}O_3$. ^{1}H NMR (500 MHz, acetone-d_6) δ: 2.59 (4H, m, H_2-9 and H_2-10), 3.97 (3H, s, 4-MeO), 6.56 (1H, d, *J* = 2.5 Hz, H-1), 6.60 (1H, d, *J* = 2.5 Hz, H-3), 6.80 (2H, d, *J* = 8.0 Hz, H-6, H-8), 7.06 (1H, t, *J* = 8.0 Hz, H-7); ^{13}C NMR (125 MHz, acetone-d_6) δ: 31.6 (C-10), 31.8 (C-9), 57.2 (4-MeO), 99.8 (C-3), 109.8 (C-1), 114.6 (C-4b), 118.2 (C-6), 120.1 (C-8), 128.1 (C-7), 130.4 (C-4a), 141.3 (C-8a), 144.2 (C-10a), 154.7 (C-4), 156.3 (C-5), 158.5 (C-2).

Ephemeranthoquinone (**2**): reddish powder; HR-ESI-MS: *m/z* 279.06185 [M + Na]$^+$ calcd. for $C_{15}H_{12}O_4Na$, 279.06333 suggesting $C_{15}H_{12}O_4$. ^{1}H NMR (300 MHz, acetone-d_6) δ: 2.62 (2H, m, H_2-10), 2.69 (2H, m, H_2-9), 3.85 (3H, s, 2-OMe), 5.98 (1H, s, H-3), 6.76 (2H, m, H-6, H-8), 7.97 (1H, d, *J* = 9.3 Hz, H-5); ^{13}C NMR (75 MHz, acetone-d_6) δ: 20.0 (C-10), 27.2 (C-9), 55.7 (2-OMe), 107.4 (C-3), 113.4 (C-5), 114.8 (C-8), 121.2 (C-4b), 132.0 (C-6), 136.1 (C-4a), 136.2 (C-10a), 141.5 (C-8a), 158.6 (C-2), 159.1 (C-7), 180.8 (C-1), 187.2 (C-4).

Densifloral B (**3**): orange powder, HR-ESIMS: *m/z* 277.0473 [M + Na]$^+$ calcd. for $C_{15}H_{10}O_4Na$, 277.04768 suggesting $C_{15}H_{10}O_4$. ^{1}H NMR (300 MHz, acetone-d_6) δ: 3.92 (3H, s, 2-OMe), 6.20 (1H, s, H-3), 7.31 (1H, d, *J* = 2.4 Hz, H-8), 7.35 (1H, dd, *J* = 2.4, 9.3 Hz, H-6), 8.04 (2H, br s, H-9 and H-10), 9.49 (1H, d, *J* = 9.3 Hz, H-5); ^{13}C NMR (75 MHz, acetone-d_6) δ: 55.8 (2-OMe), 109.9 (C-8), 110.9 (C-3), 122.0 (C-5, C-6), 124.2 (C-4b), 127.5 (C-4a), 130.1 (C-10a), 132.3 (C-10), 134.2 (C-9), 139.4 (C-8a), 157.8 (C-7), 160.2 (C-2), 180.5 (C-1), 188.4 (C-4).

Moscatin (**4**): brown amorphous solid, HR-ESIMS: m/z 241.0888 [M + H]$^+$ calcd. for $C_{15}H_{13}O_3$ 241.0865, suggesting $C_{15}H_{12}O_3$. ^{1}H NMR (500 MHz, acetone-d_6) δ: 4.15 (3H, s, 4-OMe), 6.99 (1H, d, J = 2.5 Hz, H-3), 7.07 (1H, d, J = 2.5 Hz, H-1), 7.10 (1H, dd, J = 7.5, 2.0 Hz, H-6), 7.41 (1H, dd, J = 7.5, 2.0 Hz, H-8), 7.43 (1H, t, J = 7.5 Hz, H-7), 7.49 (1H, d, J = 9.0 Hz H-10), 7.63 (1H, d, J = 9.0 Hz, H-9); ^{13}C NMR (125 MHz, acetone-d_6) δ: 58.6 (4-OMe), 102.5 (C-3), 107.8 (C-1), 113.9 (C-4a), 116.9 (C-6), 119.8 (C-4b), 121.0 (C-8), 126.9 (C-10), 127.4 (C-7), 129.7 (C-9), 135.0 (C-8a), 137.1 (C-10a), 155.2 (C-5), 156.4 (C-2), 157.3 (C-4).

4,9-Dimethoxy-2,5-phenanthrenediol (**5**): brown amorphous solid. HR-ESIMS: at m/z 271.1009, [M + H]$^+$ calcd for $C_{16}H_{15}O_4$; 271.0970, suggesting $C_{16}H_{14}O_4$. ^{1}H NMR (500 MHz, acetone-d_6) δ: 4.03 (3H, s, 9-OMe), 4.11 (3H, s, 4-OMe), 6.81 (1H, d, J = 2.5 Hz, H-3), 6.92 (1H, s, H-10), 6.99 (1H, d, J = 2.5 Hz, H-1), 7.12 (1H, dd, J = 1.5, 7.5 Hz, H-6), 7.43 (1H, t, J = 7.5 Hz, H-7), 7.85 (1H, dd, J = 1.5, 7.5 Hz, H-8), 8.82 (1H, s, 2-OH), 9.43 (1H, s, 5-OH); ^{13}C NMR (125 MHz, acetone-d_6) δ: 55.9 (9-OMe), 58.5 (4-OMe), 100.3 (C-3), 102.8 (C-10), 106.9 (C-1), 110.0 (C-4a), 114.3 (C-8), 117.6 (C-6), 120.9 (C-4b), 127.9 (C-7), 129.2 (C-8a), 137.9 (C-10a), 154.9 (C-9), 155.2 (C-5), 156.3 (C-4), 157.5 (C-2).

Gigantol (**6**): brown amorphous solid. HR-ESIMS: at m/z 297.1102, [M + Na]$^+$ calculated for $C_{16}H_{18}O_4Na$; 297.1102, suggesting $C_{16}H_{18}O_4$. ^{1}H NMR (500 MHz, acetone-d_6) δ: 2.79 (4H, m, H$_2$-α, H$_2$-α′), 3.69 (3H, s, 3-OMe), 3.77 (3H, s, 3′-OMe), 6.25 (1H, t, J = 1.5 Hz, H-4), 6.29 (1H, t, J = 1.5 Hz, H-6), 6.32 (1H, t, J = 1.5 Hz, H-2), 6.65 (1H, dd, J = 8.0, 2.0 Hz, H-6′), 6.73 (1H, d, J = 8.0 Hz, H-5′), 6.79 (1H, d, J = 2.0 Hz, H-2′); ^{13}C NMR (125 MHz, acetone-d_6) δ: 37.9 (C-α′), 38.9 (C-α), 55.2 (3′-OMe), 56.1 (3-OMe), 99.7 (C-4), 106.3 (C-6), 108.9 (C-2), 112.8 (C-5′), 115.5 (C-2′), 121.5 (C-6′), 134.1 (C-1′), 145.1 (C-4′), 145.4 (C-1), 147.9 (C-3′), 159.2 (C-3), 161.8 (C-5).

Batatasin III (**7**): brown amorphous solid. HR-ESIMS: at m/z 267.10556, [M + Na]$^+$ calculated for $C_{15}H_{16}O_3Na$; 267.099715, suggesting $C_{15}H_{16}O_3$. ^{1}H NMR (500 MHz, acetone-d_6) δ: 2.79 (4H, m, H-α, H-α′), 3.70 (3H, s, 3-OMe), 6.23 (1H, t, J = 2.0 Hz, H-4), 6.30 (1H, t, J = 2.0 Hz, H-2), 6.32 (1H, br t, J = 2.0 Hz, H-6), 6.63 (1H, m, H-4′), 6.69 (1H, br d, J = 9.0 Hz, H-6′), 6.71 (1H, br d, J = 2.4 Hz, H-2′), 7.07 (1H, t, J = 8.0 Hz, H-5′); ^{13}C NMR (125 MHz, acetone-d_6) δ: 38.2 (C-α′), 38.5 (C-α), 55.2 (3-OMe), 99.8 (C-4), 106.2 (C-2), 108.8 (C-6), 113.6 (C-4′), 116.2 (C-2′), 120.4 (C-6′), 130.0 (C-5′), 144.3 (C-1′), 145.0 (C-1), 158.2 (C-3′), 159.2 (C-3), 161.8 (C-5).

Lusianthridin (**8**): brown amorphous solid. HR-ESIMS: at m/z 265.08251, [M + Na]$^+$ calculated for $C_{15}H_{14}O_3Na$; 265.084065, suggesting $C_{15}H_{14}O_3$. ^{1}H NMR (500 MHz, acetone-d_6) δ: 2.67 (4H, m, H$_2$-9 and H$_2$-10), 3.72 (3H, s, 2-OMe), 6.37 (1H, d, J = 2.5 Hz, H-1), 6.45 (1H, d, J = 2.5 Hz, H-3), 6.73 (1H, br d, J = 7.5 Hz, H-6), 6.72 (1H, br s, H-8), 8.24 (1H, d J = 7.5 Hz, H-5); ^{13}C NMR (125 MHz, acetone-d_6) δ: 30.6 (C-9), 31.3 (C-10), 55.2 (2-OMe), 101.5 (C-3), 105.8 (C-1), 113.4 (C-6), 115.0 (C-8), 115.7 (C-4a), 125.8 (C-4b), 129.8 (C-5), 139.7 (C-8a), 141.3 (C-10a), 155.7 (C-4), 155.8 (C-7), 159.1 (C-2).

4,4′,7,7′-Tetrahydroxy-2,2′-dimethoxy-9,9′,10,10′-tetrahydro-1,1′-biphenanthrene (**9**): yellow amorphous powder, HR-ESIMS: at m/z 505.1630, [M + Na]$^+$ calculated for $C_{30}H_{26}O_6Na$; 505.1627 suggesting $C_{30}H_{26}O_6$. ^{1}H NMR (300 MHz, acetone-d_6) δ: 2.31 (4H, m, H$_2$-10, H$_2$-10′), 2.51 (4H, m, H$_2$-9, H$_2$-9′), 3.60 (6H, s, 2-OMe, 2′-OMe), 6.57 (2H, s, H-3, H-3′), 6.65 (2H, d, J = 2.4 Hz, H-8, H-8′), 6.69 (2H, dd, J = 8.4, 2.4 Hz, H-6, H-6′), 8.25 (2H, d, J = 8.4 Hz, H-5, H-5′), 8.08 (2H, s, 7-OH, 7′-OH), 8.41 (2H, s, 4-OH, 4′-OH); ^{13}C NMR (75 MHz, acetone-d_6) δ: 27.0 (C-10, C-10′), 29.7 (C-9, C-9′), 54.7 (2-OMe, 2′-OMe), 98.3 (C-3, C-3′), 112.5 (C-6, C-6′), 113.8 (C-8, C-8′), 114.6 (C-4a, C-4a′), 116.5 (C-1, C-1′), 125.5 (C-4b, C-4b′), 129.3 (C-5, C-5′), 139.3 (C-8a, C-8a′), 139.7 (C-10a, C-10a′), 154.0 (C-4, C-4′), 155.1 (C-7, C-7′), 156.4 (C-2, C-2′).

Phoyunnanin E (**10**): amorphous powder, HR-ESIMS: at m/z 505.1628, [M + Na]$^+$ calculated for $C_{30}H_{26}O_6Na$; 505.1627, suggesting $C_{30}H_{26}O_6$. ^{1}H NMR (500 MHz, acetone-d_6) δ: 2.60 (4H, m, H$_2$-9 and H$_2$-10), 2.67 (4H, m, H$_2$-9′, H$_2$-10′), 3.71 (3H, s, 2-OMe), 3.73 (3H, s, 2′-OMe), 6.37 (1H, d, J = 2.5 Hz, H-1′), 6.42 (1H, d, J = 2.5 Hz, H-3′), 6.62 (1H, dd, J = 8.5, 2.5 Hz, H-6′), 6.66 (1H, s, H-3), 6.67 (1H, d, J = 2.5 Hz, H-8′), 6.69 (1H, d, J = 2.5 Hz, H-8), 6.71 (1H, dd, J = 2.5, 9.0 Hz, H-6), 8.25 (1H, d, J = 8.5 Hz, H-5′), 8.27 (1H, d, J = 9.0 Hz, H-5); ^{13}C NMR (125 MHz, acetone-d_6) δ: 23.8 (C-10), 30.7 (C-9′, C-9), 31.3 (C-10′), 55.3

(2′-OMe), 56.0 (2-OMe), 100.8 (C-3), 101.6 (C-3′), 106.0 (C-1′), 112.6 (C-6′), 113.6 (C-6), 114.2 (C-8′), 115.0 (C-8), 115.5 (C-4a′), 115.7 (C-4a), 125.6 (C-4b), 127.6 (C-4b′), 129.8 (C-5′), 130.2 (C-5), 133.9 (C-1), 134.0 (C-10a), 139.7 (C-8a′), 139.8 (C-8a), 141.6 (C-10a′), 152.0 (C-2), 152.5 (C-4), 156.1 (C-4′), 156.4 (C-7), 157.7 (C-7′), 159.6 (C-2′).

Phoyunnanin C (**11**): amorphous powder, HR-ESIMS: at m/z 505.1635, $[M + Na]^+$ calculated for $C_{30}H_{26}O_6Na$; 505.1627, suggesting $C_{30}H_{26}O_6$. ^{1}H NMR (300 MHz, acetone-d_6) δ: 2.51 (2H, m, H_2-10), 2.53 (2H, m, H_2-9), 2.73 (4H, m, H_2-9′, H_2-10′), 3.64 (3H, s, 2-OMe), 3.72 (3H, s, 2′-OMe), 6.38 (2H, br s, H-1′, H-3′), 6.57 (1H, s, H-3), 6.66 (1H, d, J = 2.7 Hz, H-8), 6.69 (1H, dd, J = 8.4, 2.7 Hz, H-6), 6.76 (1H, s, H-8′), 8.08 (1H, s, H-5′), 8.23 (1H, d, J = 8.4 Hz, H-5); ^{13}C NMR (75 MHz, acetone-d_6) δ: 27.5 (C-10), 29.6 (C-9, C-9′), 30.7 (C-10′), 54.4 (2-OMe), 54.8 (2′-OMe), 98.5 (C-3), 100.7 (C-3′), 105.1 (C-1′), 112.5 (C-6), 113.8 (C-8), 114.3 (C-8′), 114.9 (C-4a), 115.1 (C-4a′), 117.4 (C-1), 121.7 (C-6′), 124.8 (C-4b′), 125.3 (C-4b), 129.3 (C-5), 131.7 (C-5′), 137.7 (C-8a′), 139.2 (C-8a), 140.1 (C-10a), 140.5 (C-10a′), 152.8 (C-7′), 154.3 (C-4), 155.1 (C-4′), 155.2 (C-7), 156.6 (C-2), 158.3 (C-2′).

3.4. Assay for α-Glucosidase Inhibitory Activity

The assay was based on the inhibition in the sample of α-glucosidase enzyme, which can release *p*-nitrophenol (PNP) from *p*-nitrophenyl-α-D-glucoside (PNPG) by hydrolysis [48]. In this assay, acarbose was used as the positive control. For IC_{50} determination, twofold serial dilution was performed for each sample. Each experiment was accomplished in triplicate. Data are expressed as mean ± SD.

The kinetic study of enzyme inhibition was analyzed by the double reciprocal Lineweaver–Burk plot (1/V versus 1/[S]). The experiment was carried out by performing various concentrations of substrate *p*-nitrophenol-α-D-glucopyranoside (0.25, 0.5, 1.0, 2.0 mM) in the absence or presence of compound (**10**) (12 and 22 μM) and compound (**11**) (12 and 24 μM). The reaction was monitored every 5 min for a total time of 25 min and measured at 405 nm by a microplate reader. Each experiment was performed in triplicate. The K_i value was estimated by constructing a secondary plot which is plotted by the slopes of the double-reciprocal lines versus inhibitor concentration.

3.5. Assay for Anti-Adipogenic Activity

3.5.1. Cell Culture and Adipocyte Differentiation

Mouse embryonic preadipocyte 3T3-L1 cells were cultured in DMEM containing 10% FBS, 100 units/mL of penicillin/streptomycin, and 2 mmol/L of L-glutamine under humidified conditions of 5% CO_2 at 37 °C until 70–80% confluence was reached. For differentiation into adipocyte, preadipocyte 3T3-L1 cells were incubated with differentiation media composed of 10% FBS, 0.5 mM isobutylmethylxanthine, 1 μM dexamethasone, and 5 μg/mL insulin in DMEM with or without test compound for 2 days. Then, the differentiation media was replaced with culture media containing 5 μg/mL of insulin. After further incubation for 2 days, the cells were maintained in complete DMEM, which was changed every 2 days until adipocytes containing lipid droplets were observed under microscope [40].

3.5.2. Determination of Cytotoxicity

To evaluate the effect of *D. delacourii* extracts on cell viability, 3T3-L1 preadipocytes were seeded into 96-well plates at density of 2×10^3 cells/well and allowed to attach overnight at 37 °C. Then, the cells were further cultured with extracts (5 μg/mL), compounds (20 μM), or left untreated for 48 h before adding of 0.45 mg/mL MTT solution to assess cell viability. After incubation for 3 h at 37 °C and kept from light, the optical density (OD) of the purple formazan product dissolved in DMSO was measured at 570 nm using a microplate reader (Anthros, Durham, NC, USA). The relative OD ratio of treated to non-treated cells was presented as percentage cell viability [50].

The cytotoxicity of *D. delacourii* extracts was confirmed via cell death detection using costaining of Hoechst33342 and propidium iodide. After 48 h incubation with indicated treatment, the cells were further incubated with nuclear staining solution containing 2 μg/mL of Hoechst33342 and 1 μg/mL of propidium iodide for 30 min. The mode of cell death was observed under a fluorescence microscope (Olympus IX51 with DP70, Olympus Corp., Shinjuku-ku, Tokyo, Japan).

3.5.3. Quantification of Cellular Lipid Content Using Oil Red O Staining

The lipid droplets presenting in differentiated adipocytes were detected via oil red O staining. After the differentiation process, 3T3-L1 cells were fixed with 10% formalin for 45 min and further incubated with oil red O solution at room temperature for 1 h. After washing with 60% isopropanol for three times, oil red O-stained cells were captured using a Nikon Ts2 inverted optical microscope (Tokyo, Japan). For quantification, cellular oil red O was extracted using absolute isopropanol for measurement of OD at 570 nm by microplate reader (Anthros, Durham, NC, USA) [51]. The percentage of oil red O staining was calculated relative to the total protein content determined by BCA assay [52].

3.5.4. Western Blot Analysis

After the indicated treatment, 3T3-L1 cells were washed with phosphate-buffered saline (PBS, pH 7.4), then the cell membranes were broken using RIPA buffer supplemented with a protease inhibitor cocktail. After incubation on ice for 45 min, the cell lysates were centrifuged at 12,000 rpm at 4 °C for 15 min to collect the clear supernatant containing cellular protein, which was measured for total protein content using a BCA assay kit. The total protein (30 μg) from each sample was loaded and separated onto 10% sodium dodecyl sulfate-polyacrylamide gel electrophoresis (SDS-PAGE). Subsequently, the separated proteins were transferred onto nitrocellulose membranes, which were blocked with 5% skim milk in TBST buffer (Tris-buffered saline with Tween 20, pH 7.2) and further immunoblotted with primary antibodies against p-Akt (Thr308), Akt, p-GSK3β (Ser9), GSK3β, p-AMPKα (Thr172), AMPKα, p-AMPKβ1 (Ser128), AMPKβ1/2, p-ACC (Ser79), ACC, PPARγ, C/EBPα, and β-actin at 4 °C overnight. Before immersion in horseradish peroxidase (HRP)-linked secondary antibody at room temperature for 2 h, the membranes were washed with TBST for 7 min, three times. The reactive protein signals exposed with chemiluminescent substrates were captured and quantified using Chemiluminescent ImageQuant LAS 4000 (GE Healthcare Bio-Sciences AB, Björkgatan, Uppsala, Sweden).

3.6. Statistical Analysis

All data are expressed as means ± standard deviation (SD) obtained from three independent experiments. Statistical analysis was performed using GraphPad Prism 8.0.2 (GraphPad Software Inc., San Diego, CA, USA) with one-way ANOVA. Differences with p value < 0.05 were considered to be statistically significant.

4. Conclusions

In this study, 11 compounds were isolated from the ethyl acetate extract of *D. delacourii*. Two dimeric phenanthrene derivatives, phoyunnanin E (**10**) and phoyunnanin C (**11**), revealed the most potent α-glucosidase inhibition when compared with a positive control, acarbose. An enzyme kinetic study performed on them indicated non-competitive inhibitors. Regarding anti-adipogenic activity, densifloral B (**3**) showed the most potent activity when compared with a positive control, oxyresveratrol. The anti-adipogenic properties of densifloral B (**3**) were attributed to the downregulation of PPARγ and C/EBPα expression through the modulation of Akt-related pathways including the Akt/GSK3β and Akt/AMPK–ACC signals. The findings obtained from this study demonstrate the evaluation of α-glucosidase inhibitory activity and anti-adipogenic effect of the *Dendrobium delacourii* plant, which can be used for the management of diabetes and obesity.

Author Contributions: B.S. conceived, designed, and supervised the research project, as well as preparing and editing the manuscript; C.C. supervised the anti-adipogenic activity and prepared the manuscript; M.T.T. performed the experiments and prepared the manuscript; H.E.E.K. and J.Q.L.N. assisted in anti-adipogenic activity; N.C. supervised the α-glucosidase inhibition assay; K.L. provided comments and suggestions on the preparation of the manuscript. All authors have read and agreed to the published version of the manuscript.

Funding: This research is funded by Thailand Science research and Innovation Fund Chulalongkorn University (CU_FRB65_hea (57)066_33_10). C.C. is grateful to the Faculty of Pharmaceutical Sciences, Chulalongkorn University, for a research fund (Phar2564-RG008).

Institutional Review Board Statement: Not applicable.

Informed Consent Statement: Not applicable.

Data Availability Statement: All data presented in this study are available in the article.

Acknowledgments: M.T.T. is grateful to the Graduate School, Chulalongkorn University, for a CUASEAN Ph.D. scholarship.

Conflicts of Interest: The authors declare no conflict of interest.

Sample Availability: Samples of the compounds are not available from authors.

References

1. WHO. *Definition, Diagnosis and Classification of Diabetes Mellitus and Its Complication (Part 1)*; World Health Organization: Geneva, Swizerland, 1999; pp. 2–32.
2. Bischoff, H. Pharmacology of α-glucosidase inhibition. *Eur. J. Clin. Investig.* **1994**, *24*, 3–10.
3. Peng, X.; Zhang, G.; Liao, Y.; Gong, D. Inhibitory kinetics and mechanism of kaempferol on α-glucosidase. *Food Chem.* **2016**, *190*, 207–215. [CrossRef] [PubMed]
4. Nhiem, N.X.; Kiem, P.V.; Minh, C.V.; Ban, N.K.; Cuong, N.X.; Tung, N.H.; Ha, L.M.; Ha, D.T.; Tai, B.H.; Quang, T.H.; et al. α-Glucosidase inhibition properties of cucurbitane-type triterpene glycosides from the fruits of Momordica charantia. *Chem. Pharm. Bull.* **2010**, *58*, 720–724. [CrossRef] [PubMed]
5. Butala, M.A.; Kukkupuni, S.K.; Vishnuprasad, C.N. Ayurvedic anti-diabetic formulation Lodhrasavam inhibits alpha-amylase, alpha-glucosidase and suppresses adipogenic activity in vitro. *J. Ayurveda. Integr. Med.* **2017**, *8*, 145–151. [CrossRef]
6. Abbasi, F.; Brown, B.W.; Lamendola, C.; McLaughlin, T.; Reaven, G.M. Relationship between obesity, insulin resistance, and coronary heart disease risk. *J. Am. Coll. Cardiol.* **2002**, *40*, 937–943. [CrossRef]
7. Tan, H.Y.; Iris, M.Y.; Li, E.T.; Wang, M. Inhibitory effects of oxyresveratrol and cyanomaclurin on adipogenesis of 3T3-L1 cells. *J. Funct. Foods* **2015**, *15*, 207–216. [CrossRef]
8. Klein, G.; Kim, J.; Himmeldirk, K.; Cao, Y.; Chen, X. Antidiabetes and anti-obesity activity of Lagerstroemia speciosa. *Evid. Based Complement. Altern. Med.* **2007**, *4*, 401–407. [CrossRef]
9. Yun, J.W. Possible anti-obesity therapeutics from nature—A review. *Phytochemistry* **2010**, *71*, 1625–1641. [CrossRef]
10. Hwang, J.T.; Park, I.J.; Shin, J.I.; Lee, Y.K.; Lee, S.K.; Baik, H.W.; Park, O.J. Genistein, EGCG, and capsaicin inhibit adipocyte differentiation process via activating AMP-activated protein kinase. *Biochem. Biophys. Res. Commun.* **2005**, *338*, 694–699. [CrossRef]
11. Lin, J.; Della-Fera, M.A.; Baile, C.A. Green tea polyphenol epigallocatechin gallate inhibits adipogenesis and induces apoptosis in 3T3-L1 adipocytes. *Obes. Res.* **2005**, *13*, 982–990. [CrossRef]
12. Pinent, M.; Blay, M.; Blade, M.C.; Salvado, M.J.; Arola, L.; Ardevol, A. Grape seed-derived procyanidins have an antihyperglycemic effect in streptozotocin-induced diabetic rats and insulinomimetic activity in insulin-sensitive cell lines. *Endocrinology* **2004**, *145*, 4985–4990. [CrossRef] [PubMed]
13. Daval, M.; Foufelle, F.; Ferré, P. Functions of AMP-activated protein kinase in adipose tissue. *J. Physiol.* **2006**, *574*, 55–62. [CrossRef] [PubMed]
14. Luo, Z.; Zang, M.; Guo, W. AMPK as a metabolic tumor suppressor: Control of metabolism and cell growth. *Future Oncol.* **2010**, *6*, 457–470. [CrossRef] [PubMed]
15. Habinowski, S.A.; Witters, L.A. The effects of AICAR on adipocyte differentiation of 3T3-L1 cells. *Biochem. Biophys. Res. Commun.* **2001**, *286*, 852–856. [CrossRef] [PubMed]
16. Saha, A.K.; Ruderman, N.B. Malonyl-CoA and AMP-activated protein kinase: An expanding partnership. *Mol. Cell Biochem.* **2003**, *253*, 65–70. [CrossRef] [PubMed]
17. Lago, F.; Gómez, R.; Gómez-Reino, J.J.; Dieguez, C.; Gualillo, O. Adipokines as novel modulators of lipid metabolism. *Trends Biochem. Sci.* **2009**, *34*, 500–510. [CrossRef]
18. Rizzatti, V.; Boschi, F.; Pedrotti, M.; Zoico, E.; Sbarbati, A.; Zamboni, M. Lipid droplets characterization in adipocyte differentiated 3T3-L1 cells: Size and optical density distribution. *Eur. J. Histochem.* **2013**, *57*, e24. [CrossRef]

19. Kim, J.H.; Lee, S.; Cho, E.J. Flavonoids from Acer okamotoanum inhibit adipocyte differentiation and promote lipolysis in the 3T3-L1 cells. *Molecules* **2020**, *25*, 1920. [CrossRef]

20. Kim, G.S.; Park, H.J.; Woo, J.H.; Kim, M.K.; Koh, P.O.; Min, W.; Cho, J.H. Citrus aurantium flavonoids inhibit adipogenesis through the Akt signaling pathway in 3T3-L1 cells. *BMC Complement. Altern. Med.* **2012**, *12*, 31. [CrossRef]

21. Kim, H.J.; Yoon, B.K.; Park, H.; Seok, J.W.; Choi, H.; Yu, J.H.; Kim, J.W. Caffeine inhibits adipogenesis through modulation of mitotic clonal expansion and the AKT/GSK3β pathway in 3T3-L1 adipocytes. *BMB Rep.* **2016**, *49*, 111. [CrossRef]

22. Stein, S.A.; Lamos, E.M.; Davis, S.N. A review of the efficacy and safety of oral antidiabetic drugs. *Expert. Opin. Drug Saf.* **2013**, *12*, 153–175. [CrossRef] [PubMed]

23. Kang, J.G.; Park, C.Y. Anti-obesity drugs: A review about their effects and safety. *Diabetes Metab. J.* **2012**, *36*, 13–25. [CrossRef]

24. Paudel, M.R.; Bhattarai, H.D.; Pant, B. Traditionally used medicinal Dendrobium: A promising source of active anticancer constituents. In *Orchids Phytochemistry, Biology and Horticulture: Fundamentals and Applications*; Springer: Cham, Switzerland, 2020; pp. 1–26.

25. Xu, J.; Han, Q.B.; Li, S.L.; Chen, X.J.; Wang, X.N.; Zhao, Z.Z.; Chen, H.B. Chemistry, bioactivity and quality control of Dendrobium, a commonly used tonic herb in traditional Chinese medicine. *Phytochem. Rev.* **2013**, *12*, 341–367. [CrossRef]

26. Cakova, V.; Bonte, F.; Lobstein, A. Dendrobium: Sources of active ingredients to treat age-related pathologies. *Aging Dis.* **2017**, *8*, 827–849. [CrossRef] [PubMed]

27. Cheng, J.; Dang, P.P.; Zhao, Z.; Yuan, L.C.; Zhou, Z.H.; Wolf, D.; Luo, Y.B. An assessment of the Chinese medicinal Dendrobium industry: Supply, demand and sustainability. *J. Ethnopharmacol.* **2019**, *229*, 81–88. [CrossRef]

28. Chuakul, W. Ethnomedical uses of Thai Orchidaceous plants. *Mahidol J. Pharm. Sci.* **2002**, *29*, 41–45.

29. Fisch, M.H.; Flick, F.H.; Arditti, J. Structure and antifungal activity of hircinol, loroglossol and orchinol. *Phytochemistry* **1973**, *12*, 437–441. [CrossRef]

30. Majumder, P.L.; Sen, R.C. Structure of moscatin—A new phenanthrene derivative from the orchid Dendrobium moscatum. *Indian J. Chem.* **1987**, *26B*, 18–20.

31. Fan, C.; Wang, W.; Wang, Y.; Qin, G.; Zhao, W. Chemical constituents from Dendrobium densiflorum. *Phytochemistry* **2001**, *57*, 1255–1258. [CrossRef]

32. Ono, M.; Ito, Y.; Masuika, C.; Koga, H.; Nohara, T. Antioxidative constituents from Dendrobii Herba (stems of *Dendrobium* spp.). *Food Sci. Technol. Int.* **1995**, *1*, 115–120.

33. Leong, Y.W.; Kang, C.C.; Harrison, L.J.; Powell, A.D. Phenanthrenes, dihydrophenanthrenes and bibenzyls from the orchid Bulbophyllum vaginatum. *Phytochemistry* **1997**, *44*, 157–165. [CrossRef]

34. Chen, Y.; Xu, J.; Yu, H.; Qing, C.; Zhang, Y.; Wang, L.; Liu, Y.; Wang, J. Cytotoxic phenolics from Bulbophyllum odoratissimum. *Food Chem.* **2008**, *107*, 169–173. [CrossRef]

35. Yang, M.; Zhang, Y.; Chen, Y. A new (propylphenyl) bibenzyl derivative from Dendrobium williamsonii. *Nat. Prod. Res.* **2017**, *32*, 1699–1705. [CrossRef] [PubMed]

36. Guo, X.Y.; Wang, J.; Wang, N.L.; Kitanaka, S.; Yao, X.S. 9, 10-Dihydrophenanthrene derivatives from Pholidota yunnanensis and scavenging activity on DPPH free radical. *J. Asian Nat. Prod. Res.* **2007**, *9*, 165–174. [CrossRef] [PubMed]

37. Guo, X.Y.; Wang, J.; Wang, N.L.; Kitanaka, S.; Liu, H.W.; Yao, X.S. New stilbenoids from Pholidota yunnanensis and their inhibitory effects on nitric oxide production. *Chem. Pharm. Bull.* **2006**, *54*, 21–25. [CrossRef]

38. Chougale, A.D.; Ghadyale, V.A.; Panaskar, S.N.; Arvindekar, A.U. Alpha glucosidase inhibition by stem extract of Tinospora cordifolia. *J. Enzyme Inhib. Med. Chem.* **2009**, *24*, 998–1001. [CrossRef] [PubMed]

39. Ghadyale, V.; Takalikar, S.; Haldavnekar, V.; Arvindekar, A. Effective control of postprandial glucose level through inhibition of intestinal alpha glucosidase by Cymbopogon martinii (Roxb.). *Evid. Based Complement. Altern. Med.* **2012**, *2012*, 372909. [CrossRef]

40. Yang, M.H.; Chin, Y.W.; Chae, H.S.; Yoon, K.D.; Kim, J. Anti-adipogenic constituents from Dioscorea opposita in 3T3-L1 cells. *Bio. Pharm. Bull.* **2014**, *37*, 1683–1688. [CrossRef]

41. Sarjeant, K.; Stephens, J.M. Adipogenesis. *Cold Spring Harb. Perspect. Biol.* **2012**, *4*, a008417. [CrossRef]

42. Lee, H.W.; Rhee, D.K.; Kim, B.O.; Pyo, S. Inhibitory effect of sinigrin on adipocyte differentiation in 3T3-L1 cells: Involvement of AMPK and MAPK pathways. *Biomed. Pharmacother.* **2018**, *102*, 670–680. [CrossRef]

43. Ahmad, B.; Serpell, C.J.; Fong, I.L.; Wong, E.H. Molecular mechanisms of adipogenesis: The anti-adipogenic role of AMP-activated protein kinase. *Front. Mol. Biosci.* **2020**, *7*, 76. [CrossRef] [PubMed]

44. Kang, M.C.; Ding, Y.; Kim, H.S.; Jeon, Y.J.; Lee, S.H. Inhibition of adipogenesis by diphlorethohydroxycarmalol (DPHC) through AMPK activation in adipocytes. *Mar. Drugs* **2019**, *17*, 44. [CrossRef] [PubMed]

45. Shao, Y.; Yuan, G.; Zhang, J.; Guo, X. Liraglutide reduces lipogenetic signals in visceral adipose of db/db mice with AMPK activation and Akt suppression. *Drug Des. Devel. Ther.* **2015**, *9*, 1177–1184. [CrossRef] [PubMed]

46. Liu, X.; Yao, Z. Chronic over-nutrition and dysregulation of GSK3 in diseases. *Nutr. Metab.* **2016**, *13*, 49. [CrossRef]

47. Manning, B.D.; Toker, A. AKT/PKB signaling: Navigating the network. *Cell* **2017**, *169*, 381–405. [CrossRef] [PubMed]

48. Jope, R.S.; Bijur, G.N. Mood stabilizers, glycogen synthase kinase-3β and cell survival. *Mol. Psychiatry* **2002**, *7*, S35–S45. [CrossRef]

49. Chatsumpun, N.; Sritularak, B.; Likhitwitayawuid, K. New biflavonoids with α-glucosidase and pancreatic lipase inhibitory activities from Boesenbergia rotunda. *Molecules* **2017**, *22*, 1862. [CrossRef] [PubMed]

50. Ono, M.; Fujimori, K. Antiadipogenic effect of dietary apigenin through activation of AMPK in 3T3-L1 cells. *J. Agric. Food Chem.* **2011**, *59*, 13346–13352. [CrossRef]
51. Borah, A.K.; Kuri, P.R.; Singh, A.; Saha, S. Anti-adipogenic effect of Terminalia chebula fruit aqueous extract in 3T3-L1 preadipocytes. *Pharmacogn. Mag.* **2019**, *15*, 197–204.
52. He, F. BCA (bicinchoninic acid) protein assay. *Bio-protocol* **2011**, *1*, e44. [CrossRef]

Article

Chemical Characterization of Flowers and Leaf Extracts Obtained from *Turnera subulata* and Their Immunomodulatory Effect on LPS-Activated RAW 264.7 Macrophages

Jefferson Romáryo Duarte da Luz [1,2], Eder A. Barbosa [3], Thayse Evellyn Silva do Nascimento [2,4], Adriana Augusto de Rezende [1,4], Marcela Abbott Galvão Ururahy [4], Adriana da Silva Brito [5], Gabriel Araujo-Silva [6], Jorge A. López [2] and Maria das Graças Almeida [1,2,4,*]

[1] Post-Graduation Program in Health Sciences, Health Sciences Center, Federal University of Rio Grande do Norte, R. Gen. Gustavo Cordeiro de Farias, s/n—Petrópolis, Natal 59012-570, RN, Brazil; jefferson_romaryo@hotmail.com (J.R.D.d.L.); adrirezende@yahoo.com (A.A.d.R.)

[2] Multidisciplinary Research Laboratory, DACT, Health Sciences Center, Federal University of Rio Grande do Norte, R. Gen. Gustavo Cordeiro de Farias, s/n—Petrópolis, Natal 59012-570, RN, Brazil; t_hayse_13@hotmail.com (T.E.S.d.N.); jorgejal@gmail.com (J.A.L.)

[3] Laboratory of Synthesis and Analysis of Biomolecules (LSAB), Institute of Chemistry, Darcy Ribeiro University Campus, University of Brasilia, Brasília 70910-900, DF, Brazil; bioederr@gmail.com

[4] Post-Graduation Program in Pharmaceutical Sciences, Health Sciences Center, Federal University of Rio Grande do Norte, R. Gen. Gustavo Cordeiro de Farias, s/n—Petrópolis, Natal 59012-570, RN, Brazil; marcelaururahy@yahoo.com.br

[5] Faculty of Health Sciences of Trairi (FACISA/UFRN), R. Passos de Miranda, Santa Cruz 59200-000, RN, Brazil; britoas.ufrn@gmail.com

[6] Organic Chemistry and Biochemistry Laboratory, Amapá State University (UEAP), Av. Presidente Vargas, s/n, Centro, Macapá 68900-070, AP, Brazil; gabriel_ar4@yahoo.com.br

*   Correspondence: mgalmeida84@gmail.com; Tel.: +55-84-3342-9807; Fax: +55-84-3342-9833

Citation: Luz, J.R.D.d.; Barbosa, E.A.; Nascimento, T.E.S.d.; Rezende, A.A.d.; Ururahy, M.A.G.; Brito, A.d.S.; Araujo-Silva, G.; López, J.A.; Almeida, M.d.G. Chemical Characterization of Flowers and Leaf Extracts Obtained from *Turnera subulata* and Their Immunomodulatory Effect on LPS-Activated RAW 264.7 Macrophages. *Molecules* **2022**, *27*, 1084. https://doi.org/10.3390/molecules 27031084

Academic Editor: Jacqueline Aparecida Takahashi

Received: 31 December 2021
Accepted: 31 January 2022
Published: 6 February 2022

Publisher's Note: MDPI stays neutral with regard to jurisdictional claims in published maps and institutional affiliations.

Abstract: The anti-inflammatory properties of *Turnera subulata* have been evaluated as an alternative drug approach to treating several inflammatory processes. Accordingly, in this study, aqueous and hydroalcoholic extracts of *T. subulata* flowers and leaves were analyzed regarding their phytocomposition by ultrafast liquid chromatography coupled to mass spectrometry, and their anti-inflammatory properties were assessed by an in vitro inflammation model, using LPS-stimulated RAW-264.7 macrophages. The phytochemical profile indicated vitexin-2-*O*-rhamnoside as an important constituent in both extracts, while methoxyisoflavones, some bulky amino acids (e.g., tryptophan, tyrosine, phenylalanine), pheophorbides, and octadecatrienoic, stearidonic, and ferulic acids were detected in hydroalcoholic extracts. The extracts displayed the ability to modulate the in vitro inflammatory response by altering the secretion of proinflammatory (TNF-α, IL-1β, and IL-6) and anti-inflammatory (IL-10) cytokines and inhibiting the PGE-2 and NO production. Overall, for the first time, putative compounds from *T. subulata* flowers and leaves were characterized, which can modulate the inflammatory process. Therefore, the data highlight this plant as an option to obtain extracts for phytotherapic formulations to treat and/or prevent chronic diseases.

Keywords: plant; natural compounds; chromatography; anti-inflammatory

1. Introduction

Inflammation is a human body defense response to aggressive stimuli, involving biochemical, physiological, and immunological reactions to locate, inactivate, and destroy the offending agent, in addition to promoting tissue healing and repair [1,2]. During the inflammatory process, several cellular pathways play a crucial role in recovering and maintaining homeostatic balance. In this process, macrophage participation in the immune system is critical to the activation of inflammatory pathways and in the specific inflammatory mediator release [3].

Macrophages present antigens to cells, acting as central mediators in the immune system control, contributing to both the initiation and the resolution of inflammation. Activated macrophages secrete several inflammatory mediators, including cytokines such as TNF-α, IL-1β, IL-6, and IL-10, as well as PGE-2 and nitric oxide. Cytokines, such as TNF-α, IL-1β, and IL-6, activate the inflammatory response, while anti-inflammatory cytokines such as IL-10 act in the repair processes. Dysregulation of inflammatory mediator signaling interrupts this coordinated process and can lead to pathology. Accordingly, in inflammatory diseases, the cytokine network balance is severely disrupted, leading to persistent inflammation, tissue damage, and eventually organ failure [4,5].

Nonsteroidal anti-inflammatory drugs (NSAIDs) are used daily and indiscriminately in worldwide inflammatory disease treatment [6]. Overall, NSAIDs can cause side-effects due to their mechanism of action. Major adverse effects include gastrointestinal bleeding, changes in cardiovascular and renal functions, and other complications such as hemorrhage, perforation, or death [7]. Despite the diversity of anti-inflammatory drugs on the pharmaceutical market, no formulation offers low toxicity and minimal adverse effects. This situation stimulates the search for new molecules aiming at safer drugs with low side-effects for the inflammation treatment [8,9]. The mechanisms involved in the inflammatory response are fundamental for anti-inflammatory drug development [10,11].

In this context, plant phenolic compounds (e.g., flavonoids, phenolic acids, lignins) have been described regarding their several pharmacological activities, including antioxidant, anti-inflammatory, hepatoprotective, and antimicrobial effects [12]. Polyphenols are the object of studies to develop new drugs due to their direct or indirect inhibition or activation of important cellular and molecular targets, by modulating the expression of inflammatory mediators [13,14].

On this basis, studies have focused on evaluating plant biodiversity as a compound reservoir for therapeutic purposes, in addition to elucidating medicinal activities attributed by popular tradition [12,15]. In this context, *Turnera subulata* (Passifloraceae family) is a species with wide distribution in tropical and subtropical regions [16,17], used in folk medicine due to its pharmacological properties, such as anti-inflammatory, hypoglycemic, antifungal, and antioxidant activity [16,18–21]. Phytochemical studies with species of the *Turnera* genus have revealed the presence of flavanols, alkaloids, tannins, cyanogenic glycosides, fatty acids, triterpenoids, and various phenolic compounds related to their bioactivities [16,22].

Several studies have reported the use of plants from the Passifloraceae family for inflammatory process treatment [23,24]. Regarding *T. subulata*, their anti-inflammatory effects have mainly been described by inhibiting the cytokine production [25]. In a previous study, Luz et al. [26] showed the ability of *T. subulata* extracts to directly and indirectly inhibit thrombin, promoting few side-effects. These data are relevant since this coagulation factor is also related to an inflammatory response [27].

Hence, studies on the immunomodulatory potential of this plant species are relevant, in addition to proposing an association between the phytochemical profile and their pharmacological effects. Furthermore, this study is the first report on the phytochemical characterization and immunomodulatory effect of *T. subulata* flower extracts. Thus, this work chemically analyzed *T. subulata* floral and leaf extracts and evaluated their immunomodulatory effects in an in vitro model using RAW 264.7 macrophages stimulated by LPS, aiming to contribute to the search for anti-inflammatory drugs with low side-effects.

2. Results

The *T. subulata* hydroethanolic extracts displayed the highest yields (5.43% and 6.93% for flowers and leaves, respectively) compared to values of 4.16% and 5.3% obtained with flower and leaf aqueous extracts, respectively. This result was similar to those described by Antonio and Brito [18] for extracts from *T. ulmifolia* aerial parts, indicating the importance of the solvent choice for the extract yield.

      T. subulata flower and leaf extracts were analyzed by UPLC–MS/MS, and their MS/MS spectra were submitted to compound identification using the GNPS database. Although a large number of MS/MS spectra were acquired for each extract that matched to known molecules deposited in the GNPS database, only those spectra with cosine $\geq$0.85 and mass difference $\leq$0.005 Da were considered. The extract phytochemical profile indicated the presence of vitexin-2-*O*-rhamnoside or its isoforms, a glycosylated flavone derivate from apigenin, as a constituent with potential pharmacological interest in all analyzed extracts. Furthermore, other compounds were tentatively identified, such as methoxyisoflavones, pheophorbides, octadecatrienoic, stearidonic, ferulic acids, and some bulky amino acids (e.g., tryptophan, tyrosine, phenylalanine) or their isoforms (Figure 1).

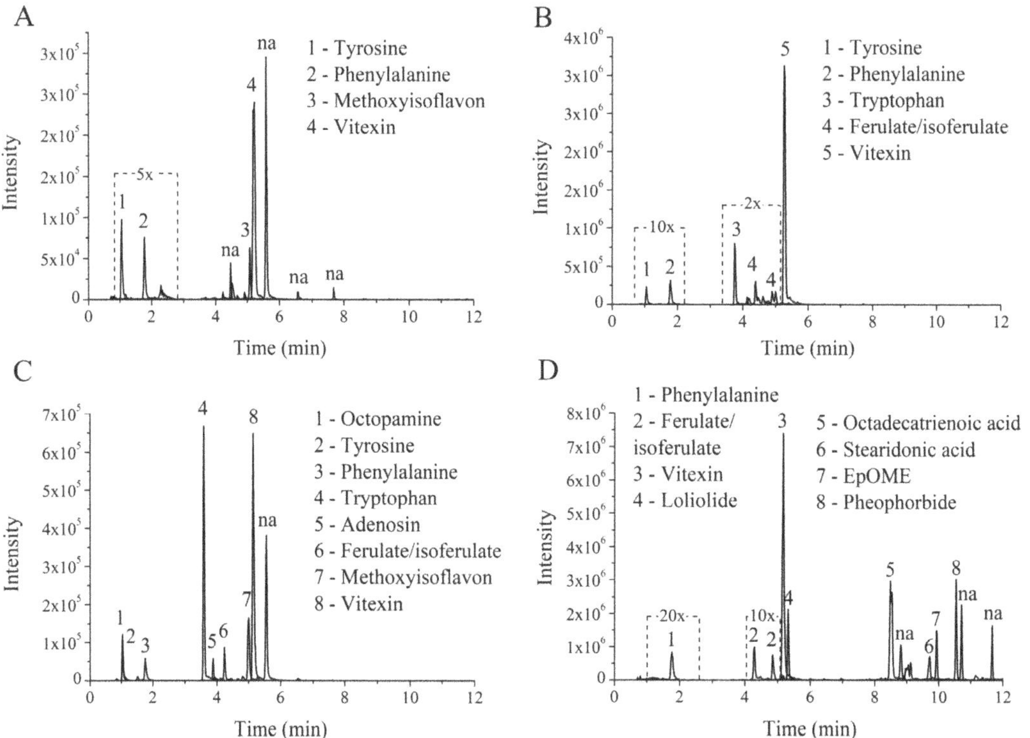

Figure 1. LC–MS/MS fingerprint of *Turnera subulata* extracts: (**A**) *T. subulata* aqueous flower extract (AFETS); (**B**) *T. subulata* aqueous leaf extract (ALETS); (**C**) *T. subulata* hydroethanolic flower extract (HEFTS); (**D**) *T. subulata* hydroethanolic leaf extract (HELTS). $2\times$, $5\times$, and $10\times$ denote the magnification applied in dotted areas of the chromatogram; na = not available.

      The extracted ion chromatograms (XICs) obtained for each of the of the structures tentatively identified by UPLC–MS/MS and GNPS analysis showed four major phytocomponents with good resolution for AFETS, (Figure 1A) five for ALETS, (Figure 1B) eight for HEFTS (Figure 1C), and eight for HELTS (Figure 1D). In all extracts, vitexin was the highest-intensity phytocomponent among the identified compounds; however, its exact sum still requires further quantification. Table 1 shows phytocomponents and their respective cosine, mass difference, and mass. The comparison between library GNPS and query spectra of the phytocomponents, as well as the draw structures identified, is presented in the Supplementary Materials.

Table 1. Phytocomponents identified in *Turnera subulata* flower and leaf extracts by LC–MS/MS analyses.

Peak.	Phytocomponents Matched with GNPS Data Base	Cosine	Mass Diff	Mass	Molecular Formula	Ion Fragments [a]	Adduct	Extract
1	L-Tyrosine	0.86	0.001	165.054	$C_9H_8O_3$	147.04, 123.05, 119.05, 95.05, 91.06	$[M - NH_3 + H]^+$	AFETS
2	Phenylalanine	0.95	0	166.086	$C_9H_{11}NO_2$	149.06, 131.05, 120.08, 103.05, 53.04	$[M + H]^+$	AFETS
3	7-*O*-beta-glucopyranosyl-4′-hydroxy-5-methoxyisoflavone	0.89	0	447.129	$C_{22}H_{22}O_{10}$	285.08, 270.05, 213.05, 152.01	$[M + H]^+$	AFETS
4	Vitexin-2-*O*-rhamnoside	0.95	0.002	579.172	$C_{27}H_{30}O_{14}$	433.11, 415.10, 397.09, 313.07, 283.06	$[M + H]^+$	AFETS
1	L-Tyrosine	0.95	0.001	182.081	$C_9H_{11}N_1O_3$	165.06, 147.04, 136.07, 123.05, 119.05	$[M + H]^+$	ALETS
2	DL-Phenylalanine	0.97	0	166.086	$C_9H_{11}NO_2$	149.06, 131.05, 120.08, 103.05, 53.04	$[M + H]^+$	ALETS
3	L-Tryptophan	0.95	0.001	205.096	$C_{11}H_{12}N_2O_2$	188.07, 159.09, 146.06, 144.08, 118.06	$[M + H]^+$	ALETS
4	Ferulate/isoferulate	0.93	0	177.054	$C_{10}H_8O_3$	149.06, 145.03, 117.03, 89.04	$M - H_2O + H$	ALETS
5	Vitexin-2-*O*-rhamnoside	0.94	0	579.17	$C_{27}H_{30}O_{14}$	433.11, 415.10, 397.09, 313.07, 283.06	$[M + H]^+$	ALETS
1	DL-Octopamine	0.95	0	136.076	$C_8H_9NO_1$	119.05, 118.06, 107.05, 91.06, 64.04	$[M - H_2O + H]^+$	HEFTS
2	L-Tyrosine	0.94	0.001	165.054	$C_9H_8O_3$	147.04, 123.05, 119.05, 95.05, 91.06	$[M - NH_3 + H]^+$	HEFTS
3	Phenylalanine	0.96	0	166.086	$C_9H_{11}NO_2$	149.06, 131.05, 120.08, 103.05, 53.04	$[M + H]^+$	HEFTS
4	L-Tryptophan	0.94	0.001	188.07	$C_{11}H_9N_1O_2$	170.06, 146.06, 144.08, 143.07, 118.07	$[M - NH_3 + H]^+$	HEFTS
5	Adenosine, 5_-*S*-methyl-5_-thio-	0.91	0.004	298.097	$C_{11}H_{15}N_5O_3S$	145.03, 136.06, 97.03, 61.01	$[M + H]^+$	HEFTS
6	Ferulate/isoferulate	0.89	0	177.054	$C_{10}H_8O_3$	149.06, 145.03, 117.03, 89.04	$[M - H_2O + H]^+$	HEFTS
7	7-*O*-beta-glucopyranosyl-4′-hydroxy-5-methoxyisoflavone	0.91	0.001	447.13	$C_{22}H_{22}O_{10}$	285.08, 270.05, 213.05, 152.01	$[M + H]^+$	HEFTS
8	Vitexin-2-*O*-rhamnoside	0.97	0.003	579.173	$C_{27}H_{30}O_{14}$	433.11, 415.10, 397.09, 313.07, 283.06	$[M + H]^+$	HEFTS
1	Phenylalanine	0.94	0	166.086	$C_9H_{11}NO_2$	149.06, 131.05, 120.08, 103.05, 53.04	$[M + H]^+$	HELTS
2	Ferulate/isoferulate	0.94	0	177.054	$C_{10}H_8O_3$	149.06, 145.03, 117.03, 89.04	$[M - H_2O + H]^+$	HELTS
3	Vitexin-2-*O*-rhamnoside	0.94	0.001	579.171	$C_{27}H_{30}O_{14}$	433.11, 415.10, 397.09, 313.07, 283.06	$[M + H]^+$	HELTS
4	Loliolide	0.94	0	197.117	$C_{11}H_{16}O_3$	179.11, 161.09, 135.12, 133.10, 107.09	$[M + H]^+$	HELTS

Table 1. *Cont.*

Peak.	Phytocomponents Matched with GNPS Data Base	Cosine	Mass Diff	Mass	Molecular Formula	Ion Fragments [a]	Adduct	Extract
5	9S-Hydroxy-10E,12Z,15Z-octadecatrienoic acid	0.86	0	277.216	$C_{18}H_{28}O_2$	259.20, 149.13, 135.12, 121.10, 93.07	$[M - H_2O + H]^+$	HELTS
6	9(10)-EpOME	0.91	0.002	279.233	$C_{18}H_{30}O_2$	173.13, 109.10, 95.09, 81.07, 67.06	$[M - H_2O + H]^+$	HELTS
7	Stearidonic acid Ethyl ester	0.86	0.001	305.248	$C_{20}H_{32}O_2$	259.20, 149.13, 135.12, 121.10, 93.07	$[M + H]^+$	HELTS
8	Pheophorbide A	0.87	0.005	593.274	$C_{35}H_{36}N_4O_5$	533.25, 460.23, 447.22, 433.24, 431.18	$[M + H]^+$	HELTS

Turnera subulata aqueous flower extract (AFETS); *T. subulata* aqueous leaf extract (ALETS); *T. subulata* hydroethanolic flower extract (HEFTS); *T. subulata* hydroethanolic leaf extract (HELTS). [a] The most intense fragment ions are described.

After determination of the chemical composition of AFETS, ALETS, HEFTS, and HELTS, their potential cytotoxic effects on RAW 264.7 macrophage cells were evaluated (Figure 2). The results showed no cytotoxic effects evaluated through cell viability assay by MTT (Figure 2A) and by Alamar Blue® (Figure 2B), after exposing murine macrophages to different concentrations of these extracts.

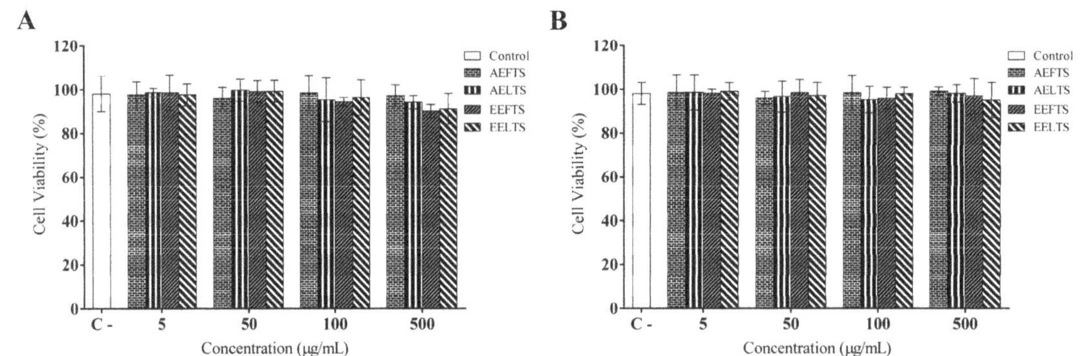

Figure 2. Cytotoxicity effects of AFETS, ALETS, HEFTS, and HELTS on RAW 264.7 murine macrophage cells: (**A**) cell viability measured by MTT assay; (**B**) cell viability measured by Alamar Blue assay. Culture medium DMEM was used as a negative control for cytotoxicity.

It is known that macrophages are activated in the presence of LPS, which promotes the secretion of a high content of proinflammatory molecules, such as cytokines. The ELISA results indicated that the treatment with AFETS, ALETS, HEFTS, and HELTS significantly reduced the secretion of inflammatory cytokines (TNF-α and IL-1β) in LPS-activated murine RAW 264.7 macrophages in a concentration-dependent manner (Figure 3). AFETS and ALETS efficiently reduced the TNF-α secretion at 100 and 500 μg/mL, reaching 50% and 65% inhibition, respectively, while HEFTS and HELTS displayed a stronger action on this cytokine secretion at the same concentrations, reaching up to 75% inhibition (Figure 3A–D). Regarding IL-1β, the aqueous extracts inhibited its secretion by approximately 50%, although the hydroethanolic extracts at 500 μg/mL exhibited the highest percentage of inhibition (70%) (Figure 4A–D).

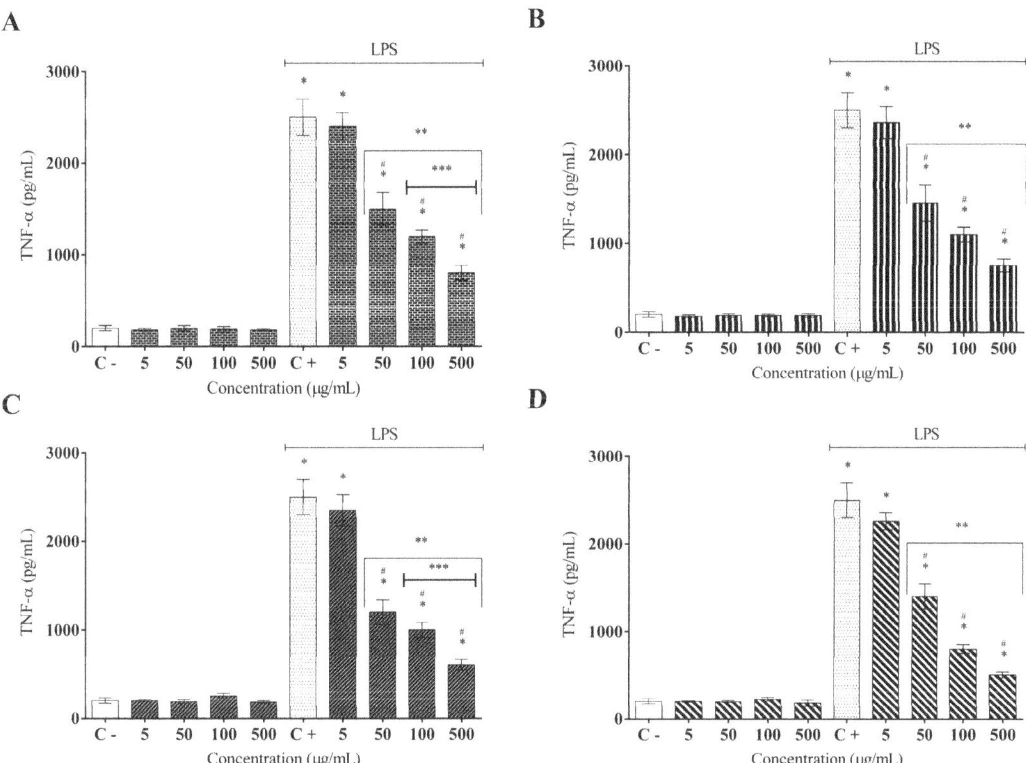

Figure 3. Effect of *Turnera subulata* extracts on the TNF-α cytokine release: (**A**) AFETS; (**B**) ALETS; (**C**) HEFTS; (**D**) HELTS. The cytokine content was released in RAW 264.7 cells and stimulated by LPS after 24 h. Release of cytokines was performed using ELISA assays. Data represent the mean ± SEM from three independent experiments. One-way ANOVA followed by the post hoc Tukey test. * $p < 0.05$ vs. the control group; # $p < 0.05$ vs. the LPS-stimulated cells; ** $p < 0.05$ between the concentrations of the extract; *** $p < 0.05$ between the higher concentrations (100 and 500 µg/mL).

Additionally, all extracts were less effective in modulating IL-6 secretion, showing no statistically significant difference for AFETS and ALETS, whereas hydroethanolic extracts (HEFTS and HELTS) stabilized their inhibitory effects by 50% at 500 µg/mL (Figure 5A–D). Overall, both aqueous and hydroethanolic extracts of *T. subulata* significantly increased the anti-inflammatory cytokine IL-10 levels (Figure 6A–D). AFETS and ALETS displayed an increase of around 15% in IL-10 secretion, while HEFTS and HELTS achieved a 20% increase.

AFETS, ALETS, HEFTS, and HELTS cell treatments were also effective in inhibiting the prostaglandin E2 production (Figure 7A–D). AFETS promoted an inhibitory effect on prostaglandin E2 secretion at the two highest concentrations tested (100 and 500 µg/mL), reaching values around 50% and 60%, respectively. The inhibition induced by ALETS at 500 µg/mL was 50%. Additionally, AFETS and HEFTS displayed an inhibitory effect, stabilizing their action at 60% at the highest concentrations, while HELTS decreased the PGE-2 secretion by around 80% at 500 µg/mL.

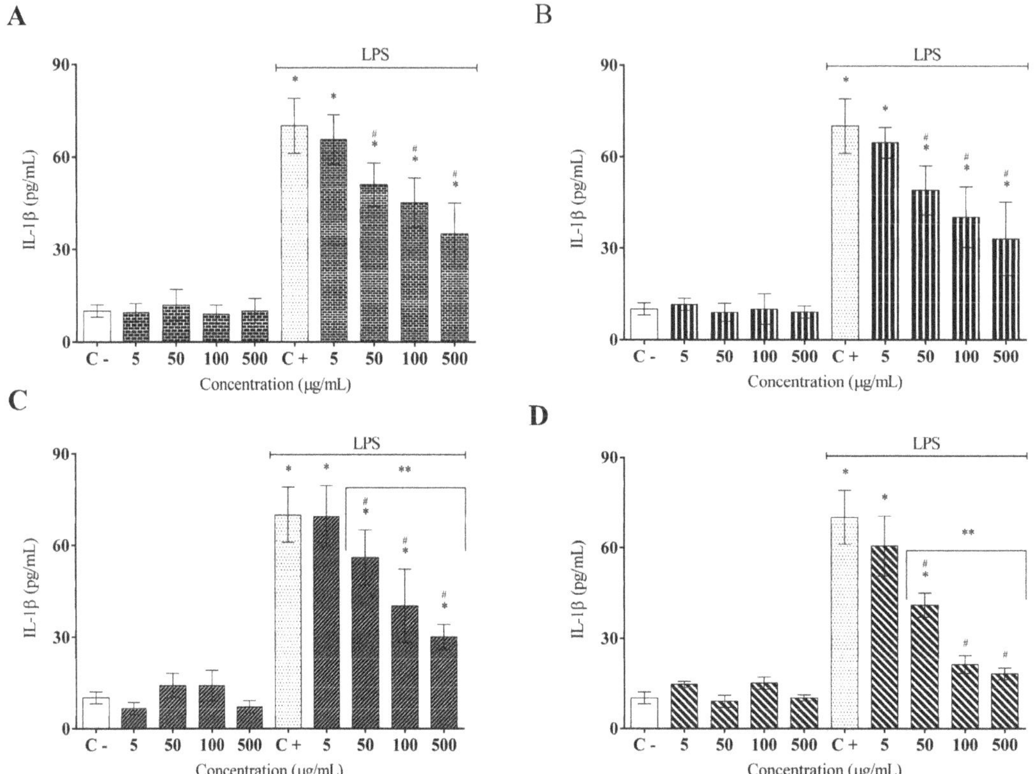

Figure 4. Effect of *Turnera subulata* extracts on the IL-1β cytokine release: (**A**) AFETS; (**B**) ALETS; (**C**) HEFTS; (**D**) HELTS. The cytokine content was released in RAW 264.7 cells and stimulated by LPS after 24 h. Release of cytokines was performed using ELISA assays. Data represent the mean ± SEM from three independent experiments. One-way ANOVA followed by the post hoc Tukey test. * $p < 0.05$ vs. the control group; # $p < 0.05$ vs. the LPS-stimulated cells; ** $p < 0.05$ between the concentrations of the extract.

Furthermore, the nitric oxide production was quantified to verify the anti-inflammatory activity of *T. subulata* extracts. The results indicated that both aqueous and hydroethanolic extracts at 100 and 500 μg/mL promoted a significant decrease ($p < 0.05$) in NO production by macrophages compared to the LPS-stimulated control group (Figure 8A–D). AFETS inhibited NO secretion by around 30% at the two highest concentrations tested (100 and 500 μg/mL), while ALETS only reached this inhibitory percentage at 500 μg/mL. HEFTS decreased NO secretion by 30% and 40% at concentrations of 100 and 500 μg/mL, respectively. No statistically significant difference was observed compared to the negative control. Regarding HELTS, the results showed that the highest concentrations tested (100 and 500 μg/mL) inhibited nitric oxide secretion by 40% and 65%, respectively, again with no statistical difference compared to the negative control.

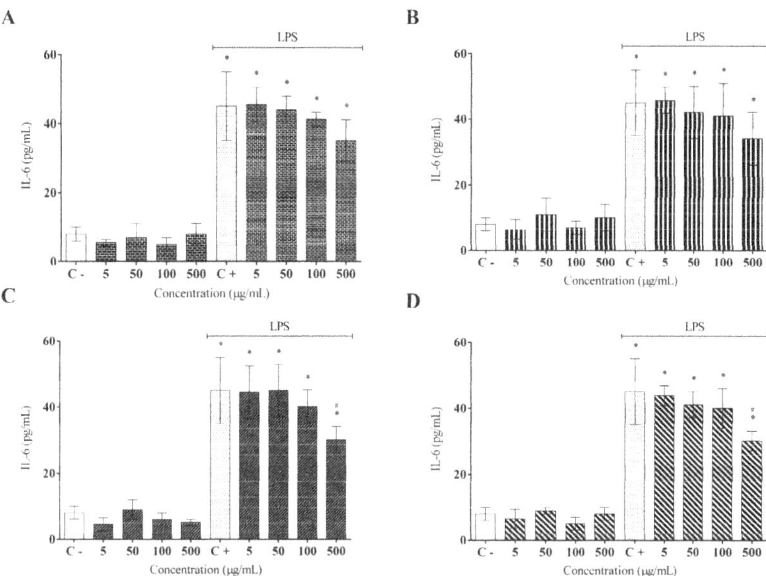

Figure 5. Effect of *Turnera subulata* extracts on the IL-6 cytokine release: (**A**) AFETS; (**B**) ALETS; (**C**) HEFTS; (**D**) HELTS. The cytokine content was released in RAW 264.7 cells and stimulated by LPS after 24 h. Release of cytokines was performed using ELISA assays. Data represent the mean $\pm$ SEM from three independent experiments. One-way ANOVA followed by the post hoc Tukey test. * $p < 0.05$ vs. the control group; # $p < 0.05$ vs. the LPS-stimulated cells.

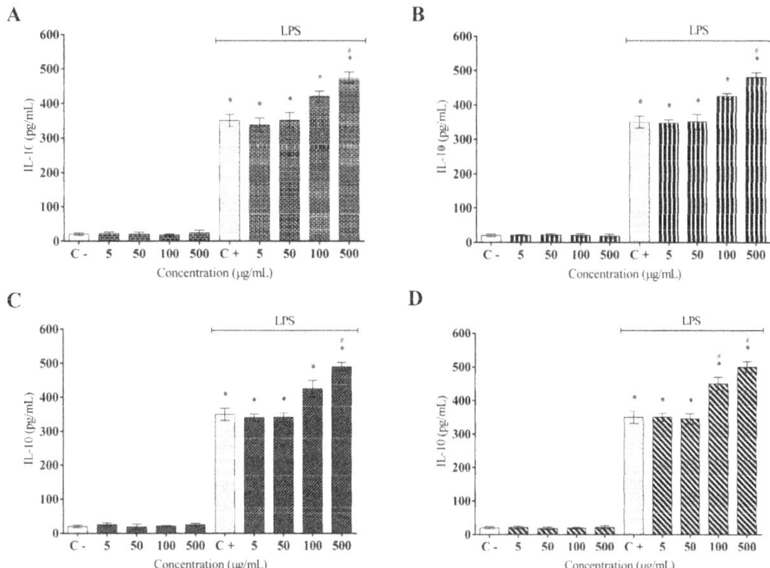

Figure 6. Effect of *Turnera subulata* extracts on the anti-inflammatory IL-10 cytokine release: (**A**) AFETS; (**B**) ALETS; (**C**) HEFTS; (**D**) HELTS. The cytokine content was released in RAW 264.7 cells and stimulated by LPS after 24 h. Release of cytokines was performed using ELISA assays. Data represent the mean $\pm$ SEM from three independent experiments. One-way ANOVA followed by the post hoc Tukey test. * $p < 0.05$ vs. the control group; # $p < 0.05$ vs. the LPS-stimulated cells.

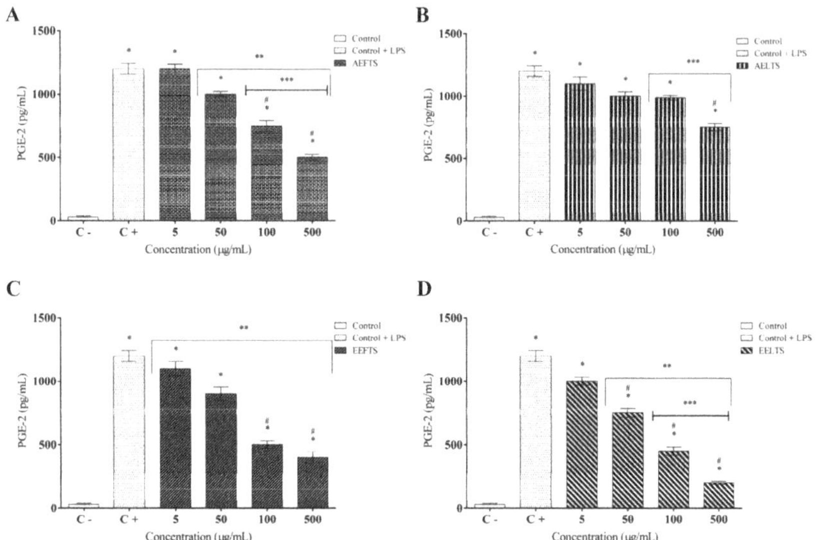

Figure 7. Inhibitory effects of *Turnera subulata* extracts on LPS-stimulated PGE-2 production in RAW 264.7 macrophages: (**A**) AFETS; (**B**) ALETS; (**C**) HEFTS; (**D**) HELTS. The level of PGE-2 in the culture medium was quantified using enzyme-linked immunoassay (ELISA) kits. Data represent the mean ± SEM from three independent experiments. One-way ANOVA followed by the post hoc Tukey test. * $p < 0.05$ vs. the control group; # $p < 0.05$ vs. the LPS-stimulated cells; ** $p < 0.05$ between the concentrations of the extract; *** $p < 0.05$ between the higher concentrations (100 and 500 μg/mL).

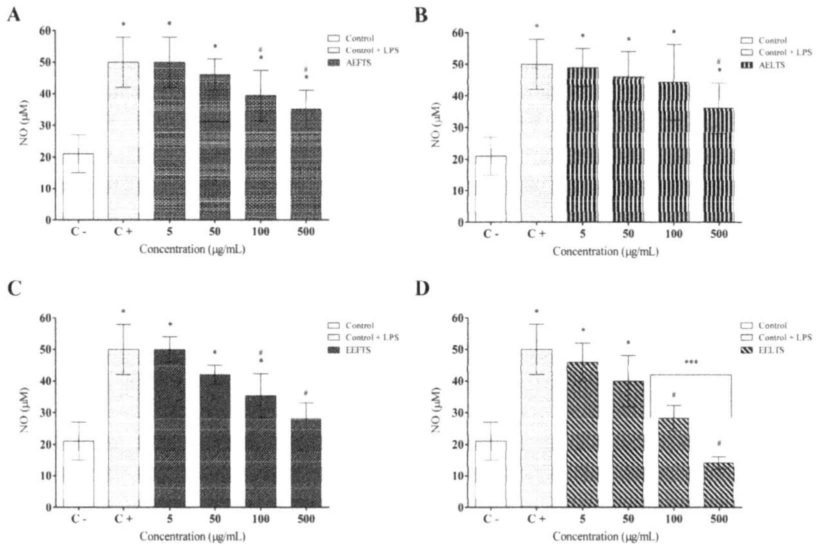

Figure 8. Inhibitory effects of *Turnera subulata* extracts on LPS-stimulated nitric oxide (NO) production in RAW 264.7 macrophages: (**A**) AFETS; (**B**) ALETS; (**C**) HEFTS; (**D**) HELTS. The level of NO in the culture medium was quantified using Griess reagent. Data represent the mean ± SEM from three independent experiments. One-way ANOVA followed by the post hoc Tukey test. * $p < 0.05$ vs. the control group; # $p < 0.05$ vs. the LPS-stimulated cells; *** $p < 0.05$ between the higher concentrations (100 and 500 μg/mL).

3. Discussion

Studies on the regulatory mechanisms of inflammatory responses have revealed that inflammation can delay healing and, in a chronic situation, can also play an important role in the onset of cancer and other chronic diseases [28]. Therefore, the use of nonsteroidal anti-inflammatory drugs for inflammation suppression is prevalent, despite several adverse effects, such as gastrointestinal bleeding and ulcers [29]. On the other hand, increasing evidence suggests that natural products, such as plant extracts, due to their phytocomposition, can decrease inflammatory processes [14,30,31].

The properties of several secondary plant metabolites have been extensively studied, mainly due to their antioxidant and anti-inflammatory properties [32–34]. Thus, this study presented the chemical profile characterization of aqueous and hydroethanolic extracts of *T. subulata* flower and leaf by mass spectrometry, as well as the evaluation of their anti-inflammatory properties in an in vitro model. At this point, it is noteworthy that only a few studies have reported the presence of phenolic compounds especially in *T. subulata* flowers, as well as their pharmacological properties [25,26,35].

Regarding the phytocomposition, it is noteworthy that phenolic compounds are the main class among the secondary metabolites present in *Turnera* genus [22]. Accordingly, *T. subulata* flower and leaf extract analyses revealed the presence of three phenolic compounds: vitexin-2-*O*-ramnhoside as a valuable chemical constituent, as well as 7-*O*-beta-glucopyranosyl-4'-hydroxy-5-methoxyisoflavone and ferulate. Although quantitative experiments need to be performed to validate the potential relevance of vitexin-2-*O*-ramnhoside in the *T. subulata* flower and leaf extracts, this phytochemical has already been characterized as the main constituent in extracts of distinct plants from the Magnoliopsida class [36–40]. These phenolic compounds are scarce in the genus *Turnera*, and this is the first report on the identification of these secondary metabolites for the genus. Studies have only identified the presence of other secondary metabolites in *Turnera* leaves, flowers, and roots [22], such as four different lutein structures and two apigenin structures [41], as well as naringenin, three apigenin coumaroyl glucosides, and five flavone aglycones [42]. The presence of these flavonoids in *T. subulata* extracts is possibly related to the climatic conditions of the Brazilian caatinga biome, a place of flower and leaf collection, where high exposure to solar radiation favors the biosynthesis of phenolic compounds such as flavones [43].

Anti-inflammatory properties observed with treatments using *T. subulata* extracts are correlated with the flavonoid presence, which points to the involvement of these compounds in inhibiting the synthesis and activity of different proinflammatory mediators [44]. Furthermore, flavonoids are reported as molecules with low cytotoxicity [44,45]. Therefore, the presence of these phytocomponents in the AFETS, ALETS, HEFTS, and HELTS could be related to the anti-inflammatory effect observed in this study. These *T. subulata* extracts exhibited an in vitro effect, evaluated by the inhibition of proinflammatory cytokines (TNF-α, IL-1β, and IL-6) in LPS-stimulated macrophages. Moreover, these extracts stimulated anti-inflammatory cytokine IL-10 secretion under the same experimental conditions.

These immunomodulatory effects may be associated with metabolites tentatively identified in the extracts, since vitexin is known to have an anti-inflammatory action by inhibiting important cytokines involved in the inflammatory cascade [46]. In a colitis experimental model of colitis, vitexin significantly inhibited the TNF-α, IL-6, and IL-1β expression, suggesting that this compound may suppress the inflammatory response to improve colitis-induced liver damage [47]. Another study showed that vitexin decreased the MCP-1 (a monocytic cytokine), IL-6, and IL-8 levels, while concomitantly increasing IL-10 expression in a murine model of septic encephalopathy [48].

Isoflavones are other metabolites with anti-inflammatory effects described in the literature. Thus, milletenol A, an isoflavone isolated from *Milletia pacchycarpa* seeds, decreased the neutrophil numbers in an inflammation model of zebrafish stimulated by copper sulfate [49]. Furthermore, isoflavones identified from *Glycine max* leaves exhibited positive effects by reducing the production of IL-1β and IL-6 in LPS-stimulated RAW 264.7 cells [50].

Regarding ferulic acid also detected in *T. subulata* extracts, studies have demonstrated its anti-inflammatory action by suppressing the expression of proinflammatory cytokines (TNF-a, IL-1B, IL-6, and IL-8) in LPS-induced primary cells isolated from bovine uterus via inhibition of the NF-κB and MAPK pathways as the main mechanism of action. Additionally, ferulic acid reduced TNF-a, IL-1b, and NF-κB secretion in an in vivo model of liver injury induced by methotrexate in rats [51,52].

During the proinflammatory response in LPS-stimulated cells, cytokines such as IL-1β, IL-6, IL-10, and TNF-α are released. The expression and release of these cytokines can be regulated by different transcription factors, such as NF-κB, and they are usually associated with metabolic processes [53]. Blockade of IL-1β action/secretion has been associated with anti-inflammatory responses [54], which may suggest that *T. subulata* extracts may also act on these metabolic pathways, since the extracts modulated the secretion of the same cytokines.

According to the chemical composition, other mechanisms of action may be associated with the immunomodulatory effects observed with aqueous and hydroethanolic extracts of *T. subulata* flowers and leaves. One mechanism suggested that the anti-inflammatory effect of this species may be due to the ability of leaf extracts to modulate MAPK signaling pathways by inhibiting ERK1/2 phosphorylation and blocking the inflammatory response in RAW 264.7 macrophages. This effect also decreases TNF-α and IL-1β secretion, increasing the blockade of RAGE and CD40 as the main mechanisms of action [25].

Inflammatory mediators, such as PGE-2 and NO, and several proinflammatory cytokines released by activated macrophages are important targets for inflammatory disease treatment, including multiple sclerosis and rheumatoid arthritis. When overexpressed, these proinflammatory mediators can lead to cell and tissue damage, resulting in various physiological disorders related to inflammation [55,56]. Therefore, compounds that inhibit the production of proinflammatory mediators such as PGE-2 and NO can be candidates as anti-inflammatory drugs [57].

The AFETS, ALETS, HEFTS, and HELTS extracts reduced the PGE-2 and nitric oxide levels by more than 60%. This study shows for the first time that *T. subulata* extracts as active natural products exhibit inhibitory effects on the secretion of these proinflammatory mediators, which may also be associated with the phenolic groups detected in the extract chemical composition. According to studies, the polyphenol combination displays synergistic effects on biological activity [58–60].

Regarding its ability to modulate cytokine secretion, the presence of vitexin may also play an important role in the *T. subulata* extract in reducing PGE-2 and NO levels. This is consistent with studies showing vitexin's capacity to decrease the PGE-2 levels in cultured chondrocytes from osteoarthritic patients [46], as well as NO production associated with the autophagic dysfunction of ischemic stroke, and to protect against cerebral endothelial permeability [61,62]. An analogous situation can be verified with isoflavones also identified in *T. subulata* extracts and their association with the modulation of NO, compared to the concentration-dependent effect exhibited by the isoflavone milletenol A, which was able to inhibit NO secretion in an in vitro assay using LPS-stimulated RAW 264.7 macrophages [50]. In the inflammatory process, the increase in PGE-2 levels is stimulated by the NO production, which is closely associated with an increase in TNF-α production [63,64]. In the present study, the TNF-α secretion was inhibited by all *T. subulata* extracts, with a correlated decrease in NO levels and, consequently, PGE-2 levels, suggesting a well-controlled mechanism of action.

In addition to the immunomodulatory effects demonstrated in this study, it is noteworthy that Luz et al. [26] showed the ability of *T. subulata* extracts to inhibit thrombin, the main protease in the coagulation cascade, which plays a role in the inflammatory response [65]. Thrombin can signal the expression of proinflammatory cytokines, such as IL-1β and IL-6, and cell adhesion molecules, promoting leukocyte activation and fibroblast proliferation, as well as contributing to leukocyte recruitment [65,66]. The studies with *T. subulata* ethanol and ethyl acetate leaf extracts demonstrated a direct inhibitory effect on thrombin and an indirect one through heparin cofactor II, corroborating the effect of the extracts on the inflammatory process [26].

Overall, these results support ethnopharmacological studies, which describe the use of flowers and leaves of the genus *Turnera* in folk medicine to treat inflammatory diseases [22]. Experimental data indicate the immunomodulatory effect of aqueous and hydroalcoholic extracts of *T. subulata* flowers and leaves through the inhibition of inflammatory cytokine secretion and an increase in anti-inflammatory cytokine IL-10 levels. Despite the scarce data in the literature regarding extracts of the species *T. subulata*, it is possible to suggest its anti-inflammatory action. Accordingly, this is a pioneering study showing the chemical characterization of its flower and leaf extracts associated with potential therapeutic effects.

4. Material and Methods

4.1. Plant Material and Extract Preparation

T. subulata flowers and leaves were collected in Natal, Rio Grande do Norte, Brazil. The species were taxonomically identified by Dr. Jomar Gomes Jardim. A voucher specimen (No. 0674/08) was deposited in the Herbarium at the Department of Botany and Zoology, Federal University of Rio Grande do Norte, Natal, RN, Brazil. The aerial parts of *T. subulata* were air-dried at 40 °C for 48 h and powdered to particle size <180 μm. Then, 300 g of powdered flowers and leaves were individually subjected to decoction in water (100 °C/10 min), filtered on Whatman filter papers, and lyophilized to obtain the aqueous flower extract (AFETS) and aqueous leaf extract (ALETS). Regarding the hydroethanolic extracts, 300 g of flowers and leaves were individually extracted by maceration with 1.5 L of ethanol/water (50:50, *v*/*v*) for 4 days at room temperature. Extracts were filtered, rotary evaporated, and lyophilized; they were denominated as hydroethanolic flower extract (HEFTS) and hydroethanolic leaf extract (HELTS).

4.2. Chemical Characterization by Ultrafast Liquid Chromatography Coupled to Mass Spectrometry (LC–MS/MS)

All extract samples were initially reconstituted in methanol (μg/mL), centrifuged (30 min/13,000 rpm), and filtered through a 0.22 mm membrane. Their respective supernatants were stored at 20 °C and then diluted in mobile phase (pure acetonitrile) to 10× and 20× for aqueous and ethanolic extracts, respectively, prior to LC–MS/MS analyses.

Extracts were analyzed by ultrafast liquid chromatography in a UPLC Eksigent UltraLC 110-XL liquid chromatograph (AB Sciex, Framingham, MA, USA) coupled to Kinetex 2.6 μm C18 100 Å column (50 × 2.1 mm) and a 5600+ TripleT spectrometer (AB Sciex, Framingham, MA, USA). After equilibrating the column with 5% acetonitrile/0.1% formic acid solution for 5 min, 2 μL of the sample was automatically injected, and the separation was performed with a linear gradient of 5% acetonitrile/0.1% formic acid ranging 5–95% over 10 min at a flow rate of 0.4 mL/min, keeping the column temperature at 40 °C. The mass spectrometer operated in positive IDA (information dependent acquisition) mode with mass range from *m*/*z* 100 to 1800 and source temperature of 650 °C. The IDA mode was configured to fragment ions from *m*/*z* 100–1250, with a load ranging from 1 to 3, and with an intensity greater than 1000 counts. The other acquisition parameters were as follows: period cycle time = 900 ms; pulser frequency = 15,392 kHz; accumulation time = 250.0 ms; curtain gas = 15,000; ion source gas 1 = 50,000; ion source gas 2 = 45,000; ion spray voltage floating = 5500. In addition to the extracts, a blank control was acquired. Before the start and every five analyses, the spectrometer was calibrated using the calibration solution (AB Sciex, Framingham, MA, USA) to obtain an accuracy of approximately 0.5 ppm (sodium iodide (2 μg/μL) and cesium iodide (50 ng/μL) in 50/50 2-propanol/water).

For data analysis, acquisition files (.WIFF) were converted to the .mzXML format using the MSConvert software (ProteoWizard 3.0, ProteoWizard, Palo Alto, CA, USA) and submitted to the GNPS platform: Global Natural Products Social Molecular Networking (http://gnps.ucsd.edu) for analysis with the Molecular-Library Search-V2 (version release_14) tool (1). Data were filtered by removing all peaks with ~17 Da referring to the m/z value of the precursor present in the MS/MS spectra, which were filtered by window choosing only the top six peaks in the 50 Da window across the spectrum. Data were then grouped with the MS-Cluster with an original mass tolerance of 0.02 Da and an ion tolerance of MS/MS fragments of 0.1 Da to create consensus spectra. Furthermore, consensus spectra containing fewer than two spectra were discarded, before analysis in the GNPS spectral libraries. The library spectra were filtered in the same way as the input data. All correspondences maintained between network and library spectra were required to have a score above 0.85 and at least four corresponding peaks. Cosine score refers to a normalized dot-product, a mathematical measure of spectral similarity between two fragmentation spectra. A cosine score of 1 represents identical spectra, whereas a cosine score of 0 denotes no similarity at all [67].

4.3. Cell Culture

Murine macrophage (RAW 264.7) cells were obtained from American Type Culture Collection (ATCC, Rockville, MD, USA) (ATCC® TIB-71™). Cells were grown in DMEM (Dulbecco's modified Eagle's medium), supplemented with 10% fetal bovine serum, and streptomycin (5000 mg/mL)/penicillin (5000 IU) at 37 °C in a humidified atmosphere with 5% CO_2.

4.4. MTT Viability Assay

RAW 264.7 macrophages were exposed to AFETS, ALETS, HEFTS, and HELTS at different extract concentrations (5, 50, 100, and 500 µg/mL) in order to assess the cytotoxicity. Cell viability was determined using the MTT assay (3-(4,5-dimethylthiazol-2-yl)-2,5-diphenyltetrazolium bromide). A total of 1×10^5 cells per well were seeded in 96-well microplates for 24 h to promote adhesion. After 24 h at 37 °C, 100 µL of MTT (5 mg/mL) was added to each well containing cells, and the plates were again incubated (37 °C/4 h). After removing the culture medium, 100 µL of DMSO was added to each well, before determining the cell viability at 570 nm in a microplate ELISA reader (Epoch-Biotek, Winooski, VT, USA). Cells grown only in DMEM medium were used as a negative control. All assays were performed in triplicate.

4.5. Alamar Blue® Viability Assay

Cell viability after challenge with AFETS, ALETS, HEFTS, and HELTS at 5, 50, 100, and 500 µg/mL was also evaluated by Alamar Blue® assay to assess their cytotoxic effect. A total of 1×10^5 cells per well were seeded in 96-well microplates for 24 h to promote adhesion. After 24 h at 37 °C of exposure to the extracts, 10% Alamar Blue®, corresponding to the volume of the medium contained in each well, was added, and the plate was again incubated (4 h/37 °C/5% CO_2). Then, the reduced Alamar Blue® was monitored at 570 nm and 600 nm in a microplate ELISA reader (Epoch-Biotek, Winooski, VT, USA). Cells grown only in DMEM medium were used as a negative control. All assays were performed in triplicate.

4.6. Cytokine Measurement of (TNF-α, IL1-β, IL-6, and IL10)

Raw 264.7 cells (1×10^5 cells per well) were plated and activated with lipopolysaccharide (LPS) from *Escherichia coli* (O55:B5), previously dissolved in DMEM (2 µg/mL). After 1 h, cells were treated with AFETS, ALETS, HEFTS, and HELTS at different concentrations (5, 50, 100, and 500 µg/mL). After 24 h, supernatants were harvested, and the levels of TNF- α, IL1-β, IL-6, and IL-10 were measured using an immunoenzymatic assay (ELISA) kit (eBioscience), according to the manufacturer's instructions. Analyses were performed in triplicate, and optical density was measured at 450 nm in a microplate ELISA reader (Epoch-Biotek, Winooski, VT, USA). Cells without LPS stimulation and LPS-stimulated cells without extract exposure were used as negative and positive control, respectively.

4.7. Measurement of Prostaglandin E_2 (PGE$_2$) and Nitric Oxide (NO) Production

Raw 264.7 cells (1×10^5 cells per well) were plated and activated with LPS from *Escherichia coli* (O55:B5), dissolved in DMEM (2 µg/mL). One hour later, cells were treated with different concentrations (5, 50, 100, and 500 µg/mL) of AFETS, ALETS, HEFTS, and HELTS. After 24 h, supernatants were collected to determine PGE$_2$ and NO, using ELISA kits according to the respective manufacturer's instructions (BD Biosciences, San Jose, CA, USA). The PGE$_2$ concentrations, expressed in pg/mL, were monitored at 405 nm and 420 nm, while the NO total concentration was assessed after addition of Griess reagent to 40 µL of supernatant and measuring the absorbance at 545 nm. All readings were performed in a microplate ELISA reader (Epoch-Biotek, Winooski, VT, USA). Cells without LPS stimulation and LPS-stimulated cells without extract exposure were used as the negative and positive control, respectively.

4.8. Statistical Analysis

The results were expressed as the mean $\pm$ SD and analyzed with one-way ANOVA and Tukey's post hoc test, considering $p < 0.05$ as statistically significant. All statistical analyses were performed using GraphPad Prism version 5.0 Software for Windows (GraphPad Software, San Diego, CA, USA).

5. Conclusions

This study demonstrated that *T. subulata* and its bioactive molecules display promising anti-inflammatory activity and no cytotoxicity toward RAW 264.7 cells. The anti-inflammatory properties of *T. subulata* extracts were suggested using an in vitro model of acute inflammation through a reduction in TNF-α, IL-1β, and IL-6 cytokines and an increase in IL-10 cytokine. The extracts also inhibited PGE$_2$ and NO production in LPS-stimulated RAW 264.7 cells. Phytochemical analyses revealed the presence of vitexin-2-*O*-rhamnoside or its isoforms, a glycosylated flavone derived from apigenin, which may be responsible for the pharmacological activities such as the immunomodulatory effect displayed by *T. subulata* extracts highlighted in this study. Overall, the results provide scientific evidence supporting the use of *T. subulata* flowers and leaves in folk medicine for inflammatory disorders, which can be applied as an alternative to assist with this treatment.

Supplementary Materials: The following supporting information can be downloaded online. Comparison between library GNPS and query spectra of phytocomponents identified in *Turnera subulata* flower and leaf extracts by LC-MS/MS analyses.

Author Contributions: Conceptualization, J.R.D.d.L. and M.d.G.A.; methodology, J.R.D.d.L. and E.A.B.; formal analysis, J.R.D.d.L., E.A.B. and M.d.G.A.; investigation, J.R.D.d.L., E.A.B. and T.E.S.d.N.; resources, J.R.D.d.L., T.E.S.d.N., A.A.d.R., A.d.S.B., E.A.B., M.d.G.A., J.A.L., G.A.-S. and M.A.G.U.; data curation, J.R.D.d.L., G.A.-S., J.A.L. and M.d.G.A.; writing—original draft preparation, J.R.D.d.L., G.A.-S., J.A.L. and M.d.G.A.; writing—review and editing, J.R.D.d.L., A.d.S.B., G.A.-S., J.A.L. and M.d.G.A.; supervision, J.R.D.d.L., G.A.-S., J.A.L. and M.d.G.A.; project administration, M.d.G.A.; funding acquisition, M.d.G.A. All authors read and agreed to the published version of the manuscript.

Funding: This research was funded by the Brazilian research funding agency National Council for Scientific and Technological Development (CNPq) (Grant No.478652/2010-0), the Banco do Nordeste (Grant No.912011), Federal University of Rio Grande do Norte (Grant No. 397/2020) and Coordenação de Aperfeiçoamento de Pessoal de Nível Superior—Brasil (CAPES)—Finance Code 001.

Institutional Review Board Statement: Not applicable.

Informed Consent Statement: Not applicable.

Acknowledgments: The authors would like to thank the CNPq for providing the postgraduate fellowship (Process No. 169246/2018-3) and the Research Center for Redox Processes in Biomedicine at the University of São Paulo (USP) for their technical assistance with cell cultures.

Conflicts of Interest: The authors declare no conflict of interest.

Sample Availability: Not available.

References

1. Kulkarni, O.P.; Lichtnekert, J.; Anders, H.J.; Mulay, S.R. The immune system in tissue environments regaining homeostasis after injury: Is "inflammation" always inflammation? *Mediators. Inflamm.* **2016**, *1*, 2856213. [CrossRef]
2. Villanueva, J.R.; Esteban, J.M.; Villanueva, L.R. Solving the puzzle: What is behind our forefathers' anti-inflammatory remedies? *J. Intercult. Ethnopharmacol.* **2016**, *6*, 128–143. [CrossRef]
3. Parisi, L.; Bassani, B.; Tremolati, M.; Gini, E.; Farronato, G.; Bruno, A. Natural killer cells in the orchestration of chronic inflammatory diseases. *J. Immunol. Res.* **2017**, *1*, 4218254. [CrossRef]
4. Oishi, Y.; Manabe, I. Macrophages in inflammation, repair and regeneration. *Int. Immunol.* **2018**, *30*, 511–528. [CrossRef] [PubMed]
5. Watanabe, S.; Alexander, M.; Misharin, A.V.; Budinger, G.R.S. The role of macrophages in the resolution of inflammation. *J. Clin. Investig.* **2019**, *129*, 2619–2628. [CrossRef] [PubMed]
6. Kim, E.Y.; Moudgil, K.D. Immunomodulation of autoimmune arthritis by pro-inflammatory cytokines. *Cytokine* **2017**, *98*, 87–96. [CrossRef]
7. Ungprasert, P.; Srivali, N.; Thongprayoon, C. Nonsteroidal anti-inflammatory drugs and risk of incident heart failure: A systematic review and meta-analysis of observational studies. *Clin. Cardiol.* **2016**, *39*, 111–188. [CrossRef]
8. Abdelwahab, D.I.; Koko, W.S.; Taha, M.M.E.; Mohan, S.; Achoui, M.; Abdulla, M.A.; Mustafa, M.R.; Ahmad, S.; Noordin, M.I.; Yong, C.L.; et al. In vitro and in vivo anti-inflammatory activities of columbin through the inhibition of cycloxygenase-2 and nitric oxide but not the suppression of NF-κB translocation. *Eur. J. Pharmacol.* **2012**, *678*, 61–70. [CrossRef] [PubMed]
9. Fokunang, C.N.; Fokunang, E.T.; Frederick, K.; Ngameni, B.; Ngadjui, B. Overview of non-steroidal anti-inflammatory drugs (nsaids) in resource limited countries. *MOJ Toxicol* **2018**, *4*, 5–13. [CrossRef]
10. Alessandri, A.L.; Sousa, L.P.; Lucas, C.D.; Rossi, A.G.; Pinho, V.; Teixeira, M.M. Resolution of inflammation: Mechanisms and opportunity for drug development. *Pharmacol. Ther.* **2013**, *139*, 189–212. [CrossRef]
11. Winand, L.; Sester, A.; Nett, M. Bioengineering of anti-inflammatory natural products. *ChemMedChem* **2021**, *16*, 767–776. [CrossRef]
12. Kisiriko, M.; Anastasiadi, M.; Terry, L.A.; Yasri, A.; Beale, M.H.; Ward, J.L. Phenolics from medicinal and aromatic plants: Characterisation and potential as biostimulants and bioprotectants. *Molecules* **2021**, *26*, 6343. [CrossRef] [PubMed]
13. Ali, S.S.; Ahmad, W.A.N.W.; Budin, S.B.; Zainalabidin, S. Implication of dietary phenolic acids on inflammation in cardiovascular disease. *Rev. Cardiovasc. Med.* **2020**, *21*, 225–240. [CrossRef]
14. Puangpraphant, S.; Cuevas-Rodríguez, E.O.; Oseguera-Toledo, M. Anti-inflammatory and antioxidant phenolic compounds. In *Current Advances for Development of Functional Foods Modulating Inflammation and Oxidative Stress*; Levy, N., Ed.; Academic Press: Cambridge, MA, USA, 2022; pp. 165–180.
15. Fitzgerald, M.; Heinrich, M.; Booker, A. Medicinal plant analysis: A historical and regional discussion of emergent complex techniques. *Front. Pharmacol.* **2020**, *10*, 1480. [CrossRef] [PubMed]
16. Kumar, S.; Taneja, R.; Sharma, A. The genus Turnera: A review update. *Pharm. Biol.* **2005**, *43*, 383–391. [CrossRef]
17. Rocha, L.; Nogueira, J.W.A.; Figueiredo, M.F.; Loiola, M.I.B. Flora of Ceará: Turneraceae. *Rodriguésia* **2018**, *69*, 1673–1700. [CrossRef]
18. Antonio, M.A.; Brito, A.R.M.S. Oral anti-inflammatory and antiulcerogenic activities of a hydroalcoholic extract and partitioned fractions of *Turnera ulmifolia* (Turneraceae). *J. Ethnopharmacol.* **1998**, *61*, 215–228. [CrossRef]
19. Galvez, J.; Gracioso, J.S.; Camuesco, D.; Vilegas, W.; Brito, A.R.M.S.; Zarzuelo, A. Intestinal antiinflammatory activity of a lyophilized infusion of *Turnera ulmifolia* in TNBS rat colitis. *Fitoterapia* **2006**, *77*, 515–520. [CrossRef]
20. Nascimento, M.A.; Silva, A.K.; França, L.C.; Quignard, E.L.J.; López, J.A.; Almeida, M.G. *Turnera ulmifolia* L. (Turneraceae): Preliminary study of its antioxidant activity. *Bioresour. Technol.* **2006**, *97*, 1387–1391. [CrossRef]
21. Brito, N.J.N.; López, J.A.; Nascimento, M.A.; Macêdo, J.B.M.; Silva, G.A.; Oliveira, C.N.; Rezende, A.U.; Brandão-Neto, J.; Schwarz, A.; Almeida, M.G. Antioxidant activity and protective effect of *Turnera ulmifolia* Linn. var. *elegans* against carbon tetrachloride-induced oxidative damage in rats. *Food Chem. Toxicol.* **2012**, *50*, 4340–4347. [CrossRef]
22. Szewczyk, K.; Zidorn, C. Ethnobotany, phytochemistry, and bioactivity of the genus Turnera (Passifloraceae) with a focus on damiana–*Turnera diffusa*. *J. Ethnopharmacol.* **2014**, *152*, 424–443. [CrossRef] [PubMed]
23. Montanher, A.B.; Zucolotto, S.M.; Schenkel, E.P.; Fröde, T.S. Evidence of anti-inflammatory effects of *Passiflora edulis* in an inflammation model. *J. Ethnopharmacol.* **2007**, *109*, 281–288. [CrossRef] [PubMed]
24. Park, J.W.; Kwon, O.K.; Ryu, H.W.; Paik, J.H.; Paryanto, I.; Yuniato, P.; Choi, S.; Oh, S.R.; Ahn, K.S. Anti-inflammatory effects of *Passiflora foetida* L. in LPS-stimulated RAW264.7 macrophages. *Int. J. Mol. Med.* **2018**, *41*, 3709–3716. [CrossRef] [PubMed]
25. Souza, N.C.; Oliveira, J.M.; Morrone, M.D.S.; Albanus, R.D.; Amarante, M.D.S.M.; Camillo, C.D.S.; Langassner, S.M.Z.; Gelain, D.P.; Moreira, J.C.F.; Dalmolin, R.J.S.; et al. *Turnera subulata*: Anti-inflammatory properties in lipopolysaccharide-stimulated RAW 264.7 macrophages. *J. Med. Food* **2016**, *19*, 922–930. [CrossRef] [PubMed]
26. Luz, J.R.D.; Nascimento, T.E.S.; Morais, L.V.F.; Cruz, A.K.M.; Rezende, A.A.; Brandão-Neto, J.; Ururahy, M.A.G.; Luchessi, A.D.; López, J.A.; Rocha, H.A.O.; et al. Thrombin inhibition: Preliminary assessment of the anticoagulant potential of *Turnera subulata* (Passifloraceae). *J. Med. Food* **2019**, *22*, 384–392. [CrossRef]

27. Neuenschwander, P.F. Coagulation cascade: Overview. In *Encyclopedia of Respiratory Medicine*, 2nd ed.; Janes, S.M., Ed.; Elsevier: Oxford, UK, 2022; pp. 479–488.
28. Furman, D.; Campisi, J.; Verdin, E.; Carrera-Bastos, P.; Targ, S.; Franceschi, C.; Ferrucci, L.; Gilroy, D.W.; Fasano, A.; Miller, G.W.; et al. Chronic inflammation in the etiology of disease across the life span. *Nat. Med.* **2019**, *25*, 1822–1832. [CrossRef]
29. Lee, M.W.; Katz, P.O. Nonsteroidal antiinflammatory drugs, anticoagulation, and upper gastrointestinal bleeding. *Clin. Geriatr. Med.* **2021**, *37*, 31–42. [CrossRef]
30. Nunes, C.D.R.; Barreto, M.B.; Pereira, S.M.F.; Cruz, L.L.; Passos, M.S.; Moraes, L.P.; Vieira, I.J.C.; Oliveira, D.B. Plants as sources of anti-inflammatory agentes. *Molecules* **2020**, *25*, 3726. [CrossRef]
31. Shin, S.A.; Joo, B.J.; Lee, J.S.; Ryu, G.; Han, M.; Kim, W.Y.; Park, H.H.; Lee, J.H.; Lee, C.S. Phytochemicals as anti-inflammatory agents in animal models of prevalent inflammatory diseases. *Molecules* **2020**, *25*, 5932. [CrossRef]
32. Yang, Y.; Saand, M.A.; Abdelaal, W.B.; Zhang, J.; Wu, Y.; Li, J.; Fan, H.; Wang, F. iTRAQ-based comparative proteomic analysis of two coconut varieties reveals aromatic coconut cold-sensitive in response to low temperature. *J. Proteom.* **2020**, *30*, 103766. [CrossRef]
33. Góral, I.; Wojciechowski, K. Surface activity and foaming properties of saponin-rich plants extracts. *Adv. Colloid. Interface Sci.* **2020**, *279*, 102145. [CrossRef] [PubMed]
34. Zaynab, M.; Fatima, M.; Abbas, S.; Sharif, Y.; Umair, M.; Zafar, M.W.; Bahadar, K. Role of secondary metabolites in plant defense against pathogens. *Microb. Pathog.* **2018**, *124*, 198–202. [CrossRef] [PubMed]
35. Freitas, C.L.A.; Santos, F.F.P.; Dantas-Junior, O.M.; Inácio, V.V.; Matias, W.F.F.; Quintans-Júnior, L.J.; Aguiar, J.J.S.; Coutinho, H.D.M. Enhancement of antibiotic activity by phytocompounds of *Turnera subulata*. *Nat. Prod. Res.* **2020**, *34*, 2384–2388. [CrossRef]
36. Salvador, M.J.; Ferreira, E.O.; Mertens-Talcott, S.U.; De Castro, W.V.; Butterweck, V.; Derendorf, H.; Dias, D.A. Isolation and HPLC quantitative analysis of antioxidant flavonoids from *Alternanthera tenella* Colla. *Z. Nat. C.* **2006**, *61*, 19–25. [CrossRef]
37. Ninfali, P.; Bacchiocca, M.; Antonelli, A.; Biagiotti, E.; Di Gioacchino, A.M.; Piccoli, G.; Stocchi, V.; Brandi, G. Characterization and biological activity of the main flavonoids from Swiss Chard (*Beta vulgaris subspecies cycla*). *Phytomedicine* **2007**, *14*, 216–221. [CrossRef]
38. Gennari, L.; Felletti, M.; Blasa, M.; Angelino, D.; Celeghini, C.; Corallini, A.; Ninfali, P. Total extract of *Beta vulgaris var. cicla* seeds versus its purified phenolic components: Antioxidant activities and antiproliferative effects against colon cancer cells. *Phytochem. Anal.* **2011**, *22*, 272–279. [CrossRef] [PubMed]
39. Zielińska-Pisklak, M.A.; Kaliszewska, D.; Stolarczyk, M.; Kiss, A.K. Activity-guided isolation, identification and quantification of biologically active isomeric compounds from folk medicinal plant *Desmodium adscendens* using high performance liquid chromatography with diode array detector, mass spectrometry and multidimentional nuclear magnetic resonance spectroscopy. *J. Pharm. Biomed. Anal.* **2015**, *102*, 54–63. [CrossRef]
40. Marincaş, O.; Feher, I.; Magdas, D.A.; Puşcaş, R. Optimized and validated method for simultaneous extraction, identification and quantification of flavonoids and capsaicin, along with isotopic composition, in hot peppers from different regions. *Food Chem.* **2018**, *267*, 255–262. [CrossRef]
41. Bezerra, A.G.; Negri, G.; Duarte-Almeida, J.M.; Smaili, S.S.; Carlini, E.A. Phytochemical analysis of hydroethanolic extract of *Turnera diffusa* Willd and evaluation of its effects on astrocyte cell death. *Einstein* **2016**, *14*, 56–63. [CrossRef]
42. Willer, J.; Jöhrer, K.; Greil, R.; Zidorn, C.; Çiçek, S.S. Cytotoxic properties of damiana (*Turnera diffusa*) extracts and constituents and a validated quantitative UHPLC-DAD assay. *Molecules* **2019**, *24*, 855. [CrossRef]
43. Rempelos, L.; Almuayrifi, A.M.; Baranski, M.; Tetard-Jones, C.; Eyre, M.; Shotton, P.; Cakmak, I.; Ozturk, L.; Cooper, J.; Volakakis, N.; et al. Effects of agronomic management and climate on leaf phenolic profiles, disease severity, and grain yield in organic and conventional wheat production systems. *J. Agric. Food. Chem.* **2018**, *66*, 10369–10379. [CrossRef] [PubMed]
44. Serafini, M.; Peluso, I.; Raguzzini, A. Flavonoids as anti-inflammatory agents. *Proc. Nutr. Soc.* **2010**, *69*, 273–278. [CrossRef] [PubMed]
45. Chang, S.K.; Alasalvar, C.; Shahidi, F. Superfruits: Phytochemicals, antioxidant efficacies, and health effects–A comprehensive review. *Crit. Rev. Food. Sci. Nutr.* **2019**, *59*, 1580–1604. [CrossRef]
46. Yang, H.; Huang, J.; Mao, Y.; Wang, L.; Li, R.; Ha, C. Vitexin alleviates interleukin-1β-induced inflammatory responses in chondrocytes from osteoarthritis patients: Involvement of HIF-1α pathway. *Scand. J. Immunol.* **2019**, *90*, e12773. [CrossRef]
47. Cao, H.; Wang, X.; Zhang, B.; Ren, M. The protective effect of vitexin in septic encephalopathy by reducing leukocyte-endothelial adhesion and inflammatory response. *Ann. Palliat. Med.* **2020**, *9*, 2079–2089. [CrossRef] [PubMed]
48. Duan, S.; Du, X.; Chen, S.; Liang, J.; Huang, S.; Hou, S.; Gao, J.; Ding, P. Effect of vitexin on alleviating liver inflammation in a dextran sulfate sodium (DSS)-induced colitis model. *Biomed. Pharmacother.* **2020**, *121*, 109683. [CrossRef]
49. Tu, Y.; Xiao, T.; Gong, G.; Bian, Y.; Li, Y. A new isoflavone with anti-inflammatory effect from the seeds of *Millettia pachycarpa*. *Nat. Prod. Res.* **2019**, *34*, 981–987. [CrossRef]
50. Xie, C.; Park, K.H.; Kong, S.S.; Cho, K.M.; Lee, D.H. Isoflavone-enriched soyben leaves attenuate ovariectomy-induced osteoporosis in rats by anti-inflammatory activity. *J. Sci. Food. Agric.* **2020**, *111*, 1499–1506. [CrossRef]
51. Yin, P.; Zhang, Z.; Li, J.; Shi, Y.; Jin, N.; Zou, W.; Gao, G.; Wang, W.; Liu, F. Ferulic acid inhibits bovine endometrial epithelial cells against LPS-induced inflammation via suppressing NF-κB and MAPK pathway. *Res. Vet. Sci.* **2020**, *126*, 164–169. [CrossRef]

52. Mahmoud, A.M.; Hussein, O.E.; Hezayen, W.G.; Bin-Jumah, M.; El-Twab, S.M.A. Ferulic acid prevents oxidative stress, inflammation, and liver injury via upregulation of Nrf2/HO-1 signaling in methotrexate-induced rats. *Environ. Sci. Pollut. Res. Int.* **2020**, *27*, 7910–7921. [CrossRef]
53. Taniguchi, K.; Karin, M. NF-κB, inflammation, immunity and cancer: Coming of age. *Nat. Rev. Immunol.* **2018**, *18*, 309–324. [CrossRef]
54. Gupta, P.; Barthwal, M.K. IL-1 β genesis: The art of regulating the regulator. *Cell. Mol. Immunol.* **2018**, *15*, 998–1000. [CrossRef] [PubMed]
55. Alunno, A.; Carubbi, F.; Giacomelli, R.; Gerli, R. Cytokines in the pathogenesis of rheumatoid arthritis: New players and therapeutic targets. *BMC Rheumatol.* **2017**, *1*, 3. [CrossRef] [PubMed]
56. Hamidzadeh, K.; Christensen, S.M.; Dalby, E.; Chandrasekaran, P.; Mosser, D.M. Macrophages and the recovery from acute and chronic inflammation. *Annu. Rev. Physiol.* **2017**, *79*, 567–592. [CrossRef]
57. Qi, Y.; Choi, S.I.; Son, S.R.; Han, H.S.; Ahn, H.S.; Shin, Y.K.; Lee, S.H.; Lee, K.T.; Kwon, H.C.; Jang, D.S. Chemical Constituents of the Leaves of *Campanula takesimana* (Korean Bellflower) and Their Inhibitory Effects on LPS-induced PGE2 Production. *Plants* **2020**, *9*, 1232. [CrossRef] [PubMed]
58. Zhang, L.; McClements, D.J.; Wei, Z.; Wang, G.; Liu, X.; Liu, F. Delivery of synergistic polyphenol combinations using biopolymer-based systems: Advances in physicochemical properties, stability and bioavailability. *Crit. Rev. Food Sci. Nutr.* **2019**, *60*, 2083–2097. [CrossRef] [PubMed]
59. Sharma, R.; Padwad, Y. Plant polyphenol-based second-generation synbiotic agents: Emerging concepts, challenges, and opportunities. *Nutrition* **2020**, *77*, 110785. [CrossRef] [PubMed]
60. Giordano, A.; Morales-Tapia, P.; Moncada-Basualto, M.; Pozo-Martínez, J.; Olea-Azar, C.; Nesic, A.; Cabrera-Barjas, G. Polyphenolic composition and antioxidant activity (ORAC, EPR and cellular) of different extracts of *Argylia radiata* vitroplants and natural roots. *Molecules* **2022**, *27*, 610. [CrossRef]
61. Jiang, J.; Dai, J.; Cui, H. Vitexin reverses the autophagy dysfunction to attenuate MCAO-induced cerebral ischemic stroke via mTOR/Ulk1 pathway. *Biomed. Pharmacother.* **2018**, *99*, 583–590. [CrossRef]
62. Babaei, F.; Moafizad, A.; Darvishvand, Z.; Mirzababaei, M.; Hosseinzadeh, H.; Nassiri-Asl, M. Review of the effects of vitexin in oxidative stress-related diseases. *Food Sci. Nutr.* **2020**, *8*, 2569–2580. [CrossRef]
63. Kelner, M.J.; Uglik, S.F. Mechanism of prostaglandin E2 release and increase in PGH2/PGE2 isomerase activity by PDGF: Involvement of nitric oxide. *Arch. Biochem. Biophys.* **1994**, *312*, 240–243. [CrossRef] [PubMed]
64. Mollace, V.; Muscoli, C.; Masini, E.; Cuzzocrea, S.; Salvemini, D. Modulation of prostaglandin biosynthesis by nitric oxide and nitric oxide donors. *Pharmacol. Rev.* **2005**, *57*, 217–252. [CrossRef] [PubMed]
65. Foley, J.H.; Conway, E.M. Cross talk pathways between coagulation and inflammation. *Circ. Res.* **2016**, *118*, 1392–1408. [CrossRef] [PubMed]
66. Palhares, L.C.G.F.; Brito, A.S.; Lima, M.A.; Nader, H.B.; London, J.A.; Barsukov, I.L.; Andrade, G.P.V.; Yates, E.A.; Chavante, S.F. A further unique chondroitin sulfate from the shrimp *Litopenaeus vannamei* with antithrombin activity that modulates acute inflammation. *Carbohydr. Polym.* **2019**, *222*, 115031. [CrossRef]
67. Wang, M.; Carver, J.J.; Phelan, V.V.; Sanchez, L.M.; Garg, N.; Peng, Y.; Nguyen, D.D.; Watrous, J.; Kapono, C.A.; Luzzatto-Knaan, T.; et al. Sharing and community curation of mass spectrometry data with global natural products social molecular networking. *Nat. Biotechnol.* **2016**, *34*, 828–837. [CrossRef]

Article

Chemical Analysis of *Eruca sativa* Ethanolic Extract and Its Effects on Hyperuricaemia

Arthur Ferrari Teixeira [1], Jacqueline de Souza [2], Douglas Daniel Dophine [1], José Dias de Souza Filho [3] and Dênia Antunes Saúde-Guimarães [1,*]

[1]   Laboratório de Plantas Medicinais (LAPLAMED), Programa de Pós-Graduação em Ciências Farmacêuticas (CiPharma), Universidade Federal de Ouro Preto, Ouro Preto 354000-000, Brazil; arthurfarmaufop@gmail.com (A.F.T.); douglasdophine@gmail.com (D.D.D.)
[2]   Laboratório de Controle de Qualidade (LCQ), Universidade Federal de Ouro Preto, Ouro Preto 354000-000, Brazil; jacsouza@ufop.edu.br
[3]   Laboratório Multiusuário de Caracterização de Moléculas (LMCM), Programa de Pós-Graduação em Ciências Farmacêuticas (CiPharma), Universidade Federal de Ouro Preto, Ouro Preto 354000-000, Brazil; jose.dsf@ufop.edu.br
*   Correspondence: saude@ufop.edu.br

Abstract: In vivo assays and chemical analyses were performed on the ethanolic extract from leaves of *Eruca sativa*. UHPLC-ESI-QTOF analysis confirmed the presence of glucosinolates and flavonol glucosides. The major flavonoid of the ethanolic extract, kaempferol-3,4′-di-*O*-β-glucoside, was isolated, a HPLC-DAD method developed and validated to quantify its content in the extract. In vivo experiments were carried out on Wistar rats with hyperuricaemia induced by potassium oxonate and uric acid. A hypouricaemic effect was observed in hyperuricaemic Wistar rats treated with ethanolic extract at dose of 125 mg/kg and kaempferol-3,4′-di-*O*-β-glucoside at dose of 10 mg/kg. The main anti-hyperuricaemic mechanism observed in the extract was uricosuric. Kaempferol-3,4′-di-*O*-β-glucoside was identified as an important component responsible for the total activity of the ethanolic extract and was considered as a good chemical and biological marker of the ethanolic extract of *E. sativa*. The obtained results indicated the potential of *E. sativa* in the treatment of hyperuricaemia and its comorbidities.

Keywords: glucosinolate; glucosylated flavonols; rocket; validation; NMR; UHPLC/ESI/QTOF; hyperuricaemia

Citation: Teixeira, A.F.; de Souza, J.; Dophine, D.D.; de Souza Filho, J.D.; Saúde-Guimarães, D.A. Chemical Analysis of *Eruca sativa* Ethanolic Extract and Its Effects on Hyperuricaemia. *Molecules* **2022**, *27*, 1506. https://doi.org/10.3390/molecules27051506

Academic Editor: Jacqueline Aparecida Takahashi

Received: 1 February 2022
Accepted: 17 February 2022
Published: 23 February 2022

Publisher's Note: MDPI stays neutral with regard to jurisdictional claims in published maps and institutional affiliations.

1. Introduction

Hyperuricaemia occurs due to the overproduction of uric acid or its insufficient excretion. The high levels of uric acid in the blood cause various damages to patients, such as the development of gout, endothelial dysfunction, hypertension, diabetes, and heart and kidney diseases [1].

Natural products are considered a great source of bioactive substances, as they have a rich chemical profile. The potential of several plant species in relation to anti-hyperuricaemic therapy has already been demonstrated [2].

Rocket, *Eruca sativa* Miller, is a vegetable widespread in several regions of the world and commonly used in food. It is consumed pure in salads, cooked with meat, processed in pesto or in the form of extracts. Rocket in Brazil is usually grown in open fields or in a hydroponic system [3]. Animal models have already shown that rocket is associated with several biological activities, such as antihypertensive, nephroprotective and antidiabetic activities [4–6]. Such biological activities are desirable since hyperuricaemia is closely related to endothelial dysfunction and is pointed out as a causal factor of diseases such as hypertension and diabetes [1].

Looking from the opposite point of view, the fact that *E. sativa* species demonstrate benefits in hypertension and diabetes may be an important lead that this plant also acts as an anti-hyperuricaemic agent. The chemical constitution of rocket has been well studied and many biological activities have been attributed to isolated compounds, such as glucosinolates and flavonoids [6,7]. However, little is known about the chemistry of *Eruca sativa* grown in Brazil. Studies on its therapeutic effects are limited, despite the fact that this plant has several ethnopharmacological uses reported in other countries.

The discovery of anti-hyperuricaemic effects in *E. sativa* would draw attention to its potential in the treatment of hyperuricaemia and its associated comorbidities.

2. Results and Discussion

2.1. Qualitative HPLC Analyses of EE from E. sativa Leaves

2.1.1. Qualitative HPLC-DAD Analysis

Fingerprinting of the ethanolic extract from leaves of *Eruca sativa* (EE) obtained by HPLC-DAD is shown in Figure 1. Peaks along the chromatogram are concentrated between Rt = 4.50 and 15.00 min, indicating a polar bias of substances in the extract. Several minor peaks were seen in this region and two major peaks (4.81 and 14.89 min) stood out. Analysis of the major peaks in the UV spectra showed behaviour characteristics of a glucosinolate and a flavonol, respectively [8,9]. In order to acquire more structural data of the major flavonoid (MF), the UV spectra of kaempferol standard was compared under the same conditions (Figure 1). No differences were observed in relation to band II, indicating that both structures contain a kaempferol aglycone moiety. In relation to band I, a hypsochromic effect of 27 nm from kaempferol to MF was noticed, suggesting substitutions of hydroxyl groups at positions 3 and 4′ [8]. This suggestion was further confirmed by HPLC-DAD analysis, by showing that flavonoid isolated from EE (YP) matched in retention time and UV spectra with the MF in EE. The relative area of the MF in YP corresponded to 96% of the chromatogram, suggesting a high degree of purity.

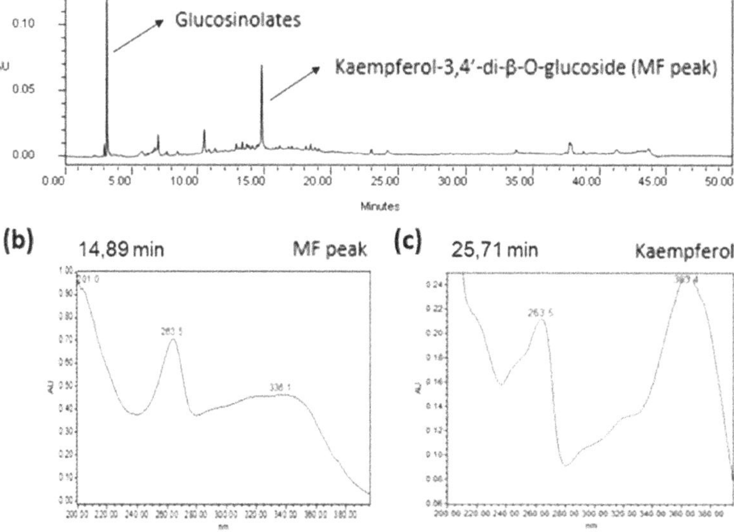

Figure 1. Chromatographic profile of ethanolic extract of *Eruca sativa* leaves and comparison of UV-spectra of MF peak and kaempferol. (**a**) Chromatogram of the ethanolic extract of *Eruca sativa* leaves obtained by HPLC-DAD at 264 nm. (**b**) UV spectra of MF peak extracted from chromatogram of Figure 1a. (**c**) UV spectra of kaempferol standard injected independently at the same chromatographic conditions of MF peak.

2.1.2. Qualitative UHPLC-ESI-QTOF Analysis

Mass spectral (MS) data obtained by UHPLC-ESI-QTOF was used to gather further knowledge about EE composition, and it is summarised in Table 1. It was possible to identify nine substances: kaempferol-3-*O*-β-glucoside, kaempferol-3,4'-di-*O*-β-glucoside, kaempferol-3-*O*-(2-sinapoyl- β-glucoside)-4'-*O*-glucoside, glucosativin glucosinolate, glucoraphanin glucosinolate, leucine, tryptophan, angustione and erucamide by comparing their molecular ions and mass fragments with the literature and mass banks available.

Table 1. Substances identified by UHPLC-ESI-QTOF in the ethanolic extract of *Eruca sativa* leaves.

Chemical Compound	RT (min)	Experimental Mass Data ESI$^+$	Reference
Glucosativin glucosinolate	1.0	328.1384; 166.0861	[10]
Leucine	1.4	182.0808; 268.1037; 132.1019	Fiocruz
Glucoraphanin glucosinolate	1.8	438.0554; 196.0457; 358.0987	[11,12]
Tryptophan	4.2	188.0702, 146.0601; 205.0968	Fiocruz
Kaempferol-3,4'-di-*O*-β-glucoside	12.6	611.1598; 449.1070; 287.0544	[13]
Angustione	14.5	197.1168; 179.1062	Fiocruz
Kaempferol-3-*O*-β-glucoside	15.1	449.1068; 287.0544	[13]
Kaempferol-3-*O*-(2-sinapoyl-β-glucoside)-4'-*O*-glucoside	16.2	817.2178; 817,2166; 655,1641; 369,1166; 207,0646	[14]
Erucamide	46.4	338,3423	Fiocruz

In the spectrum of flavonoids-*O*-glucosylated, the molecular ion [M+H]$^+$ is normally of low intensity, and the loss of a sugar molecule is evident. The product is a prominent Y0+ aglycon ion. In the case of flavonoid-di-*O*-glucosylates, an intermediate Y1+ corresponding to the loss of a sugar molecule is observed [15]. MS analyses showed the presence of a molecular ion [M+H]$^+$ of 611 m/z at 12.6 min and the formation of fragments consistent with losses of two hexose fragments, confirming the identification of a diglucoside of kaempferol. At 15.1 min, a molecular ion [M+H]$^+$ of 449 m/z and its secondary fragmentation, compatible with one loss of a hexose, allowed the identification of a monoglucoside of kaempferol. Comparison with the literature strongly suggested the identity of these flavonoids to be kaempferol-3,4'-di-*O*-β-glucoside and kaempferol-3-*O*-β-glucoside [13].

Cuyckens et al. [14] studied the fragmentation of acylated flavonol-*O*-glucosides and identified fragments of groups such as feruloyl, sinapoyl, coumaroyl and benzoyl, among others. It was found that these structures can be observed in the mass spectra in the positive mode at low-energy CID conditions. This knowledge enabled the identification of an acylated flavonol-*O*-glucoside in the EE. The molecular ion peak with m/z 817 was observed at Rt = 16.2 min. Secondary fragmentations showed the loss of a glucose fragment (m/z 162) and the formation of a fragment with m/z 655. In sequence, a loss of a fragment of m/z 286, characteristic of a kaempferol aglycone, and the formation of a fragment with m/z 369 associated with a sinapoyl-hexose group, were observed. Another loss of a hexose (m/z 162) led to the formation of a fragment with m/z 207 of the sinapoyl group. These results suggested the peak identity with Rt = 16.2 min as kaempferol-3-(2-sinapoyl-β-glucoside)-4'-*O*-glucoside.

Glucosinolate identification was sustained by the observation of fragments with m/z 328.1384 and 166.0861 characteristic of glucosativin glucosinolate (Rt = 1.0 min), and fragments of m/z 438.055, 358.0987 and 196.0457 characteristic of glucoraphanine (Rt = 1.8 min) [10–12].

Aminoacids such leucine (182 m/z, 268 m/z and 132 m/z) and tryptophan (188 m/z, 146 m/z and 205 m/z) were identified by comparison with the mass bank of Fiocruz Institute at 1.4 and 4.2 min, respectively. Previous work by Bell et al. [16] quantified amino acids in *E. sativa* varieties, finding leucine levels in the range of 2.6–3.5 µg/g. Furthermore, it is known that these amino acids (leucine and tryptophan) are biosynthetic precursors of glucosinolates [17]. Other substances, such as angustione (197 m/z and 179 m/z) and erucamide (338 m/z), an amide derived from erucic acid, were also identified by comparison with the mass bank of the Fiocruz Institute at 14.5 and 46.4 min. It was reported that erucic acid is normally found in high levels in *E. sativa* seeds [18].

2.2. Full NMR Structural Characterisation of Kaempferol-3,4′-di-O-β-Glucoside

Notably, the ^{1}H NMR spectrum profile revealed the flavonoidic nature of this compound because of the apparent AA'BB' doublets and meta-coupled AB system (J 1.76 Hz) signals in the aromatic region (Figure 2a). At δ 5.03 and δ 5.48 are seen two well resolved doublets (J 7.25 Hz and 7.50 Hz, respectively) that can be assigned to two anomeric protons (Figure 2b). Despite the lack of resolution in the 3.0–3.8 ppm region (Figure 2b) due to the water signal from the solvent, one can confirm the presence of the two pyranosidic units from the HSQC contour map by the two double correlations due to two methylenic groups, shown in Figure 2c. In the 4.2–5.6 ppm region, one can observe broad signals of hydroxyl groups due to the effect of hydrogen bonds with the solvent (Figure 2b). Moreover, a very unshielded proton signal is registered at δ 12.55, indicating the presence of a quelated hydroxy group in the molecule (Figure 2a).

As one can see, all similar pyranosidic carbon signals are paired. Figure 2c,d shows the carbon and DEPT135 spectra as the F1 projection. The ^{13}C NMR spectrum exhibits 25 signals (Table 2) that are compatible with the 27 predicted carbons in the fragments shown in Figure 2d, which indicates a structural formula containing an aglycone-type kaempferol and two sugar units.

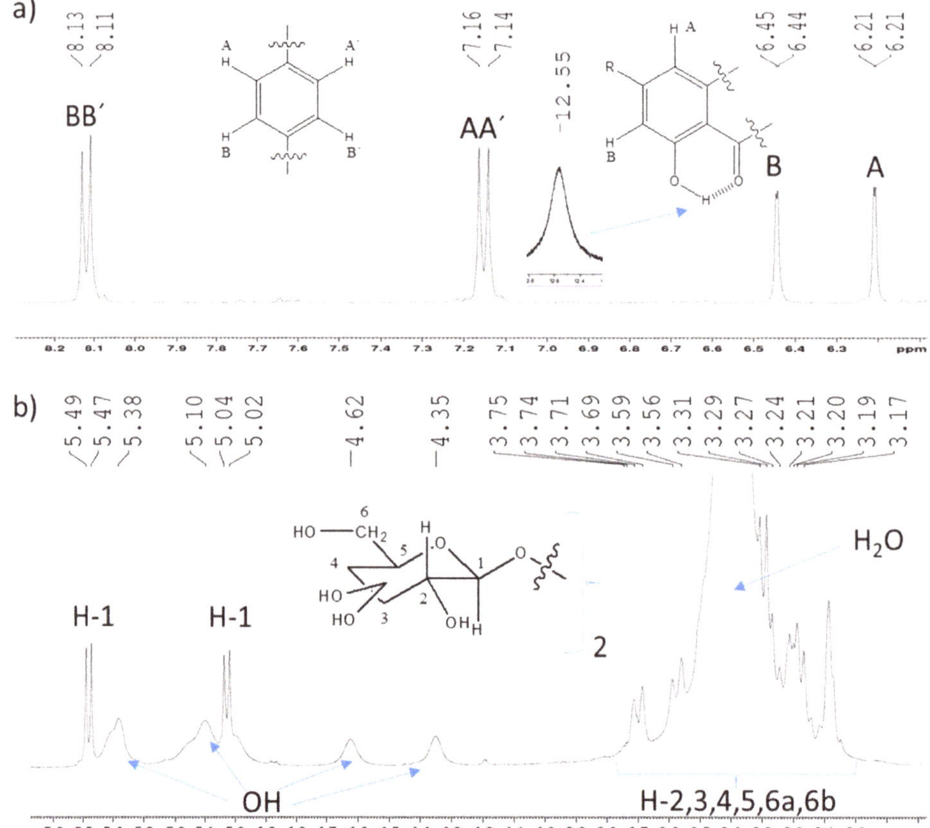

Figure 2. *Cont.*

c)

d)

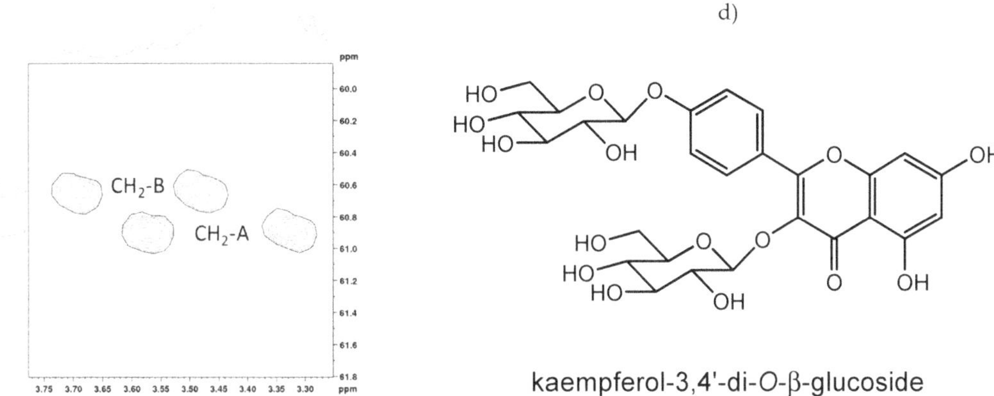

kaempferol-3,4'-di-O-β-glucoside

Figure 2. Diglucoside nature of the flavonoid. (**a**) Aromatic region; (**b**) sugar moieties region; (**c**) HSQC expansion of the pyranosidic protons focusing correlations of the two methylenic groups (DEPT135 in vertical F1 projection) and (**d**) complete chemical structure.

Table 2. NMR data of kaempferol-3,4′-di-O-β-glucoside (400 MHz, DMSO-*d6*, 300 K).

Peak	*d* C-13	C-n	*d* H HSQC	*d* H HMBC	H-n (*J*/Hz)
1	177.49	C-4	—	—	—
2	164.81	C-7	—	6.21; 6.45	—
3	161.20	C-4′	—	5.03; 8.12	—
4	159.24	C-9	—	6.45	—
5	156.55	C-5	—	6.21	—
6	155.53	C-2	—	8.12	—
7	133.72	C-3	—	5.48	—
8	130.63	C-2′,6′	8.12	8.12	H-2′,6′
9	123.76	C-1′	—	7.16	—
10	115.83	C-3′,5′	7.16	7.16	H-3′,5′
11	103.96	C-10	—	6.21; 6.45	—
12	100.86	C-1A	5.48	nd	H-1A (7.50)
13	99.97	C-1B	5.03	nd	H-1B (7.25)
14	98.93	C-6	6.21	6.45	H-6 (1.73)
15	93.85	C-8	6.45	6.21	H-8 (1.73)
16	77.62	C-5B	3.09	nd	H-5B
17	77.10	C-5A	3.39	nd	H-5A
18	76.52	C-3A	3.29	nd	H-3A
19	76.44	C-3B	3.22	nd	H-3B
20	74.21	C-2A	3.17	nd	H-2A
21	73.23	C-2B	3.27	nd	H-2B
22	69.91	C-4B	3.08	nd	H-4B
23	69.58	C-4A	3.19	nd	H-4A
24	60.87	C-6B	3.57; 3.33	nd	H-6a,b-B
25	60.63	C-6A	3.69; 3.48	nd	H-6a,b-A

nd—not defined.

The homonuclear 2 D COSY experiment produced a contour plot (Figure 3a) where the AA′BB′ and AB aromatic correlations could be promptly assigned. The 3.0–3.8 ppm region shows an extreme overlapping of the pyranosidic proton signals, and one can observe two correlations for each anomeric hydrogen (Figure 3b). This is probably due to the presence of rotamers. The chemical shifts of H-2 A (δ 3.17) and H-2B (δ 3.27) as well as those of H-5 A (δ 3.39) and H-5B (δ 3.09) are assigned to H-1 A,B and H-6 A,B, respectively. H-5B and H-4B are very close in chemical shift (δ 3.09 and 3.08, respectively) with correlations over the diagonal in the 2 D COSY contour plot. H-4B exhibits its coupling to H-3B (δ 3.22) by a discrete correlation (Figure 3b). The assignment of the correlations in this region was supported by the edited HSQC contour plots shown in Figure 3c,d.

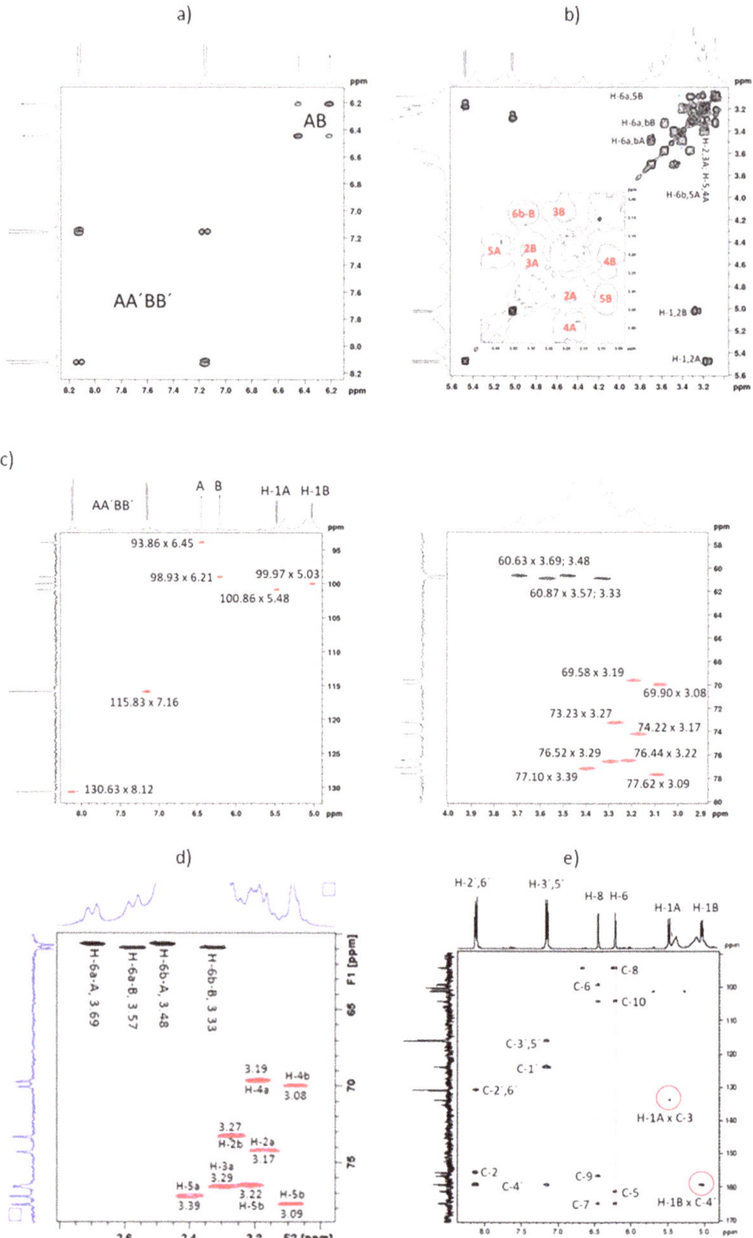

Figure 3. (**a**) COSY expanded contour plot of the aromatic region showing the AA′BB′ and AB coupling systems; (**b**) COSY expanded contour plot of the pyranosidic region; (**c**) HSQC expanded contour plot of the pyranosidic region showing the precise chemical shifts; (**d**) assignments of the pyranosidic unities and (**e**) HMBC expanded contour plot showing the kaempferol 3 and 4′ glucosilated positions.

The anomeric proton signals as doublets at δ 5.48 (H-1 A, J 7.50 Hz) and δ 5.03 (H-1B, J 7.25 Hz) are due to β-glucosides because of the magnitude of the vicinal coupling constants. The very elegant HMBC experiment determined the location of each sugar moiety A and B by the long-range coupling strategy to establish the spin systems. Figure 3e confirms the position of sugar moieties at position 3 and 4' by the presence of the correlations between the chemical shifts δ 5.48 × 133.78 (H-1 A × C4') and δ 5.03 × 159.24 (H-1B × C3). The non-hydrogenated carbons are readily assigned via three- and two-bond couplings.

To the best of our knowledge, this is the first complete assignment of the NMR data of kaempferol-3,4'-di-O-β-glucoside (Table 2). The proton signal multiplicities of the pyranosidic region could not be extracted because of the high degree of overlapping and the existence of rotamers.

2.3. Quantification of the Major Flavonoid Kaempferol-3,4'-di-O-β-glucoside in the EE of E. sativa Leaves by the Validated HPLC-DAD Method

Concentration of kaempferol-3,4'-di-O-β-glucoside quantified in sample solutions (n = 3) was 0.00716 ± 0.00021 mg/mL, which was above the LQ (0.0059 mg/mL). Therefore, it was calculated that the EE tested in rats contained 3.5 ± 0.283 g rutin equivalents/kg EE. In comparison, Martínez-Sánchez et al. [19] reported a content of kaempferol-3,4'-di-O-β-glucoside of 0.978 g/kg of fresh plant, while other flavonoids presented contents 10-fold lower. Other work carried out with several varieties of *E. sativa* showed, by means of statistical analysis (principal component analysis—PCA), that there is a strong tendency in varieties of *E. sativa* to accumulate flavonols with a kaempferol scaffold, especially kaempferol-3,4'-di-O-β-glucoside, with some cultivars presenting contents of 1.1 g/kg of kaempferol-3,4'-di-O-β-glucoside. Kaempferol-3-O-glucoside is found in lower amounts than kaempferol-3,4'-di-O-β-glucoside in most varieties, demonstrating that the biosynthesis of the diglucosylated kaempferol is favoured in this species [20]. That fact justifies choosing kaempferol-3,4'-di-O-β-glucoside as a chemical marker in the EE of *E. sativa* leaves.

2.4. Effects of Ethanolic Extract of E. sativa Leaves on Hyperuricaemia-Induced Rats

The serum uric acid level in the hyperuricaemic control group (8 ± 1.115 mg/dL) was significantly higher than in the normal group (2 ± 0.1630 mg/dL), attesting that the model was effective in increasing uric acid levels in the blood (Figure 4a). The hypouricaemic effect was associated with groups that had serum uric acid levels lower than in the hyperuricaemic group. Positive controls (allopurinol, probenecid and benzbromarone) were effective in reducing serum hyperuricaemia to normal levels, as the concentrations with these treatments showed significant differences with the hyperuricaemic control group and no differences from the normal control group. Animals treated with EE at a dose of 125 mg/kg showed serum uric acid levels similar to the normal control. Treatments with EE at doses of 40 and 10 mg/kg were statistically equivalent and demonstrated a hypouricaemic effect; however, this effect was not able to reduce uricaemia to normal levels. The flavonoid isolated from the EE (kaempferol-3,4'-di-O-β-glucoside) at a dose of 10 mg/kg was effective in reducing the uricaemia of hyperuricaemic rats to normal levels. The anti-hyperuricaemic activity of EE (125 mg/kg) and kaempferol-3,4'-di-O-β-glucoside (10 mg/kg) were statistically equal to positive controls (allopurinol, probenecid and benzbromarone).

Allopurinol is an inhibitor of xanthine oxidase (uricostatic activity), and acts by decreasing the production of uric acid in the liver [21]. The residual activity of xanthine oxidase was evaluated, and treatment with allopurinol promoted an inhibition of 54.72% in relation to the hyperuricaemic group. None of the other treatments managed to promote xanthine oxidase inhibition, suggesting that EE and kaempferol-3,4'-di-O-β-glucoside do not promote uricostatic activity.

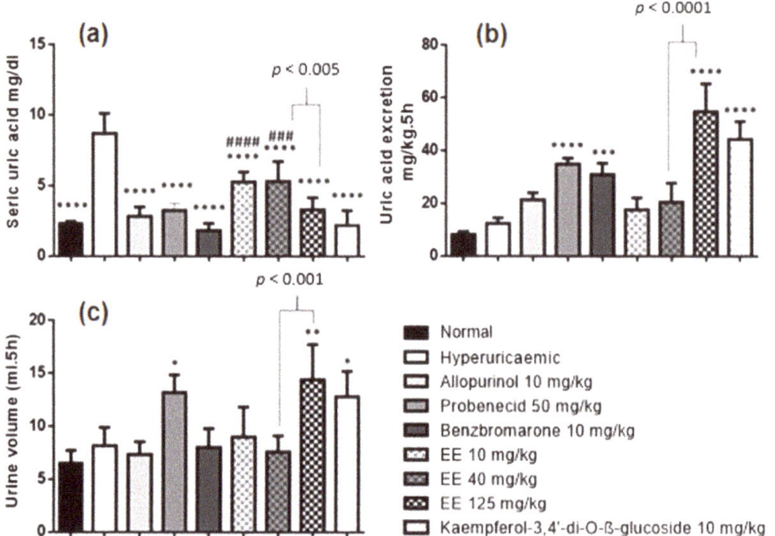

Figure 4. Biological effects in hyperuricaemic Wistar rats after treatments with E. sativa ethanolic extracts and positive controls allopurinol, probenecid and benzobromarone (**a**) Seric concentrations of uric acid (**b**) Excretion of uric acid (**c**) Urine volume excreted after 5 h of the treatment. *** $p < 0.001$ vs. hyperuricaemic control group; ### $p < 0.001$ vs. normal control group; #### $p < 0.0001$ vs. normal control group; **** $p < 0.0001$ vs. hyperuricaemic control group; *** $p < 0.001$ vs. hyperuricaemic control group; ** $p < 0.01$ vs. hyperuricaemic control group; * $p < 0.05$ vs. hyperuricaemic control group ($n = 6$, One-way ANOVA followed by the Dunnet test).

The uricosuric effect of the treatments were evaluated, and the results are illustrated in Figure 4b. Treatments with probenecid (34 ± 2.304 mg/kg 5 h) and benzbromarone (30 ± 4.253 mg/kg 5 h) were effective in exerting a uricosuric effect in the rats. Treatments with EE and with kaempferol-3,4'-di-O-β-glucoside exerted uricosuric effects. The amount of uric acid excreted by the rats when receiving EE at a dose of 125 mg/kg was 54 ± 10.66 mg/kg 5 h (Figure 4b), which was statistically more effective than the values obtained in the treatments with probenecid and benzbromarone (Tukey test, $p \leq 0.05$). Treatment with kaempferol-3,4'-di-O-β-glucoside showed 44.00 ± 6.842 mg/kg 5 h a better uricosuric effect than benzbromarone (Tukey test $p \leq 0.05$), but equivalent to probenecid. Other treatments (EE at 40 and 10 mg/kg) did not exert an uricosuric effect. These data suggest that at doses of 125 mg/kg of EE and 10 mg/kg of kaempferol-3,4'-di-O-β-glucoside, the uricosuric effect is the main mechanism responsible for the hypouricaemic effect. However, it was impossible to explain why a hypouricaemic effect was seen on treatment with EE at 10 and 40 mg/kg, given that these doses did not promote uricosuric and uricostatic effects. This suggests that other mechanisms are contributing to the anti-hyperuricaemic activity of these treatments. Furthermore, it was seen that kaempferol-3,4'-di-O-β-glucoside, the major component of EE, plays a crucial role in the overall anti-hyperuricaemic activity of the EE and can be seen as a biological marker for uricosuric activity.

Additionally, the treatments that exerted an uricosuric effect (EE 125 mg/kg; kaempferol-3, 4'-di-O-β-glucoside), also showed a diuretic effect in relation to the hyperuricaemic and normal groups (Figure 4c). This is particularly convenient because one of the problems that can appear with the use of uricosuric substances is urolithiasis, i.e., when supersaturation of uric acid takes place leading to precipitation of urate crystals in the kidneys, urethra or bladder [22]. Therefore, the diuretic effects observed in treatments with EE of *E. sativa* and kaempferol-3,4′-di-*O*-β-glucoside have a positive and synergic effect because they aid in urinary excretion and minimise possible episodes of urolithiasis.

It was observed that the water balance was negative in hyperuricaemic animals. The fact that these animals were dehydrated could have underestimated the results obtained by treatments that showed a uricosuric response in this model. Such a proposition may indicate that the uricosuric and diuretic responses may have been greater than actually shown.

Flavonoids have been associated with anti-hyperuricaemic effects. It has been reported that myricetin, quercetin and kaempferol aglycones inhibit xanthine oxidase in vitro, whereas rutin, the glycosylated form of quercetin, does not [23]. A study by Haidari et al. [24] showed that hyperuricaemic Wistar rats treated with kaempferol (5 mg/kg) had a significant reduction in serum uric acid levels and hepatic xanthine oxidase and xanthine dehydrogenase activities were inhibited. Similar results were found by Zhu et al. [25], who explored the anti-hyperuricaemic effects in mice of flavonoids isolated from *Biota orientalis* by different routes of administration. When administered orally, quercetin, rutin and *B. orientalis* extract demonstrated significant inhibition of xanthine oxidase and xanthine dehydrogenase. In contrast, when these same treatments were administered intraperitoneally, quercetin and the extract demonstrated a subtle reduction in hyperuricaemia, while rutin had no hypouricaemic effect. It is known that the pharmacokinetics of flavonoids varies by adding sugars to the structure [26]. In the present study, kaempferol-3,4′-di-*O*-β-glucoside administered intraperitoneally to rats demonstrated a potent hypouricaemic effect at a dose of 10 mg/kg. This suggests that hexoses in positions 3 and 4′ give a different pharmacological profile to kaempferol-3,4′-di-*O*-β-glucoside, since the treatment was remarkably effective intraperitoneally and one of the mechanisms involved was uricosuric. Previous studies [27] have suggested that the presence of hydroxyl groups at positions C3, C5, C7 and C-4′ of the chemical structure of flavonoids are important for XO-inhibiting and anti-hyperuricaemic activity. This theory may explain why kaempferol-3,4′-di-*O*-β-diglucoside showed no effect on hepatic xanthine oxidase activity.

Uricosuric drugs may go beyond their hypouricaemic effect. Endothelial dysfunction is closely linked to hyperuricaemia and the development of this condition predisposes to several other morbidities such as hypertension, systemic inflammation, atherosclerosis, cardiovascular diseases and diabetes [1]. A uricosuric effect may be beneficial in preventing the development of endothelial dysfunction. Kang et al. [28] demonstrated through in vitro models the role of uric acid in inducing a decrease in nitric oxide and stimulating the growth of vascular smooth muscle cells (VSMCs), which are crucial factors in the pathogenesis of endothelial dysfunction. In this study, pre-treatments with probenecid lessened the negative effects caused by uric acid. Probenecid has the ability to inhibit the entry of uric acid into VSMCs. If the EE of *E. sativa* acts by the same uricosuric mechanism, they can be potential candidates for preventing endothelial dysfunction. Kang et al. [28] also suggested that the harmful action of uric acid occurs through C-reactive protein, and that uric acid in concentrations of 6–12 mg/dL in the blood has the ability to regulate its expression. The findings of Mazzali et al. [29] stimulated the discussion about uricosuric treatments for hypertension, as it was observed that uric acid induces hypertension via activation of the renin-angiotensin system and inhibition of nitric oxide synthase 1 (ONS1) expression. Benefits in type II diabetes could also be explored, with uric acid being identified as having a bidirectional relationship with insulin resistance [30]. Anzai et al. [31] proposed a model for the renal reabsorption of uric acid, where URAT1 and GLUT9 act together. URAT1 mediates the uptake of urate from the urinary lumen into the renal tubule, which is then transported from the intracellular medium to the blood by GLUT9. Most uricosuric drugs show an inhibitory effect on URAT1. Probenecid and benzbromarone are potent URAT1 inhibitors, but moderately inhibit GLUT 9. Thiazide diuretics are able to increase uric acid reabsorption via URAT1, and while URAT1 is influenced by organic anions, GLUT9 has a high affinity for hexoses [31]. Kaempferol-3,4′- di-*O*-β-glucoside has two hexoses in its structure, which may be related to its uricosuric effect via GLUT9 inhibition.

A dose-dependent effect was observed in relation to the hypouricaemic and uricosuric effect when the doses were reduced from 40 mg/kg to 125 mg/kg in the EE. Although treatment with 40 mg/kg of EE did not promote an uricosuric effect, a less pronounced hypouricaemic effect was observed. When reducing the dose to 10 mg/kg of EE, there was no change in the effect, so the treatment with 10 mg/kg was considered equivalent to that of 40 mg/kg of EE. This indicates that there are other anti-hyperuricaemic mechanisms in addition to the uricosuric and uricostatic pathways by which *E. sativa* extracts promote the hypouricaemic effect. These findings create room for pursuing new alternatives for the treatment of hyperuricaemia that do not act by uricostatic and/or uricosuric routes. Other mechanisms include supplementation of uricase analogues, inhibition of previous pathways to the formation of xanthine and hypoxanthine and increased intestinal extra-renal excretion [32,33]. Uricases tested on humans are known to have immunogenotoxicity problems when administered intravenously. However, a recent study has shown that uricase administered orally to pigs is able to increase intestinal elimination of uric acid [34]. These findings draw attention to a new type of thinking in relation to the discovery of new anti-hyperuricaemic drugs. Products with active uricases, such as plants, could become anti-hyperuricaemic herbal medicines. As it turns out, researchers in Iraq have been able to isolate uricase from *E. sativa* seeds [35]. However, there are no reports of in vivo testing of uricases from plants, and further studies are needed.

Data from ethnopharmacological studies conducted with *E. sativa* in other countries report the use of the plant in folk and aboriginal medicine in the treatment of hypertension and diabetes [6,7]. In a complementary way, Hetta et al. [4] reinforced the proposition that *E. sativa* exhibits an antidiabetic action. Some works written in ancient Greece reported diuretic and renal benefits of the rocket plant. Such reports are confirmed by studies that indicated the nephroprotective activity of *E. sativa* against toxicity induced by pesticides [5]. From an ethnopharmacological point of view, several studies have shown similarities in the medicinal uses and biological activities of *E. sativa*, suggesting that, in general, the activities found for this plant are similar in different cultivars; however, this cannot be considered as a rule. Bell et al. [20] drew attention to the fact that *Eruca* species are still in an evolutionary process and, as a result, their phytochemical profiles may vary in composition and concentration. Interferents such as light, environmental stresses, temperature and genetic factors can also influence the chemical composition between varieties. The discovery of the uricosuric effect of kaempferol-3,4'-di-*O*-β-glucoside gives room for thinking towards the standardisation of the anti-hyperuricaemic activity of the EE of *E. sativa* leaves based on its flavonoid content. Therefore, doses of EE would be adjusted in accord with kaempferol-3,4'-di-*O*-β-glucoside concentration since it is a biological and a chemical marker.

3. Materials and Methods

3.1. Reagents, Chemicals and Analytical Instruments

Reagents: ethanol PA (Quemis®, Jundiaí, Brazil), hexane PA (Neon®, Suzano, Brazil), chloroform PA (Synth®, Diadema, Brazil), ethyl acetate PA (Qhemis®, Jundiaí, Brazil), methanol PA (Dynamic®, São Paulo, Brazil), acetonitrile (JTBaker®, Radnor, PA, USA), methanol (JTBaker®, Radnor, PA, USA), deuterated dimethylsulfoxide, diphenylbory-loxyethylamine (Sigma-Aldrich, St. Louis, MI, USA), polyethylene glycol (Sigma-Aldrich), potassium oxonate (Sigma-Aldrich), uric acid (Sigma-Aldrich), Tween 80 U.S.P (Synth®, Diadema, Brazil), distilled water, saline solution (Sequiplex, Aparecida de Goiânia, Brazil), dimethyl sulfoxide (DMSO; Vetec, Duque de Caxias, Brazil), hydrochloric acid PA (Chomoline, Diadema, Brazil), monobasic potassium phosphate (Vetec), dibasic sodium phosphate (Vetec), dibasic sodium phosphate hepatic hydrate (Vetec), phosphoric acid 85% PA (Vetec), Brilliant Blue G-250 (Sigma-Aldrich), anaesthetics (injectable Dopalen; Ceva Santé Animale, Libourne, France), xylazine (injectable Dopaser; Hertape Calier, Juatuba, Brazil), benzbromarone (Sigma-Aldrich), probenecid (Sigma-Aldrich) and allopurinol (Sigma-Aldrich). For the determination of uric acid, the monoreagent uric acid kit was supplied by Bioclin (Belo Horizonte, Brazil). One-millilitre syringes (BD Plastipak, Jersey City, NJ,

USA) and hypodermic needles (BD Precision Glide, Jersey City, NJ, USA) were used to administer solutions, treatments and anaesthesia to rats. Reverse-phase silica gel (KP-C18-HS Biotage®,Uppsala, Sweden), Phenomenex, Luna® C18, Aschaffenburg, Deutschland (5 μm, 250 × 4.6 mm), Shimadzu Shim, Canby, OR, USA, pack XR-ODS III C18 (2.2 μm; 2 × 150 mm), Millex® filtration device (0.45 μm) (Merk, Darmstadt, Germany), Millipore polytetrafluoroethylene (Merk, Darmstadt, Germany) membrane (PTFE, 0.45 μm), Milli-Q® (Millipore®) filtration system (Merk, Darmstadt, Germany), plates and CCD 10 × 10 Biotage ® KP-SIL were used (Biotage®, Uppsala, Sweden).

Instruments: Shimadzu® and Quimis® balances, M-560 Buchi® melting-point meter, R-210 rotary evaporator (Buchi®, Switzerland), MARCONI® (Piracicaba, SP, Brazil) greenhouse model MA035, MARCONI® knife mill, ultraviolet (UV) light at 254/366 nm (Blak-Ray), Bruker Avance III 400 MHz spectrometer (Madison, WI, USA), Waters® model 2695 high-performance liquid chromatograph, Waters 2996 diode array detector (DAD), UHPLC-ESI-QTOF system (Oswaldo Cruz Foundation, René Rachou Research Center), Cary® 50 Bio Varian UV-vis spectrophotometer, Australia—quartz cubes (10 × 10 mm, m; 10 × 10 mm, 600 μL) were used in the absorbance readings. Additionally used were Sigma Laboratory Centrifuges® 3K30 centrifuge, Germany, and a Microcen® 16 centrifuge from Herolab. The biological materials collected were stored in the Brastemp® Flex freezer at –20 °C or in the freezer at –80 °C. Unique ultrasonic, Shimadzu and Quimis scales, Motion® II vortex (Logen Scientific) and Digimed DM20 pH meter were used.

3.2. Plant Material

Seeds of *E. sativa* Miller of the variety Selecta (Broad Leaf) were obtained from the company Horticeres. The cultivation was carried out in a hydroponic system in semi-covered non-acclimatised greenhouses in the municipality of Mario Campos, MG, Brazil. Supplementation was made using fertilizers based on N, K, P and Fe. Collection of plant material was made on 18 June 2018. Leaves of the rocket were washed with water, dried in an oven with air circulation at 38 °C and pulverised in a knife mill.

3.3. Preparation of the Ethanolic Extract (EE) from E. sativa Leaves

Dried and pulverised leaves (370.0 g) were extracted by continuous percolation with ethanol until the plant material was depleted. The solvent was eliminated in a rotary evaporator and the material kept under vacuum to obtain 66.0 g of crude dry extract.

3.4. HPLC Analyses

3.4.1. Qualitative HPLC-DAD Analysis

The method was developed in order to obtain a chromatographic profile that would detect the greatest possible number of peaks with better resolution of the chromatographic signal baseline; the purity of the peaks was also considered. The method used by Martínez-Sánchez et al. [13] was taken as a starting point for the development of the analytical condition for the fingerprint of the EE. Analyses were conducted using a Phenomenex column, Luna® C18 (5 μm, 250 × 4.6 mm). The analysis temperature was set to 30 °C, flow rate at 1.0 mL/min and volume injection of 20 μL in all injections. Several conditions were tested to select the mobile phase (gradient elution with different proportions of water/methanol, water/acetonitrile, and acidified water/acidified acetonitrile). Thus, the mobile phase was composed of solvent A (ultrapure water/formic acid 0.1%) and solvent B (acetonitrile/formic acid 0.1%). The HPLC gradient was started with 100% of solvent A, reaching 80% A at 10.0 min, 50% A at 25.0 min and 0.0% A at 40.0 min, finally returning to the initial conditions at 45.0 min, which were maintained until 50 min. The detector was adjusted to read from 200–400 nm. Empower® software was used for data processing and all chromatograms were extracted at 264 nm.

EE diluted in methanol was analysed by HPLC-DAD and the absorption spectra of its major peaks were extracted and analysed. In addition, a kaempferol standard was injected separately under the same conditions at a concentration of 0.5 mg/mL.

3.4.2. UHPLC-ESI-QTOF Analysis

The method used was adapted from Martínez-Sánchez et al. [13]. EE (10 mg) was diluted in 500 µL of DMSO. Then, 100 µL of this solution was diluted in 300 µL of methanol. The sample injection volume was 5 µL. The analysis was conducted in an ultra-efficient liquid chromatograph using a column (2.2 µm; 2 × 150 mm; C18) coupled to an ESI-Q-q-TOF mass detector adjusted in positive ion mode. Ionisation was conducted via electrospray ionisation (ESI) and fragmentation was by collision-induced dissociation (CID). The detection range was set from 100 to 1500 m/z. The nebulisation was carried out at 3 Bar, the drying temperature was 200 °C and the capillary voltage was adjusted to 4500 V. The mobile phase was composed of solvent A (ultrapure water/formic acid 0.1%) and solvent B (acetonitrile/formic acid 0.1%). The chromatographic run started with 0% of solvent B, reaching 10% B in 20 min, 50% B in 25 min and finally reaching 100% B in 40 min. The proportion of 100% B was maintained for 5 min, returning to initial conditions in 1 min and maintained for another 9 min. The oven temperature was 40 °C during analysis and the flow of 400 µL/min was constant. The compounds were identified by comparing the results obtained with the available literature and mass spectra libraries.

3.5. Isolation of the Major Flavonoid in EE

Fractionation of EE was conducted using the method proposed by Nazif et al. [36] with adaptations. EE (55.0 g) was dissolved in distilled water at 60 °C. The solution was kept refrigerated overnight and then filtered on paper resulting in an insoluble residue (IE; 21.678 g) and an aqueous filtrate (Aq1). Aq1 was subjected to liquid–liquid partition with chloroform followed by ethyl acetate. Each solvent extraction was evaporated separately in a rotary evaporator in vacuo furnishing fractions EECl (0.349 g) and EEaE (0.273 g), respectively. The second aqueous fraction (Aq2) was filtered to yield an inorganic solid IS (7.78 g) and a filtrate (Aq3). Aq3 was submitted to open-column chromatography packed with reverse-phase (C18) silica gel; the elution was performed by using decreasing proportions of distilled water and methanol, starting with 100% water. Fractionation was monitored by silica gel TLC and revelation with NP-PEG [37]. Fraction (C1) eluted with water/methanol (75:25 *v/v*) was revealed to have an intense orange fluorescence spot when sprayed with NP-PEG. Subsequent chromatographic fractionations were performed employing the same method to yield a yellow powder (YP, 15 mg, flavonoid isolated from EE) with melting range at 217.3–218.9 °C. A DMSO-d6 solution of YP was analysed by NMR spectroscopy by ^{1}H NMR, ^{13}C NMR, COSY, HSQC and HMBC and the structure was confirmed to be kaempferol-3,4′-di-*O*-β-glucoside.

3.6. Validation of a Quantitative Method of the Major Flavonoid in EE by HPLC-DAD

Validation was carried out using parameters based on RDC 166/17 and international legislation [38,39]. Linearity, precision, intermediate precision and accuracy tests were conducted and are summarised in Table 3.

Table 3. Linearity, repeatability, intermediate precision and accuracy of the HPLC-DAD quantification method of kaempferol-3,4′-di-*O*-β-glucoside in the ethanolic extract of *E. sativa* on days 1 and 2.

Day	Linearity (*n* = 3)	Concentration Level	RSD % (*n* = 3)	Accuracy % (*n* = 3)
1	Y = 3.364e + 0.007X − 44233 R^2 = 0.9973	Lowest Medium Highest	1.0 3.5 1.7	102.7–99.0 100.6–94.5 98.5–96.9
2	Y = 3.422e + 0.007X − 53351 R^2 = 0.9946	Lowest Medium Highest	5.4 6.9 2.4	101.0–92.7 104.0–93.5 99.4–96.9

Firstly, selectivity was evaluated considering the degradation of products in alkali and extreme acid conditions (*n* = 3). EE methanolic solutions (4 mg/mL) were added to an equal volume of a 1.0 mol/L HCl solution and kept at room temperature. After 21 h, the solutions were neutralised with 0.5 mol/L NaOH to produce the final concentration of 1 mg/mL of extract. For the alkali condition, EE methanolic solutions were submitted to 1.0 mol/L NaOH solutions for 18 h and neutralised with 0.5 mol/L HCl solutions. The percentage peak degradation of kaempferol-3,4′-di-*O*-β-glucoside was 11.5 ± 0.019% in acid conditions, whereas in alkali conditions 46.6 ± 8.735% degradation was found. Peak purity analyses of the chromatographic signal were evaluated by Empower® software. In both conditions, the kaempferol-3,4′-di-*O*-β-glucoside peak was pure in the range of 200–400 nm.

Due to a lack of the reference standard (kaempferol-3,4′-di-*O*-β-glucoside) to validate a quantitative method, a standard of similar UV absorption characteristics (rutin) was chosen. EE solutions in methanol (1 mg/mL) containing increasing concentrations of rutin (0.01, 0.02, 0.03, 0.04 and 0.05 mg/mL) were prepared in triplicate on two different days. Resolution of kaempferol-3,4′-di-*O*-β-glucoside and rutin peaks were adequate and symmetry of both peaks was less than 1.5. The peak assigned to rutin was pure in all runs. Using Prisma® software, statistical treatments were applied to the obtained results in order to evaluate linearity. In both days, the average curve generated presented r > 0.99 (two-tailed test). On day 1, the curve obtained was y = 3.364e + 0.007x − 44,233 (R^2 = 0.9973) and on day 2, the curve was y = 3.422e + 0.007x − 53,351 (R^2 = 0.9946). The residue analysis followed a random distribution and confirmed that the variance of the experimental errors was homoscedastic. No outliers were evidenced. Experimental errors analysis made by Shapiro–Wilk test confirmed the normality of the experimental errors on the two days (Day: $p < 1$–0.88; Day 2: $p < 0.94$). Dependency error analysis (0.8182 < 1000) carried out on Prisma® confirmed the independence of the results. Three concentrations in the curves of each day were selected (lowest, medium, and highest) to evaluate precision (repeatability and intermediary precision) and accuracy. Precision was expressed by relative standard deviation (RSD) of the response and the accuracy as the percentage deviation from the nominal concentration. In the first day of analysis, RSD of the responses were 1.0% (lowest concentration), 3.5% (medium concentration) and 1.7% (highest concentration). On the second day, RSD of responses were 5.4, 6.9 and 2.4%. The limit of acceptance for precision was considered to be 7.3%, according to the analytical quality manual recommended by the Ministério da Agricultura e Pecuária [40]. In all concentrations, deviations from the nominal concentration were <3%. The limits of detection (LD) and quantification (LQ) were obtained from the ratio of the standard deviation of the intercept with the *y*-axis (n = 6) with the slope of the average analytical curve obtained. This ratio was multiplied by 3 to obtain LD (0.0019 mg/mL) and by 10 to obtain LQ (0.0059 mg/mL). Ultimately, the validated method was selective, linear, precise and accurate. With a 95% confidence interval, it was seen that there were no statistical differences between the slope coefficients of the curves generated on the two days and their intersections with the *y*-axis. Thus, the mean analytical curve of days 1 and 2, y = 3.39333e + 0.007x − 48791.8, was assumed for all data.

3.7. Quantification of Kaempferol-3,4′-di-O-β-glucoside by HPLC-DAD

Three stock solutions of EE at 4 mg/mL in methanol were prepared. From these, solutions with a theoretical concentration of 2 mg/mL in methanol were obtained. Kaempferol-3,4′-di-*O*-β-glucoside present in the solutions was quantified using the HPLC-DAD method, developed and validated. The mean analytical curve obtained was used to determine the concentration in the sample solutions. Results were then expressed in g of rutin equivalent/kg EE.

3.8. Anti-Hyperuricaemic Assay

3.8.1. Animals

In vivo experiments were conducted in adult and male Wistar rats weighing 180–280 g and provided by the Central Bioterium of UFOP. The experimental protocol was approved by the UFOP Animal Use Ethics Committee (CEUA-UFOP, protocol number 7504230419) and were in accord with the NIH guide for the care and use of laboratory animals published by the US National Institute of Health [41]. The 12 h/12 h light/dark cycle was maintained ad libitum.

3.8.2. In Vivo Hyperuricaemic Model

To evaluate the anti-hyperuricaemic activity, the model described by Murugaiyah and Chan [42], Ferrari et al. [43] and Bernardes et al. [44] was used. Animals were initially deprived of water and food for 12 h. Hyperuricaemia was achieved by treating animals with a single dose of potassium oxonate (200 mg/kg, i.p.) and uric acid (1 g/kg, i.g.). Thirty minutes after hyperuricaemia induction, treatments [EE, doses of 10, 40 and 125 mg/kg diluted in vehicle DMSO/Tween 80/water (1:1:8)] and drugs of reference [allopurinol 10 mg/kg, benzbromarone 10 mg/kg and probenecid 50 mg/kg diluted in the vehicle ethanol/Tween 80/water (10:20:70)] were administered. Each group was formed by six animals. The normal control group (with normal uricemia) did not receive potassium oxonate and uric acid. Instead, it was given saline intraperitoneally and vehicle orally. The hyperuricaemic control (model control) received only the vehicle used to dilute treatments after hyperuricaemia induction. After administration of treatments, the animals were placed in individual metabolic cages and received 100 mL of water ad libitum. Urine was collected in graduated tubes and water consumption measured for 5 h. The collected urine was used for the quantification of uric acid. At the end of the experiment, the animals were anaesthetised with a combination of ketamine and xylazine (240 and 60 mg/kg, respectively), intraperitoneally. Blood samples were collected from the abdominal aorta and a thoracoabdominal laparotomy was carried out. The blood was kept at room temperature until coagulation and then centrifuged at $3000 \times g$ for 15 min. The resulting supernatant was again centrifuged at $3000 \times g$ for 15 min to obtain the serum. Serum and collected urine were stored at $-20\ °C$ for further analysis. The livers (3.0 g) were removed, weighed and stored at $-80\ °C$ for later preparation of the homogenates.

3.8.3. Uric Acid Assay

Serum and urine samples had their uric acid measured by a colorimetric technique using a kit (Bioclin, Brazil), following the manufacturer's instructions.

3.8.4. Liver Homogenate Preparation

Livers were thawed and crushed in a 50 mM phosphate buffer solution (5 mL). During crushing, the samples were kept on ice. The crushed livers were centrifuged at $3000 \times g$ at $4\ °C$ for 15 min. The upper lipid layer was discarded, and the intermediate supernatants collected and centrifuged at $10,000 \times g$ at $4\ °C$ for 1 h. The intermediate supernatant was collected to obtain the homogenates used for the quantification of total proteins and for the evaluation of the residual activity of xanthine oxidase.

3.8.5. Total Protein Assay

Measurement of total proteins was performed using the Bradford method [45]. An analytical curve using increasing masses of albumin was constructed; the equation of the line obtained was y = 0.0064x + 0.0352; R^2 = 0.992. A 50 mM phosphate buffer solution was used to zero the equipment. The blank was prepared by adding 100 µL of phosphate buffer to 5 mL of Bradford's reagent, followed by homogenisation in a vortex apparatus. The samples were prepared by adding 100 µL of homogenate to 5 mL of Bradford's reagent and then homogenizing in a vortex apparatus. The absorbances were recorded at 595 nm on a spectrophotometer after 2 min.

3.8.6. Hepatic Xanthine Oxidase Activity Assay

Synthesis of uric acid from xanthine was monitored spectrophotometrically according to the method described by Hall et al. [46]. To prevent the conversion of uric acid to allantoin, 100 µL of each homogenate was added to test tubes with 5.4 mL of 1 mM potassium oxonate solution. The solution was then pre-incubated at $37\ °C$ for 15 min.

Hence, 1.2 mL of the 250 mM xanthine solution was added and incubated for 30 min at 37 °C. At 0 and 30 min after adding the xanthine, 0.5 mL of HCl solution (0.6 mol/L) was added to stop the reaction. The solutions were centrifuged at 3000 g for 5 min and the supernatants collected to read the absorbance in a spectrophotometer at 295 nm. The blank was adjusted using the 1 mM potassium oxonate solution. The quantification of the uric acid formed was made by the difference between the absorbance values at times 0 and 30 min, which was compared with the analytical curve ($y = 1.6629x - 0.0045$; $R^2 = 0.998$). With the total protein values, it was possible to express the residual activity of xanthine oxidase as nanomoles of uric acid formed per minute per milligram of protein.

3.8.7. Statistical Analysis

Results processing of the animal model was made on GraphPad® Prism software, version 6.01 (USA). Standard deviations among groups did not present a Gaussian distribution when applied the test of Shapiro–Wilk. The variance was performed using One-Way analysis of variance (ANOVA) and means were significantly different among the groups ($p < 0.0001$). Equal variances tests were performed using the Brown–Forsythe test and standard deviations among the groups were not significantly different [47]. Post-tests were performed using the Dunnett test and group pairs were compared using the Tukey test. p-values ≤ 0.05 were considered statistically significant. The results were expressed as means and standard deviation.

4. Conclusions

Analysis carried out on the ethanolic extract of *E. sativa* showed the presence of flavonoids and glucosinolates previously reported in other studies with *E. sativa*. Kaempferol-3,4'-di-*O*-β-glucoside was determined as the main flavonoid in the ethanolic extract, and it was considered a chemical and biological marker for uricosuric activity. Antihyperuricaemic activity of extracts were mainly but not only due to uricosuric action. Other mechanisms are still to be elucidated. Finally, the results obtained so far, associated with data from ethnopharmacological studies, drew attention to the potential of *E. sativa* to control hyperuricaemia and the comorbidities associated with this pathology.

Author Contributions: Conceptualisation, A.F.T., J.d.S. and D.A.S.-G.; methodology, A.F.T.; software, A.F.T. and J.D.d.S.F.; validation, A.F.T.; formal analysis, A.F.T., D.D.D. and J.D.d.S.F.; writing—original draft preparation, A.F.T. and J.D.d.S.F.; writing—review and editing, D.A.S.-G. and J.d.S.; supervision, D.A.S.-G. and J.d.S.; project administration, D.A.S.-G.; funding acquisition, D.A.S.-G. All authors have read and agreed to the published version of the manuscript.

Funding: The authors acknowledge CAPES, CNPq, FAPEMIG (APQ-01160-15 and APQ-00820-19) and UFOP (Processo N.: 23109.003209/201698 and Processo N.: 23109.004079/2019-53) for the financial support.

Institutional Review Board Statement: The study was conducted according to the guidelines of the Declaration of Helsinki, and approved by Ethics Committee of CEUA-UFOP, protocol number 7504230419 approved on 28 June 2019.

Informed Consent Statement: Not applicable.

Data Availability Statement: Not applicable.

Acknowledgments: IKA is kindly acknowledged for the laboratory amelioration and Bioclin for the reagents supplied. We thank the NMR facility—LAREMAR at the Chemistry Department in the Federal University of Minas Gerais, Brazil; Oswaldo Cruz Foundation, René Rachou Research Center, for the UHPLC/ESI/QTOF analyses; and Laboratório Multiusuário de Caracterização de Moléculas (LMCM) da Universidade Federal de Ouro Preto, Minas Gerais, Brazil.

Conflicts of Interest: The authors declare no conflict of interest.

Sample Availability: Samples of the compounds are not available from the authors.

References

1. Gustafsson, D.; Unwin, R. The pathophysiology of hyperuricaemia and its possible relationship to cardiovascular disease, morbidity and mortality. *BMC Nephrol.* **2013**, *14*, 164. [CrossRef] [PubMed]
2. Müller, C.M.D.S.; Coelho, G.B.; de Paula Michel Araújo, M.C.; Saúde-Guimarães, D.A. Lychnophora pinaster ethanolic extract and its chemical constituents ameliorate hyperuricemia and related inflammation. *J. Ethnopharmacol.* **2019**, *242*, 112040. [CrossRef] [PubMed]
3. Amorim, H.C.; Henz, G.P.; Mattos, L.M. *Boletim de Pesquisa e Desenvolvimento. Identificação Dos Tipos De Rúcula Comercializados No Varejo do Distrito Federal*; Embrapa Hortaliças: Brasília, Brazil, 2007; Volume 34, ISBN 1677-2229.
4. Hetta, M.H.; Owis, A.I.; Haddad, P.S.; Eid, H.M. The fatty acid-rich fraction of Eruca sativa (rocket salad) leaf extract exerts antidiabetic effects in cultured skeletal muscle, adipocytes and liver cells. *Pharm. Biol.* **2017**, *55*, 810–818. [CrossRef] [PubMed]
5. Meligi, N.; Hassan, H.F. Protective effects of Eruca sativa (rocket) on abamectin insecticide toxicity in male albino rats. *Environ. Sci. Pollut. Res.* **2017**, *24*, 9702–9712. [CrossRef]
6. Salma, U.; Khan, T.; Shah, A.J. Antihypertensive effect of the methanolic extract from Eruca sativa Mill., (Brassicaceae) in rats: Muscarinic receptor-linked vasorelaxant and cardiotonic effects. *J. Ethnopharmacol.* **2018**, *224*, 409–420. [CrossRef]
7. Jouad, H.; Haloui, M.; Rhiouani, H.; EL Hilaly, J.; Eddouks, M. Ethnobotanical survey of medicinal plants used for the treatment of diabetes, cardiac and renal diseases in the North centre region of Morocco (Fez–Boulemane). *J. Ethnopharmacol.* **2001**, *77*, 175–182. [CrossRef]
8. Mabry, T.J.; Markham, K.R.; Thomas, M.B. *The Systematic Identification of Flavonoids*; Springer: Berlin/Heidelberg, Germany, 1970; ISBN 978-3-642-88460-3.
9. Kiddle, G.; Bennett, R.N.; Botting, N.P.; Davidson, N.E.; Robertson, A.A.B.; Wallsgrove, R.M. High-performance liquid chromatographic separation of natural and synthetic desulphoglucosinolates and their chemical validation by UV, NMR and chemical ionisation-MS methods. *Phytochem. Anal.* **2001**, *12*, 226–242. [CrossRef]
10. Bennett, R.N.; Mellon, F.A.; Botting, N.P.; Eagles, J.; Rosa, E.; Williamson, G. Identification of the major glucosinolate (4-mercaptobutyl glucosinolate) in leaves of Eruca sativa L. (salad rocket). *Phytochemistry* **2002**, *61*, 25–30. [CrossRef]
11. Cataldi, T.R.I.; Rubino, A.; Lelario, F.; Bufo, S.A. Naturally occurring glucosinolates in plant extracts of rocket salad (*Eruca sativa* L.) identified by liquid chromatography coupled with negative ion electrospray ionization and quadrupole ion-trap mass spectrometry. *Rapid Commun. Mass Spectrom.* **2007**, *21*, 2374–2388. [CrossRef]
12. Pasini, F.; Verardo, V.; Caboni, M.F.; D'Antuono, L.F. Determination of glucosinolates and phenolic compounds in rocket salad by HPLC-DAD–MS: Evaluation of Eruca sativa Mill. and Diplotaxis tenuifolia L. genetic resources. *Food Chem.* **2012**, *133*, 1025–1033. [CrossRef]
13. Martínez-Sánchez, A.; Llorach, R.; Gil, M.I.; Ferreres, F. Identification of New Flavonoid Glycosides and Flavonoid Profiles To Characterize Rocket Leafy Salads (Eruca vesicaria and Diplotaxis tenuifolia). *J. Agric. Food Chem.* **2007**, *55*, 1356–1363. [CrossRef] [PubMed]
14. Cuyckens, F.; Shahat, A.A.; van den Heuvel, H.; Abdel-Shafeek, K.A.; El-Messiry, M.M.; Nasr, M.M.S.-E.; Pieters, L.; Vlietinck, A.J.; Claeys, M. The Application of Liquid Chromatography-Electrospray Ionization Mass Spectrometry and Collision-Induced Dissociation in the Structural Characterization of Acylated Flavonol O-Glycosides from the Seeds of Carrichtera Annua. *Eur. J. Mass Spectrom.* **2003**, *9*, 409–420. [CrossRef]
15. Cavaliere, C.; Foglia, P.; Pastorini, E.; Samperi, R.; Laganà, A. Identification and mass spectrometric characterization of glycosylated flavonoids inTriticum durum plants by high-performance liquid chromatography with tandem mass spectrometry. *Rapid Commun. Mass Spectrom.* **2005**, *19*, 3143–3158. [CrossRef]
16. Bell, L.; Methven, L.; Signore, A.; Concha, M.J.O.; Wagstaff, C. Analysis of seven salad rocket (Eruca sativa) accessions: The relationships between sensory attributes and volatile and non-volatile compounds. *Food Chem.* **2016**, *218*, 181–191. [CrossRef]
17. Mithen, R.F.; Dekker, M.; Verkerk, R.; Rabot, S.; Johnson, I.T. The Nutritional Significance, Biosynthesis and Bioavailability of Glucosinolates in Human Foods. *J. Sci. Food Agric.* **2000**, *80*, 967–984. [CrossRef]
18. Orhan, I.E.; Kartal, M.; Sekeroglu, N.; Esiyok, D.; Şener, B.; Ugur, A.; Süntar, I.; Aslan, S. Variations in fatty acid compositions of the seed oil of Eruca sativa Mill. caused by different sowing periods and nitrogen forms. *Pharmacogn. Mag.* **2010**, *6*, 305–308. [CrossRef]
19. Martínez-Sánchez, A.; Gil-Izquierdo, A.; Gil, M.I.; Ferreres, F. A Comparative Study of Flavonoid Compounds, Vitamin C, and Antioxidant Properties of Baby LeafBrassicaceaeSpecies. *J. Agric. Food Chem.* **2008**, *56*, 2330–2340. [CrossRef] [PubMed]
20. Bell, L.; Concha, M.J.O.; Wagstaff, C. Identification and quantification of glucosinolate and flavonol compounds in rocket salad (Eruca sativa, Eruca vesicaria and Diplotaxis tenuifolia) by LC–MS: Highlighting the potential for improving nutritional value of rocket crops. *Food Chem.* **2014**, *172*, 852–861. [CrossRef] [PubMed]
21. Pacher, P.; Nivorozhkin, A.; Szabó, C. Therapeutic Effects of Xanthine Oxidase Inhibitors: Renaissance Half a Century after the Discovery of Allopurinol. *Pharmacol. Rev.* **2006**, *58*, 87–114. [CrossRef]
22. Perez-Ruiz, F.; Hernandez-Baldizon, S.; Herrero-Beites, A.M.; González-Gay, M.A. Risk factors associated with renal lithiasis during uricosuric treatment of hyperuricemia in patients with Gout. *Arthritis Care Res.* **2010**, *62*, 1299–1305. [CrossRef]
23. Selloum, L.; Reichlb, S.; Müllerb, M.; Sebihi, L.; Arnhold, J. Effects of Flavonols on the Generation of Superoxide Anion Radicals by Xanthine Oxidase and Stimulated Neutrophils. *Arch. Biochem. Biophys.* **2001**, *395*, 49–56. [CrossRef]

24. Haidari, F.; Keshavarz, S.A.; Shahi, M.M.; Mahboob, S.-A.; Rashidi, M.-R. Effects of Parsley (Petroselinum crispum) and its Flavonol Constituents, Kaempferol and Quercetin, on Serum Uric Acid Levels, Biomarkers of Oxidative Stress and Liver Xanthine Oxidoreductase Aactivity in Oxonate-Induced Hyperuricemic Rats. *Iran. J. Pharm. Res.* **2011**, *10*, 811–819. [CrossRef] [PubMed]
25. Zhu, J.X.; Wang, Y.; Kong, L.D.; Yang, C.; Zhang, X. Effects of Biota orientalis extract and its flavonoid constituents, quercetin and rutin on serum uric acid levels in oxonate-induced mice and xanthine dehydrogenase and xanthine oxidase activities in mouse liver. *J. Ethnopharmacol.* **2004**, *93*, 133–140. [CrossRef] [PubMed]
26. Boyle, S.P.; Dobson, V.L.; Duthie, S.; Hinselwood, D.C.; Kyle, J.; Collins, A. Bioavailability and efficiency of rutin as an antioxidant: A human supplementation study. *Eur. J. Clin. Nutr.* **2000**, *54*, 774–782. [CrossRef] [PubMed]
27. Yuan, L.; Bao, Z.; Ma, T.; Lin, S. Hypouricemia effects of corn silk flavonoids in a mouse model of potassium oxonated-induced hyperuricemia. *J. Food Biochem.* **2021**, *45*, e13856. [CrossRef]
28. Kang, D.-H.; Park, S.-K.; Lee, I.-K.; Johnson, R. Uric Acid—Induced C-Reactive Protein Expression: Implication on Cell Proliferation and Nitric Oxide Production of Human Vascular Cells. *J. Am. Soc. Nephrol.* **2005**, *16*, 3553–3562. [CrossRef]
29. Mazzali, M.; Hughes, J.; Kim, Y.-G.; Jefferson, J.A.; Kang, D.-H.; Gordon, K.L.; Lan, H.Y.; Kivlighn, S.; Johnson, R. Elevated Uric Acid Increases Blood Pressure in the Rat by a Novel Crystal-Independent Mechanism. *Hypertension* **2001**, *38*, 1101–1106. [CrossRef]
30. Li, C.; Hsieh, M.-C.; Chang, S.-J. Metabolic syndrome, diabetes, and hyperuricemia. *Curr. Opin. Rheumatol.* **2013**, *25*, 210–216. [CrossRef]
31. Anzai, N.; Ichida, K.; Jutabha, P.; Kimura, T.; Babu, E.; Jin, C.J.; Srivastava, S.; Kitamura, K.; Hisatome, I.; Endou, H.; et al. Plasma Urate Level Is Directly Regulated by a Voltage-driven Urate Efflux Transporter URATv1 (SLC2A9) in Humans. *J. Biol. Chem.* **2008**, *283*, 26834–26838. [CrossRef]
32. Bantia, S.; Parker, C.; Harman, L.; Papac, D.; Hollister, A. OP0106 Effect of BCX4208 add-on therapy to allopurinol 300 mg on plasma hypoxanthine and xanthine concentrations in gout patients. *Ann. Rheum. Dis.* **2013**, *71*, 89. [CrossRef]
33. Hosomi, A.; Nakanishi, T.; Fujita, T.; Tamai, I. Extra-Renal Elimination of Uric Acid via Intestinal Efflux Transporter BCRP/ABCG2. *PLoS ONE* **2012**, *7*, e30456. [CrossRef]
34. Szczurek, P.; Mosiichuk, N.; Woliński, J.; Yatsenko, T.; Grujic, D.; Lozinska, L.; Pieszka, M.; Święch, E.; Pierzynowski, S.G.; Goncharova, K. Oral uricase eliminates blood uric acid in the hyperuricemic pig model. *PLoS ONE* **2017**, *12*, e0179195. [CrossRef]
35. Al-Helaly, L.; Mohammad, F. Uricase Isolated from Seeds of Leek (Allium Ampeloprasum), Celery (Apium Graveolens) and Arugula (Eruca sativa). *Iraqi Natl. J. Chem.* **2014**, *55*, 340–356.
36. Nazif, N.M.; Amira, A.E.H.; Wafaa, A.M.T.; Rasmia, A.H. Chemical Composition and Cytotoxic Activity of Eruca sativa L. Seeds Cultivated in Egypt. *Asian J. Chem.* **2010**, *22*, 2407–2416.
37. Wagner, H.; Bladt, S. *Plant Drug Analysis: A Thin Layer Chromatography Atlas*, 2nd ed.; Springer: Berlin/Heidelberg, Germany, 2009; ISBN 978-3-642-00574-9.
38. Brasil RDC N. 166, de 25 de Julho de 2017. Agência Nacional de Vigilância Sanitária. 2017; pp. 1–21. Available online: http://antigo.anvisa.gov.br/documents/10181/2721567/RDC_166_2017_COMP.pdf/d5fb92b3-6c6b-4130-8670-4e3263763401 (accessed on 31 January 2022).
39. Validation of Analytical Procedures Text and Methodoloy Q2 (R1). ICH Harmonised Tripartite Guideline. In Proceedings of the International Conference on Harmonisation of Technical Requirements for Registration of Pharmaceuticals for Human Use, 1995. Available online: https://www.ich.org/ (accessed on 31 January 2022).
40. *Brasil Manual de Garantia da Qualidade Analítica*, 1st ed.; Ministério da Agricultura, Pecuária e Abastecimento: Brasilia, Brazil, 2011; pp. 40–42.
41. National Institutes of Health, Public Health Service. *Guide for the Care and Use of Laboratory Animals*; NIH Publication: Bethesda, MD, USA, 1985; pp. 23–86.
42. Murugaiyah, V.; Chan, K.-L. Mechanisms of antihyperuricemic effect of Phyllanthus niruri and its lignan constituents. *J. Ethnopharmacol.* **2009**, *124*, 233–239. [CrossRef]
43. Ferrari, F.C.; Lima, R.D.C.L.; Filha, Z.S.F.; Barros, C.H.; Araújo, M.C.D.P.M.; Saúde-Guimarães, D.A. Effects of Pimenta pseudo-caryophyllus extracts on gout: Anti-inflammatory activity and anti-hyperuricemic effect through xantine oxidase and uricosuric action. *J. Ethnopharmacol.* **2016**, *180*, 37–42. [CrossRef] [PubMed]
44. Bernardes, A.C.F.P.F.; Coelho, G.B.; Araújo, M.C.D.P.M.; Saúde-Guimarães, D.A. In vivo anti-hyperuricemic activity of sesquiterpene lactones from Lychnophora species. *Rev. Bras. Farm.* **2019**, *29*, 241–245. [CrossRef]
45. Bradford, M.A. Rapid and Sensitive Method for the Quantitation of Microgram Quantities of Protein Utilizing the Principle of Protein-Dye Binding. *Anal. Biochem.* **1976**, *72*, 248–254. [CrossRef]
46. Hall, I.H.; Scoville, J.P.; Reynolds, D.J.; Simlot, R.; Duncan, P. Substituted cyclic imides as potential anti-gout agents. *Life Sci.* **1990**, *46*, 1923–1927. [CrossRef]
47. De Almeida, A.; Elian, S.; Nobre, J. Modifications and Alternatives to the Tests of Levene and Brown & Forsythe for Equality of Variances and Means. *Rev. Colomb. Estad.* **2008**, *31*, 241–260.

Article

Seasonal Chemical Evaluation of *Miconia chamissois* Naudin from Brazilian Savanna

Juliana de Freitas Ferreira [1], Manuel Humberto Mera López [1], João Victor Dutra Gomes [1], Diegue H. Nascimento Martins [1], Christopher William Fagg [2], Pérola Oliveira Magalhães [1], Noel William Davies [3], Dâmaris Silveira [1] and Yris Maria Fonseca-Bazzo [1,*]

1 Department of Pharmacy, Health Sciences School, University of Brasília, Brasilia 70910-900, Brazil; julianafferreiraunb@gmail.com (J.d.F.F.); manuel.humberto@aluno.unb.br (M.H.M.L.); dutra.joaovictor@gmail.com (J.V.D.G.); diegue.hen@gmail.com (D.H.N.M.); perolamagalhaes@unb.br (P.O.M.); damaris@unb.br (D.S.)
2 Department of Botany, Institute of Biological Science, Ceilândia Campus, School of Pharmacy, University of Brasília, Brasilia 70910-900, Brazil; acaciafagg@gmail.com
3 Central Science Laboratory, University of Tasmania, Hobart, TAS 7005, Australia; noel.davies@utas.edu.au
* Correspondence: yrisfonseca@unb.br

Abstract: *Miconia chamissois* Naudin is a species from the Cerrado, which is being increasingly researched for its therapeutic potential. The aim of this study was to obtain a standardized extract and to evaluate seasonal chemical variations. Seven batches of aqueous extracts from leaves were produced for the standardization. These extracts were evaluated for total solids, polyphenol (TPC) and flavonoid content (TFC), vitexin derivative content, antioxidant activity; thin-layer chromatography (TLC), and high-performance liquid chromatography (HPLC) profiles were generated. For the seasonal study, leaves were collected from five different periods (May 2017 to August 2018). The results were correlated with meteorological data (global radiation, temperature, and rainfall index). Using chromatographic and spectroscopic techniques, apigenin C-glycosides (vitexin/isovitexin) and derivatives, luteolin C-glycosides (orientin/isoorientin) and derivatives, a quercetin glycoside, miconioside B, matteucinol-7-O-β-apiofuranosyl (1 → 6) -β-glucopyranoside, and farrerol were identified. Quality parameters, including chemical marker quantification by HPLC, and biological activity, are described. In the extract standardization process, all the evaluated parameters showed low variability. The seasonality study revealed no significant correlations ($p < 0.05$) between TPC or TFC content and meteorological data. These results showed that it is possible to obtain extracts from *M. chamissois* at any time of the year without significant differences in composition.

Keywords: *Miconia chamissois* Naudin; seasonality; standardized extract; Cerrado

Citation: Ferreira, J.d.F.; López, M.H.M.; Gomes, J.V.D.; Martins, D.H.N.; Fagg, C.W.; Magalhães, P.O.; Davies, N.W.; Silveira, D.; Fonseca-Bazzo, Y.M. Seasonal Chemical Evaluation of *Miconia chamissois* Naudin from Brazilian Savanna. *Molecules* **2022**, *27*, 1120. https://doi.org/10.3390/molecules27031120

Academic Editor: Jacqueline Aparecida Takahashi

Received: 29 December 2021
Accepted: 1 February 2022
Published: 8 February 2022

Publisher's Note: MDPI stays neutral with regard to jurisdictional claims in published maps and institutional affiliations.

1. Introduction

Cerrado, with a predominantly dry and hot climate, is recognized as the richest savanna in the world, home to 11,627 species of native plants already cataloged [1]. Cerrado comprises several phytophysiognomies and has a prevalence of some botanical families with commercial, cultural, and social importance (food, ethnobotany, religious, and others), as well as the potential for ecological restoration [2].

The richness of the Cerrado is such that the range and potential of bioactive compounds produced by Cerrado species can be considered greater than those of the Amazon Forest, representing an interesting field of investigation with medicinal plants and conservation of natural resources [3].

Melastomataceae is the sixth most abundant botanical family of angiosperms in Brazil, comprising more than 1300 species, with *Miconia*, *Leandra*, *Tibouchina*, *Microlicia*, and *Clidemia* among the most diverse genera. In all of Brazil, 267 species from the *Miconia* genus have been described and are distributed in these phytogeographic domains: Amazon,

Caatinga, Cerrado, Atlantic Forest, Pampa, and Pantanal [4]. In the Cerrado, which principally covers the Midwest region of Brazil, 68 *Miconia* species have already been identified [2]. In particular, *Miconia chamissois* Naudin is a native species from Brazil found in the Caatinga, Cerrado, and Atlantic Forest [4].

Some species from Melastomataceae are used in traditional medicine as *Miconia cinnmonifolia* (DC.) Naudin and *Miconia albicans* (Sw.) Triana. *Miconia cinnmonifolia* (DC.) Naudin is used in folk medicine for the treatment of cold/fever and rheumatism. *Miconia albicans* (Sw.) Triana. is used for eupepsia. The *Tibouchina* genus is used therapeutically for pain relief. *Ossaea quinquenervia* (Mill.) Cogn. is also used in the treatment of symptomatic malarial fever by indigenous tribes of Panama [5,6].

Miconia chamissois Naudin is popularly known as "Folha de Bolo," "Sabiazeira," or "Pixirica" and has been used as food by inhabitants of Mato Grosso do Sul, Brazil [7,8]. The main phytochemical constituents of this species are anthraquinones, triterpenoids, saponins, tannins, steroids, flavonoids (rutin, isoquercitrin, and vitexin), alkaloids, and coumarins [9–11]. Recently, Gimenez et al. (2020) reported the presence of miconioside B, matteucinol 7-*O*-β-apiofuranosyl (1→6)-β-glucopyranoside, ursolic acid, and oleanolic acid on the chloroform partition and its sub-fraction of *M. chamissois* leaf extracts [7]. Many of these compounds are used therapeutically to treat inflammatory and neurodegenerative diseases, cancer, and other pathological processes that involve the presence of free radicals [12].

Some studies have shown the biological potential of *M. chamissois* Naudin extracts, such as antioxidant activity, *in vitro* enzyme inhibition activity against tyrosinase and alpha-amylase, antimicrobial activity [11], *in vitro* inhibition of MMP-2 and MMP-9 activities [10], and antineoplastic potential in human cervical cancer cell lines [13]. Silva et al. (2020) also demonstrated that matteucinol, isolated from *M. chamissois* exhibited selective cytotoxic against glioblastoma cell lines [14].

There has been an increase in research aimed at discovering pharmacologically active native plant species. The few studies published on the biological activity and chemical composition of *M. chamissois* have supported its therapeutic importance.

The secondary metabolism of plants is affected by temperature, ultraviolet radiation, rain index, soil nutrients, altitude, and other environmental factors [15]. *Miconia chamissois* is widely distributed throughout the Brazilian territory and therefore, is exposed to different environmental conditions.

Due to the complexity of plant species and chemical variation through seasonality, chemical elucidation and the definition of chemical markers are important for assessing the quality of the final product and ensure patient safety. Moreover, standardization of extracting extracts can reduce variability [16,17]. Thus, this study aimed to obtain a standardized extract and to evaluate the seasonal variability of *M. chamissois*.

2. Materials and Methods

2.1. Plant Materials

Miconia chamissois Naudin leaves were collected from Lago do Cedro, Brasília, Federal District, Brazil (coordinates 15°53′48.0″ S, 47°56′36.1″ W). A sample was deposited at the University of Brasília herbarium (UB) under the exsiccate number CW Fagg 2358. To evaluate seasonal chemical variability, leaves were collected in May 2017 (autumn), November 2017 (spring), February 2018 (summer), May 2018 (autumn), and August 2018 (winter).

Access to genetic heritage was approved by the Genetic Heritage Management Council (CGEN), registered at National System for the Management of Genetic Heritage and Associated Traditional Knowledge (SISGEN) under the number A215A9A.

2.2. Extraction Process

The leaves were dried at 37 °C in a circulating air oven for 30 h. The moisture content at the end of this process was 10.10% ± 0.02. The dried leaves were powdered in a knife mill to obtain a powder with 30 mesh (0.59 mm) using a sieve integrated into the mill. This powder was extracted in water by infusion at 70 °C to 50 °C at a ratio of 1:10. The aqueous extract was then lyophilized (VirTis SP Scientific Advantage Plus XL-70 Benchtop Freeze Dryer) and stored at −20 °C.

The yield of aqueous extract from *Miconia chamissois* Naudin (AEMC) was calculated as the weight percentage of the dried powdered leaves.

The total solids content was determined using an infrared moisture detector (Gehaka® model IV2000, São Paulo, Brazil) from 2 mL of the sample. The analysis was performed in triplicate, and the results are expressed as the weight percentage of the dried leaf material.

2.3. Extraction Process Standardization

To evaluate the standardization and reproducibility of the extraction process, seven equal batches of aqueous extracts from leaves were prepared (B1–B7). Leaves collected in May 2017 at Lago do Cedro, Brasília, Federal District, Brazil (coordinates 15°53′48.0″ S, 47°56′36.1″ W) were used to prepare the seven batches. Total solids, polyphenol and flavonoid contents, TLC and HPLC profiles, vitexin derivate content, and antioxidant activity were evaluated from these batches.

2.4. Chemical Composition

2.4.1. TLC Assay

Thin-layer chromatography was performed using silica gel (Sobernt Technologies®, 200 μm, 20 × 20 cm). The eluent solutions were ethyl acetate, formic acid, acetic acid, and deionized water (100:11:11:26). Chromatographic spots were visualized using both physical and chemical protocols (ultraviolet lamp at 254 nm and natural product/polyethylene glycol (NP/PEG) reagent (2% diphenylboryloxyethylamine methanolic solution—solution A and 5% polyethylene glycol ethanolic solution 4000—solution B) was used as the detection reagent [18].

2.4.2. Polyphenol and Flavonoid Contents

Total polyphenol (TPC) and total flavonoid (TFC) contents in AEMCs were determined using a modified colorimetric method proposed by Kumazawa et al. (2004) [19]. The TPC assay was performed in a 96-well plate, with 50 μL of standard or sample, 50 μL of 10% calcium carbonate (Na_2CO_3), and 50 μL of 1N Folin reagent. One hour after the addition of Folin reagent to the reaction medium, absorbance was measured at 760 nm using a Perkin Elmer EnSpire plate reader. The results were expressed in μg equivalents of gallic acid (μg EGA/mg or percentage of total polyphenol content (%)). The TFC assay was performed in a 96-well plate by adding 100 μL of 2% aluminum chloride ($AlCl_3$) solubilized in 40% ethanol and 100 μL of standard/sample. One hour after the addition of 2% aluminum chloride ($AlCl_3$) to the reaction medium, absorbance was measured at 420 nm using a Perkin Elmer EnSpire plate reader. The results were expressed in μg equivalents of quercetin per mg of extract (μg QE/mg) or percentage of total flavonoid content (%). The analyses were performed in triplicate.

2.4.3. HPLC-UV/DAD Assay

Chromatographic analysis by HPLC was performed using a Hitachi LaChrom Elite® HPLC System (L-2130 pump, L2200 auto-sample, L-2300 column oven, and L-2455 DAD detector). Separation was carried out on a C18 column (5 μm, 250 × 4.6 mm; LiChroCART® 150-4.6 Purospher® RP18e) at 25 °C; the flow rate was set at 0.6 mL/min, the injection volume was 10 μL, and detection was performed at 354 nm. The mobile phase consisted of phosphoric acid 1% (A) (Sigma-Aldrich®) and acetonitrile (B) (Tedia®) with gradient elution performed at 0 min. 90% (A) and 10% (B); 40 min. 70% (A) and 30% (B); 50 min. 50% (A) and 50% (B). The data were analyzed using EZChrom Elite software, version 3.3.2 SP1. All solvents used were of HPLC grade (Sigma-Aldrich® and Tedia®). The water used was obtained from a Millipore Mili-Q system. The methodology was described by Leite et al. (2014) [20]. For analysis, 25 mg of AEMC was weighed and dissolved in 5 mL of ultra-purified water/ HPLC grade methanol (6:4), which was then filtered (Hydrophyllic, 33 mm, membrane 0.45 μm). Commercial standards were used an attempt to identify the compounds present in the AEMC by comparison of retention times and ultraviolet spectra. The standards obtained were caffeic acid, chlorogenic acid, ferulic acid, kaempferol, catechin, epicatechin, gallic acid, rosmarinic acid, myricitrin, trigonelline hydrochloride, ellagic acid, isoquercitrin, hesperetin, quercetin, resveratrol, vitexin, isovitexin and myricetin acquired from Sigma-Aldrich®, hyperoside acquired from Hwi Analytik Gmbh®, and rutin from Chromadex®.

Vitexin equivalents (VE) were determined with a standard curve generated using vitexin in the range of 8 to 100 μg/mL. The sample was analyzed at a concentration of 5 mg/mL in a methanol and water solution (6:4). The data from the standard curve were used to estimate the vitexin equivalents in the sample using linear regression of the data. Analyses were performed for all seven batches.

2.4.4. UHPLC-MS/MS Assay

UHPLC analyses were performed using a Waters Acquity H-series UPLC coupled to a Waters Acquity PDA detector in series with a Xevo triple quadrupole mass spectrometer. A Waters Acquity UPLC BEH C18 column (1.7 μm, 2.1 mm × 100 mm) was used, with mobile phases A = 0.1% formic acid and B = acetonitrile. The column was held at 35 °C with a flow rate of 0.35 mL/min, with 100% A and 0% B with a linear gradient of 40% A and 60% B at 30 min, followed by 4 min re-equilibration to the original conditions. The PDA was monitored continuously over a range of 230–500 nm. The injection volume was 15 μL. The mass spectrometer was operated in several different modes for separate injections. Initially, positive ion full scan positive ion electrospray 'survey scans' were acquired over the range m/z 100 to 1500 every 0.3 s, with a cone voltage of 30 V. Scanwave daughter scans at 10 V and 20 V collision energy (CE) at 2000 m/z per second were automatically acquired from the strongest ions. To unequivocally determine molecular weights, full negative ion electrospray spectra were subsequently acquired from m/z 100 to 1000 every 0.4 s using a cone voltage ramp, followed by targeted MS/MS scans with a cone voltage of 30 V and collision energy of 40 V from the relevant major $[M - H]^-$ ions. The ion source temperature was 130 °C, the desolvation gas was nitrogen at 950 L/h, the desolvation temperature was 450 °C, and the capillary voltage was 2.7 KV in all cases. For this assay, a mixture of seven batches of aqueous extracts from leaves was used.

2.5. Antioxidant Activity

2.5.1. DPPH Assay

Antioxidant activity was evaluated by reducing the DPPH (2,2-Diphenyl-1-picrylhydrazyl) radical using an adapted methodology described by Blois (1958) [21]. Ascorbic acid was used as a positive control, and a standard curve was generated over the concentration range of 10 to 1000 μg/mL. Different AEMC samples were evaluated at a concentration of 3 μg/mL. The results were also expressed as inhibition percentage (%) and were determined using the following equation:

AI (%) = 100 − (Sample Abs − Sample Blank) ∗ 100/Control Abs
AI = antioxidant inhibition (%)
Sample Abs = sample absorbance
Sample Blank = solvent
Control Abs = inhibition

2.5.2. Phosphomolybdenum Method

Antioxidant activity was determined using the phosphomolybdenum method, as described by Pietro et al. (1999) [22]. Ascorbic acid was used as a positive control, and a standard curve was generated over the concentration range of 10–300 μg/mL. The AEMC was evaluated at a concentration of 125 μg/mL. The assay was performed by adding

1.0 mL of the reagent solution to 0.1 mL of the sample or the standard. The reagent solution consisted of 28 mM phosphate, 4 mM molybdate, and 0.6 M sulfuric acid. After a reaction time of 90 min in a water bath at 95 °C, the absorbance was measured at 695 nm using a Shimadzu® UV-1800 (Software UVProve 2.33) spectrophotometer. The data obtained from the standard curve were used to estimate the equivalents of ascorbic acid content in the sample using linear regression.

2.5.3. Lipid Peroxidation Assay

The thiobarbituric acid reactive substances assay (TBARS) was performed according to the methodology proposed by Hazzit et al. (2009), Badmus et al. (2011), and Seneviratne et al. (2016) [23–25] with adaptations. An emulsion of egg (10%) in 20 mM potassium phosphate buffer (pH 7.4) was prepared as a lipid source for the test. For the positive control, α-tocopherol was used at concentrations of 7.81 µg/mL to 1000 µg/mL in methanol. AEMC was evaluated over the concentration range of 3.09 µg/mL to 1000 µg/mL prepared in water:methanol solution (6:4). The percentage of lipid peroxidation inhibition was determined using the following equation:

LPI = [(Control Abs − Sample Abs)/Control Abs] ∗ 100.

Where:

LPI = lipid peroxidation inhibition (%).

Control Abs = control absorbance (solvent).

Sample Abs = sample absorbance.

2.6. Seasonal Study

To evaluate the seasonal chemical variability, *M. chamissois* leaves were collected in May 2017 (P1), November 2017 (P2), February 2018 (P3), May 2018 (P4), and August 2018 (P5). The development stage of the plants was observed, in which P1 was flowering/fruiting, P2 was fruiting, P3 was vegetative period, P4 was flowering/fruiting, and P5 was fruiting. The aqueous extract from these leaves was prepared by infusion following the method described above.

The meteorological data for 2017 and 2018 were provided by the Automatic Agrometeorological Station of Agroclimatology Laboratory from the University of Brasília. The data were limited to Lago do Cedro, Brasília, Federal District, Brazil, and included meteorological parameters, such as global radiation ($MJm^2 d^{-1}$), maximum temperature (°C), minimum temperature (°C), and rainfall index (mm) from the period of 17 January to 18 August.

Meteorological data were correlated with total solids content, TPC and TFC contents, vitexin derivate content, and antioxidant activity by DPPH assay. Pearson's linear correlation coefficient (r) was used to determine the correlation level.

2.7. Statistical Analysis

Microsoft Office Excel® 2016 software and GraphPad Prism® Version 5.01 were used for statistical analysis. The results are expressed as the average plus standard deviation and relative deviation standard. ANOVA tests followed by Kruskal-Wallis or Dunn's multiple comparison tests were used in different assays. Pearson's correlation was used for the seasonal study, using linear correlations evaluated according to Callegari-Jacques considering $|r| = 0$–0.3 (weak), $|r| = 0.3$–0.6 (moderate), and $|r| = 0.6$–0.9 (strong) [26].

3. Results and Discussion

Herbal medicines have a complex chemical constitution, and their pharmacological effect are often due to synergy between these compounds. Several factors can affect the chemical composition of a plant extracts, such as growth, harvest, drying, the extraction process, and storage conditions [27]. To secure a constant composition of herbal preparations, and consequently their efficacy and safety, it is necessary to ensure their pharmaceutical quality by standardizing the process for these products [20].

In this study, a method to obtain a standardized extract of *M. chamissois* Naudin was developed, and the seasonal chemical variability was evaluated.

3.1. Chemical Composition and Standardization

Seven AEMC batches were prepared according to the method described above. The reproducibility of the extraction process was evaluated in relation to total solid, polyphenol, and flavonoid contents, TLC and HPLC profiles, vitexin derivate content, and biological activity by antioxidant activity.

Total solid contents found batches 1 at 7 were (2.97% $\pm$ 0.003 (B1); 2.13% $\pm$ 0.002 (B2); 2.73% $\pm$ 0.003 (B3); 2.73% $\pm$ 0.006 (B4); 2.73% $\pm$ 0.004 (B5); 2.63% $\pm$ 0.003 (B6); 2.97% $\pm$ 0.003 (B7). The average of the total solid of the seven batches was 2.70% $\pm$ 0.003, and the relative standard deviation (RSD) found was 0.11%. The results showed no significant difference ($p < 0.05$) between batches for the total solids in analysis by Kruskal-Wallis with Dunn's test of multiple comparisons.

The extraction yields for batches 1 to 7 were 21.49%, 23.00%, 21.48%, 22.30%, 22.88%, 20.08% and 22.83%, with an average of 22.00% $\pm$ 1.06 (RSD = 4.8%).

In the TLC analysis, the same chemical profile was observed for the seven batches, with five main spots. The retention factor (R_f) of these spots was 0.17, 0.33, 0.43, 0.54, and 0.61 (Figure 1).

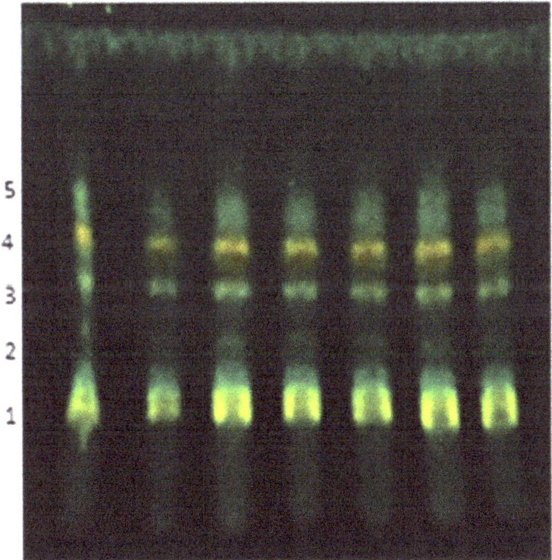

Figure 1. Thin-layer Chromatography from seven batches of aqueous extract of *M. chamissois* Naudin (AEMC). Each column represents one extraction batch. The leaves were collected in May 2017. The stationary phase was silica gel, and the mobile phase was a mixture of ethyl acetate, formic acid, acetic acid, and deionized water (100:11:11:26). Chromatographic spots were visualized using an ultraviolet lamp at 254 nm and the NP/PEG reagent. 1: R_f 0.17, 2: R_f 0.33, 3: R_f 0.43, 4: R_f 0.54 and 5: R_f 0.61.

The polyphenol and flavonoid analysis results are expressed as the average of the seven batches. The TP content found in AEMC was 19.65 $\pm$ 0.43 µg QE/mg (RSD = 2.19%) and the TF content was 2.49 % $\pm$ 0.15 (RSD = 5.55%).

Pearson (r) linear correlation was performed to assess the correlation between total polyphenol content and extractable solids content. A weak negative correlation (r = −0.20; $p = 0.37$) was not significant between the evaluated parameters. In general, the weak correlation indicates that the parameters evaluated are unlikely to be comparable and associate; that is, there was no correlation.

Another Pearson (r) linear correlation was carried out to correlate the total flavonoid and total solids content. The results showed a strong negative non-significant correlation ($r = -0.87$; $p = 0.02$), which suggested a tendency of an inversely proportional relationship between the total flavonoid content and the total solids content.

Gontijo et al. (2019) found 3.56 ± 0.17 µg equivalent of rutin/mg of extract in aqueous extract of leaves of *M. latecrenata* (DC.) Naudin prepared by infusion (1:20) [28]. In *M. albicans* (Su.) Triana fruits 510.96 ± 8.64 mg QE 100 g $^{-1}$ of total flavonoids was measured in July 2015, after a dry winter in the region where the fruits were collected and where the average annual temperature was 21.8 °C [29].

For other genera of Melastomataceae, *Bellucia*, in aqueous dry bark extract decoction lifting 1:10 (p/v) of *Bellucia dichotoma* Cogn 0.14 ± 0.030 g/100 g of flavonoids was obtained [30], and no flavonoids were detected in the aqueous extract of *Bellucia grossularioides* (L.) Triana fruits [31].

Compound detection by HPLC/DAD was performed by comparing the spectra obtained and retention times (t_R) of seven batches of AEMC and standards. The chromatogram profiles showed 12 peaks at 354 nm (Table 1 and Figure 2). The results in the Table 1 show the mean values of retention time and peak area. The chromatogram presented in Figure 2 is representative of the analysis of one of the batches (B4).

Table 1. Peak characteristics of *M. chamissois* Naudin aqueous extract from leaves (AEMC) analyzed by HPLC/DAD at 354 nm.

Peak	Rt (min.)	Area	λ max (nm)	λ min (nm)
1	16.44 ± 0.13	$605,461 \pm 75,626$	267; 397	393; 261
2	-	-	252; 350; 354	331; 351; 361
3	20.92 ± 0.08	$3,053,209 \pm 247,842$	269; 349	304; 247
4	22.68 ± 0.07	$3,165,255 \pm 196,166$	268; 256; 349	306; 246; 260
5	-	-		
6	24.89 ± 0.08	$7,057,969 \pm 298,667$	27; 336	247; 304
7	27.56 ± 0.07	$1,089,016 \pm 112,057$	269	250
8	-	-		
9	29.58 ± 0.09	$1,169,432 \pm 146,103$	256; 354	319; 241
10	-	-		
11	41.36 ± 0.12	$3,326,752 \pm 1,134,026$	28; 363	319; 249
12	50.76 ± 0.06	$3,460,879 \pm 1,572,479$	282; 363	319; 249

Detection at 354 nm, C18 column, flow rate of 0.6 mL/min, eluent: phosphoric acid 1%, and acetonitrile in gradient system.

In the identification of AEMC compounds by HPLC/DAD, the results showed similar UV spectra between peak 6 (t_R 24.89 min) and vitexin (0.9959) and isovitexin (0.9951) standards (Figure 2A–C); the retention times were similar. This is in agreement with the data reported by Gomes et al. (2021) [11], who found a similarity between the peak at 24.8 min of the aqueous extract of *M. chamissois* Naudin leaves and vitexin (0.9968) and isovitexin (0.9963). Peak 6 was considered to be closely related to vitexin and isovitexin.

To quantify the compound corresponding to peak 6, a linear regression of the standard curve generated using vitexin as the standard ($y = 104,860x - 109,129$, $r = 0.99$) was used to determine the vitexin equivalent (VE) content. The data analysis of seven batches showed 13.67 ± 0.57 µg VE/mg of AEMC (RSD = 4.17%).

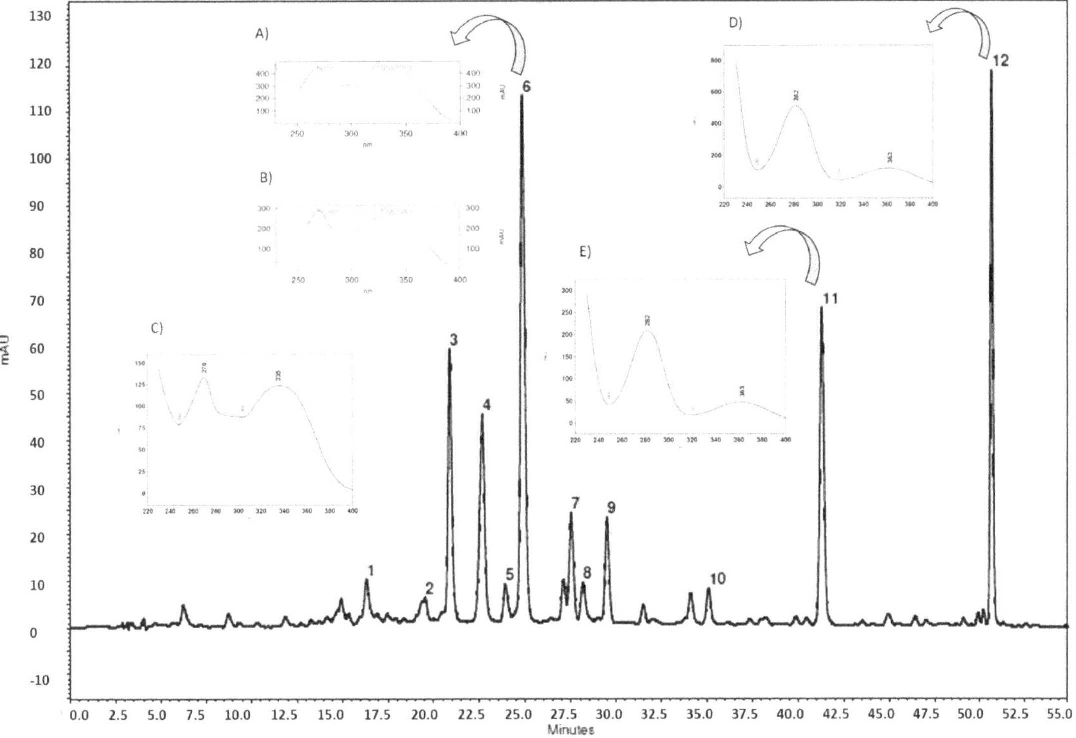

Figure 2. Chromatographic profile of aqueous extract *M. chamissois* Naudin leaves (B4) by HPLC/DAD at 354 nm. Detection at 354 nm, C18 column, flow rate of 0.6 mL/min, eluent: phosphoric acid 1%, and acetonitrile in gradient system. (**A**) similarity of peak 6 (t_R 24.89 min, λmax: 270, 336) and vitexin standard compound (similarity index: 0.9959); (**B**) similarity of peak 6 (t_R 24.89 min) and isovitexin standard compound (similarity index: 0.9951), (**C**) peak 6 (t_R 24.89 min, λmax: 270, 336); strongly suggestive a vitexin or isovitexin related compound (**D**) peak 11 (t_R 41.27 min; λmax: 282, 363) miconioside B; (**E**) peak 12 (t_R 50.76 min; λmax: 282, 363) matteucinol-7-*O*-β-apiofuranosyl (1 → 6)-β-glucopyranoside.

In addition, peaks 11 and 12 showed a flavanone profile, suggesting the presence of miconioside B (t_R 41.36 min) and matteucinol-7-*O*-β-apiofuranosyl(1 → 6)-β-glucopyranoside (t_R 50.76 min), respectively (Figure 2D,E). These compounds were also identified in *M. chamissois* by Silva et al. (2020) and Gimenez et al. (2020) [7,14].

Vitexin (apigenin-8-C-glucoside) and isovitexin (apigenin-6-C-glucoside) are chemical markers for several species of *Passiflora* sp. and are present in some species of the Melastomataceae family such as *Melastoma dodecandrum* Lour. [32], *Clidemia sericea* D. Don [33], and *Dissotis rotundifolia* (Sm.) Triana [34].

For the standardization process of AEMC, the biological activity by antioxidant activity using the DPPH assay was also evaluated. The seven batches of AEMC showed 59.34% ± 7.92 (RSD = 13.35%) of inhibition of the reduction of DPPH radical at a concentration of 3 μg/mL. The results of batches B1 to B7 showed significant differences; batch B1 was significantly different from B2 and B6, B1 from B5, B4 from B5, and B5 from B6 using ANOVA test of multiple comparisons (Dunn's) followed by the Kruskal-Wallis test.

Pearson correlation between antioxidant activity (DPPH assay) and total solids content or total polyphenol content, and flavonoid content was evaluated. The results revealed a weak positive correlation (r = 0.15; *p* = 0.42) between polyphenols and antioxidant activity and a moderate negative correlation between total flavonoids and antioxidant activity ($r = -0.59$; $p = 0.14$), with no significant difference ($p < 0.05$) between the chemical assays and DPPH antioxidant activity. There was a moderate positive correlation ($r = 0.76$; $p = 0.06$), but no significant correlation was found between antioxidant activity and total solids content.

It is possible to observe that the DPPH inhibition values for AEMC is higher than that obtained for other Melastomataceae species, considering the final concentration of 3 µg/mL with 59.34% $\pm$ 7.92 inhibition (RSD = 13.35%). DPPH inhibition by *Osbeckia aspera* var. *aspera*, *O. reticulata*, and *O. virgata* were 61.8%, 62.8%, and 60.6%, respectively, at a concentration of 100 µg/mL [35]. For *M. albicans* ethanol extract (50%) of leaves, using the same concentration of 3 µg/mL, DPPH inhibition was 36.91% $\pm$ 0.93 [36]. In addition, *Memecylon terminale* Dalz was also able to scavenge DPPH radicals with an IC_{50} value of 43 µg/mL [37].

Although the results demonstrated low variability in chemical composition, a larger variability in biological activity was observed among the seven batches tested. Nevertheless, considering the DPPH assay as a bioanalytical method, the precision should not exceed 15% (RSD) [38]. The variability in the antioxidant activity of the AEMC batches was within the allowed range.

Biological activity of several herbal medicinal products is due to the synergy between its many constituents [39]. Thus, evaluating the biological activity is an important tool for determining the quality of herbal products. Oxidative stress is present in the mechanism of many diseases, such as neurodegenerative [40], cardiovascular [41], and liver diseases [42], and in the aging process [43]. Therefore, antioxidant activity can be an important biological assay to ensure the efficacy and quality of herbal products.

UHPLC-MS/MS Analysis

To better characterize the chemical composition of AEMC, a UHPLC-MS/MS analysis was performed. Ten of the peaks detected matched the chromatographic profile obtained by HPLC/DAD, aiming to identify flavonoids (Figure 3). The data obtained are listed in Table 2.

Table 2. Data from UHPLC/UV/MS/MS analysis of *M. chamissois* Naudin. aqueous leaf extract.

Peak (n°)	t_R (min)	UV Max (nm)		$[M-H]^-$ (m/z)	MS/MS Fragments	$[M+H]^+$ (m/z)	MS/MS Fragments	MW	Compound Identity or Partial Identity
1	1.87	-	-	-	-	-	-	-	-
2	2.19	-	-	-	-	-	-	-	-
3	2.98	241	-	935	-	-	-	936	-
4	3.71	269	350	609	489, 327, 309, 298	611	449, 431, 413, 383, 353	610	a mixed O,C glycoside of luteolin eg 2″-O-hexosyl orientin
5	4.16	255	349	609	357, 327, 309, 297	611	449, 431, 413, 353, 329	610	a mixed O,C glycoside of luteolin eg 2″-O-hexosyl isoorientin
6	4.6	270	338	593	293	595	433, 415, 367, 337, 313	594	a mixed O,C glycoside of apigenin eg 2″-O-hexosyl vitexin
7	4.71	267	338	593	293	595	433, 415, 337, 313	594	a mixed O,C glycoside of apigenin eg 2″-O-hexosyl isovitexin
8	5.04	269	337	431	283	433	415, 397, 367, 337, 313	432	isovitexin
9	5.25	255	354	463	301	465	303	464	isoquercitrin or hyperoside
10	5.97	276	-	599	-	-	-	600	-
11	6.61	268	341	583	447	585	449	584	?
12	9.07	281	363	593	299, 179, 135	595	301, 181, 147	594	miconioside B
13	13.02	281	362	607	313, 192	609	315, 181, 161	608	matteucinol-7-O-β-apiofuranosyl (1→6)-β-glucopyranoside
14	15.86	297	249	299	179, 135,119	301	181, 147	300	farrerol

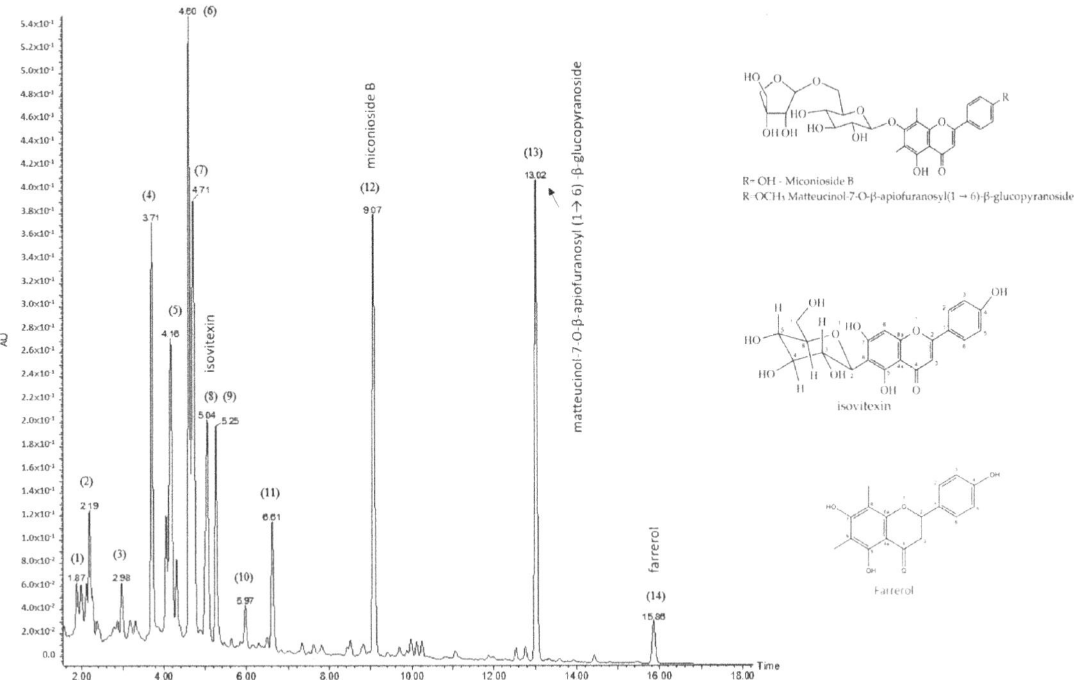

Figure 3. Chromatogram profile of *M. chamissois* Naudin aqueous extract from leaves (AEMC) analyzed UHPLC/UV/MS/MS at 354 nm. Detection at 354 nm, C18 column, flow rate of 0.3 mL/min, eluent: formic acid 1%, and acetonitrile in gradient system. The mass spectrometer was operated in several different modes for separate injections. The ion source temperature was 130 °C, the desolvation gas was nitrogen at 950 L/h, the desolvation temperature was 450 °C, and the capillary voltage was 2.7 KV.

The identities of compounds present in AEMC were based on the results obtained compared with data in the literature or with similarity to compounds in the Metlin MS database, and it was not possible to confirm all of them due to the complexity of the sample, and the lack of reference MS data for some relatively obscure flavonoids. In particular, the reference MS data available for the mixed C and O glycosides of apigenin and luteolin are very limited.

Peak 4 had a molecular weight (MW) of 610 with MS/MS fragments at m/z 489 [M−H-120], 429 [M−H-180], 327 [M−H- 282 (=162 + 120)], 309 [M−H-300 (=180 + 120)], and 298 [M−H-311]. A neutral loss of 282 is characteristic of 2″-O-hexosyl-6-C-hexosyl glycosides, [44,45]. In terms of the formal flavonoid glycoside fragmentation notation, the Z_1 ion was 429, and the subsequent $^{0,2}X_0$ ion was 309. These glycosides are characterized by this intense ion at the mass of the aglycone + 23. The data strongly suggested presence of an O-hexosyl-C-hexosyl luteolin, such as compounds like 2″-O-glucosyl-8-C-glucosyl luteolin (2″-O-glucosyl orientin) or 2″-O-glucosyl-6-C-glucosyl luteolin (2″-O-glucosyl isoorientin).

Peak 5 was also MW 610 with MS/MS fragment ions at m/z 429 [M−H-180], 357 [M−H-252 (=162 + 90)], 327 [M−H-282(=162 + 120)], 309 [(M−H-300 (=180 + 120)], and 297 [M−H-312 (=162 + 150)]. These fragment ions were consistent with the presence of two C6 sugar moieties of mixed C and O diglycosides of luteolin based on reference data and isomeric with peak 4 (Supplementary Materials) [45–49].

Peaks 6 and 7 had MWs of 594 with an intense fragment ion at m/z 293 [M−H-300(=180 + 120)], while in positive ion mode, a fragment ion at m/z 433 [M+H-162] suggested the presence of mixed C and O diglycosides of apigenin (i.e., *O*-glycosides of vitexin and/or isovitexin) (Supplementary Materials) [49,50]. These peaks together appear to correspond to peak 6 from the HPLC-UV/DAD analysis, identified as a vitexin/isovitexin derivative; we observed slightly better separation by UHPLC compared to that by HPLC.

It is noteworthy that derivatives of kaempferol, quercetin, luteolin, and apigenin have already been identified in other Melastomataceae genera, such as *Huberia* sp., *Pleroma pereirae* (kaempferol and quercetin) [51], *Melastoma* sp. [52], and *Melastoma decemfidum* Roxb. (kaempferol) [53], *Osmbeckia parvifolia, Arnbeckia parvifolia, Medinilla septentrionalis* (W.W.Sm.) H.L.Li [54], *Melastoma malabathricum* L., and *M. hirta* [55].

Peak 8 with MW 432 and a prominent fragment at m/z 283 [M−H-148] was characterized as isovitexin, similar to the Metlin MS database and the literature [56,57] (Supplementary Materials). Isovitexin has also been identified in other species of the Melastomataceae family, such as *Dissotis rotundifolia* (Sm.) Triana [34], and *Clidemia sericea*. D. Don [33].

Peak 9 had a MW of 464 with a prominent fragment in positive ion mode at m/z 303 [M+H-162], indicating loss of a hexose unit, characteristic of isoquercitrin, isoquercetin, or hyperoside [58].

Peak 12 had a MW of 594; $[M - H]^-$ at m/z 593, MS/MS gave m/z 299 (neutral loss of 294), 179, 135; $[M + H]^+$ at m/z 595 with neutral loss 294 to m/z 301. A neutral loss of 294 is characteristic of the loss of both pentose (132) and hexose (162) glycoside subunits. These data and the UV spectrum indicated that it was the farrerol glycoside miconioside B, identified in *Miconia prasina* (Sw.) DC. [59], *M. trailii* [60] and *M. chamissoi*s [7,14] (Supplementary Materials).

Peak 13 had a MW of 608, and negative ion MS/MS gave m/z 313 ([M−H-294), indicating loss of both a pentose and hexose unit. Positive ion MS gave an intense fragment at m/z 315 [M+H-294], also indicating losses of a pentose and hexose. These data and the UV data corresponded exactly to matteucinol 7-*O*-β-apiofuranosyl (1 → 6)-β-glucopyranoside (Supplementary Materials), previously identified as *M. chamissois* [7,14].

Peak 14 had MW 300, and MS/MS gave fragments at m/z 180, 135, and 119. This and the UV data correspond to farrerol (Supplementary material), identified in *M. prasina* [59]. Yin, Jintuo et al. (2019) presented data that support to our findings [61]. Several of the very minor peaks appeared to have a farrerol aglycone component, based on the prominent m/z 301 ion in positive ion MS. Neither rutin or quercetin could be detected.

3.2. Antioxidant Activity

To better characterize the antioxidant activity of AEMC, two additional methods were used: the phosphomolybdenum method and the lipid peroxidation assay.

For the phosphomolybdenum assay, the standard curve of ascorbic acid yielded the regression equation y = 0.02928x + 0.0604 (r = 0.98). AEMC showed 548.8 µg EqAA/mg.

Gomes et al. (2021) found in *Miconia chamissois* Naudin equivalents of ascorbic acid content of 0.8 ± 0.01 µg/mg and equivalents of BHT content of 0.9 ± 0.05 µg/mg [11]. Murguran and Parimelazhagan (2014) found for several crude extracts from whole plant of *O. parvifolia* a phosphomolybdenum inhibition of 47.7 ± 3.2 (*n*-Hexane), 55.1 ± 7.1 (ethylacetate), 163.9 ± 10.4 (methanol) and 157.9 ± 17.5 (ethanol) mg AAE/g extract [62].

For TBARS, AEMC did not inhibit the formation of reactive species. For the dried extract (50% ethanol v/v) of *Miconia albicans* leaves, an IC_{50} value of 1338.34 µg/mL [36] was observed. *Osbeckia parvifolia* Arn. ethyl acetate extract of whole plants showed 59.6% TBARS inhibition [62]. Leaves of other species, such as *Melastomastrum capitatum* A. Fern. & A. Fern. showed TBARS inhibition of 86.86% ± 3.63 and 39.93% ± 1.07 for ethanol and aqueous extracts, respectively [63].

3.3. Seasonality Study

Leaves of *M. chamissois* were collected in May 2017 (P1), November 2017 (P2), February 2018 (P3), May 2018 (P4) and August 2018 (P5) to evaluate the chemical seasonal variability. The results are shown in Table 3.

Table 3. Results from seasonal study of *M. chamissois* Naudin leaves.

Collection Period	Yield (%)	Solids Content (%)	TPC (µg GA/mg)	TFC (µg QE/mL)	Antioxidant Activity (%)
P1	22.01	2.70 ± 0.003	20.73 ± 0.81	8.28 ± 0.46	56.22 ± 2.44
P2	19.4	2.83 ± 0.002	19.28 ± 0.49	7.70 ± 0.23	49.38 ± 4.95
P3	21.7	2.70 ± 0.002	18.58 ± 0.23	7.58 ± 0.09	47.97 ± 4.34
P4	17.94	2.70 ± 0.002	18.60 ± 0.36	7.73 ± 0.30	52.39 ± 6.09
P5	22.78	3.03 ± 0.002	19.11 ± 0.31	6.91 ± 0.21	56.60 ± 0.70

The meteorological data are presented in Table 4, and the correlation results are presented in Table 5.

Table 4. Meteorological data during the collection months of *M. chamissois* Naudin.

	Global Radiation (MJm^{-2}d^{-1})	Rainfall Index (mm)	Temperature Max ($^\circ$C)	Temperature Min ($^\circ$C)
Total solids	r = 0.363; p = 0.548	r = −0,092; p = 0.442	r = 0.430; p = 0.235	r = 0.198; p = 0.749
TPC	r = −0.775; p = 0.124	r = −0.190; p = 0.377	r = 0.230; p = 0.355	r = 0.209; p = 0.375
TFC	r = −0.753; p = 0.142	r = 0,10; p = 439	r = −0.233; p = 0.353	r = −0.043; p = 0.473
Antioxidant activity	r = 0.174; p = 0.780	r = −0.688; = p = 0.100	r = 0.012; p = 0,492	r = −0.257; p = 0.338

The data represent the monthly average.

Table 5. Correlation index (r) from seasonal study of *M. chamissois* Naudin leaves.

Collection Months	Global Radiation (MJm^{-2}d^{-1})	Rainfall Index (mm)	Temperature Max ($^\circ$C)	Temperature Min ($^\circ$C)
May/2017 (P1)	332.05	45.21	22.88	17.19
Nov/2017 (P2)	337.65	241.81	23.79	18.62
Feb/2018 (P3)	363.89	160.78	23.32	20.49
May/2018 (P4)	350.38	27.69	21.02	17.42
Aug/2018 (P5)	360.6	22.86	23.38	18.64

Data represent the correlation index (r) and statistical significance by Pearson's linear correlation test.

According to the data obtained, there were no significant correlations ($p < 0.05$) between the TPC and TFC and meteorological data, with a strong negative correlation between global radiation, suggesting that the production of these metabolites is inversely proportional to the radiation index. Another strong negative correlation was observed between the antioxidant activity and rainfall index, suggesting that the antioxidant activity is inversely proportional to the rainfall index.

A moderate positive correlation between total solids and global radiation and a weak correlation between the other parameters were also observed.

Notably, the harvests were carried out in different reproductive periods, with P3 being the only harvest in which the species did not show flowering, fruiting or maturation; on 17 May, 17 November and 18 August, the highest total polyphenol content and lowest radiation were observed. Gobbo-Neto and Lopes (2007) understand the complexity of seasonality studies. In addition to meteorological and environmental parameters (water availability, radiation, temperature, soil, and altitude), the metabolic and hormonal conditions inherent in the development of the plant species must be attributed [15].

Interest in studies that assess changes in the secondary metabolism of plants has grown in recent decades and contributes to understanding the composition, adaptive capacity, and bioactive screening of species. When it comes to species from Cerrado, the interest is even greater because of the complexity and biological richness of this biome.

Studies of biological material from fauna and flora are very complex due to the object of the investigation itself, and the complexity increases when dealing with species from the Cerrado because of the diversity present in this biome. Zanatta et al. (2021) also pointed out that there may be a gap in the physiological response of plants. There will not necessarily be a positive correlation between composition and biological activity, especially with species from the Cerrado [64].

Melastomacatecae has adaptive capacity under different environmental conditions and, thus, can be present in all vegetation formations in the Cerrado [65]. Ishino et al. (2021) still emphasized the changes caused by anthropogenic issues and consequently provoked biological changes in the fauna and flora to guarantee protection. In a study of seasonality, some native plants may be stress-tolerant and may not show differences in physical and physical characteristics [66].

4. Conclusions

A standardized AEMC extract was obtained, and chemical marker quantification by HPLC was performed. Biological activity using the DPPH assay was proposed as an important quality parameter. In the extract's standardization process of the extract, all the parameters evaluated showed low variability. Moreover, AEMC showed high reactivity and potential antioxidant activity via the phosphomolybdenum complex. The chemical composition confirmed the presence of polyphenolic compounds already identified in the species, contributing to the chemical elucidation of the species that have been the object of study by various research groups. Using different chromatographic techniques, luteolin glycosides, apigenin glycosides, a quercetin glycoside, miconioside B, matteucinol-7-*O*-β-apiofuranosyl (1 → 6)-β-glucopyranoside and farrerol were identified. Several of the main flavonoids were mixed C,O-diglycosides of apigenin and luteolin. The seasonal evaluation is of great value considering that there was no correlation between composition and activity versus meteorological parameters, indicating the temporal adaptability of the species. These results showed that it is possible to obtain extracts from *M. chamissois* at any time of the year without significant differences in composition.

Supplementary Materials: The following are available online. Figure S1: UHPLC-UV/MS/MS data of peak 4; Figure S2: UHPLC-UV/MS/MS data of peak .5; Figure S3: UHPLC-UV/MS/MS data of peak 6; Figure S4: UHPLC-UV/MS/MS data of peak 7; Figure S5: UHPLC-UV/MS/MS data of peak 8; Figure S6: UHPLC-UV/MS/MS data of peak 12; Figure S7: UHPLC-UV/MS/MS data of peak 13; Figure S8: UHPLC-UV/MS/MS data of peak 14; Figure S9: Chemical structures of the flavonoids from *M. chamissois* Naudin aqueous extract from leaves (AEMC); Figure S10: Chromatographic profile of aqueous extract *M. chamissois* Naudin leaves in different batches (B1–B7) by HPLC/DAD at 354 nm.

Author Contributions: Conceptualization, D.S. and Y.M.F.-B.; Data curation, J.d.F.F.; Formal analysis, N.W.D.; Funding acquisition, P.O.M.; Investigation, J.d.F.F., M.H.M.L., D.H.N.M. and C.W.F.; Methodology, M.H.M.L., J.V.D.G., D.H.N.M. and C.W.F.; Project administration, Y.M.F.-B.; Supervision, Y.M.F.-B.; Writing—original draft, J.d.F.F.; Writing—review & editing, C.W.F., P.O.M., N.W.D., D.S. and Y.M.F.-B. All authors have read and agreed to the published version of the manuscript.

Funding: This research project was funded in part by the Coordenação de Aperfeiçoamento de Pessoal de Nível Superior—Brasil (CAPES)—Finance Code 001. Federal District Research Support Foundation (FAP-DF)—Grant number 0193.0000542/2018-63. National Council for Scientific and Technological Development (CNPq)—Grant number: 010337/2014-8.

Institutional Review Board Statement: Not Applicable.

Informed Consent Statement: Not Applicable.

Data Availability Statement: All data included in this study are available upon request by contact with the corresponding author.

Acknowledgments: The authors are grateful to the Agroclimatology Laboratory from University of Brasília which provided meteorological data. The authors are also grateful to the National Council for Scientific and Technological Development (CNPq), Federal District Research Support Foundation (FAP-DF), and University of Brasília (UnB) for financial support. We would like to thank Editage for English language editing.

Conflicts of Interest: The authors state no conflict of interest.

Sample Availability: Samples of the compounds are available from the authors.

References

1. Brasil. O Bioma Cerrado. Available online: https://antigo.mma.gov.br/biomas/cerrado (accessed on 10 October 2020).
2. Goldenberg, R.; Baumgratz, J.F.A.; Souza, M.L.D.R. Taxonomia de Melastomataceae no Brasil: Retrospectiva, perspectivas e chave de identificação para os gêneros. *Rodriguésia* **2012**, *63*, 145–161. [CrossRef]
3. Neto, G.G.; de Morais, R.G. Recursos medicinais de espécies do Cerrado de Mato Grosso: Um estudo bibliográfico. *Acta Bot. Bras.* **2003**, *17*, 561–584. [CrossRef]
4. Goldenberg, R.; Caddah, M.K. Miconia in Lista de Espécies da Flora do Brasil. Available online: http://floradobrasil.jbrj.gov.br/jabot/floradobrasil/FB9683 (accessed on 18 November 2021).
5. Cruz, A.V.D.M.; Kaplan, M.A.C. Uso medicinal de espécies das famílias Myrtaceae e Melastomataceae no Brasil. *Floresta Ambiente* **2004**, *11*, 47–52.
6. Calderon, A.I.; Simithy, J.; Gupta, M.P. Antimalarial natural products drug discovery in Panama. *Pharm. Biol.* **2012**, *50*, 61–71. [CrossRef] [PubMed]
7. Gimenez, V.M.; e Silva, M.L.; Cunha, W.R.; Januario, A.H.; Costa, E.J.X.; Pauletti, P.M. Influence of environmental, geographic, and seasonal variations in the chemical composition of *Miconia* species from Cerrado. *Biochem. Syst. Ecol.* **2020**, *91*, 104049. [CrossRef]
8. Bortolotto, I.M.; Damasceno-Junior, G.A.; Pott, A. Lista preliminar das plantas alimentícias nativas de mato grosso do sul, Brasil. *Iheringia* **2018**, *73*, 101–116. [CrossRef]
9. Pinto, G.F.D.S.; Kolb, R.M. Seasonality affects phytotoxic potential of five native species of Neotropical savanna. *Botany* **2016**, *94*, 81–89. [CrossRef]
10. Alves, N.R. Estudo dos extratos de três espécies do gênero *Miconia* sobre a inibição das MMPs 2 e 9 e sobre o crescimento tumoral in vitro. Ph.D. Thesis, Universidade Federal de São João Del-Rei, Divinópolis, Brazil, 2016.
11. Gomes, L.F.; Martins, D.H.N.; da Silva, S.M.M.; de Barros, Y.Y.; de Souza, P.M.; de Freitas, M.M.; Fagg, C.W.; Simeoni, L.A.; Magalhães, P.O.; Silveira, D.; et al. Propiedades biológicas y caracterización fitoquímica del extracto acuoso de *Miconia* chamissois Naudin. *BLACPMA* **2021**, *20*, 427–442. [CrossRef]
12. Cunha, G.O.S.; da Cruz, D.C.; Menezes, A.C.S. An Overview of *Miconia* Genus: Chemical Constituents and Biological Activities. *Pharmacogn. Rev.* **2021**, *13*, 77–88. [CrossRef]
13. Rosa, M.N.; e Silva, L.R.V.; Longato, G.B.; Evangelista, A.F.; Gomes, I.N.F.; Alves, A.L.V.; de Oliveira, B.G.; Pinto, F.E.; Romão, W.; de Rezende, A.R.; et al. Bioprospecting of Natural Compounds from Brazilian Cerrado Biome Plants in Human Cervical Cancer Cell Lines. *Int. J. Mol. Sci.* **2021**, *22*, 3383. [CrossRef]
14. Silva, A.G.; Silva, V.A.O.; Oliveira, R.J.S.; de Rezende, A.R.; Chagas, R.C.R.; Pimenta, L.P.S.; Romão, W.; Santos, H.B.; Thomé, R.G.; Reis, R.M.; et al. Matteucinol, isolated from *Miconia chamissois*, induces apoptosis in human glioblastoma lines via the intrinsic pathway and inhibits angiogenesis and tumor growth in vivo. *Investig. New Drugs* **2019**, *38*, 1044–1055. [CrossRef]
15. Gobbo-Neto, L.; Lopes, N.P. Plantas medicinais: Fatores de influência no conteúdo de metabólitos secundários. *Química Nova* **2007**, *30*, 374–381. [CrossRef]
16. World Health Organization. *WHO Expert Committee on Specifications for Pharmaceutical Preparations: Fiftieth Report*; WHO Technical Report Series No. 996; World Health Organization: Geneva, Switzerland, 2016; pp. 1–358. ISBN 978-92-4-069548-1.
17. Süntar, I. Importance of ethnopharmacological studies in drug discovery: Role of medicinal plants. *Phytochem. Rev.* **2020**, *19*, 1199–1209. [CrossRef]
18. Wagner, H.; Bladt, S. *Plant Drug Analysis-A thin Layer Chromatography Atlas*; Springer Science & Business Media: München, Germany, 1996; ISBN 3-540-58676-8.
19. Kumazawa, S.; Hamasaka, T.; Nakayama, T. Antioxidant activity of propolis of various geographic origins. *Food Chem.* **2004**, *84*, 329–339. [CrossRef]
20. Leite, C.F.M.; Leite, B.H.M.; de Carvalho Barros, I.M.; Gomes, S.M.; Fagg, C.W.; Simeoni, L.A.; Silveira, D.; Fonseca, Y.M. Determination of rutin in Erythroxylum suberosum extract by liquid chromatography: Applicability in standardization of herbs and stability studies. *BLACPMA* **2014**, *13*, 135–143.
21. Blois, M.S. Antioxidant Determinations by the Use of a Stable Free Radical. *Nature* **1958**, *181*, 1199–1200. [CrossRef]
22. Prieto, P.; Pineda, M.; Aguilar, M. Spectrophotometric Quantitation of Antioxidant Capacity through the Formation of a Phosphomolybdenum Complex: Specific Application to the Determination of Vitamin E. *Anal. Biochem.* **1999**, *269*, 337–341. [CrossRef]
23. Hazzit, M.; Baaliouamer, A.; Veríssimo, A.R.; Faleiro, M.L.; Miguel, M.G. Chemical composition and biological activities of Algerian *Thymus* oils. *Food Chem.* **2009**, *116*, 714–721. [CrossRef]
24. Badmus, J.A.; Adedosu, T.O.; Fatoki, J.O.; Adegbite, V.A.; Adaramoye, O.A.; Odunola, O.A. Lipid peroxidation inhibition and antiradical activities of some leaf fractions of *Mangifera indica*. *Acta Pol. Pharm.* **2011**, *68*, 23–29.
25. Seneviratne, K.N.; Prasadani, W.C.; Jayawardena, B. Phenolic extracts of coconut oil cake: A potential alternative for synthetic antioxidants. *Food Sci. Technol.* **2016**, *36*, 591–597. [CrossRef]
26. Callegari-Jacques, S.M. *Bioestatística: Princípios e Aplicações*; Artmed: Porto Alegre, Brazil, 2003; ISBN 978-85-363-1144-9.

27. Bauer, R. Quality Criteria and Standardization of Phytopharmaceuticals: Can Acceptable Drug Standards Be Achieved? *Ther. Innov. Regul. Sci.* **1998**, *32*, 101–110. [CrossRef]

28. Gontijo, D.C.; Gontijo, P.C.; Brandão, G.C.; Diaz, M.A.N.; de Oliveira, A.B.; Fietto, L.G.; Leite, J.P.V. Antioxidant study indicative of antibacterial and antimutagenic activities of an ellagitannin-rich aqueous extract from the leaves of *Miconia latecrenata*. *J. Ethnopharmacol.* **2019**, *236*, 114–123. [CrossRef]

29. Pasta, P.C.; Durigan, G.; Moraes, I.C.F.; Ribeiro, L.F.; Haminiuk, C.W.I.; Branco, I.G. Physicochemical properties, antioxidant potential and mineral content of *Miconia albicans* (Sw.) Triana: A fruit with high aluminium content. *Braz. J. Bot.* **2019**, *42*, 209–216. [CrossRef]

30. de Moura, V.M.; Bezerra, A.N.S.; Mourão, R.H.V.; Lameiras, J.L.V.; Raposo, J.D.A.; de Sousa, R.L.; Boechat, A.L.; de Oliveira, R.B.; Chalkidis, H.D.M.; Dos-Santos, M.C. A comparison of the ability of *Bellucia dichotoma* Cogn. (Melastomataceae) extract to inhibit the local effects of *Bothrops atrox* venom when pre-incubated and when used according to traditional methods. *Toxicon* **2014**, *85*, 59–68. [CrossRef]

31. Lizcano, L.J.; Bakkali, F.; Ruiz-Larrea, M.B.; Ruiz-Sanz, J.I. Antioxidant activity and polyphenol content of aqueous extracts from Colombian Amazonian plants with medicinal use. *Food Chem.* **2010**, *119*, 1566–1570. [CrossRef]

32. Tong, Y.; Jiang, Y.; Chen, X.; Li, X.; Wang, P.; Jin, Y.; Cheng, K. Extraction, Enrichment, and Quantification of Main Antioxidant Aglycones of Flavonoids and Tannins from *Melastoma Dodecandrum* Lour.: Guided by UPLC-ESI-MS/MS. *J. Chem.* **2019**, *2019*, 2793058. [CrossRef]

33. Montenegro, H.; Gonzalez, J.; Ortega-Barria, E.; Cubilla-Rios, L. Antiprotozoal Activity of Flavonoid Glycosides Isolated from *Clidemia sericea*. and *Mosquitoxylon jamaicense*. *Pharm. Biol.* **2007**, *45*, 376–380. [CrossRef]

34. Rath, G.; Touré, A.; Nianga, M.; Wolfender, J.L.; Hostettmann, K. Characterization of C-glycosylflavones from *Dissotis rotundifolia* by liquid chromatography—UV diode array detection—tandem mass spectrometry. *Chromatographia* **1995**, *41*, 332–342. [CrossRef]

35. Lawarence, B.; Murugan, K. Comprehensive Evaluation of Antioxidant Potential of Selected Osbeckia species and their in vitro Culture, Purification and Fractionation. *Pharmacogn. J.* **2017**, *9*, 674–682. [CrossRef]

36. Lima, T.C.; Matos, S.S.; Carvalho, T.F.; Silveira-Filho, A.J.; Couto, L.P.S.M.; Quintans-Júnior, L.J.; Quintans, J.S.S.; Silva, A.M.O.; Heimfarth, L.; Passos, F.R.S.; et al. Evidence for the involvement of IL-1β and TNF-α in anti-inflammatory effect and antioxidative stress profile of the standardized dried extract from *Miconia albicans* Sw. (Triana) Leaves (Melastomataceae). *J. Ethnopharmacol.* **2020**, *259*, 112908. [CrossRef]

37. Peethambar, S.K.; Puttaswamy, R.; Vinayaka, K.S.; Padukone, S.; Achur, R.N. Pharmacological and gross behavioral studies on Memecylon terminale Dalz, a medicinal plant from Western Ghats in southern India. *World J. Pharm. Sci.* **2013**, *1*, 61–92.

38. International Council for Harmonization. ICH Guideline M10 on Bioanalytical Method Validation-Draft Version. Available online: https://www.ema.europa.eu/en/documents/scientific-guideline/draft-ich-guideline-m10-bioanalytical-method-validation-step-2b_en.pdf (accessed on 20 November 2021).

39. Houghton, P.; Mukherjee, P.K. *Evaluation of Herbal Medicinal Products*; Pharmaceutical Press: London, UK, 2009; pp. 85–94. ISBN 978-0-85369-751-0.

40. Emerit, J.; Edeas, M.; Bricaire, F. Neurodegenerative diseases and oxidative stress. *Biomed. Pharmacother.* **2004**, *58*, 39–46. [CrossRef]

41. Dhalla, N.S.; Temsah, R.M.; Netticadan, T. Role of oxidative stress in cardiovascular diseases. *J. Hypertens.* **2000**, *18*, 655–673. [CrossRef] [PubMed]

42. Li, S.; Tan, H.-Y.; Wang, N.; Zhang, Z.-J.; Lao, L.; Wong, C.-W.; Feng, Y. The Role of Oxidative Stress and Antioxidants in Liver Diseases. *Int. J. Mol. Sci.* **2015**, *16*, 26087–26124. [CrossRef] [PubMed]

43. Luo, J.; Mills, K.; Le Cessie, S.; Noordam, R.; van Heemst, D. Ageing, age-related diseases and oxidative stress: What to do next? *Ageing Res. Rev.* **2020**, *57*, 100982. [CrossRef] [PubMed]

44. Ferreres, F.; Gil-Izquierdo, A.; Andrade, P.B.; Valentão, P.; Tomás-Barberán, F.A. Characterization of C-glycosyl flavones O-glycosylated by liquid chromatography–tandem mass spectrometry. *J. Chromatogr. A* **2007**, *1161*, 214–223. [CrossRef] [PubMed]

45. Vukics, V.; Guttman, A. Structural characterization of flavonoid glycosides by multi-stage mass spectrometry. *Mass Spectrom. Rev.* **2010**, *29*, 1–16. [CrossRef]

46. Llorent-Martínez, E.J.; Spínola, V.; Gouveia, S.; Castilho, P.C. HPLC-ESI-MS^n characterization of phenolic compounds, terpenoid saponins, and other minor compounds in *Bituminaria bituminosa*. *Ind. Crop. Prod.* **2015**, *69*, 80–90. [CrossRef]

47. Wojakowska, A.; Perkowski, J.; Góral, T.; Stobiecki, M. Structural characterization of flavonoid glycosides from leaves of wheat (Triticum aestivum L.) using LC/MS/MS profiling of the target compounds. *J. Mass Spectrom.* **2013**, *48*, 329–339. [CrossRef]

48. Yang, X.; Yang, L.; Xiong, A.; Li, D.; Wang, Z. Authentication of *Senecio scandens* and *S. vulgaris* based on the comprehensive secondary metabolic patterns gained by UPLC–DAD/ESI-MS. *J. Pharm. Biomed. Anal.* **2011**, *56*, 165–172. [CrossRef]

49. Cherfia, R.; Zaiter, A.; Akkal, S.; Chaimbault, P.; Abdelwahab, A.B.; Kirsch, G.; Chaouche, N.K. New approach in the characterization of bioactive compounds isolated from *Calycotome spinosa* (L.) Link leaves by the use of negative electrospray ionization LITMS^n, LC-ESI-MS/MS, as well as NMR analysis. *Bioorg. Chem.* **2020**, *96*, 103535. [CrossRef]

50. Vukics, V.; Ringer, T.; Kéry, Á.; Bonn, G.K.; Guttman, A. Analysis of heartsease (*Viola tricolor* L.) flavonoid glycosides by micro-liquid chromatography coupled to multistage mass spectrometry. *J. Chromatogr. A* **2008**, *1206*, 11–20. [CrossRef]

51. Mimura, M.R.M.; Salatino, A.; Salatino, M.L.F. Distribution of flavonoids and the taxonomy of *Huberia* (Melastomataceae). *Biochem. Syst. Ecol.* **2004**, *32*, 27–34. [CrossRef]

52. Zheng, W.-J.; Ren, Y.-S.; Wu, M.-L.; Yang, Y.-L.; Fan, Y.; Piao, X.-H.; Ge, Y.-W.; Wang, S.-M. A review of the traditional uses, phytochemistry and biological activities of the *Melastoma* genus. *J. Ethnopharmacol.* **2021**, *264*, 113322. [CrossRef]
53. Sarju, N.; Samad, A.A.; Ghani, M.A.; Ahmad, F. Detection and quantification of Naringenin and kaempferol in *Melastoma decemfidum* extracts by GC-FID and GC-MS. *Acta Chromatogr.* **2012**, *24*, 221–228. [CrossRef]
54. Han, Q.-H.; Mu, Y.-X.; Gong, X.; Zhang, N.; Zhang, C.-H.; Li, M.-H. Chemical constituents of *Medinilla septentrionalis* (W. W. Sm.) H. L. Li (Melastomataceae). *Biochem. Syst. Ecol.* **2019**, *86*, 103901. [CrossRef]
55. Azahar, N.F.; Gani, S.S.A.; Zaidan, U.H.; Bawon, P.; Halmi, M.I.E. Optimization of the Antioxidant Activities of Mixtures of Melastomataceae Leaves Species (*M. malabathricum Linn Smith*, *M. decemfidum*, and *M. hirta*) Using a Simplex Centroid Design and Their Anti-Collagenase and Elastase Properties. *Appl. Sci.* **2020**, *10*, 7002. [CrossRef]
56. Kazuno, S.; Yanagida, M.; Shindo, N.; Murayama, K. Mass spectrometric identification and quantification of glycosyl flavonoids, including dihydrochalcones with neutral loss scan mode. *Anal. Biochem.* **2005**, *347*, 182–192. [CrossRef]
57. Sun, Y.; Zhang, X.; Xue, X.; Zhang, Y.; Xiao, H.; Liang, X. Rapid Identification of Polyphenol C-Glycosides from *Swertia franchetiana* by HPLC—ESI-MS—MS. *J. Chromatogr. Sci.* **2009**, *47*, 190–196. [CrossRef]
58. Zhou, C.; Liu, Y.; Su, D.; Gao, G.; Zhou, X.; Sun, L.; Ba, X.; Chen, X.; Bi, K. A Sensitive LC–MS–MS Method for Simultaneous Quantification of Two Structural Isomers, Hyperoside and Isoquercitrin: Application to Pharmacokinetic Studies. *Chromatographia* **2011**, *73*, 353–359. [CrossRef]
59. Tarawneh, A.H.; Leon, F.; Ibrahim, M.A.; Pettaway, S.; McCurdy, C.R.; Cutler, S.J. Flavanones from *Miconia prasina*. *Phytochem. Lett.* **2014**, *7*, 130–132. [CrossRef] [PubMed]
60. Zhang, Z.; ElSohly, H.N.; Li, X.-C.; Khan, S.I.; Broedel, S.E.; Raulli, R.E.; Cihlar, R.L.; Walker, L.A. Flavanone Glycosides from *Miconia trailii*. *J. Nat. Prod.* **2003**, *66*, 39–41. [CrossRef] [PubMed]
61. Yin, J.; Ma, Y.; Liang, C.; Wang, H.; Sun, Y.; Zhang, L.; Jia, Q. A Complete Study of Farrerol Metabolites Produced in Vivo and in Vitro. *Molecules* **2019**, *24*, 3470. [CrossRef] [PubMed]
62. Murugan, R.; Parimelazhagan, T. Comparative evaluation of different extraction methods for antioxidant and anti-inflammatory properties from *Osbeckia parvifolia* Arn.–An in vitro approach. *J. King Saud. Univ. Sci.* **2014**, *26*, 267–275. [CrossRef]
63. Taiwo, J.E.; Shemishere, U.B.; Omoregie, E.S. Phytochemical constituents and in vitro antioxidant activities of *Melastomastrum capitatum* and *Solanum macrocarpon* leaf extracts. *Niger. J. Life Sci.* **2017**, *7*, 128–149.
64. Zanatta, A.C.; Vilegas, W.; Edrada-Ebel, R. UHPLC-(ESI)-HRMS and NMR-Based Metabolomics Approach to Access the Seasonality of *Byrsonima intermedia* and *Serjania marginata* From Brazilian Cerrado Flora Diversity. *Front. Chem.* **2021**, *9*, 534. [CrossRef]
65. Albuquerque, L.B.; Aquino, F.G.; Costa, L.C.; Miranda, Z.J.G.; Sousa, S.R. Espécies de Melastomataceae Juss. com potencial para restauração ecológica de mata ripária no cerrado. *Polibotánica* **2013**, *35*, 1–19.
66. Ishino, M.N.; de Sibio, P.R.; Rossi, M.N. Edge effect and phenology in *Erythroxylum tortuosum* (Erythroxylaceae), a typical plant of the Brazilian Cerrado. *Braz. J. Biol.* **2012**, *72*, 587–594. [CrossRef]

Article

Phytochemicals of Avocado Residues as Potential Acetylcholinesterase Inhibitors, Antioxidants, and Neuroprotective Agents

Geisa Gabriela da Silva [1], Lúcia Pinheiro Santos Pimenta [2], Júlio Onésio Ferreira Melo [3], Henrique de Oliveira Prata Mendonça [3], Rodinei Augusti [2] and Jacqueline Aparecida Takahashi [2,*]

[1] Department of Food Science, Faculty of Pharmacy, Universidade Federal de Minas Gerais, Av. Antônio Carlos, 6627, Belo Horizonte 31270-901, Brazil; geisambio@gmail.com

[2] Chemistry Department, Universidade Federal de Minas Gerais, Av. Antônio Carlos, 6627, Belo Horizonte 31270-901, Brazil; lpimenta@qui.ufmg.br (L.P.S.P.); augusti.rodinei@gmail.com (R.A.)

[3] Exact and Biological Sciences Department, Campus Sete Lagoas, Universidade Federal de São João del-Rei, Sete Lagoas 36307-352, Brazil; onesiomelo@gmail.com (J.O.F.M.); hp.quimico@hotmail.com (H.d.O.P.M.)

* Correspondence: jat@qui.ufmg.br; Tel.: +55-31-34095754

Abstract: Avocado (*Persea americana*) is a widely consumed fruit and a rich source of nutrients and phytochemicals. Its industrial processing generates peels and seeds which represent 30% of the fruit. Environmental issues related to these wastes are rapidly increasing and likely to double, according to expected avocado production. Therefore, this work aimed to evaluate the potential of hexane and ethanolic peel (PEL-H, PEL-ET) and seed (SED-H, SED-ET) extracts from avocado as sources of neuroprotective compounds. Minerals, total phenol (TPC), total flavonoid (TF), and lipid contents were determined by absorption spectroscopy and gas chromatography. In addition, phytochemicals were putatively identified by paper spray mass spectrometry (PSMS). The extracts were good sources of Ca, Mg, Fe, Zn, ω-6 linoleic acid, and flavonoids. Moreover, fifty-five metabolites were detected in the extracts, consisting mainly of phenolic acids, flavonoids, and alkaloids. The in vitro antioxidant capacity (FRAP and DPPH), acetylcholinesterase inhibition, and in vivo neuroprotective capacity were evaluated. PEL-ET was the best acetylcholinesterase inhibitor, with no significant difference ($p > 0.05$) compared to the control eserine, and it showed neither preventive nor regenerative effect in the neuroprotection assay. SED-ET demonstrated a significant protective effect compared to the control, suggesting neuroprotection against rotenone-induced neurological damage.

Keywords: avocado biomass; Alzheimer's disease; neuroprotective effect; avocado seed; avocado peel; paper spray mass spectrometry

Citation: da Silva, G.G.; Pimenta, L.P.S.; Melo, J.O.F.; Mendonça, H.d.O.P.; Augusti, R.; Takahashi, J.A. Phytochemicals of Avocado Residues as Potential Acetylcholinesterase Inhibitors, Antioxidants, and Neuroprotective Agents. *Molecules* **2022**, *27*, 1892. https://doi.org/10.3390/molecules27061892

Academic Editor: Simona Rapposelli

Received: 28 January 2022
Accepted: 11 March 2022
Published: 15 March 2022

Publisher's Note: MDPI stays neutral with regard to jurisdictional claims in published maps and institutional affiliations.

1. Introduction

Originated from Central America, the avocado is the fruit of the *Persea americana* Mill species from the Lauraceae family. It is rich in minerals, vitamins, proteins, fibers, and unsaturated lipids, the latter known to prevent cardiovascular diseases [1,2]. Moreover, avocado seeds display significant antioxidant, anti-inflammatory, and anti-cancer effects which have been associated with elevated percentages of hydrocarbons, sterols, and unsaturated fatty acids. Pulp has the same effects, although at a lower intensity [3]. Finally, the peel is rich in phenolic compounds and minerals, and has significant antioxidant activity, comparable to the seeds and far superior to the pulp [4,5].

In 2020, Mexico was the largest avocado producer, accounting for 2393 thousand tons per year, followed by the Dominican Republic, Peru, Colombia, Indonesia, and Brazil [6]. Since the inedible parts of avocado, mainly constituting peels and seeds, are discarded by the industry, restaurants, or at home, with the growth of avocado production, the amount of residues is also increasing, causing environmental problems [7]. These parts can reach up to 30% of the fruit. Furthermore, it is predicted that avocado production will double shortly [8]. In this sense, changes in the management of this fruit are necessary to accompany the dynamics of economic demands, with minimal environmental impacts [8–10]. Because avocado solid waste retains useful bioactive natural compounds, this residue has been targeted by several interesting studies to recover phytochemicals of industrial interest [11–14].

The antioxidant activity of avocado inedible parts can be explored in the development of neuroprotective agents against oxidative-related diseases like Alzheimer's [4,5,11,14]. Alzheimer's disease is one of the most concerning neurodegenerative disorders, affecting over 55 million people, mainly the elderly [15]. This disease can disrupt the synapses between neurons by the action of acetylcholinesterase (AChE), an enzyme that prevents acetylcholine from establishing communication between neurons. There is still no cure for this pathology, but palliative drugs have been very useful in decreasing symptoms and improve the quality of life of patients with Alzheimer's disease. However, these drugs are usually expensive. The monthly cost of galantamine (30 pills) almost reaches USD 200, while drugs for the treatment of other diseases such as diabetes and hypertension cost less than USD 20 [16]. Therefore, the development of new drugs to treat neurodegenerative diseases from avocado biowaste could widen access to treatment while generating economic gains for the industry.

In this context, this work aimed to validate the nutritional and functional value of avocado residues by carrying out chemical and biological assays on seeds and peels of the Fortuna variety. The mineral, fatty acid, phenolic, and flavonoid contents were determined using spectroscopic and chromatographic methodologies. The phytochemical profile was determined by Paper Spray Mass Spectrometry (PSMS). Biological assays targeted antioxidant potential, AChE inhibition, and neuroprotective efficacy.

2. Results and Discussion

2.1. Residues Weight and Extracts Yields

Avocado fruits were separated into parts and weighted, resulting in 69.9% pulp, 10.2% peel, and 19.9% seed. According to the FAO [17], the world production of avocado is estimated to reach 9.2 tons by 2028. In this scenario, the volume of residues can reach approximately two tons. The peels were extracted by maceration with hexane and ethanol, providing two extracts, PEL-H (1.41 g/100 g of sample) and PEL-ET (16.94 g/100 g of sample). The same procedure was repeated with the seeds to furnish hexane (SED-H, 0.94 g/100 g of sample) and ethanol (SED-ET, 28.58 g/100 g of sample) extracts. The high yields of ethanol extracts reflect greater extraction power than that of hexane. The substitution of organic solvents with an environmentally favorable solvent, such as ethanol, makes the recycling of avocado waste important in terms of green chemistry and sustainability.

2.2. Mineral Contents in Avocado Residues

The mineral contents were quantified by atomic absorption spectroscopy, and the results are presented in Table 1.

Table 1. Mineral content in avocado fresh peels and seeds.

Minerals (mg/100 g of Sample)	Peels [1]	Seeds [1]
Ca	26.78 ± 2.06	41.14 ± 8.50
Cu	0.20 ± 0.74	0.48 ± 0.01
Fe	0.72 ± 2.06	1.04 ± 7.40
Mg	23.87 ± 3.09	31.41 ± 1.82
Mn	4.23 ± 10.34	1.80 ± 6.48
Zn	0.67 ± 7.02	1.11 ± 0.57

[1] Values represent mean standard deviations ($n = 5$).

According to the results, both avocado peels and seeds are good sources of minerals, especially calcium and magnesium, along with minor amounts of copper, iron, manganese, and zinc. Calcium, together with iron and zinc, are among the most difficult nutrients to obtain using local foods, according to research targeting women and young children from Southeast Asia [18]. In the U.S., mineral supplementation is advised during pregnancy [19]. Women are likely to be more affected by mineral deficiency, especially during childhood, although in many countries like Brazil, the prevalence of mineral deficiency is underreported [20]. Additionally, each ton of an agro-industrial residue containing avocado peels and seeds would correspond to the FAO/WHO's daily recommended intake of calcium (1024 mg) for over two thousand individuals for one month, which is very considerable given calcium deficiency in some groups. The mineral content of agro-industrial residues is variable. Peels and seeds of avocado have lower mineral content than cocoa honey, a cocoa byproduct [21], but higher than pea peels [22]. Minerals and other nutrient and bioactive compounds present in agro-industrial residues can be processed in different ways, such as snack crackers and dry soup [22].

2.3. Fatty Acid Composition

After chemical derivatization to fatty acid methyl esters (FAMEs), the extracts were evaluated by gas chromatography, allowing identification of the fatty acids presented in Table 2.

Table 2. Fatty acids composition of hexane and ethanolic extracts of avocado peels and seeds.

Fatty Acids	Fatty Acids Contents (%)			
	Peels		Seeds	
	PEL-H	**PEL-ET**	**SED-H**	**SED-ET**
Miristic 14:0	0.7	1.7	0.7	1.5
Palmitic 16:0	42.5	47.9	23.0	22.2
Palmitoleic 16:1	2.7	1.8	2.9	3.2
Stearic 18:0	7.0	22.2	4.1	14.7
Oleic 18:1	18.2	2.5	17.3	16.2
Linoleic 18:2	4.5	0.7	34.8	27.4
Linolenic 18:3	1.0	0.4	3.0	1.6
Total saturated fatty acids	50.2	71.8	27.8	38.4
Total unsaturated fatty acids	26.4	5.4	58	48.4
Total	76.6	77.2	85.8	86.8

SED-H: seed hexane extract; SED-ET: seed ethanolic extract; PEL-H: peel hexane extract, PEL-ET: peel ethanolic extract.

Linoleic acid, a ω-6 fatty acid, was the component predominant in the seed extracts (27.4 and 34.8%) and palmitic acid, the major component (42.5–47.9%), in the seed extracts (22.2–23%). Oleic acid was distributed in all samples, but its content was lower in the peel ethanolic extract. Stearic acid was detected in all samples, but mainly in the ethanol extracts (14.7% in the seeds/22.2% in the peels). The composition of the different extracts revealed that seeds had a higher content of unsaturated fatty acids, such as oleic and linoleic, and small amounts of saturated fatty acids, such as palmitic and stearic.

The human body cannot synthesize linoleic acid, an essential fatty acid; therefore, it must be acquired through feeding. Linoleic acid plays a fundamental role in fetal neurological function and infant growth because it is the precursor of arachidonic acid. The uptake of this fatty acid decreases the risk of developing heart disease and it is beneficial for inflammatory processes [23]. Furthermore, palmitic and stearic acids are undesirable in food because saturated fatty acids usually contribute to increased blood cholesterol. However, stearic acid is considered an exception, because a rapid in vivo enzymatic dehydrogenation reaction catalyzed by stearoyl-CoA Δ9-desaturase converts stearic acid to oleic acid (18:1 Δ9) in the body [24]. The fatty acid composition found in avocado residue extracts suggests that they could be better used in combination with other oils, to balance the final composition. For instance, a sample of olive oil studied by Orsavova et al. [25] presented 16.4% of linoleic acid and 16.5% of palmitic acid (~1:1 proportion). A 1:1 mixture of this olive oil and the oil present in the SED-H would increase the proportional quantity of linoleic acid to 26% and the amount of palmitic acid to 19.75%. In the end, such mixed oil, composed of 50% oil recovered from waste biomass, would benefit from an improved content of linoleic acid.

The health claims for avocado fruits are closely related to the high content of fatty acids that are essential for human health [26]. Therefore, the essential fatty acids detected in the residues already consist of valuable metabolites that could be used as food additives or might be extracted, adding value to avocado waste biomass, which is a promising and inexpensive alternative source of linoleic acid.

2.4. Determination of the Total Phenolic and Flavonoid Contents

Avocado is a rich source of polyphenols, including hydroxybenzoic acid, caffeoylquinic acid derivatives, and cinnamic acids [5,27]. Flavonoids, such as quercetin, quercetin glycosides, naringenin, catechin, and epigallocatechin, were described in avocado seeds of the "Hass" variety [5]. In the current study, the total phenolic content (TPC) was estimated using the Folin–Ciocalteu method for each extract and expressed in mg of gallic acid equivalents (GAE)/g of extract. The total flavonoid content (TFC) was determined using the aluminum chloride colorimetric method and expressed as miligram of quercetin per gram of extract. Table 3 presents the results of the polyphenol and flavonoid contents according to the previous methodologies mentioned.

Table 3. Concentrations of phenols and flavonoids in different extracts of avocado peels and seeds.

Assays	Extracts			
	PEL-H	PEL-ET	SED-H	SED-ET
Total phenolic content (TPC) *	26.33 ± 0.48 [g]	35.40 ± 0.60 [d]	32.48 ± 2.00 [e]	32.15 ± 0.39 [fe]
Total flavonoid content (TFC) **	1243.78 ± 32.33 [j]	694.058 ±1.490 [l]	1199.04 ± 49.39 [k]	640.72 ± 9.30 [l]

SED-H: seed hexane extract; SED-ET: seed ethanolic extract; PEL-H: peel hexane extract, PEL-ET: peel ethanolic extract. * (TPC) = expressed as mg gallic acid/g extract; ** (TFC) = expressed in mg quercetin/g of extract. Data are expressed as the mean ± SD of five replicates. Different letters in the columns indicate statistical difference according to Tukey's test ($p < 0.05$).

The TPC values was found in the range of 26.33–35.40 mg of gallic acid/g of extract, and the TFC values ranged from 640.72 to 1199.04 mg of quercetin/g of extract. The peel ethanolic extract (PEL-ET) presented the highest content of total phenolic compounds (35.40 ± 0.599 mg of gallic acid/g of ethanol extract), which was significantly different from all other extracts. Similarly, the TPC reported for peels from another variety of avocados using the same methodology (Folin-Ciocalteau) was 47.9 ± 2.7 mg of gallic acid/g of ethanol extract [28]. Conversely, peel ethanolic extracts from different varieties of *P. Americana*, including the Fortuna variety, presented lower phenolic compounds contents [29].

The peel extracts presented significantly higher flavonoid content than the extracts of the corresponding seeds. Higher percentages of flavonoid compounds were expected in the ethanol extracts, given the polar nature of the most representative flavonoids. However, our results showed that the higher percentages were present in the hexane extracts. This discrepancy can be partially explained by the presence of vitamin E in avocado residues [2]. Vitamin E is a phenolic non-polar compound, known to be present in substantial levels in avocado. Probably, its aliphatic long side chain may have contributed to the results. In addition, the percentages were determined by a colorimetric method that, although being a widely used methodology, is not a specific analysis. Flavonoid compounds have also been identified in the peels and seeds of the Hass and Fuerte varieties using HPLC-DAD [30].

2.5. Total Antioxidant Capacity and Antioxidant Activity Determined by DPPH Radical Scavenging and Ferric Reducing Power

The results regarding the antioxidant functional properties are shown in Table 4. In the total antioxidant activity assay (expressed as mmol of ascorbic acid/g extract) ($R^2 = 0.9339$), the values ranged from 630.23 to 770.01 and all extracts from seeds (SED-H and SED-ET) and peel hexane extracts (PEL-H) showed no significant differences.

Table 4. Total antioxidant capacity, antioxidant activity determined by DPPH radical scavenging and ferric reducing power, and acetylcholinesterase inhibition of avocado peels and seeds.

Assays	Extracts			
	PEL-H	PEL-ET	SED-H	SED-ET
Total antioxidant capacity *	26.33 ± 0.48 [g]	35.40 ± 0.60 [d]	32.48 ± 2.00 [e]	32.15 ± 0.39 [fe]
DPPH scavenging (%)	7.77 ± 1.44 [o]	52.20 ± 1.05 [m]	35.89 ± 1.59 [n]	37.60 ± 1.67 [n]
Ferric reducing power (%) **	4.81 ± 1.37 [g]	1.11 ± 0.25 [i]	4.07 ± 1.21 [gh]	2.38 ± 0.24 [hi]
AChE inhibition (%)	70.8 ± 9.7 [rq]	85.6 ± 11.1 [pq]	65.0 ± 8.9 [s]	78.0 ± 6.8 [qr]

SED-H: seed hexane extract; SED-ET: seed ethanolic extract; PEL-H: peel hexane extract, PEL-ET: peel ethanolic extract, DPPH: 2,2-Diphenyl-1-picrylhydrazyl. * (TPC) = expressed as mg gallic acid/g extract; ** (TFC) = expressed in mg quercetin/g of extract. Data are expressed as the mean $\pm$ SD of five replicates. Different letters in the columns indicate statistical differences according to Tukey's test ($p < 0.05$). Controls: Ferric reducing power (ascorbic acid): $98.89 \pm 9.36\%$; DPPH capture assay (ascorbic acid): $79.16 \pm 0.37\%$; AChE inhibition (eserine): $91.5 \pm 1.6\%$.

Peel and seed hexane extracts presented similar behavior in promoting the reduction of ferricyanide ion to ferrocyanide, forming Prussia's blue. The reducing power of both seed and peel ethanolic extracts was lower. The extract capacities of scavenging DPPH radicals were, in general, low. Only the PEL-ET showed moderate activity presenting over 50% of inhibition, with an IC_{50} 92.557 µg/mL, being, therefore, the most promising extract in this assay. The high antioxidant activity may be explained by the high content of phenolic compounds, which was observed for peel and seed ethanolic extracts.

2.6. Acetylcholinesterase Inhibition

The hexane (SED-H and PEL-H) and ethanolic (SED-ET and PEL-ET) extracts were submitted to acetylcholinesterase inhibition assay. All extracts inhibited acetylcholinesterase activity up to 65%, with the ethanolic extracts more active than the hexane ones (Table 4). This range of extract activities is promising when compared with the activity presented by the control, a pure compound. The extract is a complex mixture and the active constituents are usually minor constituents in the extracts. In this way, the inhibitory capacity of the peel ethanolic extract from avocado ($85.6 \pm 11.1\%$) over acetylcholinesterase is outstanding, presenting no significant difference ($p > 0.05$) compared to the activity of the pure standard, eserine ($91.5 \pm 1.6\%$). Comparing the whole picture, the extract PEL-ET demonstrated a direct relationship between the amount of total phenolic compounds and AChE inhibition, corroborating the influence of antioxidant phenolic compounds in the acetylcholinesterase activity. By contrast, an inverse relationship was perceived for the hexane and ethanol extracts of the seeds. The principal component analysis (PCA) score biplot allowed a general qualitative visualization of the results obtained in the biological screening (Figure 1).

The PCA biplot shows the close relationship of the ethanolic extracts with the presence of phenolic compounds, DPPH assay, and acetylcholinesterase inhibition. The peel ethanolic extract (PEL-ET) demonstrated the closest association with the acetylcholinesterase enzyme inhibition assay. Concerning the hexane extracts, the seed extract (SED-H) was mainly related to the total antioxidant activity, while the peel extract (PEL-H) is associated with both ferric reduction power and total flavonoid contents. The hexane extracts are more distant from the AChE assay and present lower activity for acetylcholinesterase inhibition, as observed in the study. Therefore, only the ethanolic extracts were submitted to the in vivo *D. melanogaster* assay to evaluate the neuroprotective response.

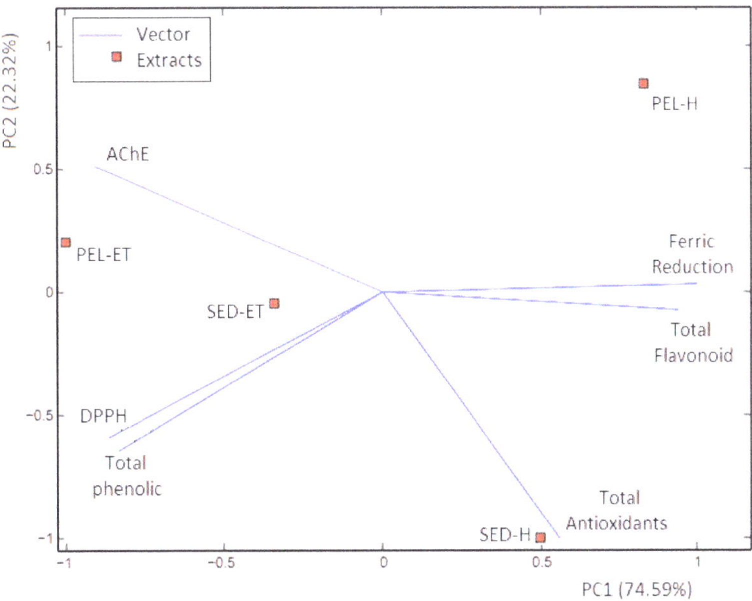

Figure 1. Principal component analysis biplot, PC1 versus PC2, correlating antioxidant responses, total phenolic content, total flavonoid content, and acetylcholinesterase (AChE) inhibitory activity. DPPH: 2,2-Diphenyl-1-picrylhydrazyl; SED-H: seed hexane extract; SED-ET: seed ethanolic extract; PEL-H: peel hexane extract, and PEL-ET: peel ethanolic extract.

2.7. Activity of P. americana Extracts in Neuroprotection Using D. melanogaster Model

D. melanogaster flies have been used as an in vivo model for the evaluation of neurological damage related to neurodegenerative conditions. In this model, neurological damage can be induced in the flies by in vitro feeding with chemicals such as rotenone, preventing these insects from flying. The *D. melanogaster* model was used by Jiménez et al. [31] to demonstrate the neuroprotective effect of methanol extract prepared from *Solanum ovalifolium*, and by Siima et al. [32] to show the effect of flavonoids and polyketides in rescuing locomotor capacity in a Parkinson's disease-related study [33]. Test compounds can be administered before or after flies' exposure to rotenone to evaluate protective and regenerative effects, respectively. In the first alternative, the test compound avoids some of the damages that rotenone would cause (preventive action). In the second, damages due to rotenone exposure are established in the flies before test compound administration, and the capacity of the target compound to revert the neurodegenerative effects is evaluated (regenerative action).

In the present work, test compounds were administered before and after exposure to rotenone, to evaluate both effects (Figure 2). The ethanol extracts (PEL-ET and SED-ET) were evaluated by this model and the results were submitted to the Shapiro–Wilk statistical normality test. The p-value found (>0.05) indicated that the distribution was not suitable. Therefore, the Kruskal–Wallis nonparametric test was conducted, identifying a significant difference; subsequently, the Dunn's test was used, which identified a significant difference $p < 0.05$ between the samples. Overall, a high mortality rate was observed during the trial in which flies were first exposed to rotenone (Figure 2). In this condition, rotenone caused 60.4–72.9% of flies to die, showing high acute toxicity. Under this condition, peel extract (PEL-ET) did not statistically differ ($p > 0.05$) from the control, in which flies were not exposed either to the extracts or to rotenone. Therefore PEL-ET showed neither preventive nor regenerative effect in this model.

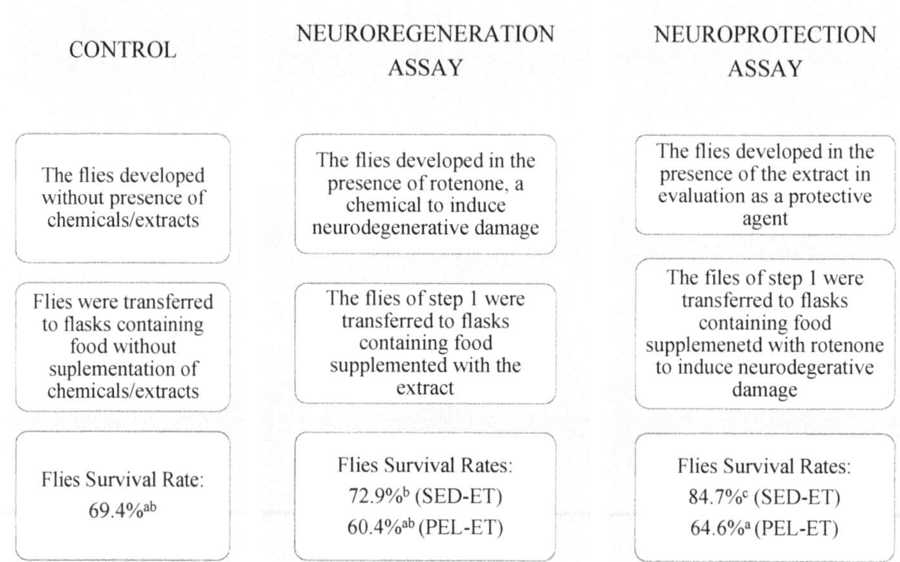

Figure 2. Outline and results of negative geotaxis assay of peels and seeds ethanol extracts in comparison with the control. Different letters indicate statistical difference by Dunn's test ($p < 0.05$). SED-H: seed hexane extract; SED-ET: seed ethanolic extract; PEL-H: peel hexane extract, and PEL-ET: peel ethanolic extract.

SED-ET showed a more significant result (Figure 2). Comparing the seed and peel results in this assay, only SED-ET demonstrated a protective effect in this treatment. In addition, the performance of SED-ET was significant compared to the control, presenting a higher number of flies capable of flying above the limit established, which suggests that the seed extract provided a neuroprotective effect against rotenone-induced neurological damage (Figure 2). This flies model has been leading to the identification of new therapeutic targets for research in Parkinson's [34], Alzheimer's [35], and other neurodegenerative diseases. The regenerative activity has been the main target of drugs to individuals in the advanced neurodegeneration stage. However, protection against degeneration can be helpful to patients in the early stages of neurodegenerative diseases.

2.8. Identification of Metabolites by Paper Spray Mass Spectrometry

Paper spray mass spectrometry (PSMS) was applied to identify the natural products present in the ethanol extracts. This quick and efficient technique for chemical profile determination is commonly used in the clinic and in forensic research to identify specific ions through direct sample analysis [36–39]. This study showed no difference in the chemical profile of peel and seed samples. PS ($-$) MS spectrum of SED-ET is presented in Figure 3. Fifty-five compounds were tentatively identified in the ethanol extracts by PSMS analyzing their MS and MS2 spectra, together with information previously reported in the literature. In the negative mode, 41 metabolites were identified, mainly flavonoids and organic acids, most of them with biological activities already reported [12].

The presence of compounds such as caffeic acid (m/z 179 [M $-$ H]$^-$), catechin (m/z 289 [M $-$ H]$^-$), rutin (m/z 609 [M $-$ H]$^-$), procyanidin isomers, and flavonoids validated the antioxidant activity of the seed extract and were in accordance with previous reports [39,40]. Corroborating the experiments, some of the metabolites detected in the current work were already reported in the seeds and peels of avocado (vanillin, caffeic acid, ferulic acid, synaptic acid, kaempferol, catechin, quercetin-3-glucoside, quercetin-O-arabinosyl-glucoside, rutin, dimers A and B, and trimer A of procyanidin) [12]. Based on the antioxidant effect, Segovia et al. [41] described the role of avocado seed extract as an additive to protect oil mixtures from oxidation. The presence of quinic acid derivatives in the extract also contributed to validating the antioxidant activity and, therefore, highlighted the beneficial properties of this residue as food ingredients for industrial use [42]. Table 5 presents the compounds tentatively identified in the extracts in negative mode.

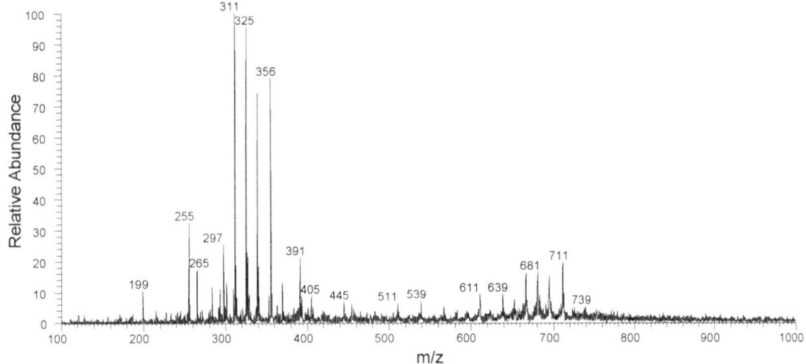

Figure 3. PS (−) MS full scan of ethanol extracts from seed of avocado.

Table 5. Compounds tentatively identified in avocado seed and peel by (−) PSMS.

m/z	Compound	Chemical Structure	Class	MS/MS	Extract	Reference
151	Vanillin	$C_8H_8O_3$	Aldehyde	136, 107, 93	S, P	[43]
179	Caffeic acid	$C_9H_8O_4$	Phenolic acid	151, 135	S	[43]
191	Quinic acid	$C_7H_{12}O_6$	Phenolic acid	93, 111	S, P	[43]
197	Syringic acid	$C_9H_{10}O_5$	Phenolic acid	153, 141, 125	S	[43]
209	5-Hydroxyferulic acid	$C_{10}H_{10}O_5$	Phenolic acid	191, 165, 118	S	[44]
223	Sinapic acid	$C_{11}H_{12}O_5$	Phenolic acid	179, 151, 85	S	[43]
269	Apigenin	$C_{15}H_{10}O_5$	Flavonoid	252, 223, 197	S, P	[43]
285	Kaempferol	$C_{15}H_{10}O_6$	Flavonoid	255, 224, 213	S, P	[43]
289	Catechin	$C_{15}H_{13}O_6$	Flavonoid	274, 245, 217, 199	S, P	[44]
301	Quercetin	$C_{15}H_{10}O_7$	Flavonoid	272, 265, 123	S, P	[43]
315	Hydroxytyrosol hexoside	$C_{14}H_{20}O_8$	Phenolic glycoside	297, 269, 243	S, P	[44]
325	p-coumaroyl hexose	$C_{15}H_{17}O_8$	Phenolic glycoside	261, 197, 183, 170	S, P	[44]
337	3-O-p-coumaroylquinic acid	$C_{16}H_{18}O_8$	Phenolic compound	293, 237, 183	S, P	[44]
341	Caffeic acid-hexoside	$C_{15}H_{18}O_9$	Phenolic glycoside	280, 185, 183, 179	S, P	[44]
353	5-O-caffeoylquinic acid	$C_{16}H_{18}O_9$	Phenolic compound	309, 211, 191, 183	S, P	[43]
417	Kaempferol-O-pentoside	$C_{20}H_{18}O_{10}$	Flavonoid glycoside	404, 344, 289	S, P	[44]
419	Cyanidin 3-O-pentatoside	$C_{20}H_{19}O_{10}$	Flavonoid glycoside	395, 361, 292, 287	S, P	[44]
431	Vitexin	$C_{21}H_{20}O_{10}$	Flavonoid glycoside	362, 351, 311, 196	S	[44]
433	Peonidin 3-O-pentoside	$C_{21}H_{21}O_{11}$	Flavonoid glycoside	420, 389, 301, 205	S, P	[44]
435	Phloridzin	$C_{21}H_{24}O_{10}$	Flavonoid glycoside	426, 416, 369	S	[43]

Table 5. *Cont.*

m/z	Compound	Chemical Structure	Class	MS/MS	Extract	Reference
447	Kaempferol-O-hexoside	$C_{21}H_{19}O_{11}$	Flavonoid glycoside	420, 403, 352, 301	S, P	[44]
449	Dihydroquercetin-3,5-rhamnoside	$C_{21}H_{22}O_{11}$	Flavonoid glycoside	430, 303, 298, 286	S	[44]
451	Cinchonain	$C_{24}H_{20}O_9$	Flavonoid	424, 414, 377	S, P	[44]
461	Isorhamnetin-O-coumaroyl	$C_{22}H_{22}O_{11}$	Flavonoid	461, 417, 216	S, P	[44]
463	Quercetin-3-hexoside	$C_{21}H_{20}O_{12}$	Flavonoid glycoside	464, 384, 316, 300	S	[44]
473	Quercetin-3-O-hexoside	$C_{21}H_{19}O_{12}$	Flavonoid glycoside	467, 436, 372	S, P	[45]
477	Quercetin glucuronide	$C_{21}H_{18}O_{13}$	Flavonoid	431, 262, 231	S	[44]
487	Caffeoyl hexose-deoxyhexoside	$C_{22}H_{31}O_{12}$	Flavonoid	442, 298, 173	S	[44]
491	Dimethyl ellagic acid hexoside	$C_{22}H_{22}O_{13}$	Flavonoid	343, 275, 269	S, P	[44]
563	Apigenin-C-hexoside-C-pentoside	$C_{26}H_{28}O_{14}$	Flavonoid glycoside	531, 446, 298	S	[44]
575	Procyanidin dimer A	$C_{30}H_{24}O_{12}$	Flavonoid	431, 404, 329	S	[43]
577	Procyanidin dimer B	$C_{30}H_{25}O_{12}$	Flavonoid	532, 516, 420	S	[44]
579	Luteolin 7-O-(2″-O-pentosyl)hexoside	$C_{26}H_{28}O_{15}$	Flavonoid glycoside	560, 542, 514	S, P	[44]
593	Catechin dihexoside	$C_{27}H_{29}O_{15}$	Flavonoid glycoside	574, 495, 347	S	[44]
595	Quercetin-O-pentatosyl-hexoside	$C_{26}H_{28}O_{16}$	Flavonoid glycoside	562, 558, 497	S, P	[44]
609	Rutin	$C_{27}H_{30}O_{16}$	Flavonoid glycoside	573, 564, 208	S, P	[44]
625	Quercetin-3,4′-O-diglucoside	$C_{27}H_{30}O_{17}$	Flavonoid glycoside	605, 588, 581	S, P	[43]
863	Procyanidin trimer A	$C_{45}H_{36}O_{18}$	Flavonoid	845, 826, 555	S, P	[43]
865	Procyanidin trimer B-isomer 1	$C_{45}H_{38}O_{18}$	Flavonoid	829, 735, 560	S	[43]

S: Avocado seed; P: Avocado peel.

In the positive mode, fifteen alkaloids were identified: anibine (**1**), duckeine (**2**), riparin I, II and III (**3–5**), norcanelilline (**6**), N-methylcoclaurine (**7**), anicanine (**8**), (-)-α-8-methylpseudoanibacanine (**9**), ceceline (**10**), (+)-manibacanine (**11**), cassythicine (**12**), isoboldine (**13**), reticuline (**14**), nantenine (**15**), and anibamine (**16**) (Figures 4 and 5, Table 6). PS(+)MS full scan of ethanol extract from peel of avocado can be found in Supplementary Materials, Figure S1. It is reported that, as the avocado fruit develops, the carbohydrate amount decreases and the fatty acids and secondary metabolites content increases [46,47]. In ripe avocado fruits, the alkaloids constitute a remarkable percentage of the natural compounds present in the pulp [46,47]. From the alkaloids identified, anibine (**1**), duckeine (**2**), and anicanine (**8**) were not detected in PEL-ET. All other alkaloids were detected in both extracts.

Alkaloids are used as active substances in prescriptions for the treatment of neurodegenerative diseases, e.g., galantamine. Therefore, the alkaloids detected in the extracts may contribute to the acetylcholinesterase inhibition and neuroprotective effect. More specifically, the alkaloids riparin (**3–5**) show antioxidant, antinociceptive, anti-inflammatory, and neuroprotective effects [49–51]. The aporphine alkaloid nantenine (**14**) has a potential role in acetylcholinesterase inhibition, since nantenine is present in *Unopsis stipitate*, a plant traditionally used for cognitive disorders [52]. Furthermore, nantenine possesses pharmaceutical interest as anticonvulsant, due to its direct inhibitory effect on calcium influx [53].

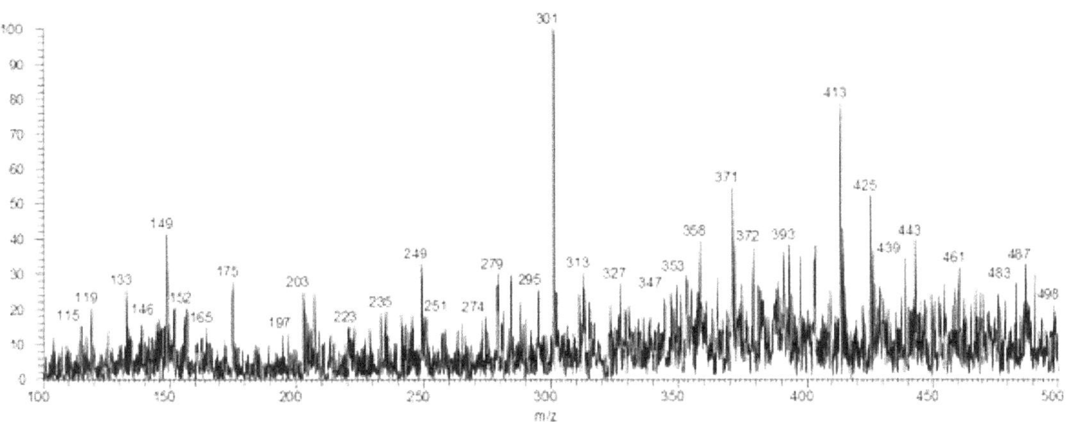

Figure 4. PS (+) MS full scan of ethanol extract from seed of avocado.

Table 6. Compounds tentatively identified in avocado seed and peel by (+) PSMS.

m/z	Compound	Chemical Structure	MS/MS	Extract	Reference
204	Anibine (**1**)	$C_{11}H_9NO_3$	183, 188, 192	S	[48]
244	Duckeine (**2**)	$C_{13}H_{11}NO_4$	226, 235, 187	S	[48]
256	Riparin I (**3**)	$C_{16}H_{17}NO_2$	241, 187, 212	S, P	[48]
270	Norcanelilline (**6**)	$C_{17}H_{19}NO_3$	214, 261, 240	S, P	[48]
272	Riparin II (**4**)	$C_{16}H_{17}NO_3$	263, 250, 254	S, P	[48]
282	Anicanine (**8**)	$C_{19}H_{23}NO_3$	200, 183, 192	S	[48]
288	Riparin III (**5**)	$C_{16}H_{17}NO_4$	270, 106, 271	S, P	[48]
296	(-)-α-8-methyl-pseudoanibacanine (**9**)	$C_{19}H_{21}NO_3$	279, 287, 239	S, P	[48]
300	N-methylcoclaurine (**7**)	$C_{18}H_{21}NO_3$	283, 291, 227	S, P	[48]
305	Ceceline (**10**)	$C_{19}H_{16}N_2O_2$	263, 273, 287	S, P	[48]
312	(+)-Manibacanine (**11**)	$C_{19}H_{21}NO_3$	116, 291, 243	S, P	[48]
325	Cassythicine (**12**)	$C_{19}H_{19}NO_4$	287, 316, 307	S, P	[45]
328	Isoboldine (**13**)	$C_{19}H_{21}NO_4$	297, 178, 310	S, P	[48]
330	Reticuline (**14**)	$C_{19}H_{23}NO_4$	187, 218, 235	S, P	[48]
339	Nantenine (**15**)	$C_{20}H_{21}NO_4$	249, 330, 321	S, P	[45]
424	Anibamine (**16**)	$C_{30}H_{50}N^+$	334, 379, 418	S, P	[48]

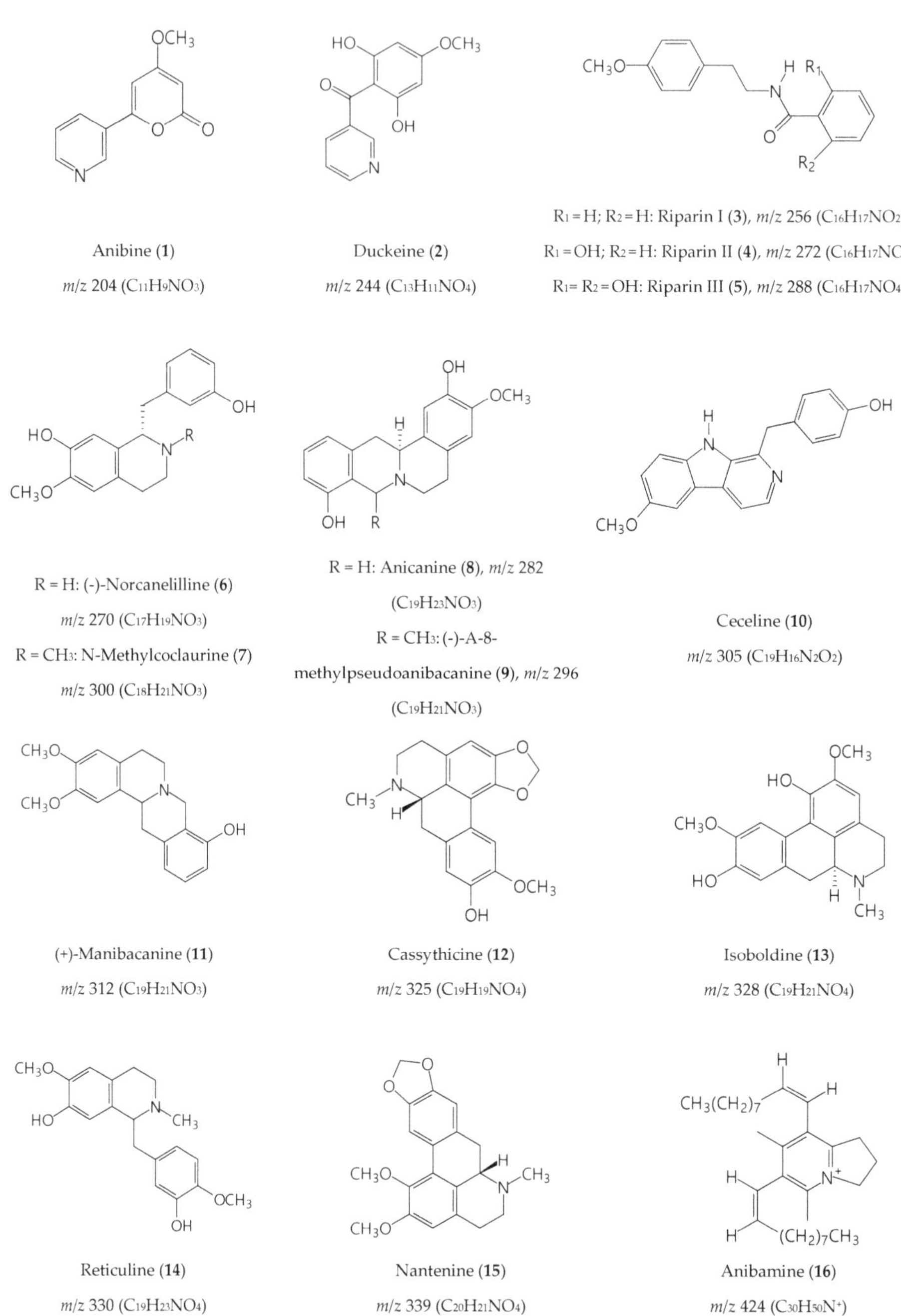

Figure 5. Chemical structures, *m/z* and chemical formulas of alkaloids putatively identified by PSMS in seed and peel of avocado.

3. Materials and Methods

3.1. Materials and Reagents

Ripe avocado (cultivar *Persea americana* var. Fortuna) fruits were purchased in Ibirité city (MG, Brazil) in the winter of 2018. P.A. grade hexane (Nox Lab Solutions, Mauá, SP, Brazil) and hydrated ethanol 96% (Emfal, Betim, MG, Brazil) were used. Ascorbic acid (Neon, Suzano, SP, Brazil), gallic acid (Neon, Suzano, SP, Brazil), quercetin (Sigma-Aldrich, St. Louis, MO, USA), and eserine (Sigma-Aldrich, St. Louis, MO, USA) were used as positive controls in the biological assays. Acetylcholinesterase iodide, acid 5′,5′-dithio-bis-(2-nitrobenzoate), bovine serum albumin, and AChE (Sigma-Aldrich, St. Louis, MO, USA) were used in the AChE inhibition assay. For cultivation of *Drosophila melanogaster* flies, bacteriological agar (Vetec, Duque de Caxias, SP, Brazil), powdered dry yeast (Fleischmann, Sorocaba, SP, Brazil), and nystatin (oral suspension from Germed, Campinas, SP, Brazil) were used. Rotenone (Sigma-Aldrich, At. Louis, MO, USA) was employed in the negative geotaxis assay to induce toxicity. The experiments were carried out in quintuplicate unless otherwise specified.

3.2. Biomass Processing and Extract Preparation

The peels (53.233 g) and seeds (103.431 g) of avocado fruits were separately removed from the pulp, sanitized under running water, and dried separately with paper towel sheets. The peels were sliced, the seeds were grated, and 3 g of each was utilized for atomic absorption spectroscopy analysis. The remaining biomasses (sliced peels and grated seeds) were separately soaked in hexane (600 mL). After 24 h, the hexane fraction was separated by filtration and another aliquot of hexane (600 mL) was added to the vegetal material. The procedure was repeated three times. The hexane fractions were combined; the solvent was removed in a rotary evaporator (60 °C), obtaining the respective hexane extracts of peels (PEL-H) and seeds (SED-H). After hexane extraction, the defatted plant material was further extracted with 96% ethanol, following the same procedure previously described, to furnish the peel (PEL-ET) and seed (SED -ET) ethanolic extracts (Figure 5) [54].

3.3. Mineral Element Analysis by Atomic Absorption

The separated aliquots of fresh peels and seeds (3 g each) were placed in a porcelain crucible and heated until complete carbonization. Subsequently, the crucibles with the carbonized plant material were placed in a muffle at 550 °C, obtaining the organic matter-free ashes. Nitric acid (2 mL) was added to an aliquot of the ashes (0.05 mg) and the volume was completed with distilled water to 5 mL. The resulting suspension was analyzed using Atomic Absorption Spectrometer AA 240 FS (Varian, Australia) to quantify the minerals Ca, Cu, Fe, Mg, Mn, and Zn [4].

3.4. Fatty Acid Gas Chromatography Analysis

Aliquots (10 mg) of the extracts were derivatized to fatty acid methyl esters (FAMEs) [55], solubilized in methanol, and injected into an HP5890 Gas Chromatograph. A Supelcowax-10 analytical column (30 m × 0.2 mm × 0.2 μm) (Supelco) was used for analysis under the following chromatographic conditions (split 1/50): 150 °C, 0 min, 10 °C /min up to 240 °C and detector at 250 °C. Hydrogen was used as carrier gas (4 mL/min) and the injection volume was 1 μL. A standard of fatty acid methyl esters (FAME C14-C22) were injected, and their retention times were used to comparatively identify the fatty acids present in the avocado residues.

3.5. Total Content of Phenol and Flavonoids Assay

The determination of total phenolic content was performed spectrophotometrically with a Folin–Ciocalteau reagent method previously described [56]. Briefly, the ethanolic extract solutions (0.5 mL) were mixed with Folin–Ciocalteau reagent (2.5 mL). The solutions were incubated for 3 min at room temperature, and 0.3 mL of saturated sodium carbonate solution was added to each one. After a further incubation period (20 min at 25 °C), absorbance was read at 760 nm using a Biospectro SP-22 Spectrophotometer. The same procedure was carried out for gallic acid (standard), which was used for constructing the calibration curve (10–100 µg/mL; $R^2 = 0.9291$). In the equation of the calibration curve established using gallic acid, "y" was replaced by the absorbance and divided by the mass of the extract/standard. The results were expressed in mg gallic acid/g extract.

The total flavonoid contents were determined by aluminum chloride spectrophotometric method as described by Yang et al. [27]. Briefly, 2.5 mL of the ethanolic extract or standard solutions were mixed with 0.3 mL sodium nitrate solution (5% w/v) with subsequent addition of aluminum chloride solution (10% w/v) (3.0 mL). The mixture was vortexed, and 2.0 mL of sodium hydroxide solution (1 M) was added and well vortexed. After 10 min, the absorbances were determined at 510 nm using a Biospectro SP-22 Spectrophotometer. Quercetin standard solutions (350–600 µg/mL) were used for constructing the calibration curve ($R^2 = 0.9887$). The result was expressed as mg of quercetin equivalent per g of extract.

3.6. Antioxidant Activity

Ethanol and hexane extracts (SED-ET, PEL-ET, SED-H, and PEL-H) and the standards (ascorbic acid and gallic acid) were dissolved in ethanol (500 µg/mL), except quercetin (1000 µg/mL). All assays were performed in quintuplicate. Microsoft Excel 2013 software was used for data analysis.

3.6.1. Total Antioxidant Capacity

A reagent solution (100 mL) was prepared mixing sulfuric acid (3.22 mL), bibasic phosphate (0.39 g), and ammonium molybdate (0.49 g). An aliquot (0.3 mL) of ethanolic extract solutions (500 µg/mL) was mixed with 3.0 mL of the reagent solution. The resulting test samples were incubated in an oven at 95 °C for 90 min. After the samples had cooled to room temperature, the absorbances were read at 695 nm on a Biospectro SP 22 Spectrophotometer. The same procedure was carried out with ascorbic acid (positive control) and a negative control was carried out with the solvent. The results were expressed in mmol of ascorbic acid per gram of extract (adapted from [57]).

3.6.2. Ferric Reducing Power Assay

An aliquot (1.0 mL) of the ethanolic extract solutions was mixed with 2.5 mL of phosphate buffer (pH 6.6, 0.2 M) and 2.5 mL of potassium ferricyanide aqueous solution (1% w/v). The mixture was homogenized and incubated in a laboratory water bath at 50 °C for 20 min. Subsequently, 2.5 mL of trichloroacetic acid solution (10% w/v) was added and the mixture was centrifuged (300 rpm, 10 min). An aliquot (2.5 mL) of the superior layer was mixed with 2.5 mL distilled water, and 0.5 mL ferric chloride solution (0.1% w/v) was added. The same procedure was conducted with the positive control. The absorbances were read at 700 nm on a Biospectro SP-22 Spectrofotometer. The results were converted into a percentage of ferric reducing power in relation to the positive control ascorbic acid [57].

3.6.3. Free Radical Capture Assay (DPPH)

Solutions of extracts (100 µL) were added to 96-well plates in quadruplicate, followed by the addition of 2,2-Diphenyl-1-picrylhydrazyl (DPPH) solution (175 µL) prepared in ethanol (0.1 mM). The materials were incubated for 30 min in the dark, after which reading was performed at 490 nm in a microplate reader. Solutions of known concentration of ascorbic acid, the positive control, were used to construct a graph of percent inhibition versus concentration (µg/mL) [58].

3.7. Acetylcholinesterase Inhibition Assay

The extracts were solubilized in dimethylsulfoxide (DMSO) (10 mg/mL) and an aliquot of each one (25 µL) was mixed with aqueous solutions of acetylcholine iodide

(15 mM) (25 µL) and 125 µL of a solution containing 5,5′-dithio-bis (2-nitrobenzoic acid) (3 mM), NaCl solution (0.1 M) and $MgCl_2 \cdot 6H_2O$ (20 mM) prepared in Tris/HCl buffer (50 mM, pH 8.0, 50 µL) in wells of 96-well plates. The first set of wavelength readings was then performed at 405 nm every 1 min (8 readings). Subsequently, a solution of acetyl-cholinesterase was prepared (0.22 U/mL) in buffer and 0.1% (w/v) of bovine serum albumin and added to the wells (25 µL each). Eserine (10 mg/mL) was used as positive control. Next, new readings were performed at the same wavelength every 1 min (10 readings) [20]. The following equation was used to calculate the percentage of inhibition (%*I*), where *Ab*: Negative control (DMSO) and *Aa*: Sample absorbance.

$$\% I = \frac{(Ab - Aa) \times 100}{Ab}$$

3.8. In Vivo Neuroprotection Activity Using Drosophila melanogaster Model

A sterile mixture containing bananas (400 g), bacteriological agar (4 g), commercial yeast (3.5 g), nystatin (2.5 mL), and propionic acid (0.6 mL) in distilled water (350 mL) was prepared and used to grow *D. melanogaster* flies for the whole experiment. The cultivations were performed in glass vessels in triplicate. Twelve 6–8-day-old male flies were grown in vessels containing (A) rotenone (neurotoxic inductor, 10 mg/mL in ethanol; 0.4 mL) or (B) extracts (5 mg/mL in water; 0.6 mL). Rotenone and extracts were mixed with the nutritive substrate. A control flask (C) was utilized. Neither rotenone nor extract were added to the control flasks. After 7 days, the flies that were in vessels A and B were anesthetized with ether and interchanged. After 15 days the flies were transferred to falcon tubes that were horizontally placed on a support. A photographic camera was positioned approximately 30 cm away. Then, the support was subjected to three hits against the bench surface, and ten photographs of the tubes containing the flies were recorded. Picture number 6 was used to account for the number of flies that were able to fly above half the tube height after hitting [31].

3.9. Paper Spray Mass Spectrometry

Pulps were analyzed using a Thermo Fisher LCQ FLEET ion-trap mass spectrometer (Thermo Scientific, San Jose, CA, USA) equipped with an ambient paper spray ionization source. Analyses were performed in triplicate in positive and negative ionization modes. For PSMS analyses, a 2 µL sample volume and 40 µL of methanol were added to a triangular chromatographic paper (equilateral, 1.5 cm lengthwise) positioned 10 mm away from the mass spectrometer inlet [37–39]. A high voltage was then applied to the paper held by a copper clamp and attached to a three-dimensional moving platform for data acquisition. Instrumental conditions of operation were: PSMS source voltage equal to +4 kV (positive mode) and −3 kV (negative mode); capillary voltage of 40 V; transfer tube temperature of 275 °C; tube lens voltage of 120 V; mass scanning range of 100 to 1000 m/z in positive and negative modes. Ions were fragmented using collision energies from 15 to 45 eV. Some representative MS can be found in Figures S2–S48. Tentative identification of the compounds was carried out using a comparison of the m/z ratios of the data obtained from literature associated with instrumental readings and subsequent fragmentation employing sequential mass spectrometry.

3.10. Statistical Analysis

GraphPad Prism 5.0 software was used for statistical analysis, applying variance analysis (ANOVA) [20]. To evaluate the differences between the means, the Tukey test with significance level of 5% was used. Kruskal–Wallis (nonparametric) and Dunn's post-test, with significance of 5%, were utilized to evaluate data from neuroprotection assay. SensoMaker 1.61 software was used for main components analyses.

4. Conclusions

The avocado residues produced during industrial processing retain nutrients and bioactive compounds independent of the extraction solvent. Concerning the minerals, avocado peels and seeds displayed significant levels of Ca, Mg, Mn, and Zn. The seeds presented high essential fatty acids content, such as linoleic, palmitic, and oleic acids. Seed and peel ethanolic extracts showed phenolic compounds correlated to their antioxidant and AChE inhibitory activities. Additionally, the seed ethanolic extract exhibited in vivo neuroprotective effect. In vivo studies are an important step to determine the real potential of phytochemicals since they overcome many problems inherent to in vitro models. Therefore, the positive in vivo results encourage research of this extract as a potential candidate to develop drugs to treat patients in the early stages of neurodegenerative illnesses such as Alzheimer's and Parkinson's diseases. The PSMS chemical profile of the residues led to the identification of fifty-five metabolites, including phenolic, hydroxycinnamic acids, flavonoids, and alkaloids, most of them already reported as pharmacologically active compounds. Quantification of the phytochemicals will be the object of future work. In addition, toxicity, bioavailability, and suitable formulations should be further investigated before using avocado residues in pharmacotherapy, but the current results pave the way to deeper studies on the in vivo potential of avocado residues as a bioresource in the development of low-cost drugs and functional foods with neuroprotective effects.

Supplementary Materials: The following supporting information can be downloaded at: https://www.mdpi.com/article/10.3390/molecules27061892/s1, Section S1: PSMS of the seed and ethanolic extracts and of the compounds identified, Figure S1: PS(+)MS full scan of ethanol extract from peel of avocado, Figure S2: Anibine PS(+)MS fragmentation and chemical structure, Figure S3: Duckeine PS(+)MS fragmentation and chemical structure, Figure S4: Riparin I PS(+)MS fragmentation and chemical structure, Figure S5: Norcanelilline PS(+)MS fragmentation and chemical structure, Figure S6: Riparin II PS(+)MS fragmentation and chemical structure, Figure S7: Anicanine PS(+)MS fragmentation and chemical structure, Figure S8: Riparin III PS(+)MS fragmentation and chemical structure, Figure S9: (-)-α-methylpseudoanibacanine PS(+)MS fragmentation and chemical structure, Figure S10: N- methylcoclaurine PS(+)MS fragmentation and chemical structure, Figure S11: Ceceline PS(+)MS fragmentation and chemical structure, Figure S12: (+)-manibacanine PS(+)MS fragmentation and chemical structure, Figure S13: Cassythicine PS(+)MS fragmentation and chemical structure, Figure S14: Isoboldine PS(+)MS fragmentation and chemical structure, Figure S15: Reticuline PS(+)MS fragmentation and chemical structure, Figure S16: Anibamine PS(+)MS fragmentation and chemical structure, Figure S17: PS(−)MS full scan of peel ethanolic extract of avocado, Figure S18: Vanillin PS(−)MS fragmentation and chemical structure, Figure S19: Caffeic acid PS(−)MS fragmentation and chemical structure, Figure S20: Quinic acid PS(−)MS fragmentation and chemical structure, Figure S21: Syrinc acid PS(−)MS fragmentation and chemical structure, Figure S22: 5-hydroxyferulic acid PS(−)MS fragmentation and chemical structure, Figure S23: Sinapic acid PS(−)MS fragmentation and chemical structure, Figure S24: Apigenin PS(−)MS fragmentation and chemical structure, Figure S25: Kaempferol PS(−)MS fragmentation and chemical structure, Figure S26: Catechin PS(−)MS fragmentation and chemical structure, Figure S27: Quercetin PS(−)MS fragmentation and chemical structure, Figure S28: Hydroxytyrosol glucoside PS(−)MS fragmentation and chemical structure, Figure S29: p-coumaroyl hexose PS(−)MS fragmentation and chemical structure, Figure S30: 3-O-p-coumaroylquinic acid PS(−)MS fragmentation and chemical structure, Figure S31: Caffeic acid hexoside PS(−)MS fragmentation and chemical structure, Figure S32: 5-O-caffeoylquinic acid PS(−)MS fragmentation and chemical structure, Figure S33: Kaempferol-O-pentoside PS(−)MS fragmentation and chemical structure, Figure S34: Cyanidin-3-O-arabinoside PS(−)MS fragmentation and chemical structure, Figure S35: Vitexin PS(−)MS fragmentation and chemical structure, Figure S36: Peonidin-3-O-pentoside PS(−)MS fragmentation and chemical structure, Figure S37: Phloridzin PS(−)MS fragmentation and chemical structure, Figure S38: Kaempferol-O-hexoside PS(−)MS fragmentation and chemical structure, Figure S39: Dihydroquercetin-3,5-rhamnoside PS(−)MS fragmentation and chemical structure, Figure S40: Isorhamnetin-O-coumaroyl PS(−)MS fragmentation and chemical structure, Figure S41: Quercetin-3-glucoside PS(−)MS fragmentation and chemical structure, Figure S42: Quercetin glucuronide PS(−)MS fragmentation and chemical structure, Figure S43: Caffeoyl hexose-deohyhexoside PS(−)MS fragmentation and chemical structure, Figure S44: Dimethyl ellagic acid hexoside PS(−)MS fragmentation and chemical structure, Figure S45: Apigenin-C-hexoside-C-pentoside PS(−)MS fragmentation and chemical structure, Figure S46: Luteolin-7-O-(2″-O-pentosyl)-hexoside PS(−)MS fragmentation and chemical structure, Figure S47: Catechin diglucopyranoside PS(−)MS fragmentation and chemical structure, Figure S48: Quercetin-3,4′-O-diglucoside PS(−)MS fragmentation and chemical structure.

Author Contributions: Conceptual idea, methodology design, writing, and editing: J.A.T.; Data collection, antioxidant assays, data analysis and interpretation, writing, and editing: G.G.d.S. and L.P.S.P.; Paper Spray analysis and data interpretation: J.O.F.M., H.d.O.P.M. and R.A. All authors have read and agreed to the published version of the manuscript.

Funding: This research was funded by CAPES (Code 001), Fundação de Amparo à Pesquisa do Estado de Minas Gerais (FAPEMIG PPM-00255-18), and Conselho Nacional de Desenvolvimento Científico e Tecnológico (CNPq Grant 304922/2018-8).

Institutional Review Board Statement: Not applicable.

Informed Consent Statement: Not applicable.

Acknowledgments: We acknowledge Lucas Xavier da Silva for drawing the graphical abstracts.

Conflicts of Interest: The authors declare no conflict of interest.

Sample Availability: Samples of the compounds are not available.

References

1. Ochoa-Zarzosa, A.; Báez-Magaña, M.; Guzmán-Rodríguez, J.J.; Flores-Alvarez, L.J.; Lara-Márquez, M.; Zavala-Guerrero, B.; Salgado-Garciglia, R.; López-Gómez, R.; López-Meza, J.E. Bioactive Molecules From Native Mexican Avocado Fruit (*Persea americana* var. drymifolia): A Review. *Plant Foods Hum. Nutr.* **2021**, *76*, 133–142. [CrossRef] [PubMed]
2. Tesfaye, T.; Ayele, M.; Gibril, M.; Ferede, E.; Limeneh, D.Y.; Kong, F. Beneficiation of avocado processing industry by-product: A review on future prospect. *Curr. Res. Green Sustain. Chem.* **2022**, *5*, 100253. [CrossRef]
3. Alkhalaf, M.I.; Alansari, W.S.; Ibrahim, E.A.; ELhalwagy, M.E.A. Anti-oxidant, anti-inflammatory and anti-cancer activities of avocado (*Persea americana*) fruit and seed extract. *J. King Saud Univ. Sci.* **2019**, *31*, 1358–1362. [CrossRef]
4. Borges, C.V.; Maraschin, M.; Coelho, D.S.; Leonel, M.; Gomez, H.A.G.; Belin, M.A.F.; Diamante, M.S.; Amorim, E.P.; Gianeti, T.; Castro, G.R.; et al. Nutritional value and antioxidant compounds during the ripening and after domestic cooking of bananas and plantains. *Food Res. Int.* **2020**, *132*, 109061. [CrossRef] [PubMed]
5. Araujo, R.G.; Rodríguez-Jasso, R.M.; Ruíz, H.A.; Govea-Salas, M.; Pintado, M.; Aguilar, C.N. Recovery of bioactive components from avocado peels using microwave-assisted extraction. *Food Bioprod. Process.* **2021**, *127*, 152–161. [CrossRef]
6. Baysal, S.S.; Ülkü, M.A. Food Loss and Waste: A Sustainable Supply Chain Perspective. ICGI-Global: Istanbul, Turkey, 2021, ISBN 9789251317549.
7. Yap, K.M.; Sekar, M.; Seow, L.J.; Gan, S.H.; Bonam, S.R.; Mat Rani, N.N.I.; Lum, P.T.; Subramaniyan, V.; Wu, Y.S.; Fuloria, N.K.; et al. *Mangifera indica* (Mango): A promising medicinal plant for breast cancer therapy and understanding its potential mechanisms of action. *Breast Cancer Targets Ther.* **2021**, *13*, 471–503. [CrossRef]
8. Béné, C.; Oosterveer, P.; Lamotte, L.; Brouwer, I.D.; de Haan, S.; Prager, S.D.; Talsma, E.F.; Khoury, C.K. When food systems meet sustainability—Current narratives and implications for actions. *World Dev.* **2019**, *113*, 116–130. [CrossRef]
9. González-Chang, M.; Wratten, S.D.; Shields, M.W.; Costanza, R.; Dainese, M.; Gurr, G.M.; Johnson, J.; Karp, D.S.; Ketelaar, J.W.; Nboyine, J.; et al. Understanding the pathways from biodiversity to agro-ecological outcomes: A new, interactive approach. *Agric. Ecosyst. Environ.* **2020**, *301*, 107053. [CrossRef]
10. Tumwesigye, K.S.; O'Brien, E.; Oliveira, J.C.; Crean, A.; Sousa-Gallagher, M.J. Engineered food supplement excipients from bitter cassava for minimisation of cassava processing waste in environment. *Future Foods* **2020**, *1–2*, 100003. [CrossRef]
11. Ortega-Arellano, H.F.; Jimenez-Del-Rio, M.; Velez-Pardo, C. Neuroprotective Effects of Methanolic Extract of Avocado *Persea americana* (var. Colinred) Peel on Paraquat-Induced Locomotor Impairment, Lipid Peroxidation and Shortage of Life Span in Transgenic knockdown Parkin Drosophila melanogaster. *Neurochem. Res.* **2019**, *44*, 1986–1998. [CrossRef]
12. Salazar-López, N.J.; Domínguez-Avila, J.A.; Yahia, E.M.; Belmonte-Herrera, B.H.; Wall-Medrano, A.; Montalvo-González, E.; González-Aguilar, G.A. Avocado fruit and by-products as potential sources of bioactive compounds. *Food Res. Int.* **2020**, *138*, 109774. [CrossRef] [PubMed]
13. Del Castillo-Llamosas, A.; del Río, P.G.; Pérez-Pérez, A.; Yáñez, R.; Garrote, G.; Gullón, B. Recent advances to recover value-added compounds from avocado by-products following a biorefinery approach. *Curr. Opin. Green Sustain. Chem.* **2021**, *28*, 100433. [CrossRef]
14. Fuloria, S.; Yusri, M.A.A.; Sekar, M.; Gan, S.H.; Rani, N.N.I.M.; Lum, P.T.; Ravi, S.; Subramaniyan, V.; Azad, A.K.; Jeyabalan, S.; et al. Genistein: A Potential Natural Lead Molecule for New Drug Design and Development for Treating Memory Impairment. *Molecules* **2022**, *27*, 265. [CrossRef] [PubMed]

15. World Health Organization (WHO). *Global Action Plan on the Public Health Response to Dementia 2017–2025*; World Health Organization: Geneva, Switzerland, 2017; p. 27.

16. Bernstein, J.J.J.; Holt, G.B.; Bernstein, J. Price dispersion of generic medications. *PLoS ONE* **2019**, *14*, e0225280. [CrossRef] [PubMed]

17. FAO; IFAD; UNICEF; WHO. *The State of Food Security and Nutrition in the World 2020*; FAO: Rome, Italy, 2020; ISBN 9789251329016. [CrossRef]

18. Ferguson, E.L.; Watson, L.; Berger, J.; Chea, M.; Chittchang, U.; Fahmida, U.; Khov, K.; Kounnavong, S.; Le, B.M.; Rojroong-wasinkul, N.; et al. Realistic Food-Based Approaches Alone May Not Ensure Dietary Adequacy for Women and Young Children in South-East Asia. *Matern. Child Health J.* **2019**, *23*, 55–66. [CrossRef]

19. Adams, J.B.; Sorenson, J.C.; Pollard, E.L.; Kirby, J.K.; Audhya, T. Evidence-based recommendations for an optimal prenatal supplement for women in the U.S., part two: Minerals. *Nutrients* **2021**, *13*, 1849. [CrossRef] [PubMed]

20. Lourenção, L.F.d.P.; de Paula, N.C.; Cardoso, M.A.; Santos, P.R.; de Oliveira, I.R.C.; Fonseca, F.L.A.; da Veiga, G.L.; Alves, B.d.C.A.; Graciano, M.M.d.C.; Pereira-Dourado, S.M. Biochemical markers and anthropometric profile of children enrolled in public daycare centers. *J. Pediatr.* 2021, *in press*. [CrossRef]

21. Guirlanda, C.P.; da Silva, G.G.; Takahashi, J.A. Cocoa honey: Agro-industrial waste or underutilized cocoa by-product? *Future Foods* **2021**, *4*, 100061. [CrossRef]

22. Mousa, M.M.H.; El-Magd, M.A.; Ghamry, H.I.; Alshahrani, M.Y.; El-Wakeil, N.H.M.; Hammad, E.M.; Asker, G.A.H. Pea peels as a value-added food ingredient for snack crackers and dry soup. *Sci. Rep.* **2021**, *11*, 22747. [CrossRef]

23. Marangoni, F.; Agostoni, C.; Borghi, C.; Catapano, A.L.; Cena, H.; Ghiselli, A.; La Vecchia, C.; Lercker, G.; Manzato, E.; Pirillo, A.; et al. Dietary linoleic acid and human health: Focus on cardiovascular and cardiometabolic effects. *Atherosclerosis* **2020**, *292*, 90–98. [CrossRef]

24. Park, W.J. *The Biochemistry and Regulation of Fatty Acid Desaturases in Animals*; Elsevier Inc.: Amsterdam, The Netherlands, 2018; ISBN 9780128112304.

25. Orsavova, J.; Misurcova, L.; Vavra Ambrozova, J.; Vicha, R.; Mlcek, J. Fatty acids composition of vegetable oils and its contribution to dietary energy intake and dependence of cardiovascular mortality on dietary intake of fatty acids. *Int. J. Mol. Sci.* **2015**, *16*, 2871. [CrossRef]

26. Guillén-Sánchez, J.; Paucar-Menacho, L.M. Oxidative stability and shelf life of avocado oil extracted cold and hot using discard avocado (*Persea americana*). *Sci. Agropecu.* **2020**, *11*, 127–133. [CrossRef]

27. Velderrain-Rodríguez, G.R.; Quero, J.; Osada, J.; Martín-Belloso, O.; Rodríguez-Yoldi, M.J. Phenolic-rich extracts from avocado fruit residues as functional food ingredients with antioxidant and antiproliferative properties. *Biomolecules* **2021**, *11*, 997. [CrossRef] [PubMed]

28. Vinha, A.F.; Moreira, J.; Barreira, S.V.P. Physicochemical Parameters, Phytochemical Composition and Antioxidant Activity of the Algarvian Avocado (*Persea americana* Mill.). *J. Agric. Sci.* **2013**, *5*, 100–109. [CrossRef]

29. Amado, D.A.V.; Helmann, G.A.B.; Detoni, A.M.; de Carvalho, S.L.C.; de Aguiar, C.M.; Martin, C.A.; Tiuman, T.S.; Cottica, S.M. Antioxidant and antibacterial activity and preliminary toxicity analysis of four varieties of avocado (*Persea americana* Mill.). *Braz. J. Food Technol.* **2019**, *22*, 04418. [CrossRef]

30. Tremocoldi, M.A.; Rosalen, P.L.; Franchin, M.; Massarioli, A.P.; Denny, C.; Daiuto, É.R.; Paschoal, J.A.R.; Melo, P.S.; De Alencar, S.M. Exploration of avocado by-products as natural sources of bioactive compounds. *PLoS ONE* **2018**, *13*, e0192577. [CrossRef] [PubMed]

31. Jimenez, E.V.; Tovar, J.; Mosquera, O.M.; Cardozo, F. Actividad Neuroprotectora de *Solanum ovalifolium* (SOlanaceae) contra la toxicidad Inducida por rotenona em *Drosophila melanogaster*. *Rev Fac Cienc. Basicas* **2017**, *13*, 27–35. [CrossRef]

32. Siima, A.A.; Stephano, F.; Munissi, J.J.E.; Nyandoro, S.S. Ameliorative effects of flavonoids and polyketides on the rotenone induced Drosophila model of Parkinson's disease. *NeuroToxicology* **2020**, *81*, 209–215. [CrossRef]

33. Kumar, S.; Behl, T.; Sehgal, A.; Chigurupati, S.; Singh, S.; Mani, V.; Aldubayan, M.; Alhowail, A.; Kaur, S.; Bhatia, S.; et al. Exploring the focal role of LRRK2 kinase in Parkinson's disease. *Environ. Sci. Pollut. Res.* **2022**, 1–15. [CrossRef] [PubMed]

34. Vos, M.; Klein, C. UnCover Cellular Pathways Underlying Parkinson's Disease. *Cells* **2021**, *10*, 579. [CrossRef]

35. Bolus, H.; Crocker, K.; Boekhoff-Falk, G.; Chtarbanova, S. Modeling neurodegenerative disorders in drosophila melanogaster. *Int. J. Mol. Sci.* **2020**, *21*, 3055. [CrossRef]

36. McBride, E.M.; Mach, P.M.; Dhummakupt, E.S.; Dowling, S.; Carmany, D.O.; Demond, P.S.; Rizzo, G.; Manicke, N.E.; Glaros, T. Paper spray ionization: Applications and perspectives. *TrAC Trends Anal. Chem.* **2019**, *118*, 722–730. [CrossRef]

37. Silva, V.D.M.; Macedo, M.C.C.; dos Santos, A.N.; Silva, M.R.; Augusti, R.; Lacerda, I.C.A.; Melo, J.O.F.; Fante, C.A. Bioactive activities and chemical profile characterization using paper spray mass spectrometry of extracts of *Eriobotrya japonica* Lindl. leaves. *Rapid Commun. Mass Spectrom.* **2020**, *34*, e8883. [CrossRef] [PubMed]

38. García, Y.M.; Ramos, A.L.C.C.; de Oliveira Júnior, A.H.; de Paula, A.C.C.F.F.; de Melo, A.C.; Andrino, M.A.; Silva, M.R.; Augusti, R.; de Araújo, R.L.B.; de Lemos, E.E.P.; et al. Physicochemical Characterization and Paper Spray Mass Spectrometry Analysis of *Myrciaria Floribunda* (H. West ex Willd.) O. Berg Accessions. *Molecules* **2021**, *26*, 7206. [CrossRef] [PubMed]

39. Do Nascimento, C.D.; de Paula, A.C.C.F.F.; de Oliveira Júnior, A.H.; Mendonça, H.d.O.P.; Reina, L.D.C.B.; Augusti, R.; Figueiredo-Ribeiro, R.d.C.L.; Melo, J.O.F. Paper spray mass spectrometry on the analysis of phenolic compounds in *rhynchelytrum repens*: A tropical grass with hypoglycemic activity. *Plants* **2021**, *10*, 1617. [CrossRef] [PubMed]

40. Weremfo, A.; Adulley, F.; Adarkwah-Yiadom, M. Simultaneous Optimization of Microwave-Assisted Extraction of Phenolic Compounds and Antioxidant Activity of Avocado (*Persea americana* Mill.) Seeds Using Response Surface Methodology. *J. Anal. Methods Chem.* **2020**, *2020*, 7541927. [CrossRef] [PubMed]
41. Segovia, F.J.; Hidalgo, G.I.; Villasante, J.; Ramis, X.; Almajano, M.P. Avocado seed: A comparative study of antioxidant content and capacity in protecting oil models from oxidation. *Molecules* **2018**, *23*, 2421. [CrossRef]
42. Mijangos-Ramos, I.F.; Zapata-Estrella, H.E.; Ruiz-Vargas, J.A.; Escalante-Erosa, F.; Gómez-Ojeda, N.; García-Sosa, K.; Cechinel-Filho, V.; Meira-Quintão, N.L.; Peña-Rodríguez, L.M. Bioactive dicaffeoylquinic acid derivatives from the root extract of *Calea urticifolia*. *Rev. Bras. Farmacogn.* **2018**, *28*, 339–343. [CrossRef]
43. Rosero, J.C.; Cruz, S.; Osorio, C.; Hurtado, N. Analysis of Phenolic Composition of Byproducts (Seeds and Peels) of Avocado (*Persea americana* Mill.) Cultivated in Colombia. *Molecules* **2019**, *24*, 3209. [CrossRef]
44. Castro-López, C.; Bautista-Hernández, I.; González-Hernández, M.D.; Martínez-Ávila, G.C.G.; Rojas, R.; Gutiérrez-Díez, A.; Medina-Herrera, N.; Aguirre-Arzola, V.E. Polyphenolic Profile and Antioxidant Activity of Leaf Purified Hydroalcoholic Extracts from Seven Mexican *Persea americana* Cultivars. *Molecules* **2019**, *24*, 173. [CrossRef]
45. Moita, I.S.; Yamaguchi, K.K.L.; Alcântara, J.M.; Silva, Y.C.; Fernandes, N.S.; Nakamura, C.V.; Veiga Junior, V.F. Phytochemical and Biological Studies on *Ocotea ceanothifolia* (Nees) Mez. *Rev. Virtual Quim.* **2019**, *11*, 1267–1276. [CrossRef]
46. Alagbaoso, C.A.; Osakwe, O.S.; Tokunbo, I.I. Changes in proximate and phytochemical compositions of *Persea americana* mill. (avocado pear) seeds associated with ripening. *J. Med. Biomed. Res.* **2017**, *16*, 28–34.
47. Setyawan, H.Y.; Sukardi, S.; Puriwangi, C.A. Phytochemicals properties of avocado seed: A review. *IOP Conf. Ser. Earth Environ. Sci.* **2021**, *733*, 012090. [CrossRef]
48. Silva, Y.C.D. Estudo de Marcadores em Espécies de Aniba (Lauraceae) Bioativas da Amazônia. 2018. Available online: https://tede.ufam.edu.br/handle/tede/6494 (accessed on 23 January 2022).
49. Nunes, G.B.L.; Costa, L.M.; Gutierrez, S.J.C.; Satyal, P.; De Freitas, R.M. Behavioral tests and oxidative stress evaluation in mitochondria isolated from the brain and liver of mice treated with riparin A. *Life Sci.* **2015**, *121*, 57–64. [CrossRef] [PubMed]
50. Rodrigues de Carvalho, A.M.; Vasconcelos, L.F.; Moura Rocha, N.F.; Vasconcelos Rios, E.R.; Dias, M.L.; Maria de França Fonteles, M.; Gaspar, D.M.; Barbosa Filho, J.M.; Chavez Gutierrez, S.J.; Florenço de Sousa, F.C. Antinociceptive activity of Riparin II from Aniba riparia: Further elucidation of the possible mechanisms. *Chem.-Biol. Interact.* **2018**, *287*, 49–56. [CrossRef] [PubMed]
51. Chaves, R.d.C.; Mallmann, A.S.V.; de Oliveira, N.F.; Capibaribe, V.C.C.; da Silva, D.M.A.; Lopes, I.S.; Valentim, J.T.; Barbosa, G.R.; de Carvalho, A.M.R.; de Fonteles, M.M.F.; et al. The neuroprotective effect of Riparin IV on oxidative stress and neuroinflammation related to chronic stress-induced cognitive impairment. *Horm. Behav.* **2020**, *122*, 104758. [CrossRef] [PubMed]
52. Singh, S.; Pathak, N.; Fatima, E.; Negi, A.S. Plant isoquinoline alkaloids: Advances in the chemistry and biology of berberine. *Eur. J. Med. Chem.* **2021**, *226*, 113839. [CrossRef] [PubMed]
53. Kaur, J.; Famta, P.; Famta, M.; Mehta, M.; Satija, S. Potential anti-epileptic phytoconstituents: An updated review. *J. Ethnopharmacol.* **2021**, *268*, 113565. [CrossRef] [PubMed]
54. Martins, B.d.A.; Sande, D.; Solares, M.D.; Takahashi, J.A. Antioxidant role of morusin and mulberrofuran B in ethanol extract of *Morus alba* roots. *Nat. Prod. Res.* **2021**, *35*, 5993–5996. [CrossRef]
55. Sande, D.; Colen, G.; dos Santos, G.F.; Ferraz, V.P.; Takahashi, J.A. Production of omega 3, 6, and 9 fatty acids from hydrolysis of vegetable oils and animal fat with *Colletotrichum gloeosporioides* lipase. *Food Sci. Biotechnol.* **2018**, *27*, 537–545. [CrossRef] [PubMed]
56. Santos, B.O.; Labanca, R.A. Development and Chemical Characterization of Pequi Pericarp Flour. *J. Braz. Chem. Soc.* **2022**, *00*, 1–11, (*in press*).
57. Bashmil, Y.M.; Ali, A.; Bk, A.; Dunshea, F.R.; Suleria, H.A.R. Screening and characterization of phenolic compounds from Australian grown bananas and their antioxidant capacity. *Antioxidants* **2021**, *10*, 1521. [CrossRef] [PubMed]
58. Seyrekoglu, F.; Temiz, H.; Eser, F.; Yildirim, C. Comparison of the antioxidant activities and major constituents of three Hypericum species (H. perforatum, H. scabrum and H. origanifolium) from Turkey. *S. Afr. J. Bot.* **2022**, *146*, 723–727. [CrossRef]

Article

The Purified Siderophore from *Streptomyces tricolor* HM10 Accelerates Recovery from Iron-Deficiency-Induced Anemia in Rats

Hassan Barakat [1,2,*] , Kamal A. Qureshi [3,4] , Abdullah S. Alsohim [5] and Medhat Rehan [5,6]

1   Department of Food Science and Human Nutrition, College of Agriculture and Veterinary Medicine, Qassim University, Buraydah 51452, Saudi Arabia
2   Department of Food Technology, Faculty of Agriculture, Benha University, Moshtohor 13736, Egypt
3   Department of Pharmaceutics, Unaizah College of Pharmacy, Qassim University, Unaizah 51911, Saudi Arabia; ka.qurishe@qu.edu.sa
4   Faculty of Biosciences and Biotechnology, Invertis University, Bareilly 243123, Uttar Pradesh, India
5   Department of Plant Production and Protection, College of Agriculture and Veterinary Medicine, Qassim University, Buraydah 51452, Saudi Arabia; a.alsohim@qu.edu.sa (A.S.A.); m.rehan@qu.edu.sa (M.R.)
6   Department of Genetics, Faculty of Agriculture, Kafrelsheikh University, Kafr El-Sheikh 33516, Egypt
*   Correspondence: haa.mohamed@qu.edu.sa or hassan.barakat@fagr.bu.edu.eg; Tel.: +966-547141277

Citation: Barakat, H.; Qureshi, K.A.; Alsohim, A.S.; Rehan, M. The Purified Siderophore from *Streptomyces tricolor* HM10 Accelerates Recovery from Iron-Deficiency-Induced Anemia in Rats. *Molecules* **2022**, *27*, 4010. https://doi.org/10.3390/molecules27134010

Academic Editor: Jacqueline Aparecida Takahashi

Received: 17 May 2022
Accepted: 19 June 2022
Published: 22 June 2022

Publisher's Note: MDPI stays neutral with regard to jurisdictional claims in published maps and institutional affiliations.

Abstract: Iron-deficiency-induced anemia is associated with poor neurological development, including decreased learning ability, altered motor functions, and numerous pathologies. Siderophores are iron chelators with low molecular weight secreted by microorganisms. The proposed catechol-type pathway was identified based on whole-genome sequences and bioinformatics tools. The intended pathway consists of five genes involved in the biosynthesis process. Therefore, the isolated catechol-type siderophore (Sid) from *Streptomyces tricolor* HM10 was evaluated through an anemia-induced rat model to study its potential to accelerate recovery from anemia. Rats were subjected to an iron-deficient diet (IDD) for 42 days. Anemic rats (ARs) were then divided into six groups, and normal rats (NRs) fed a standard diet (SD) were used as a positive control group. For the recovery experiment, ARs were treated as a group I; fed an IDD (AR), group II; fed an SD (AR + SD), group III, and IV, fed an SD with an intraperitoneal injection of 1 µg Sid Kg^{-1} (AR + SD + Sid1) and 5 µg Sid Kg^{-1} (AR + SD + Sid5) twice per week. Group V and VI were fed an iron-enriched diet (IED) with an intraperitoneal injection of 1 µg Sid Kg^{-1} (AR + IED + Sid1) and 5 µg Sid Kg^{-1} (AR + IED + Sid5) twice per week, respectively. Weight gain, food intake, food efficiency ratio, organ weight, liver iron concentration (LIC) and plasma (PIC), and hematological parameters were investigated. The results showed that ~50–60 mg Sid L^{-1} medium could be producible, providing ~25–30 mg L^{-1} purified Sid under optimal conditions. Remarkably, the AR group fed an SD with 5 µg Sid Kg^{-1} showed the highest weight gain. The highest feed efficiency was observed in the AR + SD + Sid5 group, which did not significantly differ from the SD group. Liver, kidneys, and spleen weight indicated that diet and Sid concentration were related to weight recovery in a dose-dependent manner. Liver iron concentration (LIC) in the AR + IED + Sid1 and AR + IED + Sid5 groups was considerably higher than in the AR + SD + Sid1 AR + SD + Sid5 groups or the AR + SD group compared to the AR group. All hematological parameters in the treated groups were significantly closely attenuated to SD groups after 28 days, confirming the efficiency of the anemia recovery treatments. Significant increases were obtained in the AR + SD + Sid5 and AR + IED + Sid5 groups on day 14 and day 28 compared to the values for the AR + SD + Sid1 and AR + IED + Sid1 groups. The transferrin saturation % (TSAT) and ferritin concentration (FC) were significantly increased with time progression in the treated groups associatively with PIC. In comparison, the highest significant increases were noticed in ARs fed IEDs with 5 µg Kg^{-1} Sid on days 14 and 28. In conclusion, this study indicated that Sid derived from *S. tricolor* HM10 could be a practical and feasible iron-nutritive fortifier when treating iron-deficiency-induced anemia (IDA). Further investigation focusing on its mechanism and kinetics is needed.

Keywords: *Streptomyces tricolor*; siderophore; purification; iron-deficiency-induced anemia

1. Introduction

Iron, a trace metal, is crucial in maintaining normal human metabolism and is required by most organisms. It is used in numerous living processes, including oxygen transport, redox reactions, and nucleic acid synthesis [1,2]. Iron-deficiency-induced anemia (IDA) can result in various diseases, including delays in development and behavior in children and, in particular, in pregnant women [3–6]. Over two billion people worldwide suffer from IDA, according to the FAO (Food and Agriculture Organization) [3,7]. Iron-deficiency-induced anemia and latent iron deficiency significantly increase and spread among younger individuals [8]. Iron deficiency and IDA arise when dietary iron requirements cannot meet the physiological requirements of the human body, particularly in developing countries, primarily due to insufficient iron content in food and low iron absorption efficiency. Iron-deficiency-induced anemia causes tissue hypoxia, fatigue, headaches, and palpitations. Its association with neurological and psychiatric symptoms has been reported, as epidemiological evidence has shown that restless leg syndrome is associated with iron-deficiency-induced anemia [9]. Iron deficiency symptoms are not limited to anemia alone; for example, prolonged iron deficiency is known to cause inferior papillary atrophy, angular cheilitis, etc. [10]. Iron is found in animal and plant foods in two forms, i.e., heme and non-heme iron; however, non-heme iron constitutes approximately 90% of the iron absorbed from food [11]. Though iron is absorbed in the intestinal tract [12], non-heme iron is much more poorly absorbed than heme iron [13]. Therefore, it is necessary to make iron absorption more efficient and supplement daily diets with iron.

Siderophores are low-molecular-weight compounds (200–2000 Da) secreted by microbes, fungi, and plants under limited iron conditions [14]. Under iron-deficient growth conditions, the hydrophobicity of microbial surface decreases; then, the surface protein composition will alter, leading to the limitation of biofilm formation [15]. When iron becomes unavailable in the environment, microorganisms start producing siderophores as a developed specific uptake strategy. Siderophores consider a metal-chelating agent that can be produced especially under Fe-limiting conditions, making them available for microorganisms [16,17]. They could facilitate the iron acquisition and counter iron deficiency by high-affinity iron (III) ligands [14]. Compared with other metal elements, iron is preferentially chelated by siderophores [18].

Moreover, ferric iron considers the only state that siderophores could chelate. When ferric iron is reduced to ferrous iron, siderophores will release it. This is an important strategy for microbes to combat the iron stress environment [14,19]. The role of siderophores in microbial physiological activities is to dissolve ferric iron and transport it into the cytoplasm, so siderophores have the potential to become a nutritional fortifier [20]. The majority of marine siderophores have distinctive structural features [14]. Accordingly, they contain amphipathic and photochemical properties in Fe^{3+} complexes. Depending on membrane affinity and length, siderophores could anchor a particular gradient outside the cell membrane, thereby enhancing the ability to capture iron from seawater [18,21].

A siderophore is a biological molecule with many applications, and various bacteria can produce it. It can be utilized in multiple fields such as agricultural, medicinal, and environmental applications [22–24]. In therapeutic applications and iron-overload diseases (hemosiderosis, β-thalassemia, hemochromatosis, and accidental iron poisoning), iron is needed in the body so it can be treated with siderophore-based drugs [23,24]. Because of their rapid cell division, cancer cells have a higher requirement of iron when compared to healthy cells. Additionally, their iron uptake and storage rates are higher, so iron chelators such as siderophores can be beneficial for use in cancer therapy (iron excess or iron overload). It was shown that desferrioxamines (hydroximate-type siderophores) significantly decreased tumor cell growth in patients with neuroblastoma or leukemia [25,26]. Furthermore, desferrioxamine E produced by *Streptomyces* reduced the viability of malignant melanoma cells significantly. At the same time, other siderophores (i.e., desferriexochelins, desferrithiocin, tachpyridine, O-trensox, and dexrazoxane) were used in cancer therapy as iron chelators [27,28].

The iron-conveying abilities of siderophores are used to deliver drugs. Drug delivery was shown to occur in cells by forming conjugating among antimicrobials such as albomycins and siderophores [29]. Siderophores can be potentially used as iron chelators in treating cancers, e.g., dexrazoxane, desferriexochelins, desferrithiocin, and *O*-trensox. Furthermore, siderophores can rescue non-transferrin-bound iron in blood serum [29]. Siderophores produced by *Klebsiella pneumoniae* can act as antimalarial agents and can be used to treat malaria caused by *Plasmodium falciparum* [30].

Streptomyces, Gram-positive bacteria, are famous for producing various secondary metabolites. These secondary metabolites have significant biological and clinical usages. Secondary metabolite biosynthesis is usually controlled by complex regulatory networks that respond to various biotic and abiotic stresses in the surrounding environment. *Streptomyces* are considered to be one of the most common producers of secondary metabolites (i.e., antibiotic, antibacterial, antifungal, anticancer, siderophores, etc.) [31–34].

Siderophores could potentially become exciting molecules for combating IDA. However, despite the literature showing promising potentialities related to siderophores, particular attention needs to be paid to the applicability, dose, and mechanisms of siderophores. Moreover, to the best of the authors' knowledge, the literature review mainly highlighted the purification and applications of siderophores, with only a few studies referring to in vivo applications [35,36]. The need to further verify siderophore applications motivated this work. Interestingly, the iron chelation capability of siderophores concerning malignant cancerous cells has been recently studied [37]. Suehiro et al. [38] proved that hemoglobin and hematocrit levels were significantly increased with maltobionic acid calcium salt (MBCa) intake, and recovery from iron-deficiency-induced anemia was promoted. MBCa effectively promoted the recovery of rats from subclinical iron deficiency and iron-deficiency-induced anemia. Feng et al. [35] indicated that siderophores derived from *Synechococcus* sp. PCC7002 could be practical and feasible iron-nutritive fortifiers. However, no studies have clearly demonstrated the effectiveness of using siderophores to enhance iron absorption in human food models. In anemia conditions, a lack of sufficient healthy red blood cells to carry adequate oxygen to the body's cells and low hemoglobin content causes fatigue and weakness. Iron-deficiency-induced anemia is the most common type of anemia worldwide, caused by a shortage of iron in the body. Bone marrow requires iron to produce hemoglobin. When iron levels are low, our bodies cannot produce enough hemoglobin for red blood cells. Therefore, the main objective of this study was to study the potential application of purified Sid from *S. tricolor* HM10 bacteria to accelerate recovery from iron-deficiency-induced anemia in a rat model.

2. Materials and Methods

2.1. Siderophore, Detection, Production, and Purification from S. tricolor HM10

S. tricolor HM10 strain (accession: MN527236) was used for siderophore production, as previously established by Rehan et al. [39], with minor modifications. For siderophore detection, the Chrome Azurol S (CAS) assay was applied by adding CAS solution prepared according to Schwyn and Neilands [40] to King's B agar medium (composed of peptone, 20.0; K_2HPO_4, 1.50; $MgSO_4 \cdot 7H_2O$, 1.50; agar, 15.0 (g/L), pH 7.2) with a ratio of 1:9. In Sid production, a fresh culture of *S. tricolor* HM10 was cultivated on King's B medium supplemented with $FeCl_3$ (5 µM) for 72 h at 28 °C with continuous agitation (170 rpm min^{-1}). Afterward, the culture broth was centrifugated at 10,000× g for 20 min^{-1}. Sid concentration and type were determined in clear supernatant using Arnow's Assay [41] for catechol-type siderophore and Atkin's assay [42] for hydroxamate-type siderophore. According to Clark [43], after the acidification of the collected supernatant to pH 2.0 using concentrated HCl, the Sid produced was isolated and purified. The culture supernatant was passed through a column (5 × 30 cm) packed with Amberlite XAD-2, Sid was eluted with methanol, and fractions were analyzed for the presence of Sid with Arnow's Assay. According to their hydrophobicity, the collected fractions were loaded on the Sephadex LH-20 column and eluted with methanol as the mobile phase for separation. Subsequently, TLC was used to test the collected fractions with Sid activity using a mobile phase containing methanol: ammonium acetate (60:40) [44]. Before injecting the solutions into rats, the concentration of purified Sid was determined with a standard curve generated by 2,3-Dihydroxybenzoic acid (2,3-DHBA). The injected solutions were freshly prepared at two concentrations (1 µg and 5 µg Sid) in sterilized saline buffer (pH 7.0).

2.2. Whole-Genome Sequence Analysis

The whole genome of *S. tricolor* HM10 was sequenced using Oxford Nanopore, assembled, and annotated. The genome is available in the NCBI database with accession number JAJREA000000000. In terms of searching the entire genome, the catechol-type pathway putatively coding for siderophores close to petrobactin was identified using blast search and antiSMASH [45,46].

2.3. Determination of Total Phenolic Content (TPC) and Antioxidant Activity (AOA) of Purified Sid

The TPC-purified Sid was determined by using Folin–Ciocalteu reagent according to Yawadio Nsimba et al. [47], and the TPC content was expressed as milligrams of Gallic acid equivalents (GAE) per gram (mg of GAE g^{-1} DW). The radical scavenging activity was measured spectrophotometrically based on the bleaching of DPPH radicals in purple solution according to Yawadio Nsimba et al. [39], and ABTS (2,2'-azino-bis (3-ethylbenzothiazoline-6-sulphonic acid)) radicals in a dark green solution using the method by Barakat and Rohn [48]. The final results were expressed as millimoles of Trolox Equivalents (TE) per gram (mmol of TE g^{-1}).

2.4. Bio-Evaluation of Iron Absorption Using a Rat Model

2.4.1. Experimental Animals

The study protocol was approved by the Institutional Animal Ethics Committee (IAEC) of Qassim University, KSA (21-07-06 on 26 January 2022). It was regulated by the Purpose of the Control and the Supervision of Experiments on Animals (CPCSEA) Committee under the National Committee of Bioethics (NCBE), Implementing Regulations of the Law of Ethics of Research on Living Creatures. Fifty-six female Wistar rats (6–8 weeks) weighing 150–200 g were housed in polypropylene laboratory cages (four each) in a room at 24 ± 1 °C under inverted 12 h light–dark cycle conditions. Before the feeding experiment, the rats were allowed to adapt to the AIN-93G standard diet (SD) for one week, as Suehiro et al. [38] recommended, as shown in Table 1.

Table 1. Composition of the experimental diets used in iron-deficiency anemia induction and control (g 100 g^{-1} diet).

Ingredients	Anemia-Inducing Period			Experimental Diet and Anemia Recovery Period				
	SD [a]	IDD	AR	AR + SD	AR + SD + Sid1	AR + SD + Sid5	AR + IED + Sid1	AR + IED + Sid5
Corn starch	53.43	53.45	53.45	53.43	53.43	53.43	53.41	53.41
Casein	20.00	20.00	20.00	20.00	20.00	20.00	20.00	20.00
Sucrose	10.00	10.00	10.00	10.00	10.00	10.00	10.00	10.00
Soybean oil	7.00	7.00	7.00	7.00	7.00	7.00	7.00	7.00
Cellulose powder	5.00	5.00	5.00	5.00	5.00	5.00	5.00	5.00
L-Cystine	0.30	0.30	0.30	0.30	0.30	0.30	0.30	0.30
Choline bitartrate	0.25	0.25	0.25	0.25	0.25	0.25	0.25	0.25
AIN-93 vitamin mixture	1.00	1.00	1.00	1.00	1.00	1.00	1.00	1.00
AIN-93G mineral mixture [b]	1.75	1.75	1.75	1.75	1.75	1.75	1.75	1.75
Calcium carbonate	1.25	1.25	1.25	1.25	1.25	1.25	1.25	1.25
Ferric citrate	0.02	–	–	0.02	0.02	0.02	0.04	0.04
Sid (µg Kg^{-1} Rat) [c]	–	–	–	–	1 µg	5 µg	1 µg	5 µg

[a] Standard diet (SD) used for adaptation, the iron content per 100 g of diet: 4.50 mg, iron-deficient diet (IDD); 0.14 mg, iron-enriched diet (IED); 9.00 mg, [b] iron-free mineral mixture; [c] pure Sid was intraperitoneally injected twice per week as (µg Kg^{-1} rat).

2.4.2. Experimental Diets

Table 1 shows the diet compositions utilized during the anemia induction and recovery periods. An iron-deficient diet (IDD) was created during the anemia-inducing period by slightly altering the AIN-93G diet used during the adaptation period. The rats were divided into two groups: the first (8 rats) were fed an SD for 42 days, and the second (48) were provided an IDD for 42 days. According to Suehiro et al. [38], rats administered the IDD had their serum iron level measured randomly and were deemed anemic if their plasma iron concentration (PIC) was less than 50 g dL^{-1}. The animals were given unlimited access to deionized water. Anemic rats were separated into six groups during the anemia recovery period, each having eight rats. Anemic rats (ARs) in group I was fed an IDD for another four weeks. Group II (AR + SD) received the SD, group III (AR + SD + Sid1) received the SD and received purified Sid at 1 µg Kg^{-1} rat twice per week, and group IV (AR + SD + Sid5) received the SD and received pure Sid at 5 µg Kg^{-1} rat twice each week. The iron-enriched diet (IED) was created with double the iron content of the standard diet (SD) and administered to groups V and VI. Group V (AR + IED + Sid1) received pure Sid intraperitoneally twice a week at a dose of 1 µg Kg^{-1} rat. Group VI (AR + IED + Sid5) received pure Sid intraperitoneally twice a week at a dose of 5 µg Kg^{-1} rat. The animals were given deionized water ad libitum during iron-deficient anemia induction, while ordinary tap water was offered to drink throughout the recovery period. Rats were anesthetized and slaughtered, and their spleen and liver were taken and weighed before the liver was perfused with physiological saline to eliminate blood at the end of the experiment. The relative weight of organs was calculated using the following equation:

$$\text{The relative weight of the organ} = \frac{\text{Weight of the organ}}{\text{Weight of the rat}} \times 100 \qquad (1)$$

2.4.3. Measurement of Hematological Parameters

Five hundred microliters of blood were collected in 1.5 mg mL^{-1} purple capped tubes incorporated with disodium salt of ethylenediaminetetraacetic acid (EDTA-2Na) from the tail vein of rats at the end of the 0, 2nd, and 4th week. The hemoglobin level (g dL^{-1}) and hematocrit (%) were analyzed on the day of blood collection using CBC (Mindray Vet 28000, China). The plasma iron (µg dL^{-1}) was determined using a photometric–colorimetric test based on the chromazurol B (CAB) protocol according to Garčic [49], and ferritin concentration (ng mL^{-1}) was determined using an enzyme immunoassay test kit according to White et al. [50]. The total iron-binding capacity (TIBC, µg dL^{-1}) was determined using an in vitro quantitative kit (TIBC-LS, MG, STC, USA). At the same time, serum transferrin iron saturation (TSAT %) values were calculated using the following equations:

$$\text{TSAT (\%)} = \text{PIC } (\mu g \ dL^{-1}) / \text{TIBC } (\mu g \ dL^{-1}) \times 100 \qquad (2)$$

2.5. Statistical Analysis

The statistical analysis was conducted using the SPSS program (ver. 23), applying the experimental design variance analysis using one-way ANOVA for growth parameters, food intake, feed efficiency, liver, kidney, and spleen weight, and iron concentration. In contrast, two-way ANOVA was applied for hematological parameters, iron concentration, transferrin saturation %, and ferritin concentration in plasma. The significance level was 0.05, and Tukey's test was applied following Steel et al. [51].

3. Results

3.1. Production, Isolation, Purification, and Identification of Catechol-Type Sid from S. tricolor HM10

The *S. tricolor* HM10 strain (Accession: MN527236) was used for Sid production, which was first detected on King's B agar medium (Figure 1, step 1). Later, mass production on King's B broth medium at 28 C with contentious agitation for 72 h (step 2) was carried out. Around 50–60 mg Sid L^{-1} medium could be produced using optimal growth conditions. Afterward, the culture broth was cleaned of bacterial cells (step 3). Sid detection and type determination in clear supernatant was achieved according to Arnow's method (step 4). Only 25–30 mg L^{-1} could be isolated and purified by applying the established protocol using acidified supernatant [43]. To test the purity of the isolated Sid, an appropriate sample was loaded onto a cellulose plate, and one-dimensionally separated, resulting in one single visible band, as shown in Figure 1 (step 5).

Figure 1. Production, isolation, and purification diagram of catechol-type Sid from *S. tricolor* strain HM10. 1: detection of Sid production on King's B medium supplemented with Chrome Azurol S (CAS); the orange zone refers to produced Sid, 2: production of Sid in King's B broth medium, 3: clear supernatant containing produced Sid, 4: detection of catechol-type Sid by Arnow's Assay [41], and 5: confirming the purity of purified Sid using thin-layer chromatography (TLC).

3.2. Catechol-Type Siderophore Pathway

A putative pathway for catechol-type Sid was identified by searching the sequenced genome of *S. tricolor* HM10. The proposed pathway consisted of the gene that encodes diaminobutyrate-2-oxoglutarate transaminase family protein (gene A), iron transporter (gene B), the IucA/IucC family protein for siderophore biosynthesis (gene C), ferric iron reductase FhuF-like transporter (gene D), and the acetyltransferase (GNAT) domain (gene E). The proposed pathway consisted of five genes with predicted Enterochelin-like Sid production (Figure 2).

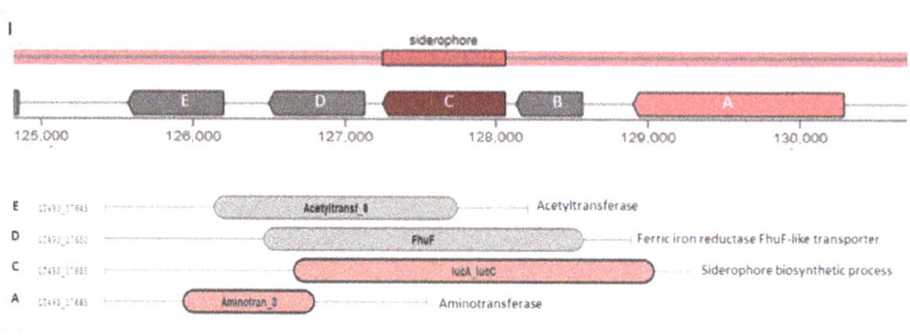

Figure 2. The proposed catechol-type Sid in *S. tricolor* HM10. (**I**) represents Sid pathway in contigs 28 in the genome, (**II**) represents genes involved in Sid production; A, aminotransferase; B, iron transporter; C, IucA/IucC family protein; D, ferric iron reductase FhuF, and E, acetyltransferase.

3.3. TPC and Relative Antioxidant Capacities of Purified Sid from S. tricolor HM10

The TPC and potential antioxidant capacities of purified Sid from *S. tricolor* levels were determined. The TPC in Sid was 452.12 ± 5.78 mg GAE g^{-1}, which presented 85.18 ± 4.21 and 97.38 ± 2.95 mmol of TE g^{-1} for DPPH-RSA- and ABTS-RSA-based antioxidant capacities, respectively.

3.4. Weight Gain, Food Intake, and Food Efficiency Ratio

Table 2 shows the final body weight, weight gain, food intake, and food efficiency ratio in the rats during the recovery period compared to the SD group. The highest body weight gain was recorded in the SD group, while the lowest body weight gain was recorded in the AR group. The AR group fed an SD with different concentrations of injected Sid exhibited significant changes in weight gain. Remarkably, the AR group fed an SD with 5 µg Kg^{-1} showed the highest weight gain. There was no significant difference with ARs provided an IED with 5 µg Kg^{-1} of Sid. However, feeding ARs with an IED containing double iron content did not significantly affect weight gain compared to AR + SD + Sid5. Notably, high Sid concentration is a key promoter in increasing weight gain. Consequently, the highest food intake was observed with the SD, followed by ARs, which were fed the SD diet and injected with 5 µg Kg^{-1} of Sid, while the lowest food intake was seen in the AR group. Feed efficiency significantly increased when feeding an SD in the SD group compared to the AR group. A significant change was noted with ARs provided an SD or IED with Sid injection. The highest feed efficiency was observed in the AR + SD + Sid5 group, which did not significantly differ from the SD group.

Table 2. Growth parameters, food intake, and feed efficiency of rats fed different formulated experimental diets with an intraperitoneal injection of Sid after a recovery period of 4 weeks ($n = 8$).

Items	Experimental Groups						
	SD	AR	AR + SD	AR + SD + Sid1	AR + SD + Sid5	AR + IED + Sid1	AR + IED + Sid5
Initial BW (g) *	178.24 ± 7.21	168.00 ± 11.67	187.00 ± 7.78	179.83 ± 6.94	176.50 ± 5.10	171.33 ± 5.77	174.83 ± 3.89
Final BW (g)	271.45 ± 6.48	185.50 ± 14.73	236.67 ± 4.24	246.33 ± 6.14	255.83 ± 5.51	238.67 ± 5.02	251.67 ± 6.95
BW gain (g)	93.21 ± 5.75 [a]	12.50 ± 4.91 [e]	49.87 ± 2.83 [d]	66.50 ± 4.09 [c]	79.33 ± 3.68 [b]	67.33 ± 3.35 [c]	76.83 ± 4.63 [b]
Food intake (g day^{-1}) [#]	19.39 ± 1.76 [a]	12.03 ± 1.01 [d]	15.78 ± 1.63 [c]	16.42 ± 1.74 [bc]	17.06 ± 1.84 [b]	15.91 ± 1.65 [c]	16.11 ± 1.69 [bc]
Feed efficiency [##]	0.172 ± 0.015 [a]	0.074 ± 0.004 [e]	0.112 ± 0.013 [d]	0.145 ± 0.015 [bc]	0.166 ± 0.017 [a]	0.151 ± 0.013 [bc]	0.148 ± 0.017 [bc]

SD: standard diet, AR: anemic rat fed iron-deficient diet, AR + SD: AR fed SD, AR + SD + Sid1: AR fed SD and intraperitoneally injected with Sid at 1 µg Kg^{-1} rat twice per week, AR + SD + Sid5: AR fed SD and intraperitoneally injected with Sid at 5 µg Kg^{-1} rat twice per week. AR + IED + Sid1: AR fed iron-enriched diet (IED) and intraperitoneally injected with Sid at 1 µg Kg^{-1} rat twice per week, AR + IED + Sid5: AR fed IED diet and intraperitoneally injected with Sid at 5 µg Kg^{-1} rat twice per week, *: initial BW was recorded at the beginning of the experiment, BW: body weight, [#]: calculated as gram diet/rate per day, [##] feed efficiency = BW gain/food intake, [a,b,c,d,e]: there is no significant difference ($p > 0.05$) between any two means within the same row with the same superscripted letters.

3.5. Liver, Kidney, and Spleen Weight and Liver Iron Concentration

The relative weights of the liver, kidneys, and spleen and liver iron concentration (mg g^{-1}) are shown in Table 3. The relative weights of the liver and kidneys varied significantly across all groups. The AR group had the highest liver weight among all groups, with a significant difference. Compared to the SD group, the AR + SD and AR + IED groups with 1 or 5 µg Kg^{-1} of Sid showed no significant differences. Consequently, ARs had significantly higher relative liver weights than AR groups fed an SD or IED and injected with 1 or 5 µg Kg^{-1} of Sid. Diet and Sid concentration were clearly related to the relative weight of the organ, and reductions in relative organ weights took place in a dose-dependent manner. The SD and AR groups demonstrated that an IDD affected the relative kidney weight. However, there was no significant difference in relative kidney weight between ARs fed an SD or IED with 1 and 5 µg Kg^{-1} of Sid groups and SD groups. The SD group had the highest relative spleen weight, which differed significantly from the AR group. ARs fed an SD or IED with 1 or 5 µg Kg^{-1} Sid did not display significantly different results from the SD and AR + SD groups. The LIC was significantly lower in the AR group than in the SD group. Among iron-deficient groups, the values observed in the AR + IED + Sid1 and AR + IED + Sid5 groups were considerably higher than those observed in the AR + SD + Sid1 and AR + SD + Sid5 groups or in the AR + SD group.

Table 3. The relative weight of liver, kidneys, and spleen and iron concentration in the liver of rats fed different formulated experimental diets with an intraperitoneal injection of Sid after a recovery period of 4 weeks ($n = 8$).

Items	Experimental Groups						
	SD	AR	AR + SD	AR + SD + Sid1	AR + SD + Sid5	AR + IED + Sid1	AR + IED + Sid5
Relative liver weight (%)	3.45 ± 0.12 [bc]	3.85 ± 0.14 [a]	3.56 ± 0.06 [b]	3.56 ± 0.05 [b]	3.62 ± 0.09 [b]	3.42 ± 0.07 [c]	3.60 ± 0.05 [b]
Relative kidney weight (%)	0.70 ± 0.08 [b]	0.89 ± 0.02 [a]	0.79 ± 0.05 [b]	0.73 ± 0.01 [b]	0.72 ± 0.02 [b]	0.72 ± 0.02 [b]	0.73 ± 0.03 [b]
Relative spleen weight (%)	0.32 ± 0.02 [b]	0.39 ± 0.02 [a]	0.34 ± 0.02 [b]	0.31 ± 0.02 [b]	0.32 ± 0.01 [b]	0.30 ± 0.02 [b]	0.30 ± 0.03 [b]
LIC (mg g^{-1} liver)	98.12 ± 3.48 [a]	11.24 ± 1.89 [e]	35.24 ± 4.18 [d]	46.21 ± 3.24 [c]	48.54 ± 3.28 [c]	64.25 ± 4.19 [b]	69.24 ± 4.28 [b]

SD: standard diet, AR: anemic rat fed iron-deficient diet, AR + SD: AR fed SD, AR + SD + Sid1: AR fed SD and intraperitoneally injected with Sid at 1 µg Kg^{-1} rat twice per week, AR + SD + Sid5: AR fed SD and intraperitoneally injected with Sid at 5 µg Kg^{-1} rat twice per week. AR + IED + Sid1: AR fed iron-enriched diet (IED) and intraperitoneally injected with Sid at 1 µg Kg^{-1} rat twice per week, AR + IED + Sid5: AR fed IED diet and intraperitoneally injected with Sid at 5 µg Kg^{-1} rat twice per week, LIC: liver iron concentration, [a,b,c,d,e]: there is no significant difference ($p > 0.05$) between any two means within the same row with the same superscripted letters.

3.6. Measurement of Hematological Parameters of Blood

The hematological parameters, such as RBC (10^{12} L^{-1}), HGB (g dL^{-1}), HCT (%), MCV (fL), MCH (pg), and MCHC (g dL^{-1}), are determined through complete blood counts (CBCs) for blood samples taken at the start (day 0) and at the end of the recovery period (day 28) for different experimental groups are illustrated in Table 4. Immediately after inducing anemia, rats were divided into six boxes with eight rats each, and 0.35 mL of fresh blood was taken from the tail vein. The RBC, HGB, HCT, MCV, MCH, and MCHC were determined through CBCs. The results on day 0 presented the means of the hematological parameters for each group. However, no considerable variation could be seen among AR divided groups regarding RBC, HGB, HCT, MCV, MCH, and MCHC. Of course, the SD group differed significantly from other AR groups. After 14 days, RBC, HGB, HCT, MCV, MCH, and MCHC values were improved considerably due to the attenuation of induced anemia compared to SD or AR groups. Quick enhancement was seen in ARs fed an IED with Sid injection compared to ARs fed an SD with Sid injection.

Table 4. Hematological parameters of rats fed different formulated experimental diets with an intraperitoneal injection of Sid at 1 and 5 µg Kg^{-1} at starting and recovery period up to 28 days ($n = 8$).

Hematological Parameters	Days	Experimental Groups						
		SD	AR	AR + SD	AR + SD + Sid1	AR + SD + Sid5	AR + IED + Sid1	AR + IED + Sid5
RBC [10^{12} L^{-1}]	0	6.26 ± 0.08 aC	2.68 ± 0.17 bA	2.57 ± 0.10 bC	2.30 ± 0.41 bC	2.67 ± 0.15 bC	2.57 ± 0.15 bC	2.69 ± 0.24 bC
	14	6.57 ± 0.27 aB	2.47 ± 0.24 dA	3.27 ± 0.31 cB	4.18 ± 0.75 bB	4.77 ± 0.59 bB	4.61 ± 0.09 bB	5.37 ± 0.81 bB
	28	8.20 ± 0.31 aA	2.41 ± 0.15 cA	6.52 ± 0.19 bA	7.21 ± 0.25 aA	7.80 ± 0.54 aA	7.23 ± 0.30 aA	7.37 ± 0.27 aA
HGB [g dL^{-1}]	0	15.90 ± 0.35 aA	7.06 ± 1.06 abA	7.88 ± 2.09 abB	7.52 ± 1.01 bC	6.52 ± 0.47 bC	6.43 ± 0.20 bC	6.87 ± 0.25 bC
	14	16.13 ± 0.47 aA	6.91 ± 0.92 eA	8.98 ± 0.87 dB	9.23 ± 0.84 dB	12.24 ± 1.18 bcB	11.08 ± 0.79 cB	13.01 ± 0.64 bB
	28	16.60 ± 0.22 aA	6.36 ± 0.95 dA	13.58 ± 1.42 bcA	15.43 ± 0.74 abA	16.18 ± 0.85 aA	14.88 ± 0.30 bA	15.15 ± 0.51 abA
HCT [%]	0	36.05 ± 2.86 aA	13.69 ± 1.06 bA	11.63 ± 0.46 bC	10.12 ± 1.81 bC	11.93 ± 0.58 bC	11.60 ± 0.87 bC	14.70 ± 0.84 bC
	14	36.78 ± 1.13 aA	13.74 ± 0.79 dA	13.15 ± 0.91 dB	13.14 ± 0.97 dB	14.79 ± 1.47 cB	15.27 ± 0.59 bcB	17.25 ± 1.51 bB
	28	38.60 ± 0.48 aA	12.33 ± 0.95 cA	21.30 ± 0.57 bA	24.00 ± 0.94 aA	25.10 ± 1.07 aA	23.10 ± 0.72 abA	23.20 ± 0.78 abA
MCV [fL]	0	91.30 ± 0.74 aA	15.38 ± 0.25 bA	65.13 ± 0.04 bA	74.72 ± 0.77 bA	84.95 ± 0.69 bA	65.08 ± 0.38 bA	68.28 ± 0.58 abA
	14	90.77 ± 0.92 aA	13.91 ± 0.54 eA	55.78 ± 0.25 bB	48.17 ± 1.81 cB	40.80 ± 1.29 dB	49.27 ± 2.47 cB	39.75 ± 3.19 dB
	28	92.08 ± 0.67 aA	13.84 ± 0.22 bA	52.88 ± 1.96 aC	33.40 ± 0.83 aC	32.23 ± 1.64 aC	32.32 ± 1.49 aC	31.48 ± 1.87 aC
MCH [pg]	0	29.08 ± 2.80 aA	7.00± 0.52 bA	8.47 ± 1.93 bC	6.67 ± 0.20 bC	6.76 ± 0.14 bC	6.94 ± 0.20 bC	7.11 ± 0.47 bC
	14	28.75 ± 0.19 aA	7.08 ± 0.49 dA	12.57 ± 2.01 bcB	14.17 ± 0.93 bcB	15.34 ± 1.28 bB	15.07 ± 0.68 bB	15.08 ± 1.27 bB
	28	30.25 ± 0.89 aA	6.30 ± 0.47 cA	20.80 ± 1.13 bA	21.38 ± 0.41 bA	20.55 ± 0.79 bA	20.72 ± 0.49 bA	20.55 ± 1.05 bA
MCHC [g dL^{-1}]	0	31.71 ± 1.27 aA	6.85 ± 0.42 aB	8.43 ± 1.92 aB	6.82 ± 0.16 aB	6.82 ± 0.35 aB	6.94 ± 0.32 aB	5.84 ± 0.22 aB
	14	30.91 ± 2.28 aA	6.71 ± 0.64 eAB	12.80 ± 1.52 dA	15.98 ± 0.86 cA	19.72 ± 0.98 bA	16.74 ± 1.65 cA	21.76 ± 2.15 bA
	28	33.75 ± 1.66 aA	6.17 ± 0.38 bA	31.60 ± 0.99 aA	32.20 ± 1.03 aA	34.10 ± 1.46 aA	31.42 ± 1.51 aA	32.55 ± 1.39 aA

SD: standard diet, AR: anemic rat fed iron-deficient diet, AR + SD: AR fed SD, AR + SD + Sid1: AR fed SD and intraperitoneally injected with Sid at 1 µg Kg^{-1} rat twice per week, AR + SD + Sid5: AR fed SD and intraperitoneally injected with Sid at 5 µg Kg^{-1} rat twice per week. AR + IED + Sid1: AR fed iron-enriched diet (IED) and intraperitoneally injected with Sid at 1 µg Kg^{-1} rat twice per week, AR + IED + Sid5: AR fed IED diet and intraperitoneally injected with Sid at 5 µg Kg^{-1} rat twice per week, RBC: red blood cells, HGB: hemoglobin, HCT: hematocrit, MCV: mean corpuscular volume, MCH: mean corpuscular hemoglobin, MCHC: mean corpuscular hemoglobin concentration, a,b,c,d,e: there is no significant difference ($p > 0.05$) between any two means within the same row that have the same superscripted letters, A,B,C: there is no significant difference ($p > 0.05$) between any two means for the same attribute within the same column with the same superscripted letters.

After 28 days, a non-significant difference was found in the SD group regarding all hematological parameters except RBC. After 28 days, a non-significant reduction in all hematological parameters was found in the AR group except the MCHC value compared to results recorded on day 0. On the contrary, significant changes were found in all parameters in all experimental groups, confirming the efficiency of the anemia recovery treatments.

The RBC count exhibited substantial increases in the group of ARs fed SD. There was no significant difference between ARs that provided an SD or IED with 1 and 5 µg Kg^{-1} Sid and the SD group. The group of ARs provided an SD with Sid injection at 1 and 5 µg Kg^{-1}, which resulted in significant HBG increases compared to AR + SD or AR groups. Accordingly, HCT % exhibited a considerable increase in the blood of ARs fed an SD with 1 and 5 µg Kg^{-1} than the group of ARs fed an SD. Although there was a significant difference between SD- and AR-treated groups, the HCT % in the treated group indicated high iron absorption rates depending on the treatment. The MCV of RBC showed a significant difference between the AR group and AR-treated groups. Most of the monitored hematological parameters observed when providing an SD in the presence of Sid injection were efficiently better than those observed when providing an SD without Sid injection or even an IED with Sid injection.

In contrast, feeding an SD or IED with Sid helped improve the RBC status and attenuate the MCV compared with the SD group. Consequently, the MCH and MCHC values underwent a considerable improvement in hemoglobin content in the RBC of ARs fed an SD with 1 and 5 µg Kg^{-1} than the group of ARs fed an SD, which indicated a normal status compared to the SD group.

3.7. Plasma Iron Concentration, Transferrin Saturation %, and Ferritin Concentration in Plasma

Figure 3 shows the plasma iron levels, transferrin saturation %, and ferritin concentration in different experimental groups. At the beginning of the recovery period, the PIC values of the SD and iron-deficient diet groups were 143.83 µg dL^{-1} and (17.68–26.30 µg dL^{-1}), respectively. Significant increases were seen during the recovery period when ARs were fed an SD or an SD or IED with 1 or 5 µg Kg^{-1} Sid. The PIC values in the AR + SD group were significantly lower than in the other four groups on days 14 and 28. Significant increases were observed in AR + SD + Sid5 and AR + IED + Sid5 groups on day 14 and day 28 compared to the values for the AR + SD + Sid1 and AR + IED + Sid1 groups. The TSAT levels in the SD and AR groups were 59.93% and (10.34–19.84%), respectively, at the beginning of the recovery period. No significant difference was recorded in the SD group until the end of the recovery period. On the contrary, the TSAT in the AR group was significantly decreased with time progression. At the beginning of the recovery period, ferritin concentration in the SD and AR blood serum was 71.92 and (12.48–19.56) ng mL^{-1} (Figure 3).

No significant difference was recorded in the AR group until the end of the recovery period. On the contrary, the FC in the SD group was significantly increased with time progression. In comparison, the most significant increases were noticed in ARs fed an IED with 5 µg Kg^{-1} Sid on days 14 and 28.

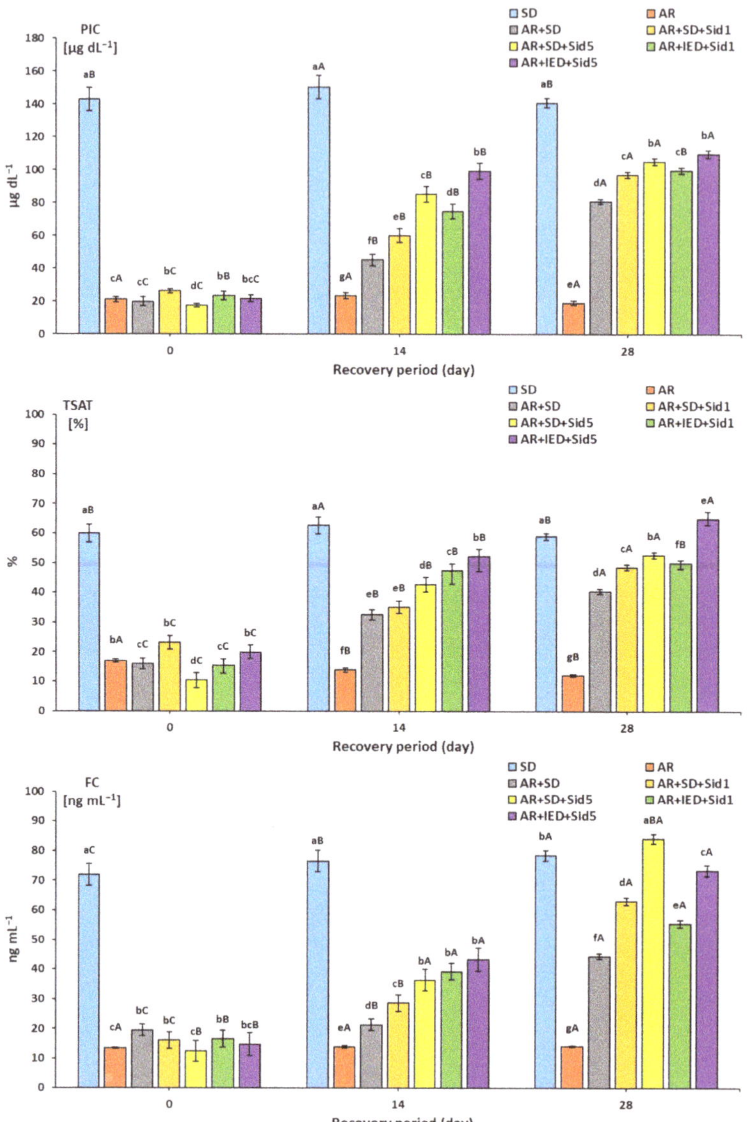

Figure 3. Plasma iron concentration (PIC), transferrin saturation % (TSAT), and ferritin concentration (FC) in rats fed different formulated experimental diets during a recovery period of up to 28 days ($n = 8$). SD: standard diet, AR: anemic rat fed iron-deficient diet, AR + SD: AR fed SD, AR + SD + Sid1: AR fed SD and intraperitoneally injected with Sid at 1 µg Kg^{-1} rat twice per week, AR + SD + Sid5: AR fed SD and intraperitoneally injected with Sid at 5 µg Kg^{-1} rat twice per week. AR + IED + Sid1: AR fed iron-enriched diet (IED) and intraperitoneally injected with Sid at 1 µg Kg^{-1} rat twice per week, AR + IED + Sid5: AR fed IED diet and intraperitoneally injected with Sid at 5 µg Kg^{-1} rat twice per week, [a,b,c,d,e,f,g]: bars for the same recovery day not sharing similar letters are significantly different ($p > 0.05$), [A,B,C]: bars for each treatment not sharing similar letters during the recovery period are significantly different ($p > 0.05$).

4. Discussion

Siderophores are secondary metabolites produced by various organisms to scavenge iron, thereby increasing its bioavailability and absorption into the cell. They are characterized by a high affinity for ferric iron, forming powerful iron-chelating complexes [14,18]. Siderophores recently received considerable attention, and many studies regarding their application in medicine and agriculture have been carried out [14,19–24]. Different synthesized siderophores have been identified, including catecholate-type, hydroxamate-type, carboxylate-type, and mixed-type siderophores, as reviewed by Sah and Singh [52]. The production and isolation of catecholate-type Sid from *S. tricolor* HM10 were previously established [39]. In the current research, a catechol-type Sid was successfully isolated and purified from the *S. tricolor* HM10 strain in significant quantities as 25–30 mg Sid L^{-1} medium under optimal conditions. The obtained results agree with the produced and purified siderophores in previous studies [43,53–55].

The proposed catechol-type siderophore pathway was identified based on genome sequencing and bioinformatic tools, i.e., antiSMASH online software. The intended pathway contains five genes similar to the petrobactin biosynthetic gene cluster from *Bacillus anthracis*. This pathway is composed of *asb* locus (*asbABCDEF*) and activates catecholate siderophore production [56]. The absB protein demonstrated substrate positions in the binding pocket and reaction flexibility [57]. *Bacillus anthracis* can produce two catechol-type siderophores (petrobactin and bacillibactin) under iron-limited conditions. Bacillibactin siderophores are highly sensitive to iron concentration, whereas petrobactin is less dependent on Fe [58]. Catechol-type siderophores are formed from 3,4-dihydroxybenzoyl spermidine (3,4-DHB-SPD) via the condensation process, followed by an activated form of citrate to produce 4-dihydroxybenzoyl spermidinyl citrate (3,4-DHB-SPD-CT). Both previously produced components condensate to form a mature siderophore [56]. Our study identified the expected catechol-type siderophore molecular weight to be 658.33 (M). This siderophore is similar close to Enterochelin (MW = 668.5), which can be produced from *E. coli* [59] and *Pseudomonas aeruginosa* [60]. Iron plays an essential role in maintaining normal human metabolism and is necessary for most organisms [2]. As is evidently known, iron-deficiency-induced anemia could cause a low hemoglobin concentration and a deficiency of iron enzymes, which initiates systemic dysfunction and abnormal macrophage secretion, further damaging immune function [61–63]. IDA could lead to numerous pathologies, particularly delays in development and behavior among children, girls, and pregnant women [3–5]. Interestingly, the isolated Sid in the current study expressed valuable TPC with potential antioxidant capacities against DPPH and ABTS radical scavenging activity, adding additional features to the Sid. Indeed, phenolic content and relative antioxidant capacity revealed that catecholate-type Sid has potential antioxidant activity [64,65].

The rats' final body weight, weight gain, food intake, and food efficiency ratio compared to the SD group during the recovery period were carefully investigated. During the recovery period, the AR group fed an SD and ARs fed an SD with different concentrations of injected Sid exhibited significant changes in weight gain. It could be noticed that Sid injection improved the weight gain in a dose-dependent manner in the presence of available iron in the diet. The increases in the weight gain might have been due to improved metabolism and increased fat and muscle mass, which was observed in the feed efficiency ratio [35]. Remarkably, the AR group fed an SD with 5 µg Kg^{-1} from Sid showed the highest weight gain. Although AR groups were fed an IED containing double iron content, the weight gain was not significantly changed. It might be because a diet containing high inorganic salt such as ferric citrate had side effects on the stomach and affected appetite and digestion [66]. Feed efficiency significantly increased with the feeding of an SD in the SD group compared to the AR group. A significant change was noted with ARs fed an SD or IED with Sid injection. The highest feed efficiency was observed in the AR + SD + Sid5 group, which did not significantly differ from the SD group. Notably, Sid concentration is a key promoter in increasing weight gain [35].

Consequently, the relative weights of the liver, kidneys, and spleen were linked to growth and body weight gain. Weight gained in the organs might have been due to improved metabolism and increased fat and muscle mass, as observed and indicated [36,39]. Consequently, ARs with an SD or IED injected with 1 μg Kg^{-1} of Sid had significantly lower relative liver weights than in the AR group, reflecting the improvement in body weight. Indeed, the higher relative weight may explain the higher organ weight and lower body weight. Diet and Sid concentration were clearly related to weight recovery in a dose-dependent manner, similarly noted in [35,38]. IDA resulted in a negatively skewed iron balance, causing a reduction in LIC and increased serum ferritin levels. Subsequently, decreasing iron intake will cause a decrease in the TSAT level and an increase in the TIBC level, resulting in "latent iron deficiency" [38]. In addition, as the iron deficiency state progresses, subjective symptoms of anemia appear, and the HGB level and HCT% decrease, resulting in "iron deficiency anemia" [67]. Indeed, LIC was measured as an index to determine the possible iron deficiency level in the body. On day 0, the PIC concentration confirmed that the iron-deficient diet group induced the latent iron deficiency condition. Among the iron-deficient groups, the values observed in the AR + IED + Sid1 and AR + IED + Sid5 groups were considerably higher than those observed in the AR + SD + Sid1 and AR + SD + Sid5 groups or in the AR + SD group. This may have indicated that iron was more efficiently absorbed and stored in the liver in the presence of both high iron content in the diet and Sid [35,38]. This relationship is directly related to serum transferrin iron saturation (TSAT %), which increased associatively (Figure 2) and was confirmed [38].

Surprisingly, a noticeable improvement in the hematological parameters such as RBC, HGB, HCT, MCV, MCH, and MCHC was observed. Immediately after inducing anemia, the RBC, HGB, HCT, MCV, MCH, and MCHC were monitored every two weeks up to 4 weeks. Significant changes were found in all parameters in all experimental groups, confirming the efficiency of the anemia recovery treatments. After 2 weeks, the remarkable improvements in HGB, HCT%, MCV, and MCHC in treated AR were evident in the efficiency of Sid in the presence of iron in the diet to attenuate anemia. This may be because Sid could anchor a specific gradient outside the cell membrane, depending on membrane affinity and length, thereby enhancing the ability to capture iron from surrounding matrixes [20,21]. The above results clarified that providing an SD to AR groups with Sid injection at 1 and 5 μg Kg^{-1} resulted in more significant improvements in the hematological parameters than feeding an SD alone or even an IED with Sid injection. Accordingly, considerable increases in HGB, HCT%, MCV, MCH, and MCHC values promoted recovery from iron-deficiency-induced anemia. Iron is mainly absorbed in the duodenum; there are two types of iron in the diet: non-heme and heme iron, each with a different absorption mechanism. Non-heme iron is taken up into intestinal epithelial cells by the divalent metal transporter 1 (DMT1), though the iron in the intestinal tract would then need to be solubilized [68]. However, acidified digesta by gastric acid is gradually neutralized by pancreatic and intestinal juice in the duodenum, and poorly soluble salts are formed with coexisting phosphoric acid as the pH increases, which results in poor iron absorption [69]. The solubility of the iron citrate used in the tested diet in this study was low, and insoluble salts were formed with an increase in pH. However, the results indicated that the presence of Sid might contribute to enhanced iron absorption [35]. Siderophores have low molecular weights and facilitate the iron acquisition and counter iron deficiency by high-affinity iron (III) ligands [14]. As Sid preferentially chelates iron, ferric iron could be chelated by Sid, increasing iron availability. Later on, when ferric iron is reduced to ferrous iron, Sid will release it. This is a crucial strategy for increasing iron absorption [14,70]. The role of siderophores is to dissolve ferric iron and transport it into the cytoplasm, so siderophores have the potential to become a nutritional fortifier [20].

As previously hypothesized, Sid could anchor a specific gradient outside the cell membrane, depending on membrane affinity and length, thereby enhancing the ability to capture iron from surrounding matrixes [20,21]. Iron is an indispensable nutrient; however, excess iron works as a pro-oxidant to produce hydroxyl radicals. Therefore, iron absorption has a regulatory system [71,72]. There are two iron transporters in epithelial cells, namely DMT1, an importer of iron, and feroportin (Fpn), an exporter of iron. The expression of hepcidin, a peptide hormone produced in the liver, is promoted during iron deficiency. Blood hepcidin binds to Fpn, which accelerates the internalization of the complex and the degradation of Fpn. Thus, the exportation of iron from enterocytes to ferritin is reduced. An inverse relationship between hepcidin expression and iron absorption and the expression of duodenal iron transporters was reported in rats [73]. These might be reasonable explanations for the significant increases observed in the PIC of ARs fed an SD or IED with 1 or 5 µg Kg^{-1} Sid. Likewise, the TSAT was associated significantly with serum iron (SI), as evidently presented [35,38]. Significant increases in ferritin concentration were noticed in ARs fed an IED with 5 µg Kg^{-1} Sid on days 14 and 28. On the other hand, indigestible saccharides, such as water-soluble dietary fibers and oligosaccharides, are reportedly fermented in the large intestine by intestinal bacteria to produce short-chain fatty acids such as acetic acid, propionic acid, n-butyric acid, or organic acids such as lactic acid and succinic acid. These acids acidify the lumen, increase the number of minerals solubilized in the intestine, and promote iron absorption in the intestine [74–76].

Indeed, in small species, intraperitoneal injection is used instead of intravenous access, and it can be used to safely administer large volumes of fluid. A higher affinity between siderophores and iron than between transferrin and iron is recognized. The catecholate-type siderophore has stronger iron affinities than transferrin in comparison to other categories of siderophores. A siderophore–iron complex is expected to enter the cell through simple diffusion (according to the low iron concentration inside the cells regarding anemia induced by iron deficiency) via the endocytosis of the transferrin receptor (TfR1) Siderocalin/Lipocalin 2/Neutrophil Gelatinase Associated Lipocalin, which is an innate immune system protein with bacteriostatic activity, or ATP-binding cassette (ABC) transporters, the energy-dependent efflux transporters. Red pulp macrophages (RPMs) and Kupffer cells (KCs) express complete machinery for RBC clearance and heme iron recycling [77–82].

Noticeably, at present, we did not delineate the detailed mechanism of these changes, and further experiments are needed to elucidate the exact underlying mechanisms. Combing our results with clearly reviewed studies, we suggest that the mechanism of action of Sid is to be effective in chelating iron in the entire intestine and binding to the cell membrane and will then find a way to transfer iron elements into the cell. However, to further confirm this hypothesis, we need to design a signaling experiment to elucidate iron pathways. In addition, the possibility of excess iron absorption needs to be examined using an iron-repleted diet and prolonged feeding conditions.

5. Conclusions

Isolated Sid from *S. tricolor* HM10 was evaluated through an anemia-induced rat model to study its potential to accelerate recovery from IDA. This study revealed that Sid significantly improved weight gain and iron-deficiency-induced anemia was effective during recovery from anemia and could be a practical iron-nutritive fortifier. Combining our findings with previously mentioned studies, we suggest that Sid's mechanism of action is effective in chelating iron in the entire intestine, binding to the cell membrane, and then finding a way to transfer iron elements into the cell. Noticeably, at present, we did not delineate the detailed mechanism of these changes, and the design of a signaling experiment to confirm this hypothesis is highly recommended. In addition, the possibility of excess iron absorption needs to be examined using an iron-repleted diet and prolonged feeding conditions.

Author Contributions: Conceptualization, M.R. and H.B.; methodology, H.B. and K.A.Q.; validation, A.S.A. and M.R.; formal analysis, H.B.; investigation, M.R.; data curation, H.B., M.R. and A.S.A.; writing—original draft preparation, M.R., K.A.Q. and H.B.; writing—review and editing, H.B. and M.R. All authors have read and agreed to the published version of the manuscript.

Funding: Qassim University, Deanship of Scientific Research, project (CAVM-2019-2-2-I-5457), the academic year 1440 AH/2019 AD.

Institutional Review Board Statement: The study was approved by the committee of Research Ethics, Deanship of Scientific Research, Qassim University (21-07-06 on 26 January 2022), which is governed by the Control and Supervision of Experiments on Animals (CPCSEA) Committee of the National Committee of BioEthics (NCBE), which implements regulations related to the ethics of research on living creatures.

Informed Consent Statement: Not applicable.

Data Availability Statement: The data presented in this study are available on request from the corresponding author.

Acknowledgments: The authors gratefully acknowledge Qassim University, represented by the Deanship of Scientific Research, for the financial support for this research under the number (CAVM-2019-2-2-I-5457) during the academic year 1440 AH/2019 AD.

Conflicts of Interest: The authors declare no conflict of interest.

Sample Availability: Not applicable.

Abbreviations

2,3-DHBA	2,3-Dihydroxybenzoic acid
ABC	ATP-binding cassette transporters
ABTS	2,2′-azino-bis(3-ethylbenzothiazoline-6-sulfonic acid)
AOA	Antioxidant activity
AR	Anemic rat
AR + IED + Sid1	Anemic rat fed iron-enriched diet with an intraperitoneal injection of 1 μg Sid Kg^{-1}
AR + IED + Sid5	Anemic rat fed iron-enriched diet with an intraperitoneal injection of 5 μg Sid Kg^{-1}
AR + SD + Sid1	Anemic rat fed standard diet with an intraperitoneal injection of 1 μg Sid Kg^{-1}
AR + SD + Sid5	Anemic rat fed standard diet with an intraperitoneal injection of 5 μg Sid Kg^{-1}
ARs	Anemic rats
CAB	Chromazurol B
CAS	Chrome Azurol S
CPCSEA	Purpose of the Control and the Supervision of Experiments on Animals
DPPH	1,1-diphenyl-2-picryl hydrazine
dw	Dry weight
EDTA-2Na	Ethylenediaminetetraacetic acid
FC	Ferritin concentration
GAE	Gallic acid equivalent
HCT	Hematocrit
HGB	Hemoglobin
IAEC	Institutional Animal Ethics Committee
IDA	Iron-deficiency-induced anemia
IDD	Iron-deficient diet
IED	Iron-enriched diet
KC	Kupffer cell
KSA	Kingdom of Saudi Arabia
LIC	Lever's iron concentration
MBCa	Maltobionic acid calcium salt
MCH	Mean corpuscular hemoglobin
MCHC	Mean corpuscular hemoglobin concentration
MCV	Mean corpuscular volume
NCBE	National Committee of Bioethics
NCBI	National Center for Biotechnology Information
NRs	Normal rats
PIC	Plasma iron concentration
RBC	Red blood cells
RPM	Red pulp macrophage

SD	Standard diet
Sid	Siderophore
TE	Trolox Equivalents
TIBC	Total iron-binding capacity
TLC	Thin-layer chromatography
TPC	Total phenolic compounds
TSAT	Transferrin saturation %

References

1. Hentze, M.W.; Muckenthaler, M.U.; Galy, B.; Camaschella, C. Two to Tango: Regulation of Mammalian Iron Metabolism. *Cell* **2010**, *142*, 24–38. [CrossRef] [PubMed]
2. Lv, C.; Zhao, G.; Lönnerdal, B. Bioavailability of Iron from Plant and Animal Ferritins. *J. Nutr. Biochem.* **2015**, *26*, 532–540. [CrossRef]
3. Camaschella, C. Iron deficiency. *Blood* **2019**, *133*, 30–39. [CrossRef] [PubMed]
4. Bresgen, N.; Eckl, P.M. Oxidative Stress and the Homeodynamics of Iron Metabolism. *Biomolecules* **2015**, *5*, 808–847. [CrossRef] [PubMed]
5. McClung, J.P.; Murray-Kolb, L.E. Iron Nutrition and Premenopausal Women: Effects of Poor Iron Status on Physical and Neuropsychological Performance. *Annu. Rev. Nutr.* **2013**, *33*, 271–288. [CrossRef] [PubMed]
6. Coad, J.; Pedley, K. Iron Deficiency and Iron Deficiency Anemia in Women. *Scand. J. Clin. Lab. Investig. Suppl.* **2014**, *74*, 82–89. [CrossRef]
7. Kassebaum, N.J.; Jasrasaria, R.; Naghavi, M.; Wulf, S.K.; Johns, N.; Lozano, R.; Regan, M.; Weatherall, D.; Chou, D.P.; Eisele, T.P.; et al. A systematic Analysis of Global Anemia Burden from 1990 to 2010. *Blood* **2014**, *123*, 615–624. [CrossRef]
8. Kusumi, E.; Shoji, M.; Endou, S.; Kishi, Y.; Shibata, T.; Murashige, N.; Hamaki, T.; Matsumura, T.; Yuji, K.; Yoneyama, A.; et al. Prevalence of Anemia among Healthy Women in 2 Metropolitan Areas of Japan. *Int. J. Hematol.* **2006**, *84*, 217–219. [CrossRef]
9. Algarín, C.; Nelson, C.A.; Peirano, P.; Westerlund, A.; Reyes, S.; Lozoff, B. Iron-deficiency Anemia in Infancy and Poorer Cognitive Inhibitory Control at Age 10 Years. *Dev. Med. Child. Neurol.* **2013**, *55*, 453–458. [CrossRef]
10. Lee, G.R. Iron deficiency and iron-deficiency anemia. In *Wintrobe's Clinical Hematology*; John Wiley & Sons, Ltd.: London, UK, 1999.
11. Hallberg, L.; Björn-Rasmussen, E.; Howard, L.; Rossander, L. Dietary Heme Iron Absorption. *Scand. J. Gastroenterol.* **1979**, *14*, 769–779. [CrossRef]
12. Terato, K.; Hiramatsu, Y.; Yoshino, Y. Studies on iron absorption. II. Transport Mechanism of Low Molecular Iron Chelate in Rat Intestine. *Am. J. Dig. Dis.* **1973**, *18*, 129–134. [CrossRef] [PubMed]
13. Björn-Rasmussen, E.; Hallberg, L.; Isaksson, B.; Arvidsson, B. Food Iron Absorption in Man. Applications of the Two-pool Extrinsic Tag Method to Measure Heme and Nonheme iron Absorption from the Whole Diet. *J. Clin. Investig.* **1974**, *53*, 247–255. [CrossRef] [PubMed]
14. Butler, A.; Theisen, R.M. Iron(III)–Siderophore Coordination Chemistry: Reactivity of Marine Siderophores. *Coord. Chem. Rev.* **2010**, *254*, 288–296. [CrossRef] [PubMed]
15. Simões, L.C.; Simões, M.; Vieira, M.J. Biofilm Interactions between Distinct Bacterial Genera Isolated from Drinking Water. *Appl. Environ. Microbiol.* **2007**, *73*, 6192–6200. [CrossRef] [PubMed]
16. Vaughn, C.B.; Weinstein, R.; Bond, B.; Rice, R.; Vaughn, R.W.; McKendrick, A.; Ayad, G.; Rockwell, M.A.; Rocchio, R. Ferritin Content in Human Cancerous and Noncancerous Colonic Tissue. *Cancer Investig.* **1987**, *5*, 7–10. [CrossRef]
17. Wandersman, C.; Delepelaire, P. Bacterial iron sources: From siderophores to hemophores. *Annu. Rev. Microbiol.* **2004**, *58*, 611–647. [CrossRef]
18. Neilands, J.B. Microbial Iron Compounds. *Annu. Rev. Biochem.* **1981**, *50*, 715–731. [CrossRef]
19. Homann, V.V.; Edwards, K.J.; Webb, E.A.; Butler, A. Siderophores of *Marinobacter aquaeolei*: Petrobactin and its Sulfonated Derivatives. *BioMetals* **2009**, *22*, 565–571. [CrossRef]
20. Sandy, M.; Butler, A. Microbial Iron Acquisition: Marine and Terrestrial Siderophores. *Chem. Rev.* **2009**, *109*, 4580–4595. [CrossRef]
21. Ito, Y.; Butler, A. Structure of Synechobactins, New Siderophores of the Marine Cyanobacterium *Synechococcus* sp. PCC 7002. *Limnol. Oceanogr.* **2005**, *50*, 1918–1923. [CrossRef]
22. Saha, M.; Sarkar, S.; Sarkar, B.; Sharma, B.K.; Bhattacharjee, S.; Tribedi, P. Microbial Siderophores and their Potential Applications: A review. *Environ. Sci. Pollut. Res.* **2016**, *23*, 3984–3999. [CrossRef] [PubMed]
23. Albelda-Berenguer, M.; Monachon, M.; Joseph, E. Chapter Five—Siderophores: From Natural Roles to Potential Applications. In *Advances in Applied Microbiology*; Gadd, G.M., Sariaslani, S., Eds.; Academic Press: Cambridge, MA, USA, 2019; Volume 106, pp. 193–225.
24. Kurth, C.; Kage, H.; Nett, M. Siderophores as Molecular Tools in Medical and Environmental Applications. *Org. Biomol. Chem.* **2016**, *14*, 8212–8227. [CrossRef] [PubMed]
25. Buss, J.L.; Torti, F.M.; Torti, S.V. The Role of Iron Chelation in Cancer Therapy. *Curr. Med. Chem.* **2003**, *10*, 1021–1034. [CrossRef] [PubMed]

26. Lovejoy, D.B.; Richardson, D.R. Iron Chelators as Anti-Neoplastic Agents: Current Developments and Promise of the PIH Class of Chelators. *Cur. Med. Chem.* **2003**, *10*, 1035–1049. [CrossRef] [PubMed]
27. Nakouti, I.; Sihanonth, P.; Palaga, T.; Hobbs, G. Effect of a siderophore producer on animal cell apoptosis: A possible role as anticancer agent. *Int. J. Pharma Med. Biol. Sci.* **2013**, *2*, 1–5.
28. Miethke, M.; Marahiel, M.A. Siderophore-Based Iron Acquisition and Pathogen Control. *Microbiol. Mol. Biol. Rev.* **2007**, *71*, 413–451. [CrossRef]
29. AlMatar, M.; Albarri, O.; Makky, E.A.; Var, I.; Köksal, F. A glance on the Role of Bacterial Siderophore from the Perspectives of Medical and Biotechnological Approaches. *Curr. Drug Targets* **2020**, *21*, 1326–1343. [CrossRef]
30. Ali, S.S.; Vidhale, N. Bacterial Siderophore and their Application: A review. *Int. J. Curr. Microbiol. Appl. Sci.* **2013**, *2*, 303–312.
31. Lee, N.; Kim, W.; Chung, J.; Lee, Y.; Cho, S.; Jang, K.-S.; Kim, S.C.; Palsson, B.; Cho, B.-K. Iron Competition Triggers Antibiotic Biosynthesis in *Streptomyces coelicolor* during Coculture with *Myxococcus xanthus*. *ISME J.* **2020**, *14*, 1111–1124. [CrossRef]
32. Terra, L.; Ratcliffe, N.; Castro, H.C.; Vicente, A.C.; Dyson, P. Biotechnological Potential of Streptomyces Siderophores as New Antibiotics. *Curr. Med. Chem.* **2021**, *28*, 1407–1421. [CrossRef]
33. Imbert, M.; Béchet, M.; Blondeau, R. Comparison of the main Siderophores Produced by some Species of Streptomyces. *Curr. Microbiol.* **1995**, *31*, 129–133. [CrossRef]
34. Armin, R.; Zühlke, S.; Grunewaldt-Stöcker, G.; Mahnkopp-Dirks, F.; Kusari, S. Production of Siderophores by an Apple Root-Associated *Streptomyces ciscaucasicus* Strain GS2 Using Chemical and Biological OSMAC Approaches. *Molecules* **2021**, *26*, 3517. [CrossRef] [PubMed]
35. Feng, X.; Jiang, S.; Zhang, F.; Wang, R.; Zhao, Y.; Zeng, M. Siderophore (from *Synechococcus* sp. PCC 7002)-Chelated Iron Promotes Iron Uptake in Caco-2 cells and Ameliorates Iron Deficiency in Rats. *Mar. Drugs* **2019**, *17*, 709. [CrossRef] [PubMed]
36. Zhang, X.G.; Wang, N.; Ma, G.D.; Liu, Z.Y.; Wei, G.X.; Liu, W.J. Preparation of S-iron-enriched Yeast using Siderophores and its Effect on Iron Deficiency Anemia in Rats. *Food Chem.* **2021**, *365*, 130508. [CrossRef] [PubMed]
37. Khan, A.; Singh, P.; Srivastava, A. Synthesis, Nature and Utility of Universal Iron Chelator—Siderophore: A review. *Microbiol. Res.* **2018**, *212–213*, 103–111. [CrossRef] [PubMed]
38. Suehiro, D.; Kawase, H.; Uehara, S.; Kawase, R.; Fukami, K.; Nakagawa, T.; Shimada, M.; Hayakawa, T. Maltobionic acid Accelerates Recovery from Iron Deficiency-Induced Anemia in Rats. *Biosci. Biotechnol. Biochem.* **2020**, *84*, 393–401. [CrossRef]
39. Rehan, M.; Alsohim, A.S.; Abidou, H.; Rasheed, Z.; Al Abdulmonem, W. Isolation, Identification, Biocontrol Activity, and Plant Growth Promoting Capability of a Superior *Streptomyces tricolor* Strain HM10. *Pol. J. Microbiol.* **2021**, *70*, 245–256. [CrossRef]
40. Schwyn, B.; Neilands, J.B. Universal Chemical Assay for the Detection and Determination of Siderophores. *Anal. Biochem.* **1987**, *160*, 47–56. [CrossRef]
41. Arnow, L.E. Colorimetric Determination of the Components of 3, 4-dihydroxyphenylalanine-tyrosine Mixtures. *J. Biol. Chem* **1937**, *118*, 531–537. [CrossRef]
42. Atkin, C.L.; Neilands, J.B.; Phaff, H.J. Rhodotorulic Acid from Species of *Leucosporidium, Rhodosporidium, Rhodotorula, Sporidiobolus*, and *Sporobolomyces*, and a New Alanine-Containing Ferrichrome from *Cryptococcus melibiosum*. *J. Bacteriol.* **1970**, *103*, 722–733. [CrossRef]
43. Clark, B.L. Characterization of a Catechol-Type Siderophore and the Detection of a Possible Outer Membrane Receptor Protein from *Rhizobium leguminosarum* Strain IARI 312. Master's Thesis, East Tennessee State University, Ann Arbor, MI, USA, 2004.
44. Storey, E.P. Isolation, Purification, and Chemical Characterization of the Dihydroxamate-type Siderophore, "Schizokinen", Produced by *Rhizobium leguminosarum* IARI 917. Master's Thesis, East Tennessee State University, Ann Arbor, MI, USA, 2005.
45. Zhang, Z.; Schwartz, S.; Wagner, L.; Miller, W. A Greedy Algorithm for Aligning DNA Sequences. *J. Comput. Biol.* **2000**, *7*, 203–214. [CrossRef] [PubMed]
46. Blin, K.; Shaw, S.; Kloosterman, A.M.; Charlop-Powers, Z.; van Wezel, G.P.; Medema, M.H.; Weber, T. AntiSMASH 6.0: Improving Cluster Detection and Comparison Capabilities. *Nucleic Acids Res.* **2021**, *49*, W29–W35. [CrossRef] [PubMed]
47. Yawadio Nsimba, R.; Kikuzaki, H.; Konishi, Y. Antioxidant Activity of Various Extracts and Fractions of *Chenopodium quinoa* and *Amaranthus* spp. seeds. *Food Chem.* **2008**, *106*, 760–766. [CrossRef]
48. Barakat, H.; Rohn, S. Effect of Different Cooking Methods on Bioactive Compounds in Vegetarian, Broccoli-based Bars. *J. Funct. Foods* **2014**, *11*, 407–416. [CrossRef]
49. Garčic, A. A highly Sensitive, Simple Determination of Serum Iron Using Chromazurol B. *Clin. Chim. Acta* **1979**, *94*, 115–119. [CrossRef]
50. White, D.; Kramer, D.; Johnson, G.; Dick, F.; Hamilton, H. Estimation of Serum Ferritin by Using Enzyme Immunoassay Method. *Am. J. Clin. Pathol.* **1986**, *72*, 346–351.
51. Steel, R.G.; Torrie, J.H.; Dickey, D.A. *Principles and Procedures of Statistics: A Biological Approach*; McGraw-Hill: New York, NY, USA, 1997.
52. Sah, S.; Singh, R. Siderophore: Structural and Functional Characterisation-A Comprehensive Review. *Agriculture* **2015**, *61*, 97–114. [CrossRef]
53. Schwabe, R.; Senges, C.H.R.; Bandow, J.E.; Heine, T.; Lehmann, H.; Wiche, O.; Schlömann, M.; Levicán, G.; Tischler, D. Cultivation Dependent Formation of Siderophores by *Gordonia rubripertincta* CWB2. *Microbiol. Res.* **2020**, *238*, 126481. [CrossRef]
54. Rachid, D.; Ahmed, B. Effect of Iron and Growth Inhibitors on Siderophores Production by *Pseudomonas fluorescens*. *Afr. J. Biotechnol.* **2005**, *4*, 697–702.

55. Sayyed, R.Z.; Chincholkar, S.B. Purification of Siderophores of *Alcaligenes faecalis* on Amberlite XAD. *Bioresour. Technol.* **2006**, *97*, 1026–1029. [CrossRef]
56. Lee, J.Y.; Janes, B.K.; Passalacqua, K.D.; Pfleger, B.F.; Bergman, N.H.; Liu, H.; Håkansson, K.; Somu, R.V.; Aldrich, C.C.; Cendrowski, S. Biosynthetic Analysis of the Petrobactin Siderophore Pathway from Bacillusanthracis. *J. Bacteriol.* **2007**, *189*, 1698–1710. [CrossRef] [PubMed]
57. Nusca, T.D.; Kim, Y.; Maltseva, N.; Lee, J.Y.; Eschenfeldt, W.; Stols, L.; Schofield, M.M.; Scaglione, J.B.; Dixon, S.D.; Oves-Costales, D. Functional and Structural Analysis of the Siderophore Synthetase AsbB Through Reconstitution of the Petrobactin Biosynthetic Pathway from *Bacillus anthracis*. *J. Biol. Chem.* **2012**, *287*, 16058–16072. [CrossRef] [PubMed]
58. Lee, J.Y.; Passalacqua, K.D.; Hanna, P.C.; Sherman, D.H. Regulation of Petrobactin and Bacillibactin Biosynthesis in *Bacillus anthracis* under Iron and Oxygen Variation. *PLoS ONE* **2011**, *6*, e20777. [CrossRef] [PubMed]
59. Gehring, A.M.; Bradley, K.A.; Walsh, C.T. Enterobactin Biosynthesis in *Escherichia coli*: Isochorismate Lyase (Entb) is a Bifunctional Enzyme that is Phosphopantetheinylated by Entd and then Acylated by Ente using Atp and 2,3-Dihydroxybenzoate. *Biochemistry* **1997**, *36*, 8495–8503. [CrossRef]
60. Gasser, V.; Baco, E.; Cunrath, O.; August, P.S.; Perraud, Q.; Zill, N.; Schleberger, C.; Schmidt, A.; Paulen, A.; Bumann, D.; et al. Catechol Siderophores Repress the Pyochelin Pathway and Activate the Enterobactin Pathway in *Pseudomonas aeruginosa*: An Opportunity for Siderophore–Antibiotic Conjugates Development. *Environ Microbiol.* **2016**, *18*, 819–832. [CrossRef]
61. Wu, H.; Zhu, S.; Zeng, M.; Liu, Z.; Dong, S.; Zhao, Y.; Huang, H.; Lo, Y.M. Enhancement of Non-heme Iron Absorption by Anchovy (*Engraulis japonicus*) Muscle Protein Hydrolysate Involves a Nanoparticle-mediated Mechanism. *J. Agric. Food Chem.* **2014**, *62*, 8632–8639. [CrossRef]
62. Young, M.F.; Griffin, I.; Pressman, E.; McIntyre, A.W.; Cooper, E.; McNanley, T.; Harris, Z.L.; Westerman, M.; O'Brien, K.O. Utilization of Iron from an Animal-Based Iron Source Is Greater Than That of Ferrous Sulfate in Pregnant and Nonpregnant Women. *J. Nutr.* **2010**, *140*, 2162–2166. [CrossRef]
63. Martinez-Torres, C.; Cubeddu, L.; Dillmann, E.; Brengelmann, G.L.; Leets, I.; Layrisse, M.; Johnson, D.G.; Finch, C. Effect of Exposure to Low Temperature on Normal and Iron-deficient Subjects. *Am. J. Physiol. Regul. Integr. Comp. Physiol.* **1984**, *246*, R380–R383. [CrossRef]
64. Achard, M.E.S.; Chen, K.W.; Sweet, M.J.; Watts, R.E.; Schroder, K.; Schembri, M.A.; McEwan, A.G. An Antioxidant Role for Catecholate Siderophores in Salmonella. *Biochem. J.* **2013**, *454*, 543–549. [CrossRef]
65. Huo, Y.; Kang, J.P.; Ahn, J.C.; Kim, Y.J.; Piao, C.H.; Yang, D.U.; Yang, D.C. Siderophore-producing Rhizobacteria Reduce Heavy Metal-induced Oxidative Stress in *Panax ginseng* Meyer. *J. Ginseng Res.* **2021**, *45*, 218–227. [CrossRef]
66. Zhao, G. Phytoferritin and its Implications for Human Health and Nutrition. *Biochim. Biophys. Acta Gen. Subj.* **2010**, *1800*, 815–823. [CrossRef] [PubMed]
67. Bainton, D.F.; Finch, C.A. The Diagnosis of Iron Deficiency Anemia. *Am. J. Med.* **1964**, *37*, 62–70. [CrossRef]
68. Gunshin, H.; Mackenzie, B.; Berger, U.V.; Gunshin, Y.; Romero, M.F.; Boron, W.F.; Nussberger, S.; Gollan, J.L.; Hediger, M.A. Cloning and Characterization of a Mammalian Proton-coupled Metal-ion Transporter. *Nature* **1997**, *388*, 482–488. [CrossRef] [PubMed]
69. Ward, F.W.; Coates, M.E. Gastrointestinal pH Measurement in Rats: Influence of the Microbial Flora, Diet and Fasting. *Lab. Anim.* **1987**, *21*, 216–222. [CrossRef] [PubMed]
70. Nemeth, E.; Tuttle, M.S.; Powelson, J.; Vaughn, M.B.; Donovan, A.; Ward, D.M.; Ganz, T.; Kaplan, J. Hepcidin Regulates Cellular Iron Efflux by Binding to Ferroportin and Inducing Its Internalization. *Science* **2004**, *306*, 2090–2093. [CrossRef]
71. Ganz, T.; Nemeth, E. Hepcidin and Iron Homeostasis. *Biochim. Biophys. Acta-Mol. Cell Res.* **2012**, *1823*, 1434–1443. [CrossRef]
72. Frazer, D.M.; Wilkins, S.J.; Becker, E.M.; Vulpe, C.D.; McKie, A.T.; Trinder, D.; Anderson, G.J. Hepcidin Expression Inversely Correlates with the Expression of Duodenal Iron Transporters and Iron Absorption in Rats. *Gastroenterology* **2002**, *123*, 835–844. [CrossRef]
73. Asvarujanon, P.; Ishizuka, S.; Hara, H. Promotive Effects of Non-digestible Disaccharides on Rat Mineral Absorption Depend on the Type of Saccharide. *Nutrition* **2005**, *21*, 1025–1035. [CrossRef]
74. Ohta, A.; Ohtsuki, M.; Baba, S.; Takizawa, T.; Adachi, T.; Kimura, S. Effects of Fructooligosaccharides on the Absorption of Iron, Calcium and Magnesium in Iron-deficient Anemic Rats. *J. Nutr. Sci. Vitam.* **1995**, *41*, 281–291. [CrossRef]
75. Laparra, J.M.; Díez-Municio, M.; Herrero, M.; Moreno, F.J. Structural Differences of Prebiotic Oligosaccharides Influence their Capability to Enhance iron Absorption in Deficient Rats. *Food Funct.* **2014**, *5*, 2430–2437. [CrossRef]
76. Turner, P.V.; Brabb, T.; Pekow, C.; Vasbinder, M.A. Administration of Substances to Laboratory Animals: Routes of Administration and Factors to Consider. *J. Am. Assoc. Lab. Anim. Sci.* **2011**, *50*, 600–613. [PubMed]
77. Correnti, C.; Strong, R.K. Mammalian Siderophores, Siderophore-binding Lipocalins, and the Labile Iron Pool. *J. Biol. Chem.* **2012**, *287*, 13524–13531. [CrossRef] [PubMed]
78. Courbon, G.; Francis, C.; Gerber, C.; Neuburg, S.; Wang, X.; Lynch, E.; Isakova, T.; Babitt, J.L.; Wolf, M.; Martin, A.; et al. Lipocalin 2 Stimulates Bone Fibroblast Growth Factor 23 Production in Chronic Kidney Disease. *Bone Res.* **2021**, *9*, 35. [CrossRef] [PubMed]
79. Page, M.G.P. The Role of Iron and Siderophores in Infection, and the Development of Siderophore Antibiotics. *Clin. Infect. Dis.* **2019**, *69*, S529–S537. [CrossRef]

80. Clifton, M.C.; Rupert, P.B.; Hoette, T.M.; Raymond, K.N.; Abergel, R.J.; Strong, R.K. Parsing the Functional Specificity of Siderocalin/Lipocalin 2/NGAL for Siderophores and Related Small-molecule Ligands. *J. Struct. Biol. X* **2019**, *2*, 100008. [CrossRef]
81. Golonka, R.; Yeoh, B.S.; Vijay-Kumar, M. The Iron Tug-of-War Between Bacterial Siderophores and Innate Immunity. *J. Innate Immun.* **2019**, *11*, 249–262. [CrossRef]
82. Nairz, M.; Theurl, I.; Swirski, F.K.; Weiss, G. "Pumping iron"-how Macrophages Handle Iron at the Systemic, Microenvironmental, and Cellular Levels. *Pflug. Arch.* **2017**, *469*, 397–418. [CrossRef]

Article

Isolation and Characterization of Bacteriocin-Producing *Lacticaseibacillus rhamnosus* XN2 from Yak Yoghurt and Its Bacteriocin

Yonghua Wei [1], Jinze Wang [2], Zhe Liu [2], Jinjin Pei [2,*], Charles Brennan [3] and A.M. Abd El-Aty [4,5]

[1] College of Food Science, Shanxi Normal University, Taiyuan 030031, China; weiyonghua1983@126.com
[2] Qinba State Key Laboratory of Biological Resources and Ecological Environment, Qinling-Bashan Mountains Bioresources Comprehensive Development C.I.C., Shaanxi Province Key Laboratory of Bio-Resources, College of Bioscience and Bioengineering, Shaanxi University of Technology, Hanzhong 723001, China; 15029368245@163.com (J.W.); l1115711913@163.com (Z.L.)
[3] College of Food Science and Engineering, Royal Melbourne Institute of Technology, Melbourne 3046, Australia; charles.brennan@rmit.edu.au
[4] Department of Pharmacology, Faculty of Veterinary Medicine, Cairo University, Giza 12211, Egypt; abdelaty44@hotmail.com
[5] Department of Medical Pharmacology, Faculty of Medicine, Atatürk University, Erzurum 25240, Turkey
* Correspondence: jinjinpeislg@163.com

Citation: Wei, Y.; Wang, J.; Liu, Z.; Pei, J.; Brennan, C.; Abd El-Aty, A.M. Isolation and Characterization of Bacteriocin-Producing *Lacticaseibacillus rhamnosus* XN2 from Yak Yoghurt and Its Bacteriocin. *Molecules* **2022**, *27*, 2066. https://doi.org/10.3390/molecules27072066

Academic Editor: Jacqueline Aparecida Takahashi

Received: 2 December 2021
Accepted: 18 March 2022
Published: 23 March 2022

Publisher's Note: MDPI stays neutral with regard to jurisdictional claims in published maps and institutional affiliations.

Abstract: Lactic acid bacteria (LAB) produce antimicrobial substances that could potentially inhibit the growth of pathogenic and food spoilage microorganisms. *Lacticaseibacillus rhamnosus* XN2, isolated from yak yoghurt, demonstrated antibacterial activity against *Bacillus subtilis*, *B. cereus*, *Micrococcus luteus*, *Brochothrix thermosphacta*, *Clostridium butyricum*, *S. aureus*, *Listeria innocua* CICC 10416, *L. monocytogenes*, and *Escherichia coli*. The antibacterial activity was estimated to be 3200 AU/mL after 30 h cultivation. Time-kill kinetics curve showed that the semi-purified cell-free supernatants (CFS) of strain XN2 possessed bactericidal activity. Flow cytometry analysis indicated disruption of the sensitive bacteria membrane by semi-purified CFS, which ultimately caused cell death. Interestingly, sub-lethal concentrations of semi-purified CFS were observed to reduce the production of α-haemolysin and biofilm formation. We further investigated the changes in the transcriptional level of *luxS* gene, which encodes signal molecule synthase (Al-2) induced by semi-purified CFS from strain XN2. In conclusion, *L. rhamnosus* XN2 and its bacteriocin showed antagonistic activity at both cellular and quorum sensing (QS) levels. Finally, bacteriocin was further purified by reversed-phase high-performance liquid chromatography (RP-HPLC), named bacteriocin XN2. The amino acid sequence was Met-Lue-Lys-Lys-Phe-Ser-Thr-Ala-Tyr-Val.

Keywords: *Lacticaseibacillus rhamnosus*; yak yoghurt; antibacterial activity; purification; bacteriocin

1. Introduction

A yak (*Bos mutus*) is a long-haired bovid commonly farmed in the Himalayan region. It is distributed in an area above the altitude of 3000 m in the Tibet Plateau in China [1]. Homemade and naturally fermented yak milk yoghurt is one of the favorite foods for the Tibetan people. The yoghurt is prepared by naturally fermenting yak milk in a custom-made, specially treated tung-made big jar for at least 7–10 days at ambient temperatures around 15 °C to produce acidity, alcohol, and flavor to the desired level [2]. A characteristic common feature of this yak yoghurt is the presence of alcohol in addition to lactic acid [3]. These yak milk yoghurts in the Qinghai-Tibet Plateau area may be considered a good reservoir for lactic acid bacteria. Ding et al., (2011) investigated the microorganisms among the yoghurts from Gansu province in China [1]. Among the isolates, 164 isolates (51.41% of the total) were classified under *Lactobacilli*, and 155 (48.59%) belonged to *cocci*. All the isolates were classified into six genera (*Lactobacillus*, *Lactococcus*, *Leuconostoc*, *Streptococcus*,

Enterococcus, and *Weissella*) and 21 species. Luo et al. (2017) indicated that the average counts of lactic acid bacteria of the naturally fermented yak milk were higher than that of yoghurt and suggested new strains of lactic acid bacteria and yeasts may be isolated from the yak yoghurts, as well as for new functional products [2].

Many researchers are screening and isolating bacteriocin-producing lactic acid bacteria from naturally fermented yoghurt due to concerns of potential damage to health by artificial chemical preservatives [4,5]. Bacteriocins from lactic acid bacteria are peptides secreted by some lactic acid bacteria with antimicrobial activity against other microorganisms, including food spoilage and pathogens. They are generally considered safe [6]. Todorov et al. isolated bacteriocins ST461BZ and ST462BZ produced by *L. rhamnosus* isolated from boza [7]. Srinivasan et al. reported that *L. rhamnosus* L34 isolated from the feces of Thai Breastfed infants could produce bacteriocins [8]. However, there is no report on screening bacteriocins producing lactic acid bacteria from naturally fermented yak yoghurts. This study aimed to screen and isolate bacteriocin-producinger lactic acid bacteria (LAB) from the naturally fermented yak yoghurt in the Qinghai-Tibet plateau in China.

Most of the research regarding the antibacterial mechanism action of bacteriocins has focused on their mechanism of action against sensitive bacteria cells. For example, they damage the integrity of the cell membrane and bind to the DNA [4,5]. However, some researchers denoted that bacteriocins may also inhibit biofilm formation and the production of secondary metabolites. For instance, *Lactiplantibacillus paraplantarum*, isolated from cheese, produces bacteriocin FT259, potentially influencing *Listeria monocytogenes* biofilm formation [9]. Chopra et al. [10] investigated a new bacteriocin with the potential to prevent biofilm formation. This study aimed to characterize the antibacterial activity and the mode of action of the bacteriocin-producing strain isolated from naturally fermented yak yoghurt and its bacteriocin. As biofilm formation may be regulated by the quorum sensing (QS) system [11,12], we also investigated whether bacteriocins were also able to regulate the QS system of sensitive bacteria in this study.

2. Results and Discussion

2.1. Screening of Bacteriocin Production by LABs Isolated from Yak Yoghurt

Among the isolates from Qinghai yak yoghourt, strain XN2 could inhibit the growth of both gram-positive and gram-negative bacteria (the diameter of the inhibition zone was 15.1 ± 1.2 mm for *S. aureus* CICC10384, and 7.2 ± 0.5 mm for *E. coli* CICC10302). After eliminating the antibacterial activity caused by organic acids or H_2O_2, strain XN2 was selected as the bacteriocin producer. Based on carbohydrate utilization profile using API 50 CHL kits and 16S rRNA sequence analyses, the strain XN2 was identified as *L. rhamnosus*, referred to as *L. rhamnosus* XN2.

Naturally fermented yak yoghurts from different locations may contain different microorganisms. For instance, Ding et al. (2011) reported that *Lactobacillushelveticus*, *Leuconostoc mesenteroides* subsp. *mesenteroides*, *Streptococcus thermophilus*, *Lactobacillus casei*, and *Lactococcus lactis subsp. lactis* were the predominant populations in the yak milk products from Gansu Province in China [1]. This study isolated the functional strain XN2 from yak yoghurts in Qinghai province. The use of *L. rhamnosus* in food has been well documented in probiotics and starter culture strain, and it is generally accepted as a safe and probiotic lactic acid bacteria [7,8,13].

The activity spectrum is listed in Table 1. The semi-purified CFS of stain XN2 deployed the inhibitory activities towards both gram-positive (*Bacillus subtilis*, *B. cereus*, *B. megaterium*, *Micrococcus luteus*, *S. aureus*, *Listeria innocua*, and *L. monocytogenes*) and gram-negative bacteria (*E. coli*) (Table 1).

Table 1. Effect of enzymes, temperature, and pH on bacteriocin XN2.

	Testing Condition	Antibacterial Activities of Semi-Purified CFS of Stain XN2
	Lipase, α-Amylase,	+
	Proteinase K, Papain, α-Chymotrypsin, Trypsin, Pepsin	−
	Catalase	+
Effect of temperature and storage on the activity of semi-purified CFS of stain XN2	60 °C, 80 °C, and 100 °C	+
	37 °C for 14 d	+
	2 months at 4 °C	+
Effect of pH on the activity of semi-purified CFS of stain XN2	pH 2–8	+
	pH 9–10	−
	Bacillus subtilis CICC 10034	+
	B. cereus CICC 2155	+
	Micrococcus luteus CICC 10209	+
	Brochothrix thermosphacta CICC 10509	+
The activity of semi-purified CFS of stain XN2 against gram-positive bacteria	*Clostridium butyricum* CICC 10350	+
	Staphylococcus aureus CICC 10384	+
	S. aureus CICC 10201	+
	Methicillin-resistant S. aureus *	+
	Listeria innocua CICC 10416	+
	L. monocytogenes CICC 21529	+
	Escherichia coli CICC 10302	+
	E. coli CGMCC 3373	+
The activity of semi-purified CFS of stain XN2 against gram-negative bacteria	*E. coli* CICC 10300	+
	Pseudomonas aeruginosa CICC 21636	−
	Enterobacter cloacae CICC 21539	−
	Salmonella paratyphi β CICC 10437	−
The activity of semi-purified CFS of stain XN2 against fungi	*Aspergillus niger* CICC 2124	−
	Candida albicans CICC 1965	−
	Saccharomyces cerevisiae CICC 1002	−

CICC: China Center of Industrial Culture Collection. *: Methicillin-resistant *S. aureus* provided by a local hospital in Xi'an, Shaanxi, China. +: There is a clear inhibition circle. −: there is no inhibition circle.

It is worth noting that the semi-purified CFS of strain XN2 could inhibit gram-negative (*E. coli*), while most bacteriocins from LAB belonged to Class I and Class II, only inhibiting gram-positive bacteria [14,15]. As well as bacteriocin from strain XN2, some bacteriocins were also found to have the ability to interact with intracellular enzyme systems and nucleic acid, besides damaging the integrity of the bacterial cell membrane. The determination and characterization of these bacteriocins with broad antibacterial spectra require the attention of more and more researchers [16,17]. One of the bottlenecks of applying bacteriocins in the food industry or medical clinics is that the antibacterial spectra of bacteriocins is usually narrow. The ability of the strain XN2 and its bacteriocin, which was active against both gram-positive and negative, might improve its potential.

2.2. Bacteriocin Characterization of XN2 Strain

2.2.1. Growth and Bacteriocin Production Curve

Bacteriocin production began after 8 h, and the stable high level was recorded after 30 h of growth in MRS broth (Figure 1). Similar results were also reported for other LAB bacteriocin-producing strains, such as *Companilactobacillus crustorum* MN047 [17], *Lactobacillus plantarum* ST16Pa [18], and *Pediococcus acidilactici* HW01 [19]. Bacteriocin might be the secondary metabolites of lactic acid bacteria.

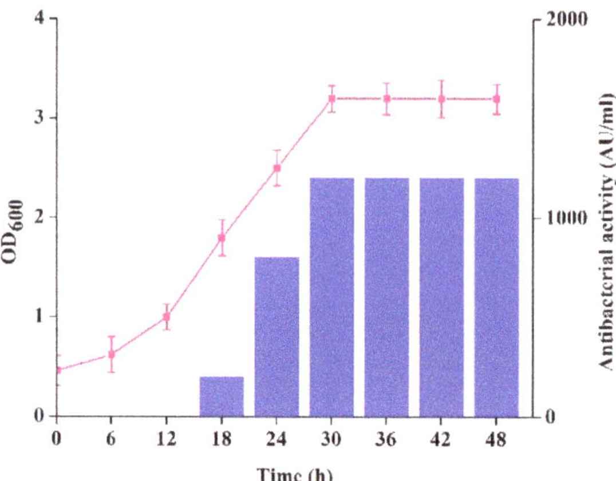

Figure 1. Production of bacteriocin by strain XN2. Square: growth of strain XN2; column: antibacterial activity of the CFS of strain XN2. The data are shown as mean ± standard deviation. The critical limit is 5%. $p < 0.05$, the differences are significant.

2.2.2. Bacteriocin Stability

Treatment with proteinase resulted in inactivation, whereas treatment with catalase, α-amylase, and lipase did not affect the antibacterial activity (Table 1). Similar results were observed for bacteriocin RC20975 produced by *L. rhamnosus* RC20975 [13] and bacteriocin BacC1 produced by *Enterococcus faecium* C1 [20]. The antibacterial activity was stable under heat treatment and over a pH range from 2–8 (Table 1). However, bacteriocin RC20975, similar to Sonorensin, enterocin E50-52, and pediocin PA-1 [10,13,21] were quickly inactivated under alkaline conditions

2.3. Kill Kinetics Curve of Semi-Purified CFS of Strain XN2 against S. aureus

Treatment with 50 µg/mL semi-purified CFS of strain XN2 did not affect the population of *S. aureus*. However, treatment with 100 µg/mL semi-purified CFS of strain XN2 inhibited the growth of *S. aureus*. On the other hand, the population of *S. aureus* was drastically decreased within 1 h of treatment with 200 or 400 µg/mL semi-purified CFS (Figure 2).

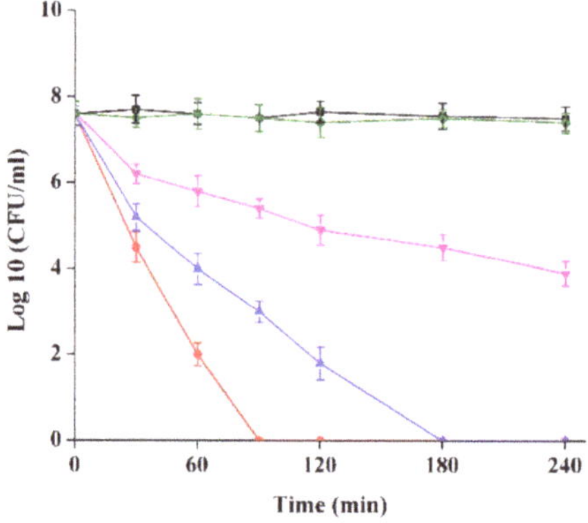

Figure 2. *S. aureus* in the PBS treated with 0 µg/mL (♦), 50 µg/mL (■), 100 µg/mL (▼), 200 µg/mL (▲) and 400 µg/mL (●) of the semi-purified CFS of strain XN2. The data are shown as mean ± standard deviation. The critical limit is 5%. $p < 0.05$, the differences are significant.

The bactericidal activity against *S. aureus* implied that the inhibitory effect of the semi-purified CFS was concentration dependent. Similar action was reported for Enterocin FH99 isolated from *Enterococcus faecium* FH99 (active against *L. monocytogenes*) by Kaur et al. (2013) [22], bacteriocin RC20975 produced by *L. rhamnosus* RC20975 (active against *Alicyclobacillus. Spp.*) by Yue et al. (2013) [13], and Plantaricin GZ1-27 produced by *L. plantarum* GZ1-27 (against *Bacillus cereus*) by Du et al. (2018) [16].

2.4. Effect of Semi-Purified CFS of Strain XN2 on the Integrity of S. aureus

Living cells cannot allow the passage of *propidium iodide* (PI) across the intact cell membrane. However, increase in permeability would increase the concentration and fluorescence intensity of PI within the cells. Herein, an extension of the incubation time led to a significant increase in the amount of PI passed through the cells; the finding was proved by the cell membrane rupture (Figure 3). These results indicate that the semi-purified CFS of strain XN2 can increase the cell permeability, ultimately resulting in cell death.

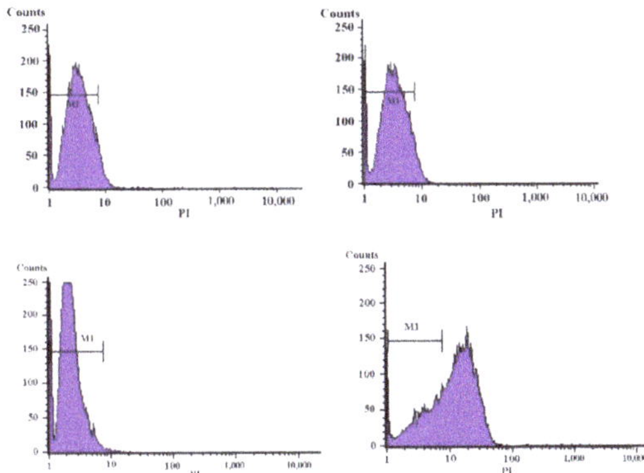

Figure 3. Effects of the semi-purified CFS of strain XN2 on the membrane integrity of *S. aureus*, tested by fluorescent staining and flow cytometry. Left-up: *S. aureus* cells treated with 2 MIC of the semi-purified CFS of strain XN2 for 0 min; right-up: for 30 min; left-down: for 60 min; and right-down: for 240 min.

The great diversity of chemical structures of bacteriocins from different lactic acid bacteria determines the diversity of their functions and modes of action [17,23]. However, researchers believe that the alteration of cell membrane permeability is the principal mechanism of action of bacteriocins isolated from LAB [24,25]. Similar mode of action was also seen in Bacteriocin RC20975 from *Lacticaseibacillus rhamnosus* CICC20975, bacteriocin BacC1 from *Enterococcus faecium* C1, and bacteriocin SLG10 from *Lactobacillus plantarum* SLG10 [13,20].

2.5. Effect of Semi-Purified CFS of Strain XN2 on α-Haemolysin Secreted by S. aureus

Compared with the control, the hemolytic activity of *S. aureus* CICC 10384 decreased to 90.05% and 27.34%, following treatment with 1/2 MIC and 2 MIC semi-purified CFS of strain XN2, respectively (Figure 4).

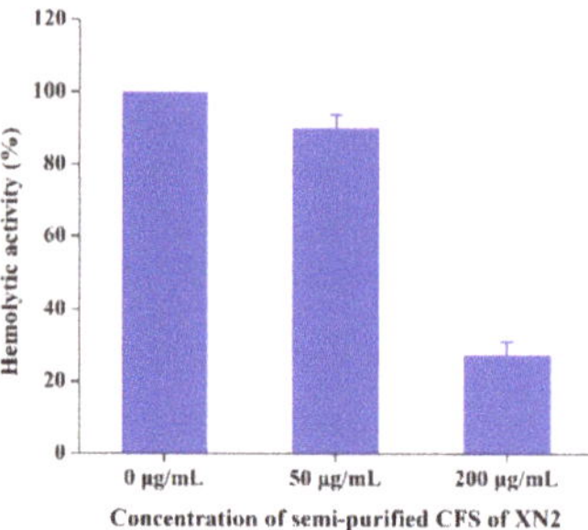

Figure 4. Effects of semi-purified CFS of strain XN2 on the secretion of α-haemolysin by *S. aureus*. The data are shown as mean ± standard deviation. The critical limit is 5%. $p < 0.05$, the differences are significant.

These results could justify the ability of the semi-purified CFS of strain XN2 to inhibit the secretion of α-haemolysin in a dose-dependent manner, even at the sub-MIC. α-haemolysin is one of the major virulence factors produced by *S. aureus* [26]. Semi-purified CFS of strain XN2 reduced the secretion of α-haemolysin, which meant that bacteriocin from stain XN2 had good potential characteristics to control *S. aureus* in the future.

2.6. Effect of Semi-Purified CFS of Strain XN2 on the Biofilm Formation of S. aureus

In the control group, *S. aureus* (Figure 5A) was able to form a thin biofilm. It should be noted that samples treated with 1/2 MIC semi-purified CFS of strain XN2 would have fewer bacteria attached to the carrier surface (Figure 5B). The inhibition percentage of biofilm formation of *S. aureus* by the same concentration of semi-purified CFS of strain XN2 was 45.8 ± 3.1. The results were consistent. For the sample treated with 2 MIC semi-purified CFS of strain XN2, it could be seen that there were fewer cells in the biofilms, the biofilm was looser (Figure 5C), and the inhibition precent of biofilm formation was 90.3 ± 5.9. This finding suggested that bacteriocin XN2 could inhibit the formation of biofilm.

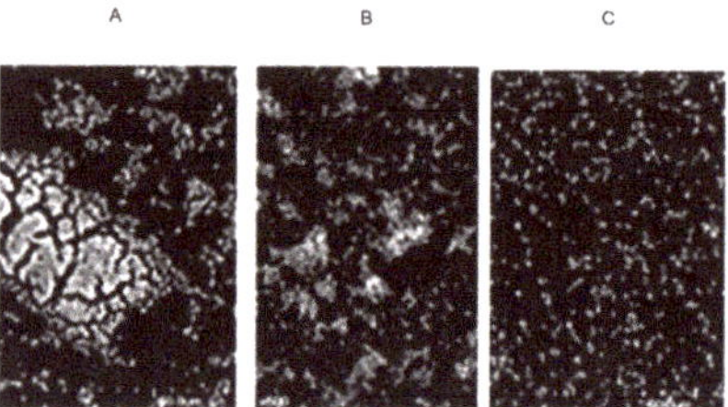

Figure 5. Effect of semi-purified CFS of strain XN2 on the biofilm formation of *S. aureus*; (**A**): control (*S. aureus* cells treated with no semi-purified CFS); (**B**): with 1/2 MIC semi-purified CFS; and (**C**): with 2 MIC semi-purified CFS.

It is well known that only a tiny proportion of bacteria are present in the form of free-floating plankton [9]. The formation of biofilm communities is one of the most critical strategies for microbial survival under a particular ecological niche [11]. As strain XN2 and its bacteriocin could inhibit biofilm formation of food pathogens, such as *S. aureus*, they might be used in the food industry. The biofilm formation and α-haemolysin secretion may be regulated by the QS system [12]. Thus, it can be assumed that strain XN2 and its bacteriocin can also regulate the QS system of sensitive bacteria. Therefore, we further investigated the effect of semi-purified CFS of strain XN2 on the QS system of *S. aureus*.

2.7. Effect of Semi-Purified CFS of Strain XN2 on the QS System of S. aureus

With 1/2 MIC or 2 MIC semi-purified CFS of strain XN2, the gene expression of *luxS* was increased 2 and 5.6 fold (Figure 6). The result proved that strain XN2 or its bacteriocin could affect the QS system of *S. aureus* at a molecular level.

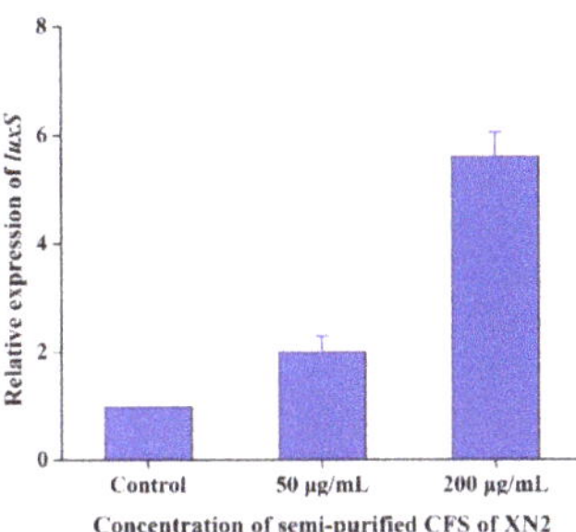

Figure 6. Effect of semi-purified CFS of strain XN2 on the relative expression of *luxS* of *S. aureus*. The data are shown as mean $\pm$ standard deviation. The critical limit is 5%. $p < 0.05$, the differences are significant.

It was reported that the expression of gene *luxS* is one of the vital targets for QS inhibitors. Therefore, the effect of bacteriocin on the *luxS* gene expression was tested to determine the encoding of the protein LuxS, the enzyme for the synthesis of AI-2, and the signal molecule of the QS system. It was generally accepted that the formation of biofilm and production of some secondary compounds, for instance, α-haemolysin, were regulated by the QS system. When the bacterial population was big enough, inferred by the amount of signal molecules, the processing of the biofilm formation and production of secondary compounds, like α-haemolysin, starts. The results in this study were complementary with the findings that semi-purified CFS of strain XN2 can affect the secretion of α-haemolysin and the formation of biofilm aforementioned above. Most of the previous research has focused on bacteriocin's inhibition/killing mechanism on sensitive bacteria via increasing the membrane permeability and eventually cell death [14,18]. There are few reports investigating the effect of bacteriocin on the QS system of sensitive bacteria [9]. In the present study, a bacteriocin from XN2 has dual activities, damaging the integrity of *S. aureus* cells and affecting its QS system.

2.8. Purification of Bacteriocin

After $(NH_4)_2SO_4$ precipitation, the specific activity of bacteriocin improved to 262.44 IU/mg. After passing through the Sephadex G-50 column, fractions 33 to 37 showed antibacterial activity. A single sharp peak in the RP-HPLC spectrum at retention time of 16.58 min showed the antibacterial activity.

The three-step method used in this study for isolation and purification of bacteriocin XN2 was frequently used and was utilized for successful purification of diverse bacteriocins from culture supernatants, such as bacteriocin RC20975 [13], Plantaricin GZ127 [16] and bacteriocin BacC1 [20]. The industrial production of nisin, the most well-known bacteriocin, was also obtained by the standard three steps. The standard purification method would make this bacteriocin easier to obtain.

N-sequencing indicated that the amino acid sequence of the bacteriocin was Met-Leu-Lys-Lys-Phe-Ser-Thr-Ala-Try-Val (MLKKFSTAYV). Similarity analysis revealed none in the database (http://web.expasy.org/blast/, accessed on 5 February 2019). Therefore, this bacteriocin was designated as a novel peptide, bacteriocin XN2. Bacteriocin XN2 falls into the category of Class IId bacteriocins because it does not contain lanthionine or YGNGVXC (characteristics of Class IIa bacteriocins).

The mode of action of small-sized bacteriocins has recently attracted much research attention. Due to the small size, bacteriocins might not cause "pore formation" on sensitive bacteria. They might inhibit cell wall/membrane synthesis through binding with precursors of cell wall/membrane synthesis, destroying the integrity of the wall/membrane, and interacting with enzyme system or DNA in cells [27].

3. Materials and Methods

3.1. Isolation and Identification of Bacteriocin-Producing Stain

Traditional Natural fermented Qinghai yak yoghurts (a total of 40 bottles from 20 different local markets (GaoYuanZhiBao Co.Ltd, Xining, China)) were used to isolate bacteriocin-producing LAB. One loop of the yoghurt was inoculated into 20 mL MRS medium and incubated anaerobically at 37 °C for 24 h. Then, one loop was streaked onto a De Man, Rogosa, Sharpe (MRS) agar (Oxoid, Basingstoke, UK) plate. Gram-positive, catalase-negative, and oxidase-negative bacterial strains were chosen as LAB [27]. Afterwards, cell-free culture supernatants (CFS) from LAB were produced by centrifugation (Avanti J-E, Beckman, CA, USA) at $5000 \times g$ for 15 min at 4 °C, followed by 0.22 μm micro-filtration (Millipore, MA, USA), according to Yue et al. [13]. The pH of the CFS was adjusted to pH 6.5 to eliminate the antibacterial effect of organic acid. Catalase (Solarbiio, Beijing, China) was added to eliminate the effect of peroxide, as described by Yue et al. [13]. Furthermore, the agar well diffusion method was utilized for the antibacterial activity screening of CFS. Gram-positive, *S. aureus* CICC10384, and Gram-negative bacteria, *E. coli* CICC 10302, were used as indicator bacteria [28]. The indicator strains were pre-cultured at 30 °C to mid-exponential with anaerobic culture. The well with 10 mM phosphate buffer (pH 6.5) was used as a negative control. The well with 2 MIC Ampicillin (Solarbiio, Beijing, China) in buffer was considered as a positive control. The strain in which CFS showed a clear inhibition zone with both indicator strains was chosen and isolated. Primary strain identification was performed using a commercial kit (API 50 CHL, BioMerieux, Montalieu Versie, France), which recognizes the carbohydrate fermentation pattern of bacteria [29]. Finally, the bacterial strain genotype was identified by 16S rRNA gene sequence analysis using the same 27F (5′-AGTTTGATCMTGGCTCAG-3′) and 1492R primers (5′-GGTTACCTTGTTACGACTT-3′) as reported earlier [7]. Sequence similarity searches were compared with NCBI (www.ncbi.nlm.nlh.gov, accessed on 9 January 2018) database.

3.2. Bacteriocins Production by Strain XN2

A 12 h old culture of strain XN2 was inoculated (5%, *v/v*) into MRS broth and incubated at 30 °C. Cell growth, antibacterial activity, and pH value of the culture were tested at 6 h intervals [13]. The growth of strain XN2 was evaluated by testing the optical density at 600 nm (OD_{600}). Antibacterial activity was tested with *S. aureus* CICC10384 as the sensitive strain. The pH values of the culture were monitored with a pH meter (Lei-ci SJ-5, INESA Scientific instrument Co., Ltd., Shanghai, China).

3.3. Semi-Purification of CFS of Strain XN2

The CFS of strain XN2 was semi-purified using ammonium sulfate (($NH_4)_2SO_4$) precipitation and size-exclusion chromatography (SEC) [13]. After 30 h cultivation, the CFS of strain XN2 was recovered by centrifugation at $10,000 \times g$ for 30 min at 4 °C. Afterwards, CFS was concentrated to 1/5th of the initial volume using a rotavapor (RV-8V, IKA, Staufen, Germany). Subsequently, $(NH_4)_2SO_4$ was slowly added up to the concentration of 60% (*v/v*). After stirring overnight at 4 °C, the precipitate was collected and resuspended in 20 mM Na_2HPO_4-citric acid buffer (pH 3.6). The biologically active solution was added onto the Sephadex G-50 column (80 × 2.0 cm, Sigma, Santa Clara, CA, USA) and eluted by Na_2HPO_4-citric acid buffer at a flow rate of 0.5 mL/min. The highest active fraction was collected and freeze-dried. The Bradford protein assay method was used [30] to quantify protein in a fraction. The antibacterial activity was assessed by the agar well diffusion method using *S. aureus* CICC10384 as an indicator strain [7].

3.4. Antimicrobial Spectrum and Stability Testing

The antibacterial spectrum was determined using indicator strains, including *Bacillus subtilis* CICC 10034, *B. cereus* CICC 2155, *Micrococcus luteus* CICC 10209, *Brochothrix thermosphacta* CICC 10509, *Clostridium butyricum* CICC 10350, *Staphylococcus aureus* CICC 10384, *S. aureus* CICC 10201, *Listeria innocua* CICC 10416, *L. monocytogenes* CICC 21529, *Escherichia coli* CICC 10302, *E. coli* CGMCC 3373, *E. coli* CICC 10300, *Pseudomonas aeruginosa* CICC 21636, *Enterobacter cloacae* CICC 21539, *Salmonella paratyphi* β CICC 10437, *Aspergillus niger* CICC 2124, *Candida albicans* CICC 1965, *Saccharomyces cerevisiae* CICC 1002 All the strains are purchsed from China Center of Industrial Culture Collection(CICC). To determine the influence of pH or enzymes on antibacterial activity, the pH of the samples was adjusted between 2 and 10 with 1 M HCl and 1 M NaOH solutions, or the samples were treated with different enzymes (Lipase, α-Amylase, Proteinase K, Papain, α-Chymotrypsin, Trypsin, Pepsin, and Catalase) (Solarbiio, Beijing, China), with the final concentrations of 1.0 mg/mL, as described by Du et al. [18]. The effect of temperature on antibacterial activity was tested, as shown in Table 1. The antibacterial activity was tested as described above, with S. aureus CICC 10384 as indicator strain [7]; samples without any treatments were used as controls.

3.5. Time-Killing Kinetics

Overnight-cultured *S. aureus* CICC 10384 was washed with and suspended in PBS (10 mM, pH 7.0), adjusted to the OD_{600} of 0.5, and treated with 50, 100, 200, and 400 μg/mL of semi-purified CFS of strain XN2. At fixed time points, the plate count method determined viable cells [13].

3.6. Flow Cytometry (FCM) Analysis

The mid-logarithmic phase of *S. aureus* CICC10384 at a concentration of 10^4 CFU/ mL was co-cultured with 2 MIC semi-purified CFS of strain XN2 at 37 °C for 240 min. Cells were then labeled with propidium iodide (PI) and subjected to flow cytometry analysis (Becton Dickinson, NJ, USA), as described by Pei et al. (2018) [27,28,31]. Untreated *S. aureus* CICC 10384 cells were used as a control.

3.7. α-Haemolysin Secretion

S. aureus CICC 10384 was inoculated in LB broth containing semi-purified CFS of strain XN2 at 1/2 MIC and 2 MIC (control without semi-purified CFS of strain XN2) and was cultured at 30 °C for 24 h. Following filtration with 0.22 um filter membrane, the α-haemolysin content was tested according to Zhou et al. (2017) [26]. Fresh fibrin removed from rabbit blood was washed with calcium chloride buffer three times. A mixture of 875 μL calcium chloride buffer, 100 μL sample supernatant, and 25 μL red blood cells was incubated for 30 min at 37 °C, centrifuged at 5500× *g* 1 min, then tested for the D_{543} value.

3.8. Biofilm Formation

The mid-logarithmic phase culture of *S. aureus* CICC 10384 was diluted to approximately 10^7 CFU/mL. Each 6-well cell culture plate hole was filled with 1 mL of the suspension. The cell culture plate was pre-positioned with a sterile cover glass (22 mm × 22 mm). Plates were cultured at 30 °C after adding 0 (control), 1/2 MIC, or 2 MIC of semi-purified CFS of strain XN2. After 25 h of cultivation (visible and mature biofilm can be seen in control samples), the cover slides were taken out from each hole and washed with distilled water three times. After fixation with 2.5% glutaraldehyde and washing with PBS, the samples were then dehydrated with ethanol (100% repeated thrice, 90, 80, 70, and 50%) for 15 min each. The gold powder was used to cover the sample surface under vacuum conditions. Samples were observed under a scanning electron microscope.

The percent of biofilm inhibition was assessed by the crystal violet staining method according to the protocol adopted by Saporito et al. (2018) [32]. Briefly, after stained by crystal violet (0.1% *w/v* in water) for 10 min at room temperature (a washing step with PBS removed the excess dye) and re-dissolved by adding 96% ethanol for 10 min, OD595 of the samples in a new plate were recorded (SpectraMax 190, San Jose, CA, USA). The percent of biofilm inhibition was calculated by comparing the optical density values for the treated samples and the untreated control [32].

3.9. RT-PCR

To verify the effect of semi-purified CFS of strain XN2 on the quorum sensing (QS) system of *S. aureus* CICC 10384, RT-PCR was developed for the detection of transcriptional changes observed in the *luxS* gene (encoding the synthase of QS signal molecular (Al-2) of *S. aureus*) and induced by 0 (control), 1/2 MIC or 2 MIC of semi-purified CFS of strain XN2. One loop of stain *S. aureus* CICC 10384 was added onto 10 mL sterilized LB broth and cultured at 30 °C for 12 h. The OD_{600} value was adjusted to 0.5 and inoculated into the fresh LB broth containing 0 μg/mL (control), 1/2 MIC, or 2 MIC semi-purified CFS of strain XN2 with the inoculum size of 3%, *v/v*. After co-culture in LB broth at 30 °C for 24 h, the gene expression of *luxS* was investigated.

The total RNA from *S. aureus* CICC 10384 was extracted by the Trizol method (Cui et al., 2012) [33]. The reverse transcription reaction system (10 μL) contained: 2 μL of total RNA, 0.5 μL of dNTP, 0.5 μL of random primers, and 4 μL of distilled water without RNAase. All samples were heated at 70 °C for 5 min and then placed immediately in an ice bath. Then, 2 μL of 5× reverse transcription buffer, 0.5 μL of RNAase inhibitor, and 0.5 μL of MMLV reverse transcription enzyme were added immediately. The reverse transcription reaction procedure was as follows: 30 °C for 10 min, 42 °C for 1 h, 70 °C for 15 min, and 4 °C forever. The DNA templates were stored at −20 °C until use.

PCR reaction contained: 0.2 μL of upstream primer (10 μM), 0.2 μL of downstream primer (10 μM), 5 μL of 2× Ultra SYBR Mixture, 3.6 μL of double-distilled H_2O, and 1 μL of 5× DNA templates. The PCR reaction conditions were: 95 °C for 10 min, 95 °C for 15 s, 60 °C for 1 min, and 40 cycles. Primers were designed according to the gene sequence of *luxS* Primer Premier 5.0 software (version 5.0, Primer Premier, Canada, 2017) as (5′-GAGATCTTATGCCATCAGTAGAAAG-3′; 5′-GGTCACCTTTATCCAAACACTTTCTC-3′).

3.10. Bacteriocin Purification

The semi-purified CFS of strain XN2 was additionally purified using an HPLC system with *photodiode array detector* (PDA) (UltiMate 3000, Dionex, Sunnyvale, CA, USA) and HC-C_{18} column (5 μm, 250 mm × 4.6 mm, Agilent Technologies, Palo Alto, CA, USA) [13] (Yue et al., 2013). The bacteriocin was eluted by linearly increasing the ACN content in ACN/water (Milli-Q) mixture from 10 to 95% over 40 min, at a flow rate of 0.5 mL/min with an injection volume of 1 mL. The amount of proteins was quantified according to the Bradford assay method [30]. The antibacterial activity was tested using an agar well diffusion assay, using *S. aureus* CICC10384 as an indicator strain [7]. The amino acids sequence were tested by N-sequence by Shengo bioengineering Co. Ltd. (Shanghai, China).

3.11. Statistical Analyses

SPSS 18.0 software (IBM SPSS Statistics, Amund, NY, USA) was used to analyze the data [16]. All data were the average representation of three measurements and shown as mean ± standard deviation. T-test was used to calculate whether there were significant differences in statistics for the group of data. $t > t_0$ ($p < 0.05$) was considered that the differences (between the control and the testing groups, between testing groups) are significant.

4. Conclusions

In this study, one bacteriocin-producing strain, *Lacticaseibacillus rhamnosus* XN2, was successfully isolated from the yak yoghurt produced in the Xining and Qinghai Provinces in China. The bacteriocin, bacteriocin XN2, exhibited antibacterial activities against *Bacillus subtilis, B. cereus, Micrococcus luteus, Brochothrix thermosphacta, Clostridium butyricum, S. aureus, Listeria innocua CICC 10416, L. monocytogenes*, and *Escherichia coli*. The bacteriocin was able to disrupt the cell membrane (ultimately causing cell death), inhibit the secretion of α-haemolysin, and regulate the QS system of *S. aureus*. The amino acid sequence of the bacteriocin XN2 was YGNGVFSVIK.

Author Contributions: Conceptualization, Y.W. and J.P.; methodology, J.W. and Z.L.; software, Y.W., J.W. and Z.L.; validation, Y.W.; formal analysis, J.W. and Z.L.; investigation, J.W. and Y.W.; resources, Y.W.; data curation, J.W.; writing—original draft preparation, Y.W.; writing—review and editing, J.P.; visualization, J.P.; supervision, A.M.A.E.-A. and C.B.; project administration, writing—review and editing; J.P.; funding acquisition, J.P. All authors have read and agreed to the published version of the manuscript.

Funding: This study was funded by the National Natural Science Foundation of China (31801563), a special support plan for high-level talents in Shaanxi Province (Jinjin Pei) and foreign expert project of the Ministry of science and technology.

Institutional Review Board Statement: Not applicable.

Informed Consent Statement: Not applicable.

Data Availability Statement: The data presented in this study are available on request from the corresponding author.

Conflicts of Interest: The authors declare no conflict of interest.

Sample Availability: Samples of the compounds are available from the authors.

References

1. Ding, W.; Shi, C.; Chen, M.; Zhou, J.; Long, R.; Guo, X. Screening for lactic acid bacteria in traditional fermented Tibetan yak milk and evaluating their probiotic and cholesterol-lowering potentials in rats fed a high-cholesterol diet. *J. Funct. Foods* **2017**, *32*, 324–332. [CrossRef]
2. Luo, F.; Feng, S.; Sun, Q.; Xiang, W.; Zhao, J.; Zhang, J.; Yang, Z. Screening for bacteriocin-producing lactic acid bacteria from kurut, atraditional naturally-fermented yak milk from Qinghai–Tibet plateau. *Food Control* **2011**, *22*, 50–53. [CrossRef]
3. Ji, X.; Li, X.; Ma, Y.; Li, D. Differences in proteomic profiles of milk fat globule membrane in yak and cow milk. *Food Chem.* **2017**, *221*, 1822–1827. [CrossRef] [PubMed]
4. Duan, X.; Duan, S.; Wang, Q.; Ji, R.; Cao, Y.; Miao, J. Effects of the natural antimicrobial substance from Lactobacillus paracasei FX-6 on shelf life and microbial composition in chicken breast during refrigerated storage. *Food Control* **2020**, *109*, 106906. [CrossRef]
5. Xu, C.; Fu, Y.; Liu, F.; Liu, Z.; Ma, J.; Jiang, R.; Song, C.; Jiang, Z.; Hou, J. Purification and antimicrobial mechanism of a novel bacteriocin produced by *Lacticaseibacillus rhamnosus* 1.0320—Sciencedirect. *LWT* **2021**, *137*, 110338. [CrossRef]
6. Mirkovic, N.; Kulas, J.; Miloradovic, Z.; Miljkovic, M.; Tucovic, D.; Miocinovic, J.; Jovcic, B.; Mirkov, I.; Kojic, M. Lactolisterin BU-producer Lactococcus lactis subsp. lactis BGBU1-4: Bio-control of *Listeria monocytogenes* and *Staphylococcus aureus* in fresh soft cheese and effect on immunological response of rats. *Food Control* **2020**, *111*, 107076. [CrossRef]
7. Todorov, S.D.; Prévost, H.; Lebois, M.; Dousset, X.; LeBlanc, J.G.; Franco, B.D. Bacteriocinogenic *Lacticaseibacillus rhamnosus* ST16Pa isolated from papaya (*Carica papaya*)—From isolation to application: Characterization of a bacteriocin. *Food Res. Int.* **2011**, *44*, 1351–1363. [CrossRef]
8. Srinivasan, R.; Kumawat, D.K.; Kumar, S.; Saxena, A.K. Purification and characterization of a bacteriocin from *Lactobcillus rhamnosus* L34. *Ann. Microbiol.* **2013**, *3*, 37–392.
9. Winkelströter, L.K.; Tulini, F.L.; De Martinis, E.C. Identification of the bacteriocin produced by cheese isolate *Lactiplantibacillus paraplantarum*, FT259 and its potential influence on *Listeria monocytogenes* biofilm formation. *LWT-Food Sci. Technol.* **2015**, *64*, 586–592. [CrossRef]
10. Chopra, L.; Singh, G.; Kumar, J.K.; Sahoo, D.K. Sonorensin: A new bacteriocin with potential of an anti-biofilm agent and a food biopreservative. *Sci. Report.* **2015**, *5*, 13412. [CrossRef]
11. Al-Gburi, A.; Zehm, S.; Netrebov, V. Subtilosin Prevents Biofilm Formation by Inhibiting Bacterial Quorum Sensing. *Probiotics Antimicrob. Proteins* **2016**, *9*, 81–90. [CrossRef]
12. Mhatre, E.; Monterrosa, R.G.; Kovács, Á.T. From environmental signals to regulators: Modulation of biofilm development in Grampositive bacteria. *J. Basic Microbiol.* **2014**, *54*, 616–632. [CrossRef] [PubMed]
13. Pei, J.; Li, X.; Han, H.; Tao, Y. Purification and characterization of plantaricin SLG1, a novel bacteriocin produced by Lb. plantarum isolated from yak cheese. *Food Control* **2018**, *84*, 111–117. [CrossRef]
14. Yue, T.; Pei, J.; Yuan, Y. Purification and Characterization of Anti-Alicyclobacillus Bacteriocin Produced by *Lacticaseibacillus rhamnosus*. *J. Food Prot.* **2013**, *76*, 1575–1581. [CrossRef]
15. Pei, J.; Feng, Z.; Ren, T.; Sun, H.; Han, H.; Jin, W.; Dang, J.; Tao, Y. Purification, characterization and application of a novel antimicrobial peptide from *Andrias davidianus* blood. *Lett. Appl. Microbiol.* **2018**, *66*, 38–43. [CrossRef]
16. Song, Y.L.; Kato, N.; Matsumiya, Y.; Liu, C.X.; Kato, H.; Watanabe, K. Identification of Lactobacillus species of human origin by a commercial kit, API50CHL. *J. Assoc. Rapid Method Autom. Microbiol.* **1999**, *10*, 77–82.

17. Bradford, M.M. A rapid and sensitive method for the quantification of microgram quantities of protein utilizing the principle of protein-dye binding. *Anal. Biochem.* **1976**, *72*, 248–254. [CrossRef]

18. Du, H.; Yang, J.; Lu, X.; Lu, Z.; Bie, X.; Zhao, H.; Zhang, C.; Lu, F. Purification, characterization, and mode of action of action of Plantaricin GZ1-27, a novel bactriocin against *Bacillus cereus*. *J. Agric. Food Chem.* **2018**, *66*, 4716–4724. [CrossRef]

19. Pei, J.; Feng, Z.; Ren, T.; Jin, W.; Li, X.; Chen, D.; Dang, J. Selectivelyscreen the antibacterial peptide from the hydrolysates of highland barley. *Eng. Life Sci.* **2018**, *18*, 28–54. [CrossRef]

20. Zhou, Y.; Chen, C.; Pan, J.; Deng, X.; Wang, J. Epigallocatechin gallate can attenuate human alveolar epithelial cell injury induced by alpha-haemolysin. *Microb. Pathog.* **2017**, *115*, 222–226. [CrossRef]

21. Saporito, P.; Mouritzen, M.V.; Løbner-Olesen, A.; Jenssen, H. LL-37 fragments have antimicrobial activity against *Staphylococcus epidermidis* biofilms and wound healing potential in HaCaT cell line. *J. Pept. Sci.* **2018**, *24*, e3080. [CrossRef] [PubMed]

22. Cui, Y.; Zhao, Y.; Tian, Y.; Zhang, W.; Li, X.; Jiang, X. The molecular mechanism of action of bactericidal gold nanoparticles on Escherichia coli. *Biomaterials* **2012**, *33*, 2327–2333. [CrossRef] [PubMed]

23. Cavicchioli, V.Q.; Camargo, A.C.; Todorov, S.D.; Nero, L.A. Novel bacteriocinogenic *Enterococcus hirae* and *Pediococcus pentosaceus* strains with antilisterial activity isolated from Brazilian artisanal cheese. *J. Dairy Sci.* **2017**, *100*, 2526–2535. [CrossRef] [PubMed]

24. Rahmeh, R.; Akbar, A.; Kishk, M.; Onaizi, T.A.; Al-Shatti, A.; Shajan, A.; Akbar, B.; Al-Mutairi, S.; Yateem, A. Characterization of semipurified enterocins produced by *Enterococcus faecium* strains isolated from raw camel milk. *J. Dairy Sci.* **2018**, *101*, 4944–4952. [CrossRef]

25. Hwanhlem, N.; Ivanova, T.; Biscola, V.; Choiset, Y.; Haertlé, T. Bacteriocin producing Enterococcus faecalis isolated from chicken gastrointestinal tract originating from Phitsanulok, Thailand: Isolation, screening, safety evaluation and probiotic properties. *Food Control* **2017**, *78*, 187–195. [CrossRef]

26. Sabo, S.S.; Converti, A.; Ichiwaki, S.; Oliveira, R.P. Bacteriocin production by *Lactobacillus plantarum* ST16Pa in supplemented whey powder formulations. *J. Dairy Sci.* **2019**, *102*, 87–99. [CrossRef]

27. Ahn, H.; Kim, J.; Kim, W.J. Isolation and characterization of bacteriocin-producing *Pediococcus acidilactici* HW01 from malt and its potential to control beer spoilage lactic acid bacteria. *Food Control* **2017**, *80*, 59–66. [CrossRef]

28. Goh, H.F.; Philip, K. Isolation and mode of action of bacteriocin BacC1 produced by nonpathogenic *Enterococcus faecium* C1. *J. Dairy Sci.* **2015**, *98*, 5080–5090. [CrossRef]

29. Tiwari, S.K.; Noll, S.K.; Cavera, V.L.; Chikindas, M.L. Improved antimicrobial activities of synthetic-hybrid bacteriocins designed from enterocin E50-52 and pediocin PA-1. *Appl. Environ. Microbiol.* **2015**, *81*, 1661–1667. [CrossRef]

30. Kaur, G.; Singh, T.; Malik, R. Antibacterial efficacy of Nisin, Pediocin 34 and Enterocin FH99 against *Listeria monocytogenes* and cross resistance of its bacteriocin resistant variants to common food preservatives. *Braz. J. Microbiol.* **2013**, *14*, 63–71. [CrossRef]

31. Heredia-Castro, P.Y.; Méndez-Romero, J.I.; Hernández-Mendoza, A.; Acedo-Félix, E.; González-Córdova, A.F.; Vallejo-Cordoba, B. Antimicrobial activity and partial characterization of bacteriocin-like inhibitory substances produced by Lactobacillus spp. isolated from artisanal Mexican cheese. *J. Dairy Sci.* **2015**, *98*, 8285–8293. [CrossRef]

32. Panagiota, K.; Kyriakou, E.B.; Kristiansen, P.E.; Kaznessis, Y.N. Interactions of a class IIb bacteriocin with a model lipid bilayer, investigated through molecular dynamics simulations. *Biochim. Biophys. Acta (BBA)—Biomembr.* **2016**, *1858*, 824–835.

33. Snyder, A.B.; Worobo, R.W. Chemical and genetic characterization of bacteriocins: Antimicrobial peptides for food safety. *J. Sci. Food Agric.* **2014**, *94*, 28–44. [CrossRef]

Article

Antioxidant, Antibacterial and Dyeing Potential of Crude Pigment Extract of *Gonatophragmium triuniae* and Its Chemical Characterization

Ajay C. Lagashetti [1,2], Sanjay K. Singh [1,2,*], Laurent Dufossé [3,*], Pratibha Srivastava [2,4] and Paras N. Singh [1,2]

[1] National Fungal Culture Collection of India (NFCCI), Biodiversity and Palaeobiology Group, MACS' Agharkar Research Institute, G.G. Agarkar Road, Pune 411004, India; lagashettiajay@gmail.com (A.C.L.); pnsingh@aripune.org (P.N.S.)
[2] Faculty of Science, Savitribai Phule Pune University, Pune 411007, India
[3] CHEMBIOPRO Chimie et Biotechnologie des Produits Naturels, ESIROI Département Agroalimentaire, Université de la Réunion, F-97490 Sainte-Clotilde, Ile de La Réunion, France
[4] Bioprospecting Group, MACS' Agharkar Research Institute, G.G. Agarkar Road, Pune 411004, India; psrivastava@aripune.org
* Correspondence: sksingh@aripune.org (S.K.S.); laurent.dufosse@univ-reunion.fr (L.D.); Tel.: +91-20-25325103 (S.K.S.); +33-66-8731906 (L.D.)

Abstract: Filamentous fungi synthesize natural products as an ecological function. In this study, an interesting indigenous fungus producing orange pigment exogenously was investigated in detail as it possesses additional attributes along with colouring properties. An interesting fungus was isolated from a dicot plant, *Maytenus rothiana*. After a detailed study, the fungal isolate turned out to be a species of *Gonatophragmium* belonging to the family *Acrospermaceae*. Based on the morphological, cultural, and sequence-based phylogenetic analysis, the identity of this fungus was confirmed as *Gonatophragmium triuniae*. Although this fungus grows moderately, it produces good amounts of pigment on an agar medium. The fermented crude extract isolated from *G. triuniae* has shown antioxidant activity with an IC_{50} value of 0.99 mg/mL and antibacterial activity against Gram-positive bacteria (with MIC of 3.91 µg/mL against *Bacillus subtilis*, and 15.6 µg/mL and 31.25 µg/mL for *Staphylococcus aureus* and *Micrococcus luteus*, respectively). Dyeing of cotton fabric mordanted with $FeSO_4$ using crude pigment was found to be satisfactory based on visual observation, suggesting its possible use in the textile industry. The orange pigment was purified from the crude extract by preparative HP-TLC. In addition, UV-Vis, FTIR, HRMS and NMR (^{1}H NMR, ^{13}C NMR), COSY, and DEPT analyses revealed the orange pigment to be "1,2-dimethoxy-3*H*-phenoxazin-3-one" ($C_{14}H_{11}NO_4$, *m/z* 257). To our understanding, the present study is the first comprehensive report on *Gonatophragmium triuniae* as a potential pigment producer, reporting "1,2-dimethoxy-3*H*-phenoxazin-3-one" as the main pigment from the crude hexane extract. Moreover, this is the first study reporting antioxidant, antibacterial, and dyeing potential of crude extract of *G. triuniae*, suggesting possible potential applications of pigments and other bioactive secondary metabolites of the *G. triuniae* in textile and pharmaceutical industry.

Keywords: *Gonatophragmium triuniae*; pigments; bioactivity; dyeing; chemical characterization

Citation: Lagashetti, A.C.; Singh, S.K.; Dufossé, L.; Srivastava, P.; Singh, P.N. Antioxidant, Antibacterial and Dyeing Potential of Crude Pigment Extract of *Gonatophragmium triuniae* and Its Chemical Characterization. *Molecules* **2022**, *27*, 393. https://doi.org/10.3390/molecules27020393

Academic Editor: Jacqueline Aparecida Takahashi

Received: 13 December 2021
Accepted: 5 January 2022
Published: 8 January 2022

Publisher's Note: MDPI stays neutral with regard to jurisdictional claims in published maps and institutional affiliations.

1. Introduction

Ascomycetous filamentous fungi are known to produce bio-pigments extensively used as colourants, additives, colour intensifiers, antimicrobial, antioxidants, etc., in different industries such as food, beverages, cosmetics, textiles, and pharmaceuticals [1–3]. Natural pigments/colours are gaining increased attention currently because of the adverse effects of synthetic colours on humans and the environment. Most of the synthetic colours have been found to be toxic, allergic, and carcinogenic to human beings, and hazardous to the environment [4–6]. This has increased the need for safe, natural, eco-friendly pigments as an alternative to synthetic pigments. Increasing demand for natural pigments necessitates a greater need to explore colours or pigments from safe, natural sources, especially from microbes (bacteria, fungi, algae, lichens, and actinomycetes).

Fungi are currently emerging as a better and excellent source of natural pigments. Many researchers have reported fungi of different taxonomic groups exhibiting the production of pigments of diverse chemicals classes such as carotenoids, flavins, ubiquinones, anthraquinones, quinines, phenazines, etc. [1,7–10]. Due to the additional attributes of these "mycopigments" such as antimicrobial, anticancer, antioxidant, cytotoxicity, activity, etc. in addition to colouring property, they are being extensively used for a wide range of applications in food, textiles, medicines, paints, cosmetics, and electronics [8,9]. Some fungal pigments such as azaphilones, astaxanthin, Arpink Red, riboflavin, β-carotene isolated from *Monascus, Xanthophyllomyces dendrorhous, Penicillium oxalicum, Ashbya gossypii,* and *Blakeslea trispora*, respectively, are already in the market for their commercial and industrial applications [11]. Numerous studies have reported the dyeing potential of fungal pigments and suggested their possible use in the textile industry for dyeing different types of textile fabrics like cotton, silk, wool, etc. [7,8,12,13].

Literature indicates that extensive studies have been done worldwide on production, optimization, and applications of pigments from common ascomycetous fungi like *Monascus, Talaromyces, Aspergillus, Penicillium, Fusarium,* etc. [8,9]. Besides these conventional ones, several other genera, such as *Epicoccum, Trichoderma, Alternaria, Chaetomium,* etc., are reported to have good pigment production potential [7,14–18]; however, several genera of filamentous ascomycetes are still unknown and unexplored for their pigment production potential and exploitations. These unexplored fungi might prove to be a hidden treasure of novel bio-active pigments having a variety of applications.

Taking this view into account, we have planned the present research in which we have isolated an uncommon fungus, *Gonatophragmium triuniae,* from infected leaves of the *Maytenus rothiana*, a plant endemic to central Maharashtra (Western Ghats). Interestingly, it was found that this rare fungus produces a very good extracellular orange pigment on solid media [potato dextrose agar medium (PDA)] as well as in liquid media [potato dextrose (PD) broth]. For the characterization of pigment, pure culture of *G. triuniae* was subjected to flask level fermentation in PD broth, and pigments were extracted from the culture filtrate using Hexane and dried. The dried Hexane extract was then examined for its antimicrobial and antioxidant properties and also assessed for its dyeing potential on cotton fabric using two mordants (Alum & $FeSO_4$). Finally, by preparative thin-layer chromatography (TLC), we purified an orange pigment from the crude pigment extract and identified it as "1,2-dimethoxy-3*H*-phenoxazin-3-one" based on ultra-violet (UV); Fourier transform infrared (FTIR); high-resolution mass (HRMS) spectroscopy; and ^{1}H & ^{13}C NMR, COSY, and DEPT analysis.

Several natural and synthetic phenoxazines are well known for their bioactivity and dyeing properties. These phenoxazines were found to exhibit antioxidant, antibacterial, anti-proliferative, and anti-tumoral activities [19]. 3*H*-phenoxazin-3-one and its derivatives exhibiting numerous biological activities (antimicrobial, anticancer, antitumor, antiviral, antitubercular, anticoccidial, antineoplastic, phytotoxic, and cell growth-stimulating) have been reported from different microorganisms such as actinomycetes, lichens, and fungi [20]. Phenoxazine class of pigments such as Phenoxazone, Pycnosanguin, Cinnabarine, *O*-acetyl cinnabarine, 2-Amino-9-formylphenoxazone-1-carbonic acid, and 9-Hydroxymethyl-2-methylaminophenoxazone-1-carbonic acid methyl ester have been described from the fungus *Pycnoporus sanguineus* [21]. Similarly, Chandrananimycins A-C belonging to class 3*H*-phenoxazin-3-one, exhibiting antitumor activity against colon cancer (CCL HT29); breast cancer (LCL H460, CNCL SF268, MACL MCF-7); lung cancer (LXFA 526L, LXFL 529L); melanoma (MEXF 514L); and kidney tumor cells (PRCL PC3M, RXF 631L) has also been reported from *Actinomadura* sp. [22,23]. One of the recent studies evaluated the antibacterial activity of the synthesized derivatives of 3*H*-phenoxazin-3-one [20]. Based on the history of Phenoxazin class of pigments, their promising bioactivity, and their dyeing property, we may consider present compound 1,2-dimethoxy-3*H*-phenoxazin-3-one isolated from *G. triuniae* NFCCI 4873 as a good colourant as well as a potential candidate for application in the textile and pharmaceutical industry.

2. Results and Discussion

2.1. Morphological Identification

Leaf lesions, amphigenous, necrotic spots single or irregular in concentric rings, later spots unite to form large spots. Margin: greyish-white; center: white. Colonies hypophyllus, velvety, brown. Mycelium superficial. Hyphae branched, septate, pale olivaceous to subhyaline, smooth-walled, up to 6.5 μm wide. Stroma and hyphopodia are absent. Conidiophores arising from superficial hyphae, dichotomously branched, multi-septate (6–8), the width of conidiophore gradually decreasing towards the length; macronematous to mononematous, erect, smooth-walled, highly geniculate, nodose, basal half part of conidiophore olivaceous brown and subhyaline to light olivaceous towards the apex, 55–145 × 3.22–7 μm. Conidiogenous cells integrated, polyblastic, terminal to intercalary, swollen towards the apex, variable in size: 10–20 μm long, bearing 10–15 loci, scars thickened and darkened, dentate or plate-like about 1 μm diameter. Conidia solitary, dry, holoblastic, acropleurogenous, clavate, cylindrical, straight to slightly curved, 0–1 septate, smooth-walled, subhyaline to light olivaceous, base narrowly truncate, apex obtuse, hilum thickened and darkened, 6–15 × 2–3.6 μm (Figure 1).

Colonies on Potato Dextrose Agar (PDA), slow-growing, reach 26–29 mm diameter after 4 weeks of incubation at 25 °C; front view of colony light yellow (4A4), circular, raised, aerial mycelium slightly cottony, margin smooth. Reverse dark brown (7F8) with diffusible yellowish orange (5B8) pigment in entire media. Mycelium branched, septate, smooth-walled, with frequent anastomosis, hyaline, sterile. Colonies on Potato Carrot Agar (PCA), reach 27–28 mm diameter after 4 weeks at 25 °C; front view of colony grey (6C1), circular, raised, slightly cottony with margin smooth. Reverse dark brown (6F8) and with diffusible yellowish orange (5B8) pigment in entire media. Mycelium branched, septate, smooth-walled, anastomosis, hyaline, sterile (Figure 1).

2.2. Molecular Identification and Phylogeny

Mega Blast analysis of ITS sequence of *G. triuniae* NFCCI 4873 showed 100% identity with type strain *G. triuniae* CBS138901; whereas LSU sequence showed 99.74% identity with *G. epiloblii* CPC 34889 and 99.61% similarity with *G. triuniae* CBS 138901. A phylogenetic tree was constructed based on combined ITS & LSU rDNA sequence data of a total of 19 genetically-related isolates, which shows that our isolate was clustered with *G. triuniae* CBS138901 with a very strong bootstrap value of 98.4 (Figure 2). Therefore, based on combined morphological and molecular phylogenetic analysis, the present isolate was identified as *G. triuniae*.

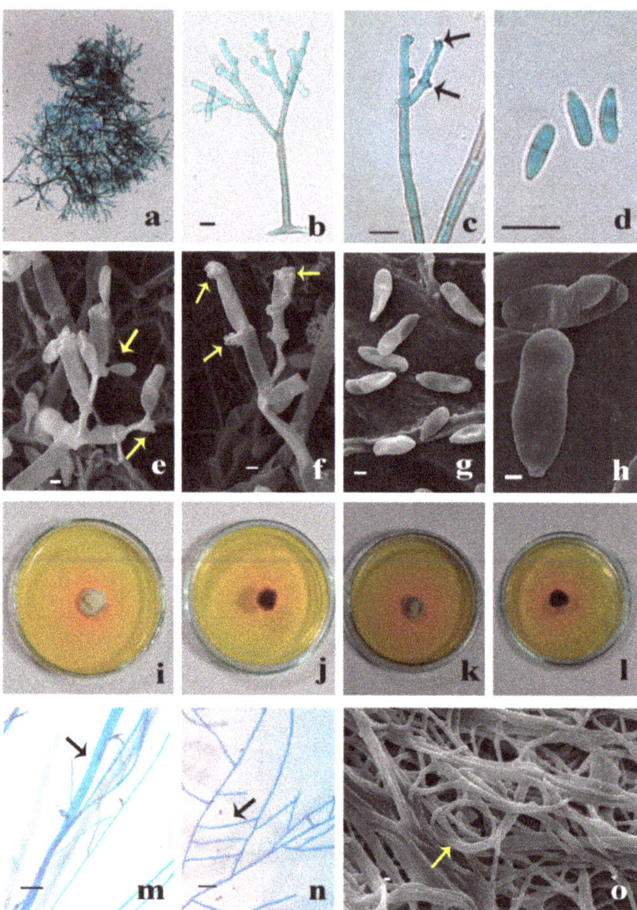

Figure 1. *Gonatophragmium triuniae* (NFCCI 4873): (**a**) Numerous conidiophores in low magnification; (**b**) Single dichotomously branched conidiophore; (**c**) Conidiophore and conidiogenous cells bearing dark scars (showing arrows); (**d**) Numerous conidia in high magnification; (**e**) SEM of conidiophores with attached conidia; (**f**) SEM of conidiophore bearing intercalary and terminal conidiogenous cells with dark scars (showing arrows); (**g**,**h**) SEM image of conidia; (**i**–**l**) Colonies of *G. triuniae* NFCCI 4873 on PDA and PCA (front and reverse view); (**m**,**n**) Hyaline sterile hyphal bundle and anastomosis in-vitro culture (showing arrows); (**o**) SEM of sterile hyphal bundles with coiling (showing arrow).

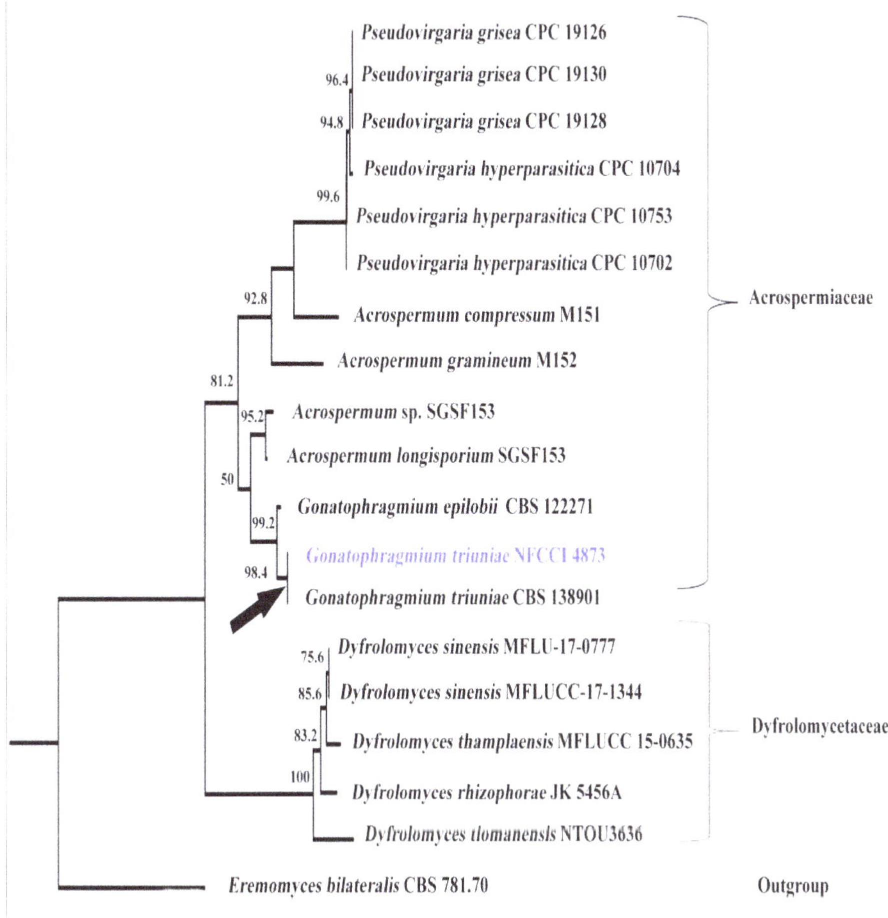

Figure 2. Phylogenetic tree of *G. triuniae* NFCCI 4873 based on the combined ITS & LSU rDNA sequence data. Digits on the nodes represent the likelihood bootstrap values.

2.3. Analysis of Pigment Production on Different Media

After 4 weeks of incubation, *G. triuniae* NFCCI 4873 produced yellowish orange (5B8) pigment on potato dextrose agar (PDA), potato carrot agar (PCA), Sabouraud dextrose agar (SDA), and Czapek Yeast Extract Agar (CYA), which was completely diffused in media. In contrast, no pigment production was observed on cornmeal agar (CMA) and Czapek Dox agar (CZA) (Figure 3).

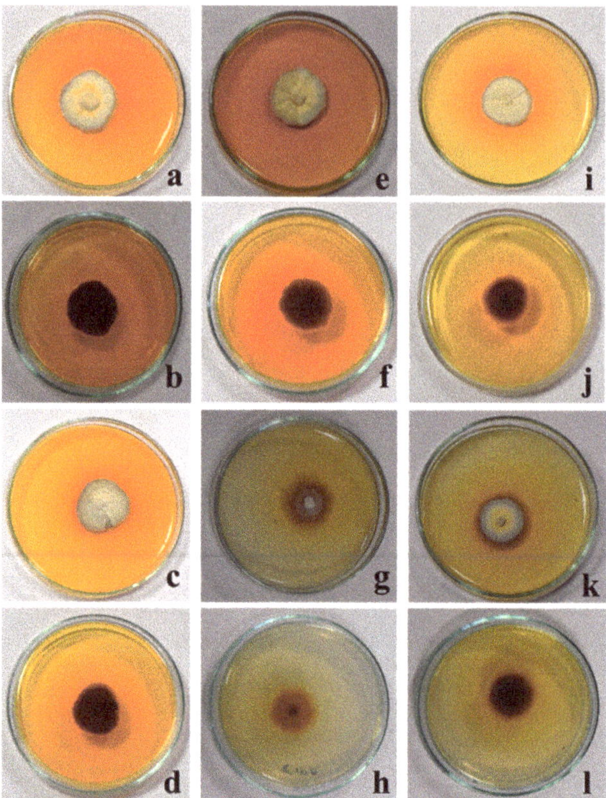

Figure 3. Studies on pigment production by *G. triuniae* (NFCCI 4873) on different media: (**a**,**b**) *G. triuniae* on PDA (front and reverse view); (**c**,**d**) *G. triuniae* on PCA (front and reverse view); (**e**,**f**) *G. triuniae* on SDA (front and reverse view); (**g**,**h**) *G. triuniae* on CMA (front and reverse view); (**I**,**j**) *G. triuniae* on CYA (front and reverse view; and (**k**,**l**) *G. triuniae* on CZA (front and reverse view).

2.4. Pigment Production in Liquid Media

G. triuniae NFCCI 4873 started pigment production earlier in PD broth of Hi-media compared to natural PD broth and natural PC broth. However, pigment production increased in natural PD broth and PC broth after 28 days of incubation compared to the PD broth of Hi-media (Figure 4). Scanning of coloured culture filtrate of *G. triuniae* NFCCI 4873 from three different media at a wavelength ranging from 390–760 nm shows that pigment production was higher in natural PD broth than PD broth (Hi-media) and natural PC broth (Figure 5). This clearly shows that natural potato dextrose broth supports the pigment production by *G. triuniae* NFCCI 4873 and suggests it as a good media for optimum pigment production.

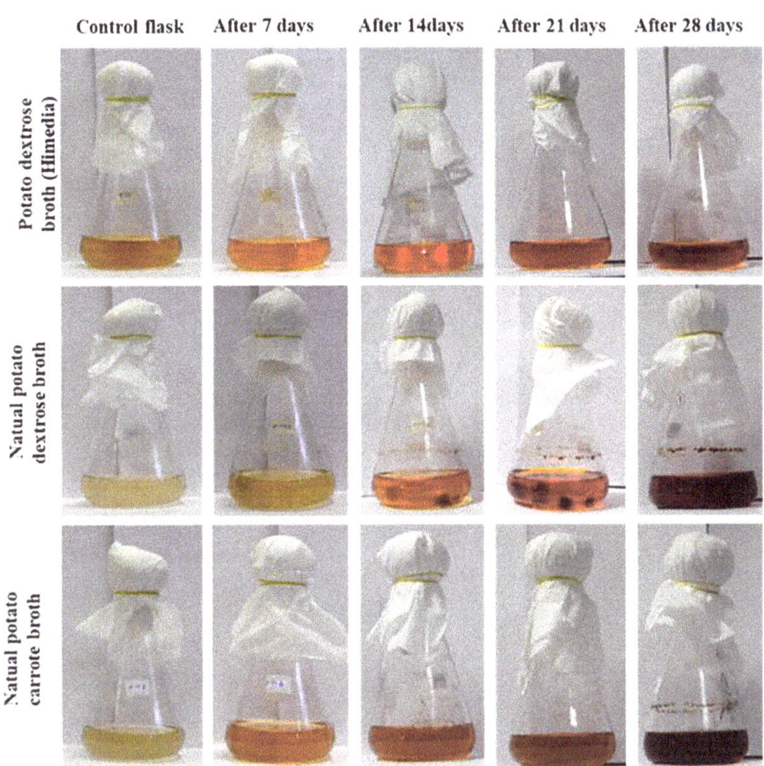

Figure 4. Analysis of pigment production by *G. triuniae* NFCCI 4873 in different liquid media.

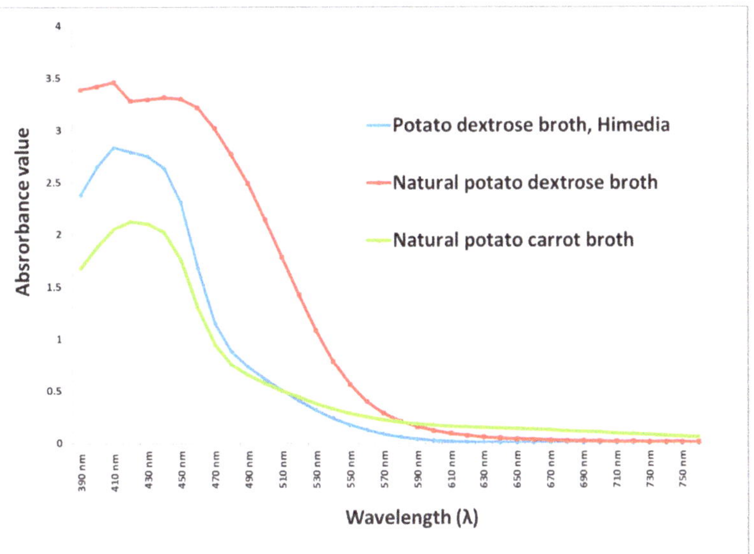

Figure 5. UV–Vis spectroscopy scanning (390–760 nm) of culture filtrates of different media.

2.5. Fermentation and Extraction of Pigments

Upon filtration of 6-L fermentation broth of *G. triuniae* NFCCI 4873, approximately 4 L of coloured culture filtrate and 75 g of dry fungal biomass were obtained. Pigments from the coloured culture filtrate were extracted with Hexane, and the concentration of hexane extract in a rota evaporator under reduced pressure yielded 526.26 mg of crude hexane extract. This dried crude hexane extract was then used for subsequent analysis, testing, and purification (Figure 6).

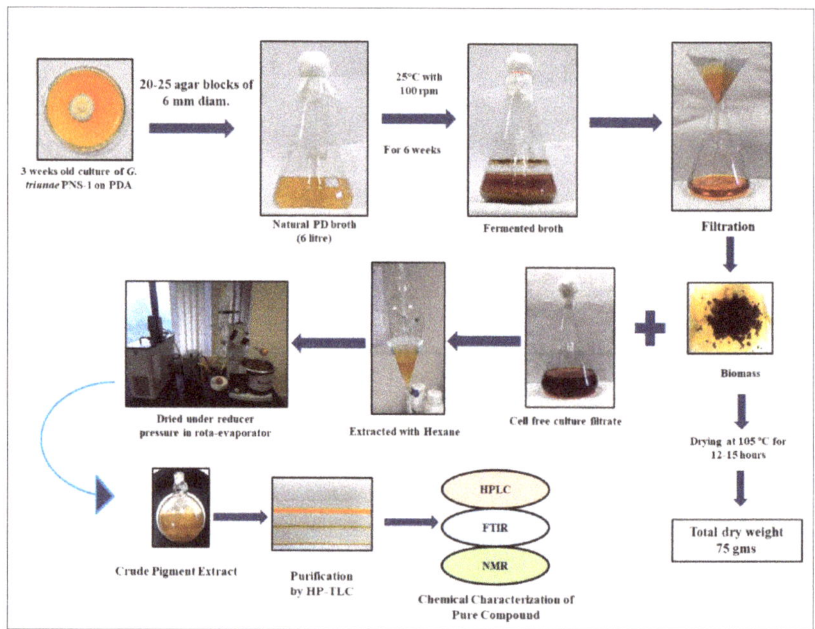

Figure 6. Schematic representation of fermentation, extraction, and purification of the compound from a pure culture of *G. triuniae* NFCCI 4873.

2.6. UV–Vis Spectroscopy Analysis of Hexane Extract

Dried hexane extract dissolved in methanol showed maximum absorption at 220 nm (λ_{max}) upon scanning at a wavelength ranging from 190–760 nm (Figure 7).

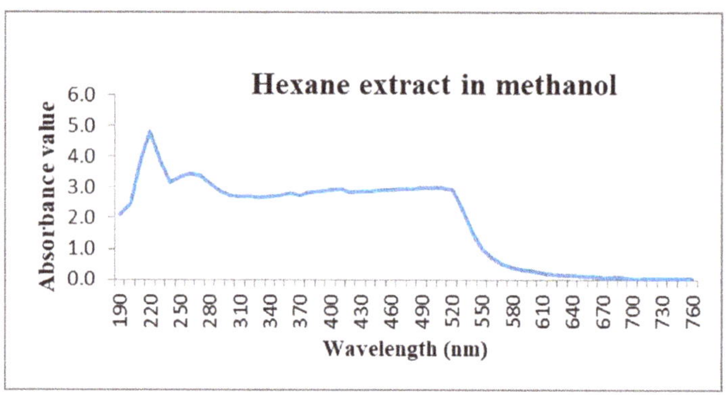

Figure 7. UV–Vis spectrum of hexane extract dissolved in methanol of *G. triuniae* NFCCI 4873.

2.7. Antagonistic Activity Testing

The dual culture assay shows that *G. triuniae* NFCCI 4873 retarded the growth of fungal plant pathogens (*Colletotrichum gloeosporioides*, *Fusarium oxysporum*, and *Fusarium solani*). Among them, *G. triuniae* showed 30% inhibition of radial growth of *F. solani*, followed by 23% inhibition of radial growth of *C. gloeosporioides* and 16% inhibition of radial growth of *F. oxysporum*. This indicates that our isolate has the potential to inhibit the growth of other fungal plant pathogens (Figure 8).

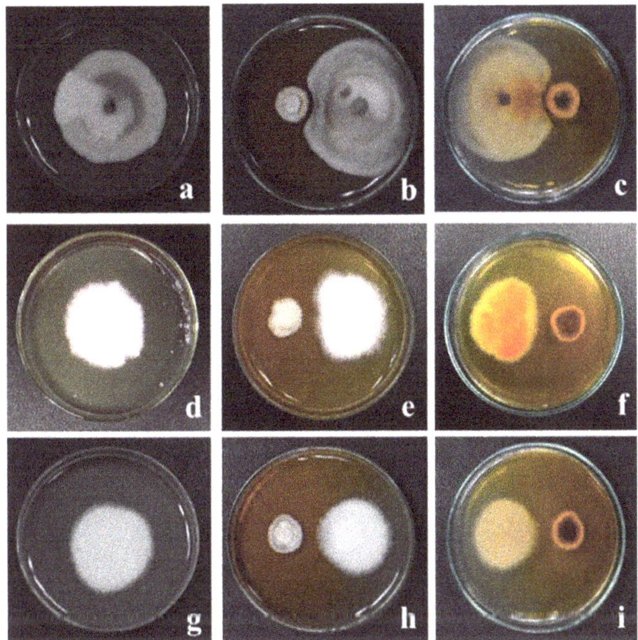

Figure 8. Antagonistic activity of *G. triuniae* NFCCI 4873: (**a**) *C. gloeosporioides* on PDA (control), (**b**,**c**) *G. triuniae* against *C. gloeosporioides* on PDA (front and reverse view), (**d**) *F. oxysporum* on PDA (control), (**e**,**f**) *G. triuniae* against *F. oxysporum* on PDA (front and reverse view), (**g**) *F. solani* on PDA (control), and (**h**,**i**) *G. triuniae* against *F. oxysporum* on PDA (front and reverse view).

2.8. Antioxidant Activity Testing of G. triuniae NFCCI 4873

From the result of in-vitro antioxidant activity, it was observed that hexane extract of *G. triuniae* NFCCI 4873 showed satisfactory dose-dependent DPPH radical scavenging activity with an IC_{50} value of 0.99 mg/mL when tested with standard ascorbic acid (IC_{50} value of 0.24 mg/mL). This suggests the promising antioxidant potential of crude hexane extract of *G. triuniae* NFCCI 4873 (Figure 9).

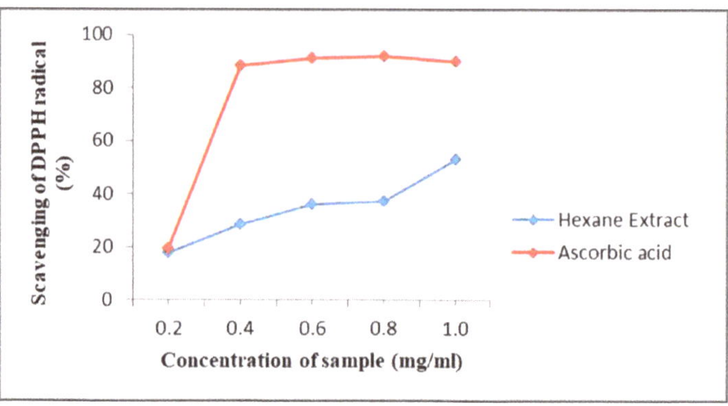

Figure 9. Scavenging effects of hexane extract of *G. triuniae* NFCCI 4873 on DPPH radical.

2.9. Antimicrobial Activity Testing by Disc Diffusion

A crude hexane extract of *G. triuniae* showed promising antibacterial activity in the disc diffusion assay (Figure 10). The mean diameters of the inhibition zones of test strains for the crude hexane extract of *G. triuniae* NFCCI 4873 are shown in Table 1. From the results of disk diffusion assay, it was observed that crude hexane extract *G. triuniae* NFCCI 4873 showed promising antimicrobial activity mainly against Gram-positive bacteria (*B. subtilis*, *S. aureus*, and *M. luteus*), displaying an average zone diameter of 17.33 mm, 18.67 mm, and 17.33 mm respectively (Table 1). However, it showed very little activity against Gram-negative bacterium *R. planticola* with an average zone diameter of 12.33 mm; whereas no activity was observed against *E. coli* and *P. aeruginosa*. In general, *S. aureus* was more sensitive to the crude hexane extract, followed by *B. subtilis* and *M. luteus,* which show similar sensitivity to the crude hexane extract. The study revealed that crude hexane extract of *G. triuniae* NFCCI 4873 has potential antibacterial compounds that can be used in medicines for their possible applications as an antibiotic.

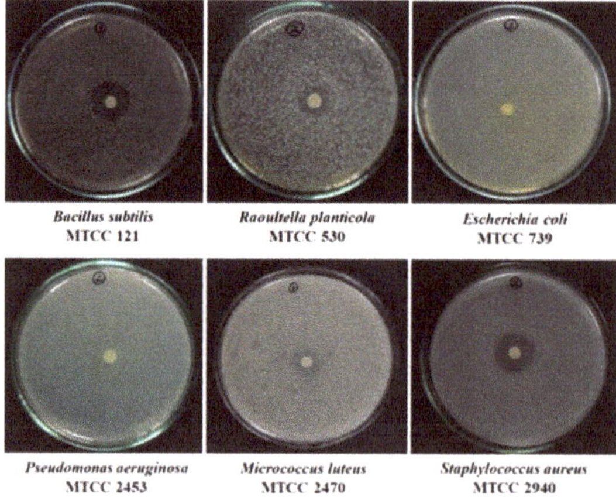

Figure 10. Antimicrobial activity of hexane extract of *G. triuniae* NFCCI 4873 against different test bacteria.

Table 1. The zone of inhibition of test organisms against sample and standards with control.

Samples	The Diameter of the Zone of Inhibition in Millimeters (Values are Average of Three Readings)					
	E. coli MTCC 739	*B. subtilis* MTCC 121	*P. aeruginosa* MTCC 2453	*S. aureus* MTCC 2940	*R. planticola* MTCC 530	*M. luteus* MTCC 2470
Hexane extract of *G. triuniae* NFCCI 4873	-	17.33 ± 1.53	-	18.67 ± 0.58	12.33 ± 0.58	17.33 ± 1.53
Streptomycin	23.33 ± 2.89	18 ± 0	9 ± 0	-	19.67 ± 0.58	38.66 ± 1.15
Ampicillin	18.33 ± 2.89	30.67 ± 2.08	-	14.33 ± 1.15	13.67 ± 0.58	54 ± 0
Chloramphenicol	23.33 ± 2.89	29 ± 1.00	-	26 ± 1.00	19 ± 1.00	26.67 ± 1.53
Ciprofloxacin hydrochloride	43.33 ± 2.89	42 ± 1.00	40 ± 0	34 ± 1.00	30.33 ± 0.58	36.67 ± 1.53
Vancomycin hydrochloride	-	23.33 ± 0.58	-	20 ± 1.00	18.67 ± 0.58	26.67 ± 0.58
Milli-Q water	-	-	-	-	-	-
Methanol	-	-	-	-	-	-

2.10. MIC and MBC of Crude Pigment

The MIC of the crude hexane extract was found to be 3.91 µg/mL against *B. subtilis* and 15.6 µg/mL against *S. aureus*; whereas it was 31.25 µg/mL for *M. luteus*. The results of MIC show that *B. subtilis* had the lowest MIC, while the highest MIC obtained was for *M. luteus* (Figure 11).

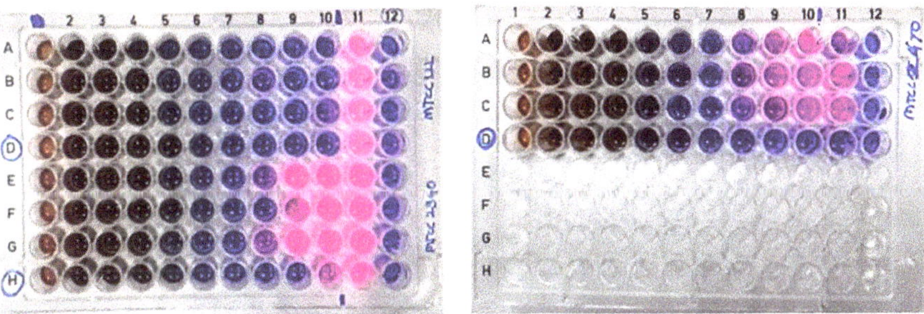

Figure 11. Microtitre plate showing MIC of crude pigment of *G. triuniae* NFCCI 4873.

The MBC of the crude hexane extract of *G. triuniae* NFCCI 4873 was found to be 1 mg/mL against *B. subtilis* MTCC 121 and>1 mg/mL against *S. aureus* MLS 16 MTCC 2940; whereas it was 0.25 mg/mL against *M. luteus* MTCC 2470.

2.11. Dyeing of Cotton Fabric

Results of the dyeing experiment showed that cotton fabrics mordanted with different mordants (FeSO$_4$ and Alum) show more pigment uptake than un-mordanted fabric. Among the two mordants used, cotton fabrics mordanted with FeSO$_4$ have shown more pigment uptake than cotton fabrics mordanted with different concentrations of Alum (Figures 12 and 13). This clearly shows that pigments of *G. triuniae* NFCCI 4873 have potential applications in the textile industry for dyeing different textile fabrics. Moreover, previous studies have already reported phenoxazines and their derivatives as promising textile dyes. Their intense colours and chemical nature make them excellent vat dyes. Besides this, these dyes act as good colourants for paint, ink, papers, candles, soap, and plastic materials [24]. This confirms that the main pigment "1,2-dimethoxy-3*H*-phenoxazin-3-one" isolated and characterized from the present study may have promising dyeing potential in the textile industry.

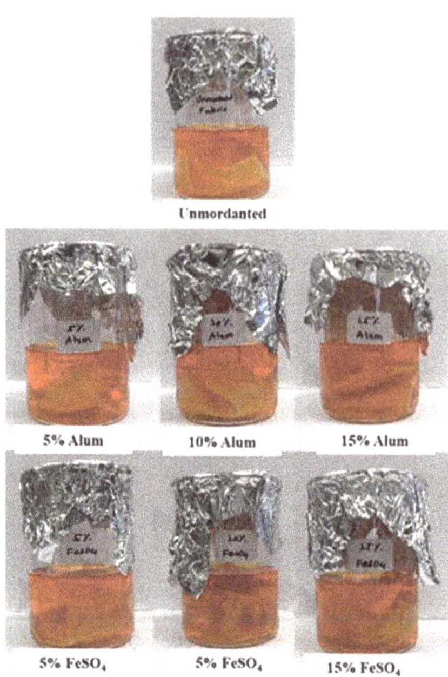

Figure 12. Dyeing of mordanted and un-mordanted cotton fabrics with crude pigment extract of *G. triuniae* NFCCI 4873.

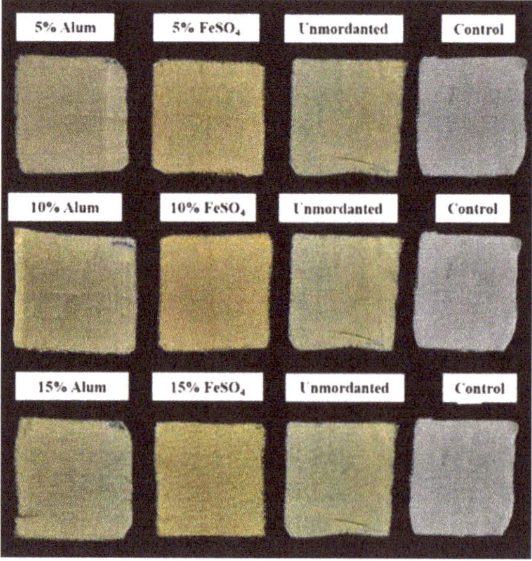

Figure 13. Cotton fabrics dyed with crude pigment extract of *G. triuniae* NFCCI 4873.

2.12. GC-MS Analysis of Crude Hexane Extract

GC-MS analysis of crude hexane extract showed the peaks along with the mass of the organic molecules present in the extract. The chromatogram showed 25 peaks corresponding to the molecules (Figure S7). The detailed data of compounds present in the extract, their molecular weight, and respective retention time is presented in Table 2. Crude hexane extract majorly showed the presence of fatty acids [n-hexadecanoic acid and octadecanoic acid] and their derivatives, esters (glycidyl palmitate and dibutyl phthalate), and alkanes (hexatriacontane and undecane).

Table 2. Compounds identified from hexane extract by GC–MS analysis.

No.	Compound Name	Retention Time	Molecular Weight	Relative Content (%)
1	Undecane	8.361	156	0.34
2	Unknown	17.817	283	0.21
3	1,2-benzenedicarboxylic acid, bis(2-methylpropyl) ester	18.243	278	0.40
4	alpha-D-Mannopyranose, 5TMS derivative	18.594	540	0.19
5	Dibutyl phthalate	18.717	278	0.98
6	n-Hexadecanoic acid	19.116	256	3.56
7	Dibutyl phthalate	19.189	278	0.75
8	Phthalic acid, butyl nonyl ester	19.386	348	0.30
9	alpha-D-Mannopyranose, 1,2,3,4,6-pentakis-O-(trimethylsilyl)-	19.459	540	0.28
10	Hexadecanoic acid, trimethylsilyl ester	19.854	328	1.75
11	9-Octadecenoic acid, methyl ester, (E)-	20.401	296	0.21
12	9-Octadecenoic acid, (E)-	20.801	282	17.62
13	Octadecanoic acid	20.989	284	2.85
14	Unknown	21.380	355	0.25
15	9-Octadecenoic acid, (E)-, TMS derivative	21.422	354	1.37
16	9,12-Octadecadienoic acid (Z,Z)-	21.550	280	0.47
17	Unknown	21.655	327	1.13
18	Hexatriacontane	21.747	506	2.71
19	Unknown	21.820	383	0.51
20	Glycidyl palmitate	22.149	312	0.53
21	Unknown	22.355	340	0.65
22	Unknown	22.697	355	28.08
23	Unknown	22.854	411	30.35
24	Unknown	23.126	397	2.97
25	9-octadecenoic acid, 1,2,3-propanetriyl ester,	23.990	884	1.56

2.13. Purification of Pigment

Purification of crude pigment extract of *G. triuniae* NFCCI 4873 by preparative TLC yielded 40 mg of purified orange compound (band 2), which was named PNS-1-OR (Figure 14). This purified pigment was used for further chemical characterization for identification.

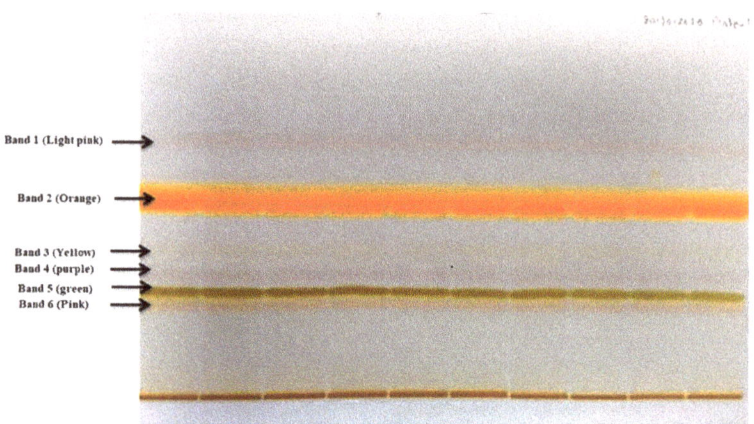

Figure 14. Separation of compounds in the crude pigment of *G. triuniae* NFCCI 4873 on TLC plate using HP-TLC.

2.14. Chemical Characterization of Pure Compound (PNS-1-OR)

HRMS spectrum of the pigment gave a sodium adduct $[M + Na]^+$ at *m/z* 280 ($C_{14}H_{11}NO_4Na$) and molecular ion peak of $C_{14}H_{11}NO_4$ at *m/z* 258 [M+1], suggesting $C_{14}H_{11}NO_4$ as its molecular formula (Figure S2). The IR spectra showed a significant carbonyl ($-C=O$) peak at 1647 cm^{-1}, ($-C=N$) 2331 cm^{-1}, and aromatic C-H stretching at 2943 cm^{-1} (Figure S1).

1,2-dimethoxy-3*H*-phenoxazin-3-one: This compound is orange solid. mp 145−150 °C; UV (MeOH) λ_{max} (log ε) 220 nm; IR (KBr) v_{max} 2943, 2331, 1647 cm^{-1}; ^{1}H and ^{13}C NMR data, see Table 3; HR-MS *m/z* 258 [M+1]$^+$ (calcd. for $C_{14}H_{11}NO_4$, *m/z* 257).

Table 3. The ^{1}H NMR (500 MHz, CDCl$_3$) and ^{13}C NMR (125 MHz, CDCl$_3$) spectroscopic data for 1,2-dimethoxy-3*H*-phenoxazin-3-one isolated from *G. triuniae* NFCCI 4873 along with the reported one [25].

Position	1,2-dimethoxy-3*H*-phenoxazin-3-one from *G. triuniae* NFCCI 4873		1,2-dimethoxy-3*H*-phenoxazin-3-one Reported from *A. viticola*	
	δ_C	δ_H (*J* in Hz)	δ_C	δ_H (*J* in Hz)
1	145.2 C	-	145.1 C	-
2	145.8 C	-	145.9 C	-
3	181.8 C=O	-	181.8 C=O	-
4	104.6 CH	6.23, s	104.7 CH	6.23, s
4a	147.2 C	-	147.3 C	-
5a	143.4 C	-	143.5 C	-
6	115.9 CH	7.33, dd (8.1, 1.07)	116.0 CH	7.33, dd (8.2)
7	132.2 CH	7.54, td (7.7, 1.37)	132.2 CH	7.53, td (8.2)
8	125.3 CH	7.39, td (7.71, 1.37)	125.3 CH	7.39, td (8.2)
9	130.3 CH	7.92, dd (7.93, 1.53)	130.3 CH	7.92, dd (8.2)
9a	132.6 C		132.7 C	-
10a	147.2 C		147.8 C	-
1-OMe	62.2 CH$_3$	4.12, s	62.3 CH$_3$	4.12, s
2-OMe	61.1 CH$_3$	4.14, s	61.2 CH$_3$	4.14, s

The ^{1}H NMR, ^{13}C NMR details are shown in Table 3 along with the reported data and were found in agreement with the natural product, 1,2-dimethoxy-3*H*-phenoxazin-3-one (Figure 15) reported from *Acrospermum viticola*, a leaf spot fungus of Mulberry [25].

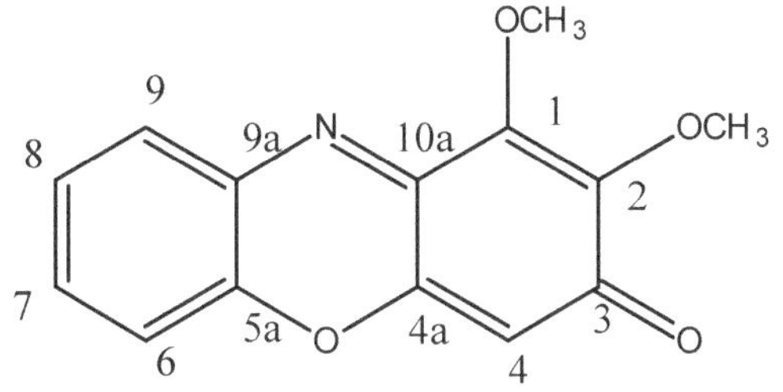

Figure 15. Natural product 1,2-dimethoxy-3*H*-phenoxazin-3-one.

The pigment is isolated as an orange solid: yield (40 mg, 20%), mp 148–150 °C. The ^{1}H NMR shows the distribution of protons signals between 1.0 and 8.0 ppm. Analysis of the ^{1}H NMR (Figure S3) and 2D-COSY spectra (Figure S4) revealed a sequence of 11 total hydrogens at δ 4.12 (s, 3H, OCH$_3$), 4.14 (s, 3H, OCH$_3$), 6.23 (s, 1H, 4-H), 7.33 (dd, 1H, *J* = 8.1, 1.07 Hz, 6-H), 7.39 (td, 1H, *J* = 7.71, 1.37 Hz, 8-H), 7.54 (td, 1H, *J* = 7.7, 1.37 Hz, 7-H), and 7.92 (dd, 1H, *J* = 7.9, 1.53 Hz, 9-H). The ^{1}H NMR of **1a** showed two singlets at δ 4.11 (s, 3H, OCH$_3$) and δ 4.14 (s, 3H, OCH$_3$), revealing two methoxy groups at positions one and two. A singlet was observed at δ 6.23 (s, 1H, C-**4**) next to the carbonyl group at position four. The characteristic signals for aromatic protons were observed at δ 7.33 as a doublet for (dd, *J* = 8.09 Hz, 1H, 6-H), triplet at δ 7.39 (td, *J* = 7.71Hz, 1H, 8-H), triplet at δ 7.54 (td, *J* = 7.70, 1H, 7-H), and doublet at δ 7.92 (dd, *J* = 7.93 Hz, 1H, 9-H). The ^{1}H COSY spectrum showed the correlation of a 6-H proton at δ 7.33 with a 7-H proton at δ 7.54 only. The 8-H proton at δ 7.39 showed a correlation with 7-H and 9-H at δ 7.54 and 7.92, respectively. The 7-H proton at δ 7.54 displayed a correlation with 6-H and 8-H at δ 7.33 and 7.39, respectively. The 9-H proton at 7.92 displayed a correlation with 8-H at δ 7.39.

The ^{13}C spectrum of PNS-1-OR was observed between 60 ppm and 200 ppm. The ^{13}C NMR spectra (Figure S5) showed 14 carbons at δ_C 61.1 (OCH$_3$), 62.2 (OCH$_3$), 104.6 (C-**4**), 115.9 (C-**6**), 125.3 (C-**8**), 130.3 (C-**9**), 132.2 (C-**7**), 132.6 (C-**9a**), 143.4 (C-**5a**), 145.2 (C-**1**), 145.8 (C-**2**), 147.2 (C-**4a**, **10a**), and 181.8 (C-**3**). The DEPT 135 analysis (Figure S6) showed the carbons that are attached to hydrogens. Therefore, peaks observed at δ 61.22 and 62.26 belong to two methoxy carbon. The peak detected at δ 104.65 corresponds to C-**4**. The displayed peaks at δ 115.98, 125.375, 130.38, and 132.21 belong to C-**6**, C-**8**, C-**9**, and C-**7**, respectively.

The elemental analysis of CHN for the formula $C_{14}H_{11}NO_4$ was obtained as C 64.97, H 4.29, and N 5.44%, which was found in agreement with the calculated values C 65.37, H 4.31, and N 5.44%. It was confirmed that the isolated pigment is 1,2-dimethoxy-3*H*-phenoxazine-3-one.

3. Materials and Methods

3.1. Collection and Isolation of Fungus

Infected leaves of *Maytenus rothiana* were collected in sterile paper bags from the Western Ghat region (Mahabaleshwar), Maharashtra, India. Collected samples were transported to the laboratory and stored in a refrigerator at 4 °C till their processing. Collected leaves (infected with fungus) were used to isolate fungus. The lower leaf surface was found to be colonized by fungus. For the in-vitro culture of fungus, spore mass was lifted with the help of a fine needle from the infected leaf surface and suspended in 1 mL sterile distilled water incorporated with Tween 20. Then, with a micropipette's help, 200 µL of spore suspension was spread on a 2% Neutral agar plate using a spreader, and the plate was incubated at 25 °C overnight. On the next day, the plate was observed under the CX-21 compound microscope, and germinated single spores with agar block were picked up with the help of a sterile needle and transferred on sterile potato dextrose agar (PDA) plates. Plates were incubated at 25 °C for 7 days. The pure growing colonies were further sub-cultured on fresh PDA plates and slants. Slants were stored in a refrigerator at 4 °C till further use.

3.2. Morphological Identification and Deposition of Fungal Culture

Necrotic lesions on the infected leaves were marked. Scrape mount slides were prepared in lactophenol cotton blue and observed under a microscope; based on the literature, the fungus was identified as *Gonatophragmium* sp. Similarly, slides were prepared in lactophenol cotton blue mount from axenic culture. Microphotographs were taken using Carl Zeiss AXIO-10 microscope, and scanning electron microscopic (SEM) images were taken using images ZEISS EVO MA 15 Scanning electron microscope at 20 KV.

A pure culture of *G. triuniae* was inoculated on potato dextrose agar (PDA) and potato carrot agar (PCA) to study cultural and microscopic characters. Plates were incubated at 25 °C for 14 days, and cultural and microscopic characters were noted upon completion of incubation. Slide culture [26,27] and grass leaf technique [28] were used to get the sporulation, but no sporulation was observed.

Live and pure fungal culture of *G. triuniae* were deposited in the National Fungal Culture Collection of India (NFCCI), Agharkar Research Institute, Pune, under the accession number NFCCI 4873. Voucher culture *G. triuniae* NFCCI 4873 was deposited in Ajrekar Mycological Herbarium (AMH), Agharkar Research Institute, Pune, with accession number AMH 10289.

3.3. Molecular Identification and Phylogeny

For the identification and authentication of fungal culture up to species level, molecular characterization was done. The fungal genomic DNA was isolated following the standard protocol [29]. Then by polymerase chain reaction (PCR), the ITS (internal transcribed spacer) and LSU (large subunit) regions of rDNA were amplified from the extracted genomic DNA using the primers ITS-4 & ITS-5 [30] and LR-0R & LR-7 [31], respectively. FavorPrepTM PCR Purification Kit (Favorgen, Biotech Corporation, Taiwan) was used to purify PCR products. Purified PCR products were subjected for sequencing by the BigDye Terminator v3.1 Cycle Sequencing Kit (Applied Biosystems, Waltham, MA, USA) and ABI Avant 3100 automated DNA sequencer (Applied Biosystems, USA). The manually edited sequences of ITS and LSU regions of rDNA of our fungal isolate were deposited in the nucleotide sequence database of NCBI (Gene Bank Accession Numbers: ITS- MW193329 and LSU- MW144438).

The ITS and LSU rDNA sequences of our fungal isolate were subjected to Mega BLASTn sequence homology searches. Based on the BLASTn search results, genetically related species, including genus *Gonatophragmium*, *Acrospermum*, *Pseudovirgaria*, and *Dyfrolomyces*, were chosen to construct the phylogenetic tree (Table 4). Phylogenetic analysis of *G. triuniae* NFCCI 4873 was performed based on a combined ITS and LSU rDNA sequence data of a total of 19 fungal cultures. The *Eremomyces bilateralis* CBS 781.70 was chosen as an out-group. With the help of the MUSCLE algorithm, multiple sequence alignment was performed in MEGA 7 [32]. Phylogenetic tree of *G. triuniae* NFCCI 4873 was constructed based on combined data of ITS & LSU rDNA sequences in IQ-TREE multicore version 1.6.11 [33] using the Maximum Likelihood method with best-fit model TN+F+G4. Selection of the best-fit model was done using the ModelFinder employed in IQ-TREE.

3.4. Analysis of Pigment Production on Different Media

A pure culture of *G. triuniae* NFCCI 4873 was inoculated on different media such as potato dextrose agar (PDA), potato carrot agar (PCA), Sabouraud dextrose agar (SDA), Czapek Dox agar (CZA), cornmeal agar (CMA), and Czapek Yeast Extract Agar (CYA) in duplicates and incubated at 25 °C for 28 days to assess the pigment production potential. After incubation, the colour of the pigment diffused in media was recorded using the Methuen handbook of colour [34].

3.5. Analysis of Pigment Production in Liquid Media

The culture of *G. triuniae* was also tested for its pigment production ability in different liquid media such as potato dextrose broth (PDB, Hi-media), natural potato dextrose broth (n-PDB), and natural potato carrot broth (n-PCB). Four agar blocks of a pure culture of *G. triuniae* (6 mm diameter) from 20-days-old PDA culture plate were inoculated in a 250 mL Erlenmeyer flask containing 100 mL media (each media in the separate flask), and flasks were incubated at 25 °C with 150 rpm. All tests were performed in duplicates. All flasks were observed intermittently after every week from the date of inoculation for pigment production, and observations were noted down. After 4 weeks of incubation, culture broths were filtered, and absorption spectrum analyses of the coloured filtrates were performed using a UV–VIS spectrophotometer (Shimadzu UV-2450). The absorbance of coloured culture filtrates was recorded in the visible light range from 390–760 nm with a 10 mm optical pathlength and 0.1 nm resolution.

Table 4. GenBank accession numbers of taxa used for phylogenetic analysis.

Sr. No.	Fungal Culture	Strain	GenBank Accession No.	
			ITS	LSU
1	*Gonatophragmium triuniae*	NFCCI 4873	MW193329	MW144438
2	*Gonatophragmium triuniae*	CBS 138901	NR_137932	NG_058117
3	*Gonatophragmium epilobii*	CBS 122271	MH863183	MH874728
4	*Acrospermum* sp.	SGSF153	MK335823	MK754265
5	*Acrospermum longisporium*	MFLU 17-2849	-	NG_064506
6	*Acrospermum gramineum*	M152	-	EU940085
7	*Acrospermum compressum*	M151	-	EU940084
8	*Pseudovirgaria grisea*	CPC 19130	JF957607	JF957612
9	*Pseudovirgaria hyperparasitica*	CPC 10702	EU041765	EU041822
10	*Pseudovirgaria hyperparasitica*	CPC 10704	EU041766	EU041823
11	*Pseudovirgaria hyperparasitica*	CPC 10753	EU041767	EU041824
12	*Pseudovirgaria grisea*	CPC 19126	JF957605	JF957610
13	*Pseudovirgaria grisea*	CPC 19128	JF957606	JF957611
14	*Dyfrolomyces sinensis*	MFLU 17-0777	-	NG_064507
15	*Dyfrolomyces sinensis*	MFLUCC 17-1344	-	MG836699
16	*Dyfrolomyces tiomanensis*	NTOU3636	-	KC692156
17	*Dyfrolomyces rhizophorae*	JK 5456A	-	GU479799
18	*Dyfrolomyces thamplaensis*	MFLUCC 15-0635	-	KX925435
19	*Eremomyces bilateralis*	CBS 781.70	NR_145364	NG_059206

3.6. Fermentation and Extraction of Pigments

G. triuniae culture was subjected to flask scale fermentation in a total of 6 L (four flasks containing 1.5 L of media) of natural potato dextrose broth (PD broth). Each flask was inoculated with 20–25 mycelial disks (6 mm diameter) of *G. triuniae* from 3-weeks-old PDA culture plate using a cork borer and incubated at 25 °C with 100 rpm for 4–6 weeks. After incubation, the coloured culture broth was filtered through pre-weighed blotting paper, and culture filtrate was collected in a separate flask. Later, the pigments from the culture filtrates were extracted thrice with an equal volume of Hexane. With the help of a separating funnel, the Hexane part was separated from the culture filtrate. The separated hexane part was evaporated to dryness under reduced pressure in a rota evaporator (Heidolph, Schwabach, Germany). The resulting concentrated hexane extract was used for further experiments.

Finally, biomass collected in a pre-weighed blotting paper was dried at 105 °C for 12–15 h and weighed to measure the yield of biomass concentration [35].

3.7. UV–VIS Spectroscopy Analysis of Hexane Extract

UV-Visible spectroscopic analyses of the crude hexane extracts were performed using a UV-VIS spectrophotometer (Shimadzu UV-2450). The crude hexane extract was evaporated to dryness and then dissolved in methanol solvent, and absorbance was recorded in the range of 190–760 nm with a 0.1 nm resolution and 10 mm optical path length.

3.8. Antagonistic Activity of G. triuniae NFCCI 4873

Antagonistic activity of *G. triuniae* against three fungal pathogens (*Colletotrichum gloeosporioides*, *Fusarium oxysporum* & *Fusarium solani*) was evaluated by dual culture technique [36]. Mycelial disks of 6 mm diameter were excised from the edge of actively growing culture of *G. triuniae* and fungal pathogens and inoculated on opposite ends of PDA plates equidistant from the periphery (each pathogen separately with test culture). For control, PDA plates were inoculated with a pathogen without test culture. Plates were then incubated at 25 °C for 14 days. Experiments were performed in duplicates. After completion of incubation, the radial growth of each fungal pathogen on PDA plates was measured, and percentage inhibition of radial growth (PIRG) of fungal pathogens was calculated relative to the control plate using the following formula:

$$PIRG = [(R_1 - R_2)/R_1] \times 100$$

where R_1 is the radial growth of the fungal pathogen in the control plate, and R_2 is the radial growth of the pathogen in the presence of test culture (*G. triuniae*) [37].

3.9. In-vitro Antioxidant Activity of Crude Pigment

In-vitro antioxidant activity of hexane extracts was tested using DPPH radical scavenging method [38]. Different concentrations (0.2, 0.4, 0.6, 0.8, and 1.0 mg/mL) of the extract and standard solution (Ascorbic acid, Sigma, USA) were used, and for that, dilutions were prepared in methanol. For the assay, 10 μL of extract or standard solution was added to 200 μL of 0.1 mM DPPH in methanol solution in a 96-well microtitre plate (Thermofisher, Waltham, MA, USA). All reactions were performed in triplicates. The plate was then incubated at 37 °C for 30 min in the dark. After incubation, the absorbance of the solution in each well was measured at 490 nm using a Synergy HT Multi-detection microplate reader [BioTek, Winooski, VT, USA]. The percentage of radical scavenging activity of the hexane extract was calculated by the following formula:

$$\text{DPPH radical scavenging activity (\%)} = [(OD\ control - OD\ sample)/OD\ control] \times 100$$

where OD means optical density or absorbance value. The IC_{50} value (concentration of sample required to scavenge 50% of free radicals) of hexane extract was determined.

3.10. Antimicrobial Activity of Crude Hexane Extract

Crude pigment sample was screened against a panel of test organisms, including *Escherichia coli* (MTCC 739), *Bacillus subtilis* (MTCC 121), *Staphylococcus aureus* MLS 16 (MTCC 2940), *Pseudomonas aeruginosa* (MTCC 2453), *Raoultella planticola* (MTCC 530), and *Micrococcus luteus* (MTCC 2470). The test strains were procured from Microbial Type Culture Collection (MTCC), CSIR-Institute of Microbial Technology (IMTECH), Chandigarh, India. Ciprofloxacin hydrochloride (1 mg/mL), vancomycin hydrochloride (1 mg/mL), chloramphenicol (1 mg/mL), streptomycin (1 mg/mL), and ampicillin (1 mg/mL) against bacteria, were used as positive controls.

3.11. Antimicrobial Activity by Disc Diffusion Method

The antimicrobial activity of the hexane extract of *G. triuniae* was tested using the disk diffusion method. The standardized microbial inoculum of the test strain was prepared in a saline solution (~10^6–10^8 CFU/mL) from the 24 h old culture plates. The culture media used was Muller-Hinton agar (MHA). A direct colony suspension method was used for the inoculum preparation in which well-isolated colonies from 24 h old culture plates were selected and suspended in sterile saline. Then saline suspensions of the test cultures were adjusted to achieve turbidity equivalent to a 0.5 McFarland standard. This was done by adjusting the turbidity of the cell suspensions between 0.08 and 0.12 AU in a UV-Vis Spectrophotometer (Shimadzu UV-2450) at 625 nm as recommended by CLSI guidelines. This results in a suspension with approximately 1–2 × 10^8 colony-forming units (CFU)/mL for *Escherichia coli* ATCC®a 25922. The resulted suspension was added in sterile molten Muller-Hinton agar (0.5 mL/100 mL of media). The final concentration of the cells in the media was 0.5–1.0 × 10^8 CFU/mL. Then media was poured in Petri plates and allowed to solidify [39].

Sterile Whatman filter paper discs impregnated with known amounts of the test sample (1mg/mL of dried hexane extract dissolved in methanol) were placed on the surface of an agar plate that was inoculated with a standardized suspension of microorganisms that were to be tested. Standard antibiotics like ampicillin (1 mg/mL), ciprofloxacin hydrochloride (1 mg/mL), streptomycin (1 mg/mL), vancomycin hydrochloride (1 mg/mL), and chloramphenicol (1 mg/mL) were used as positive controls. Paper discs impregnated with only methanol and sterile water were used as negative controls. Plates were left at room temperature for 1–2 h for the diffusion of samples and then incubated at 37 °C for 24 h. After completion of incubation, all plates were observed for the zone of inhibition. The diameters of the zone of inhibition were measured in millimeters, including the disk diameter (6 mm). All experiments were performed in triplicates.

3.12. Determination of Minimum Inhibitory Concentration (MIC) and Minimum Bactericidal Concentration (MBC)

3.12.1. Preparation of Pigment Stock

Forty milligrams of crude pigment was dissolved in 1 mL of DMSO in a 2 mL Eppendorf tube, and a tube was mixed well. Then this 1 mL solution was then diluted 10 times in sterile Muller–Hinton broth (MHB), and the resultant solution was stored in the refrigerator at 4 °C until further use.

3.12.2. Preparation of Standardized Inoculum

The standard inoculum (0.5 McFarland) of Gram-positive bacteria (*Bacillus subtilis*, *Staphylococcus aureus* MLS 16, and *Micrococcus luteus*) was prepared by the direct colony suspension method as recommended by CLSI guidelines in which the OD_{625} value was adjusted to the equivalent of 10^8 CFU/mL in a Shimadzu UV-2450 UV–VIS spectrophotometer [39].

3.12.3. MIC and MBC Experiment

The minimum inhibitory concentration (MIC) of the crude pigment extract was assessed using the standard method [39]. The MIC of the crude pigment extract was determined by the broth microdilution method in a 96-well plate as per the CLSI recommended protocol. All the wells of the microtitre plate from columns 2–11 were added with 50 µL of sterile Muller–Hinton broth (MHB). The last (12th) column was added with 100 µL of MHB as sterility control. One hundred microlitres of pigment solution (2 mg/mL) was added to the first column of the 96-well microtitre plate. Then, using a micropipette, a two-fold serial dilution was performed by transferring 50 µL of the pigment solution from the first well to the succeeding well and up to the 10th well, and the final 50 µL of the solution was discarded. Standardized inoculums of bacteria were diluted 1:150 times in sterile MH broth to get 10^6 CFU/mL concentrations of bacteria. Then, 50 µL of microbial suspension

(10^6 CFUs/mL) was added in each well from well 2–11; while in rows D and H, from well 1–10, there was no addition of bacterial suspension, which were treated as pigment control. Plates were incubated at 37 °C for 20 h. Each test was performed in triplicates.

After incubation at 37 °C for 20 h, 30 µL of 0.01% resazurin was added to each well, and then plates were further incubated for 2 h at 37 °C for the change of colour. The well-containing lowest concentration of pigment showing no colour change was considered as the MIC value.

After 20 h of incubation at 37 °C, 10 µL solution from each well (2–11) of the 96-well microtitre plate was plated on Muller–Hinton agar plate, and plates were incubated at 37 °C for 24 h. After completion of incubation, plates were observed for the growth of bacteria. The lowest concentrations that completely kill the bacteria and do not show growth on the MHA plate were considered minimum bactericidal concentration (MBC).

3.13. Dyeing of Cotton with a Crude Pigment of G. triuniae NFCCI 4873

3.13.1. Textile Fabric

Raw cotton fabric was collected from the local market, Pune. Cotton fabric was then cut into 10 × 10 cm pieces (each weighing 1 gm) and used for the dyeing experiment.

3.13.2. Scouring

Twelve cotton fabric pieces of 10 × 10 cm (12 g) were pre-soaked in milli-Q water and then cooked in 2.5 L of Milli-Q water containing 20 mL of non-ionic detergent (Triton-X-100) for 1 h at 70 °C in a water bath to remove oil and dirt. Scoured cotton fabrics were then rinsed thoroughly with running water and air-dried [40].

3.13.3. Mordanting

Scoured cotton fabric pieces were mordanted by the pre-mordanting technique [40,41]. Cotton fabric pieces (10 × 10 cm) were mordanted with different concentrations (5%, 10%, and 15% *w/w* of fabric) of Alum and $FeSO_4$ for 45 min at 70 °C with a 1:20 material to liquid ratio (MLR). After completing mordanting, fabric pieces were rinsed in running water and finally allowed to air dry.

3.13.4. Preparation of Dye Bath

One hundred milligrams of dried crude pigment of *G. triuniae* NFCCI 4873 was re-dissolved in 500 mL of Milli-Q water using a magnetic stirrer. The resultant coloured solution was used as a dye bath for dyeing cotton fabric.

3.13.5. Dyeing

Unmordanted and pre-mordanted cotton fabric pieces were dyed with a crude pigment solution of *G. triuniae* NFCCI 4873. Cotton fabric pieces (10 × 10 cm) were dyed at a material to liquor ratio (MLR) of 1:50 at 70 °C for 45 min in a water bath. The pH of the dye bath was not controlled. Dyed fabric pieces were then treated with 1% acetic acid and washed thoroughly in running water. The dyed fabric pieces were then rinsed with cold water and dried overnight in the shed [40].

3.14. GC–MS Analysis

The dried hexane extract was subjected to derivatization using *N*-Methyl-*N*-(trimethylsilyl) trifluoroacetamide [MSTFA] as a silylating agent. For derivatization, about 5 mg of sample was dissolved in 100 µL of pyridine in a glass vial followed by the addition of 100 µL of MSTFA. The solution was mixed well, heated at 60 °C for 20 min, and cooled to room temperature. After derivatization, the sample was subjected to gas chromatography-mass spectrometric analysis using GCMS-TQ8030 (Shimadzu, Nakagyo-ku, Kyoto, Japan). A fused silica column [RTX-5MS (30 m × 0.25 mm × 0.25 µm)] was used. The temperature of the column was programmed from 50 to 280 °C at a rate of 10 °C/min. The injection port temperature was set at 280 °C, and a split ratio of 1:40 was used for the analysis. Helium was used as the carrier gas at a flow rate of 1.0 mL/min. Electron ionization source of 70 eV and a mass range of m/z 35–800 U was used for MS detection. The resultant MS peaks in the GC-MS chromatogram were identified by comparing and matching the mass and mass fragmentation pattern with the reference mass and mass fragmentation pattern in the NIST05 MS library.

3.15. Purification of a Compound by HP-TLC

Two-hundred milligrams of crude pigment extract of *G. triuniae* NFCCI 4873 was dissolved in 4 mL of acetone, and compounds present in crude extract were separated on silica plates using HP-TLC (CAMAAG). Hexane:ethyl Acetate (40:60) was used as the mobile phase; thin aluminium silica plates (Merck) were used as the stationary phase. After separating compounds through HP-TLC, the orange band separated on plates was cut, and the compound attached to silica was extracted with acetone. Acetone extracts were collected in a round bottom flask and dried using a rota evaporator under reduced pressure using Heidolph rota evaporator. Finally, the weight of the dried, purified pigment was measured and recorded.

3.16. Chemical Characterization of Purified Orange Compound (PNS-1-OR)

The high-resolution mass spectrum (HRMS) of the pure orange pigment labeled as PNS-1-OR was recorded on a Bruker IMPACT HD. FTIR spectrum of pigment was recorded on a Shimadzu-IRAffinity-1 FTIR spectrophotometer in the frequency range 4000–400 cm^{-1}. Pure compound was subjected to ^{1}H (500 MHz) and ^{13}C (125 MHz) NMR in Bruker 500 MHz NMR instrument using CDCl$_3$ as a solvent for dissolving sample and TMS as internal standard. Chemical shifts (δ-values) are given in parts per million (ppm), and the coupling constants (J-values) are given in hertz (Hz).

4. Conclusions

Many fungi of different taxonomic groups producing a wide variety of pigments of different colours and chemical classes have been reported by researchers across the world. Among them, some fungal pigments find their application in different industries possessing promising colouring properties. The present study also reports one of the unconventional fungi, i.e., *G. triuniae*, showing very good pigment production potential. Moreover, this is the first experimental work reporting pigments and other secondary metabolites from the fungus *G. triuniae*. Based on the results of the antibacterial activity of the crude pigment extract, we conclude that the crude pigment extract shows the presence of antibiotic compounds, exhibiting antibacterial activity against Gram-positive bacteria. In addition to this, the DPPH radical scavenging activity of the crude pigment extract confirmed the presence of antioxidant compounds in the crude pigment extract. Such bioactivities (antibacterial and antioxidant) of the crude pigment extract have elevated the *G. triuniae* as a promising source of bioactive compounds for their possible use in the medicine and pharmaceutical industry. Besides this, the dyeing property of the crude pigment extract revealed the potential use of pigments of *G. triuniae* in the textile industry for dyeing different types of fabrics.

The purification of crude pigment extract of *G. triuniae* finally yielded into a major orange-colored phenoxazine class pigment, which was characterized and identified as 1,2-dimethoxy-3*H*-phenoxazin-3-one (C$_{14}$H$_{11}$NO$_4$, M.W. 257), based on UV-Vis, FTIR, HRMS, and NMR analysis. Although this pigment was already described from fungus *A. viticola*, this is the first study reporting the phenoxazine class of pigment from fungus *G. triuniae*. Considering the previous studies describing dyeing potential and bioactivity of phenoxazines, we may finally conclude that the orange pigment "1,2-dimethoxy-3*H*-phenoxazin-3-one" is a promising colourant and possible bioactive compound of *G. triuniae*, having future applications in the textile and pharmaceutical industry.

Supplementary Materials: FTIR (Figure S1), HR-MS (Figure S2), ^{1}H NMR (Figure S3), COSY (Figure S4), ^{13}C NMR (Figure S5), and DEPT-135 (Figure S6) spectra for compound 1,2-dimethoxy-3*H*-phenoxazin-3-one and GC-MS chromatogram (Figure S7) of hexane extract (PDF).

Author Contributions: Conceptualization, A.C.L., S.K.S.; Formal analysis, A.C.L., P.S.; Methodology, A.C.L.; Supervision, S.K.S., L.D. and P.N.S.; Writing—original draft, A.C.L.; Writing—review & editing, S.K.S., L.D., P.S. and P.N.S. All authors have read and agreed to the published version of the manuscript.

Funding: This research received no external funding.

Institutional Review Board Statement: Not applicable.

Informed Consent Statement: Not applicable.

Data Availability Statement: Not applicable.

Acknowledgments: We thank Prashant Dhakephalkar, MACS-Agharkar Research Institute, Pune, for providing the necessary facilities and encouragement to carry out the research work. A.C. Lagashetti acknowledges CSIR (Council of Scientific and Industrial Research), New Delhi, for granting Senior Research Fellowship (SRF), and S.P. Pune University, Pune, for granting permission to register for the Ph.D. degree. We acknowledge technical support and help by S.B. Gaikwad, N.S. Gaikwad, and S. Bagale. Laurent Dufossé deeply thanks the Conseil Régional de La Réunion, Réunion island, Indian Ocean, for continuous financial support of research activities dedicated to microbial pigments.

Conflicts of Interest: The authors declare no conflict of interest.

References

1. Rao, M.P.; Xiao, M.; Li, W.J. Fungal and Bacterial Pigments: Secondary Metabolites with Wide Applications. *Front. Microbiol.* **2017**, *8*, 1113.
2. Mansour, R. Natural Dyes and Pigments: Extraction and Applications. In *Handbook of Renewable Materials for Coloration & Finishing*; Scrivener Publishing LLC: Beverly, MA, USA, 2018; Volume 9, pp. 75–102.
3. Ranaweera, S.J.; Ampemohotti, A.A.L.T.; Arachchige, U.S. Advantages, and Considerations for the Applications of Natural Food Pigments in the Food Industry. *J. Res. Technol. Eng.* **2020**, *1*, 8–18.
4. Arora, S. Textile Dyes: Its Impact on Environment and Its Treatment. *J. Bioremediation Biodegrad.* **2014**, *5*, 1000e146. [CrossRef]
5. Ardila-Leal, L.D.; Poutou-Piñales, R.A.; Pedroza-Rodríguez, A.M.; Quevedo-Hidalgo, B.E. A Brief History of Colour, The Environmental Impact of Synthetic Dyes and Removal by Using Laccases. *Molecules* **2021**, *26*, 3813. [CrossRef] [PubMed]
6. Manzoor, J.; Sharma, M. Impact of Textile Dyes on Human Health and Environment. In *Impact of Textile Dyes on Public Health and the Environment*; IGI Global: Hershey, PA, USA, 2020; pp. 162–169.
7. Caro, Y.; Venkatachalam, M.; Lebeau, J.; Fouillaud, M.; Dufossé, L. Pigments and Colorants from Filamentous Fungi. In *Fungal Metabolites*; Merillon, J.-M., Ramawat, K.G., Eds.; Springer International Publishing: Cham, Switzerland, 2017; pp. 499–568.
8. Lagashetti, A.C.; Dufossé, L.; Singh, S.K.; Singh, P.N. Fungal Pigments and Their Prospects in Different Industries. *Microorganisms* **2019**, *7*, 604. [CrossRef]
9. Ramesh, C.; Vinithkumar, N.V.; Kirubagaran, R.; Venil, C.K.; Dufossé, L. Multifaceted Applications of Microbial Pigments: Current Knowledge, Challenges and Future Directions for Public Health Implications. *Microorganisms* **2019**, *7*, 186. [CrossRef] [PubMed]
10. Khan, A.A.; Alshabi, A.M.; Alqahtani, Y.S.; Alqahtani, A.M.; Bennur, R.S.; Shaikh, I.A.; Muddapur, U.M.; Shakeel Iqubal, S.M.; Mohammed, T.; Dawoud, A.; et al. Extraction and Identification of Fungal Pigment from *Penicillium europium* Using Different Spectral Studies. *J. King Saud Univ.–Sci.* **2021**, *33*, 101437. [CrossRef]
11. Dufossé, L. Microbial Production of Food Grade Pigments. *Food Technol. Biotechnol.* **2006**, *44*, 313–323.
12. Kumar, A.; Vishwakarma, H.S.; Singh, J.; Dwivedi, S.; Kumar, M. Microbial Pigments: Production and Their Applications in Various Industries. *Int. J. Pharm. Chem. Biol. Sci.* **2015**, *5*, 203–212.
13. Sajid, S.; Akbar, N. Applications of Fungal Pigments in Biotechnology. *Pure Appl. Biol.* **2018**, *7*, 922–930. [CrossRef]
14. Mapari, S.A.; Meyer, A.S.; Thrane, U. Evaluation of *Epicoccum nigrum* for Growth, Morphology, and Production of Natural Colorants in Liquid Media and on a Solid Rice Medium. *Biotechnol. Lett.* **2008**, *30*, 2183–2190. [CrossRef]
15. Sibero, M.T.; Triningsih, D.W.; Radjasa, O.K.; Sabdono, A.; Trianto, A. Evaluation of Antimicrobial Activity and Identification of Yellow Pigmented Marine Sponge-Associated Fungi from Teluk Awur, Jepara, Central Java. *Indones. J. Biotechnol.* **2016**, *21*, 1–11. [CrossRef]
16. Heo, Y.M.; Kim, K.; Kwon, S.L.; Na, J.; Lee, H.; Jang, S.; Kim, C.H.; Jung, J.; Kim, J.J. Investigation of Filamentous Fungi Producing Safe, Functional Water-Soluble Pigments. *Mycobiology* **2018**, *46*, 269–277. [CrossRef]
17. Wang, W.; Liao, Y.; Chen, R.; Hou, Y.; Ke, W.; Zhang, B.; Gao, M.; Shao, Z.; Chen, J.; Li, F. Chlorinated Azaphilone Pigments with Anti-microbial and Cytotoxic Activities Isolated from the Deep Sea-Derived Fungus *Chaetomium* sp. NA–S01–R1. *Mar. Drugs* **2018**, *16*, 61. [CrossRef] [PubMed]

18. Kalra, R.; Conlan, X.A.; Goel, M. Fungi as A Potential Source of Pigments: Harnessing Filamentous Fungi. *Front. Chem.* **2020**, *8*, 369. [CrossRef] [PubMed]

19. Polak, J.; Wlizło, K.; Pogni, R.; Petricci, E.; Grąz, M.; Szałapata, K.; Osińska-Jaroszuk, M.; Kapral-Piotrowska, J.; Pawlikowska-Pawlęga, B.; Jarosz-Wilkołazka, A. Structure and Bioactive Properties of Novel Textile Dyes Synthesised by Fungal Laccase. *Int. J. Mol. Sci.* **2020**, *21*, 2052. [CrossRef] [PubMed]

20. Sudhakar, C.; Reddy, M.K.; Rajyalakshmi, K.; Raju, K.R. Antibacterial Activity of Substituted 2-Bromo-1, 4-Dimethoxy-3*H*-Phenoxazin-3-Ones. *J. Chem. Pharm. Res.* **2016**, *8*, 571–574.

21. Achenbach, H.; Blümm, E. Investigation of the Pigments of *Pycnoporus sanguineus*-Pycnosanguin and New Phenoxazin-3-Ones. *Arch. Pharm.* **1991**, *324*, 3–6. [CrossRef]

22. Maskey, R.P.; Li, F.; Qin, S.; Fiebig, H.H.; Laatsch, H. Chandrananimycins A approximately C: Production of Novel Anticancer Antibiotics from A Marine *Actinomadura* sp. Isolate M048 by Variation of Medium Composition and Growth Conditions. *J. Antibiot.* **2003**, *56*, 622–629. [CrossRef]

23. Mondal, A.; Bose, S.; Banerjee, S.; Patra, J.K.; Malik, J.; Mandal, S.K.; Kilpatrick, K.L.; Das, G.; Kerry, R.G.; Fimognari, C.; et al. Marine Cyanobacteria and Microalgae Metabolites—A Rich Source of Potential Anticancer Drugs. *Mar. Drugs* **2020**, *18*, 476. [CrossRef]

24. Onoabedje, E.A.; Okoro, U.C.; Knight, D.W. Rapid Access to New Angular Phenothiazine and Phenoxazine Dyes. *J. Heterocycl. Chem.* **2017**, *54*, 206–214. [CrossRef]

25. Kinjo, J.E.; Yokomizo, K.; Awata, Y.; Shibata, M.; Nohara, T.; Teramine, T.; Takahashi, K. Structures of Phytotoxins, AV–Toxins C, D, and E, Produced by Zonate Leaf Spot Fungus of Mulberry. *Tetrahedron Lett.* **1987**, *28*, 3697–3698. [CrossRef]

26. Riddell, R.W. Permanent Stained Mycological Preparations Obtained by Slide Culture. *Mycologia* **1950**, *42*, 265–270. [CrossRef]

27. Senanayake, I.C.; Rathnayaka, A.R.; Marasinghe, D.S.; Calabon, M.S.; Gentekaki, E.; Lee, H.B.; Hurdeal, V.G.; Pem, D.; Dissanayake, L.S.; Wijesinghe, S.N.; et al. Morphological Approaches in Studying Fungi: Collection, Examination, Isolation, Sporulation, and Preservation. *Mycosphere* **2020**, *11*, 2678–2754. [CrossRef]

28. Srinivasan, M.C.; Chidambaram, P.; Mathur, S.B.; Neergaard, P. A Simple Method for Inducing Sporulation in Seed–Borne Fungi. *Trans. Br. Mycol. Soc.* **1971**, *56*, 31–35. [CrossRef]

29. Aamir, S.; Sutar, S.; Singh, S.K.; Baghela, A. A Rapid and Efficient Method of Fungal Genomic DNA Extraction, Suitable for PCR Based Molecular Methods. *Plant Pathol. Quar.* **2015**, *5*, 74–81. [CrossRef]

30. White, T.J.; Bruns, T.D.; Lee, S.; Taylor, J. Amplification and Direct Sequencing of Fungal Ribosomal RNA Genes for Phylogenetics. In *PCR Protocols, A Guide to Methods and Applications*; Innis, M.A., Gelfand, D.H., Sninsky, J.J., White, T.J., Eds.; Academic Press: San Diego, CA, USA, 1990; pp. 315–322.

31. Vilgalys, R.; Hester, M. Rapid Genetic Identification and Mapping of Enzymatically Amplified Ribosomal DNA from Several *Cryptococcus* Species. *J. Bacteriol.* **1990**, *172*, 4238–4246. [CrossRef]

32. Kumar, S.; Stecher, G.; Tamura, K. MEGA7, Molecular Evolutionary Genetics Analysis Version 7.0 for Bigger Datasets. *Mol. Biol. Evol.* **2016**, *33*, 1870–1874. [CrossRef]

33. Nguyen, L.T.; Schmidt, H.A.; Von Haeseler, A.; Minh, B.Q. IQ–TREE: A Fast and Effective Stochastic Algorithm for Estimating Maximum-Likelihood Phylogenies. *Mol. Biol. Evol.* **2015**, *32*, 268–274. [CrossRef]

34. Kornerup, A.; Wanscher, J.H. *Methuen Handbook of Colour*, 3rd ed.; Eyre Methuen Ltd.: London, UK, 1978; pp. 1–252.

35. Velmurugan, P.; Kim, M.J.; Park, J.S.; Karthikeyan, K.; Lakshmanaperumalsamy, P.; Lee, K.J.; Park, Y.J.; Oh, B.T. Dyeing of Cotton Yarn with Five Water Soluble Fungal Pigments Obtained from Five Fungi. *Fibers Polym.* **2010**, *11*, 598–605. [CrossRef]

36. Morton, D.J.; Stroube, W.H. Antagonistic and Stimulatory Effects of Soil Microorganisms Upon *Sclerotium rolfsii*. *Phytopathology* **1955**, *45*, 417–420.

37. Rabha, A.J.; Naglot, A.; Sharma, G.D.; Gogoi, H.K.; Veer, V. In Vitro Evaluation of Antagonism of Endophytic *Colletotrichum gloeosporioides* Against Potent Fungal Pathogens of *Camellia sinensis*. *Indian J. Microbiol.* **2014**, *54*, 302–309. [CrossRef] [PubMed]

38. Jinesh, V.K.; Jaishree, V.; Badami, S.; Shyam, W. Comparative Evaluation of Antioxidant Properties of Edible and Non–Edible Leaves of *Anethum graveolens* Linn. *Indian J. Nat. Prod. Resour.* **2010**, *1*, 168–173.

39. Balouiri, M.; Sadiki, M.; Ibnsouda, S.K. Methods for In Vitro Evaluating Anti-microbial Activity: A Review. *J. Pharm. Anal.* **2016**, *6*, 71–79. [CrossRef] [PubMed]

40. Devi, S.; Karuppan, P. Reddish Brown Pigments from *Alternaria alternata* for Textile Dyeing and Printing. *Indian J. Fibre Text. Res.* **2015**, *40*, 315–319.

41. Janani, L.; Hillary, L.; Phillips, K. Mordanting Methods for Dyeing Cotton Fabrics with Dye from *Albizia coriaria* Plant Species. *Int. J. Sci. Res. Publ.* **2014**, *4*, 1–6.

Article

Antimicrobial Polyketide Metabolites from *Penicillium bissettii* and *P. glabrum*

Melissa M. Cadelis [1,2,*], Natasha S. L. Nipper [1], Alex Grey [2], Soeren Geese [2], Shara J. van de Pas [2], Bevan S. Weir [3], Brent R. Copp [1,†] and Siouxsie Wiles [2,*,†]

[1] School of Chemical Sciences, University of Auckland, Waipapa Taumata Rau, Private Bag 92019, Auckland 1142, New Zealand; natasha.nipper@gmail.com (N.S.L.N.); b.copp@auckland.ac.nz (B.R.C.)
[2] Bioluminescent Superbugs Lab, School of Medical Sciences, University of Auckland, Waipapa Taumata Rau, Private Bag 92019, Auckland 1142, New Zealand; alex.grey@auckland.ac.nz (A.G.); s.geese@auckland.ac.nz (S.G.); s.vandepas@auckland.ac.nz (S.J.v.d.P.)
[3] Manaaki Whenua, Landcare Research, Private Bag 92170, Auckland 1142, New Zealand; WeirB@landcareresearch.co.nz
* Correspondence: m.cadelis@auckland.ac.nz (M.M.C.); s.wiles@auckland.ac.nz (S.W.)
† These authors contributed equally to this work.

Abstract: Screening of several fungi from the New Zealand International Collection of Microorganisms from Plants identified two strains of *Penicillium*, *P. bissettii* and *P. glabrum*, which exhibited antimicrobial activity against *Escherichia coli*, *Klebsiella pneumoniae*, and *Staphylococcus aureus*. Further investigation into the natural products of the fungi, through extraction and fractionation, led to the isolation of five known polyketide metabolites, penicillic acid (**1**), citromycetin (**2**), penialdin A (**3**), penialdin F (**4**), and myxotrichin B (**5**). Semi-synthetic derivatization of **1** led to the discovery of a novel dihydro (**1a**) derivative that provided evidence for the existence of the much-speculated open-chained form of **1**. Upon investigation of the antimicrobial activities of the natural products and derivatives, both penicillic acid (**1**) and penialdin F (**4**) were found to inhibit the growth of Methicillin-resistant *S. aureus*. Penialdin F (**4**) was also found to have some inhibitory activity against *Mycobacterium abscessus* and *M. marinum* along with citromycetin (**2**).

Keywords: antimicrobial; penicillium; metabolite; fungi; natural products

Citation: Cadelis, M.M.; Nipper, N.S.L.; Grey, A.; Geese, S.; van de Pas, S.J.; Weir, B.S.; Copp, B.R.; Wiles, S. Antimicrobial Polyketide Metabolites from *Penicillium bissettii* and *P. glabrum*. *Molecules* **2022**, *27*, 240. https://doi.org/10.3390/molecules27010240

Academic Editor: Jacqueline Aparecida Takahashi

Received: 24 November 2021
Accepted: 27 December 2021
Published: 31 December 2021

Publisher's Note: MDPI stays neutral with regard to jurisdictional claims in published maps and institutional affiliations.

1. Introduction

Manaaki Whenua's International Collection of Microorganisms from Plants (ICMP) contains over 10,000 fungal cultures, most of which originate from Aotearoa New Zealand and the South Pacific [1]. These fungal cultures include a diverse range of fungal species, many of which are unique to Aotearoa New Zealand and most of which have not been rigorously tested for antimicrobial activity or investigated for novel natural products. The most well-known and significant example of antibiotic discovery from microbes was the discovery of penicillin from a *Penicillium rubens* culture by Alexander Fleming in 1928 [2,3]. The structure of Fleming's penicillin was proposed in 1942 by Edward Abraham and confirmed by Dorothy Hodgkin in 1945. While the original penicillins were active mostly against Gram-positive bacteria, numerous semisynthetic antibiotics were produced in the subsequent years with an improved spectrum of activity [2].

The *Penicillium* genus is large with more than 350 currently accepted species divided into 28 sections [4]. Section *Aspergilloides* contains many common cosmopolitan species isolated from soil, food, and plants, including *Penicillium glabrum* ICMP 5686 isolated from sheep dung in New Zealand [5]. Section *Lanata-Divaricata* contains many common soil-inhabiting fungi, including *Penicillium bissetti* ICMP 21333 isolated from tree roots in New Zealand [6]. Investigation of ICMP 5685 and ICMP 21333 led to the isolation of five known polyketide secondary metabolites. Structure elucidation of secondary metabolites **1–5** with

NMR spectroscopic and mass spectrometric data identified the natural products as penicillic acid (**1**) [7,8], citromycetin (**2**) [9], penialdin A (**3**) [10], penialdin F (**4**) [11], and myxotrichin B (**5**) [12] (Figure 1). The antimicrobial activities of **2–4** have been previously investigated and were found to be not active when tested against *S. aureus*, *E. coli*, *Bacillus subtilis*, and *Acinetobacter* sp. at concentrations less than 10 μg/mL [9,10,13]. However, **3** has been reported to exhibit anti-inflammatory activity [14], antioxidant activity [15], and moderate anti-diabetic activities [16]. Semisynthetic derivatization of penicillic acid (**1**), including hydrogenation and acetylation, resulted in the discovery of a novel keto acid derivative **1a** and two other known derivatives, **1b** and **1c**. The structure of keto acid derivative **1a** was suspected to be a hydrogenation product of the open chain tautomer of penicillic acid (**1**), the existence of which has somewhat been disagreed upon. Herein, we report the isolation, derivatization, structure elucidation, and bioactivities of polyketide secondary metabolites from *Penicillium* and provide evidence for the existence of the open-chain tautomer of penicillic acid (**1**).

Penicillic acid (**1**) Citromycetin (**2**) Penialdin A (**3**)

Penialdin F (**4**) Myxotrichin B (**5**)

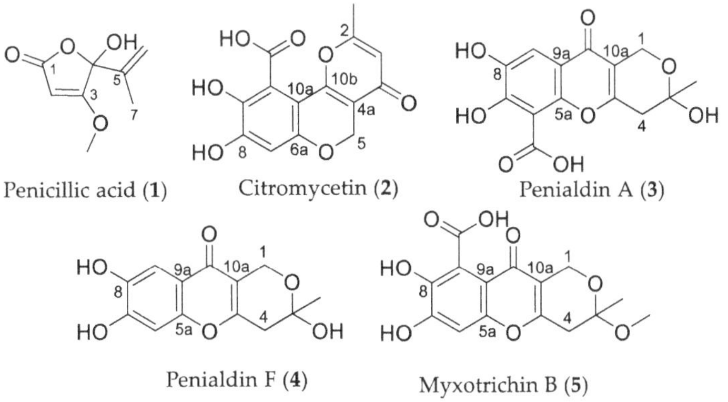

Figure 1. Structures of polyketide metabolites isolated from *Penicillium*.

2. Results and Discussion

2.1. Evaluation of Extract Bioactivity

During initial whole cell screening, *Penicillium bissettii* (ICMP 21333) showed antibacterial activity against *E. coli*, *K. pneumoniae*, and antibiotic-sensitive and resistant *S. aureus*, with zones of inhibition of up to 35 mm in diameter (Figure 2).

From the crude extract of *Penicillium bissettii* ICMP 21333, a polyketide metabolite **1** was isolated via flash column chromatography. HRESI mass spectrometric data for **1** identified a sodiated adduct observed at *m/z* 193.0494 [M + Na]$^+$ corresponding to a molecular formula of $C_8H_{11}O_4$. The ^{1}H NMR spectrum of **1** exhibited one broad exchangeable singlet (δ_H 7.87, OH), one sharp olefinic singlet (δ_H 5.41), two broad olefinic signals (δ_H 5.11 and 5.30), a methoxy singlet (δ_H 3.85), and a methyl singlet (δ_H 1.65). The ^{13}C NMR spectrum showed eight signals at δ_C 179.4, 170.0, 140.2, 115.3, 102.4, 89.4, 59.8, and 17.1. HSQC NMR data connected the two olefinic signals (δ_H 5.11 and 5.30, H_2-6) to a carbon signal at δ_C 115.3 (C-6), giving evidence for the presence of a terminal alkene. The HMBC spectrum identified correlations between these terminal alkene protons (δ_H 5.11 and 5.30, H_2-6) and the methyl carbon at δ_C 17.1 (C-7) as well as the quaternary carbons at δ_C 102.4 (C-4) and 140.2 (C-5), suggesting the presence of a fragment with a methyl group adjacent to the terminal alkene. HMBC correlations between the other olefinic proton (δ_H 5.41, H-2) and both the carbonyl carbon at δ_C 179.4 (C-1) and quaternary carbon at δ_C 102.4 (C-4), and the correlation between the methoxy singlet (δ_H 3.85) and the other quaternary carbon at δ_C 170.0 (C-3), identified the presence of an α,β-unsaturated γ-lactone ring. Additional HMBC correlations between both the methyl (δ_H 1.65, H_3-7) and olefinic (δ_H 5.41, H-2) protons and the quaternary carbon signal at δ_C 102.4 (C-4) connected the lactone ring to the terminal alkene fragment. The mass spectrometric and ^{1}H NMR data were consistent with data previously reported for penicillic acid (**1**) (Figure 1) [7,8].

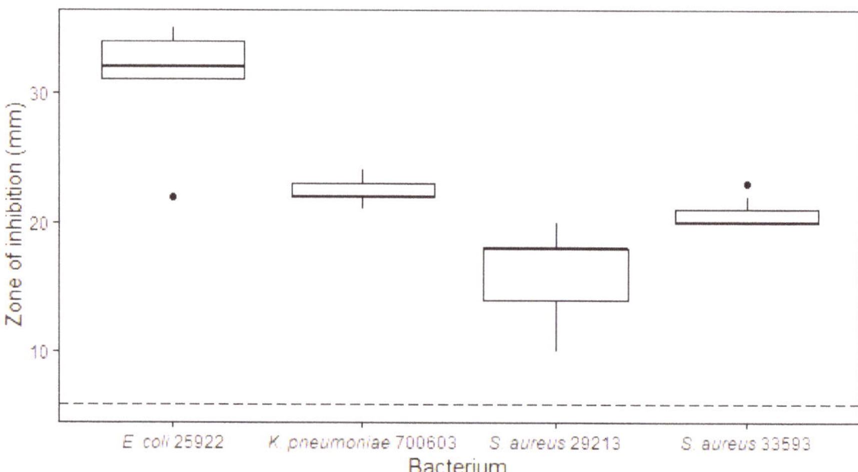

Figure 2. Antibacterial activity of *Penicillium bissettii* ICMP 21333 against *Escherichia coli*, *Klebsiella pneumoniae*, and *Staphylococcus aureus*. The antibacterial activity of *P. bissettii* ICMP 21333 was measured by the production of zones of inhibition (mm) when fungal plugs were incubated on lawns of *E. coli* ATCC 25922, *K. pneumoniae* ATCC 700603, *S. aureus* ATCC 29213 (antibiotic-sensitive), and *S. aureus* ATCC 33593 (antibiotic-resistant). Boxes are upper and lower quartiles with median shown. The whiskers extend up to 1.5× the inter-quartile range, and any dots beyond those bounds are outliers. The dotted line represents the diameter of the fungal plugs. Experiments were performed on three separate occasions with three technical replicates per experiment.

From the crude extracts of *P. glabrum*, four metabolites were isolated after purification by flash chromatography. Mass spectrometric analysis of metabolite **2** afforded a molecular formula of $C_{14}H_{10}O_7$ from a sodiated adduct at *m/z* 313.0328 [M + Na]$^+$. The ^{1}H NMR spectrum showed the presence of four singlets corresponding to a methyl group (δ_H 2.35), a methylene group (δ_H 5.03), and two olefinic protons (δ_H 6.22 and 6.53). Analysis of the HMBC spectrum identified correlations between the olefinic proton at δ_H 6.53 (H-7) and four quaternary carbons at δ_C 104.7 (C-10a), 141.9 (C-9), 153.3 (C-8), and 158.1 (C-6a) and between the methylene protons at δ_H 5.03 (H$_2$-5) and three quaternary carbons at δ_C 111.1 (C-4a), 152.5 (C-10b), and 158.1 (C-6a), indicating the presence of a chromene. The other olefinic proton at δ_H 6.22 (H-3) exhibited HMBC correlations to the methyl carbon at δ_C 19.1 (Me), a carbonyl carbon at δ_C 177.2 (C-4), and two quaternary carbons at δ_C 166.2 (C-2) and 111.1 (C-4a), indicating the presence of a γ-pyrone fused to the chromene ring. Substructure searching through the literature led to the identification of this metabolite as citromycetin (**2**) [9]. The other three metabolites showed similar ^{1}H and ^{13}C NMR spectra to **2**. Comparison of mass spectrometric and NMR data to the literature led to the identification of these metabolites as penialdin A (**3**) [10], penialdin F (**4**) [11], and myxotrichin B (**5**) [12].

2.2. Derivatisation of Penicillic Acid

Penicillic acid (**1**) was first reported in 1913 during an investigation between the incidence of pellagra and mould spoiled corn where they isolated an unknown fungal metabolite [17]. The structure of this metabolite was not determined until 1936 when Birkinshaw, Oxford, and Raistrick, who had been working on Alsberg and Black's original cultures of *Penicillium puberculum*, were able to isolate the same compound in significantly higher yields from cultures of *P. cyclopium* [18]. They used this material to complete their investigation on the structure, which they reported as a substituted γ-keto acid that tautomerized to a γ-hydroxy lactone (Figure 3) [10].

Penicillic acid (**1**)

Figure 3. Tautomerization of penicillic acid (**1**).

There has been some disagreement on the existence of the open chain γ-keto acid tautomer [18–21]. The chemical studies carried out by Birkinshaw et al. provided evidence for the open-chain form [18]. They observed esterification of anhydrous penicillic acid (**1**) with diazomethane and the reaction of a cold aqueous solution of penicillic acid with one mole equivalent of free hydroxylamine, both of which they suggested provided evidence for the presence of the carboxyl group. However, a study by Raphael presented UV absorption spectra of both penicillic acid (**1**) and the methyl ester in acid and alkaline aqueous solutions in which a maximal absorbance was observed at ~220 nm for both compounds, which was consistent with the closed-ring lactone but not with the open-chain keto acid/keto ester [20]. Similar data for dihydropenicillic acid (**1b**), a derivative formed via hydrogenation of the natural product, were also obtained suggesting that penicillic acid exists purely as the lactone and that the keto acid tautomer is not required to explain the chemical properties observed by Birkinshaw et al. [18,20]. A study by Shaw also presented UV absorbance data for penicillic acid (**1**) in both acid and alkaline solutions [21]. Results comparable to Raphael's were observed for the acid solution, but a shift in the absorption maxima to ~295 nm was observed in 0.1N NaOH solution, consistent with a change to the keto acid structure [20,21].

To investigate the existence of the open chain γ-keto acid tautomer, in the present study, we proceeded to prepare dihydropenicillic acid (**1b**) (Figure 4) via catalytic hydrogenation of penicillic acid (**1**) in methanol, using palladium on carbon under a hydrogen atmosphere. The crude product was characterized via mass spectrometry and NMR. The sodiated adduct observed at m/z 195.0622 for $[M + Na]^+$ in the (+)-HRESI mass spectrum of the crude product **1a** confirmed the addition of one mole of H_2 to penicillic acid (**1**), giving a molecular formula of $C_8H_{13}O_4$. The ^{1}H NMR spectrum (Figure S4) of **1a** exhibited one sharp olefinic singlet (δ_H 5.19), one methoxy singlet (δ_H 3.72), and one dimethyl singlet (δ_H 1.07) (Table 1). The absence of the two olefinic signals of the terminal alkene in penicillic acid (**1**) and the presence of a dimethyl signal in the ^{1}H NMR spectrum, suggested successful hydrogenation of the alkene at C-5/C-6. Of particular note was the absence of the expected septet signal of the H-5 proton and the expected splitting of the dimethyl signal into a doublet was also not observed. The HMBC NMR data of the crude product identified a correlation between the dimethyl singlet and a carbon signal at δc 39.3, consistent with an alkyl carbon at the C-5 position. Also present in the HMBC data was a correlation to a carbon signal at δc 217.1, consistent with the presence of a ketone, suggesting that **1a** was the ring opened keto acid structure as shown. This explained the absence of the ^{1}H NMR signal for H-5 (α-keto proton) and the lack of splitting of the dimethyl signal, as it was speculated that the acidic α-keto proton had exchanged with deuterium from the CD_3OD NMR solvent. This was confirmed by the ^{1}H NMR spectrum recorded in non-exchangeable solvent $CDCl_3$, which showed all expected signals: one sharp olefinic singlet (δ_H 5.12), one methoxy singlet (δ_H 3.82), one septet (δ_H 2.39, J = 7.0 Hz), and one dimethyl doublet (δ_H 1.01, J = 7.0 Hz) (Figure S6). Purification of **1a** was achieved using silica-gel-column chromatography. Changes were observed in the ^{1}H NMR spectrum of the purified product (Figure S7) when compared to the crude product, necessitating further NMR characterization of the purified product **1b**.

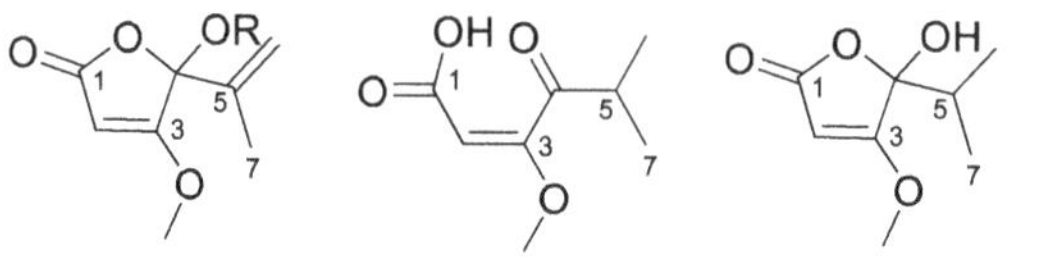

1 R = H
1c R = Ac

1a

1b

Figure 4. Structures of penicillic acid (**1**) and its derivatives (**1a–1c**).

Table 1. ^{1}H and ^{13}C NMR (CD$_3$OD) signals observed for crude (**1a**) and purified (**1b**) products.

Position	1a		1b	
	δ_H	δ_C	δ_H	δ_C
1		174.6		173.5
2	5.19	97.2	5.20	89.2
3		171.2		182.1
4		217.1		107.2
5	Not observed	39.3	2.12	34.6
6	1.07	17.5	1.04	16.1
6′	1.07	17.5	0.88	16.4
7	3.72	57.8	3.93	59.9

The ^{1}H NMR spectrum of the purified hydrogenation product **1b** showed one sharp olefinic singlet (δ_H 5.20), one methoxy singlet (δ_H 3.93), one septet (δ_H 2.12, J = 6.9 Hz), and two broad methyl doublets (δ_H 1.04, J = 6.8 Hz and δ_H 0.88, J = 6.8 Hz). The presence of the septet and the splitting of the methyl signals into doublets indicated that H-5 (δ_H 2.12, J = 6.9 Hz) of the purified product **1b** was not exchangeable with the NMR solvent, suggesting a change to the ring-closed lactone structure. The presence of two nonequivalent methyl signals was also consistent with the ring-closed form, with the chiral center at C-4 (δ_C 107.2) inducing non-equivalence in the pendant gem dimethyl groups. A change to the ring-closed structure was confirmed by HMBC data of the purified material, which exhibited a correlation between the two methyl proton signals and a carbon signal at δ_C 107.2 as well a correlation between the H-5 (δ_H 2.12, J = 6.9 Hz) proton signal and the same carbon signal at δ_C 107.2. The shift in this carbon signal was consistent with the shift observed for the hemiacetal carbon of penicillic acid (**1**) (δ_C 106.8). The HMBC data of the purified product also identified a second weaker set of correlations from a proton signal partially obscured by the broad methyl signal at δ_H 1.04, to carbons at δ_C 17.5, 39.3, and 217.1 (data not shown). This set of correlations suggested that some of the keto acid tautomer was present as the correlations were consistent with those observed in HMBC of the crude product. However, the expected second methoxy signal at δ_H 3.72 in the ^{1}H NMR spectrum was absent. The NMR data for the crude and purified products (**1a** and **1b**) clearly show the existence of the two forms.

Acetylated penicillic acid (**1c**) was prepared via acetylation with acetic anhydride in pyridine over 18 h, under a nitrogen atmosphere. The crude material was purified by diol-bonded silica-gel-column chromatography. The sodiated adduct observed at m/z 235.0578 [M + Na]$^+$ in the (+)-HRESI mass spectrum was consistent with the loss of a proton and the addition of an acetyl group, giving a molecular formula of C$_{10}$H$_{12}$O$_5$. The ^{1}H NMR spectrum (Figure S9) identified the introduction of new methyl protons (δ_H 2.12, s), which in turn exhibited a correlation in the HMBC spectrum to a carbon at δ_C 167.7, which is consistent with the introduction of an acetyl group. The mass spectrometric values and ^{1}H NMR data obtained were in agreement with the literature [22].

2.3. ^{13}C NMR Study of Penicillic Acid (**1**) Biosynthesis

A sodium [1-^{13}C] acetate labelling study was conducted to confirm the biosynthetic origin of penicillic acid (**1**). Cultures of *P. bissettii* ICMP 21333 were grown on PDA agar supplemented with 1g/L 1-^{13}C sodium acetate. Extraction and fractionation as conducted previously afforded penicillic acid (**1**) (132 mg). The ^{13}C NMR spectrum (Figure S3) of the ^{13}C-labelled penicillic acid (**1**) was acquired and compared to the spectrum observed for non-labelled compound. The signals observed for C-1, C-3, and C-5 showed increased intensity, consistent with ^{13}C enrichment. An indication of the extent of ^{13}C enrichment was given by the relative peak integrals when compared to the spectrum of unlabeled penicillic acid (**1**). The relative peak integrals for labelled penicillic acid (**1**) clearly show enhancement for peaks at δc 181.2 (C-1, 12.3×), 173.2 (C-3, 10.1×), and 141.7 (C-5, 8.5×) (Table 2). The observed labelling pattern (Figure 5) was consistent with previous studies utilizing ^{14}C-labelled acetate [23,24].

Table 2. Relative Integral Values for ^{13}C NMR Peaks Observed for ^{13}C-labelled Penicillic Acid (**1**).

Position	Δc	Relative Integral (Labelled/Unlabelled)
C-1	181.2	12.3
C-2	90.1	1.0
C-3	173.2	10.1
C-4	Not observed	
C-5	141.7	8.5
C-6	116.7	1.1
C-7	17.5	1.0
C-8	60.4	0.9

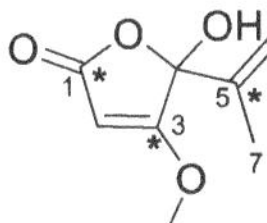

Figure 5. Labelling pattern observed in ^{13}C NMR spectrum of penicillic acid (**1**) produced by cultures supplemented with 1-^{13}C sodium acetate.

2.4. Bioactivity

Penicillic acid (**1**) and its derivatives **1b** and **1c**, as well as citromycetin (**2**) and penialdin A (**3**) and F (**4**), were screened for antimicrobial activity. Compounds **1**–**4** were tested in growth-inhibition assays against Methicillin-resistant *S. aureus* (MRSA), *E. coli*, *K. pneumoniae* (multi-drug resistant strain), *Acinetobacter baumannii*, *Pseudomonas aeruginosa*, *Candida albicans*, and *Cryptococcus neoformans* by the Community for Open Antimicrobial Drug Discovery (COADD) at The University of Queensland. The COADD test compound activity as an inhibition of microbial growth at a single concentration of 32 μg/mL is presented in Table 3.

At the concentration tested, penicillic acid (**1**) showed some inhibition of the growth of MRSA (53%) and *E. coli* (26%), as expected based on previous literature [25,26]. However, neither of the penicillic acid derivatives (**1b** and **1c**) showed any activity. Loss of activity on hydrogenation or acetylation of **1** suggests that the unsaturation at C-5 to C-6 is of importance as is the ability to convert to the open-chain form. Penialdin F (**4**) exhibited the strongest antimicrobial activity, inhibiting the growth of MRSA by over 71% (Table 3) at the concentration tested. This was surprising as **4** has been previously reported to be not active against wild-type and multidrug-resistant strains of *E. faecalis*, *E. faecium*, and *S. aureus* [13]. However, the inhibition was not complete, suggesting an MIC >32 μg/mL, which is considerably higher than antibiotics used for clinical treatment of antibiotic-sensitive *S. aureus* infections, which range from <0.001-~4 μg/mL [27] depending on the antibiotic. None of the other natural products showed strong enough inhibitory activity against any of the organisms, as expected. Citromycetin (**2**) has been previously investigated for antimicrobial activity (*E. coli*, *Bacillus subtilis*, and *Septoria nodurum*) and antiparasitic activity (*Haemonchus contortus*) and was found not active while also being non-cytotoxic against NS-1 cells [12]. Penialdin A (**3**) has also been reported as not active at a concentration of 10 μg/mL against *S. aureus*, *E. coli*, *B. subtilis*, and *Acinetobacter* sp. [10].

Table 3. Antimicrobial and antifungal activities of natural products **1–4** and derivatives **1b** and **1c**.

Compound	Percentage Inhibition at 32 µg/mL						
	S. a [a]	*E. c* [b]	*K. p* [c]	*P. a* [d]	*A. b* [e]	*C. a* [f]	*C. n* [g]
1	53.07 ****	26.11 ***	18.06 *	−1.15	6.12	6.50	2.15
1b	−1.05	0.34	4.32	2.75	−26.55	2.49	18.33 **
1c	17.41 *	5.15	9.34	−10.06	−11.88	2.56	−7.04
2	14.22	11.16	13.49	4.70	−4.31	−1.86	1.11
3	−7.23	13.60	−7.65	−4.08	−24.24	−2.20	20.72 **
4	71.52 ****	9.29	15.62 *	0.52	12.55	8.84	6.73

All values are presented as the mean ($n = 2$). Negative values represent increases in growth compared to the no-compound controls. Data were analysed using a two-way ANOVA with Dunn's multiple comparison test. The adjusted p values of those compounds that were significantly different to the relevant no inhibition control are given by * (* $p < 0.05$; ** $p < 0.005$; *** $p < 0.0005$; **** $p < 0.0001$). [a] *Staphylococcus aureus* ATCC 43300 (MRSA) with vancomycin (MIC 1 µg/mL) used as a positive control; [b] *Escherichia coli* ATCC 25922 with colistin (MIC 0.125 µg/mL) used as a positive control; [c] *Klebsiella pneumoniae* ATCC 700603 with colistin (MIC 0.25 µg/mL) as a positive control; [d] *Pseudomonas aeruginosa* ATCC 27853 with colistin (MIC 0.25 µg/mL); [e] *Acinetobacter baumanii* ATCC 19606 with colistin (MIC 0.25 µg/mL) as a positive control; [f] *Candida albicans* ATCC 90028 with fluconazole (MIC 0.125 µg/mL) as a positive control; [g] *Cryptococcus neoformans* ATCC 208821 with fluconazole (MIC 8 µg/mL) as a positive control.

In addition to the assays performed by COADD, the compounds citromycetin (**2**) and penialdin F (**4**) were evaluated for their bioactivity against luminescent derivatives of *Mycobacterium abscessus* and *M. marinum* at a maximum concentration of 64 µg/mL using inhouse assays. Luminescence is frequently used as rapid readout of potential antimycobacterial activity [28–30]. Citromycetin and another compound of the penialdin family, penialdin C, have previously been shown to have activity against the Mycobacterial species *M. smegmatis*, with minimum inhibitory concentrations (MIC) of 31.2 and 15.6 µg/mL, respectively [31]. We found that citromycetin (**2**) and penialdin F (**4**) both displayed some activity against *M. abscessus* BSG301 and *M. marinum* BSG101, decreasing bacterial light output compared to the no-compound control (Figure 6). However, at the concentration tested, neither compound caused a reduction in light production of at least 1 log, which is how we define the MIC indicating their MIC against *M. abscessus* and *M. marinum* is greater than 64 µg/mL.

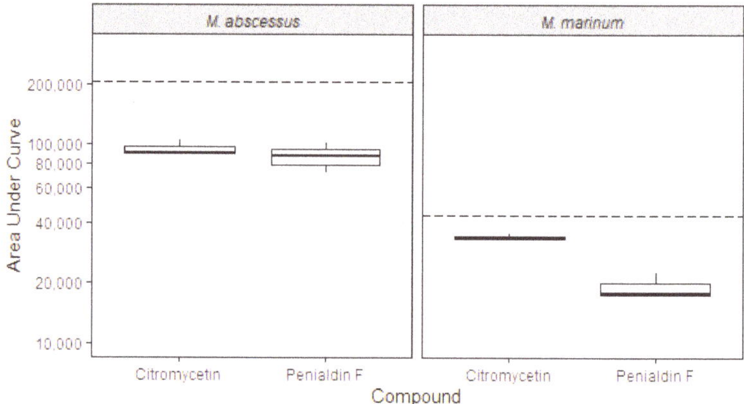

Figure 6. Effect of citromycetin (**2**) and penialdin F (**4**) on the luminescence of *Mycobacterium abscessus*

BSG301 and *M. marinum* BSG101. The antibacterial activity of citromycetin (**2**) and penialdin F (**4**) were measured by changes in bacterial luminescence. Data are presented as box and whisker plots of the area under the curve values calculated from luminescence readings taken over 72 h. Boxes are upper and lower quartiles with median shown. The whiskers extend up to 1.5 × the inter-quartile range. The dotted line represents the no-compound control.

3. Materials and Methods

3.1. General Experimental Procedures

Infrared spectra were recorded on a Perkin–Elmer spectrometer 100 Fourier Transform infrared spectrometer equipped with a universal ATR accessory. Mass spectra were acquired on a Bruker micrOTOF Q II spectrometer (Bruker Daltonics, Bremen, Germany). Melting points were recorded on an electrothermal melting point apparatus and were uncorrected. ^{1}H and ^{13}C NMR spectra were recorded at 298K on a Bruker AVANCE 400 spectrometer (Bruker, Karlsruhe, Germany) at 400 and 100 MHz, respectively, using standard pulse sequences. Proto–deutero solvent signals were used as internal references (CD$_3$OD: δ_H 3.31, δ_C 49.0; CDCl$_3$: δ_H 0.00 (TMS), δ_C 77.16; DMSO-d_6: δ_H 2.50, δ_C 39.52). For ^{1}H NMR, the data are quoted as position (δ), relative integral, multiplicity (s = singlet, d = doublet, t = triplet, q = quartet, p = pentet, m = multiplet, dd = doublet of doublets, ddd = doublet of doublets of doublets, td = triplet of doublets, and br = broad), coupling constant (*J*, Hz), and assignment to the atom. The 13C NMR data are quoted as position (δ) and assignment to the atom. Flash column chromatography was carried out using Kieselgel 60 (40-63 μm) (Merck, Munich, Germany) silica gel or Merck Diol bonded silica (40–63 μm), with Merck C$_8$ reversed–phase (40–63 μm) solid support. Gel filtration was carried out on Sephadex LH–20 (Pharmacia, Uppsala, Sweden). Thin-layer chromatography was conducted on Merck DC–plastikfolien Kieselgel 60 F254 plates. All solvents used were of analytical grade or better and/or purified according to standard procedures. Chemical reagents used were purchased from standard chemical suppliers and used as purchased.

3.2. Fungal Material

P. bissettii was described in 2016 from Canada [6] and later found to be present in South Korea [32] and New Zealand. The ecology of *P. bissettii* is unknown; however, it appears to be associated with tree soils or roots; the Canadian isolates were from Spruce forest soil and the Korean isolates from pine forest soil. Culture ICMP 21333 was isolated as an endophyte from the living roots of a kauri (*Agathis australis*) tree in Auckland, New Zealand in 2016. The identification of this isolate is supported by DNA sequences of the ITS (GENBANK# MW862794), BTUB (GENBANK# MZ393465), and CAL (GENBANK# MZ393460) genes. *P. glabrum* is a common cosmopolitan (GBIF 2021) filamentous ascomycete fungus. It has long been recognised as a cause of post-harvest fruit and vegetable rots since it was first described as *Citromyces glaber* in 1893. Culture ICMP 5686 was isolated from sheep dung in Palmerston North, New Zealand in July 1976. The identification of this isolate is supported by DNA sequences of the ITS (GENBANK# MW862782), BTUB (GENBANK# MZ393464), and CAL (GENBANK# MZ393462) genes. Cultures of *P. bissettii* and *P. glabrum* were grown at room temperature on Potato Dextrose Agar (PDA) (Fort Richard, New Zealand) plates. For extraction, the cultures were grown on 40 PDA plates until the fungus covered an area of 75% and were freeze dried for 5 days.

3.3. Antimicrobial Testing of Fungal Cultures

The ICMP fungal cultures were screened for antimicrobial activity after growth on PDA. Briefly, cultures of *E. coli* ATCC 25922, *K. pneumoniae* ATCC 700603, and *S. aureus* ATCC 29213 (antibiotic-sensitive (oxacillin)) and ATCC 33593 (antibiotic-resistant (gentamicin and methicillin)) were grown overnight in Tryptic Soy Broth (TSB) (Fort Richard, New Zealand) at 37 °C with shaking at 200 rpm. Cultures were diluted in TSB to an optical density at 600 nm of 0.1 and spread onto PDA plates for *E. coli* and *K. pneumoniae* and Mueller Hinton Agar (MHA) (Fort Richard, New Zealand) for *S. aureus* to produce a confluent bacterial lawn. Plugs were removed from fungal cultures using a 6 mm Paramount biopsy punch (Capes Medical, New Zealand) when the fungal cultures had grown to cover 75% of the plates. These were placed onto the inoculated PDA and MHA plates and incubated overnight at 37 °C. Any zones of inhibition were measured the next day using a ruler, and the diameter was recorded. Experiments were performed three times with three technical replicates per experiment.

3.4. Extraction and Isolation

Freeze-dried cultures of *P. bissettii* (17.49 g) were extracted with MeOH (1 L) for 4 h followed by CH_2Cl_2 (1 L) overnight. Concentration of the combined organic extracts afforded 2.46 g of crude extract. The crude extract was subjected to C_8 flash column chromatography (H_2O/MeOH), which afforded 5 fractions (A–E). Penicillic acid (**1**) was isolated from fraction C (1:1, H_2O/MeOH) and further purified by Sephadex LH20 (MeOH) to give white crystals (25 mg).

The dry cultures of *P. glabrum* (20.86 g) were extracted with MeOH (1 L) for 4 h followed by CH_2Cl_2 (1 L) overnight. Concentration of the combined organic extracts afforded 2.03 g of crude extract. The crude extract was subjected to C_8 flash column chromatography (H_2O/MeOH), which afforded 5 fractions (A–E). Subsequent purification of fraction B (4:1, H_2O/MeOH) by Sephadex LH20 (MeOH) afforded 5 fractions (B1–B5). Citromycetin (**2**) was isolated from fraction B4 after further purification by Diol-bonded silica column chromatography (*n*-Hexane/EtOAc, gradient) to give a yellow oil (2.9 mg). Purification of fraction C (1:1, H_2O/MeOH) by Sephadex LH20 (MeOH) afforded 4 fractions (C1–C4). Penialdin A (**3**) (3.5 mg), penialdin F (**4**) (4.6 mg), and myxotrichin B (**5**) (0.75 mg) were isolated from fraction C2 after further purification by Diol-bonded silica column chromatography (*n*-Hexane/EtOAc, gradient).

3.4.1. Penicillic acid (1)

^{1}H NMR (CD_3OD, 400 MHz) δ 5.41 (1H, br s, H-6), 5.27 (1H, s, H-2), 5.17 (1H, br s, H-6), 3.92 (3H, s, OMe), 1.75 (3H, dd, *J* = 1.0, 1.0 Hz, H_3-7); ^{13}C NMR (CD_3OD, 100 MHz) 181.2 (C-1), 173.0 (C-3), 141.5 (C-5), 116.6 (C-6), 106.8 (C-4), 90.0 (C-2), 60.5 (C-8), 17.5 (C-7); ^{1}H NMR (DMSO-d_6, 400 MHz) δ 7.87 (1H, br s, OH), 5.41 (1H, s, H-2), 5.30 (1H, br s, H-6), 5.11 (1H, br s, H-6), 3.85 (3H, s, OMe), 1.65 (3H, s, H_3-7); ^{13}C NMR (DMSO-d_6, 100 MHz) δ 179.4 (C-1), 170.0 (C-3), 140.2 (C-5), 115.3 (C-6), 102.4 (C-4), 89.4 (C-2), 59.8 (C-8), 17.1 (C-7); (+)-HRESIMS *m/z* 193.0474 [M + Na]$^+$ (calcd for $C_8H_{10}NaO_4$, 193.0470).

3.4.2. Citromycetin (2)

^{1}H NMR (CD_3OD, 400 MHz) δ 6.53 (1H, s, H-7), 6.22 (1H, s, H-3), 5.03 (2H, s, H_2-5), 2.35 (3H, s, Me); ^{13}C NMR (CD_3OD, 100 MHz) δ 177.2 (C-4), 167.0 ($\underline{C}$OOH), 166.1 (C-2), 158.1 (C-6a), 153.3 (C-8), 152.5 (C-10b), 141.9 (C-9), 120.0 (C-10), 113.3 (C-3), 111.1 (C-4a), 104.7 (C-10a), 104.5 (C-6a), 62.7 (C-5), 19.1 (Me); (+)-HRESIMS *m/z* 313.0328 [M + Na]$^+$ (calcd for $C_{14}H_{10}NaO_7$, 313.0319).

3.4.3. Penialdin A (3)

^{1}H NMR (CD_3OD, 400 MHz) δ 7.09 (1H, s, H-9), 4.65 (2H, s, H_2-1), 2.99 (1H, d, *J* = 17.8 Hz, H_2-4_A), 2.77 (1H, d, *J* = 17.8 Hz, H_2-4_B), 1.57 (3H, s, Me); ^{13}C NMR (CD_3OD, 100 MHz) δ 175.2 (C-10), 168.0 ($\underline{C}$OOH), 160.8 (C-4a), 154.8 (C-8), 152.6 (C-5a), 144.1 (C-7), 119.1 (C-6), 114.6 (C-10a), 111.8 (C-9a), 105.6 (C-9), 94.3 (C-3), 56.1 (C-1), 37.2 (C-5), 27.2 (Me); (+)-HRESIMS *m/z* 287.0529 [M + Na]$^+$ (calcd for $C_{13}H_{12}NaO_6$, 287.0526).

3.4.4. Penialdin F (4)

^{1}H NMR (CD_3OD, 400 MHz) δ 7.36 (1H, s, H-9), 6.86 (1H, s, H-6), 4.62 (2H, s, H_2-1), 2.91 (1H, d, *J* = 17.0 Hz, H_2-4_A), 2.68 (1H, d, *J* = 17.0 Hz, H_2-4_B), 1.55 (3H, s, Me); ^{13}C NMR (CD_3OD, 100 MHz) δ 174.9 (C-10), 160.8 (C-4a), 151.9 (C-5a, C-8), 144.3 (C-7), 115.7

(C-9a), 114.6 (C-10a), 106.7 (C-9), 101.9 (C-6), 94.3 (C-3), 46.1 (C-1), 37.1 (C-5), 27.2 (Me); (+)-HRESIMS *m/z* 331.0428 [M + Na]$^+$ (calcd for $C_{14}H_{12}NaO_8$, 331.0424).

3.4.5. Myxotrichin B (5)

^{1}H NMR (CD$_3$OD, 400 MHz) δ 7.09 (1H, s, H-6), 4.68 (1H, d, *J* = 15.0 Hz, H$_2$-1$_A$), 4.40 (1H, d, *J* = 15.0 Hz, H$_2$-1$_B$), 3.33 (3H, s, OMe), 3.02 (1H, d, *J* = 17.9 Hz, H$_2$-4$_A$), 2.81 (1H, d, *J* = 17.9 Hz, H$_2$-4$_B$), 1.54 (3H, s, Me); ^{13}C NMR (CD$_3$OD, 100 MHz) δ 175.5 (C-10), 167.8 (COOH), 160.4 (C-4a), 154.1 (C-7), 151.5 (C-5a), 144.4 (C-7), 118.4 (C-9), 113.8 (C-10a), 111.8 (C-9a), 105.1 (C-6), 98.2 (C-3), 57.0 (C-1), 48.3 (OMe), 37.2 (C-5), 21.3 (Me); (+)-HRESIMS *m/z* 345.0588 [M + Na]$^+$ (calcd for $C_{15}H_{14}NaO_8$, 345.0581).

3.4.6. (E)-3-Methoxy-5-methyl-4-oxohex-2-enoic acid (1a)

To a solution of penicillic acid (**1**) (5 mg, 0.03 mmol) in anhydrous MeOH (1 mL) was added Pd/C (1 mg, 0.003 mmol), and the reaction was stirred under an H$_2$ atmosphere for 18 h. The reaction mixture was filtered through celite and washed with MeOH (50 mL). The solvent was removed under reduced pressure to give the crude product as a pale-yellow oil (3.43 mg, 66%).

^{1}H NMR (CD$_3$OD, 400 MHz) δ 5.19 (1H, s, H-2), 3.72 (3H, s, OMe), 3.31 (1H, obs by solvent, H-5), 1.07 (6H, s, H$_3$-6, H$_3$-7); ^{13}C NMR (CD$_3$OD, 100 MHz) δ 217.2 (C-4), 174.6 (C-1), 171.2 (C-3), 97.2 (C-2), 57.8 (OMe), 39.2 (C-5), 17.7 (C-6, C-7); ^{1}H NMR (CDCl$_3$, 400 MHz) δ 5.12 (1H, s, H-2), 3.82 (3H, s, OMe), 2.39 (1H, septet, *J* = 7.0 Hz, H-5) 1.01 (6H, d, *J* = 7.0 Hz, H$_3$-6, H$_3$-7); (+)-HRESIMS *m/z* 195.0622 [M + Na]$^+$ (calcd for $C_8H_{12}NaO_4$, 195.0630).

3.4.7. 5-Hydroxy-5-isopropyl-4-methoxyfuran-2(5H)-one (1b)

The product was purified by silica-gel column chromatography (10% MeOH:CH$_2$Cl$_2$) to give the product as white crystals (1.72 mg, 33%).

^{1}H NMR (CD$_3$OD, 400 MHz) δ 5.20 (1H, s, H-2), 3.93 (3H, s, OMe), 2.12 (1H, septet, *J* = 6.8 Hz, H-5), 1.04 (3H, d, *J* = 6.4 Hz H$_3$-6), 0.88 (3H, d, *J* = 6.4 Hz, H$_3$-7); ^{13}C NMR (CD$_3$OD, 100 MHz) δ 182.1 (C-3), 173.5 (C-1), 107.2 (C-4), 89.2 (C-2), 59.9 (OMe), 34.6 (C-5), 16.4 (C-7), 16.1 (C-6); ^{1}H NMR (CDCl$_3$, 400 MHz) δ 5.06 (1H, s, H-2), 3.91 (3H, s, OMe), 2.21 (1H, p, *J* = 6.8 Hz, H-5), 1.12–1.02 (3H, m, H$_3$-6), 0.96–0.88 (3H, m, H$_3$-7).

3.4.8. 3-Methoxy-5-oxo-2-(prop-1-en-2-yl)-2,5-dihydrofuran-2-yl acetate (1c)

To a solution of penicillic acid (**1**) (5 mg, 0.03 mmol) in pyridine (1 mL) was added acetic anhydride (0.05 mL, 0.52 mmol) dropwise, and the reaction was stirred for 18 hr under a nitrogen atmosphere. The reaction mixture was quenched with 10% HCl and extracted with CH$_2$Cl$_2$ (3 × 10 mL) and ethyl acetate (1 × 10 mL). The combined organic layers were dried with MgSO$_4$ and the solvent removed under reduced pressure. The product was purified by diol-bonded silica-gel column chromatography, eluting with CH$_2$Cl$_2$, to give the product as a brown oil (1.96 mg, 30%).

^{1}H NMR (CDCl$_3$, 400 MHz) δ 5.38 (1H, br s, H-6), 5.17 (2H, br s, H-2, H-6), 3.91 (3H, s, OMe), 2.12 (3H, s, OAc), 1.83 (3H, d, *J* = 1.0 Hz, H$_3$-7); ^{13}C NMR (CDCl$_3$, 100 MHz) δ 178.1 (C-1), 168.9 (C-3), 167.7 (OAc), 138.4 (C-5), 116.1 (C-6), 101.6 (C-4), 89.8 (C-2), 59.8 (OMe), 21.3 (OAc), 17.3 (C-7); (+)-HRESIMS *m/z* 235.0578 [M + Na]$^+$ (calcd for $C_{10}H_{12}NaO_5$, 235.0580).

3.5. Sodium [1-^{13}C] Acetate Incorporated Acetate Fermentation of P. bissettii

Extraction and Isolation

Cultures of *Penicillium bissettii* were grown on 17 PDA plates supplemented with 1g/L 1-^{13}C sodium acetate until the fungus covered 75% of the plate. The cultured plates were freeze dried (11.00 g) and extracted with methanol and dichloromethane to give 2.28 g of crude extract. The crude extract was purified via C$_8$ reverse-phased column chromatography (H$_2$O/MeOH), and penicillic acid (**1**) (132 mg) was isolated from fraction C (1:1, H$_2$O/MeOH).

3.6. Antimicrobial Assays of Pure Compounds

Antimicrobial evaluation of the pure compounds against *S. aureus* ATCC 43300 (MRSA), *E. coli* ATCC 25922, *Pseudomonas aeruginosa* ATCC 27853, *Klebsiella pneumoniae* ATCC 700603, *Acinetobacter baumannii* ATCC 19606, *Candida albicans* ATCC 90028, and *Cryptococcus neoformans* ATCC 208821 was undertaken at the Community for Open Antimicrobial Drug Discovery at The University of Queensland (Queensland, Australia) according to their standard protocols [33]. For antimicrobial assays, the tested strains were cultured in either Luria broth (LB) (In Vitro Technologies, USB75852, Victoria, Australia), nutrient broth (NB) (Becton Dickson 234000, New South Wales, Australia), or MHB at 37 °C overnight. A sample of culture was then diluted 40-fold in fresh MHB and incubated at 37 °C for 1.5–2 h. The compounds were serially diluted 2-fold across the wells of 96-well plates (Corning 3641, nonbinding surface), with compound concentrations ranging from 0.015 to 64 µg/mL, plated in duplicate. The resultant mid-log-phase cultures were diluted to the final concentration of 1×10^6 CFU/mL; then, 50 µL were added to each well of the compound containing plates giving a final compound concentration range of 0.008–32 µg/mL and a cell density of 5×10^5 CFU/mL. All plates were then covered and incubated at 37 °C for 18 h. Inhibition of bacterial growth was determined measuring absorbance at 600 nm (OD600), using a Tecan M1000 Pro monochromator plate reader. The percentage of growth inhibition was calculated for each well, using the negative control (media only) and positive control (bacteria without inhibitors) on the same plate as references.

For the antifungal assay, fungi strains were cultured for 3 days on Yeast Extract Peptone Dextrose (YPD) agar at 30 °C. A yeast suspension of 1×10^6 to 5×10^6 CFU/mL was prepared from five colonies. These stock suspensions were diluted with yeast nitrogen base (YNB) (Becton Dickinson, 233520, New South Wales, Australia) broth to a final concentration of 2.5×10^3 CFU/mL. The compounds were serially diluted 2-fold across the wells of 96-well plates (Corning 3641, nonbinding surface), with compound concentrations ranging from 0.015 to 64 µg/mL and final volumes of 50 µL, plated in duplicate. Then, 50 µL of the fungi suspension that was previously prepared in YNB broth to the final concentration of 2.5×10^3 CFU/mL were added to each well of the compound-containing plates, giving a final compound concentration range of 0.008–32 µg/mL. Plates were covered and incubated at 35 °C for 36 h without shaking. *C. albicans* MICs were determined by measuring the optical density at 530 nm. For *C. neoformans*, resazurin was added at 0.006% final concentration to each well and incubated for a further 3 h before MICs were determined by measuring the optical density at 570–600 nm.

Colistin and vancomycin were used as positive bacterial inhibitor standards for Gram-negative and Gram-positive bacteria, respectively. Fluconazole was used as a positive fungal inhibitor standard for *C. albicans* and *C. neoformans*. The antibiotics were provided in 4 concentrations, with 2 above and 2 below its MIC value, and they were plated into the first 8 wells of column 23 of the 384-well NBS plates. The quality control (QC) of the assays was determined by the antimicrobial controls and the Z'-factor (using positive and negative controls). Each plate was deemed to fulfil the quality criteria (pass QC) if the Z'-factor was above 0.4, and the antimicrobial standards showed a full range of activity, with full growth inhibition at their highest concentration and no growth inhibition at their lowest concentration.

Antimicrobial evaluation against *M. abscessus* and *M. marinum* was undertaken using inhouse assays. *M. abscessus* BSG301 and *M. marinum* BSG101 [28] are stable bioluminescent derivatives transformed with the integrating plasmid pMV306G13ABCDE [34]. This allows light production to be used as a surrogate for bacterial viability [28–30]. Mycobacterial cultures were grown shaking at 200 rpm in Middlebrook 7H9 broth (Fort Richard, Auckland, New Zealand) supplemented with 10% Middlebrook ADC enrichment media (Fort Richard, Auckland, New Zealand), 0.4% glycerol (Sigma-Aldrich, St. Louis, MO, USA), and 0.05% tyloxapol (Sigma-Aldrich, St. Louis, MO, USA). *M. abscessus* was grown at 37 °C and *M. marinum* at 28 °C. Cultures were grown until they reached the stationary phase (approximately 3–5 days for *M. abscessus* BSG301 and 7–10 days for *M. marinum* BSG101) and then diluted in Mueller Hinton broth II (MHB) (Fort Richard, Auckland, New Zealand) supplemented with 10% Middlebrook ADC enrichment media and 0.05% tyloxapol to give an optical density at 600 nm (OD600) of 0.001, which is the equivalent of ~10^6 bacteria per mL. Pure compounds were dissolved in DMSO and added in duplicate to the wells of a black 96-well plate (Nunc, Thermo Scientific, Waltham, MA, USA) at doubling dilutions with a maximum concentration of 128 μg/mL. Then, 50 μL of diluted bacterial culture was added to each well of the compound containing plates giving final compound concentrations of 0–64 μg/mL and a cell density of ~5×10^5 CFU/mL. Rifampicin (Sigma-Aldrich, St. Louis, MO, USA) was used as positive control at 1000 μg/mL for *M. abscessus* and 10 μg/mL for *M. marinum*. Between measurements, plates were covered, placed in a plastic box lined with damp paper towels, and incubated with shaking at 100 rpm at 37 °C for *M. abscessus* and 28 °C for *M. marinum*. Bacterial luminescence was measured at regular intervals over 72 h using a Victor X-3 luminescence plate reader (PerkinElmer, Boston, MA, USA) with an integration time of 1 s. We defined the MIC as causing a 1-log reduction in light production, as previously described [29]. More detailed protocols are available at protocols.io (accessed on 19 November 2021) [35,36].

3.7. Statistical Analysis

Statistical analysis was performed as described using GraphPad Prism version 8.4.3.

4. Conclusions

Investigation of the secondary metabolites of two strains of *Penicillium* led to the isolation of five known natural products. Antimicrobial screening of selected natural products identified penicillic acid (**1**) and penialdin F (**4**), which exhibited good activity against MRSA with the latter also exhibiting weak antimycobacterial activity against *M. abscessus* and *M. marinum* along with citromycetin (**2**). To the best of our knowledge, this is the first instance where these compounds have been evaluated for antimycobacterial activity. Derivatisation of **1** through hydrogenation led to the synthesis of a novel dihydro (**1a**) derivative, which provided evidence to the existence of an open-chained γ-keto acid tautomer, with purification attempts leading to ring closure of **1a**, forming the lactone **1b**. Although none of the natural products or their derivatives exhibited significant antimicrobial activity to be of interest as lead compounds for antibiotic development, the ICMP collection remains a vast untapped resource for novel bioactive natural products worthy of further exploration.

Supplementary Materials: The following are available online, Figures S1–S10: ^{1}H and ^{13}C NMR spectra of **1**, **1a**, and **1c** and the ^{1}H NMR spectrum of **1b** as well as the ^{13}C NMR spectrum of 1-^{13}C labelled **1**.

Author Contributions: Conceptualization and supervision, S.W. and B.R.C.; methodology and validation, M.M.C., N.S.L.N., S.G., S.J.v.d.P. and A.G.; formal analysis, investigation, and data curation, S.G., N.S.L.N. and M.M.C.; resources, B.S.W.; writing—original draft preparation, M.M.C.; writing—review and editing, M.M.C., B.R.C., and S.W.; funding acquisition, S.W. All authors have read and agreed to the published version of the manuscript.

Funding: This research was funded by Cure Kids, NZ Carbon Farming, the Maurice Wilkins Centre for Molecular Biodiscovery, and donations from the New Zealand public.

Institutional Review Board Statement: Not applicable.

Informed Consent Statement: Not applicable.

Acknowledgments: We gratefully acknowledge the iwi Te Kawerau ā Maki for permission to collect samples of kauri roots on their whenua (land). We would like to thank Michael Schmitz and Tony Chen for their assistance with the NMR and mass spectrometric data and the Community for Antimicrobial Drug Discovery (CO-ADD), funded by the Wellcome Trust (UK) and The University of Queensland (Australia), for carrying out the biological activity and toxicity testing of the pure compounds.

Conflicts of Interest: The authors declare no conflict of interest.

Sample Availability: Samples of compounds **1–4** are available from the authors.

References

1. Johnston, P.R.; Weir, B.S.; Cooper, J.A. Open Data on Fungi and Bacterial Plant Pathogens in New Zealand. *Mycology* **2017**, *8*, 59–66. [CrossRef]
2. Lobanovska, M.; Pilla, G. Penicillin's Discovery and Antibiotic Resistance: Lessons for the Future? *Yale J. Biol. Med.* **2017**, *90*, 135–145. [PubMed]
3. Houbraken, J. Fleming's Penicillin Producing Strain Is Not Penicillium Chrysogenum but *P. Rubens*. *Ima Fungus* **2011**, *2*, 87–95. [CrossRef] [PubMed]
4. Visagie, C.M.; Houbraken, J.; Frisvad, J.C.; Hong, S.-B.; Klaassen, C.H.W.; Perrone, G.; Seifert, K.A.; Varga, J.; Yaguchi, T.; Samson, R.A. Identification and Nomenclature of the Genus Penicillium. *Stud. Mycol.* **2014**, *78*, 343–371. [CrossRef]
5. Houbraken, J.; Visagie, C.M.; Meijer, M.; Frisvad, J.C.; Busby, P.E.; Pitt, J.I.; Seifert, K.A.; Louis-Seize, G.; Demirel, R.; Yilmaz, N.; et al. A Taxonomic and Phylogenetic Revision of Penicillium Section Aspergilloides. *Stud. Mycol.* **2014**, *78*, 373–451. [CrossRef]
6. Visagie, C.M.; Renaud, J.B.; Burgess, K.M.N.; Malloch, D.W.; Clark, D.; Ketch, L.; Urb, M.; Louis-Seize, G.; Assabgui, R.; Sumarah, M.W.; et al. Fifteen New Species of *Penicillium*. *Pers. Int. Mycol. J.* **2016**, *36*, 247–280. [CrossRef]
7. Pohland, A.E.; Schuller, P.L.; Steyn, P.S.; Van Egmond, H.P. Physicochemical data for some selected mycotoxins. *Pure Appl. Chem.* **1982**, *54*, 2219–2284. [CrossRef]
8. Li, H.-J.; Cai, Y.-T.; Chen, Y.-Y.; Lam, C.-K.; Lan, W.-J. Metabolites of Marine Fungus *Aspergillus* sp. Collected from Soft Coral Sarcophyton tortuosum. *Chem. Res. Chin. U.* **2010**, *26*, 415–419.
9. Capon, R.J.; Stewart, M.; Ratnayake, R.; Lacey, E.; Gill, J.H. Citromycetins and Bilains A–C: New Aromatic Polyketides and Diketopiperazines from Australian Marine-Derived and Terrestrial *Penicillium* Spp. *J. Nat. Prod.* **2007**, *70*, 1746–1752. [CrossRef]
10. Jouda, J.-B.; Kusari, S.; Lamshöft, M.; Mouafo Talontsi, F.; Douala Meli, C.; Wandji, J.; Spiteller, M. Penialidins A–C with Strong Antibacterial Activities from *Penicillium* Sp., an Endophytic Fungus Harboring Leaves of *Garcinia nobilis*. *Fitoterapia* **2014**, *98*, 209–214. [CrossRef]
11. Cheng, X.; Yu, L.; Wang, Q.; Ding, W.; Chen, Z.; Ma, Z. New Brefeldins and Penialidins from Marine Fungus *Penicillium Janthinellum* DT-F29. *Nat. Prod. Res.* **2018**, *32*, 282–286. [CrossRef]
12. Yuan, C.; Wang, H.-Y.; Wu, C.-S.; Jiao, Y.; Li, M.; Wang, Y.-Y.; Wang, S.-Q.; Zhao, Z.-T.; Lou, H.-X. Austdiol, Fulvic Acid and Citromycetin Derivatives from an Endolichenic Fungus, *Myxotrichum* Sp. *Phytochem. Lett.* **2013**, *6*, 662–666. [CrossRef]
13. Kumla, D.; Pereira, J.A.; Dethoup, T.; Gales, L.; Freitas-Silva, J.; Costa, P.M.; Lee, M.; Silva, A.M.S.; Sekeroglu, N.; Pinto, M.M.M.; et al. Chromone Derivatives and Other Constituents from Cultures of the Marine Sponge-Associated Fungus *Penicillium Erubescens* KUFA0220 and Their Antibacterial Activity. *Mar. Drugs* **2018**, *16*, 289. [CrossRef] [PubMed]
14. Liu, Y.; Ruan, Q.; Jiang, S.; Qu, Y.; Chen, J.; Zhao, M.; Yang, B.; Liu, Y.; Zhao, Z.; Cui, H. Cytochalasins and Polyketides from the Fungus *Diaporthe* Sp. GZU-1021 and Their Anti-Inflammatory Activity. *Fitoterapia* **2019**, *137*, 104187. [CrossRef]
15. Fu, J.; Hu, L.; Shi, Z.; Sun, W.; Yue, D.; Wang, Y.; Ma, X.; Ren, Z.; Zuo, Z.; Peng, G.; et al. Two Metabolites Isolated from Endophytic Fungus *Coniochaeta* Sp. F-8 in *Ageratina Adenophora* Exhibit Antioxidative Activity and Cytotoxicity. *Nat. Prod. Res.* **2021**, *35*, 2840–2848. [CrossRef]
16. Gou, X.; Tian, D.; Wei, J.; Ma, Y.; Zhang, Y.; Chen, M.; Ding, W.; Wu, B.; Tang, J. New Drimane Sesquiterpenes and Polyketides from Marine-Derived Fungus *Penicillium* Sp. TW58-16 and Their Anti-Inflammatory and α-Glucosidase Inhibitory Effects. *Mar. Drugs* **2021**, *19*, 416. [CrossRef] [PubMed]
17. Alsberg, C.L.; Black, O.F. *Contributions to the Study of Maize Deterioration: Biochemical and Toxicological Investigations of Penicillium Puberulum and Penicillium Stoloniferum*; G.P.O.: Washington, DC, USA, 1913; Volume 270.
18. Birkinshaw, J.H.; Oxford, A.E.; Raistrick, H. Studies in the Biochemistry of Micro-Organisms. *Biochem. J.* **1936**, *30*, 394–411. [CrossRef]
19. Frisvad, J.C. A Critical Review of Producers of Small Lactone Mycotoxins: Patulin, Penicillic Acid and Moniliformin. *World Mycotoxin, J.* **2018**, *11*, 73–100. [CrossRef]
20. Raphael, R.A. Compounds Related to Penicillic Acid; Synthesis of Dihydropenicillic Acid. *J. Chem. Soc.* **1947**, 805–808. [CrossRef] [PubMed]
21. Shaw, E. A Synthesis of Protoanemonin. The Tautomerism of Acetylacrylic Acid and of Penicillic Acid. *J. Am. Chem. Soc.* **1946**, *68*, 2510–2513. [CrossRef]
22. Shan, R.; Stadler, M.; Anke, H.; Sterner, O. The Reactivity of the Fungal Toxin Papyracillic Acid. *Tetrahedron* **1997**, *53*, 6209–6214. [CrossRef]
23. Birch, A.J.; Blance, G.E.; Smith, H. Studies in Relation to Biosynthesis. Part XVIII. Penicillic Acid. *J. Chem. Soc.* **1958**, 4582–4583. [CrossRef]
24. Bentley, R.; Keil, J.G. Tetronic Acid Biosynthesis in Molds. II. Formation of Penicillic Acid in *Penicillium Cyclopium*. *J. Biol. Chem.* **1962**, *237*, 867–873. [CrossRef]
25. Olivigni, F.J.; Bullerman, L.B. A Microbiological Assay for Penicillic Acid. *J. Food. Prot.* **1978**, *41*, 432–434. [CrossRef]

26. Geiger, W.B.; Conn, J.E. The Mechanism of the Antibiotic Action of Clavacin and Penicillic Acid. *J. Am. Chem. Soc.* **1945**, *67*, 112–116. [CrossRef]
27. EUCAST: Clinical Breakpoints and Dosing of Antibiotics. Available online: https://www.eucast.org/clinical_breakpoints/ (accessed on 23 December 2021).
28. Dalton, J.P.; Uy, B.; Okuda, K.S.; Hall, C.J.; Denny, W.A.; Crosier, P.S.; Swift, S.; Wiles, S. Screening of Anti-Mycobacterial Compounds in a Naturally Infected Zebrafish Larvae Model. *J. Antimicrob. Chemother.* **2017**, *72*, 421–427. [CrossRef] [PubMed]
29. Dalton, J.P.; Uy, B.; Phummarin, N.; Copp, B.R.; Denny, W.A.; Swift, S.; Wiles, S. Effect of Common and Experimental Anti-Tuberculosis Treatments on Mycobacterium tuberculosis Growing as Biofilms. *Peer J.* **2016**, *4*, e2717. [CrossRef]
30. Grey, A.B.J.; Cadelis, M.M.; Diao, Y.; Park, D.; Lumley, T.; Weir, B.S.; Copp, B.R.; Wiles, S. Screening of Fungi for Antimycobacterial Activity Using a Medium-Throughput Bioluminescence-Based Assay. *Front. Microbiol.* **2021**, *12*, 739995. [CrossRef]
31. Jouda, J.B.; Mawabo, I.K.; Notedji, A.; Mbazoa, C.D.; Nkenfou, J.; Wandji, J.; Nkenfou, C.N. Anti-mycobacterial activity of polyketides from *Penicillium* sp. endophyte isolated from Garcinia nobilis against Mycobacteriumsmegmatis. *Int. J. Mycobacteriol.* **2016**, *5*, 192–196. [CrossRef]
32. Oh, S.-Y.; Park, M.S.; Lim, Y.W. The Influence of Microfungi on the Mycelial Growth of Ectomycorrhizal Fungus *Tricholoma Matsutake*. *Microorganisms* **2019**, *7*, 169. [CrossRef]
33. Blaskovich, M.A.T.; Zuegg, J.; Elliott, A.G.; Cooper, M.A. Helping Chemists Discover New Antibiotics. *ACS Infect. Dis.* **2015**, *1*, 285–287. [CrossRef] [PubMed]
34. Andreu, N.; Zelmer, A.; Fletcher, T.; Elkington, P.T.; Ward, T.H.; Ripoll, J.; Parish, T.; Bancroft, G.J.; Schaible, U.; Robertson, B.D.; et al. Optimisation of Bioluminescent Reporters for Use with Mycobacteria. *PLoS ONE* **2010**, *5*, e10777. [CrossRef] [PubMed]
35. Wiles, S.; Grey, A. Bioluminescence-Based Minimum Inhibitory Concentration (MIC) Testing of Pure Compounds Isolated from Fungi against *Mycobacterium abscessus*. 2021. Available online: Dx.doi.org/10.17504/protocols.io.bumcnu2w (accessed on 30 December 2021).
36. Wiles, S.; Grey, A. Bioluminescence-Based Minimum Inhibitory Concentration (MIC) Testing of Pure Compounds Isolated from Fungi against *Mycobacterium marinum*. 2021. Available online: Dx.doi.org/10.17504/protocols.io.3x7gprn (accessed on 30 December 2021).

Review

Recently Discovered Secondary Metabolites from *Streptomyces* Species

Heather J. Lacey [1,2,*] and Peter J. Rutledge [1,*]

[1] School of Chemistry, The University of Sydney, Camperdown, Sydney, NSW 2006, Australia
[2] Microbial Screening Technologies, Smithfield, Sydney, NSW 2164, Australia
* Correspondence: hlac5959@uni.sydney.edu.au (H.J.L.); peter.rutledge@sydney.edu.au (P.J.R.); Tel.: +61-2-9351-5020 (P.J.R)

Abstract: The *Streptomyces* genus has been a rich source of bioactive natural products, medicinal chemicals, and novel drug leads for three-quarters of a century. Yet studies suggest that the genus is capable of making some 150,000 more bioactive compounds than all *Streptomyces* secondary metabolites reported to date. Researchers around the world continue to explore this enormous potential using a range of strategies including modification of culture conditions, bioinformatics and genome mining, heterologous expression, and other approaches to cryptic biosynthetic gene cluster activation. Our survey of the recent literature, with a particular focus on the year 2020, brings together more than 70 novel secondary metabolites from *Streptomyces* species, which are discussed in this review. This diverse array includes cyclic and linear peptides, peptide derivatives, polyketides, terpenoids, polyaromatics, macrocycles, and furans, the isolation, chemical structures, and bioactivity of which are appraised. The discovery of these many different compounds demonstrates the continued potential of *Streptomyces* as a source of new and interesting natural products and contributes further important pieces to the mostly unfinished puzzle of Earth's myriad microbes and their multifaceted chemical output.

Keywords: biosynthesis; macrolides; natural products; peptides; polyketides; secondary metabolites; *Streptomyces*; structure elucidation; terpenoids

Citation: Lacey, H.J.; Rutledge, P.J. Recently Discovered Secondary Metabolites from *Streptomyces* Species. *Molecules* **2022**, *27*, 887. https://doi.org/10.3390/molecules27030887

Academic Editor: Jacqueline Aparecida Takahashi

Received: 23 December 2021
Accepted: 22 January 2022
Published: 28 January 2022

Publisher's Note: MDPI stays neutral with regard to jurisdictional claims in published maps and institutional affiliations.

1. Introduction

Analysis of *Streptomyces* derived natural product discoveries from 1947 to 1997 indicates that, in terms of the number of publications, the discipline peaked in the late 1960s before tapering off through the 1970s and 1980s and significantly decreasing in the 1990s [1]. Factors contributing to this decline include the development of combinatorial chemistry and associated high-throughput screening (HTS) techniques, a reduction in funding for traditional drug discovery pipelines, and the greater profitability of quality-of-life drugs. Had new bioinformatics technologies and methods such as genome mining not emerged to reinvigorate the field, the declining trend could have seen a cessation of work to exploit *Streptomyces* as a source of new bioactives by 2020 [1]. This is in spite of studies that conservatively estimate that the genus can synthesise some 150,000 more antimicrobial compounds than those currently known, suggesting that *Streptomyces* is far from an exhausted resource [1,2].

A Clarivate Web of Science search was undertaken to gauge the recent level of documented drug discovery research activity in the genus [3], using the terms *Streptomyces*, *isolation*, *new compound* and *novel compound* in the All Fields parameter of the Web of Science search engine. A review of the number of publications dedicated to novel natural product discovery from the *Streptomyces* genus over the last five years reveals a reasonably consistent level of publication on the subject (Figure 1). Comparing the raw numbers of this survey with the Watve study (2001) shows that the current levels of research output (in terms of publication numbers) are comparable to those at the start and end of the Golden Age of natural products [1]. This suggests that *Streptomyces* natural products research publications did not dissipate by 2020 as earlier trajectories may have indicated, nor has there been a major upswing in publications per year in the last five years [1,3]. A proportion of the publications returned in the Web of Science search (on average 7% per year) were excluded from this count because the research they describe did not yield verified novel compounds as isolated secondary metabolites of a *Streptomyces* species. Similarly, reports of partial characterisation were also excluded as the natural products described therein could not be verified as novel. This indicates that a literature count with strict eligibility criteria likely underestimates the true interest in *Streptomyces* drug discovery by focusing on verified novel compound elucidation, rather than simply research activity.

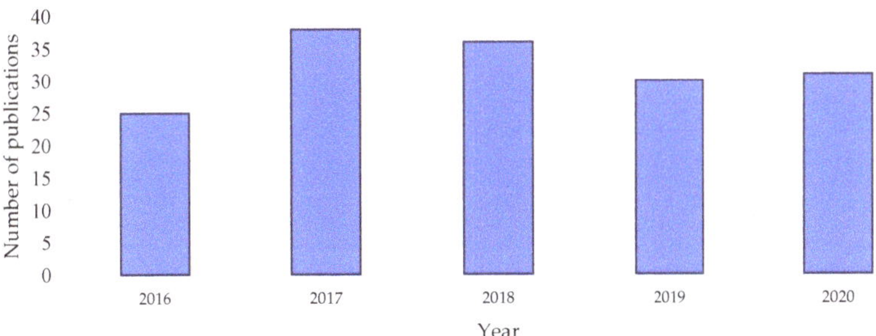

Figure 1. Drug discovery publications 2016–2020 focused on the isolation of *Streptomyces* natural products [3].

An insight analysis was conducted on scientific reports pertaining to the discovery of new *Streptomyces* secondary metabolites published in 2020. The search returned 43 references relevant to the search terms. Filtering resulted in 32 research papers that met the requirements of the review.

Each of these reports described a unique *Streptomyces* species or strain, and a total of 74 novel compounds were reported across the 32 papers surveyed. Further analysis revealed that marine and soil environments were the equal most common isolation locations, with 56% of these publications citing a *Streptomyces* strain isolated from either a marine environment or terrestrial soil (Figure 2). Notably, a significant proportion (19%, n = 6) of publications did not specify the environmental source of the *Streptomyces* strain; where this occurred, the publication often focused on high performance liquid chromatography-UV-diode array detection (HPLC-DAD) [4], LC-MS [5], or genome mining [6,7] as a drug discovery guidance strategy.

The preliminary analysis of novel secondary metabolites produced by these *Streptomyces* strains also revealed that the largest proportion (39%) of compounds was isolated from marine-derived *Streptomyces*, followed by terrestrial soil (27%) (Figure 3). A further eleven secondary metabolites were isolated from unspecific environments. The compounds reported possess a wide range of chemical scaffold variability and include cyclic and linear peptides, terpenoids, macrolactams, macrolides, glycosides, polyaromatics, and linear polyketides.

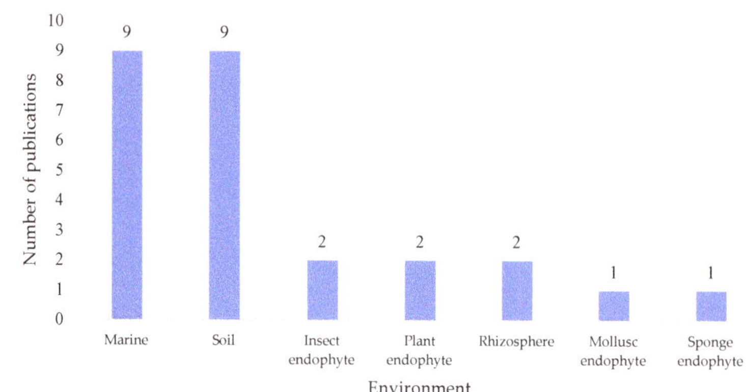

Figure 2. Number of publications for each of the isolation environments of secondary-metabolite-producing *Streptomyces* strains reported in 2020. A further 6 publications did not specify the environmental source of the strain under study.

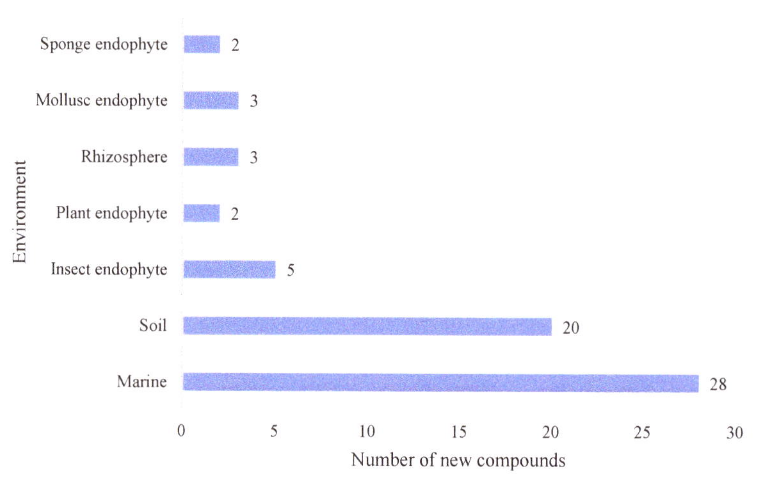

Figure 3. The environmental distribution of *Streptomyces* secondary metabolites isolated in 2020. A further 11 compounds originated from organisms of unspecified origin.

Though *Streptomyces* species no longer generate the same level of research output as at their peak in the 1960s, the genus has continued to be the subject of interest and research activity. New genome mining strategies in addition to classic bioprospecting and bioassay-guided isolation have contributed to the plethora of unique chemical scaffolds that continue to characterise both the *Streptomyces* genus and natural products more generally, as an important source of drug leads.

2. Cyclic Peptides

In 2020, ten novel cyclic peptides were reported across seven publications [4,6,8–12]. The size of the peptide core ranged from two to seven amino acids, though three of these compounds, ulleungamide C (**1**) [11] and viennamycins A and B (**2** and **3**) (Figure 4) [8], are depsipeptides, and two more, NOS-V1 (**4**) and NOS-V2 (**5**) (Figure 5), possess the typical di-macrocycle composition of a thiopeptide nosiheptide [6].

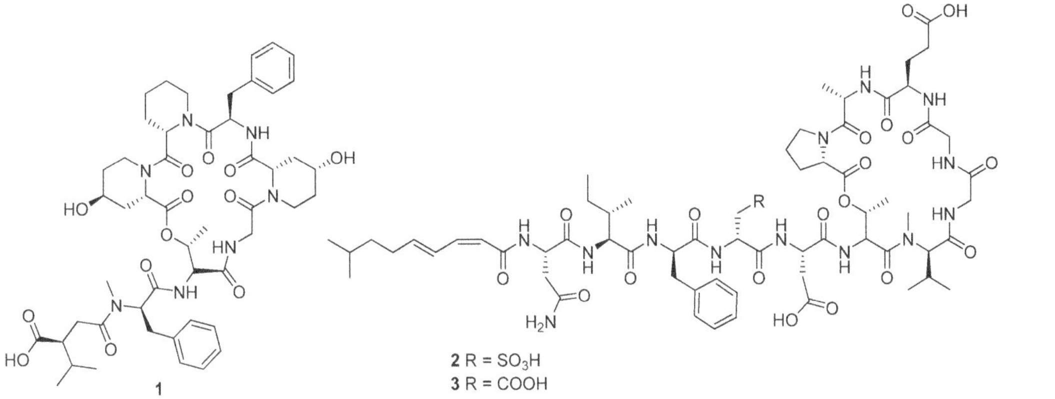

Figure 4. The molecular structures for ulleungamide C (**1**) and viennamycins A and B (**2** and **3**) [8,11].

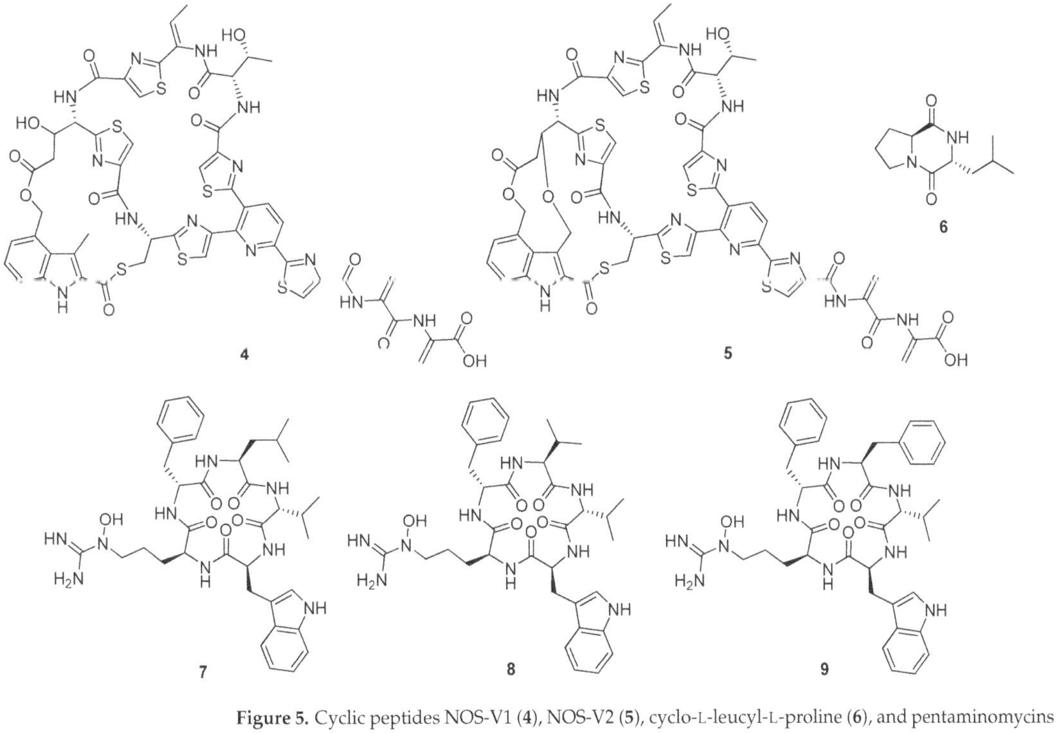

Figure 5. Cyclic peptides NOS-V1 (**4**), NOS-V2 (**5**), cyclo-L-leucyl-L-proline (**6**), and pentaminomycins C–E (**7–9**) [6,9,10].

These compounds originate from both non-ribosomal peptide synthetases (NRPSs) and ribosomal synthesis with post-translational modification. The most common bioactivity observed for this group of compounds was antibacterial activity, followed by cytotoxicity and antifungal proliferation. Pentaminomycin C (**7**) showed activity against Gram-positive bacteria down to 16 μg mL^{-1} [4], but no activity against Gram-negative bacteria. In comparison, NOS-V1 (**4**) and NOS-V2 (**5**) showed activity against Gram-positive and Gram-negative cultures with a more potent effect against Gram-positive than Gram-negative bacteria [6].

The cyclic dipeptide L-leucyl-L-proline (**6**) showed activity against a range of bacterial and fungal targets in inhibition zone studies [10]. The antibacterial activity of the viennamycins **2** and **3** was identified in tests against *S. aureus* 209P and *B. cereus* IP5812, in which these compounds were found to have comparable activity to daptomycin and vancomycin, though no activity was identified against Gram-negative bacteria [8]. Ulleungamide C **1** displayed cytotoxic activity in vitro. In contrast, in vitro studies observed the protective effects of pentaminomycins C (**7**) and D (**8**) (Figure 5) against menadione-induced cytotoxicity [9].

3. Peptide Derivatives and Linear Peptides

Three linear peptidic compounds were reported in three publications during 2020. Spongiicolazolicins A (**10**) and B (**11**) (Figure 6) are ribosomally synthesised and post-translationally modified azole-containing peptides isolated from the marine *Streptomyces* sp. CWH03 (NBRC 114659) [13]. Comprising twenty-nine (spongiicolazolicin A) and twenty-eight (spongiicolazolicin B) amino acids, the structures were deduced from 2D NMR, MS/MS experiments, and the sequence of the biosynthetic gene cluster (BGC). Despite studies of similar compounds, such as microcin B17 [14], finding them to have antibacterial bioactivity, biological assay data did not indicate similar activity for the spongiicolazolicins against Gram-positive or Gram-negative species.

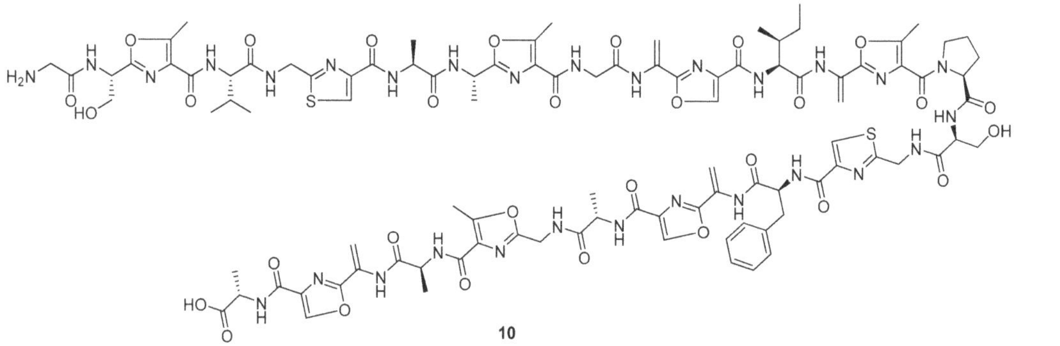

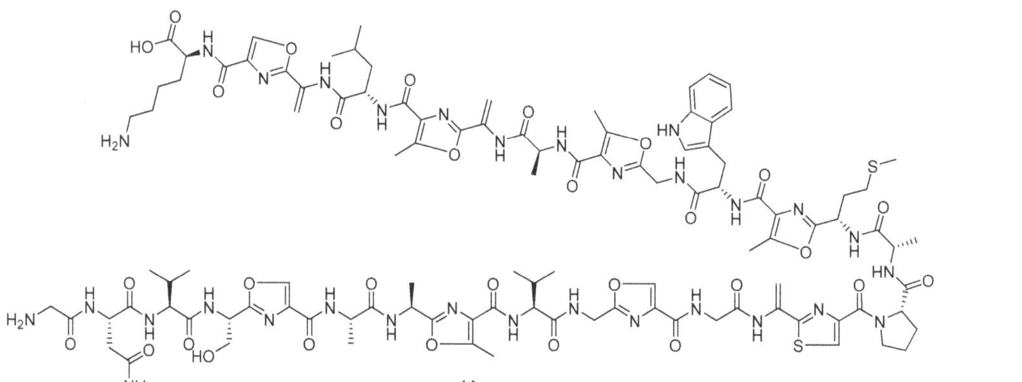

Figure 6. The structures of spongiicolazolicins A (**10**) and B (**11**), ribosomally synthesised and post-translationally modified azole-containing peptides [13].

Legonimide (**12**, Figure 7) is an indole alkaloid composed of modified phenylalanine and tryptophan amino acids, and was isolated from soil-derived *Streptomyces* sp.

CT37 [15]. The authors proposed that this compound is biosynthesised via a condensation reaction between phenyl-acetyl-CoA and indole-3-acetamide. The structure of legonimide was determined from a combination of 1D and 2D NMR data, and biological profiling showed that the compound has moderate antifungal activity against *Candida albicans* (MIC 21.5 µg mL^{-1}). The final peptide derivative reported in the review period, 2,3-dihydroxy-1-(indolin-3-yl)butan-1-one (**13**, Figure 7), was isolated from *S. triticiradicis* sp. strain NEAU-H2^{T} [16], collected from the rhizosphere of wheat (*Triticum aestivum* L.). This strain displayed antifungal effects against the mycelial growth of phytopathogenic fungi, though this activity could not be linked to the novel compound in subsequent tests [16].

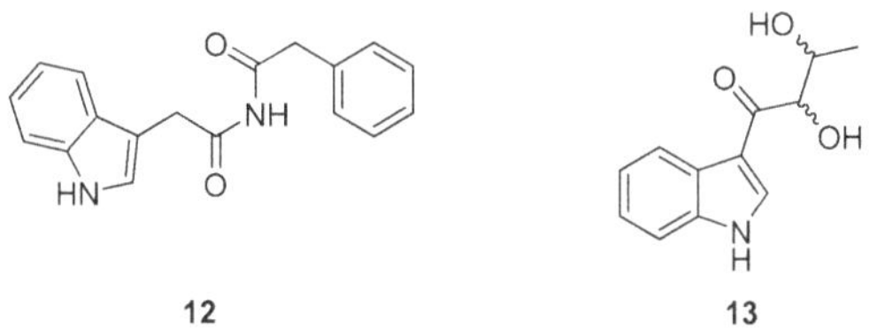

Figure 7. The structures of legonimide (**12**) and 2,3-dihydroxy-1-(indolin-3-yl)butan-1-one (**13**) [15,16].

4. Linear Polyketides

An HTS experiment looking for inhibitors of coenzyme A biosynthesis led to the isolation and spectroscopic characterisation of six new phenolic lipids, adipostatins E–J (**14–19**), from the marine-derived *S. blancoensis* (Figure 8) [17]. These novel compounds showed activity against both Gram-positive and Gram-negative bacteria: *Enterococcus coli*, *E. faecalis*, *Bacillus subtilis*, *B. anthracis*, *Listeria monocytogenes*, and *Streptococcus pneumoniae*. In contrast, the compounds had low to no effect on *Staphylococcus aureus*, *Salmonella enterica*, *Shigella flexneri*, or *Klebsiella pneumoniae*. The differential in vitro activity of adipostatins E–J suggests that the biological activity is mediated by the chain length and terminal branching.

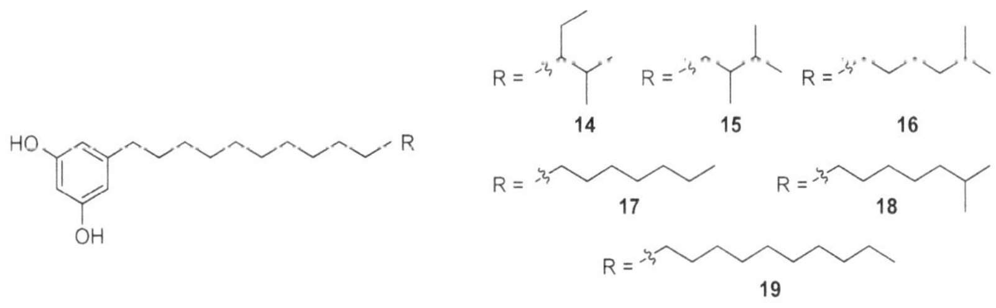

Figure 8. Molecular structures of adipostatins E–J (**14–19**) [17].

Similarly, bioactivity screening for tyrosinase inhibitors led to the serendipitous identification of trichostatic acid B (**20**) from *Streptomyces* sp. CA-129531 as a co-metabolite of the target molecules (Figure 9) [18]. Notably, trichostatic acid B did not display any cytotoxic activity up to 50 mg mL^{-1}. Nor was any antityrosinase activity detected for trichostatic acid B in a mushroom tyrosinase bioassay, despite trichostatin A (**21**) showing strong inhibitory activity [18], suggesting that the hydroxylamine and/or tertiary methyl group is important to its biological activity. Two additional bioassay-driven natural product discoveries were chresdihydrochalcone (**22**) and chresphenylacetone (**23**, Figure 9), isolated from *S. chrestomyceticus* BCC 24770 [19]. Their molecular structures were spectroscopically characterised, and chresdihydrochalcone was assessed for haemolytic, cytotoxic, and antibacterial activity. Notably, the compound showed moderate activity against methicillin-resistant *S. aureus* (MRSA, MIC 25 µg mL^{-1}) and a mutant *E. coli* strain, and stronger activity against *S. aureus* (MIC 18 µg mL^{-1}) and *B. subtilis* (MIC 10 µg mL^{-1}).

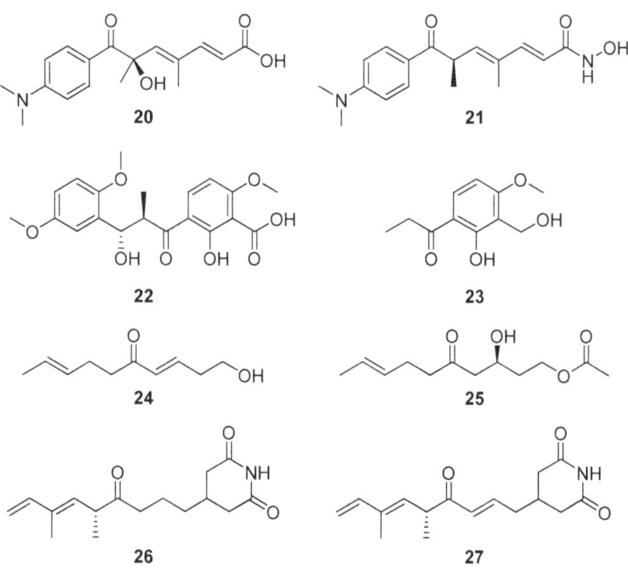

Figure 9. The molecular structures of trichostatic acid B (**20**), trichostatin A (**21**), chresdihydrochalcone (**22**), chresphenylacetone (**23**), (3*E*,8*E*)-1-hydroxydeca-3,8-dien-5-one (**24**), (*S*,*E*)-3-hydroxy-5-oxodec-8-en-1-yl acetate (**25**), 3-(5,7-dimethyl-4-oxo-6*E*,8-nonadienyl)-glutarimide (**26**), and 3-(5,7-dimethyl-4-oxo 2*E*,6*E*,8-nonatrienyl)-glutarimide (**27**) [5,18–20].

From a group of six compounds isolated from *Streptomyces* sp. MJM3055 and spectroscopically characterised, two compounds, (3*E*,8*E*)-1-hydroxydeca-3,8-dien-5-one (**24**) and (*S*,*E*)-3-hydroxy-5-oxodec-8-en-1-yl acetate (**25**), were identified as novel natural products via a Jurkat cell antiproliferation screening study (Figure 9) [20]. Using LCMS-guided screening of *Streptomyces* sp. W3002, Lee et al. purified and characterised the new deoxy-streptimidone compounds **26** and **27** (Figure 9) along with two glutarimide ring-opened derivatives of **27** [5]. The new compounds, 3-(5,7-dimethyl-4-oxo-6*E*,8-nonadienyl)-glutarimide (**26**) and 3-(5,7-dimethyl-4-oxo 2*E*,6*E*,8-nonatrienyl)-glutarimide (**27**), possess the characteristic glutarimide ring and alkyl chain and showed no cytotoxic activities against human cervical carcinoma, human hepatoma, or myeloid leukemia up to 100 μg mL^{-1} [21]. Analysis of the putative BGC of both glutarimide derivatives revealed that apart from module five, which contains the C-terminal thioesterase domain, the other four of the first five modules showed a strong similarity to those of the BGC that encodes 9-methylstreptimidone in *S. himastatinicus* [22]. It was proposed that the instability of the dehydratase accessory reactions catalysed by module three was responsible for the absence of the hydroxyl group from the alkyl chain (when these new structures are compared to 9-methylstreptimidone), while an oxidoreductase transforms **26** to **27** [5].

5. Terpenoids

Ten novel terpenoid natural products were isolated from four *Streptomyces* species in 2020. Notably, half of these compounds are napyradiomycin derivatives. Napyradiomycins are a large group of meroterpenoids that incorporate a halogenated semi-naphthoquinone moiety. This group of over fifty compounds is divided into three sub-groups based on side-chain characteristics: type A napyradiomycins possess a linear terpenoid chain, type B have a cyclohexane ring, and type C possess a 14-membered ring [23]. In the course of reassessing secondary metabolite biosynthesis by marine-derived *Streptomyces*, Carretero-Molina et al. identified four compounds from *Streptomyces* sp. CA-271078: one each of types A (**28**) and C (**31**), and two of which were type B napyradiomycins (**29** and **30**) (Figure 10) [23].

These compounds were spectroscopically characterised and their stereochemistry was inferred from biosynthetic relationships with previously reported napyradiomycins. These compounds all displayed weak antibacterial activity against MRSA and cytotoxicity to human liver carcinoma cell lines.

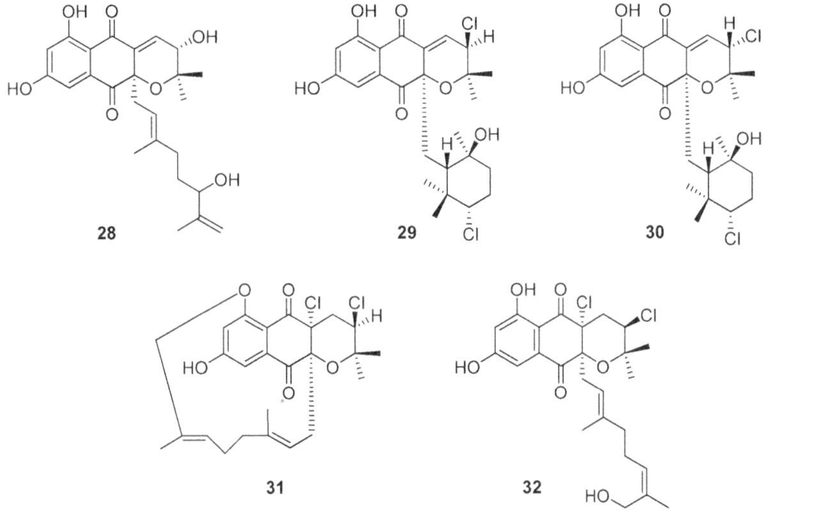

Figure 10. The molecular structures of four unnamed napyradiomycins (**28–31**) and 16*Z*-19-hydroxynapyradiomycin A1 (**32**) first reported in 2020 [23,24].

The fifth compound was also a type A napyradiomycin, and one of ten isolated from marine *Streptomyces* sp. YP127: 16*Z*-19-hydroxynapyradiomycin A1 (**32**) [24]. This structure was flagged as a potential drug lead through an Nrf-2-activating efficacy bioassay, which indicated that this compound would modulate the activity of Nrf 2, an essential component in the expression of antioxidant defence genes.

Three structurally related naphthoquinones were isolated from the marine-derived *Streptomyces* sp. B9173: flaviogeranins B1 (**33**), B (**34**), and D (**35**), the latter two of which are meroterpenoids (Figure 11). These compounds differ in their prenylation pattern and presence of amino and methyl groups on the hydroxynaphthoquinone core. Flaviogeranins B1 and B showed weak to moderate activity against *S. aureus* and *M. smegmatis* as well as human alveolar basal epithelial adenocarcinoma and human cervical cancer cells. In contrast, flaviogeranin D displayed stronger activity against *S. aureus* (MIC 9.2 µg mL^{-1}) and *M. smegmatis* (MIC 5.2 µg mL^{-1}), the alveolar basal epithelial adenocarcinoma (IC$_{50}$ 0.6 µg mL^{-1}) and human cervical cancer cell lines in vitro (IC$_{50}$ 0.4 µg mL^{-1}).

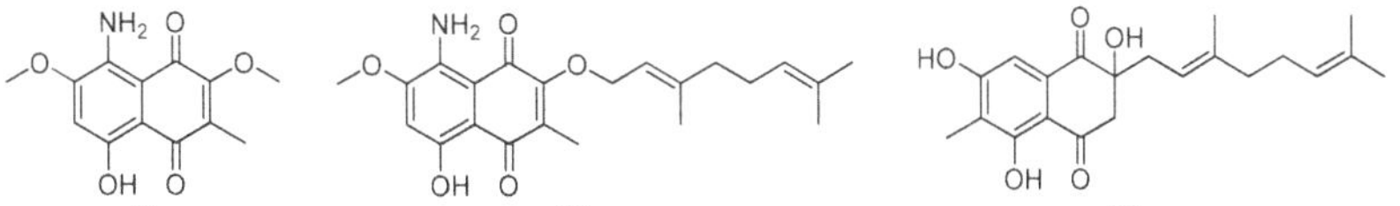

Figure 11. The molecular structures of flaviogeranins B1 (**33**), B (**34**), and D (**35**) [25].

The final terpenoids isolated from the *Streptomyces* genera are two cardinane sesquiterpenes (**36** and **37**) isolated from marine *Streptomyces* sp. ST027706 (Figure 12), a plant endophyte taken from the mangrove *Bruguiera gymnorrihiza* [26]. Cardinanes have primarily been isolated from plant and fungal sources, with only two previous examples of the class being isolated from *Streptomyces* [27,28]. Thus, this finding raises the possibility that bacterial fermentation could be used to obtain cardinane-derived pharmaceuticals.

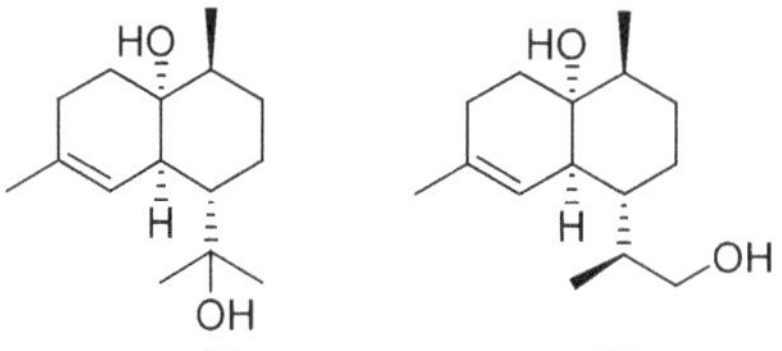

Figure 12. Molecular structures of two cardinanes (**36** and **37**) isolated from an endophytic *Streptomyces* sp. [26].

6. Polyaromatics

In total, eleven polyaromatic compounds were isolated from six *Streptomyces* species in 2020. Through herbicidal bioassay screenings, Kim et al. identified strong activity associated with *Streptomyces* strain KRA17-580 against the weed *Digitaria ciliaris* [29]. Two compounds were spectroscopically characterised, a cinnoline-4-carboxamide, 580-H1 (**38**), and a cinnoline-4-carboxylic acid, 580-H2 (**39**, Figure 13). In further phytotoxicity bioassays, the carboxamide proved to be less potent than the carboxylic acid, and electrolyte leakage assays were used to verify that the observed herbicidal activity was caused by destruction of the cell membrane [29].

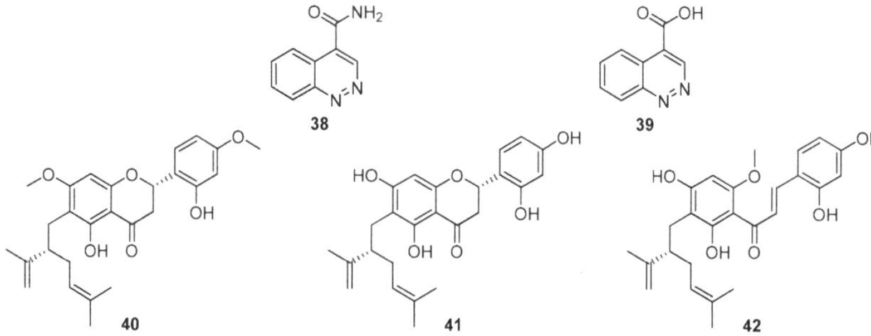

Figure 13. The structure of 580-H1 (**38**), 580-H2 (**39**), (2*S*,2″*S*)-6-lavandulyl-7,4′-dimethoxy-5,2′-dihydroxylflavanone (**40**), (2*S*,2″*S*)-6-lavandulyl-5,7,2′,4′-tetrahydroxylflavanone (**41**), and (2″*S*)-5′-lavandulyl-2′-methoxy-2,4,4′,6′-tetrahydroxylchalcone (**42**) [29,30].

Flavonoids possess a generic structure in which two aromatic rings (A and B) are connected by a heterocyclic pyran. Investigations into bacterial endophytes of the sea sponge *Halichondria panicea* led to the discovery of *Streptomyces* sp. G248 and three lavandulylated flavonoid secondary metabolites: (2*S*,2″*S*)-6-lavandulyl-7,4′-dimethoxy-5,2′-dihydroxylflavanone (**40**), (2*S*,2″*S*)-6-lavandulyl-5,7,2′,4′-tetrahydroxylflavanone (**41**), and (2″*S*)-5′-lavandulyl-2′-methoxy-2,4,4′,6′-tetrahydroxylchalcone (**42**, Figure 13) [30]. Notably, (2*S*,2″*S*)-6-lavandulyl-5,7,2′,4′-tetrahydroxylflavanone showed strong antibacterial activity (IC$_{90}$ 1 µg mL^{-1}) against *E. faecalis*, *S. aureus*, *B. cereus*, *P. aeruginosa*, and *C. albicans*, suggesting that the diol on ring B and the presence of an intact pyran are essential to antibacterial activity [30].

While investigating expression of the nybomycin BGC from *S. albus* subsp. *chlorinus* in heterologous host *S. albus* Del14, Rodriquez Estevez et al. discovered the production of a new metabolite, benzanthric acid (**43**, Figure 14) [31]. This new structure comprises a benzoic acid fused to an anthranilate moiety, and a ^{13}C feeding study indicated that anthranilic acid is incorporated into **43**, but not into the nybomycin scaffold (**44**). Notably, benzanthric acid was not identified in either the origin strain (*chlorinus*) or the unmodified host strain (Del14). Given the absence of a benzoic acid supply from the nybomycin BGC, Rodriquez Estevez et al. proposed that the benzoic acid is supplied by the host strain, possibly through a phenylalanine degradation pathway, and so too the enzyme that catalyses reaction between the anthranilate moiety and benzoic acid. This report indicated that despite the simultaneous production of nybomycin and benzanthric acid in this strain, there are substantial differences in the biosynthetic routes to the two compounds. The new benzanthric acid compound was tested for phytotoxic activity against *Agrostis stolonfera*, though no activity was identified. This study demonstrates an unexpected interplay between a target BGC and host strain, once integrated.

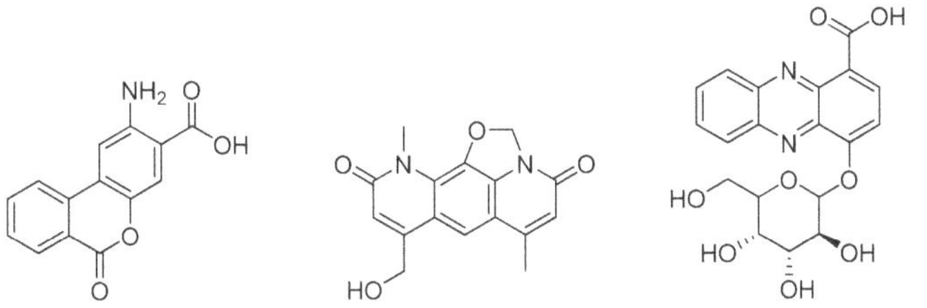

Figure 14. Structures of benzanthric acid (**43**), nybomycin (**44**), and 4-(3*S*,4*R*,5*S*)-3,4,5-trihydroxy-6-(hydroxymethyl) tetrahydro-2Hpyran-2-yloxy)phenazine-1-carboxylic acid (**45**) [31,32].

The second fused tricyclic compound isolated in 2020 was 4-((3*S*,4*R*,5*S*)-3,4,5-trihydroxy-6-(hydroxymethyl) tetrahydro-2Hpyran-2-yloxy)phenazine-1-carboxylic acid (**45**), a phenazine that features a benzoic acid moiety and glucose pendant (Figure 14) [32]. The producing strain *Streptomyces* sp. strain UICC B-92 is an endophyte of the large crane fly *Nephrotoma altissima*. The isolated compound was found to be active against Gram-positive bacteria *S. aureus* and *B. cereus* in disk diffusion assays.

In a successful attempt to highlight the value of chemical exploration of less investigated ecological niches, Voitsekhovskaia et al. isolated anthraquinone derivatives baikalomycins A–C (**46–48**, Figure 15) from the mollusc (*Benedictia baicalensis*) endophyte *Streptomyces* sp. strain IB201691-2A [33]. Baikalomycin A was identified as an aquayamycin derivative modified with β-D-amicetose and two additional hydroxyl groups on C-6a and C-12a of the aglycone. Baikalomycin B **47** is a disaccharide derivative of the same aglycone that incorporates epimeric α-L-amicetose as the *O*-linked second sugar. In baikalomycin C, the A ring of the parent aquayamycin structure is opened to afford an anthraquinone (from rings B, C, and D) with a 3-hydroxy-3-methyl butanoic acid side chain at C4a of the parent aglycone, and a disaccharide comprising β-D-amicetose and α-L-aculose at C9. The stereochemistry within the A ring of baikalomycins A (**46**) and B (**47**) and (by extrapolation) the sidechain of **48** was assigned by comparison with previously isolated compounds with similar biosynthetic origin, while the stereochemistry in the B ring of **46** and **47** remains uncertain. The putative type II polyketide synthase (PKS) BGC possesses typical genes responsible for angucycline core assembly, genes necessary for biosynthesis of the deoxy sugar, and three genes encoding the glycosyltransferase enzymes. Weak antibacterial and moderate cytotoxic activity were identified for baikalomycin C, while baikalomycins A and B only showed weak or no cytotoxic activity against various cell lines.

Two further tetracyclic metabolites were isolated from an endophytic *Streptomyces* species: gardenomycins A and B (**49** and **50**) [34]. *S. bulli* GJA1 was extracted from the bark of *Gardenia jasminoides*, and the molecular structures of the secondary metabolites deduced from spectroscopic analysis. Notably, gardenomycins A and B possess an unprecedented ether bridge between their B and C rings. Kim et al. proposed that an ether bridge was formed from ring opening of an α-oriented epoxide on ring C by a hydroxyl group on ring B [34]. These compounds did not display antiproliferative or antivirulence effects in bioassays.

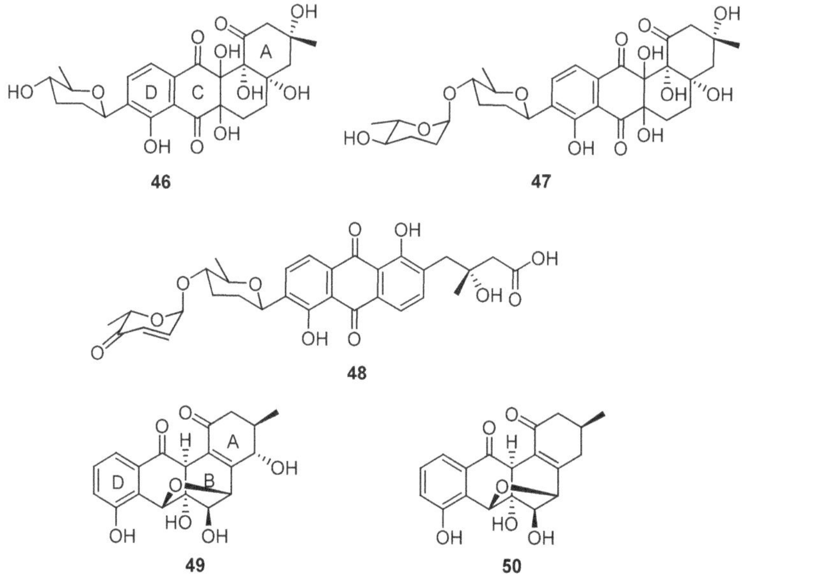

Figure 15. The structures of baikalomycins A–C (46–48) and gardenomycins A and B (49 and 50) [33,34].

7. Macrocycles

Macrocycle discoveries fall into two subgroups: macrolides (macrocyclic esters) and macrolactams (macrocyclic amides). Five macrolides (51–55, Figure 16) were identified while reactivating the quiescent pladienolide B biosynthetic pathway in a laboratory strain of *S. platensis* AS6200 [35]. Unsurprisingly, the laboratory strain showed high similarity (>95% nucleotide similarity with the published *pld* BGC [36]). One notable difference was that the PKS was encoded by five genes in AS6200, while the previously reported PKS was encoded by four genes [35,36]. The molecular structure of these compounds was deduced from spectroscopic data, and the compounds were tested in antibacterial, antifungal, antiprotozoal, and antiproliferation bioassays. The results indicated that all the congeners have less cytotoxic activity than the parent compound pladienolide B.

Six new analogues of oxazole-bearing macrodiolide conglobatin (56) were isolated from the previously unreported Australian *Streptomyces* strain MST-91080 (Figure 17) [37]. Conglobatins B–E (57–62) vary in the distribution of methyl groups around the macrocyclic core, proposed to arise via variable addition of methylmalonyl-CoA extender units during their biosynthesis. Three of the new conglobatins (B1 56, C1 57, and C2 58) displayed enhanced cytotoxicity against NS-1 myeloma cells, IC_{50} = 0.084, 1.05, and 0.45 μg mL^{-1} respectively, an increase in potency relative to the parent compound (IC_{50} = 1.39 μg mL^{-1}).

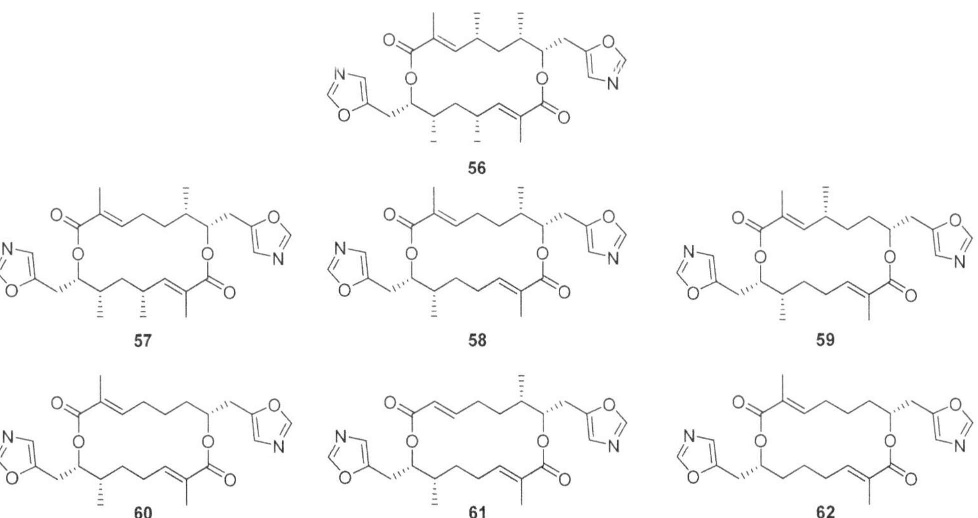

Figure 16. The structures of macrolide natural products **51–55** identified while reactivating the pladienolide B biosynthetic pathway in *S. platensis* AS6200 [35].

Figure 17. Structures of conglobatins **56–62** isolated from *Streptomyces* sp. MST-91080; the parent compound conglobatin (**56**) was previously reported as a secondary metabolite of *Streptomyces conglobatus* ATCC 31005; the others, conglobatins B1 (**57**), C1 (**58**), C2 (**59**), D1 (**60**), D2 (**61**), and E (**62**), were reported for the first time in 2020 [37,38].

Five macrolactams were described in two separate reports (Figure 18). Hashimoto et al. identified a gene cluster containing a type-I PKS with glutamate mutase genes in the genome of *S. rochei* IFO12908 [7]. The BGC was expressed in a host strain derived from *S. avermitilis* and produced a polyene macrolactam **63,** which was subsequently named JBIR-156; its stereochemistry was predicted from the biosynthetic pathway. Given the previous reports of cytotoxicity for related macrolactam structures, the compound was tested on a range of mammalian cells and showed strong activity against human ovarian adenocarcinoma (IC$_{50}$ 9.5 µM) and T lymphoma Jurkat (IC$_{50}$ 3.5 µM) cell lines. The final four compounds are polycyclic tetramate macrolactams and were uncovered using a mixture of gene- and growth-condition-manipulation techniques on a cryptic BGC in the genome of *S. somaliensis* SCSIO ZH66 [39]. Somamycin A (**64**) was identified spectroscopically as a polycyclic tetramate macrolactam, with the remaining compounds identified as the 10-*epi*-hydroxymaltophilin (B, **65**), 10-*epi*-maltophilin (C, **66**), and dihydro 10-*epi*-maltophilin (D, **67**) analogues. These compounds were tested against two plant fungal pathogens and mammalian cell lines, showing antifungal activity and cytotoxicity, with somamycin displaying the strongest antifungal activity against *Fusarium oxysporum* and *Alternaria brassicae* (1.56 and 3.12 µg mL^{-1}, respectively) and human cancer cell lines HCT116 and K562 (IC$_{50}$ 0.6 and 1.5 µM, respectively) [39].

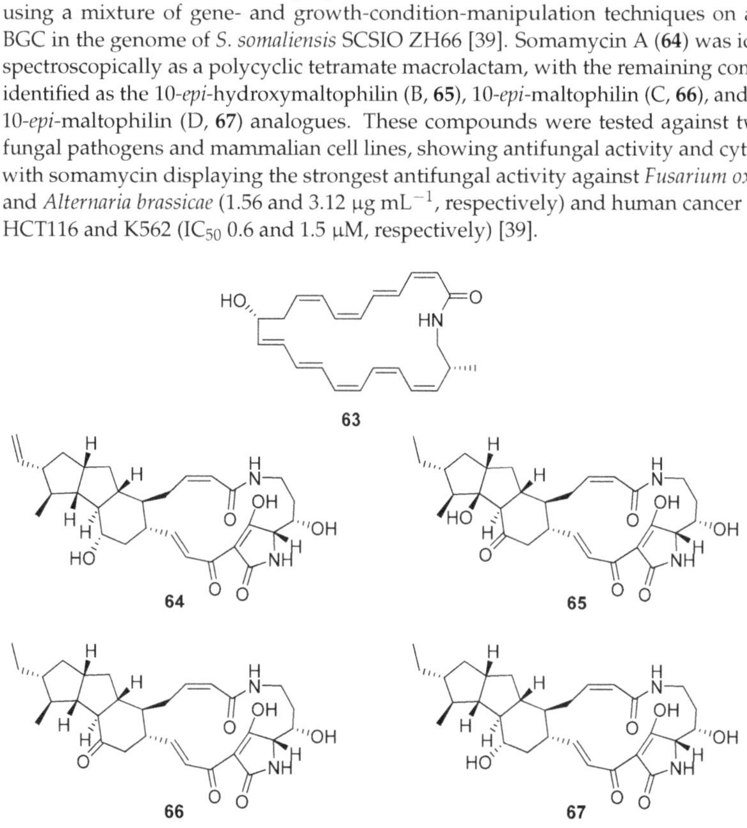

Figure 18. The molecular structures of JBIR-156 (**63**) and somamycins A–D (**64–67**) [7,39].

8. Furans

Five *Streptomyces*-derived secondary metabolites containing modified benzofurans were isolated in 2020 from *S. pratensis* strain KCB-132: (±)-pratenone A (**68** and **69**), and furamycins I–III (**70–72**, Figure 19) [40]. The core of these compounds is a benzofuranone ring system linked to naphthalene. Pratenone A is distinguished by the presence of a 3-methylnaphthalen-1,7-diol moiety, furamycins I and II possess 3,7-dihydroxy-3-methyl-3,4-dihydronaphthalen-1(2H)-one ring systems, and furamycin III is characterised by a methoxy group in place of the carbonyl on the benzofuran(one) portion. These compounds were tested for antibacterial and cytotoxic activity. Furamycin III showed moderate activity against pancreatic cancer cells (IC$_{50}$ 26.0 µg mL^{-1}) and hepatocellular carcinoma (IC$_{50}$ 42.7 µg mL^{-1}) in vitro. Only pratenone A displayed antibacterial activity against *S. aureus* (MIC 8 µg mL^{-1}) [40].

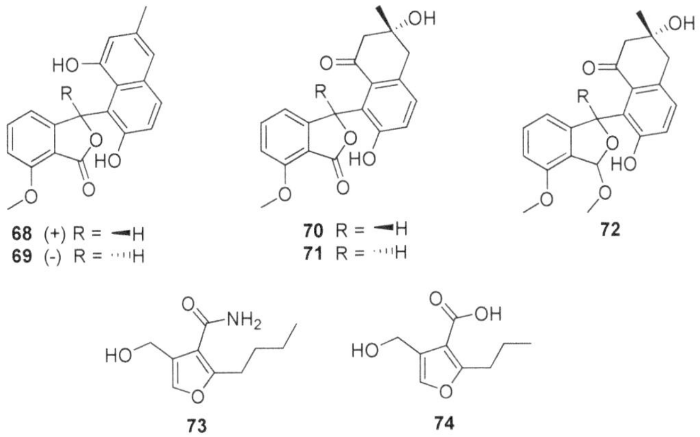

Figure 19. The structures of (±)-pratenone A (**68**, **69**), furamycin I (**70**), furamycin II (**71**), furamycin III (**72**), 2-alkyl-4-hydroxymethylfuran carboxamide (AHFA, **73**), and methylenomycin furan (MMF, **74**) [40,41].

The final furanic compound reported was 2-alkyl-4-hydroxymethylfuran carboxamide (AHFA, **73**), isolated from a soil-derived *Streptomyces* sp. RK44 [41]. AHFA is a homolog of methylenomycin furan (MMF, **74**) and is distinguished by the presence of an amide moiety at C-3 rather than a carboxylic acid (Figure 19). MMFs share a common furan ring structure and are autoregulators that act in the signalling pathway that stimulates production of methylenomycin biosynthesis [42]. Weak antiproliferative activity was observed (EC_{50} 89.6 µM) when AHFA was tested on a melanoma cell line, though no antibacterial or antiplasmodial activities were detected when tested.

9. Conclusions

A review of the natural products isolated from *Streptomyces* in 2020 reveals a diversity of novel chemical structures and a thriving and multifaceted area of drug discovery research. The level of research is on par with the end of the Golden Age of antibiotics on a publications per year basis. During the year, over seventy novel compounds were isolated from *Streptomyces* species using a combination of HPLC-DAD, genomics, and bioassays to guide the isolation of new compounds.

The *Streptomyces* species investigated derive from a wide range of environments, with marine locales being the most investigated, followed by terrestrial soil. Unsurprisingly, this resulted in most new natural products being isolated from marine and soil environments as well. A diverse range of compounds spanning many different chemical classes have been identified, including cyclic and linear peptides, linear polyketides, terpenoids, polyaromatics, macrocycles, and furans. Interestingly, the biological activity identified for these compounds was generally limited, and not tested across a consistent set of pathogenic species or cell lines, leaving an incomplete activity profile for the compound set.

Overall, the discovery of these many and different compounds demonstrates the continued potential of *Streptomyces* as a source of new and interesting natural products and contributes important pieces to the largely unfinished puzzle of Earth's myriad microbes and their multifaceted chemical output.

Author Contributions: H.J.L. wrote the first draft of the manuscript. H.J.L. and P.J.R. edited and revised the manuscript together, wrote additional sections, and created the figures. All authors have read and agreed to the published version of the manuscript.

Funding: Heather Lacey is the recipient of an Australian Government Research Training Program Scholarship and receives financial support from WaterNSW.

Institutional Review Board Statement: Not applicable.

Informed Consent Statement: Not applicable.

Data Availability Statement: Not applicable.

Acknowledgments: We acknowledge and pay respect to the Gadigal people of the Eora Nation, the traditional owners of the land on which we research, teach, and collaborate at the University of Sydney.

Conflicts of Interest: The authors declare no conflict of interest.

References

1. Watve, M.G.; Tickoo, R.; Jog, M.M.; Bhole, B.D. How many antibiotics are produced by the genus *Streptomyces*? *Arch. Microbiol.* **2001**, *176*, 386–390. [CrossRef] [PubMed]
2. Antoraz, S.; Santamaría, R.I.; Díaz, M.; Sanz, D.; Rodríguez, H. Toward a new focus in antibiotic and drug discovery from the *Streptomyces* arsenal. *Front. Microbiol.* **2015**, *6*, 461. [CrossRef] [PubMed]
3. Clarivate Web of Science. Available online: https://www.webofscience.com (accessed on 8 July 2021).
4. Kaweewan, I.; Hemmi, H.; Komaki, H.; Kodani, S. Isolation and structure determination of a new antibacterial peptide pentaminomycin C from *Streptomyces cacaoi* subsp. *Cacaoi*. *J. Antibiot.* **2020**, *73*, 224–229. [CrossRef] [PubMed]
5. Lee, B.; Son, S.; Lee, J.K.; Jang, M.; Heo, K.T.; Ko, S.-K.; Park, D.-J.; Park, C.S.; Kim, C.-J.; Ahn, J.S. Isolation of new streptimidone derivatives, glutarimide antibiotics from *Streptomyces* sp. W3002 using LC-MS-guided screening. *J. Antibiot.* **2020**, *73*, 184–188. [CrossRef]
6. Bai, X.; Guo, H.; Chen, D.; Yang, Q.; Tao, J.; Liu, W. Isolation and structure determination of two new nosiheptide-type compounds provide insights into the function of the cytochrome P450 oxygenase NocV in nocathiacin biosynthesis. *Org. Chem. Front.* **2020**, *7*, 584–589. [CrossRef]
7. Hashimoto, T.; Kozone, I.; Hashimoto, J.; Ueoka, R.; Kagaya, N.; Fujie, M.; Sato, N.; Ikeda, H.; Shin-Ya, K. Novel macrolactam compound produced by the heterologous expression of a large cryptic biosynthetic gene cluster of *Streptomyces rochei* IFO12908. *J. Antibiot.* **2020**, *73*, 171–174. [CrossRef] [PubMed]
8. Bekiesch, P.; Zehl, M.; Domingo-Contreras, E.; Martín, J.; Pérez-Victoria, I.; Reyes, F.; Kaplan, A.; Rückert, C.; Busche, T.; Kalinowski, J.R. Viennamycins: Lipopeptides produced by a *Streptomyces* sp. *J. Nat. Prod.* **2020**, *83*, 2381–2389. [CrossRef]
9. Hwang, S.; Le, L.T.H.L.; Jo, S.-I.; Shin, J.; Lee, M.J.; Oh, D.-C. Pentaminomycins C–E: Cyclic pentapeptides as autophagy inducers from a mealworm beetle gut bacterium. *Microorganisms* **2020**, *8*, 1390. [CrossRef]
10. Saadouli, I.; Zendah El Euch, I.; Trabelsi, E.; Mosbah, A.; Redissi, A.; Ferjani, R.; Fhoula, I.; Cherif, A.; Sabatier, J.-M.; Sewald, N.; et al. Isolation, characterization and chemical synthesis of large spectrum antimicrobial cyclic dipeptide (L-leu-L-pro) from *Streptomyces misionensis* V16R3Y1 bacteria extracts. A novel ^{1}H NMR metabolomic approach. *Antibiotics* **2020**, *9*, 270. [CrossRef]
11. Son, S.; Jang, M.; Lee, B.; Jang, J.-P.; Hong, Y.-S.; Kim, B.Y.; Ko, S.-K.; Jang, J.-H.; Ahn, J.S. A pipecolic acid-rich branched cyclic depsipeptide ulleungamide C from a *Streptomyces* species induces G0/G1 cell cycle arrest in promyelocytic leukemia cells. *J. Antibiot.* **2021**, *74*, 181–189. [CrossRef]
12. Vasilchenko, A.S.; Julian, W.T.; Lapchinskaya, O.A.; Katrukha, G.S.; Sadykova, V.S.; Rogozhin, E.A. A novel peptide antibiotic produced by *Streptomyces roseoflavus* strain INA-Ac-5812 with directed activity against Gram-positive bacteria. *Front. Microbiol.* **2020**, *11*, 1–13. [CrossRef] [PubMed]
13. Suzuki, M.; Komaki, H.; Kaweewan, I.; Dohra, H.; Hemmi, H.; Nakagawa, H.; Yamamura, H.; Hayakawa, M.; Kodani, S. Isolation and structure determination of new linear azole-containing peptides spongiicolazolicins A and B from *Streptomyces* sp. CWH03. *Appl. Microbiol. Biotechnol.* **2021**, *105*, 93–104. [CrossRef] [PubMed]
14. Collin, F.; Maxwell, A. The microbial toxin microcin B17: Prospects for the development of new antibacterial agents. *J. Mol. Biol.* **2019**, *431*, 3400–3426. [CrossRef] [PubMed]
15. Fang, Q.; Maglangit, F.; Mugat, M.; Urwald, C.; Kyeremeh, K.; Deng, H. Targeted isolation of indole alkaloids from *Streptomyces* sp. CT37. *Molecules* **2020**, *25*, 1108. [CrossRef]
16. Yu, Z.; Han, C.; Yu, B.; Zhao, J.; Yan, Y.; Huang, S.; Liu, C.; Xiang, W. Taxonomic characterization, and secondary metabolite analysis of *Streptomyces triticiradicis* sp. nov.: A novel actinomycete with antifungal activity. *Microorganisms* **2020**, *8*, 77. [CrossRef]
17. Gómez-Rodríguez, L.; Schultz, P.J.; Tamayo-Castillo, G.; Dotson, G.D.; Sherman, D.H.; Tripathi, A. Adipostatins E–J, new potent antimicrobials identified as inhibitors of coenzyme-A biosynthesis. *Tetrahedron. Lett.* **2020**, *61*, 151469. [CrossRef]
18. Georgousaki, K.; Tsafantakis, N.; Gumeni, S.; Gonzalez, I.; Mackenzie, T.A.; Reyes, F.; Lambert, C.; Trougakos, I.P.; Genilloud, O.; Fokialakis, N. Screening for tyrosinase inhibitors from actinomycetes; identification of trichostatin derivatives from *Streptomyces* sp. CA-129531 and scale up production in bioreactor. *Bioorg. Med. Chem. Lett.* **2020**, *30*, 126952. [CrossRef]
19. She, W.; Ye, W.; Shi, Y.; Zhou, L.; Zhang, Z.; Chen, F.; Qian, P.-Y. A novel chresdihydrochalcone from *Streptomyces chrestomyceticus* exhibiting activity against Gram-positive bacteria. *J. Antibiot.* **2020**, *73*, 429–434. [CrossRef]
20. Qiu, Y.; Yoo, H.M.; Cho, N.; Yan, P.; Liu, Z.; Cheng, J.; Suh, J.-W. Secondary metabolites isolated from *Streptomyces* sp. MJM3055 and their cytotoxicity against Jurkat cells. *Nat. Prod. Comm.* **2020**, *15*, 1–6. [CrossRef]

21. Blin, K.; Shaw, S.; Steinke, K.; Villebro, R.; Ziemert, N.; Lee, S.Y.; Medema, M.H.; Weber, T. antiSMASH 5.0: Updates to the secondary metabolite genome mining pipeline. *Nucleic Acids Res.* **2019**, *47*, W81–W87. [CrossRef]
22. Wang, B.; Song, Y.; Luo, M.; Chen, Q.; Ma, J.; Huang, H.; Ju, J. Biosynthesis of 9-methylstreptimidone involves a new decarboxylative step for polyketide terminal diene formation. *Org. Lett.* **2013**, *15*, 1278–1281. [CrossRef] [PubMed]
23. Carretero-Molina, D.; Ortiz-López, F.J.; Martín, J.; Oves-Costales, D.; Díaz, C.; de la Cruz, M.; Cautain, B.; Vicente, F.; Genilloud, O.; Reyes, F. New napyradiomycin analogues from *Streptomyces* sp. strain CA-271078. *Mar. Drugs* **2020**, *18*, 22. [CrossRef] [PubMed]
24. Choi, J.W.; Kim, G.J.; Kim, H.J.; Nam, J.-W.; Kim, J.; Chin, J.; Park, J.-H.; Choi, H.; Park, K.D. Identification and evaluation of a napyradiomycin as a potent Nrf2 activator: Anti-oxidative and anti-inflammatory activities. *Bioorg. Chem.* **2020**, *105*, 104434. [CrossRef] [PubMed]
25. Shen, X.; Wang, X.; Huang, T.; Deng, Z.; Lin, S. Naphthoquinone-based meroterpenoids from marine-derived *Streptomyces* sp. B9173. *Biomolecules* **2020**, *10*, 1187. [CrossRef]
26. Ding, L.; Görls, H.; Hertweck, C. Plant-like cadinane sesquiterpenes from an actinobacterial mangrove endophyte. *Magn. Reson. Chem.* **2021**, *59*, 34–42. [CrossRef]
27. Hu, Y.; Chou, W.K.; Hopson, R.; Cane, D.E. Genome mining in *Streptomyces clavuligerus*: Expression and biochemical characterization of two new cryptic sesquiterpene synthases. *Chem. Biol.* **2011**, *18*, 32–37. [CrossRef]
28. Pollak, F.C.; Berger, R.G. Geosmin and related volatiles in bioreactor-cultured *Streptomyces citreus* CBS 109.60. *Appl. Environ. Microbiol.* **1996**, *62*, 1295–1299. [CrossRef]
29. Kim, H.J.; Bo, A.B.; Kim, J.D.; Kim, Y.S.; Khaitov, B.; Ko, Y.-K.; Cho, K.M.; Jang, K.-S.; Park, K.W.; Choi, J.-S. Herbicidal characteristics and structural identification of the potential active compounds from *Streptomyces* sp. KRA17-580. *J. Agric. Food Chem.* **2020**, *68*, 15373–15380. [CrossRef]
30. Cao, D.D.; Do, T.Q.; Doan Thi Mai, H.; Vu Thi, Q.; Nguyen, M.A.; Le Thi, H.M.; Tran, D.T.; Chau, V.M.; Cong Thung, D.; Pham, V.C. Antimicrobial lavandulylated flavonoids from a sponge-derived actinomycete. *Nat. Prod. Res.* **2020**, *34*, 413–420. [CrossRef]
31. Rodríguez Estévez, M.; Gummerlich, N.; Myronovskyi, M.; Zapp, J.; Luzhetskyy, A. Benzanthric acid, a novel metabolite from *Streptomyces albus* Del14 expressing the nybomycin gene cluster. *Front. Chem.* **2020**, *7*, 896. [CrossRef]
32. Pratiwi, R.H.; Hidayat, I.; Hanafi, M.; Mangunwardoyo, W. Isolation and structure elucidation of phenazine derivative from *Streptomyces* sp. strain UICC B-92 isolated from *Neesia altissima* (Malvaceae). *Iran J. Microbiol.* **2020**, *12*, 127. [PubMed]
33. Voitsekhovskaia, I.; Paulus, C.; Dahlem, C.; Rebets, Y.; Nadmid, S.; Zapp, J.; Axenov-Gribanov, D.; Rückert, C.; Timofeyev, M.; Kalinowski, J. Baikalomycins AC, New Aquayamycin-type angucyclines isolated from Lake Baikal derived *Streptomyces* sp. IB201691-2A. *Microorganisms* **2020**, *8*, 680. [CrossRef] [PubMed]
34. Kim, J.W.; Kwon, Y.; Bang, S.; Kwon, H.E.; Park, S.; Lee, Y.; Deyrup, S.T.; Song, G.; Lee, D.; Joo, H.-S. Unusual bridged angucyclinones and potent anticancer compounds from *Streptomyces bulli* GJA1. *Org. Biomol. Chem.* **2020**, *18*, 8443–8449. [CrossRef] [PubMed]
35. Booth, T.J.; Kalaitzis, J.A.; Vuong, D.; Cromble, A.; Lacey, E.; Piggott, A.M.; Wilkinson, B. Production of novel pladienolide analogues through native expression of a pathway-specific activator. *Chem. Sci.* **2020**, *11*, 8249–8255. [CrossRef]
36. Machida, K.; Arisawa, A.; Takeda, S.; Tsuchida, T.; Aritoku, Y.; Yoshida, M.; Ikeda, H. Organization of the biosynthetic gene cluster for the polyketide antitumor macrolide, pladienolide, in *Streptomyces platensis* Mer-11107. *Biosci. Biotech. Bioch.* **2008**, *72*, 2946–2952. [CrossRef]
37. Lacey, H.J.; Booth, T.J.; Vuong, D.; Rutledge, P.J.; Lacey, E.; Chooi, Y.-H.; Piggott, A.M. Conglobatins B–E: Cytotoxic analogues of the C2-symmetric macrodiolide conglobatin. *J. Antibiot.* **2020**, *73*, 756–765. [CrossRef]
38. Westley, J.W.; Liu, C.-M.; Evans, R.H.; Blount, J.F. Conglobatin, a novel macrolide dilactone from *Streptomyces conglobatus* ATCC 31005. *J. Antibiot.* **1979**, *32*, 874–877. [CrossRef]
39. Hou, L.; Liu, Z.; Yu, D.; Li, H.; Ju, J.; Li, W. Targeted isolation of new polycyclic tetramate macrolactams from the deepsea-derived *Streptomyces somaliensis* SCSIO ZH66. *Bioorg. Chem.* **2020**, *101*, 103954. [CrossRef]
40. Zhang, S.M.; Zhang, L.; Kou, L.J.; Yang, Q.L.; Qu, B.; Pescitelli, G.; Xie, Z.P. Isolation, stereochemical study, and racemization of (±)-pratenone A, the first naturally occurring 3-(1-naphthyl)-2-benzofuran-1 (3H)-one polyketide from a marine-derived actinobacterium. *Chirality* **2020**, *32*, 299–307. [CrossRef]
41. Fang, Q.; Maglangit, F.; Wu, L.; Ebel, R.; Kyeremeh, K.; Andersen, J.H.; Annang, F.; Pérez-Moreno, G.; Reyes, F.; Deng, H. Signalling and bioactive metabolites from *Streptomyces* sp. RK44. *Molecules* **2020**, *25*, 460. [CrossRef]
42. Corre, C.; Song, L.; O'Rourke, S.; Chater, K.F.; Challis, G.L. 2-Alkyl-4-hydroxymethylfuran-3-carboxylic acids, antibiotic production inducers discovered by *Streptomyces coelicolor* genome mining. *Proc. Natl. Acad. Sci. USA* **2008**, *105*, 17510–17515. [CrossRef] [PubMed]

Review

Pentacyclic Triterpenoids Isolated from Celastraceae: A Focus in the ^{13}C-NMR Data

Karen Caroline Camargo, Mariana Guerra de Aguilar, Acácio Raphael Aguiar Moraes, Raquel Goes de Castro, Daiane Szczerbowski, Elizabeth Luciana Marinho Miguel, Leila Renan Oliveira, Grasiely Faria Sousa *, Diogo Montes Vidal * and Lucienir Pains Duarte *

Departamento de Química, Instituto de Ciências Exatas, Universidade Federal de Minas Gerais, Avenida Presidente Antônio Carlos, 6627, Pampulha, Belo Horizonte 31270-901, MG, Brazil; karen_camargo345@hotmail.com (K.C.C.); marianag.a9@gmail.com (M.G.d.A.); acacioramoraes@gmail.com (A.R.A.M.); raquelgooes@hotmail.com (R.G.d.C.); daianeszcz@gmail.com (D.S.); elizabethmmiguel@yahoo.com.br (E.L.M.M.); leila.renan97@hotmail.com (L.R.O.)
* Correspondence: grasielysousa@ufmg.br (G.F.S.); vidal@qui.ufmg.br (D.M.V.); lucienir@ufmg.br (L.P.D.); Tel.: +55-31-3409-5728 (G.F.S.); +55-31-3409-5750 (D.M.V.); +55-31-3409-5722 (L.P.D.)

Abstract: The Celastraceae family comprises about 96 genera and more than 1.350 species, occurring mainly in tropical and subtropical regions of the world. The species of this family stand out as important plant sources of triterpenes, both in terms of abundance and structural diversity. Triterpenoids found in Celastraceae species display mainly lupane, ursane, oleanane, and friedelane skeletons, exhibiting a wide range of biological activities such as antiviral, antimicrobial, analgesic, anti-inflammatory, and cytotoxic against various tumor cell lines. This review aimed to document all triterpenes isolated from different botanical parts of species of the Celastraceae family covering 2001 to 2021. Furthermore, a compilation of their ^{13}C-NMR data was carried out to help characterize compounds in future investigations. A total of 504 pentacyclic triterpenes were compiled and distinguished as 29 aromatic, 50 dimers, 103 friedelanes, 89 lupanes, 102 oleananes, 22 quinonemethides, 88 ursanes and 21 classified as others.

Keywords: Celastraceae; triterpenes; quinonemethide; ^{13}C-NMR

Citation: Camargo, K.C.; de Aguilar, M.G.; Moraes, A.R.A.; de Castro, R.G.; Szczerbowski, D.; Miguel, E.L.M.; Oliveira, L.R.; Sousa, G.F.; Vidal, D.M.; Duarte, L.P. Pentacyclic Triterpenoids Isolated from Celastraceae: A Focus in the ^{13}C-NMR Data. *Molecules* **2022**, *27*, 959. https://doi.org/10.3390/molecules27030959

Academic Editor: Vassilios Roussis

Received: 24 December 2021
Accepted: 24 January 2022
Published: 31 January 2022

Publisher's Note: MDPI stays neutral with regard to jurisdictional claims in published maps and institutional affiliations.

1. Introduction

The Celastraceae family comprises approximately 96 genera, reaching about 1350 species distributed in the tropical and subtropical regions of the world [1,2]. Species of this family stand out for producing compounds with several pharmacological activities, such as antitumor [3,4], anti-inflammatory [5], antimicrobial [6–8], antioxidant [9] antiviral [10], analgesic [5,11], antiulcerogenic [12], hepatoprotective [13], hypoglycemic [13,14], immunomodulatory [15], among others. Considering the chemical composition, species of the Celastraceae family are rich in pentacyclic triterpenes (PCTTs). PCTTs show a range of biological properties, characterizing these plants as research targets aiming to obtain new bioactive compounds or prototypes of new drugs [16–22].

PCTTs are structurally diverse compounds and are therefore classified according to their main skeletal structure. The main classes found in Celastraceae family possess friedelane, oleanane, lupane, ursane and quinonemethide skeletons. Quinonemethides are chemomarkers of this family are found exclusively in these species [13]. These PCTTs can occur as alcohols, ketones, carboxylic acids, lactones, aldehydes, epoxides, esters, or even glycosylated derivatives. Furthermore, these PCTTs can be sub-classified as *seco*, generally due to the opening of one of their rings, the most common being the ring 'A' opening between carbons 3 and 4, and sub-classified as *nor* when there is a lack of any of the methyl groups that constitute the basic skeleton.

This review aims to present the PCTTs reported for species of the Celastraceae family in the 21st century, exhibiting from which species they were isolated and contributing to the chemical characterization process of these compounds listing their ^{13}C NMR data. The information about the PCTTs was obtained from SciFinder, Scopus, and Web of Science, using as key search terms: "Celastraceae and triterpenes", "Celastraceae and compounds", "Celastraceae and phytochemistry" and "Celastraceae and metabolites". Articles with only ethnopharmacological information and data from *in vitro* and *in vivo* tests involving extracts or isolated substances were excluded. The period covering from January 2001 to September 2021 was considered since the group has already developed a free online database (in Portuguese) for the previous years [23]. This review reports a total of 504 pentacyclic triterpenoids, 29 aromatics (A), 50 dimers (D), 103 friedelanes (F), 89 lupanes (L), 102 oleananes (O), 22 quinonemethides (Q), 88 ursanes (U) and 21 classified as others. Table S1 (Supplementary Material) summarizes all these PCTTs, as well as the plant species and parts from which they were isolated.

2. Pentacyclic Triterpenoids (PCTTs)

PCTTs consist of 30 carbon atoms (six isoprene units) distributed over five fused rings (named A, B, C, D and E). This ring arrangement yields five six-membered rings or four six-membered rings fused to a 5-membered ring, numbered as shown in Figure 1 [24].

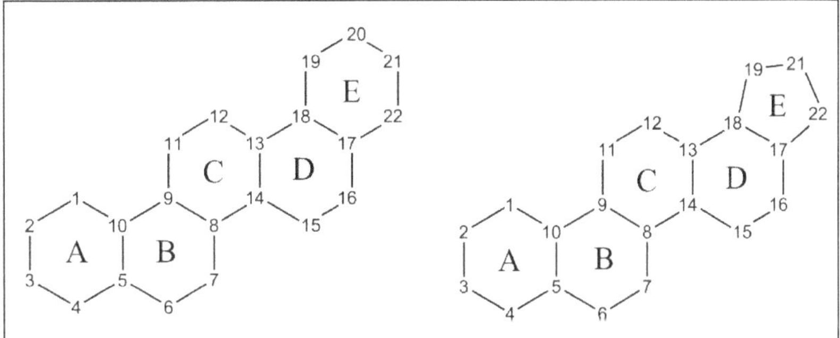

Figure 1. Core rings A, B, C, D, and E found in PCTTs.

As terpenes, the biosynthesis of PCTTs starts by the coupling of active isoprene units. Initially, there is an electrophilic condensation of IPP (isopentenyl diphosphate), with DMAPP (dimethylallyl diphosphate), yielding the precursor of monoterpenes, geranyl diphosphate (GPP). The addition of IPP to GPP generates farnesyl diphosphate (FPP), which is the precursor of sesquiterpenes. Then a tail-tail condensation of two FPP molecules leads to squalene, after the release of a diphosphate unit and a 1,3-alkyl shift (Figure 2) [25].

The biosynthesis of PCTTs continues with the oxidation of squalene, catalyzed by squalene epoxidase, forming 2,3-oxidosqualene. This intermediary assumes the "chair-chair-chair-boat" conformation and after a sequence of cyclizations yields the dammarenyl cation, which then undergoes a rearrangement forming the baccharenyl cation. From the baccharenyl cation, the key step in PCTTs biosynthesis occurs, characterized by the formation of the lupanyl cation (Figure 3) [25,26]. Through a sequence of carbocation rearrangements (1,2-shifts), involving hydride, methyl, and ring-opening shifts, the lupanyl cation yields the different PCTTs skeletons, which then could oxidize, reduce, and isomerize, leading to the formation of the different currently known PCTTs [25,26].

The most powerful spectroscopic method in the structural elucidation of PCTTs is [13]C Nuclear Magnetic Resonance (NMR). Comparison of experimental ^{13}C NMR chemical shifts with literature data is a useful tool in identifying the basic skeleton of these compounds. Through this data, it is possible to make predictions about the influence of a functional group on the chemical displacement of carbons from its basic skeleton [27]. According to Mahato & Kundu [27], for example, the introduction of a hydroxyl group in the PCTT structure induces a deshielding of about 34–50 ppm of the α carbon, 2–10 ppm of the β carbons and 0–9 ppm of the γ carbons. The effect of the hydroxyl presence on the ^{13}C NMR chemical shift of the α-carbon, is related to its configuration, and with the number of γ-gauche-type, and 1,3-diaxial-type interactions with the carbon atoms of the triterpene skeleton [27].

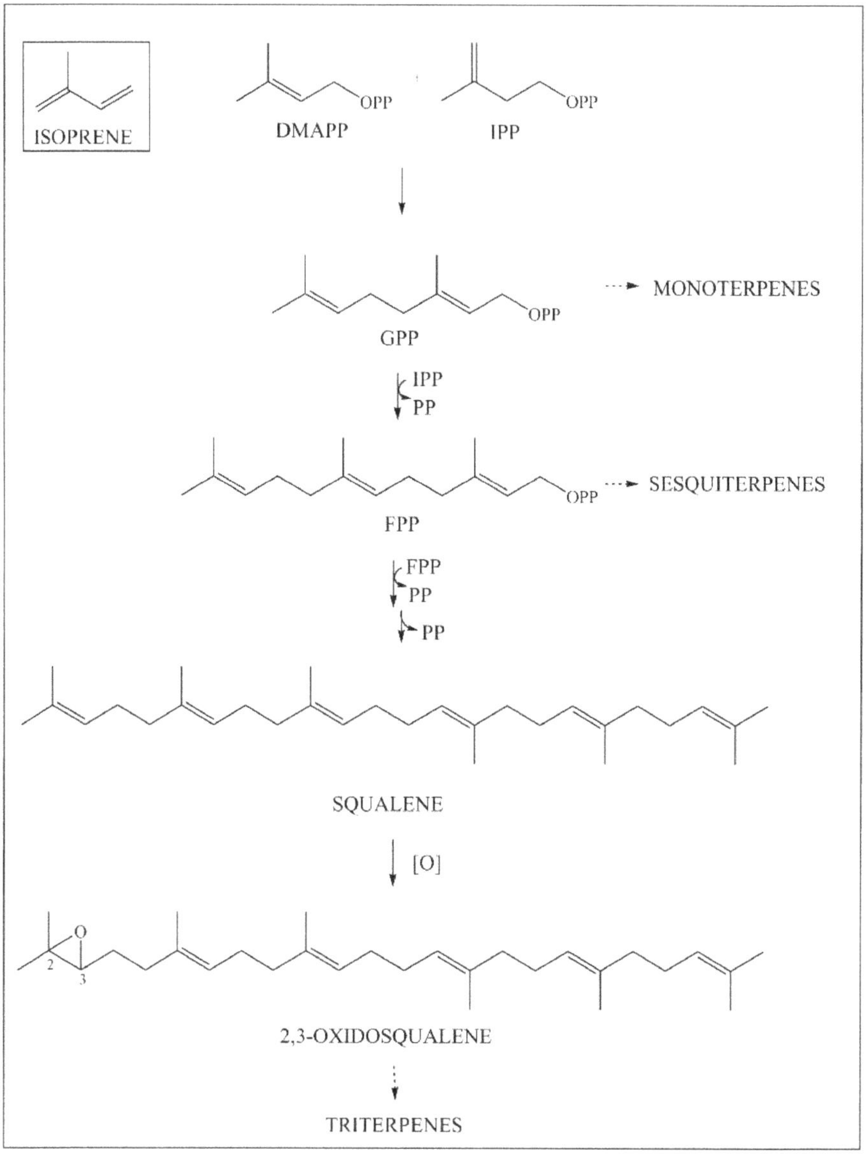

Figure 2. Simplified biosynthetic route of 2,3−oxidosqualene, the direct precursor of triterpenes, from isoprene. DMAPP: dimethylallyl diphosphate; IPP: isopentenyl diphosphate; GPP: geranyl diphosphate; FPP: farnesyl diphosphate; PP: diphosphate [26,28].

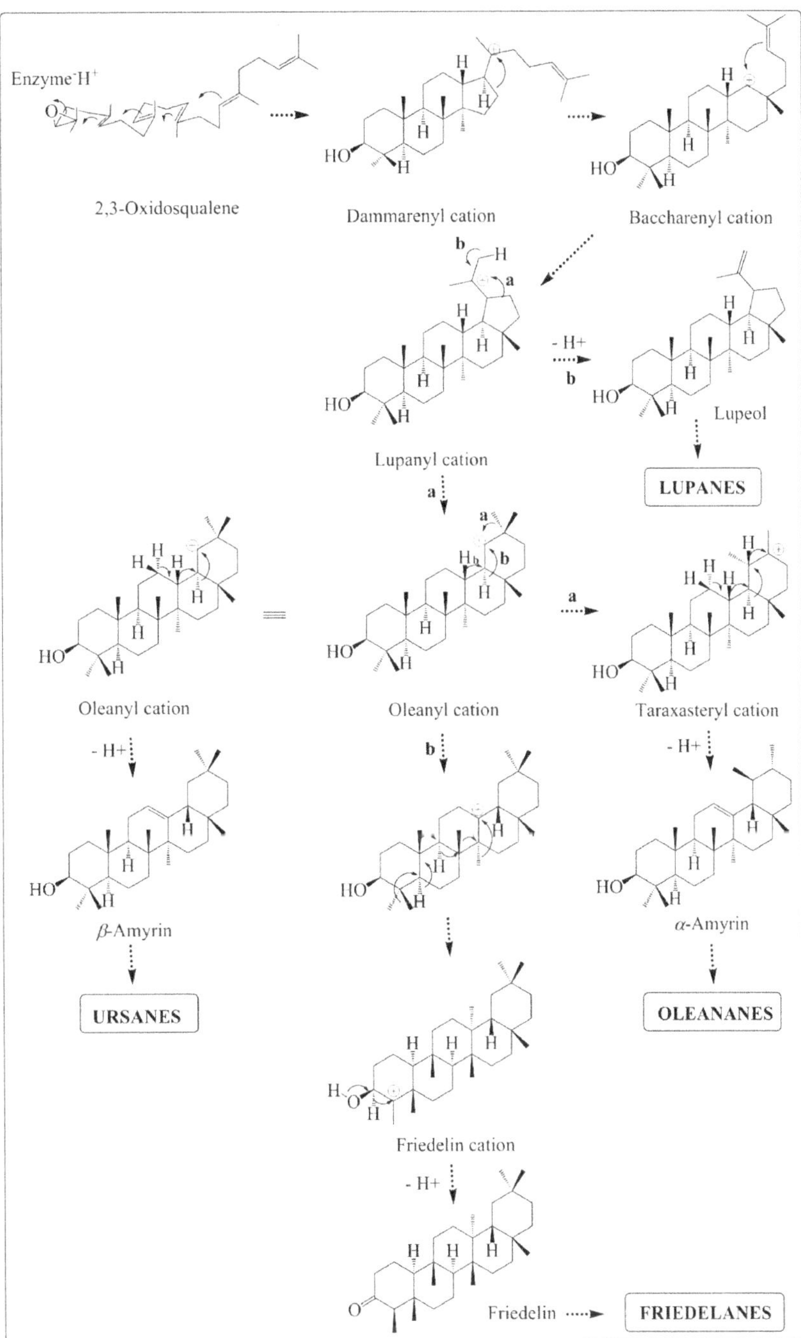

Figure 3. Simplified terpenoid biosynthetic route for the formation of the main pentacyclic triterpene skeletons isolated from Celastraceae species. "a" and "b" indicate two possible biosynthetic pathways [26,28].

2.1. Friedelanes

Compounds presenting a friedelane skeleton, together with the oleananes, are the most abundant PCTTs in the Celastraceae family, being found in the leaves, branches, roots and other parts of these plants [13]. These systems are formed by five six-membered rings fused. Rings A/B, B/C and C/D have *trans* configuration (H-10α and H-8α), while ring D/E is *cis* (H-18β). They have eight methyl groups; six attached to distinct carbons, at positions 4 (Me 23β), 5 (Me 24β), 9 (Me 25β), 13 (Me 27α), 14 (Me 26β) and 17 (Me 28β), and two geminal methyl groups at carbon 20 (Me 29α and 30β) [22,29]. In this work, 103 PCTTs of friedelan skeleton (**F**) are reported, compounds **F1**–**F103** (Figure 4).

	R_1	R_2	R_3	R_4	R_5	R_6	R_7	R_8	R_9
F1	H	=O	OHβ	H	H	H	H	CH_3	CH_3
F2	H	=O	H	OHα	H	H	H	CH_3	CH_2OH
F3	H	=O	H	OHα	H	=O	H	CH_3	CH_3
F4	H	=O	H	OHα	H	H	H	CH_3	CH_3
F5	H	=O	H	OHβ	H	H	H	CH_3	CH_3
F6	H	=O	H	H	OHα	H	H	CH_2OH	CH_3
F7	H	=O	H	H	OHα	H	H	CH_3	CH_3
F8	=O	=O	H	H	H	$OCOCH_3$	H	CH_2OH	CH_3
F10	=O	=O	H	H	H	OHα	H	CH_3	CH_3
F11	H	=O	H	H	H	OHα	H	CH_3	CH_3
F13	H	=O	H	H	H	OHβ	H	CH_3	CH_2OH
F14	H	=O	H	H	H	OHβ	H	CH_3	CH_3
F15	OHα	=O	H	H	H	H	H	CH_3	CH_2OH
F18	OHβ	=O	H	H	H	H	H	CH_3	CH_3
F19	=O	=O	H	H	H	H	OHα	CH_3	CH_3
F20	H	=O	H	H	H	H	OHα	CH_3	CH_3
F21	H	=O	H	H	=O	H	OHα	CH_3	CH_3
F22	H	=O	H	H	H	H	OHβ	CH_3	CH_3
F26	=O	=O	H	H	H	H	H	CH_2OH	CH_3
F33	=O	=O	H	H	H	H	H	CH_3	CH_2OH
F34	H	=O	H	H	H	=O	H	CH_3	CH_2OH
F35	H	=O	H	H	H	H	H	CH_3	CH_2OH
F46	=O	H	H	H	H	H	H	CH_3	CHO
F49	H	=O	H	H	H	H	H	CH_3	$COOCH_3$
F54	=O	OHβ	H	H	H	H	H	CH_3	CH_2OH
F73	H	=O	H	H	H	H	H	CH_3	COOH
F80	H	OHβ	H	H	H	H	H	CH_3	CH_2OH
F82	=O	=O	H	H	H	H	=O	CH_3	CH_3
F83	=O	=O	H	H	H	H	H	CH_3	CH_3
F84	H	=O	=O	H	H	H	H	CH_3	CH_3
F85	H	=O	H	=O	H	H	H	CH_3	CH_3
F86	H	=O	H	H	=O	H	H	CH_3	CH_3
F87	H	=O	H	H	H	=O	H	CH_3	CH_3
F89	H	=O	H	H	H	H	H	CH_3	CH_3
F90	H	OHβ	OHβ	H	H	H	H	CH_3	CH_3
F93	H	OHβ	H	H	H	H	H	CH_3	CH_3
F94	H	=O	H	H	H	H	=O	CH_3	CH_3

(**a**)

Figure 4. *Cont.*

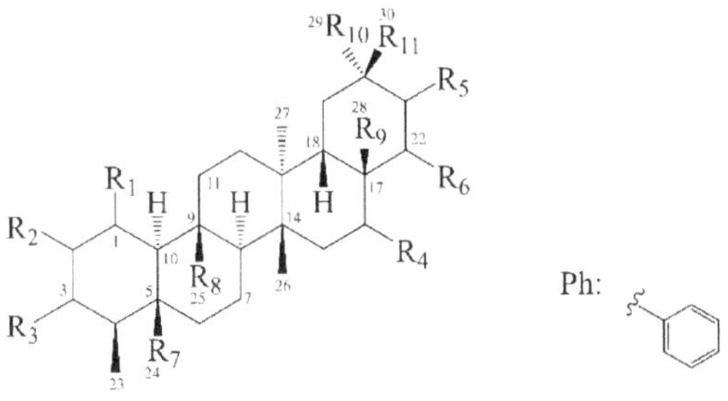

Ph:

	R_1	R_2	R_3	R_4	R_5	R_6	R_7	R_8	R_9	R_{10}	R_{11}
F9	H	H	=O	OHα	H	H	CH_3	CH_3	CH_2OH	CH_3	CH_3
F12	H	H	=O	OHβ	H	H	CH_3	CH_3	CH_2OH	CH_2OH	CH_3
F16	OHβ	H	=O	H	H	H	CH_3	CH_2OH	CH_3	CH_3	CH_3
F17	OHβ	H	=O	H	H	H	CH_3	CH_3	CH_3	CH_3	CH_2OH
F24	H	H	=O	H	H	H	CH_3	CH_2OH	CH_3	CH_3	CH_3
F25	H	H	=O	H	=O	H	CH_3	CH_2OH	CH_3	CH_3	CH_3
F28	H	H	=O	H	H	H	CH_3	CH_3	CH_2OH	CH_2OH	CH_3
F29	H	H	=O	H	H	H	CH_3	CH_3	CH_2OH	CH_3	CH_2OH
F30	=O	H	=O	H	H	H	CH_3	CH_3	CH_2OH	CH_3	CH_3
F31	H	H	=O	H	H	H	CH_3	CH_3	CH_2OH	CH_3	CH_3
F32	H	H	=O	H	H	H	CH_3	CH_3	CHO	CH_2OH	CH_3
F37	H	OHα	=O	H	H	H	CH_3	CH_3	CH_3	CH_3	CH_3
F40	H	H	=O	H	=O	H	CH_3	OCOPh	CH_3	CH_3	CH_3
F41	=O	H	=O	H	H	H	CH_3	CH_3	CH_3	CH_3	CH_2OH
F42	H	H	=O	H	H	H	CH_3	CH_3	CH_3	CH_3	CH_2OH
F47	H	H	=O	H	H	H	CH_3	CH_3	CHO	CH_3	CH_3
F48	H	H	=O	H	H	H	CH_3	CH_3	CH_3	CH_3	CHO
F50	H	=O	OHα	H	H	H	CH_3	CH_2OH	CH_3	CH_3	CH_3
F51	H	=O	OHα	H	H	H	CH_3	CH_3	CH_3	CH_3	CH_3
F63	OHβ	H	=O	H	H	H	CH_3	CH_3	CH_3	CH_3	COOH
F65	H	OHα	OHα	H	=O	OHβ	COOH	CH_3	CH_3	H	CH_3
F66	H	OHα	=O	H	H	H	CH_3	CH_3	CH_3	COOH	CH_3
F67	H	OHα	=O	H	H	H	CH_3	CH_3	CH_3	CH_3	COOH
F68	H	OHβ	=O	H	H	H	CH_3	CH_3	CH_3	CH_3	COOH
F71	H	H	=O	H	H	H	CH_3	CH_3	CH_3	CH_3	COOH
F72	H	H	=O	H	H	H	CH_3	CH_3	COOH	CH_3	CH_3
F74	H	=O	OHα	H	H	H	CH_3	CH_3	CH_3	COOH	CH_3
F76	H	H	OHα	H	H	H	CH_3	CH_3	CH_3	COOH	CH_3
F78	H	=O	OHβ	H	H	H	CH_3	CH_3	CH_3	COOH	CH_3
F92	H	H	OHβ	H	H	H	CH_3	CH_3	CH_2OH	CH_2OH	CH_3

(**b**)

Figure 4. *Cont.*

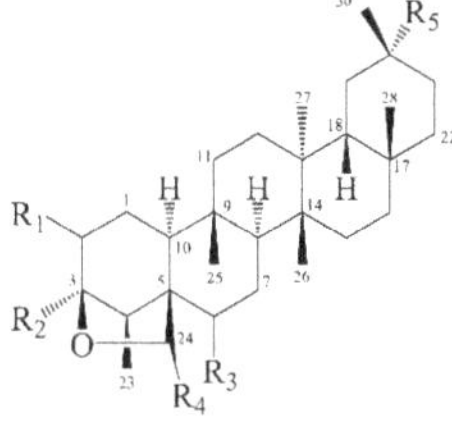

	R₁	R₂	R₃	R₄	R₅	R₆	R₇	R₈	R₉
F27	H	=O	H	CH₃	CH₃	CH₂OH	CH₃	CH₃	CH₃
F58	=O	=O	OHα	CH₃	CH₃	CH₃	CH₃	CH₃	CH₃
F59	H	=O	OHα	CH₃	CH₂OH	CH₃	CH₃	CH₃	CH₃
F60	H	=O	OHα	CH₃	CH₃	CH₃	CH₃	CH₃	CH₂OH
F61	H	=O	OHβ	CH₃	CH₃	CH₃	CH₃	CH₂OH	CH₃
F62	H	=O	OHβ	CH₃	CH₃	CH₃	COOCH₃	CH₃	CH₃
F79	H	=O	OHβ	CH₃	CH₃	CH₃	COOH	CH₃	CH₃
F88	H	=O	=O	CH₃	CH₃	CH₃	CH₃	CH₃	CH₃
F91	H	OHβ	H	CH₂OH	CH₃	CH₃	CH₃	CH₃	CH₃

	R₁	R₂	R₃	R₄	R₅
F36	OCOCH₃α	OH	H	H	COOCH₃
F52	OHα	OH	OHβ	H	COOCH₃
F53	OHα	OH	H	H	COOCH₃
F64	OHα	OH	H	H	COOH
F75	H	OH	H	H	COOH
F77	OHα	H	H	OH	COOH

	R₁	R₂
F39	H	H
F56	OHα	OHβ
F99	H	OHβ

	R₁	R₂
F43	CHO	COOCH₃
F69	CHO	COOH
F70	CH₃	COOH

	R₁	R₂	R₃
F57	OHβ	OHα	=O
F81	H	H	H

(**c**)

Figure 4. *Cont.*

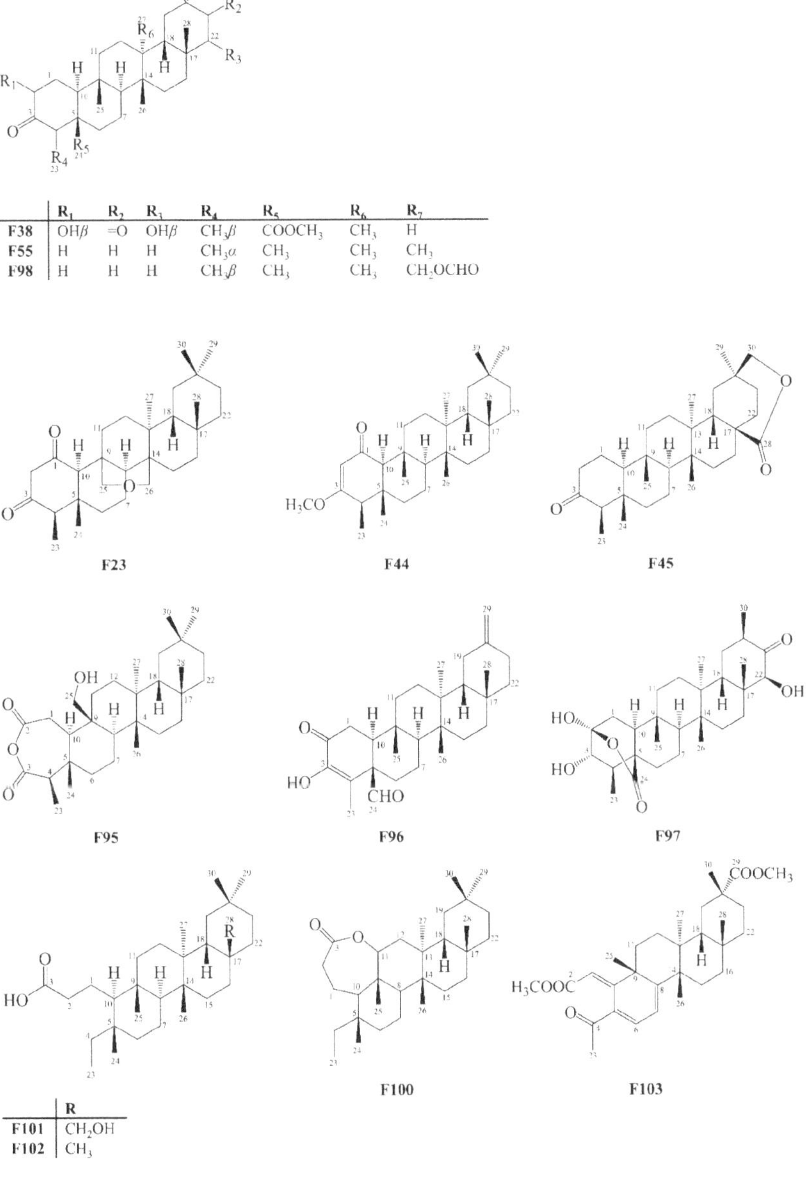

	R₁	R₂	R₃	R₄	R₅	R₆	R₇
F38	OH*β*	=O	OH*β*	CH₃*β*	COOCH₃	CH₃	H
F55	H	H	H	CH₃*α*	CH₃	CH₃	CH₃
F98	H	H	H	CH₃*β*	CH₃	CH₃	CH₂OCHO

	R
F101	CH₂OH
F102	CH₃

(d)

Figure 4. Structures of friedelane-type pentacyclic isolated from Celastraceae species (2001–2021). (a) Compounds **F1**–**F8**, **F10**, **F11**, **F13**–**F15**, **F18**–**F22**, **F26**, **F33**–**F35**, **F46**, **F49**, **F54**, **F73**, **F80**, **F82**–**F87**, **F89**, **F90**, **F93** and **F94**. (b) Compounds **F9**, **F12**, **F16**, **F17**, **F24**, **F25**, **F28**–**F32**, **F37**, **F40**–**F42**, **F47**, **F48**, **F50**, **F51**, **F63**, **F65**–**F68**, **F71**, **F72**, **F74**, **F76**, **F78** and **F92**. (c) Compounds **F27**, **F36**, **F39**, **F43**, **F52**, **F53**, **F56**, **F57**, **F58**–**F62**, **F64**, **F69**, **F70**, **F75**, **F77**, **F79**, **F81**, **F88**, **F91** and **F99**. (d) Compounds **F23**, **F38**, **F44**, **F45**, **F55**, **F95**–**F98**, **F100**–**F103**.

An important observation in the ^{13}C NMR data of 3-oxo friedelanes is the shielding of methyl group 23, which has a chemical shift value around δ_C 7.0 ppm. This occurs since this methyl is found in a cone region, generated by the π electrons of the carbonyl group at C-3, which promotes a region of shielding magnetic anisotropy [30].

2.2. Quinonemethides and Aromatics

Quinonemethides are compounds isolated exclusively in species of the Celastraceae family, and can also be found in the form of dimers or trimers [31]. Hypotheses about their origin assume that they are formed from friedelane derivatives, which are transported from the leaves to the roots, where they are converted into quinonemethides [32]. They are characterized as 24-*nor*-triterpenoids, due to the absence of methyl 24, and also they have functional oxygenated groups attached to carbons 2 and 3 [33]. Aromatic skeleton PCTTs are a subgroup of quinonemethides, which are characterized by the aromaticity of the A ring. Between 2001 and 2021 about 22 quinonemethides (**Q**), **Q1–Q22** (Figure 5), and 29 aromatics analogues (**A**), **A1–A29** (Figure 6) were isolated from Celastraceae species.

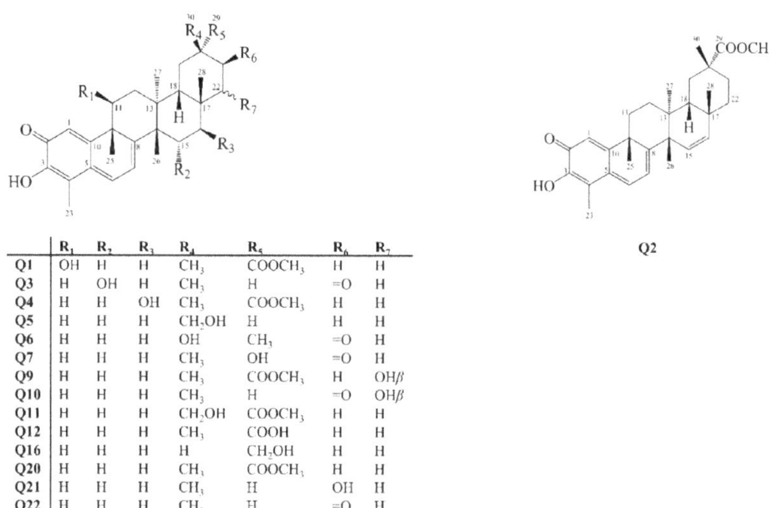

	R₁	R₂	R₃	R₄	R₅	R₆	R₇
Q1	OH	H	H	CH₃	COOCH₃	H	H
Q3	H	OH	H	CH₃	H	=O	H
Q4	H	H	OH	CH₃	COOCH₃	H	H
Q5	H	H	H	CH₂OH	H	H	H
Q6	H	H	H	OH	CH₃	=O	H
Q7	H	H	H	CH₃	OH	=O	H
Q9	H	H	H	CH₃	COOCH₃	H	OHβ
Q10	H	H	H	CH₃	H	=O	OHβ
Q11	H	H	H	CH₂OH	COOCH₃	H	H
Q12	H	H	H	CH₃	COOH	H	H
Q16	H	H	H	H	CH₂OH	H	H
Q20	H	H	H	CH₃	COOCH₃	H	H
Q21	H	H	H	CH₃	H	OH	H
Q22	H	H	H	CH₃	H	=O	H

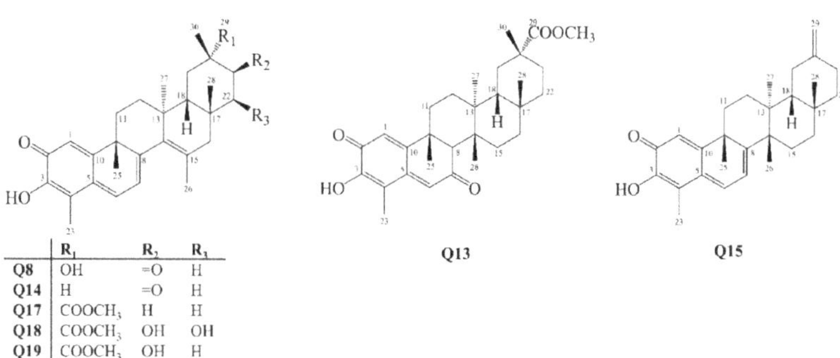

	R₁	R₂	R₃
Q8	OH	=O	H
Q14	H	=O	H
Q17	COOCH₃	H	H
Q18	COOCH₃	OH	OH
Q19	COOCH₃	OH	H

Figure 5. Structures of quinonemethide-type pentacyclic triterpenoids isolated from Celastraceae species (2001–2021).

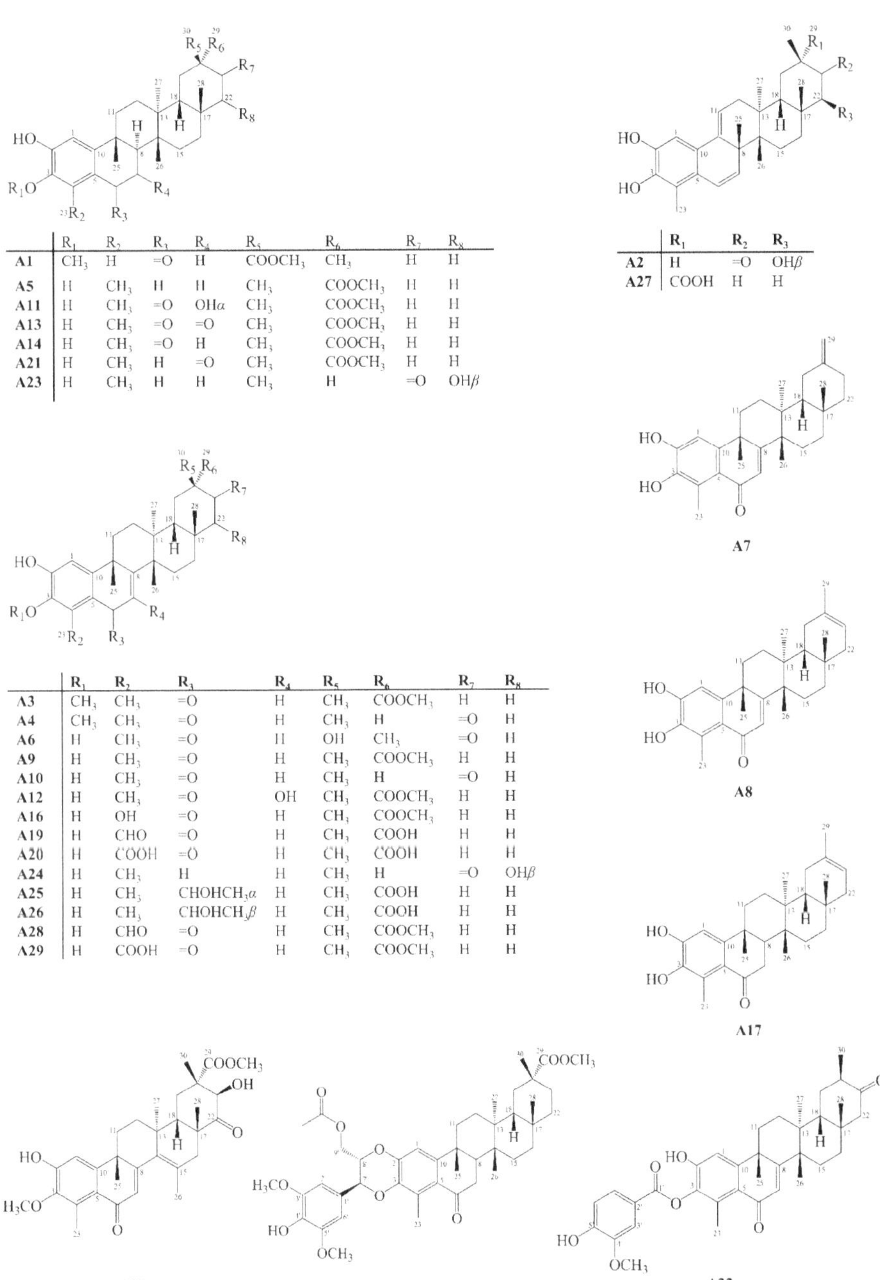

	R₁	R₂	R₃	R₄	R₅	R₆	R₇	R₈
A1	CH₃	H	=O	H	COOCH₃	CH₃	H	H
A5	H	CH₃	H	H	CH₃	COOCH₃	H	H
A11	H	CH₃	=O	OHα	CH₃	COOCH₃	H	H
A13	H	CH₃	=O	=O	CH₃	COOCH₃	H	H
A14	H	CH₃	=O	H	CH₃	COOCH₃	H	H
A21	H	CH₃	H	=O	CH₃	COOCH₃	H	H
A23	H	CH₃	H	H	CH₃	H	=O	OHβ

	R₁	R₂	R₃
A2	H	=O	OHβ
A27	COOH	H	H

	R₁	R₂	R₃	R₄	R₅	R₆	R₇	R₈
A3	CH₃	CH₃	=O	H	CH₃	COOCH₃	H	H
A4	CH₃	CH₃	=O	H	CH₃	H	=O	H
A6	H	CH₃	=O	H	OH	CH₃	=O	H
A9	H	CH₃	=O	H	CH₃	COOCH₃	H	H
A10	H	CH₃	=O	H	CH₃	H	=O	H
A12	H	CH₃	=O	OH	CH₃	COOCH₃	H	H
A16	H	OH	=O	H	CH₃	COOCH₃	H	H
A19	H	CHO	=O	H	CH₃	COOH	H	H
A20	H	COOH	=O	H	CH₃	COOH	H	H
A24	H	CH₃	H	H	CH₃	H	=O	OHβ
A25	H	CH₃	CHOHCH₃α	H	CH₃	COOH	H	H
A26	H	CH₃	CHOHCH₃β	H	CH₃	COOH	H	H
A28	H	CHO	=O	H	CH₃	COOCH₃	H	H
A29	H	COOH	=O	H	CH₃	COOCH₃	H	H

Figure 6. Structures of aromatic-type pentacyclic triterpenoids isolated from Celastraceae species (2001–2021).

In the ^{13}C NMR spectra of the quinonemethides, signals are observed in the characteristic carbonyl region, between δ_C 170–200 ppm, and in the typical olefinic carbon region, around δ_C 110–160 ppm.

2.3. Dimers

Dimers are formed from PCTTs of the quinonemethide class and its aromatic derivatives, therefore they are also restricted to the Celastraceae family. According to Bazzocchi, Núñez and Reyes [31], these triterpenes are possibly biosynthesized through a Diels-Alder reaction, in which the different possible orientations of the monomers during the reaction result in a variety of isomers.

Between the years 2001 and 2021, 50 dimers (**D**), **D1–D50**, were reported (Figure 7). Most of these dimers are formed by two triterpenes with quinonemethide skeleton or their aromatic derivatives. However, the formation of adducts can also occur from the combination of a triterpene and a sesquiterpene (**D15–D24; D37–D38**).

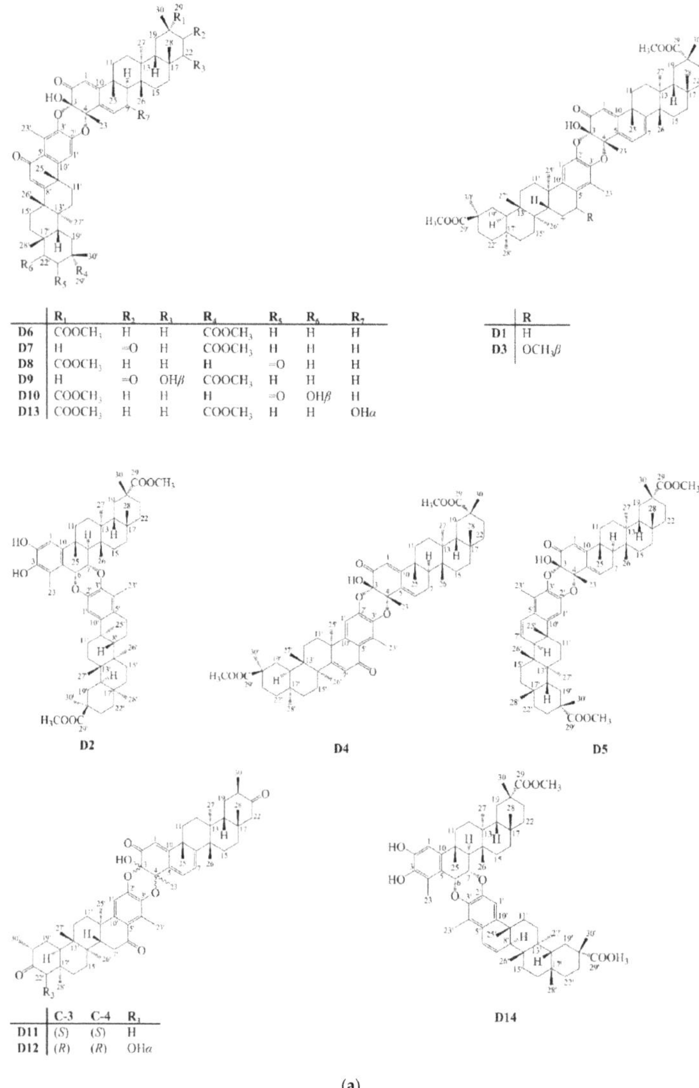

	R$_1$	R$_2$	R$_3$	R$_4$	R$_5$	R$_6$	R$_7$
D6	COOCH$_3$	H	H	COOCH$_3$	H	H	H
D7	H	=O	H	COOCH$_3$	H	H	H
D8	COOCH$_3$	H	H	H	=O	H	H
D9	H	=O	OHβ	COOCH$_3$	H	H	H
D10	COOCH$_3$	H	H	H	=O	OHβ	H
D13	COOCH$_3$	H	H	COOCH$_3$	H	H	OHα

	R
D1	H
D3	OCH$_3\beta$

D2

D4

D5

	C-3	C-4	R$_1$
D11	(S)	(S)	H
D12	(R)	(R)	OHα

D14

(**a**)

Figure 7. *Cont.*

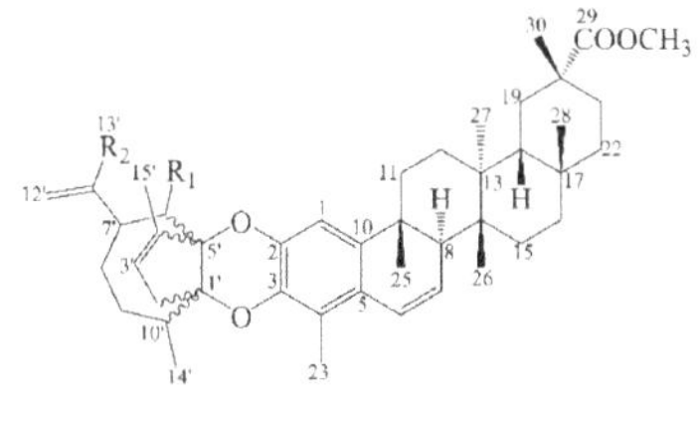

	C-1'	C-5'	R$_1$	R$_2$
D16	(S)	(S)	H	CH$_3$
D18	(R)	(R)	H	CH$_3$
D20	(R)	(R)	OH	COOH

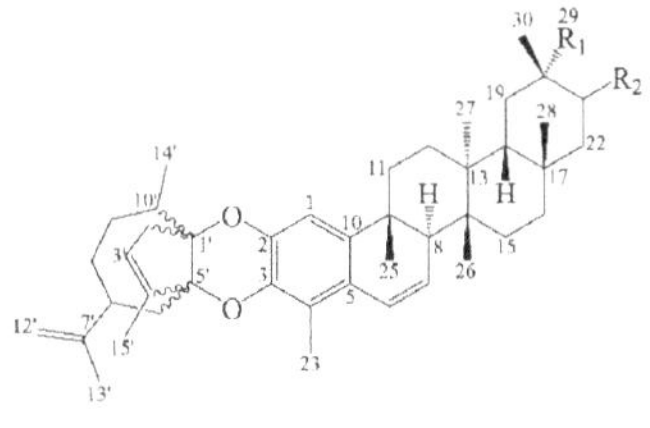

	C-1'	C-5'	R$_1$	R$_2$
D17	(R)	(R)	COOCH$_3$	H
D19	(S)	(S)	COOCH$_3$	H
D37	(S)	(S)	H	=O

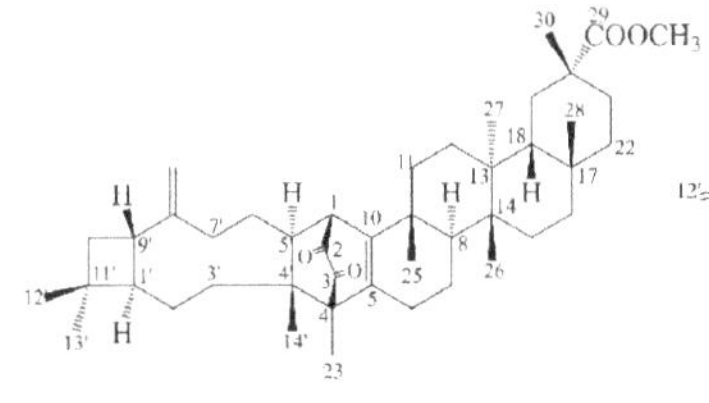

D15

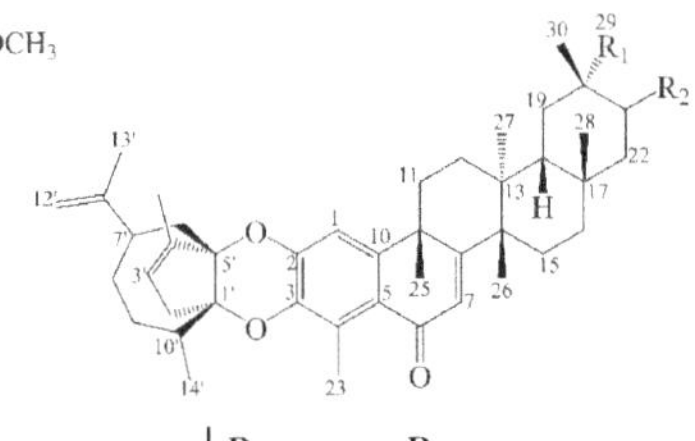

	R$_1$	R$_2$
D21	COOCH$_3$	H
D38	H	=O

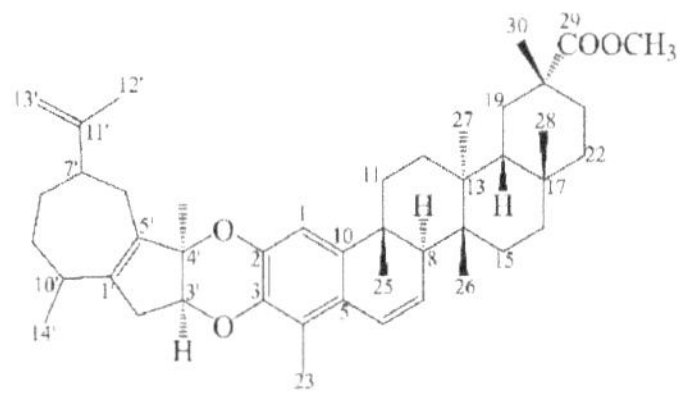

D22

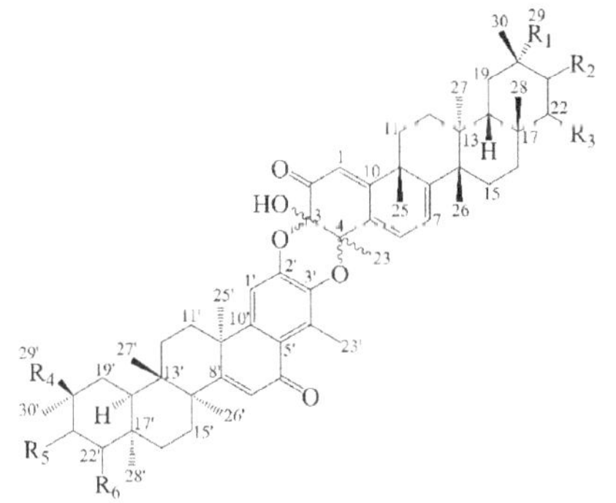

	C-3	C-4	R$_1$	R$_2$	R$_3$	R$_4$	R$_5$	R$_6$
D25	(R)	(R)	COOH	H	H	COOH	H	H
D26	(S)	(S)	COOH	H	H	COOH	H	H
D40	(R)	(R)	COOCH$_3$	H	H	COOCH$_3$	H	H
D41	(S)	(S)	H	=O	H	COOCH$_3$	H	H
D42	(S)	(S)	H	=O	OHβ	COOCH$_3$	H	H
D43	(R)	(R)	H	=O	OHβ	COOCH$_3$	H	H
D44	(S)	(S)	COOCH$_3$	H	H	H	=O	OHα
D45	(R)	(R)	COOCH$_3$	H	H	H	=O	OHα
D46	(S)	(S)	H	=O	H	COOCH$_3$	=O	H
D47	(S)	(S)	COOCH$_3$	=O	H	H	=O	H

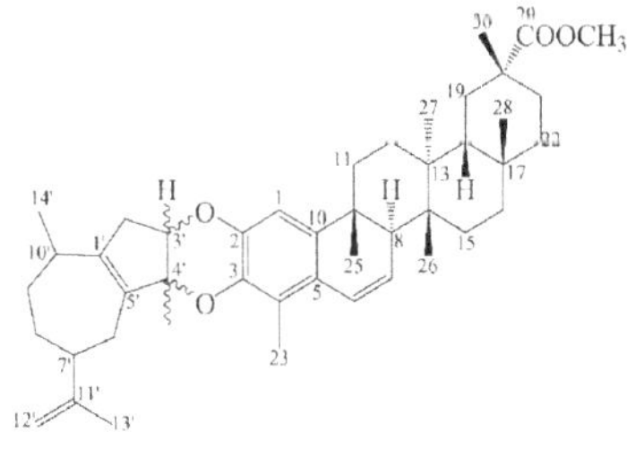

	C-3'	C-4'
D23	(R)	(S)
D24	(S)	(R)

(b)

Figure 7. *Cont.*

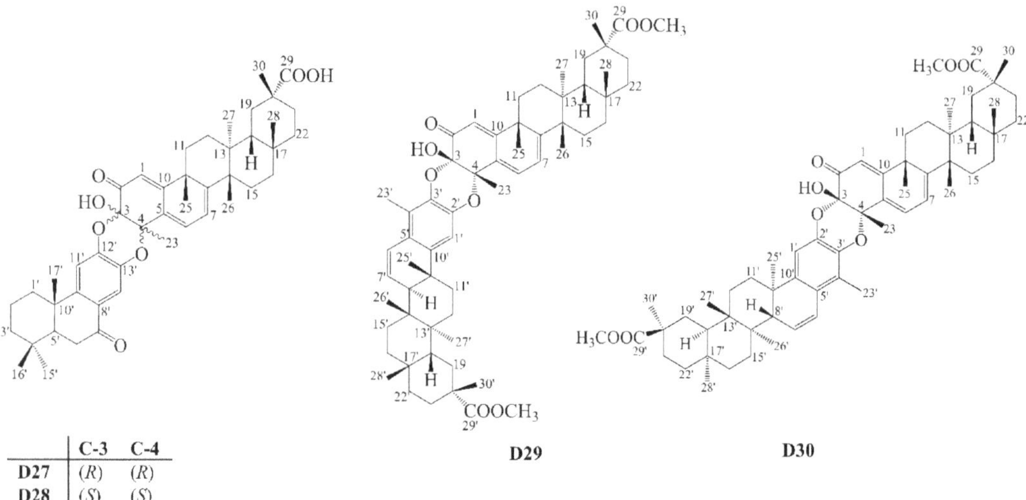

	C-3	C-4
D27	(*R*)	(*R*)
D28	(*S*)	(*S*)

D29

D30

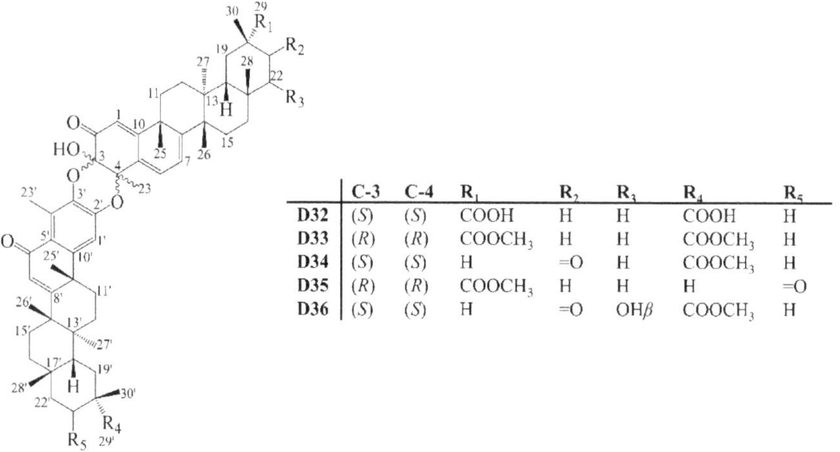

	C-3	C-4	R₁	R₂	R₃	R₄	R₅
D32	(*S*)	(*S*)	COOH	H	H	COOH	H
D33	(*R*)	(*R*)	COOCH₃	H	H	COOCH₃	H
D34	(*S*)	(*S*)	H	=O	H	COOCH₃	H
D35	(*R*)	(*R*)	COOCH₃	H	H	H	=O
D36	(*S*)	(*S*)	H	=O	OH*β*	COOCH₃	H

(**c**)

Figure 7. *Cont.*

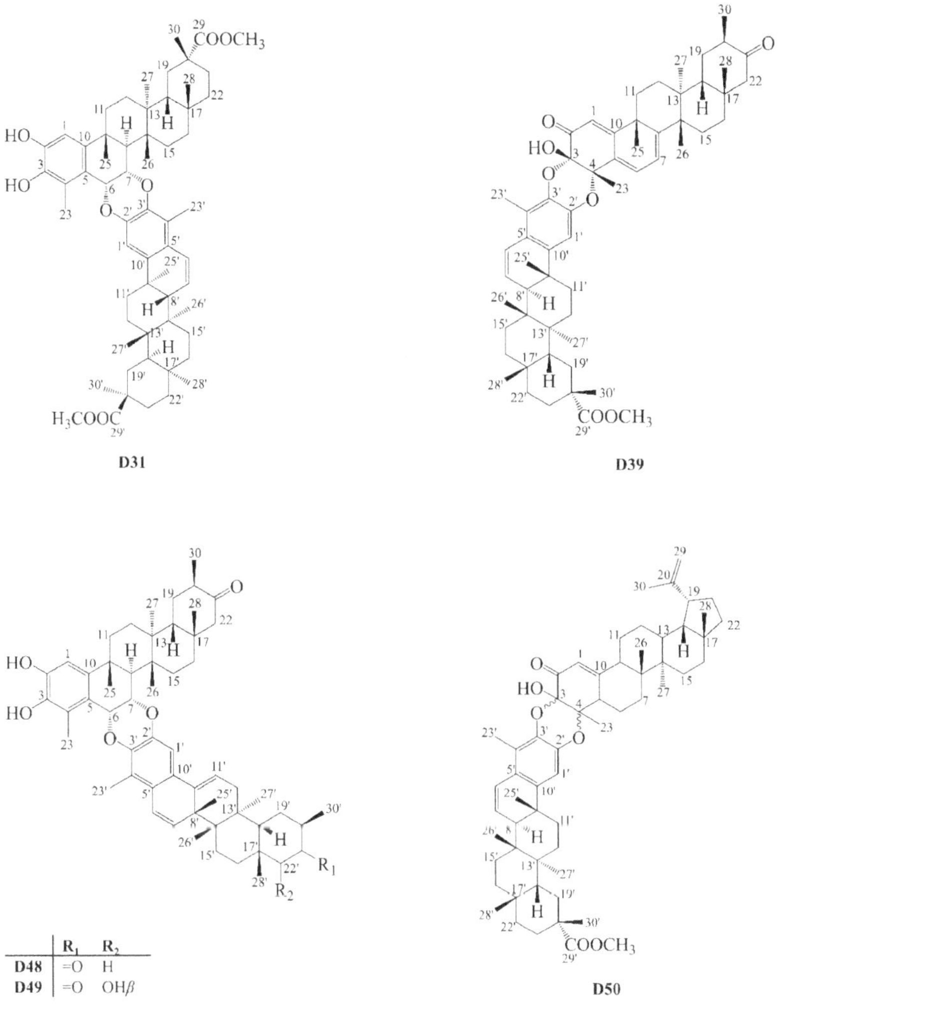

D31

D39

D48 | R₁ =O | R₂ H
D49 | =O | OHβ

D50

(**d**)

Figure 7. Structures of dimer-type pentacyclic triterpenoids isolated from Celastraceae species (2001–2021). (**a**) Compounds **D1–D14**. (**b**) Compounds **D15–D26**, **D37**, **D38** and **D40–D47**. (**c**) Compounds **D27–D30** and **D32–D36**. (**d**) Compounds **D31**, **D39** and **D48–D50**.

2.4. Lupanes

Unlike other skeletons, lupane-type PCTTs are formed by a *trans* pentacyclic ring system, in which the E ring is five-membered with an isopropenyl α substituent at carbon 19, containing a double bond between carbons 20 and 29 [19,22]. They have seven methyl groups, with two geminal ones attached to carbon 4 (Me 23α and 24β) and the others attached to carbon 8 (Me 26β), 10 (Me 25β), 14 (Me 27α), 17 (Me 28β), and 20 (Me 30), respectively. In this review, 89 pentacyclic triterpenoids of the lupane-type (**L**), **L1–L89**, were reported (Figure 8).

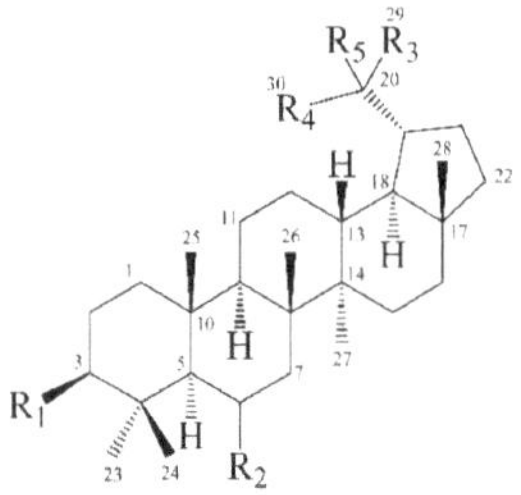

	R$_1$	R$_2$	R$_3$	R$_4$	R$_5$
L1	OHβ	H	CHO	CH_3	H
L2	OHβ	H	CH_3	CHO	H
L44	=O	OHβ	CH_3	CH_3	OH
L76	OHβ	H	CH_3	CH_3	OH
L84	OHβ	OHβ	CH_3	CH_3	OH

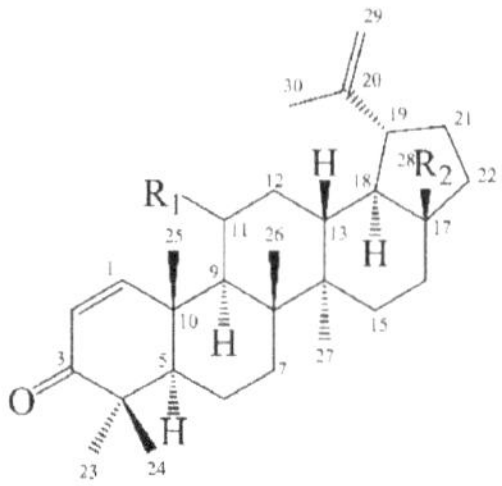

	R$_1$	R$_2$
L4	OHα	CH_3
L13	H	CH_2OH
L19	H	CHO
L61	H	CH_3

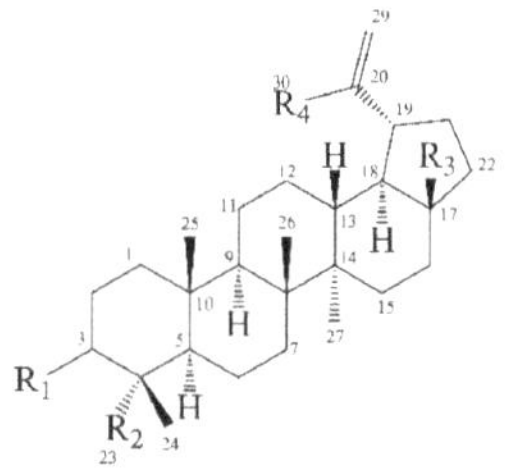

	R$_1$	R$_2$	R$_3$	R$_4$
L7	=O	CH_3	OH	CH_3
L83	OHβ	CH_2OH	CH_3	CH_2OH

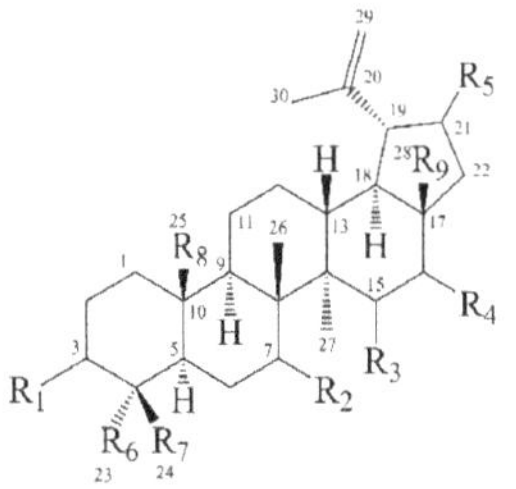

Senecioyl:

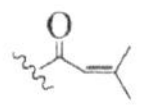

	R$_1$	R$_2$	R$_3$	R$_4$	R$_5$	R$_6$	R$_7$	R$_8$	R$_9$
L6	=O	H	H	OHβ	H	CH_3	CH_3	CH_3	CH_3
L9	=O	H	H	H	OHβ	CH_3	CH_3	CH_3	CH_3
L10	=O	H	H	H	H	CH_3	CH_2OH	CH_3	CH_2OH
L33	OHβ	OHβ	H	H	H	CH_3	CH_3	CH_3	$COOCH_3$
L51	=O	H	H	H	H	CH_2OH	CH_3	CH_3	COOH
L56	OHα	OSenecioylβ	H	H	H	CH_3	CH_3	CH_3	COOH
L67	=O	H	H	H	=O	CH_3	CH_3	CH_3	CH_3
L74	OHβ	H	OHα	H	H	CH_3	CH_3	CH_3	CH_3
L75	OHβ	H	H	OHβ	H	CH_3	CH_3	CH_3	CH_3
L77	OHβ	H	H	H	H	CH_3	CH_3	CH_2OH	CH_3

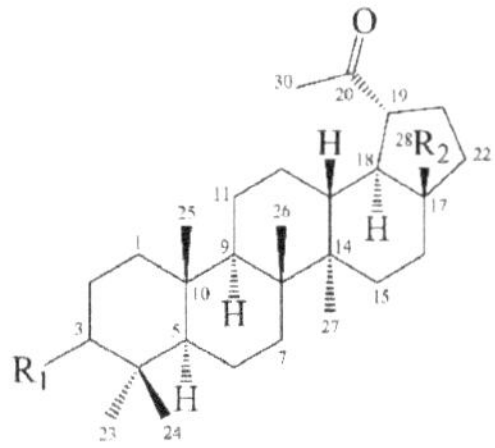

	R$_1$	R$_2$
L15	=O	CH_3
L28	OHβ	OHβ
L30	OHβ	CH_2OH
L39	OHβ	CHO
L40	OHβ	CH_3

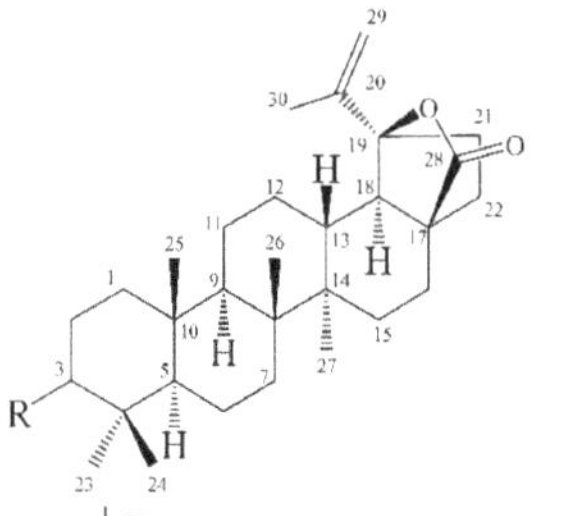

	R
L17	OHα
L87	OHβ
L88	=O

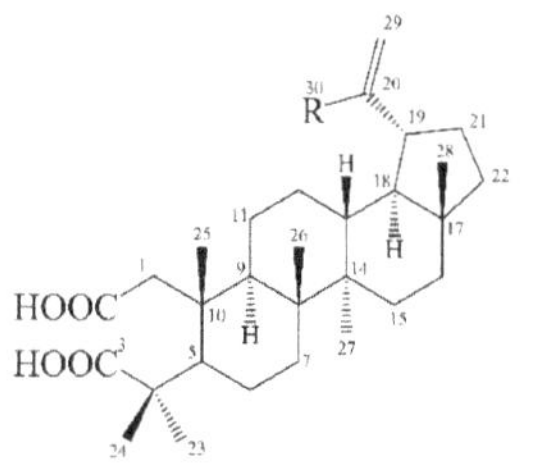

	R
L49	CH_3
L50	CH_2OH

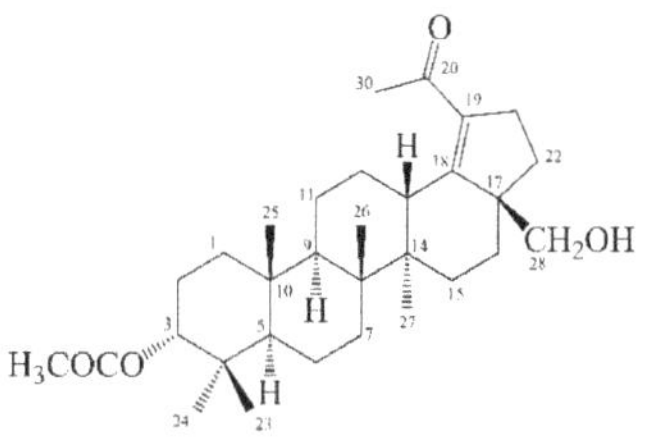

L22

(**a**)

Figure 8. *Cont.*

	R_1	R_2	R_3	R_4	R_5	R_6	R_7
L3	H	H	=O	H	OHα	OH	CH$_3$
L5	H	H	=O	H	OHα	CH$_3$	CH$_3$
L8	OHα	H	=O	H	H	CH$_3$	CH$_3$
L11	H	H	=O	H	H	CH$_2$OH	CH$_2$OH
L12	H	H	=O	H	H	CH$_2$OH	CHO
L14	H	H	=O	H	H	CH$_2$OH	CH$_3$
L16	H	H	=O	H	H	CH$_3$	CH$_2$OH
L18	H	H	=O	H	OHα	CHO	CH$_3$
L20	H	H	=O	H	H	CH$_3$	CHO
L21	H	=O	OHα	H	H	CH$_2$OH	CH$_3$
L23	H	H	OCOCH$_3\alpha$	H	H	CH$_3$	CHO
L24	H	H	OHα	H	H	CHO	CH$_3$
L25	H	=O	OHα	H	H	CH$_3$	CH$_3$
L26	H	H	(*E*) OCoumaroyl	H	H	CH$_2$OH	CH$_3$
L27	H	H	(*Z*) OCoumaroyl	H	H	CH$_2$OH	CH$_3$
L31	H	=O	OHβ	H	H	CH$_2$OH	CH$_3$
L32	H	H	OHβ	H	H	CHO	CH$_2$OH
L34	OHβ	H	OCOCH$_3\beta$	H	H	CH$_3$	CH$_3$
L35	H	H	OCaffeoylβ	H	H	CH$_2$OH	CH$_3$
L36	OHβ	H	OCaffeoylβ	H	H	CH$_3$	CH$_3$
L37	H	H	OCaffeoylβ	H	H	CH$_3$	CH$_3$
L38	H	H	OCO(CH$_2$)$_{16}$CH$_3\beta$	H	H	CH$_3$	CH$_3$
L41	H	=O	OHβ	H	H	CH$_3$	CH$_3$
L42	H	H	OHβ	H	H	CHO	CH$_3$
L43	H	H	OHβ	H	H	CH$_3$	CHO
L45	H	H	=O	OHβ	H	CH$_2$OH	CH$_3$
L46	H	H	=O	OHβ	H	CH$_3$	CH$_2$OH
L47	H	H	=O	OHβ	H	CH$_3$	CHO
L48	H	H	OCOCH$_3\beta$	H	H	CH$_3$	CH$_3$
L52	H	H	=O	OHβ	H	COOH	CH$_3$
L53	H	H	=O	H	H	COOH	CH$_3$
L54	H	H	=O	H	H	CH$_3$	COOH
L57	H	H	OHα	H	H	CH$_3$	COOH
L58	H	H	OHα	H	H	COOH	CH$_3$
L59	H	H	OHβ	H	H	COOH	CH$_3$
L60	H	H	OCH$_3\beta$	H	H	COOH	CH$_3$
L62	H	H	OHβ	H	H	CH$_3$	CH$_2$OH
L63	OHβ	H	OHα	H	H	CH$_2$OH	CH$_3$
L64	OHβ	H	OHα	H	H	CH$_3$	CH$_3$
L65	OHβ	H	OHβ	H	H	CH$_3$	CH$_3$
L66	H	OHβ	OHβ	H	H	CH$_3$	CH$_3$
L68	H	H	=O	H	H	CH$_3$	CH$_3$
L69	H	H	OHα	H	OHα	CH$_2$OH	CH$_3$
L70	H	H	OHα	H	OHα	CH$_3$	CH$_3$
L71	H	H	OHα	H	H	CH$_2$OH	CH$_3$
L72	H	H	OHα	H	H	CH$_3$	CH$_3$
L73	H	H	OHβ	H	OHα	CH$_3$	CH$_3$
L78	H	H	OHβ	H	H	CH$_2$OH	CH$_2$OH
L79	H	H	OHβ	H	H	CH$_2$OH	CH$_3$
L80	H	H	OHβ	H	H	CH$_3$	CH$_2$OH
L81	H	H	OHβ	OHβ	H	CH$_3$	CH$_3$
L82	H	H	OHβ	H	H	CH$_3$	CH$_3$
L85	H	H	OHβ	OHβ	H	CH$_2$OH	CH$_3$

(*E*)Coumaroyl:

(*Z*)Coumaroyl:

Caffeoyl:

(**b**)

Figure 8. *Cont.*

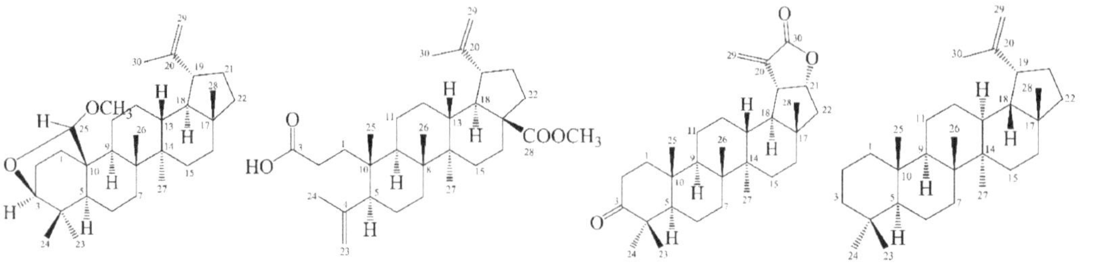

Figure 8. Structures of lupane-type pentacyclic triterpenoids isolated from Celastraceae species (2001–2021). (a) Compounds **L1**, **L2**, **L4**, **L6**, **L7**, **L9**, **L10**, **L13**, **L15**, **L17**, **L19**, **L22**, **L28**, **L30**, **L33**, **L39**, **L40**, **L44**, **L49-L51**, **L56**, **L61**, **L67**, **L74–L77**, **L83**, **L84**, **L87** and **L88**. (b) Compounds **L3**, **L5**, **L8**, **L11**, **L12**, **L14**, **L16**, **L18**, **L20**, **L21**, **L23–27**, **L31**, **L32**, **L34–L38**, **L41–L43**, **L45–L48**, **L52–L54**, **L57–L60**, **L62–L66**, **L68–L73**, **L78–L82** and **L85**. (c) Compounds **L29**, **L55**, **L86** and **L89**.

Characteristic ^{13}C-NMR signals of the class of lupanes are those in the olefinic region, which appear around δ_C 109 (C-29) and δ_C 150 ppm (C-20), and signals from the methine carbons C-5 (Hα), C- 9 (Hα), C-13 (Hβ), C-18 (Hα), and C-19 (Hβ), observed around δ_C 55, 50, 38, 48 and 47 ppm, respectively.

2.5. Oleananes

Oleanane-type triterpenoids are characterized by the presence of a double bond, most commonly between carbons 12 and 13. Rings A/B, B/C, and C/D have *trans* configuration, whereas rings D/E are *cis*. They have eight methyl groups. Geminal ones 23 (α) and 24 (β) are connected to carbon 4, and 29 (α) and 30 (β) to carbon 20. The others are connected to carbons 8 (Me 26β), 10 (Me 25β), 14 (Me 27α) and 17 (Me 28β) [19]. In this work, 102 pentacyclic triterpenoids with oleanane skeleton (**O**), **O1–O102**, were reported (Figure 9).

In the ^{13}C-NMR spectrum, the signals that characterize oleananes are those related to the double bond carbon atoms. For the most common oleananes with double bond between carbons 12 and 13, the chemical shifts are observed around δ_C 122 (C-12) and δ_C 145 ppm (C-13), except for those that have substituents close to these carbons [27].

2.6. Ursanes

Ursanes differ structurally from oleananes only by the position of methyl group 29, which is attached to carbon 19, in a β position. In the structure of ursanes, methyl group 30 is found in α position. Rings A/B, B/C and C/D have *trans* configuration, while rings D/E have *cis* configuration, like oleananes. The most common ursanes also present a double bond between carbons 12 and 13 [19]. There were 88 ursanes (**U**) isolated from Celastraceae species, triterpenoids **U1** to **U88**, were reported (Figure 10).

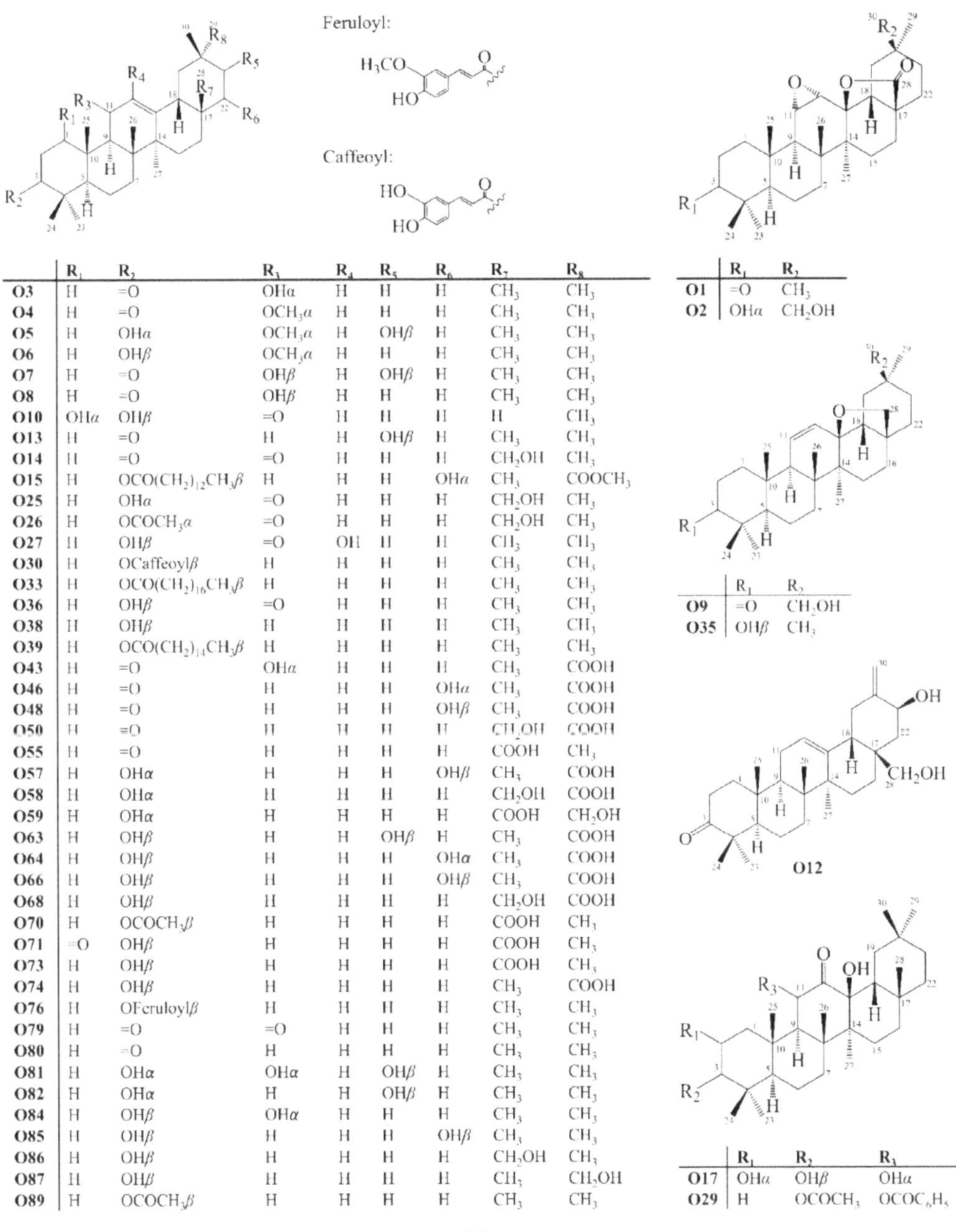

	R₁	R₂	R₃	R₄	R₅	R₆	R₇	R₈
O3	H	=O	OHα	H	H	H	CH₃	CH₃
O4	H	=O	OCH₃α	H	H	H	CH₃	CH₃
O5	H	OHα	OCH₃α	H	OHβ	H	CH₃	CH₃
O6	H	OHβ	OCH₃α	H	H	H	CH₃	CH₃
O7	H	=O	OHβ	H	OHβ	H	CH₃	CH₃
O8	H	=O	OHβ	H	H	H	CH₃	CH₃
O10	OHα	OHβ	=O	H	H	H	H	CH₃
O13	H	=O	H	H	OHβ	H	CH₃	CH₃
O14	H	=O	=O	H	H	H	CH₂OH	CH₃
O15	H	OCO(CH₂)₁₂CH₃β	H	H	H	OHα	CH₃	COOCH₃
O25	H	OHα	=O	H	H	H	CH₂OH	CH₃
O26	H	OCOCH₃α	=O	H	H	H	CH₂OH	CH₃
O27	H	OHβ	=O	OH	H	H	CH₃	CH₃
O30	H	OCaffeoylβ	H	H	H	H	CH₃	CH₃
O33	H	OCO(CH₂)₁₆CH₃β	H	H	H	H	CH₃	CH₃
O36	H	OHβ	=O	H	H	H	CH₃	CH₃
O38	H	OHβ	H	H	H	H	CH₃	CH₃
O39	H	OCO(CH₂)₁₄CH₃β	H	H	H	H	CH₃	CH₃
O43	H	=O	OHα	H	H	H	CH₃	COOH
O46	H	=O	H	H	H	OHα	CH₃	COOH
O48	H	=O	H	H	H	OHβ	CH₃	COOH
O50	H	=O	H	H	H	H	CH₂OH	COOH
O55	H	=O	H	H	H	H	COOH	CH₃
O57	H	OHα	H	H	H	OHβ	CH₃	COOH
O58	H	OHα	H	H	H	H	CH₂OH	COOH
O59	H	OHα	H	H	H	H	COOH	CH₂OH
O63	H	OHβ	H	H	OHβ	H	CH₃	COOH
O64	H	OHβ	H	H	H	OHα	CH₃	COOH
O66	H	OHβ	H	H	H	OHβ	CH₃	COOH
O68	H	OHβ	H	H	H	H	CH₂OH	COOH
O70	H	OCOCH₃β	H	H	H	H	COOH	CH₃
O71	=O	OHβ	H	H	H	H	COOH	CH₃
O73	H	OHβ	H	H	H	H	COOH	CH₃
O74	H	OHβ	H	H	H	H	CH₃	COOH
O76	H	OFeruloylβ	H	H	H	H	CH₃	CH₃
O79	H	=O	=O	H	H	H	CH₃	CH₃
O80	H	=O	H	H	H	H	CH₃	CH₃
O81	H	OHα	OHα	H	OHβ	H	CH₃	CH₃
O82	H	OHα	H	H	OHβ	H	CH₃	CH₃
O84	H	OHβ	OHα	H	H	H	CH₃	CH₃
O85	H	OHβ	H	H	H	OHβ	CH₃	CH₃
O86	H	OHβ	H	H	H	H	CH₂OH	CH₃
O87	H	OHβ	H	H	H	H	CH₃	CH₂OH
O89	H	OCOCH₃β	H	H	H	H	CH₃	CH₃

	R₁	R₂
O1	=O	CH₃
O2	OHα	CH₂OH

	R₁	R₂
O9	=O	CH₂OH
O35	OHβ	CH₃

	R₁	R₂	R₃
O17	OHα	OHβ	OHα
O29	H	OCOCH₃	OCOC₆H₅

(a)

Figure 9. *Cont.*

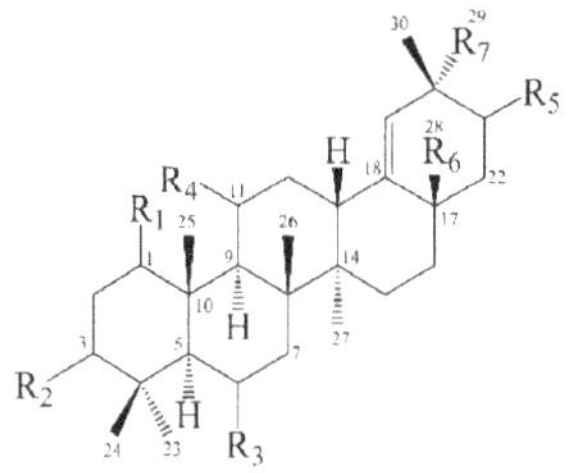

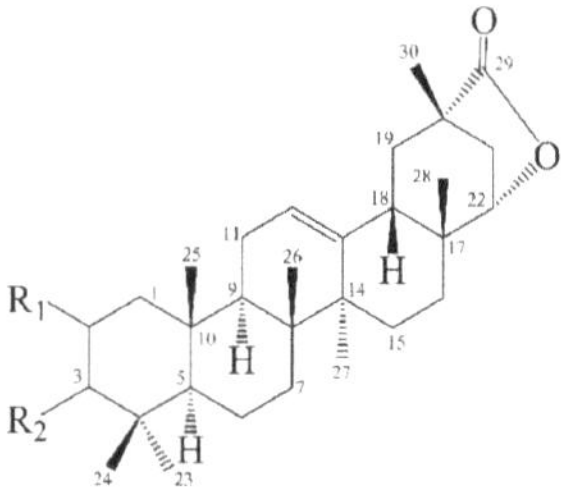

	R_1	R_2	R_3	R_4	R_5	R_6	R_7
O11	H	=O	H	H	OHα	CH$_3$	CH$_3$
O16	H	=O	H	H	H	CH$_3$	CH$_2$OH
O28	=O	OHβ	H	H	H	CH$_2$OH	CH$_3$
O32	H	OCO(CH$_2$)$_{16}$CH$_3$β	H	H	H	CH$_3$	CH$_3$
O40	H	=O	OHβ	H	H	CH$_3$	CH$_2$OH
O41	H	=O	OHβ	H	H	CH$_3$	CH$_3$
O90	H	=O	H	H	=O	CH$_3$	CH$_3$
O91	H	=O	H	H	H	CH$_3$	CH$_3$
O92	H	OHβ	H	OHα	H	CH$_2$OH	CH$_3$
O93	H	OHβ	H	OHα	H	CH$_3$	CH$_3$
O94	H	OHβ	H	H	OHα	CH$_3$	CH$_3$
O95	H	OHβ	H	H	H	CH$_3$	CH$_2$OH
O96	H	OHβ	OHβ	H	H	CH$_3$	CH$_3$
O97	H	OHβ	H	H	H	CH$_3$	CH$_3$

	R_1	R_2
O20	OHα	OHβ
O23	H	=O
O34	=O	OHβ
O37	H	OHβ

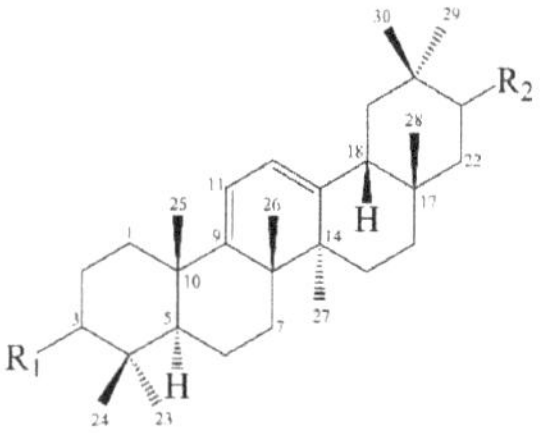

	R_1	R_2
O31	OCaffeoylβ	H
O98	=O	H
O99	OHα	OHβ
O100	OHβ	H

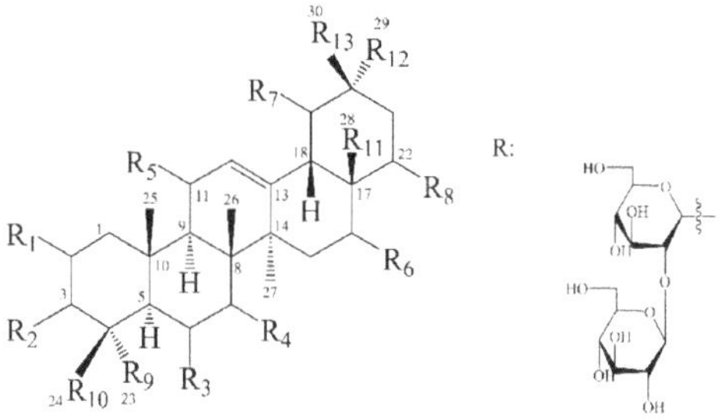

R:

	R_1	R_2	R_3	R_4	R_5	R_6	R_7	R_8	R_9	R_{10}	R_{11}	R_{12}	R_{13}
O18	OHα	OHβ	H	H	H	H	OHα	H	CH$_3$	CH$_2$OH	COOR	CH$_3$	CH$_3$
O19	OHα	OHβ	H	H	H	H	H	H	CH$_3$	CH$_2$OH	COOR	CH$_3$	CH$_3$
O22	H	=O	H	H	=O	H	H	H	CH$_3$	CH$_3$	CH$_3$	CH$_3$	CH$_2$OH
O42	H	=O	H	OHβ	H	H	H	H	CH$_3$	CH$_3$	CH$_2$OH	CH$_3$	CH$_3$
O44	H	=O	H	H	OCH$_3$α	H	H	H	CH$_3$	CH$_3$	CH$_3$	CH$_3$	COOH
O47	H	=O	H	H	H	H	H	OHα	CH$_3$	CH$_3$	CH$_3$	CH$_3$	COOH
O49	H	=O	H	H	H	H	H	H	CH$_2$OH	CH$_3$	COOH	CH$_3$	CH$_3$
O51	OHα	OHβ	H	H	H	H	H	H	CH$_3$	CH$_3$	COOH	CH$_3$	CH$_3$
O52	H	OH	H	H	OHα	H	H	H	CH$_3$	CH$_3$	CH$_3$	CH$_3$	COOH
O54	H	=O	H	H	H	H	H	H	CH$_3$	H	COOH	CH$_3$	CH$_3$
O56	H	=O	H	H	H	H	H	H	CH$_3$	CH$_3$	CH$_3$	CH$_3$	COOH
O60	H	OHα	H	H	H	H	H	H	CH$_3$	CH$_3$	COOH	CH$_3$	CH$_2$OH
O61	H	OHα	H	H	H	H	H	H	COOH	CH$_3$	CH$_3$	CH$_3$	CH$_3$
O62	H	OHβ	H	H	H	OHα	H	H	CHO	CH$_3$	COOH	CH$_3$	CH$_3$
O65	H	OHβ	H	H	H	H	H	OHα	CH$_3$	CH$_3$	CH$_3$	CH$_3$	COOH
O67	H	OHβ	H	H	H	H	H	H	CH$_2$OH	CH$_3$	COOH	CH$_3$	CH$_3$
O69	H	OHβ	OHβ	H	H	H	H	H	CH$_3$	CH$_3$	COOH	CH$_2$OH	CH$_3$
O83	H	OHα	H	H	H	H	H	H	CH$_2$OH	CH$_3$	CH$_2$OH	CH$_3$	CH$_3$
O88	H	OHβ	H	H	H	H	H	H	CH$_3$	CH$_3$	CH$_3$	CH$_3$	CH$_2$OH

(**b**)

Figure 9. *Cont.*

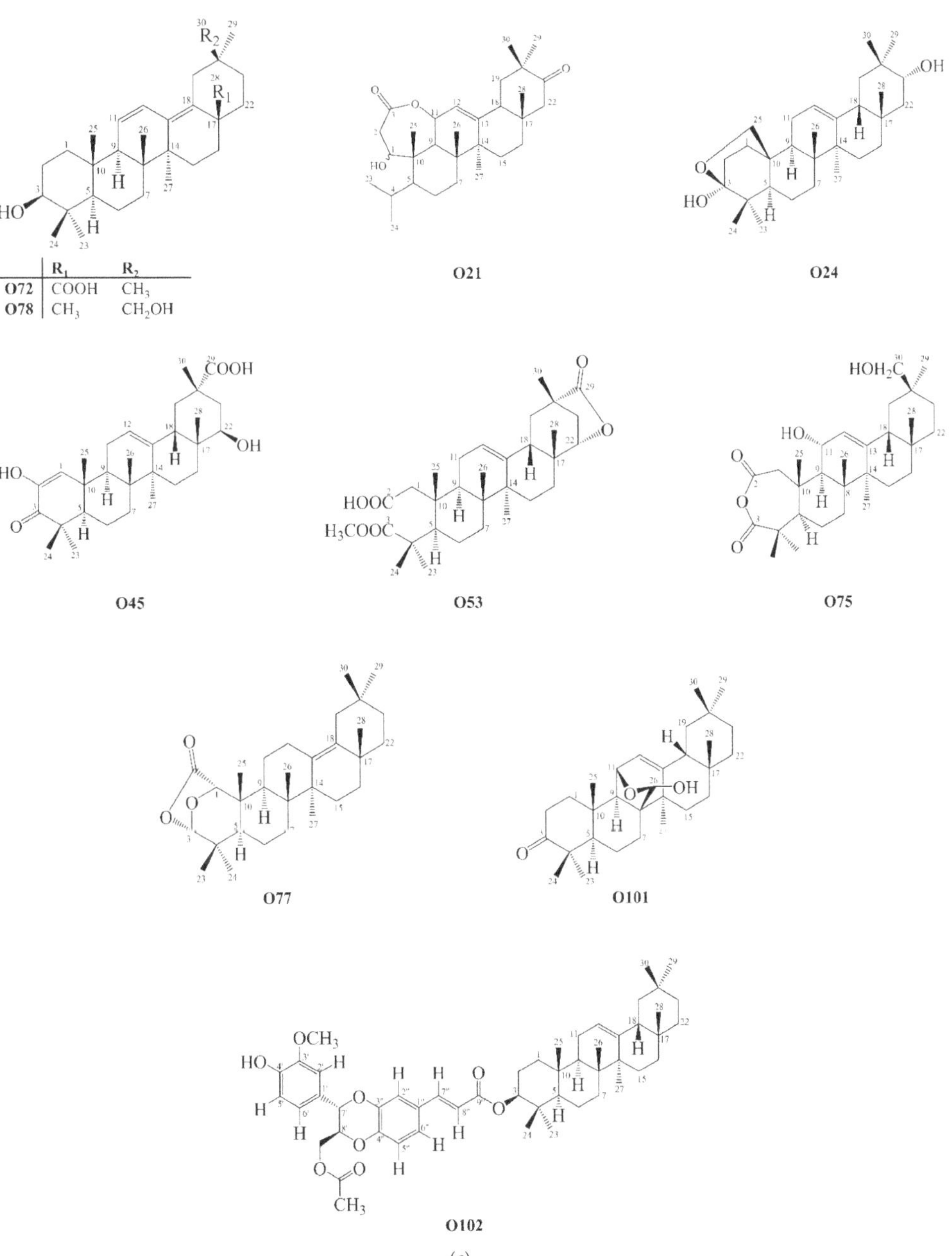

Figure 9. Structures of oleanane-type pentacyclic triterpenoids isolated from Celastraceae species (2001–2021). (**a**) Compounds **O1–O10**, **O12–O15**, **O17**, **O25–O27**, **O29**, **O30**, **O33**, **O35**, **O36**, **O38**, **O39**, **O43**, **O46**, **O48**, **O50**, **O55**, **O57–O59**, **O63**, **O64**, **O66**, **O68**, **O70**, **O71**, **O73**, **O74**, **O76**, **O79–O82**, **O84–O87** and **O89**. (**b**) Compounds **O11**, **O16**, **O18–O20**, **O22**, **O23**, **O28**, **O31**, **O32**, **O34**, **O37**, **O40–O42**, **O44**, **O47**, **O49**, **O51**, **O52**, **O54**, **O56**, **O60–O62**, **O65**, **O67**, **O69**, **O83**, **O88** and **O90–O100**. (**c**) Compounds **O21**, **O24**, **O45**, **O53**, **O72**, **O75**, **O77**, **O78**, **O101**, and **O102**.

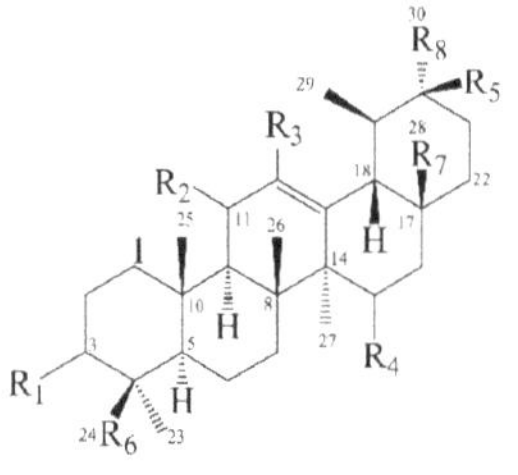

Coumaroyl:

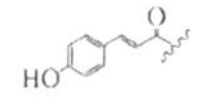

Caffeoyl:

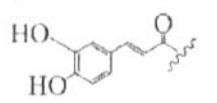

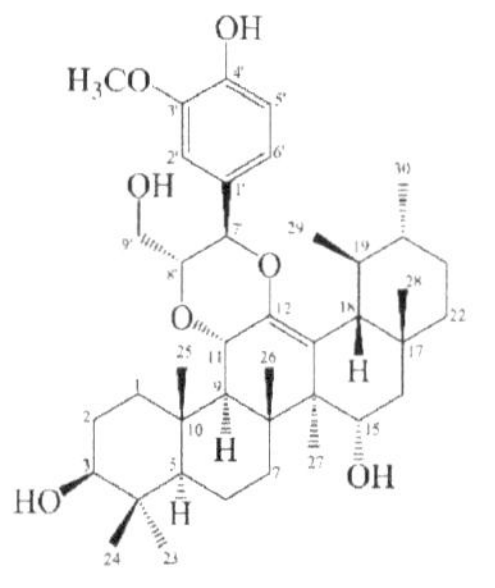

U4

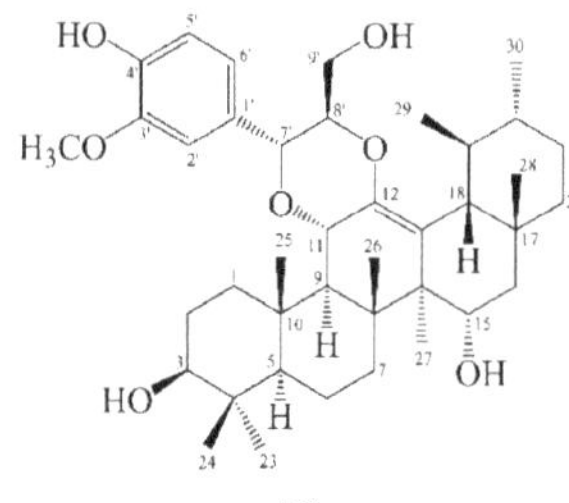

U5

U52

U53

	R_1	R_2	R_3	R_4	R_5	R_6	R_7	R_8
U1	=O	OHα	OH	OHα	OHβ	CH$_3$	CH$_3$	CH$_3$
U8	OHβ	OCH$_3\alpha$	OH	OHα	OHβ	CHO	CH$_3$	CH$_3$
U12	OHβ	OCH$_3\alpha$	OH	OHα	OHβ	CH$_3$	CH$_3$	CH$_3$
U18	OHβ	OCH$_3\alpha$	H	H	H	CH$_3$	CH$_2$OH	CH$_3$
U22	=O	=O	OH	OHα	H	CH$_2$OH	CH$_3$	COOCH$_3$
U23	=O	=O	OH	OHα	OHβ	CH$_3$	CH$_3$	CH$_3$
U24	=O	=O	OH	OHα	H	CH$_2$OH	CH$_3$	CH$_2$OH
U34	=O	H	H	H	H	CH$_3$	CH$_2$OH	CH$_2$OH
U38	OCaffeoyl	H	H	H	H	CH$_3$	CH$_2$OH	CH$_3$
U39	OCoumaroyl	H	H	H	H	CH$_3$	CH$_2$OH	CH$_3$
U40	=O	=O	H	H	H	CH$_3$	CH$_3$	CH$_2$OH
U41	OHα	=O	H	H	H	CH$_3$	CH$_3$	CH$_2$OH
U42	OHβ	OHα	OH	OHα	OHβ	CHO	CH$_3$	CH$_3$
U54	OHβ	=O	H	H	H	CH$_3$	CH$_3$	CH$_2$OH
U68	=O	H	H	H	H	CH$_3$	CH$_2$OH	COOH
U71	=O	=O	H	H	H	CH$_3$	COOH	CH$_3$
U72	OHα	=O	H	H	H	CH$_3$	COOH	CH$_3$
U76	OCOCH$_3\alpha$	H	H	H	H	CH$_3$	COOH	CH$_3$
U77	OHβ	H	H	H	H	CH$_3$	COOH	CH$_3$
U84	OHβ	H	H	H	H	CH$_3$	CH$_2$OH	CH$_3$

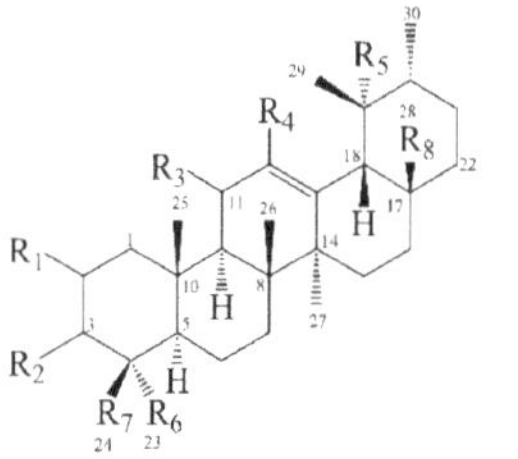

	R_1	R_2	R_3	R_4	R_5	R_6	R_7	R_8
U10	OHα	OHβ	OCH$_3\alpha$	OH	H	CH$_3$	CH$_2$OH	CH$_3$
U35	OHα	OHβ	=O	OH	H	CH$_3$	CH$_2$OH	CH$_3$
U36	OHα	OHβ	H	H	H	CH$_3$	CH$_2$OH	COOR
U37	OHα	OHβ	H	H	OHα	CH$_3$	CH$_2$OH	COOR
U55	H	OCOCH$_3\beta$	OHα	H	OHα	COOCH$_3$	CH$_3$	CH$_2$OH
U57	H	OCOCH$_3\beta$	H	H	OHα	CH$_2$OH	CH$_3$	CH$_2$OH
U64	H	=O	H	H	OHα	CH$_3$	COOH	CH$_2$OH
U69	OHα	OHβ	H	H	H	CH$_3$	CH$_3$	COOH
U70	OHα	=O	H	H	H	CH$_3$	CH$_3$	COOH
U73	H	OHβ	OHα	H	OHα	COOH	CH$_3$	COOH
U78	OHα	OHβ	H	H	OHα	CH$_3$	CH$_3$	COOR

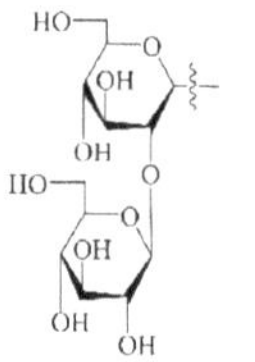

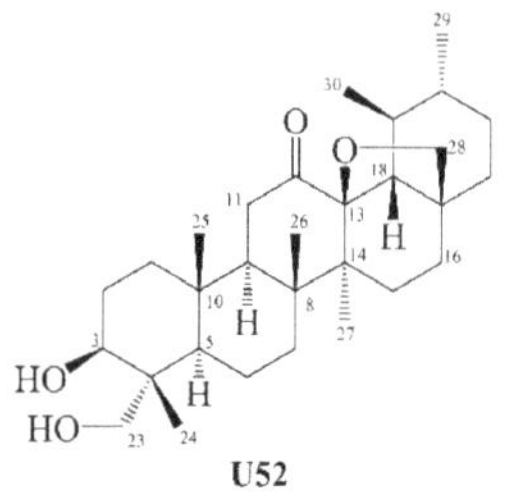

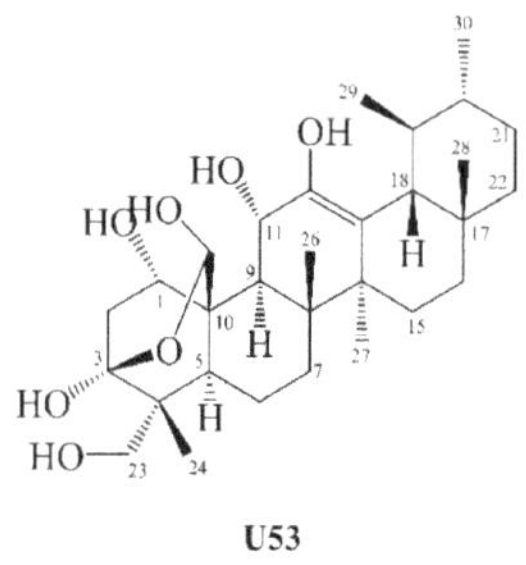

(**a**)

Figure 10. *Cont.*

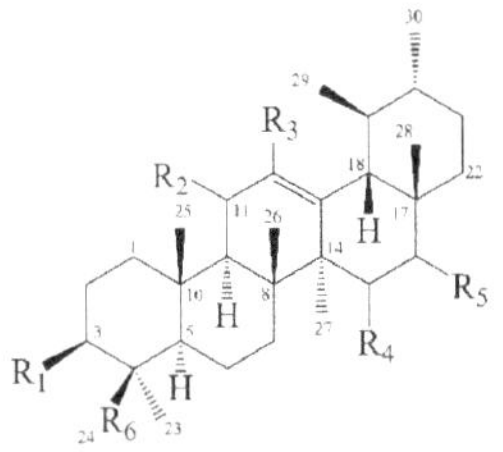

	R_1	R_2	R_3	R_4	R_5	R_6
U2	=O	OHα	OH	OHα	H	CH$_2$OH
U3	=O	OHα	OH	H	OHβ	CH$_3$
U6	=O	OHα	H	H	H	CH$_3$
U7	=O	OCH$_3\alpha$	OH	OHα	H	CH$_2$OH
U9	OHβ	OCH$_3\alpha$	OH	OHα	H	CHO
U11	=O	OCH$_3\alpha$	H	H	H	CH$_3$
U13	OHβ	OCH$_3\alpha$	OH	OHα	H	CH$_2$OH
U14	OHβ	OCH$_3\alpha$	OH	OHα	H	CH$_3$
U15	OHβ	OCH$_3\alpha$	OH	H	H	CH$_2$OH
U16	OHβ	OCH$_3\alpha$	OH	H	H	CH$_3$
U17	OHβ	OCH$_3\alpha$	H	OHα	H	CH$_3$
U19	OHβ	OCH$_3\alpha$	H	H	H	CH$_3$
U20	=O	OHβ	H	H	OHβ	CH$_3$
U21	=O	OHβ	H	H	H	CH$_3$
U25	=O	=O	OH	OHα	H	CH$_2$OH
U32	OHβ	OHα	OH	OCOC$_6$H$_5\alpha$	H	CHO
U33	OHβ	OHα	H	OCOC$_6$H$_5\alpha$	H	CH$_3$
U43	OHβ	OHα	OH	OHα	H	CHO
U44	OHβ	OHα	OH	OHα	H	COOCH$_3$
U48	OHβ	=O	OH	OHα	H	CH$_2$OH
U49	OHβ	=O	OH	OHα	H	CH$_3$
U50	OHβ	=O	OH	H	H	CH$_2$OH
U51	OHβ	=O	H	H	OHβ	CH$_3$
U58	OCOCH$_3\beta$	H	H	H	H	CH$_3$
U59	OCO(CH$_2$)$_{16}$CH$_3\beta$	H	H	H	H	CH$_3$
U61	OHβ	=O	H	H	H	CH$_3$
U62	OCO(CH$_2$)$_{14}$CH$_3\beta$	H	H	H	H	CH$_3$
U79	=O	=O	H	H	H	CH$_3$
U80	OHβ	OHα	OH	OHα	H	CH$_3$
U81	OHβ	OHα	OH	H	H	CH$_2$OH
U82	OHβ	OHα	H	OHα	H	CH$_3$
U85	OHβ	H	H	H	H	CH$_3$

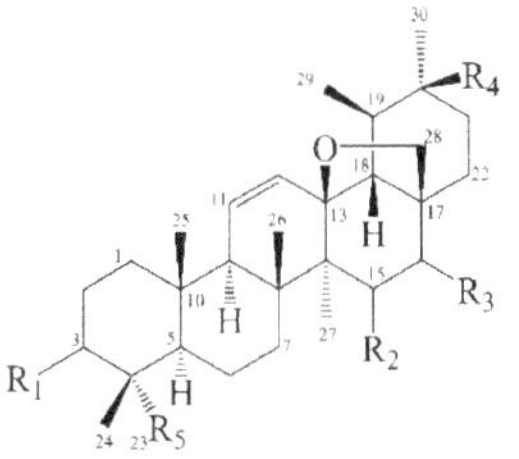

	R_1	R_2	R_3	R_4	R_5
U27	OHβ	OHα	H	H	CH$_3$
U28	OHβ	H	OHβ	H	CH$_3$
U29	OHβ	H	H	H	CH$_2$OH
U30	OHβ	H	H	H	CH$_3$
U31	=O	OHα	H	OHβ	CH$_3$

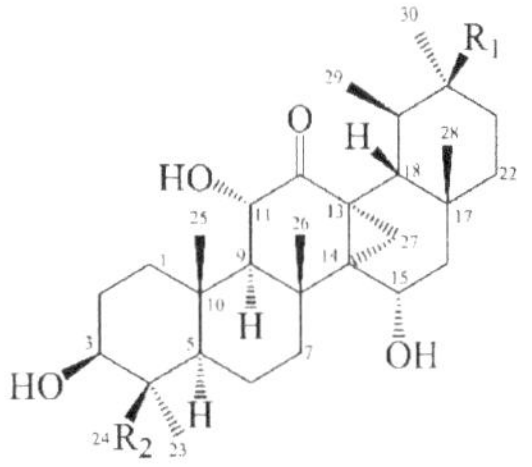

	R_1	R_2
U45	OH	CH$_3$
U46	H	CHO
U47	H	CH$_3$

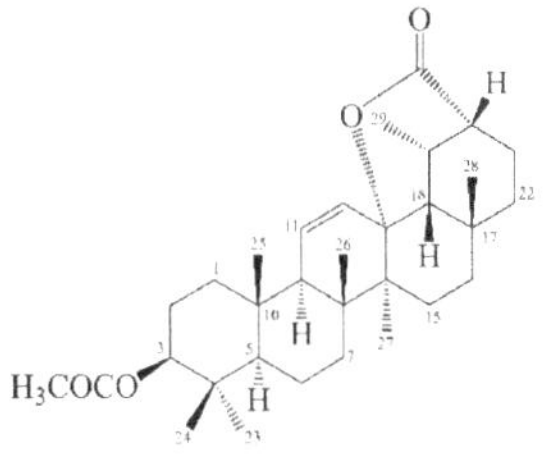

U56

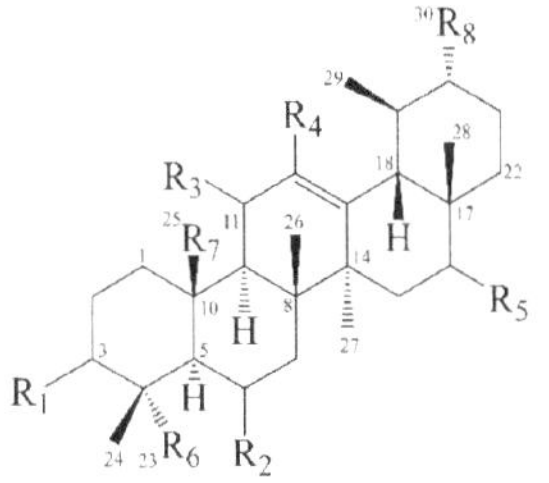

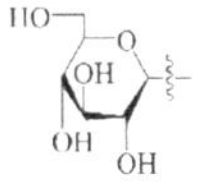

	R_1	R_2	R_3	R_4	R_5	R_6	R_7	R_8
U26	OGlcβ	H	OCH$_3\alpha$	OH	OGlcβ	CH$_2$OH	CH$_3$	CH$_3$
U63	=O	OHβ	OCH$_3\alpha$	OH	H	CH$_2$OH	CH$_3$	CH$_3$
U83	OHβ	H	H	H	H	CH$_3$	CH$_2$OH	CH$_2$OH

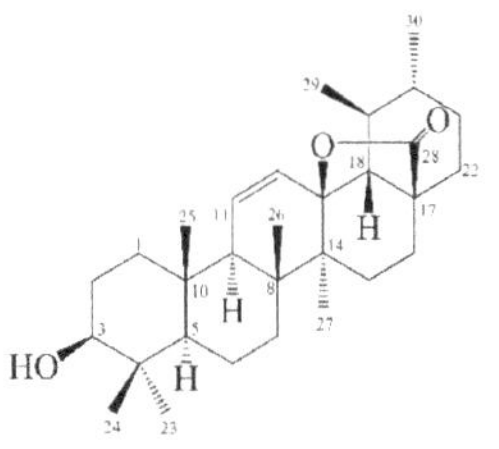

Glc:

U60

(b)

Figure 10. *Cont.*

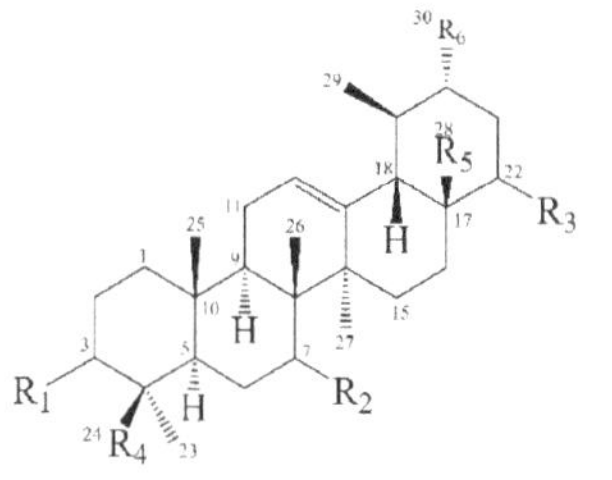

	R_1	R_2	R_3	R_4	R_5	R_6
U66	=O	H	OHα	CH$_3$	CH$_3$	COOH
U67	=O	H	OHβ	CH$_3$	CH$_3$	COOH
U74	OHβ	H	OHα	CH$_3$	CH$_3$	COOH
U75	OHβ	OHβ	H	CH$_2$OH	COOH	CH$_3$

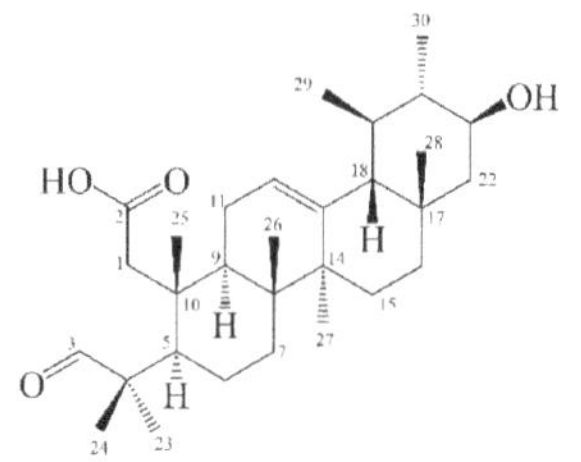

U65

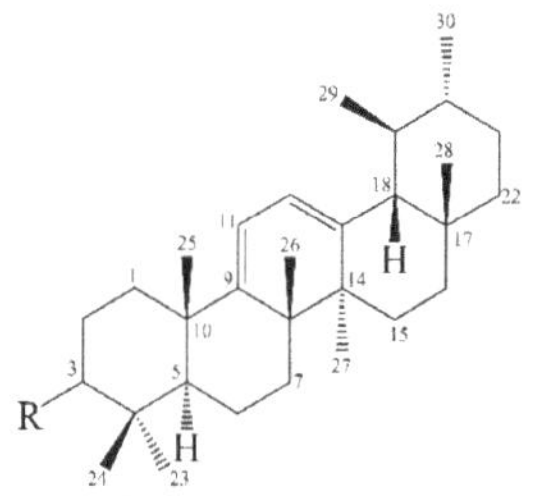

	R
U87	=O
U88	OHβ

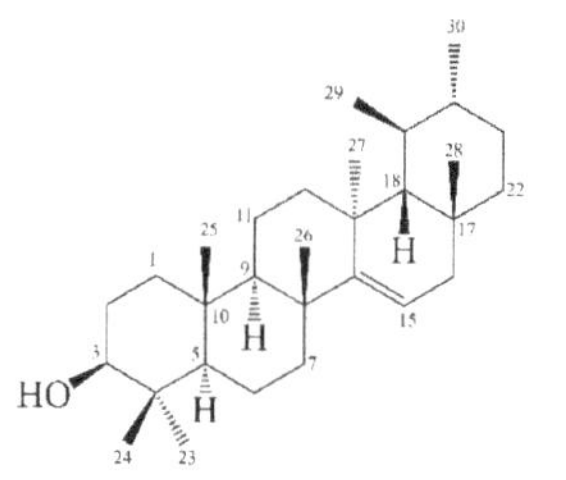

U86

(c)

Figure 10. Structures of ursane-type pentacyclic triterpenoids isolated from Celastraceae species (2001–2021). (**a**) Compounds **U1**, **U4**, **U5**, **U8**, **U10**, **U12**, **U18**, **U22–U24**, **U34–42**, **U52–U55**, **U57**, **U64**, **U68–U73**, **U76–U78** and **U84**. (**b**) Compounds **U2**, **U3**, **U6**, **U7**, **U9**, **U11**, **U13–U17**, **U19–U21**, **U25–U33**, **U43–U51**, **U56**, **U58–U63**, **U79–U83** and **U85**. (**c**) **U65–U67**, **U74**, **U75**, **U86–U88**.

^{13}C-NMR spectrum of ursanes differ from the spectrum of oleananes by the chemical shift signals of the olefinic carbon atoms, which are observed around δ_C 124 (C-12) and δ_C 139 ppm (C-13). In ursanes, the proximity of methyl group 29 with the double bond promotes a steric effect on these carbons, causing a shielding effect on C-13 and deshielding on C-12 [27,34]. This effect can be observed by comparing the ^{13}C-NMR data of **O3** and **U6**, for example. Additionally, the number of quaternary carbon signals also represents a distinction parameter between these two skeletons, since 6 signals are observed in the oleananes spectrum and 5 signals in the ursanes spectrum.

2.7. Other Triterpenoid Skeletons Isolated from Celastraceae

In addition to the PCTT types described above, other 21 types of pentacyclic structures were also isolated from Celastraceae species (Figure 11). The terpenoid skeletons are gammacerane (**OT1**), taraxane (**OT2**), hopane (**OT3**, **OT4**), glutinane (**OT6**, **OT16**, **OT18**, **OT19**), taraxerane (**OT7**, **OT21**), germanicane (**OT17**) and unidentified types (**OT5**, **OT8**, **OT9**, **OT10**, **OT11**, **OT12**, **OT13**, **OT14**, **OT15**, **OT20**).

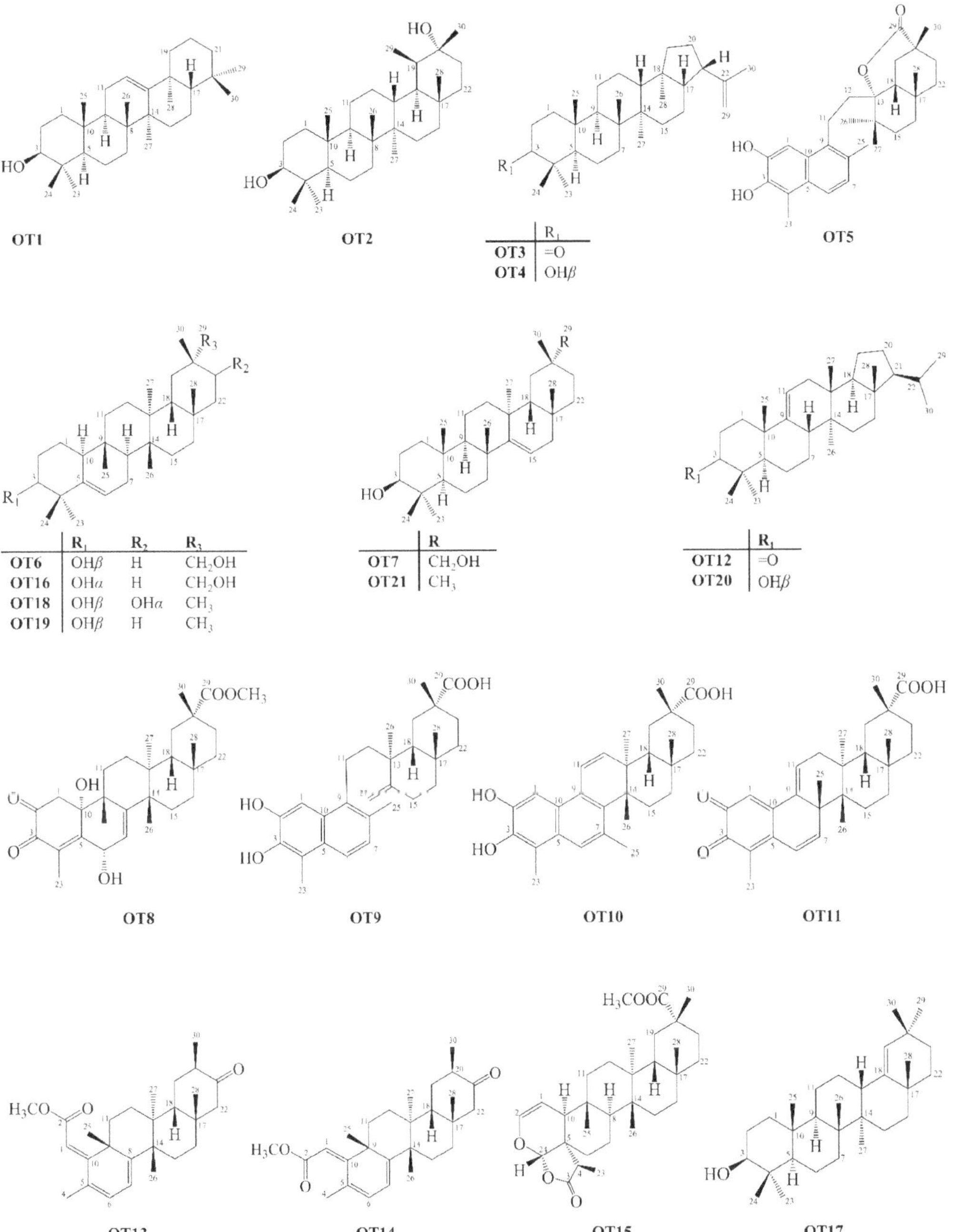

Figure 11. Structures of pentacyclic triterpenoids classified as others isolated from Celastraceae species (2001–2021).

3. ^{13}C-NMR Data of Pentacyclic Triterpenoids Isolated from Celastraceae Species (2001–2021)

Tables 1–8 list the literature ^{13}C-NMR data of the PCTTs that were isolated and characterized in the period of 2001–2021.

Table 1. ^{13}C-NMR data of friedelane-type pentacyclic triterpenoids isolated from Celastraceae species (2001–2021).

C	F1	F2	F3	F4 [a]	F6 [b]	F7	F8 [b]	F9	F10	F11
1	25.0	22.3	22.3	22.3	22.8	22.3	202.8	22.2	202.4	22.8
2	41.6	41.4	41.2	41.4	42.7	41.5	60.6	41.4	60.4	41.9
3	213.3	212.7	212.3	212.7	211.9	213.0	204.0	213.0	203.8	212.1
4	52.1	58.1	58.0	58.1	57.8	58.2	58.9	58.1	58.6	58.3
5	43.1	41.8	41.8	41.9	42.4	42.0	38.1	42.0	37.4	42.4
6	42.2	41.1	40.8	41.2	41.7	41.3	41.5	41.1	40.2	41.5
7	17.9	18.3	18.4	18.1	22.7	20.0	20.4	18.3	17.9	18.9
8	52.8	51.3	52.0	52.9	54.4	53.5	52.4	49.9	48.8	50.8
9	44.1	38.3	38.1	38.2	38.0	37.8	37.1	37.4	36.6	38.0
10	60.1	59.3	59.1	59.4	59.8	59.4	72.3	59.5	71.5	59.8
11	76.9	47.0	47.3	47.4	37.1	35.6	34.9	34.9	33.7	35.6
12	42.0	72.1	71.2	72.7	31.6	31.2	29.9	29.3	29.1	30.4
13	41.1	44.8	45.7	45.3	41.0	40.6	39.7	39.9	39.3	39.9
14	38.2	41.1	40.8	40.4	46.8	44.1	42.9	39.2	38.8	40.2
15	32.4	31.6	50.3	33.4	75.5	74.6	31.3	36.8	38.7	40.2
16	35.9	35.9	218.3	36.1	48.4	48.4	78.1	74.3	75.5	75.7
17	30.0	31.3	45.7	30.8	30.9	30.2	35.0	41.0	37.0	37.7
18	42.5	44.3	45.4	44.2	42.6	41.6	45.5	41.7	45.6	46.66
19	35.4	31.7	38.5	38.4	32.2	35.6	35.7	33.7	33.8	34.3
20	28.1	33.3	27.8	28.4	28.4	28.1	28.1	28.0	27.7	28.6
21	32.7	29.9	31.5	32.7	36.2	31.9	31.8	33.8	33.2	35.0
22	39.2	38.1	30.7	39.6	39.5	38.7	34.9	19.8	26.2	27.8
23	6.9	6.8	6.8	6.8	7.2	6.8	7.3	6.7	7.0	7.5
24	14.8	14.6	14.6	14.6	14.4	14.5	15.8	14.7	15.7	15.1
25	12.9	19.2	18.7	19.3	16.9	17.9	18.2	19.2	18.8	19.4
26	20.1	18.7	20.5	20.5	65.8	14.0	63.2	16.5	16.8	17.7
27	19.5	11.6	9.0	11.6	19.7	18.7	20.9	19.7	19.1	20.0
28	32.0	31.7	27.3	31.8	32.7	32.6	25.9	71.3	29.9	31.0
29	31.7	71.6	31.3	34.9	35.7	30.9	37.8	37.0	31.8	32.7
30	35.0	29.2	35.0	31.9	31.0	35.6	30.4	31.8	36.2	36.9
C=O							171.3			
OCH$_3$							21.3			
Ref	[35]	[36]	[37]	[38]	[39]	[40]	[41]	[42]	[43]	[44]

C	F12 [a]	F13	F14 [b]	F15 [a]	F16	F17	F18	F19	F20	F21
1	21.7	22.2	22.3	68.8	75.8	74.0	71.4	202.7	22.3	22.3
2	40.9	41.4	41.6	51.9	35.0	30.1	52.7	60.6	41.5	41.3
3	211.1	212.8	212.5	210.4	212.3	213.2	211.3	204.1	213.2	212.7
4	57.4	58.0	58.3	57.5	55.9	53.3	59.1	59.1	58.2	58.1
5	41.5	42.1	42.3	39.4	43.3	42.7	43.9	37.8	42.0	41.8
6	40.5	41.1	41.4	41.8	41.8	41.2	34.3	40.6	41.2	40.2
7	18.0	18.4	18.6	17.8	17.9	18.2	18.8	18.1	18.0	21.2
8	53.0	53.3	53.5	53.4	53.8	53.0	53.9	50.6	51.2	44.7
9	36.9	37.4	37.6	36.6	42.0	36.9	38.6	37.3	37.5	37.9
10	58.7	59.3	59.7	64.8	57.4	52.4	62.6	71.9	59.5	59.0
11	35.1	35.6	35.8	37.0	30.2	35.9	35.7	34.2	35.3	36.8
12	30.0	30.7	30.8	29.9	31.0	30.5	30.4	29.8	30.2	33.9
13	39.6	40.1	39.3	39.4	39.8	39.8	39.8	38.8	39.0	47.5
14	38.8	39.1	40.1	39.3	37.5	38.4	38.5	38.7	38.8	56.3
15	44.7	44.3	44.4	33.2	32.7	32.1	32.7	30.5	30.4	213.3

Table 1. *Cont.*

C										
16	75.7	74.4	75.6	36.0	36.0	35.4	36.2	35.9	36.1	53.8
17	40.5	36.4	32.1	30.4	29.9	30.0	30.2	32.5	32.5	34.0
18	39.9	44.1	44.8	41.9	42.7	42.7	43.0	44.2	44.3	45.3
19	29.3	30.4	35.8	30.8	35.3	28.9	35.5	36.0	35.9	38.9
20	32.7	33.1	28.0	33.2	28.1	33.4	28.2	34.3	34.4	28.0
21	27.7	27.5	32.1	27.8	32.7	28.2	33.0	74.3	74.3	71.7
22	29.5	36.4	36.0	39.7	39.3	28.1	39.3	47.0	47.0	46.3
23	6.4	6.9	6.8	7.1	6.7	6.9	6.9	7.3	6.8	6.8
24	14.0	14.6	14.7	16.0	14.7	14.3	17.4	16.0	14.6	14.9
25	17.5	18.1	18.2	18.4	63.0	17.9	19.2	18.2	17.7	18.9
26	21.1	20.1	20.1	18.6	20.1	20.0	20.2	17.8	18.2	15.4
27	19.3	21.4	21.5	21.0	18.6	18.7	18.7	19.3	19.3	12.1
28	66.7	25.4	24.9	32.1	32.2	32.2	32.2	33.1	33.2	31.5
29	74.0	74.4	30.8	74.5	35.0	28.9	31.8	24.9	31.9	31.4
30	25.4	25.7	35.5	25.9	31.7	72.0	34.9	31.8	24.9	33.8
Ref	[45]	[46]	[27]	[45]	[47]	[48]	[48]	[49]	[50]	[51]

C	F22	F23	F24	F25	F26	F27	F28 [b]	F29	F30	F31
1	22.4	202.8	24.7	24.6	202.8	22.5	22.4	22.3	202.7	22.1
2	41.6	60.5	42.6	42.5	60.6	41.6	41.6	41.5	60.6	41.3
3	213.2	230.7	212.9	212.8	204.1	213.3	211.7	212.9	204.1	213.6
4	58.4	59.7	58.5	58.5	58.9	58.4	57.9	58.3	59.0	57.8
5	42.4	37.5	42.4	42.4	38.1	42.3	42.1	42.1	37.2	41.9
6	41.4	38.5	41.8	41.7	41.7	41.5	41.2	41.4	40.6	41.0
7	18.4	17.0	17.9	18.0	20.4	18.6	18.4	18.3	18.0	18.1
8	53.4	45.2	53.7	53.9	52.0	53.8	53.2	53.5	51.5	52.2
9	37.6	37.1	42.0	42.0	37.1	37.6	37.5	37.5	37.8	37.3
10	59.6	69.0	60.7	60.7	72.4	59.7	59.2	59.6	71.9	59.1
11	35.8	35.8	29.9	30.0	35.0	37.8	35.7	35.5	33.4	35.3
12	30.8	26.5	31.2	31.4	29.9	24.1	30.5	30.0	29.7	29.9
13	39.8	38.1	39.7	39.9	39.7	45.4	40.1	39.6	39.2	39.1
14	38.4	40.0	37.7	37.9	42.0	38.4	38.2	38.4	39.1	38.0
15	32.5	34.7	32.7	33.0	20.1	32.2	29.5	31.1	31.2	31.3 *
16	38.8	36.8	36.0	35.0	35.3	36.2	32.6	29.2	29.0	29.0
17	30.0	30.8	30.1	33.1	30.4	30.3	36.6	35.1	35.1	35.1
18	42.8	44.0	42.7	41.9	43.5	43.3	39.0	38.9	39.3	39.2
19	36.0	35.4	35.3	37.0	35.4	37.1	29.8	31.5	34.4	34.4
20	33.9	28.3	28.1	42.7	28.3	28.5	33.6	33.3	28.1	27.9
21	75.8	32.5	32.7	218.7	32.9	32.6	29.2	30.2	31.4	31.4 *
22	48.6	38.9	39.2	55.0	39.1	40.1	33.0	28.3	34.3	33.2
23	6.9	7.5	7.0	7.0	7.3	7.1	7.2	6.8	7.3	6.7
24	14.8	15.7	14.7	14.7	15.7	14.9	14.7	14.7	15.9	14.5
25	18.0	67.1	63.0	62.7	17.8	18.2	18.0	18.2	18.1	18.0
26	20.8	69.9	20.1	20.9	64.0	22.3	20.1	18.6	19.1	18.9
27	18.9	19.3	18.5	18.4	19.7	63.4	19.0	19.1	19.2	19.1
28	34.5	30.0	32.1	33.6	31.7	32.8	67.1	69.0	68.0	67.0
29	34.9	31.4	35.0	28.8	34.5	35.8	73.6	28.6	34.2	32.9
30	24.1	35.1	31.7	25.0	32.0	30.6	27.5	73.4	32.8	34.2
Ref	[52]	[53]	[48]	[54]	[41]	[55]	[56]	[57]	[58]	[57]

C	F32	F34	F35	F36	F37	F38	F39 [c]	F40	F41	F42
1	22.7	22.2	22.5	25.9	32.4	31.7	28.2	24.8	202.7	22.3
2	41.4	41.3	41.6	75.4	76.9	74.4	71.0	42.6	60.6	41.5
3	213.4	212.2	212.8	105.6	212.0	209.7	199.6	212.3	204.1	212.9
4	58.6	57.9	58.6	47.4	52.7	52.4	127.1	58.6	59.1	58.3
5	42.3	42.0	42.4	46.8	42.7	54.2	158.8	42.3	37.8	42.1
6	41.9	40.8	41.7	33.3	40.8	36.7	30.4	41.6	40.6	41.4
7	18.4	18.5	18.5	19.2	17.5	19.4	20.3	18.2	18.0	18.3
8	53.2	52.3	53.7	49.8	52.7	49.6	47.3	54.0	52.0	53.1

Table 1. *Cont.*

9	38.0	37.5	37.8	37.1	38.0	37.5	37.2	41.1	37.2	37.5
10	59.6	59.0	59.9	53.2	52.7	55.7	51.8	60.3	71.9	59.6
11	35.7	35.2	36.0	34.3	35.6	34.5	32.4	30.1	34.5	35.7
12	31.0	28.9	30.8	29.0	30.2	29.4	28.9	30.7	30.1	29.4
13	37.6	39.0	38.6	39.2	39.4	40.2	39.2	39.9	39.6	39.9
14	39.3	40.6	40.3	39.1	38.9	39.4	39.5	37.9	38.3	38.2
15	28.6	50.2	33.0	29.1	32.7	28.3	27.9	33.1	32.1	32.2
16	32.8	218.4	36.2	36.2	36.5	29.6	35.1	34.9	35.8	29.7
17	48.6	45.7	30.8	30.1	29.8	44.8	37.7	33.0	30.0	30.0
18	35.9	43.2	42.3	44.5	42.4	45.4	43.4	41.8	42.6	42.9
19	29.8	30.1	39.8	30.4	35.0	31.6	31.3	36.9	29.4	30.6
20	33.7	32.7	33.4	40.6	27.8	41.4	41.9	42.7	33.4	33.4
21	28.0	27.0	30.0	30.0	32.0	214.2	213.7	218.7	28.1	28.3
22	33.7	31.1	28.1	36.6	40.8	77.5	53.3	55.0	38.1	39.9
23	7.2	6.9	7.0	7.1	6.3	7.9	10.7	7.0	7.3	6.8
24	15.0	14.6	14.9	72.4	13.8	174.2		14.8	16.0	14.7
25	17.5	17.2	18.1	16.7	17.9	16.8	17.0	65.1	18.1	18.0
26	20.5	20.3	18.6	16.2	18.4	15.5	15.0	21.5	20.0	18.6
27	19.2	15.9	20.9	17.5	19.9	19.4	17.9	18.6	18.6	19.9
28	208.9	27.4	32.3	31.9	32.1	25.4	32.3	33.8	32.1	32.2
29	24.4	74.1	75.0	179.4	31.7			28.8	28.9	29.0
30	74.7	25.8	26.1	32.1	34.7	15.0	14.7	25.0	72.0	72.1
OCH_3						51.7				
C=O								166.8		
Iso								130.2		
Orto								129.5		
Meta								128.6		
Para								133.1		
Ref	[48]	[46]	[59]	[48]	[60]	[61]	[62]	[54]	[49]	[57]

C	F43	F44	F45	F47	F49	F50	F51	F52	F53	F54
1	31.3	200.5	27.6	22.2	22.3	39.0	36.1	26.7	26.7	210.9
2	193.1	102.6	41.2	41.5	41.5	211.4	211.7	72.7	73.2	53.9
3	146.5	175.5	213.0	213.0	213.0	77.2	77.0	106.3	106.8	75.4
4	126.9	49.9	57.9	58.2	58.2	54.8	54.6	45.6	45.7	50.0
5	54.7	41.6	41.7	42.1	42.1	38.0	38.1	52.8	47.0	44.5
6	30.7	41.1	40.7	41.2	41.3	41.1	40.7	72.8	33.6	42.7
7	18.4	17.6	17.8	18.0	18.2	17.3	17.6	30.4	19.2	18.2
8	49.4	52.3	51.5	52.8	50.4	53.8	53.2	46.5	49.7	53.2
9	37.0	37.1	37.4	37.1	37.5	41.8	37.7	37.0	37.0	37.2
10	55.8	68.7	59.0	59.2	59.6	60.8	60.5	52.7	52.6	71.8
11	33.1	35.3	34.8	35.4	35.2	29.5	35.1	34.0	34.1	35.2
12	29.2	30.4	30.4	30.6	30.5	30.5	20.3	29.7	29.0	30.9
13	39.2	39.4	37.5	38.7	39.3 *	39.7	39.7	39.2	39.1	40.6
14	39.1	38.3	38.4	37.6	39.4 *	38.2	38.4	38.9	49.1	38.8
15	28.9	32.5	32.6	32.4	30.0	32.7	32.4	29.1	29.0	33.3
16	36.0	36.0	29.3	34.9	36.2	35.9	36.0	36.1	36.2	36.6
17	30.1	29.9	42.4	47.7	30.2	30.0	30.0	30.5	30.1	31.1
18	44.4	42.7	38.5	36.4	44.6	42.6	42.9	44.5	44.5	42.6
19	30.4	35.1	34.1	35.4	29.2	35.3	35.4	30.1	30.4	30.7
20	40.5	28.2	32.8	28.3	40.6	28.1	28.2	40.5	40.5	34.0
21	29.9	32.8	31.0	32.4	29.6	32.7	32.8	30.0	30.0	28.7
22	36.5	39.3	22.0	28.0	36.6	39.2	39.3	36.5	36.4	40.5
23	10.6	8.5	6.6	6.8	6.8	10.9	10.8	9.9	6.9	12.4
24	194.9	15.1	14.3	14.6	14.6	14.0	14.2	64.6	72.2	18.4
25	17.4	18.1	17.0	17.2	17.5	63.7	17.4	16.5	16.7	19.0
26	15.9	20.4	19.0	20.0	18.5	20.3	20.2	16.1	16.1	21.3
27	17.2	18.7	16.1	18.8	16.1	18.5	18.5	1.4	17.4	19.1
28	31.8	32.0	177.0	209.1	31.9	32.2	31.1	31.8	31.8	32.7

Table 1. *Cont.*

29	179.1	35.0	26.7	34.5	179.3	35.0	31.8	179.1	179.1	74.7
30	31.9	31.7	79.7	29.4	31.9	31.7	35.0	32.0	32.0	26.9
OCH$_3$	51.5	55.6			51.5					
Ref	[63]	[64]	[48]	[65]	[66]	[47]	[67]	[68]	[48]	[69]

C	F55	F56	F57	F59 [b]	F60	F61	F62	F63	F64 [b]	F65 [c]
1	21.7	29.6	146.3	35.6	21.9	22.3	22.4	74.0	28.7	26.6
2	37.1	72.4	130.1	41.1	41.2	41.7	41.7	29.7	74.2	68.5
3	216.6	202.3	200.9	212.4	212.4	212.8	212.7	231.2	108.1	71.2
4	58.7	129.5	57.7	58.1	58.2	58.8	58.8	53.3	46.8	42.8
5	39.9	157.8	49.1	42.5	42.6	41.3	41.3	42.7	47.5	49.0
6	37.4	65.4	77.2	50.4	52.2	49.0	48.9	41.2	34.0	37.2
7	17.7	29.9	28.7	69.2	68.6	68.4	68.3	18.1	19.7	19.0
8	53.5	40.9	48.2	58.2	58.6	52.9	52.6	53.1	50.4	49.2
9	37.0	38.5	36.7	38.8	39.0	37.2	37.6	36.9	37.5	44.0
10	49.4	48.9	60.3	59.3	58.9	60.0	59.8	52.4	53.3	50.0
11	35.7	33.4	45.4	36.0	35.9	37.6	31.2	32.7	34.7	33.8
12	30.5	29.5	69.3	29.9	29.6	30.5	37.5	29.5	29.5	28.8
13	39.7	39.3	40.0	40.3	40.3	40.4	39.5	39.7	39.6	39.6
14	38.3	40.6	45.6	44.2	40.0	39.1	44.9	38.0	39.3	38.8
15	32.4	28.3	49.6	27.0	29.5	32.4	32.8	30.2	29.7	27.8
16	36.0	29.8	214.2	35.5	38.3	35.8	29.1	35.3	36.7	28.9
17	30.0	45.2	47.1	30.1	30.4	30.5	38.3	30.0	30.5	36.4
18	42.7	45.8	44.7	43.3	42.3	41.6	38.0	42.4	44.8	45.0
19	35.3	32.0	39.6	21.9	35.3	29.6	35.1	31.3	30.9	30.9
20	28.1	41.7	42.3	28.3	33.5	33.1	28.5	40.2	40.7	40.8
21	32.7	214.1	218.1	32.4	27.9	27.9	35.9	28.2	30.5	213.2
22	39.2	78.1	47.2	39.2	36.1	39.4	32.5	38.2	37.4	76.7
23	13.5	11.1	9.8	6.9	6.9	6.9	6.9	6.8	8.4	11.3
24	23.1		8.7	15.9	15.9	16.1	16.1	14.4	72.1	176.5
25	18.0	17.0	19.6	18.8	18.9	19.0	18.8	17.5	16.9	15.2
26	20.4	15.8	19.9	64.2	13.1	21.8	21.4	20.9	16.8	17.3
27	18.7	19.3	8.7	20.1	20.8	18.3	18.5	17.7	18.1	18.6
28	32.1	25.9	29.1	31.2	31.8	32.1	179.2	32.0	32.1	25.5
29	35.0		28.4	34.7	29.1	74.5	34.4	31.8	181.3	
30	31.7	15.4	24.4	31.1	71.3	26.0	29.6	183.1	32.3	14.9
Ref	[70]	[71]	[72]	[39]	[73]	[48]	[48]	[48]	[74]	[75]

C	F66	F67 [+]	F68	F69 [b]	F70 [b]	F71	F72	F73	F74	F75 [b]
1	30.8	28.2	32.6	30.9	39.0	22.3	22.2	22.3	36.5	20.3
2	74.1	76.5	75.0	193.5	194.6	41.5	42.0	41.5	212.3	38.9
3	213.3	208.1	212.4	149.0	144.2	213.2	213.3	213.3	77.3	105.9
4	52.7	54.3	55.6	125.8	139.9	58.2	58.2	58.3	54.8	53.7
5	43.1	43.1	43.1	54.9	40.0	42.1	44.7	42.0	39.6	47.0
6	41.3	41.0	41.2	32.7	33.6	41.2	41.5	41.3	40.9	33.9
7	18.5	18.2	18.1	18.9	18.2	18.2	18.4	18.2	17.9	19.4
8	50.9	50.4	53.2	49.5	50.8	53.1	52.9	50.7	51.0	50.0
9	37.2	36.6	37.4	37.1	37.8	37.5	37.7	37.4	38.3	37.3
10	52.6	53.3	56.5	55.5	56.1	59.4	59.2	59.8	60.9	57.1
11	35.4	35.1	35.3	33.3	35.0	35.6	35.9	36.1	35.0	34.6
12	30.5	30.1	30.2	30.9	30.7	30.3	31.0	29.5	29.7	29.4
13	39.5	39.2	39.7	39.4	39.7	39.7	41.0	39.2	38.0	39.4
14	39.7	39.3	38.1	39.6	39.8	38.0	38.8	39.1	39.5	39.2
15	29.8	30.4	32.8	30.5	30.6	32.8	29.3	29.4	30.1	29.3
16	37.8	36.2	35.4	36.6	36.8	35.4	32.6	36.6	36.5	36.5
17	30.4	29.1	29.6	30.5	31.2	29.5	37.6	30.1	30.5	30.3
18	44.8	44.5	42.5	44.8	45.0	42.5	37.6	44.2	44.7	44.6
19	30.8	36.9	31.3	29.5	29.9	31.2	35.4	29.3	30.6	30.7
20	40.7	40.6	40.3	40.7	40.9	40.3	28.4	40.4	40.6	40.5
21	29.7	29.9	28.2	29.4	30.1	28.2	34.8	30.2	29.7	30.3

Table 1. *Cont.*

C										
22	36.7	29.4	38.2	37.4	37.2	38.2	32.6	35.2	37.1	37.2
23	6.8	6.5	6.5	10.7	10.5	6.8	6.8	6.9	11.2	8.4
24	14.2	14.1	14.7	195.8	19.1	14.6	14.6	14.6	14.6	72.9
25	18.0	18.3	17.8	17.2	18.2	17.7	17.5	18.4	18.1	16.7
26	18.5	16.1	20.9	17.9	18.5	20.9	20.7	16.3	18.2	17.9
27	16.1	17.5	17.7	16.2	17.0	17.7	18.5	18.0	16.6	16.5
28	32.1	31.8	31.8	32.3	32.0	32.0	185.0	31.8	32.2	31.9
29	181.2	31.9	183.3	181.5	181.2	31.9	29.7	184.8	32.1	181.2
30	32.1	179.4	32.0	32.1	32.3	184.7	34.5	31.5	182.4	32.2
$\underline{C}OOCH_3$		179.4								
$COO\underline{C}H_3$		51.6								
$\underline{C}OOCH_3$		169.8								
$COO\underline{C}H_3$		21.2								
Ref	[76]	[77]	[78]	[59]	[79]	[80]	[56]	[81]	[82]	[83]

C	F76	F77 [c]	F79	F80	F81	F82	F83	F84	F85	F86
1	19.3	26.6	22.4	16.0	148.2	202.7	202.8	22.2	22.3	22.3
2	36.3	67.6	41.7	35.4	130.4	60.6	60.6	40.8	41.1	41.4
3	71.9	83.1	212.7	73.0	201.5	204.0	204.2	212.1	212.1	213.1
4	52.9	43.3	58.8	49.4	57.6	59.0	59.1	57.9	58.1	58.2
5	37.4	48.6	41.3	38.0	43.7	37.8	37.8	42.2	42.2	42.0
6	40.9	33.3	48.9	41.9	39.5	40.5	40.6	41.0	41.0	40.5
7	17.4	18.7	68.3	17.7	18.1	18.1	18.1	18.4	18.6	21.3
8	52.6	49.6	52.9	53.7	51.5	52.4	52.2	53.0*	53.1	45.3
9	37.0	36.8	37.6	37.3	36.8	37.2	37.2	55.5	43.8	37.2
10	59.6	52.2	59.8	61.5	61.9	71.8	71.9	53.1*	59.3	59.3
11	35.6	33.7	31.1	35.8	34.4	34.5	34.6	214.2	51.4	34.4
12	30.9	28.7	37.8	30.9	28.8	30.3	30.2	51.2	214.2	29.4
13	37.8	38.6	39.5	40.1	39.0	39.7	39.5	44.0	55.5	42.4
14	38.6	38.6	44.6	38.5	40.5	38.1	38.2	43.8	44.0	54.2
15	32.3	29.4	32.8	32.8	49.8	32.7	32.4	31.6	35.6	214.9
16	29.2	35.9	29.2	36.1	218.7	35.0	35.9	36.1	36.2	54.0
17	44.5	29.7	38.4	30.7	45.3	33.2	30.0	29.6	29.7	33.5
18	37.6	44.0	38.4	42.0	43.9	41.8	42.7	36.4	36.6	44.0
19	34.5	29.7	35.1	29.9	35.4	37.0	35.3	35.4	31.7	34.9
20	28.2	40.6	28.5	33.3	27.6	42.8	28.2	28.3	28.4	27.9
21	35.1	29.4	35.8	28.0	31.6	218.8	32.8	33.0	33.1	33.8
22	32.2	36.3	32.4	39.7	30.7	55.0	39.3	38.9	39.1	38.6
23	9.6	14.3	6.9	11.8	6.7	7.3	7.3	6.8	6.9	6.8
24	14.3	96.5	16.1	16.6	13.7	16.0	16.0	14.5	14.6	15.0
25	17.4	16.3	18.8	18.4	19.1	17.8	18.0	18.1	18.2	17.4
26	20.3	16.1	21.7	20.8	19.8	21.3	20.3	19.0	19.9	14.7
27	18.3	17.2	18.5	18.7	16.2	18.5	18.7	19.8	19.1	18.9
28	183.7	31.7	183.3	32.3	27.4	33.5	32.0	31.8	31.9	32.2
29	29.5	179.6	34.3	75.0	35.2	28.8	31.8	31.7	31.8	33.3
30	34.2	31.6	29.7	26.0	31.1	25.0	35.0	34.2	34.3	33.4
Ref	[48]	[84]	[48]	[69]	[45]	[52]	[85]	[86]	[35]	[51]

C	F87	F88	F89	F90	F91 [b]	F92	F93 [a]	F94	F95	F96
1	22.2	21.6	22.2	18.6	17.0	16.0	16.2	41.3	27.8	31.4
2	41.4	40.8	41.4	35.4	40.3	35.5	36.1	41.6	170.4	193.2
3	212.5	210.6	212.8	72.0	74.7	72.0	71.6	213.0	177.0	146.5
4	58.2	57.8	58.1	49.4	491	49.4	49.6	59.5	47.3	127.0
5	42.1	47.0	42.0	38.3	41.4	37.9	38.1	42.2	38.9	54.7
6	41.0	56.9	41.2	41.9	36.0	41.8	42.0	35.6	34.5	30.6
7	18.6	210.2	18.1	17.5	19.3	17.7	17.7	30.7	17.1	18.3
8	52.4	63.4	53.0	52.9	53.8	53.2	53.3	53.4	52.0	48.7
9	37.7	42.4	37.4	43.9	36.8	37.1	37.2	37.5	35.3	37.0
10	59.4	59.0	59.4	62.3	61.6	61.5	61.6	58.2	46.1	56.0
11	35.4	35.5	35.5	76.6	36.5	35.5	35.7	22.4 *	32.0	33.3

Table 1. *Cont.*

C										
12	29.1	29.8	30.4	42.0	31.4	30.3	30.7	32.9	30.1	28.8
13	39.2	39.4	39.6	38.5	37.6	39.7	38.4	38.3	39.6	40.7
14	40.5	37.5	38.2	40.9	39.9	38.0	39.7	40.0	38.0	39.5
15	50.2	31.6	32.2	32.3	32.5	32.3	32.3	18.4 *	32.3	28.2
16	218.8	36.3	35.9	36.0	35.6	32.0	35.9	37.1	35.7	35.9
17	45.3	30.1	29.9	30.0	30.7	36.1	30.0	42.8	30.0	31.4
18	44.0	41.8	42.8	42.6	43.6	38.5	42.9	42.0	42.6	45.3
19	35.5	34.9	35.2	35.7	35.6	29.3	35.4	35.1	35.2	29.8
20	27.6	28.0	28.0	28.1	28.8	33.1	28.2	33.3	28.1	149.1
21	31.7	32.8	32.7	32.8	32.9	28.4	32.9	218.8	32.6	30.7
22	30.8	38.6	39.2	39.2	39.0	28.7	39.3	55.1	39.1	38.0
23	6.8	6.8	6.7	12.1	14.8	11.9	12.1	6.9	12.5	10.7
24	14.7	15.1	14.5	16.7	65.8	16.5	16.6	17.9	20.6	194.9
25	17.3	18.2	17.8	13.5	18.4	18.3	18.3	14.8	72.6	18.1
26	20.3	19.2	18.5	19.9	18.9	20.0	20.1	33.6	20.2	15.1
27	16.2	19.4	20.1	19.5	20.8	18.8	18.7	21.4	18.5	17.9
28	27.4	32.1	32.0	32.0	33.6	66.7	32.1	18.6	32.1	31.4
29	31.1	31.8	31.7	35.0	35.6	73.8	35.0	25.1	35.0	
30	35.2	34.6	34.9	31.7	32.7	26.5	31.8	28.9	31.6	107.6
Ref	[57]	[57]	[57]	[87]	[88]	[69]	[89]	[90]	[47]	[91]

C	F97[c]	F98	F99	F100	F101	F102	F103
1	26.9	22.0	28.7	19.1	21.3	21.1	116.0
2	105.0	41.2	71.5	34.5	38.0	37.2	166.1
3	77.1	212.8	200.1	175.6	176.2	177.8	
4	47.7	57.9	127.7	36.1	35.8	36.2	197.9
5	44.6	41.9	159.2	36.8	37.5	37.9	135.5
6	30.1	41.0	30.9	38.8	38.7	39.0	136.0
7	18.6	18.0	20.8	18.0	17.9	18.2	118.6
8	48.1	53.1	47.7	52.6	52.1	53.1	163.3
9	37.5	37.2	37.6	42.9	38.8	39.1	39.2
10	52.1	59.2	52.3	58.2	59.6	59.9	165.9
11	33.3	35.3	33.1	84.1	34.7	35.3	31.2
12	28.8	29.8	29.5	37.6	29.5	30.3	29.2
13	40.2	38.0	39.5	40.7	39.0	39.7	38.3
14	39.1	39.6	40.0	37.9	37.9	38.4	47.6
15	28.3	32.3	28.1	32.1	31.1	32.4	28.5
16	29.6	35.5	29.7	35.9	29.0	36.1	36.6
17	44.8	30.1	44.9	30.0	35.0	30.1	30.4
18	45.6	41.6	45.3	42.6	39.3	43.0	44.6
19	31.7	30.3	31.7	35.3	34.3	35.4	31.1
20	41.5	31.4	41.3	28.1	27.8	28.2	40.4
21	213.9	27.7	214.0	32.7	33.2	32.9	29.7
22	77.4	38.8	77.2	39.2	31.3	39.3	34.4
23	16.1	6.6	11.2	7.7	7.3	7.6	18.1
24	175.3	14.4		22.1	19.2	19.5	
25	16.6	17.6	15.6	13.6	17.8	18.0	25.6
26	15.8	18.2	19.1	19.9	18.6	20.2	24.0
27	19.3	20.7	17.5	19.3	19.0	18.7	20.3
28	26.1	31.8	25.2	32.0	67.3	32.2	31.5
29		74.3		34.9	32.6	35.0	179.1
30	15.6	25.9	14.8	31.7	33.9	31.9	32.8
C=O		161.2					
OCH$_3$							51.8
OCH$_3$							51.8
Ref	[62]	[92]	[93]	[94]	[95]	[95]	[96]

Ref: References; [+] ^{13}C-NMR data of acetylated compound; * Values bearing the same superscript are interchangeable; Solvent: CDCl$_3$; [a] CDCl$_3$ + Pyridine-d$_5$, [b] Pyridine-d$_5$; [c] DMSO-d$_6$; ^{13}C-NMR data of some compounds were not found. In these cases, the reported identification was performed by comparison of other physical data: F5 [97], F33 [98], F46 (^{1}H-NMR) [99], F48 (^{1}H-NMR) [100], F58 (IR, MS) [101] and F78 (X-ray) [102].

Table 2. ^{13}C-NMR data of quinonemethide-type pentacyclic triterpenoids isolated from Celastraceae species (2001–2021).

C	Q1	Q2	Q3	Q4	Q5	Q6 [b]	Q7	Q8	Q9	Q10
1	121.6	119.8	119.8	164.7	119.5	119.7	119.8	120.0	119.5	119.8
2	178.8	178.4	178.3	178.3	178.1	178.4	178.4	178.1	178.3	178.4
3	145.9	146.0	146.2	146.0	145.9	146.1	146.2	146.3	146.0	146.1
4	118.0	117.0	117.5	118.2	118.0	117.2	117.1	116.8	117.2	117.2
5	128.6	127.5	128.5	127.5	127.3	127.9	127.9	127.9	127.3	127.8
6	132.2	133.8	132.1	134.0	134.0	133.3	133.3	134.4	134.0	133.6
7	118.8	117.9	118.6	117.2	118.0	118.3	118.3	121.9	118.0	118.2
8	167.2	165.7	164.9	169.7	170.1	168.7	168.7	18.6	169.7	168.4
9	48.1	43.0	43.0	47.8	43.0	42.9	42.9	44.6	42.4	42.6
10	161.6	164.4	164.2	164.0	164.8	164.2	164.2	159.7	164.6	164.7
11	65.3	34.3	32.5	33.7	33.0	33.2	33.2	37.3	33.5	34.0
12	43.4	27.5	31.1	30.2	30.0	29.9	29.9	35.9	29.5	30.0
13	40.6	40.3	39.0	42.6	40.7	40.0	40.0	42.4	39.2	40.6
14	44.9	47.3	49.6	44.6	45.0	44.2	44.2	136.9	44.5	44.3
15	28.7	129.4	73.0	28.4	28.6	29.4	29.4	128.6	28.2	28.3
16	36.2	135.6	41.4	68.2	36.5 *	35.7	35.7	40.5	36.1	29.5
17	30.87	33.7	37.6	30.2	31.6	35.9	35.9	39.8	38.1	44.8
18	43.9	42.6	43.7	46.9	43.2	43.3	43.3	43.7	46.1	45.1
19	30.81	30.7	31.9	29.6	24.8 **	36.9	36.9	40.7	30.8	32.0
20	40.4	41.0	41.8	39.4	35.7	73.7	73.7	75.0	42.6	40.9
21	29.7	29.4	213.7	28.4	24.8 **	214.9	214.9	213.0	38.1	213.5
22	34.3	32.3	54.2	38.3	36.5 *	50.5	50.5	50.5	67.9	76.4
23	10.4	10.3	10.3	10.3	10.4	10.3	10.3	10.4	10.1	10.3
24										
25	34.4	37.4	41.0	38.3	38.9	38.5	38.5	29.6	38.1	39.2
26	21.4	28.8	23.6	21.7	21.6	23.3	23.3	22.1	21.6	21.6
27	18.5	18.0	23.5	19.7	21.4	19.4	19.4	23.7	19.0	20.5
28	31.4	27.4	32.9	24.2	31.4	33.2	33.2	30.4	24.0	25.0
29	178.7	178.3		178.2		29.0			178.2	
30	32.7	31.3	15.0	32.4	69.6		29.0	24.8	32.2	14.7
OCH$_3$	51.8	51.6		51.8					51.6	
Ref	[103]	[104]	[104]	[105]	[106]	[107]	[108]	[104]	[109]	[63]

C	Q11	Q12	Q13	Q14	Q15	Q16	Q17	Q18	Q19	Q20 [c]
1	119.4	120.6	119.7	120.0	119.6	119.4	119.9	120.0	119.4	119.8
2	178.4	178.3	181.1	178.1	178.3	178.4	178.0	178.1	178.8	178.6
3	146.0	147.0	161.9	146.3	146.0	146.0	146.2	146.2	146.1	146.3
4	117.1	127.5	140.9	116.8	117.1	117.2	116.7	116.7	118.6	117.4
5	127.5	120.4	117.2	128.1	127.4	127.4	127.5	127.7	127.6	127.6
6	133.9	135.4	131.7	134.4	133.9	134.2	134.9	134.5	137.8	134.4
7	118.2	118.3	200.4	122.1	117.9	118.2	121.6	121.5	122.2	118.3
8	169.8	165.3	41.9	158.6	170.1	170.7	159.7	159.0	160.2	170.4
9	42.9	39.3	58.4	44.2	42.9	43.3	44.5	44.5	45.5	43.0
10	164.7	172.6	146.6	159.7	165.0	164.4	159.7	159.9	163.5	165.1
11	33.6	28.7	32.1	36.0	33.9	33.1	37.5	37.4	37.3	33.6
12	29.8	29.3	30.2	31.8	29.7	29.4	35.6	34.8	35.2	29.7
13	39.4	39.9	39.6	42.1	41.3	40.0	43.1	42.4	43.3	39.5
14	45.0	43.1	29.2	135.6	44.8	44.0	135.3	136.0	135.5	45.1
15	28.6	29.5	27.9	127.9	28.4	29.7	128.3	126.6	129.8	28.7
16	36.4	32.4	35.9	45.6	36.0	36.5	37.8	37.9	39.0	36.4
17	30.8	45.3	31.0	36.3	31.6	30.2	33.7	38.9	35.6	30.9
18	43.6	44.2	44.2	47.8	44.9	43.7	43.9	38.7	43.6	44.4
19	25.5	33.8	28.8	37.8	30.4	25.2	33.9	33.9	34.2	30.9
20	46.2	31.0	40.5	41.7	147.9	33.0	42.6	47.4	48.0	40.5
21	25.1	34.5	29.8	213.8	30.5	22.5	28.6	69.2	68.0	29.9
22	34.0	36.3	36.5	49.8	36.9	35.2	36.1	79.4	39.0	34.8

Table 2. *Cont.*

C										
23	10.3	10.5	10.2	10.4	10.2	10.3	10.3	10.3	10.5	10.3
24										
25	38.4	38.4	38.2	28.5	38.9	37.7	29.4	29.5	29.4	38.3
26	21.7	21.5	21.5	21.3	21.3	23.4	21.9	21.8	21.9	21.7
27	18.4	18.7	18.3	23.1	19.7	17.9	24.0	24.5	24.3	18.4
28	31.5	31.5	31.5	30.2	31.1	36.2	31.5	27.0	31.1	31.7
29	177.4	182.5	178.8			69.3	179.3	179.0	178.8	179.1
30	74.1	30.7	30.8	16.0	108.2		19.8	13.7	17.5	32.7
OCH$_3$	51.9		51.4				51.8	52.7	52.3	51.7
Ref	[91]	[110]	[111]	[112]	[113]	[91]	[104]	[109]	[109]	[114]

C	Q21	Q22
1	119.8	119.8
2	178.6	178.4
3	146.3	146.0
4	117.4	117.1
5	127.8	127.7
6	134.1	133.6
7	118.3	118.1
8	170.3	168.7
9	43.2	42.7
10	165.1	164.7
11	34.0	33.8
12	30.2	29.9
13	40.7	40.6
14	45.2	44.6
15	28.9	28.5
16	37.4	35.5
17	30.6	38.2
18	44.3	43.5
19	25.0	32.0
20	31.4	41.8
21	71.2	213.6
22	44.4	52.5
23	10.4	10.2
24		
25	38.9	39.0
26	21.8	21.5
27	21.4	19.7
28	35.4	32.5
29		
30	18.6	15.1
Ref	[115]	[116]

Ref: References; *,**: Values bearing the same superscript are interchangeable; Solvent: CDCl$_3$; [b] C$_6$D$_6$:CDCl$_3$; [c] CD$_2$Cl$_2$.

Table 3. ^{13}C-NMR data of aromatic-type pentacyclic triterpenoids isolated from Celastraceae species (2001–2021).

C	A1	A3	A5	A6	A7	A8 [+]	A9 [a]	A10 [b]	A11	A12 [c]
1	109.5	109.1	108.4	108.6	125.6	105.9	110.0	108.4	106.8	109.0
2	150.8	144.2	140.9	148.0	147.7	156.0	144.0	148.8	150.0	151.7
3	152.8	132.9	139.7	140.6	140.3	146.2	126.8	141.4	143.0	143.5
4	113.2	132.0	122.0	125.4	125.1	134.1	122.7	126.1	128.0	127.9
5	141.9	154.9	126.6	122.5	122.6	123.7	151.4	122.3	119.0	119.9
6	201.2	187.2	28.3	187.6	187.7	187.7	187.3	187.9	201.2	182.8
7	43.8	126.1	18.5	126.1	108.8	126.1	126.8	126.0	74.5	147.4
8	35.4	151.9	44.1	171.2	172.1	171.4	150.7	170.8	50.3	139.1

Table 3. *Cont.*

C										
9	38.6	40.4	36.8	40.1	44.4	40.6	40.6	40.3	38.11	40.7
10	123.5	171.2	143.8	151.2	151.9	154.9	170.7	151.2	152.8	152.8
11	32.4	34.0	34.1	33.8	34.4	35.6	34.6	35.6	36.3	34.2
12	29.9	29.8	30.2	30.0	29.9	30.2	30.2	30.3	30.6	30.5
13	36.8	38.9	38.9	39.6	40.8	40.6	39.3	40.0	40.5	40.0
14	35.9	44.7	39.4	43.8	40.1	44.8	44.6	44.3	39.1	47.2
15	30.3	28.5	29.0	29.0	28.2	28.8	28.9	28.4	29.7	29.6
16	NR	36.4	36.5	35.5	36.8 **	37.5	36.6	32.1	36.2	37.8
17	35.7	30.8	30.3	36.2	31.6	30.8	30.6	38.4	29.8	31.1
18	40.4	44.3	44.5	43.3	44.8	42.7	44.4	43.5	44.1	45.4
19	29.5	30.5	30.6	36.8	30.5 *	27.9	31.1	34.3	29.8	31.9
20	39.1	40.3	40.6	73.8	148.2	132.4	44.4	42.0	41.0	41.7
21	29.6	29.7	30.0	215.0	30.5 *	119.9	30.1	214.7	29.9	30.9
22	28.2	34.8	36.2	50.2	36.8 **	35.2	35.1	52.7	34.0	36.2
23		14.7	11.3	13.6	13.6	14.4	14.8	13.7	14.0	14.0
24										
25	31.9	37.6	27.4	37.9	38.4	39.2	37.8	38.6	27.5	41.6
26	25.3	20.8	15.9	22.1	20.4	21.4	20.9	20.8	15.9	19.6
27	16.7	18.3	17.3	19.4	19.6	15.7	18.5	19.7	15.9	20.6
28	15.2	31.6	31.8	33.1	31.1	33.1	31.6	32.6	31.7	32.1
29	179.4	178.8	179.3	28.9	106.5	24.0	178.7		179.8	180.7
30	31.5	32.7	31.9				32.6	15.1	31.9	32.9
OCH$_3$	55.7	51.5				56.0				
OCH$_3$	55.1	61.1	51.5			60.7	51.5		51.4	52.2
Ref	[117]	[118]	[119]	[120]	[106]	[121]	[122]	[123]	[124]	[125]

C	A13	A14	A15	A16	A17 [+]	A18	A19 [d]	A20	A21	A22
1	107.9	107.5	109.2	107.9	104.1	110.1	116.6	112.4	107.8	108.8
2	150.0	148.4	146.6	150.0	156.3	151.7	150.7	152.6 *	141.7	148.2
3	141.4	140.9	139.5	141.4	146.1	144.7	150.0	143.5	140.4	140.7
4	129.1	127.0	130.0	129.1	135.1	134.0	117.0	120.4	122.0	125.6
5	122.8	124.5	125.0	122.8	125.1	122.4	122.3	121.8	123.9	122.5
6	182.0	201.9	200.7	182.0	201.0	185.8	186.3	185.8	43.9	187.7
7	197.0	37.9	37.7	197.0	37.7	129.3	125.2	124.8	209.8	125.9
8	60.3	42.9	42.6	60.3	42.5	160.1	174.4	177.4 **	58.3	171.3
9	38.7	37.6	37.2	38.7	38.0	42.2	40.5	41.7	38.8	40.3
10	153.6	153.2	152.2	153.6	156.2	151.2	150.4	151.7 *	142.4	151.7
11	33.5	36.6	33.2	33.5	33.8	37.1	33.7	35.0	33.7	34.3
12	27.8	30.15	29.9	27.8	29.6	35.8	29.6	30.8	29.4	30.2
13	39.5	39.7	38.8	39.5	40.2	40.7	39.4	40.7	39.1	40.1
14	39.2	39.2	39.4	39.2	40.3	134.7	45.1	46.4	38.7	44.4
15	38.7	28.8	28.5	38.7	28.2	125.9	28.7	30.0	28.4	28.4
16	35.8	36.4	36.2	35.8	38.5	38.0	26.4	37.6	36.1	35.5
17	30.2	30.7	30.3	30.2	30.7	48.8	30.5	31.6	30.3	38.2
18	43.6	45.1	44.7	43.6	42.9	46.2	44.3	45.7	43.6	43.5
19	30.6	31.1	30.5	30.6	27.6	34.2	30.8	31.9	30.6	32.0
20	40.7	41.1	40.1	40.7	133.0	50.3	40.1	41.3	40.6	41.9
21	30.3	30.1	29.8	30.3	120.2	74.3	29.8	30.9	29.8	214.0
22	35.8	33.5	36.1	35.8	35.4	214.0	34.8	36.0	35.9	52.6
23	13.8	14.0	13.2		14.3	15.1	200.2	173.1 **	11.5	13.7
24										
25	31.5	26.0	25.6	31.5	27.0	21.6	36.3	37.4	27.9	38.5
26	14.7	15.7	15.4	14.7	15.4	21.7	20.5	21.3	15.1	20.7
27	16.9	15.7	16.9	16.9	15.4	23.9	18.7	19.4	16.8	19.7
28	31.5	32.1	31.7	31.5	33.0	22.5	31.6	32.1	31.5	32.6
29	180.0	180.1	179.1	180.0	24.0	175.2	181.2	182.5 **	179.4	
30	32.5	32.9	32.2	32.5		13.7	32.6	33.2	32.3	15.1
OCH$_3$					56.0	52.7				
OCH$_3$	51.8	52.2	51.4	51.8	60.6	61.7			51.6	

Table 3. *Cont.*

1′			126.5							170.0
2′			104.2							121.3
3′			147.4							112.2
4′			135.8							146.2
5′			147.4							150.8
6′			104.2							125.2
7′			77.2							114.2
8′			75.5							
9′			62.8							
OCH_3			56.4							56.1
OCH_3			51.4							
OCH_3			20.6							
C=O			170.3							
Ref	[119]	[124]	[126]	[119]	[124]	[109]	[127]	[127]	[119]	[128]

C	A23	A24	A25 [a]	A26 [a]	A27 [d]	A28	A29
1	108.3	107.7	110.7	110.6	108.5	116.5	113.8
2	141.4	141.6	143.4	141.3	142.9	150.5	173.7
3	139.8	139.5	145.2	143.1	141.9	149.7	155.5
4	122.5	120.9	121.4	122.5	119.6	116.3	111.3
5	126.3	125.7	126.7	126.0	122.8	117.1	119.4
6	28.0	27.8	45.4	46.1	120.9	186.0	188.0
7	18.3	126.0	119.6	120.9	137.3	122.8	124.4
8	43.3	139.0	151.1	NR	43.3	149.1	153.4
9	36.7	36.2	38.3	38.1	144.1	45.3	45.6
10	143.8	140.0	142.6	145.1	128.5	125.3	152.8
11	34.2	33.0	37.6	37.7	119.4	36.4	36.6
12	30.0	29.4	31.4	30.6	32.6	28.6	28.6
13	40.0	43.5	38.8	38.7	39.4	39.7	39.7
14	39.3	58.0	44.7	44.5	40.1	40.5	40.5
15	27.9	211.4	29.9	29.7	23.3	31.0	30.9
16	29.6	47.5	37.6	37.4	36.7	34.8	34.8
17	44.9	49.4	31.1	30.9	31.1	44.1	43.0
18	45.4	44.3	45.1	44.9	46.3	44.1	44.2
19	31.7	30.7	31.3	31.1	30.7	29.8	29.8
20	41.3	40.0	40.8	40.6	38.3	30.7	30.5
21	214.2	212.4	31.4	31.2	29.2	29.8	29.8
22	77.6	77.8	35.9	35.7	37.2	33.5	34.4
23	11.5	11.6	13.4	13.8	11.1	200.3	178.7
24							
25	28.2	33.4	36.5	35.7	22.4	36.4	36.8
26	15.3	25.6	22.7	22.4	19.1	20.6	20.2
27	19.2	21.4	18.9	18.8	18.7	32.8	32.7
28	25.2	24.6	32.0	31.9	32.7	18.3	18.3
29			181.3	181.0	179.7	178.9	179.8
30	14.8	14.8	33.5	33.3	30.4	31.8	31.6
CH_3			22.7	21.4		51.6	51.6
CHOH			70.8	72.1			
Ref	[129]	[39]	[130]	[130]	[127]	[131]	[131]

Ref: References; [+] ^{13}C-NMR data of methylated compound; *,**: Values bearing the same superscript are interchangeable; NR: Not reported; Solvent: $CDCl_3$; [a] Pyridine-d_5, [b] $CDCl_3$+CD_3OD, [c] CD_3OD, [d] DMSO-d_6; ^{13}C-NMR data for A2 were not found. The reported identification was performed by comparison of ^{1}H-NMR data from Sotanaphun [132]. ^{13}C-NMR data for A4 were reported by Shirota et al. [122], but the chemical shifts were not attributed to each carbon atom.

Table 4. ^{13}C-NMR data of dimer-type pentacyclic triterpenoids isolated from Celastraceae species (2001–2021).

C	D1	D2	D3	D6	D7	D8	D9	D10	D11	D12
1	115.5	108.2	115.5	113.0	113.0	113.1	113.0	113.1	115.5	115.0
2	191.2	144.3	191.0	191.5	191.5	191.5	191.6	191.4	190.2	189.4
3	91.7	140.6	91.8	91.3	91.3	91.2	91.3	91.3	92.0	91.1
4	79.9	123.2	79.0	79.5	79.5	79.5	79.4	79.5	79.4	76.9
5	130.8	124.6	130.8	134.1	134.3	134.1	134.3	134.1	130.2	132.2
6	126.1	71.4	125.9	134.1	133.9	133.9	134.1	133.8	126.6	128.5
7	116.2	73.7	116.1	24.2	24.2	24.2	24.2	24.2	116.3	117.2
8	160.6	44.7	160.6	41.6	41.2	41.6	41.1	41.6	160.4	163.3
9	41.7	40.2	41.7	37.4	37.4	37.4	37.3	37.4	41.7	43.4
10	173.8	143.7	173.5	170.2	169.8	170.0	169.8	170.0	173.8	172.9
11	32.7	34.7	32.7	30.7	30.5	30.6	30.7	32.0	33.2	33.1
12	29.5	30.7	29.5	29.4	29.4	29.4	29.7	29.4	29.8 [1]	29.8
13	38.1	39.6	38.1	38.9	40.1[1]	38.9	40.0 [1]	40.1	39.5 [2]	39.9
14	44.5	40.00	44.5	40.1	40.2[1]	40.1	39.9 [1]	38.9	44.3	44.0
15	28.4	31.4	28.4	28.3	27.9	28.4[1]	27.7	28.3	28.3	28.5
16	36.5	36.1	36.4	36.0	35.3	36.0	29.3	36.0	35.4 [3]	35.4
17	30.5	29.8	30.2	30.2	38.1	30.2	44.7 [2]	30.1	38.3 [4]	38.2
18	44.6	43.6	44.1	44.6	43.9	44.6	45.4	44.6	43.4	43.6
19	30.9	29.9	30.9	30.5	31.8	30.5	31.7	30.5	32.1	31.9
20	40.5	40.6	40.5	40.5	42.3	40.5	41.2	40.5	42.3	41.9
21	29.9	30.7	29.9	29.9	213.8	29.9	213.8	29.7	213.6	213.6
22	34.8	37.0	34.7	36.0	53.5	36.0	77.2	36.0	52.5	52.4
23	22.4	11.1	22.5	22.7	22.7	22.7	22.8	22.7	22.3	24.6
24										
25	34.8	28.0	34.8	22.1	22.9	22.1	23.0	22.1	35.6	39.7
26	22.5	17.6	22.5	16.0	15.7	16.0	15.8	16.0	22.3	22.4
27	18.6	18.3	18.6	16.9	18.1	16.9	18.9	16.8	20.1	19.7
28	31.6	31.9	31.7	31.7	32.7	31.7	25.1	31.7	32.8 [5]	32.5
29	179.1[1]	179.4	179.0 [1]	179.0		179.0		179.0		
30	32.9	31.7	31.9	32.3	15.2	32.3	14.9	32.3	15.2 [6]	15.1
OCH$_3$	51.7[2]	51.3	51.5 [2]	51.7		51.7		51.7		
1′	110.4	109.5	110.4	110.5	110.6	110.4	110.6	110.4	109.6	108.8
2′	139.1	140.1	141.3	144.5	144.5	144.5	144.5	144.5	145.0	145.3
3′	136.2	138.9	137.0	138.3	138.3	138.4	138.2	138.4	137.6	137.4
4′	122.9	125.6	125.5	129.4	129.4	129.5	129.3	129.6	129.0	129.7
5′	127.9	123.6	127.2	123.3	123.3	123.4	123.7	123.4	126.0	125.3
6′	26.4	28.0	75.2	187.2	187.2	187.1	187.2	187.0	201.1	200.0
7′	18.5	18.5	21.8	126.3	126.3	126.3	126.3	126.3	37.6	37.4
8′	43.9	44.1	38.5	171.0	171.0	170.0	171.1	169.7	41.9	41.8
9′	36.8	36.9	37.6	40.1	40.1 [1]	39.9	40.1	39.7	37.1	37.0
10′	144.4	145.1	144.7	151.8	151.8	151.7	151.8	151.7	151.7	152.2
11′	33.9	34.3	33.8	34.2	34.3	34.4	34.3	34.6	32.9	33.2
12′	30.0	30.3	29.8	29.7	29.9	30.2	29.9	29.8	29.7 [1]	29.5
13′	38.9	39.0	38.9	39.0	39.0	40.2	39.0	40.1	39.4 [2]	39.2
14′	39.4	39.4	39.1	44.7	44.7	44.3	44.8 [2]	44.0	40.0	39.8
15′	28.9	29.1	29.0	28.5	28.6	28.3 [1]	28.6	28.2	28.0	27.7
16′	36.4	36.3	36.1	36.4	36.4	35.5	36.4	29.6	35.3 [3]	29.3
17′	30.6	30.2	29.3	30.5	30.5	38.2	30.5	44.9[1]	38.2 [4]	45.0
18′	44.1	44.4	44.4	44.3	44.3	43.5	44.3	45.0[1]	44.0	45.3
19′	30.4	29.5	30.5	31.0	31.0	32.0	31.1	30.1	31.8	31.7
20′	40.4	40.6	40.4	40.6	40.6	41.9	40.6	40.9	41.9	41.3
21′	30.3	30.7	30.3	29.9	29.7	213.7	29.7	213.6	214.1	214.0
22′	36.3	36.6	36.6	35.0	35.0	52.7	35.0	76.1	53.6	77.2
23′	10.9	10.9	10.6	13.4	13.4	13.4	13.4	13.4	13.0	13.3
24′										
25′	27.2	27.2	31.5	37.7	37.7	38.7	37.6	38.9	26.2	26.5

Table 4. *Cont.*

C	D13	D14	D15	D16	D17	D18	D19	D20	D21	D22
26′	15.8	16.1	16.3	20.9	20.9	20.8	20.8	20.9	15.0	15.1
27′	17.2	17.3	17.4	18.6	18.5	20.0	18.6	20.8	18.2	19.0
28′	31.8	31.9	26.4	31.6	31.6	32.6	31.6	25.0	32.6 [5]	25.1
29′	178.7 [1]	179.8	178.7 [1]	179.2	179.3		179.4			
30′	32.7	30.8	32.7	32.9	32.9	15.1	33.0	14.8	15.1 [6]	14.8
OCH$_3$	51.5 [2]	51.6	51.7 [2]	51.6	51.6		51.7			
OCH$_3$			55.4							
Ref	[133]	[134]	[133]	[135]	[136]	[136]	[137]	[137]	[136]	[136]
1	114.0	108.3	49.1	108.6	107.8	107.6	109.1	108.8	111.4	108.9
2	191.2	144.4	193.1	146.5	142.0	141.4	144.9	141.1	149.6	137.9
3	91.3	140.7	192.1	141.7	137.7	141.4	142.7	141.7	143.2	142.2
4	79.1	123.2	60.5	123.0	121.4	120.8	122.3	121.3	126.0	121.1
5	134.7	124.7	131.7	125.4	124.3	124.4	125.7	126.0	124.1	125.0
6	135.9	71.5	28.2	124.3	124.4	123.6	124.3	124.2	187.7	124.3
7	68.5	74.5	18.3	128.6	128.4	127.7	128.7	129.2	126.1	128.8
8	51.7	45.1	45.7	45.4	45.6	44.8	44.4	45.6	171.2	45.5
9	41.1	40.3	37.4	37.5	37.2	36.3	37.2	37.2	40.05	37.0
10	168.7	143.7	148.8	143.6	142.2	141.3	143.3	142.8	151.7	141.8
11	31.2	34.8	31.0	31.0	31.1	30.4	30.5	31.2	34.12	31.9
12	29.4	30.4	29.7	30.0	30.4	29.2	38.2	30.5	28.5	30.5
13	39.4	39.7	39.6	38.9	39.3	38.1	38.8	38.9	39.0	38.8
14	41.8	40.0	39.3	39.0	39.2	38.2	45.4	39.0	44.6	39.0
15	31.0	31.7	28.9	28.1	28.1	27.3	28.1	28.2	29.8	28.1
16	36.2	36.2	36.4	35.5	36.0	35.6	35.4	36.0	35.4	36.2
17	30.0	29.8	30.3	30.3	30.6	29.6	30.3	30.4	30.5	30.4
18	44.8	43.6	44.5	44.4	44.5	43.7	43.6	44.5	44.3	44.4
19	30.6	30.0	30.4	30.6	30.0	29.2	30.0	30.1	30.8	31.0
20	40.6^1	40.7	40.7	40.6	40.5	39.7	40.5	40.5	40.4	40.5
21	29.8	31.6	30.0	29.7	29.9	29.8	29.9	29.9	29.9	29.4
22	35.8	37.1	36.7	36.5	36.4	36.1	35.9	36.4	36.4	36.3
23	22.4	11.1	9.0	10.8	10.8	9.9	10.7	10.7	13.1	10.9
24										
25	24.0	28.0	22.3	22.2	16.8	21.6	22.1	22.3	37.7	22.3
26	16.5	17.8	16.0	17.0	22.2	16.0	16.8	16.9	20.8	16.8
27	17.4	18.3	17.2	17.5	17.4	16.6	17.3	17.4	18.3	17.4
28	31.7	32.00	31.9	31.8	31.8	31.0	31.7	31.8	31.6	31.8
29	179.0	179.4	179.2	179.3	179.1	178.3	179.1	179.1	178.8	179.1
30	32.4	32.00	31.9	32.2	32.1	31.3	32.1	32.1	32.7	32.1
OCH$_3$	51.7	51.6	51.6	51.6	51.5	50.7	51.6	51.5	51.5	51.5
1′	110.6	107.9	60.8	92.7	87.1	90.3	92.2	90.5	92.7	139.9
2′	144.2	141.6	24.4	38.6	41.8	44.6	36.3	45.8	38.4	36.0
3′	138.0	139.1	41.3	128.7	124.2	124.4	128.2	128.2	128.1	74.2
4′	129.3	122.0	39.8	140.5	151.2	141.2	140.6	140.6	140.6	90.1
5′	123.5	124.1	44.4	97.1	84.3	89.4	96.8	89.7	96.8	134.6
6′	187.2	124.3	25.1	38.1	35.2	35.2	38.9	84.6	39.1	30.0
7′	126.2	128.6	36.1	43.7	45.3	39.9	45.5	41.4	43.3	49.4
8′	171.0	45.8	151.4	36.0	38.9	35.8	32.7	29.7	34.8	29.9
9′	40.2	37.5	42.5	32.7	39.0	29.7	31.1	29.6	33.0	39.8
10′	151.9	143.8	36.2	40.1	32.9	43.0	39.4	43.2	39.3	42.6
11′	34.2	31.7	34.8	151.0	151.9	150.6	151.1	140.1	150.8	150.7
12′	29.9	30.0	21.9	108.7	108.4	108.0	108.6	119.1	108.8	108.6
13′	39.0	38.9	30.0	19.9	19.4	20.1	19.9	169.7	20.0	20.5
14′	44.7	39.1	16.7	19.1	20.3	17.8	19.1	19.0	19.1	17.1
15′	28.6	30.4	111.8	12.6	22.9	10.7	12.5	14.4	12.4	13.9
16′	36.4	36.1								
17′	30.5	30.5								
18′	44.3	44.5								

Table 4. *Cont.*

C	D23	D24	D25	D26	D27	D28	D30	D31	D32	D33
19′	30.9	29.4								
20′	40.5 [1]	40.6								
21′	29.8	31.6								
22′	35.0	36.6								
23′	13.4	10.8								
24′										
25′	37.6	22.3								
26′	20.9	17.2								
27′	18.4	17.3								
28′	31.6	31.8								
29′	179.0	179.6								
30′	32.8	31.6								
OCH$_3$	51.7	51.4								
Ref	[136]	[134]	[75]	[138]	[138]	[138]	[138]	[138]	[138]	[138]
1	107.6	108.2	116.0	114.8	116.5	114.6	115.4	108.2	115.9	114.9
2	145.8	142.4	187.8	187.3	193.3	189.6	191.0	144.4	190.8	189.6
3	136.2	138.6	92.2	90.8	93.1	91.3	91.8	140.6	92.0	90.7
4	122.7	121.6	79.1	76.9	79.2	77.2	79.1	123.0	79.4	77.2 [11]
5	124.0	124.6	130.3	131.7	130.7	131.3	130.6	124.7	130.5	132.1
6	124.2	124.2	126.6	128.8	126.7	129.8	126.1	71.3	125.9	128.7
7	128.3	128.6	116.0	117.3	115.7	117.2	116.2	74.2	116.2	116.9
8	45.6	45.6	161.8	164.4	162.4	164.8	160.8	45.0	161.2	164.5
9	37.3	37.3	41.9	43.8	42.0	43.9	41.8	40.3	41.6	44.1
10	143.0	142.7	174.4	173.4	174.8	173.8	173.8	143.9	173.4	173.3
11	31.1	31.1	32.8	32.9	33.4	32.88	32.8	34.6	32.8	32.9
12	30.1	30.0	29.2	29.4	29.3	29.3	29.5	30.3	29.2	29.9 [1]
13	39.0	38.9	38.1	38.9	38.0	38.5	38.2	39.6	37.9	38.7
14	38.9	39.0	44.6	44.3	44.6	44.4	44.5	39.9	44.6	44.4
15	28.2	28.2	28.3	28.6	28.4	28.6	28.3	28.5	28.5	28.6 [2]
16	36.4	36.0	36.3	36.2	36.2	36.2	36.4	36.1	36.3	36.4
17	30.4	30.4	30.5	30.4	30.4	30.4	30.4	29.9	30.5	30.6
18	44.5	44.5	44.0	44.1	44.1	44.1	44.2	43.9	44.2	44.4 [3]
19	30.6	30.6	30.9	30.6	31.0	30.7	30.9	30.7	30.6	30.9 [4]
20	40.5	40.5	40.0	40.3	39.7	40.1	40.5	40.7	40.3	40.5 [5]
21	30.0	29.8	29.6	29.8	29.7	29.4	29.8	29.7	29.4	29.9 [6]
22	36.0	36.4	34.6	34.8	34.3	34.4	34.8	36.9	34.7	34.8 [7]
23	11.0	10.9	22.2	24.5	22.2	24.4	22.0	11.1	22.4	24.2
24										
25	22.3	22.2	34.8	39.2	35.2	39.1	34.8	28.2	37.5	39.3
26	16.9	16.9	22.4	22.5	18.4	18.6	22.3	17.1	22.0	22.4
27	17.5	17.4	18.8	18.5	22.2	22.2	18.7	18.4	18.9	18.3 [8]
28	31.8	31.8	31.6	31.5	31.4	31.4	31.6	31.8	31.6	31.6 [9]
29	179.1	179.1	184.5	184.3	182.4	183.5	178.7	179.4	184.4	178.9 [10]
30	32.1	32.1	32.6	32.4	31.8	32.4	32.7	31.3	32.7	32.8
OCH$_3$	51.5	51.5					51.7	51.6		51.6
1′	135.4	142.7	111.2	110.8	36.2	37.9	108.8	108.2	110.4	110.7
2′	36.5	34.5	144.4	145.0	18.7	18.8	140.8	142.1	144.7	144.2
3′	74.2	78.8	137.5	137.4	40.4	41.3	136.5	138.0	138.2	138.5
4′	89.9	86.1	127.2	128.3	33.0	33.1	121.3	121.0	129.5	128.2
5′	137.7	138.0	124.4	123.6	49.8	49.2	126.0	125.7	23.0	124.0
6′	32.1	35.0	192.0	189.9	35.7	35.9	124.0	124.4	187.5	187.8
7′	49.6	46.3	126.0	126.1	197.1	198.4	129.4	128.9	126.0	126.2
8′	29.7	30.0	171.4	171.0	126.0	125.9	45.5	45.9	171.4	171.5
9′	39.6	38.5	40.3	40.0	150.3	151.3	37.4	37.6	40.0	39.9
10′	42.6	34.7	150.1	151.1	37.7	37.7	142.8	141.9	151.9	151.1
11′	150.7	151.2	34.0	33.8	112.6	112.1	30.6	31.1	34.2	34.2
12′	108.7	108.5	29.3	29.1	147.2	147.0	30.0	30.0	29.9	29.6 [1]
13′	20.4	20.4	38.9	38.5	139.8	139.4	38.9	39.0	39.0	39.0

Table 4. *Cont.*

14′	17.1	19.3	44.7	44.6	114.9	115.8	39.0	39.1	44.6	44.7
15′	14.0	22.3	28.3	28.4	32.8	32.5	28.4	29.9	28.5	28.5 [2]
16′			36.2	36.2	21.3	21.3	36.0	36.1	36.3	36.4
17′			30.4	30.3	23.3	23.2	30.6	30.5	30.4	30.6
18′			44.1	44.1			44.5	44.5	44.0	44.3 [3]
19′			30.5	30.7			31.0	30.7	31.1	30.8 [4]
20′			39.7	40.1			40.4	40.6	40.2	40.4 [5]
21′			29.5	29.7			29.7	30.0	29.3	29.7 [6]
22′			34.4	34.5			36.4	36.7	34.4	34.7 [7]
23′			12.9	13.2			10.8	11.0	13.2	12.8
24′										
25′			36.8	37.5			22.5	22.2	34.7	38.0
26′			20.7	20.9			16.9	17.1	20.9	20.9
27′			18.6	18.8			17.4	17.6	18.8	18.2 [8]
28′			31.3	31.5			31.8	31.8	32.5	31.6 [9]
29′			183.6	184.3			178.7	179.5	183.6	178.7 [10]
30′			32.4	32.2			32.1	32.1		32.8
OCH$_3$							51.6	51.5	32.4	51.6
Ref	[138]	[138]	[139]	[139]	[139]	[139]	[133]	[134]	[139]	[136]

C	D34	D35	D36	D37	D38	D39	D40	D41	D42	D43
1	116.0	114.9	116.0	109.0	111.4	115.8	114.7	115.5	115.5	115.0
2	190.4	189.5	190.4	145.1	149.7	191.1	189.4	190.2	190.3	189.5
3	91.8	90.6	91.8	142.8	143.3	91.9	91.0	92.0	92.0	91.0
4	79.4	76.9	79.3	122.5	125.2	78.8	77.3	79.3	79.4	76.8
5	130.9	132.0	130.9	125.7	122.4	131.1	131.8	130.3	130.3	132.1
6	126.1	128.7	126.2	124.6	187.5	126.1	128.8	126.6	126.6	128.7
7	116.2	116.8	116.4	128.3	126.2	116.4	117.2	116.3	116.3	117.3
8	160.2	164.6	159.9	44.9	170.2	159.5	164.4	160.3	160.1	163.0
9	41.5	44.2	41.4	39.4	39.3	41.3	43.9	41.7	41.6	43.7 [1]
10	173.2	173.4	173.2	143.3	151.7	173.4	173.2	173.7	173.8	173.0
11	33.3	32.8	33.5	31.7	34.3	33.5	32.8	33.2	33.4	33.3
12	29.8 [1]	29.5	29.9 [1]	29.9	30.2	30.0	29.8	29.9 [1]	29.9 [1]	29.9
13	39.5	38.6	39.5	38.3	39.9	39.5	38.6	39.4	39.4	39.8
14	44.2	44.4 [1]	43.9	32.8	44.3	44.1	44.4	44.3	44.0	43.61
15	28.3	28.6	28.1	27.5	28.4	28.3	28.6	28.3	28.0	28.2
16	35.5	36.4	29.5 [2]	35.3	35.6	29.7	36.4	35.4	29.5	29.5
17	38.2	30.5	44.7	35.4	38.2	39.0	30.5	38.2	44.7	45.0
18	43.4	44.1	44.9	43.5	43.5	43.8	44.2	43.4	44.9	44.8
19	32.2	30.8	32.1	31.0	32.0	32.1	30.9	32.1	32.1	31.9
20	41.9	40.4	40.8	40.1	41.9	41.8	40.5	41.9	40.8	40.9
21	213.6	29.9	213.5	214.3	213.7	213.6	29.8	213.6	213.6	213.5
22	52.5	34.7	76.5	53.9	52.6	52.5	35.0	52.5	76.5	76.4
23	22.1	24.2	22.2	10.8	13.2	22.1	24.6	22.3	22.2	24.6
24										
25	35.7	39.3	35.8	32.8	38.7	35.6	39.2	35.6	35.6	40.1
26	22.3	22.4	22.4	18.6	20.8	22.3	22.3	22.3	22.4	22.3
27	20.1	18.2	20.9	16.1	19.7	15.0	18.2	20.0	20.9	20.4
28	32.5	31.5	25.0	22.7	32.6	32.5	31.6	32.6	25.0	25.0
29		178.8					179.1			
30	15.1	32.7	14.7	15.1	15.1	20.0	32.9	15.1	14.8	14.8
OCH$_3$		51.6					51.6			
1′	110.5	110.7	110.6	92.9	92.8	108.1	110.6	111.4	111.3	110.6
2′	144.4	144.3	144.4	38.4	38.8	141.6	145.2	144.6	144.6	145.1
3′	138.3	138.5	138.3	128.2	128.1	137.6	137.5	137.6	137.6	137.5
4′	129.3	128.3	129.3	140.6	140.7	122.5	128.3	127.6	127.7	128.4
5′	123.3	123.9	123.4	96.8	96.8	125.0	123.8	124.5	124.5	123.9
6′	187.2	187.6	187.2	38.2	38.4	124.0	187.4	187.9	187.8	187.3
7′	126.3	126.1	126.2	43.6	43.3	129.2	126.1	126.1	126.2	126.1

Table 4. Cont.

8'	171.0	170.4	171.1	35.7	35.4	45.5	171.2	171.7	171.6	171.7
9'	40.1	39.7	40.8	32.7	33.0	38.2	40.0	40.0	39.9	40.0
10'	151.8	151.0	151.8	42.2	39.3	143.8	151.1	150.5	150.5	151.2
11'	34.3	34.3	34.3	151.2	150.9	36.6	34.0	34.2	34.2	34.0
12'	30.0	30.2	29.8 [1]	108.6	108.8	36.4	29.8	29.9 [1]	29.9 [1]	29.9
13'	39.0	40.2	39.0	19.9	20.0	37.5	39.0	39.0	39.0	39.0
14'	44.7	44.3 [1]	44.9	19.0	19.1	38.2	44.7	44.7	44.9	44.7
15'	28.6	28.4	28.6	12.5	12.4	28.3	28.5	28.5	28.5	28.5
16'	36.4	35.5	36.4			35.4	36.4	36.4	36.4	36.4
17'	30.5	38.2	30.5			30.4	30.5	30.5	30.5	30.5
18'	44.3	43.5	44.3			44.4	44.2	44.3	44.3	44.2
19'	31.1	32.0	31.1			29.8	30.8	30.9	30.8	30.9
20'	40.6	41.9	40.7			40.6	40.4	40.4	40.4	40.5
21'	29.7 [1]	214.7	29.7 [2]			29.8	29.5	29.9 [1]	29.8	29.7
22'	35.0	52.6	35.0			35.8	34.7	34.7	34.8	35.0
23'	13.3	12.8	13.3			10.8	13.2	13.0	13.0	13.2
24'										
25'	37.6	38.9	37.7			22.3	37.7	37.6	37.6	37.7
26'	20.9	20.8	20.8			17.5	20.9	20.8	20.8	20.9
27'	18.5	19.7	18.6			17.0	18.4	18.3	18.3	18.5
28'	31.6	32.6	31.6			31.8	31.5	31.6	31.6	31.6
29'	179.3		179.4			179.3	178.8	178.7	178.8	179.1
30'	32.9	15.1	33.0			31.8	32.7	32.7	32.7	32.9
OCH$_3$	51.6		51.8			51.5	51.4	51.6	51.6	51.4
Ref	[136]	[136]	[137]	[128]	[128]	[128]	[140]	[136]	[137]	[137]

C	D44	D45	D46	D47	D48	D49	D50
1	115.2	114.6	115.5	115.5	108.2	108.2	128.3
2	190.2	189.4	190.2	189.5	144.5	144.5	183.6
3	92.0	91.1	92.0	92.0	140.7	140.7	96.9
4	79.4	77.2	79.4	79.4	122.8	122.9	92.3
5	129.8	131.8	130.3	130.2	124.7	124.7	39.5
6	126.8	128.9	126.6	126.5	71.4	71.4	28.1
7	116.1	117.2	116.2	116.3	74.1	74.1	32.8
8	161.5	164.4	160.4	160.5	43.6	43.6	30.4
9	42.0	44.0	41.7	41.8	40.4 [1]	40.4 [1]	43.7
10	174.3	173.2	173.7	173.6	143.7	143.7	140.7
11	32.9	32.8	33.2	32.8	34.5	34.5	35.4
12	29.5	29.6 [1]	29.8	29.5	30.2	30.2	30.6
13	38.1	38.6	39.4	38.5	40.6 [1]	40.6 [1]	44.5
14	44.7	44.4	44.3	44.2	40.4 [1]	40.4 [1]	40.5
15	28.4	28.6	28.3	28.4	29.3	29.3	30.0
16	36.3	36.4	35.4	35.5	35.5	35.5	36.4
17	30.5	30.5	38.2	37.1	37.9	37.9	40.5
18	44.1	44.3	43.4	43.6	43.8	43.7	43.7
19	30.9	30.8	32.1	34.2 [1]	31.9	31.9	43.7
20	40.4	40.4	41.9	53.6	42.2	42.2	151.2
21	29.8	30.1 [2]	213.6	209.4	214.7	214.6 [2]	39.0
22	34.7	34.7	52.5	51.9	53.9	53.9	38.2
23	22.2	24.6	22.3	22.3	11.3	11.3	12.6
24							
25	34.9	39.1	35.6	35.2	29.1	29.1	
26	22.5	22.4	22.3	22.8	15.7	15.7	19.1
27	18.7	18.2	20.0	18.2	19.3	19.2	32.1
28	31.6	31.6	32.6	32.6	33.1	33.1	20.5
29	178.9	178.8		175.1			108.7
30	32.7	32.8	15.1	25.1	15.2	15.2	19.9
OCH$_3$	51.6	51.6		52.6			
1'	111.4	110.5	111.4	111.4	109.7	109.7	109.2
2'	144.7	145.2	144.7	144.6	141.5	141.4	145.0

Table 4. *Cont.*

3′	137.7	137.6	137.7	137.6	140.3	140.3	142.8
4′	127.8	128.6	127.8	127.8	121.8	121.8	122.4
5′	124.5	123.9	124.5	124.5	124.3	124.3	125.7
6′	187.7	187.2	187.7	187.7	119.7	119.8	124.4
7′	126.1	126.2	126.2	126.1	138.2	138.1	128.7
8′	170.4	169.9	170.8	170.7	43.9	43.9	45.5
9′	39.7	39.7	39.8	39.8	142.8	143.0	37.3
10′	150.4	151.1	150.3	150.4	131.8	131.8	143.3
11′	34.5	34.3	34.1	34.3 [1]	122.7	122.6	31.2
12′	30.2	30.1 [2]	29.9	30.1	37.5	37.4	30.0
13′	40.1	40.1	39.2	40.2	40.1	40.1	38.9
14′	44.0	43.9	44.3	44.3	40.7	40.4 [1]	38.9
15′	28.2	28.2	28.6	28.4	24.0	23.6	28.1
16′	29.6	29.5 [1]	35.7	35.5	35.9	29.5	36.4
17′	44.9	44.9	37.1	38.2	39.2	45.3	30.4
18′	45.0	45.0	43.9	43.5	45.7	46.7	44.5
19′	32.0	31.9	34.2	32.0	32.5	32.1	30.6
20′	40.9	40.9	53.6	41.9	42.4	41.3	40.5
21′	213.7	213.6	209.4	213.7	214.9	214.5 [2]	30.0
22′	76.5	76.7	51.9	52.6	51.2	75.4	36.0
23′	13.0	13.2	13.0	13.0	10.7	10.7	10.8
24′							
25′	38.7	38.9	38.2	38.4	22.4	22.4	22.1
26′	20.9	20.9	21.2	20.8	19.6	19.8	16.9
27′	20.6	20.7	17.8	19.7	20.6	21.4	17.4
28′	25.0	25.0	32.6	32.6	31.5	24.2	31.8
29′			175.0				179.2
30′	14.8	14.7	25.1	15.1	15.4	15.0	32.1
OCH$_3$			52.5				51.6
Ref	[137]	[137]	[137]	[137]	[141]	[141]	[142]

Ref: References; [1,2,3,4,5,6,7,8,9,10]: Values bearing the same superscript are interchangeable; [11]: Signal bearing this superscript was superimposed on solvent signals; Solvent: CDCl$_3$; ^{13}C-NMR data for D4, D5 and D29 were reported by Gonzalez et al. [143] but the chemical shifts were not attributed to each carbon atom.

Table 5. ^{13}C-NMR data of lupane-type pentacyclic triterpenoids isolated from Celastraceae species (2001–2021).

C	L1 [+]	L2	L3	L4	L5	L6	L7	L8	L9	L10
1	38.4	38.7	41.8	123.6	42.1	39.6	39.6	79.5	39.8	39.3
2	23.7	27.4	34.0	165.1	34.2	34.1	34.1	42.9	34.3	34.3
3	80.9	79.0	218.3	205.3	218.8	218.0	217.7	216.1	218.2	221.4
4	37.9	38.9	47.5	45.0	47.6	47.3	47.2	47.1	47.5	50.7
5	55.4	55.2	54.6	52.9	54.8	54.9	54.9	51.2	55.1	55.3
6	18.2	18.3	19.4	18.9	19.6	19.6	19.6	19.6	19.8	19.2
7	34.3	34.3	33.9	34.5	34.3	33.5	33.6	32.9	33.7	33.6
8	40.9	40.8	42.2	42.9	42.4	40.8	40.6	41.0	41.0	40.7
9	49.9	50.0	54.5	48.9	54.9	49.3	49.8	50.6	50.0	49.5
10	37.1	37.1	38.0	40.6	38.2	36.8	36.9	45.1	37.0	36.6
11	20.9	20.8	70.1	70.6	70.5	21.4	21.5	22.9	21.6	21.7
12	27.6 *	26.5	37.2	37.3	37.4	24.8	25.1	25.1	25.0	25.2
13	37.9	37.8	36.1	37.1	37.2	37.4	37.7	37.9	37.8	37.5
14	43.1	42.9	42.2	42.8	42.6	44.1	41.9	42.9	42.8	42.8
15	27.2	27.2	26.8	27.3	27.4	36.8	26.9	27.5	27.3	27.0
16	35.3	35.3	33.6	35.3	35.4	76.9	33.1	35.5	35.9	34.0
17	43.0	43.1	47.3	43.0	43.0	48.6	80.2	42.9	42.0	47.8
18	49.0	47.1	47.8	47.5	47.6	47.6	48.3	47.9 *	48.2	48.6
19	42.8	37.4	47.4	47.7	47.7	47.5	48.0	48.2 *	60.0	47.8
20	49.0	49.7	149.5	150.2	150.2	149.9	149.6	150.7	148.4	150.3

Table 5. *Cont.*

21	25.2 *	23.6	29.4	29.7	29.8	29.8	29.4	29.8	77.9	29.7
22	40.0	40.5	28.8	39.8	39.8	37.7	38.5	40.0	49.6	29.1
23	28.0	28.0	27.2	21.4	27.5	26.6	26.6	28.0	26.8	22.1
24	16.6	15.4	20.5	28.2	20.8	21.0	21.0	19.8	21.2	65.3
25	16.1 **	16.0	16.5	19.9	16.7	16.0	16.0	11.9	16.1	17.0
26	16.0 **	15.9	16.6	17.4	16.9	15.8	15.9	15.9	16.0	15.6
27	14.4	14.3	14.4	14.4	14.4	16.1	13.8	14.4	14.6	14.7
28	18.0	17.9	60.3	18.0	18.1	11.7		18.1	19.8	60.5
29	207.0	7.3	110.1	113.5	109.9	109.8	109.6	109.5	111.5	109.8
30	14.5	205.1	18.9	19.3	19.4	19.3	19.3	19.3	19.9	19.1
COO$\underline{C}$H$_3$	21.3									
$\underline{C}$OOCH$_3$	171.0									
Ref	[144]	[145]	[146]	[147]	[148]	[149]	[145]	[150]	[14]	[151]

C	L11	L12	L13	L14	L15	L16	L17 [a]	L18	L20	L21
1	39.6	39.6	125.2	39.8	40.2	39.7	34.1	42.4	39.7	53.5
2	34.1	34.1	159.2	34.0	35.7	34.3	26.6	34.3	34.2	211.4
3	218.2	218.0	205.0	218.0	216.5	218.3	75.1	218.6	217.9	82.5
4	47.3	47.4	44.6	47.3	47.7	47.5	38.1	47.6	47.4	45.6
5	54.9	55.0	53.5	54.8	55.4	55.1	49.2	54.9	55.1	54.6
6	19.6	19.6	19.0	19.0	20.4	19.8	18.5	19.5	19.7	18.5
7	33.5	33.5	33.7	33.5	34.2	33.8	34.5	34.2	33.7	34.0
8	40.9	42.7	41.8	40.9	41.5	41.0	41.1	42.1	40.8	41.3
9	49.7	49.6	44.5	49.7	49.8	50.0	50.8	54.7	49.7	50.4
10	36.8	36.9	39.5	36.8	37.7	37.0	37.7	38.2	36.9	43.9
11	21.4	21.4	21.2	21.3	22.2	21.7	20.8	70.1	21.5	21.0
12	26.8	27.6	25.2	25.2	28.2	26.8	26.1	37.6	27.6	25.1
13	37.3	37.1	37.5	37.4	38.0	38.3	35.2	37.6	37.9	37.2
14	42.7	40.8	43.0	42.8	43.5	43.0	41.4	42.3	42.8	42.8
15	27.0	26.9	27.0	27.0	28.2	27.6	28.2	28.8	27.4	27.1
16	33.8	29.1	29.1	29.1	34.5	35.6	22.9	28.9	35.4	29.3
17	47.8	48.0	47.7	47.7	37.6	43.2	54.0	59.3	43.3	47.8
18	49.3	52.3	48.6	48.6	50.4	49.0	55.5	59.3	47.7	48.7
19	43.5	36.5	47.8	47.3	52.8	43.9	92.2	47.2	47.7	47.8
20	154.4	157.0	150.3	150.0	211.5	154.9	141.4	149.1	157.2	150.2
21	29.1	32.8	29.7	29.7	27.8	31.9	34.4	29.8	32.7	29.7
22	31.7	33.9	33.9	33.9	40.4	40.0	29.3	32.9	40.0	33.8
23	26.7	26.6	21.4	26.6	27.1	26.8	29.2	20.7	26.6	29.1
24	21.1	21.1	27.8	21.0	21.3	21.2	22.8	27.4	21.1	16.4
25	16.0	15.9	19.2	15.8	16.1	16.1	16.5	16.8	15.9	17.0
26	15.8	15.8	16.5	15.9	14.8	16.0	15.7	16.7	15.8	14.8
27	14.7	14.6	14.6	14.7	16.4	14.6	13.6	14.1	14.4	15.6
28	60.2	60.2	60.6	60.5	18.3	17.9	178.6	206.0	17.8	60.5
29	107.2	133.2	109.9	109.7		106.9	112.3	110.7	133.0	109.9
30	65.0	194.9	19.0	19.6	29.3	65.2	19.3	19.0	195.0	19.1
Ref	[152]	[153]	[151]	[149]	[14]	[14]	[153]	[151]	[148]	[73]

C	L22	L23 [b]	L24	L25	L26	L28	L29	L30	L31	L32
1	34.0	35.4	33.6	53.4	38.6	38.6	26.3	38.6	53.5	38.7
2	22.9	23.7	25.9 *	211.5	24.0	27.4	23.1	27.3	211.4	26.9
3	78.2	80.3	76.4	82.9	81.0	78.9	76.0	78.8	82.9	78.9
4	36.7	37.7	37.5	45.6	38.3	38.9	36.4	38.8	45.6	38.8
5	50.3	51.6	49.9 **	54.6	55.6	55.3	52.4	55.1	54.6	55.3
6	18.0	19.2	18.4	18.5	18.4	18.3	20.9	18.2	18.5	18.2
7	34.4	35.3	34.4	33.8	34.2	34.3	33.0	34.1	33.8	34.3
8	40.9	42.2	41.0	41.2	41.2	40.6	43.1	40.7	43.9	40.8
9	50.7	51.5	50.5 **	50.4	50.5	50.4	45.7	50.2	50.4	50.4
10	37.2	38.3	37.3	44.0	37.3	37.2	39.8	37.1	41.3	37.1
11	20.8	21.9	20.8	21.1	21.1	21.0	23.6	20.8	21.0	20.8
12	27.2	28.8	25.6 *	24.8	25.4	27.3	25.8	27.2	25.0	27.4

Table 5. *Cont.*

C										
13	39.8	39.2	38.7	37.9	37.3	36.6	38.3	36.2	37.2	38.5
14	44.0	43.9	42.6	42.9	42.9	41.8	40.3	42.5	42.8	42.5
15	28.1	28.5	29.5	29.4	27.3	26.8	27.8	26.9	27.0	29.2
16	32.1	36.4	28.8	35.4	29.3	32.8	35.6	28.8	29.1	31.7
17	54.3	44.3	59.3	42.9	48.0	80.3	43.0	47.8	47.7	59.3
18	143.8	52.0	48.0 ***	48.2	49.0	49.2	48.3	49.6	48.6	48.6
19	139.2	38.3	47.5 ***	47.29	48.0	52.2	47.9	52.0	47.7	43.2
20	207.2	158.3	149.8	150.7	150.7	212.2	150.9	213.5	48.6	154.1
21	34.8	33.4	30.0	29.8	30.0	27.5	29.9	27.6	29.7	28.9
22	35.1	40.9	33.2	29.9	34.4	38.7	40.0	33.9	33.9	32.9
23	27.7	28.3	28.2	29.2	28.2	28.0	29.6	27.9	29.3	28.0
24	21.7	22.2	22.2	16.4	16.9	15.4	24.0	15.4	16.3	15.3
25	16.5	16.5	15.9 ****	17.0	16.4	16.2	101.4	16.0	17.0	15.9
26	16.1	16.7	16.1 ****	15.6	16.2	16.0	16.2	15.9	15.6	16.1
27	15.7	15.1	14.2	14.5	15.0	13.2	14.7	14.6	14.7	14.3
28	66.0	18.3	205.6	18.0	60.8		18.0	60.5	60.5	206.3
29		134.8	110.1	109.5	109.9		109.3		109.9	107.4
30	30.7	197.3	19.0	19.3	19.3	29.9	19.3	29.4	19.1	65.0
OCH$_3$	21.4	21.3					54.6			
C=O	170.8	173.4			167.7					
1′					126.7					
2′					130.1					
3′					115.8					
4′					159.3					
5′					130.1					
6′					130.1					
7′					116.3					
8′					144.5					
Ref	[154]	[154]	[155]	[73]	[156]	[145]	[54]	[157]	[158]	[145]

C	L33 [a]	L35 [c]	L36	L37	L38	L39	L40	L41	L42	L43
1	39.3	38.4	78.7	38.4	38.4	38.6	38.6	53.5	38.7	38.7
2	28.3	23.8	4.2	23.9	23.8	27.3	27.4	211.5	27.3	27.4
3	78.0	81.3	77.6	55.4	80.6	78.9	78.9	82.9	78.9	79.0
4	39.3	38.1	NR	38.1	37.9	38.8	38.8	45.6	38.8	38.9
5	53.2	55.4	53.1	55.4	55.4	55.2	55.2	54.6	55.5	55.3
6	30.3	18.2	17.8	18.2	18.2	18.2	18.3	18.5	18.2	18.3
7	74.4	34.1	34.0	34.2	34.3	34.2	34.2	33.8	34.3	34.3
8	47.2	41.9	41.4	40.9	40.9	40.7	40.7	41.2	40.8	40.8
9	51.1	50.3	51.4	50.3	50.4	50.3	50.2	50.4	50.4	50.2
10	37.6	37.1	43.5	37.1	37.1	37.1	37.2	44.0	37.1	37.1
11	21.3	21.8	23.8	21.0	21.0	20.7	20.9	21.1	20.7	20.9
12	26.3	25.1	25.0	25.1	25.1	27.5	27.3	24.9	25.5	27.6 [A]
13	39.3	37.3	38.1	38.1	38.1	37.2	37.0	37.9	38.7	37.7
14	44.0	42.7	37.6	43.0 *	42.9	42.3	42.7	42.9	42.5	42.7
15	34.0	27.0	27.5	27.4	27.5	29.0	27.4	27.4	29.2 *	27.3
16	32.8	29.2	35.6	35.6	35.6	28.7	35.0	35.5	28.8 *	35.4
17	26.7	48.2	42.9	42.8 *	43.0	59.3	43.0	43.0	59.3	43.3
18	49.7	48.8	48.3	48.3	48.3	48.0	49.7	48.2	48.0 **	51.2 [B]
19	47.7	47.7	48.0	47.7	48.0	51.1	52.6	47.9	47.5 **	36.7 [C]
20	151.1	150.4	150.8	151.0	151.0	211.8	212.9	150.8	149.7	157.0 [B]
21	30.9	29.7	29.8	29.9	29.9	27.6	27.6	29.8	29.8	32.6 [B]
22	37.2	34.0	40.0	40.0	40.0	32.4	39.8	39.9	33.2	39.9
23	28.5	28.0	27.8	28.0	28.0	28.0	28.0	29.3	27.9	28.0
24	16.4	16.7	16.2	16.7	16.6	15.4	15.4	16.4	15.4	15.4
25	16.3	16.2	12.0	16.2 **	16.2	16.1	16.1	17.0	15.9 ***	16.1
26	10.9	16.0	16.2	16.0 **	16.0	15.8	15.9	15.6	16.1 ***	15.9
27	15.1	14.8	14.4	14.5	14.5	14.2	14.5	14.5	14.2	14.4
28	176.6	60.6	17.9	18.0	18.0	206.1	18.0	18.0	205.6	17.8
29	110.0	109.8	109.4	109.4	109.4	30.2	29.2	109.5	110.1	132.9 [B]

Table 5. *Cont.*

C										
30	19.5	19.1	19.2	19.3	19.3			19.3	19.0	195.1
1′		127.3	167.6	127.1	173.7					
2′		115.3	115.7	115.1	34.9					
3′		144.2	145.1	144.9	25.2					
4′		146.8	127.4	147.3	29.2					
5′		114.2	114.4	113.9	29.3					
6′		122.3	144.0	122.0	29.4					
7′		144.9	146.6	144.9	29.6					
8′		115.8	115.5	115.7	29.7					
9′–14′		168.0	122.4	167.7	29.7					
15′					29.5					
16′					31.9					
17′					22.7					
18′					14.1					
Ref	[153]	[159]	[160]	[159]	[161]	[145]	[145]	[162]	[155]	[163]

C	L44	L45	L46	L47	L48	L49	L50[a]	L51	L52	L53
1	42.5	42.5	42.1	42.1	38.4	40.9	42.6	38.4	40.1	39.8
2	34.4	34.4	34.4	34.5	27.7	178.5	174.4	36.2	34.5	34.3
3	216.8	216.8	216.7	216.8	81.0	187.5	182.4	217.2	216.3	218.3
4	48.9	48.9	48.9	49.0	37.8	45.6	46.9	47.0	42.2	47.0
5	56.6	56.6	56.5	56.6	55.4	48.2	48.4	49.6	56.5	55.1
6	69.7	69.7	69.6	69.7	18.2	21.3	21.6	19.9	69.8	19.8
7	42.1	42.1	42.2	42.2	34.2	33.7	33.8	33.5	41.9	37.2
8	40.7	40.7	40.0	40.0	40.9	41.8	41.1	40.8	37.5	40.8
9	50.6	50.6	50.6	50.5	50.4	41.7	42.0	52.4	50.9 *	50.0
10	36.8	36.8	36.7	36.7	37.1	40.7	42.3	36.6	34.5	37.1
11	21.8	21.8	21.3	21.3	21.0	19.2	22.2	21.7	21.3	21.5
12	28.7	28.7	26.7	29.7	25.1	24.9	27.3	26.0	25.2	25.7
13	36.8	36.8	37.1	36.8	38.1	37.9	38.6	38.6	37.5	38.6
14	43.9	43.9	42.9	43.2	42.8	43.2	43.5	42.7	42.9	42.6
15	27.7	27.7	27.4	27.4	27.5	27.5	27.9	31.1	29.9	29.8
16	35.6	35.6	35.3	35.3	35.6	35.5	35.6	32.6	32.2	32.2
17	44.6	44.6	43.1	43.0	43.0	43.2	43.2	56.4	56.9	56.5
18	48.5	48.5	48.9	50.6	48.3	48.4	48.9	47.5	49.5 *	49.3
19	50.1	50.1	43.7	42.1	48.0	48.0	43.8	49.6	47.0	47.5
20	73.4	73.4	154.6	157.2	151.0	150.9	156.5	151.1	150.3	150.5
21	29.1	29.1	31.7	32.7	30.0	29.8	32.2	30.0	30.7	30.7
22	40.2	40.2	39.8	39.8	40.0	39.9	40.0	37.3	37.0	33.8
23	24.9	24.9	25.0	25.0	27.1	29.8	27.8	68.0	25.7	26.8
24	23.7	23.7	23.7	23.7	16.2	21.3	24.8	17.2	21.3	21.2
25	17.5	17.5	17.0	16.9	16.5	20.8	20.4	16.0	17.3 **	16.1
26	17.0	17.0	17.1	17.1	16.0	15.9	16.3	15.9	17.1 **	16.0
27	15.2	15.2	14.8	14.7	14.5	14.6	15.0	14.6	15.0	14.8
28	19.2	19.2	17.7	17.8	18.0	18.0	17.9	178.6	181.9	181.2
29	31.7	31.7	106.9	133.2	109.4	109.4	105.9	109.6	109.8	109.9
30	25.2	25.2	65.0	195.1	19.3	19.2	64.3	19.3	19.5	19.5
C=O					171.0					
OCH$_3$					21.3					
Ref	[164]	[164]	[146]	[146]	[165]	[73]	[153]	[166]	[167]	[14]

C	L54	L55	L56	L57 [a]	L58 [d]	L59 [a]	L61	L62	L63	L64
1	39.6	33.9	34.0	34.0	34.0	38.7	159.9	38.7	75.9	75.9
2	34.1	28.1	26.2	26.7	25.1	28.4	125.1	27.4	36.4	36.4
3	218.2	179.1	75.0	75.3	76.9	78.2	205.6	79.0	76.6	76.9
4	47.3	147.5	38.1	38.2	38.5	39.4	44.6	38.8	37.4	37.5
5	54.9	50.4	46.4	49.3	54.9	56.0	53.4	55.3	47.8	47.9
6	19.6	24.5	26.1	18.7	18.0	18.9	19.0	18.3	18.4	18.5
7	33.6	32.8	76.7	34.7	33.9	34.9	33.7	34.3	34.0	34.1
8	40.7	40.9	46.3	41.2	41.2	41.2	41.7	40.8	42.9	41.7
9	49.6	40.4	51.2	50.5	49.9	51.0	44.4	50.4	51.2	51.3

Table 5. *Cont.*

10	36.8	39.2	37.8	37.7	36.7	37.6	39.5	37.1	37.4	43.7
11	21.5	21.4	20.8	21.2	20.5	21.3	21.2	21.0	23.7	23.9
12	29.7	25.4	26.5	28.2	27.1	26.2	25.1	26.7	25.2	25.2
13	37.8	38.3	39.1	38.2	37.6	39.6	38.2	38.0	36.9	37.7
14	42.8	42.8	44.4	43.1	42.6	42.9	43.0	43.0	41.7	42.9 *
15	27.3	30.6	33.0	27.8	31.7	31.3	27.4	27.4	27.1	27.5
16	35.4	32.1	33.4	35.9	36.4	33.0	35.5	35.5	29.2	35.7
17	43.1	56.5	56.3	43.5	55.5	56.7	43.1	42.8	43.6	43.0 *
18	50.9	49.4	49.6	51.3	46.7	47.9	48.1	48.9	48.8	48.1 **
19	40.7	46.9	48.0	41.4	48.7	49.9	47.3	43.8	47.8	48.4 **
20	146.3	150.4	151.4	149.2	150.4	151.4	150.8	154.8	150.3	150.8
21	32.9	29.7	31.2	33.5	30.1	30.4	29.8	31.8	29.7	29.8
22	39.7	36.9	37.7	40.2	38.3	37.7	40.0	39.8	34.0	40.0
23	26.6	113.4	29.0	29.4	28.1	28.8	27.8	28.0	27.7	27.6
24	21.0	23.2	22.4	22.7	15.8	16.4	21.4	16.1	21.9	22.0
25	15.9	20.1	16.1	16.3	15.9	16.5	19.2	16.0	11.7	11.5
26	15.8	15.9	12.1	16.4	16.0	16.5	16.4	15.4	16.2	16.1
27	14.4	14.6	15.1	14.8	14.4	15.0	14.4	14.5	14.8	14.6
28	17.9	176.6	179.0	18.2	177.3	178.9	18.0	17.7	60.6	18.1
29	124.9	109.7	109.9	122.4	109.6	110.0	109.5	107.6	109.8	109.5
30	171.2	19.3	19.5	170.3	19.0	19.6	19.3	65.0	19.0	19.3
OCH$_3$		51.3			51.3					
1′			166.3							
2′			117.6							
3′			156.4							
4′			27.0							
5′			20.2							
Ref	[146]	[118]	[153]	[154]	[168]	[169]	[170]	[145]	[146]	[150]

C	L65	L66	L67	L68	L69	L70	L71	L72	L73	L74
1	79.0	44.5	39.5	39.6	35.5	35.4	33.5	32.5	39.0	38.9
2	37.5	71.3	34.1	34.2	25.6	25.6	28.2	25.3	27.5	27.4
3	75.7	78.5	217.8	218.1	75.9	75.9	76.0	76.4	78.6	78.9
4	38.9	38.2	47.3	47.3	37.8	37.8	38.8	37.6	39.4	38.8
5	53.1	55.2	54.9	54.9	48.9	48.9	48.5	48.2	55.6	54.9
6	18.0	18.1	19.6	19.7	18.0	18.3	21.1	18.4	18.1	18.5
7	34.1	34.2	33.2	33.6	35.1	35.1	34.2	34.4	35.3	37.8
8	41.3	40.9	40.9	40.8	42.6	42.7	40.9	41.1	41.1	42.5
9	51.4	50.9	49.6	49.8	55.6	55.6	50.4	50.5	55.7	51.0
10	43.5	36.9	36.8	36.9	39.1	39.1	37.3	37.3	37.7	37.4
11	23.8	21.1	21.2	21.5	70.5	70.6	20.8	21.2	70.5	21.0
12	25.0	25.2	25.3	25.2	37.6	37.6	25.2	25.3	27.7	25.2
13	38.0	38.0	37.3	38.2	36.3	37.1	37.1	38.0	37.7	37.6
14	42.8	42.9	42.7	42.9	42.9	42.7	42.7	43.0	42.6	47.9
15	27.4	27.3	26.9	27.4	27.0	27.3	27.1	27.6	27.5	69.7
16	35.5	35.6	34.8	35.5	33.8	35.4	29.2	35.8	35.5	46.5
17	42.9	43.0	37.8	43.0	47.8	43.0	48.1	43.2	43.0	43.0
18	48.3	48.3	47.0	48.3	48.2	47.7	47.8	48.2	47.7	48.1
19	47.9	48.0	59.0	48.0	47.6	47.7	48.0	48.2	47.4	47.4
20	150.8	151.0	143.4	150.6	149.8	150.2	150.4	151.2	150.2	150.4
21	29.7	29.9	217.7	29.9	29.1	29.8	29.8	29.9	29.9	30.1
22	39.9	40.0	55.4	40.0	29.3	39.8	34.0	40.2	39.9	39.7
23	27.8	29.6	26.6	26.7	22.3	22.3	28.1	28.2	28.3	27.9
24	14.9	17.1	21.0	21.0	28.7	28.7	16.0	22.4	15.6	15.4
25	11.9	17.1	15.9	16.0	16.3	16.2	16.1	16.4	16.1	16.1
26	16.2	16.0	15.7	15.8	17.2	17.2	16.1	16.2	17.3	16.6
27	14.4	14.5	14.5	14.5	14.1	14.6	14.8	14.7	14.5	8.0
28	18.0	18.0	18.7	18.0	60.2	18.0	60.1	18.2	18.1	19.2
29	109.4	109.3	115.0	109.4	110.2	109.8	109.6	109.6	109.8	109.7
30	19.2	19.3	20.8	19.3	19.1	19.3	18.9	19.5	19.4	19.4

Table 5. *Cont.*

Ref	[171]	[172]	[40]	[145]	[151]	[147]	[173]	[174]	[175]	[176]
C	**L75**	**L76** [c]	**L77**	**L78**[c]	**L79**	**L80**	**L81**	**L82**	**L83**	**L84**
1	38.7	38.7	34.2	38.2	38.7	38.9	40.7	38.7	38.4	40.7
2	27.3	27.5	27.2	30.6	27.0	27.5	29.7	27.4	27.0	27.6
3	78.3	78.9	79.0	82.3	79.0	78.8	79.1	78.8	76.8	79.1
4	38.8	38.7	38.9	40.9	38.9	38.8	39.6	38.3	41.9	39.6
5	55.6	55.2	55.4	59.5	55.3	55.4	55.6	55.2	49.9	55.5
6	18.2	18.3	18.2	22.0	18.3	18.3	69.0	18.3	18.4	69.1
7	34.1	34.5	34.1	37.5	34.2	33.4	42.1	34.2	34.0	42.4
8	40.8	41.3	40.8	42.7	40.9	40.9	39.9	40.9	40.8	40.4
9	49.4	50.2	50.4	54.5	50.4	50.4	51.9	50.3	50.4	50.9
10	36.8	37.0	42.6	32.2	37.1	37.2	36.7	37.1	37.0	36.6
11	21.4	21.4	20.8	24.7	20.8	21.0	21.1	20.9	21.4	21.6
12	24.8	27.4	25.3	30.8	25.1	26.7	25.3	25.1	26.6	28.8
13	37.4	37.4	37.9	41.3	37.3	38.0	37.2	38.0	37.9	36.6
14	44.1	44.6	42.9	44.8	42.7	42.8	43.0	42.8	42.8	43.8
15	36.8	27.7	27.2	33.0	27.3	27.5	27.5	27.4	27.4	27.5
16	76.9	35.5	35.3	38.0	29.2	35.5	35.5	35.5	35.4	35.5
17	48.6	43.5	42.8	46.4	47.8	43.0	43.1	42.9	43.0	44.6
18	47.6	48.3	48.2	53.3	48.7	48.9	48.4	48.2	48.8	48.4
19	47.4	49.9	47.8	47.4	47.8	43.8	48.0	47.9	43.8	49.9
20	150.0	73.5	150.4	150.7	150.5	155.2	150.9	150.6	154.7	73.5
21	29.8	29.0	29.7	35.4	29.8	31.8	29.9	29.8	31.7	29.2
22	37.7	40.2	40.0	33.1	33.9	39.9	40.0	39.9	39.8	40.2
23	28.0	28.0	28.0	31.2	27.9	28.1	27.6	28.0	72.1	27.7
24	15.0	15.4	15.6	19.3	15.3	15.4	16.6	15.4	11.2	16.9
25	16.0	16.1	60.7	18.6	16.1	16.0	17.7	16.1	16.5	17.8
26	16.0	16.1	16.2	19.2	15.9	16.1	16.9	15.9	16.0	17.2
27	16.0	14.8	14.8	17.8	14.7	14.6	14.9	14.5	14.6	15.2
28	11.6	19.2	18.1	62.9	60.6	17.7	18.0	18.0	17.7	19.2
29	109.6	24.7	109.9	109.8	109.7	106.4	109.4	109.2	106.8	24.8
30	19.0	31.5	19.2	67.8	19.0	64.6	19.3	19.3	65.0	31.6
Ref	[149]	[177]	[178]	[157]	[179]	[148]	[151]	[180]	[145]	[146]

C	**L85**	**L86**
1	40.7	39.5
2	27.5	34.0
3	79.1	217.8
4	39.6	47.2
5	55.6	54.7
6	68.9	19.6
7	42.0	33.4
8	40.0	40.8
9	51.0	49.2
10	36.7	36.8
11	21.0	21.2
12	25.3	26.5
13	36.4	37.7
14	42.9	43.0
15	27.1	27.7
16	33.6	34.5
17	47.7	43.8
18	48.8	52.2
19	47.7	45.0
20	150.4	139.1
21	29.7	82.2
22	29.1	47.7
23	27.6	26.7
24	16.9	21.0
25	17.7	15.9

Table 5. *Cont.*

26	16.9	15.7
27	15.1	14.1
28	60.4	19.3
29	109.7	124.7
30	19.1	171.2
Ref	[151]	[181]

Ref: References; [+] ^{13}C-NMR data of acetylated compound, *, **, ***, **** Values bearing the same superscript are interchangeable; NR: Not reported; Solvent: CDCl$_3$; [a] Pyridine-d$_5$; [b] CD$_3$COOD; [c] CD$_3$OD; [d] DMSO-d$_6$; [A] Slightly broadened peak, measured at 25 °C; [B] Broad peak, measured at 35°C; [C] Very broad peak, determined from HSQC and HMBC spectra at 35°. ^{13}C-NMR data for L19, L27 [159], L34, L60, L87 [182,183], L88 [182,183] and L89 were not found. These compounds were identified on the references cited here by comparison of spectroscopic data with those reported in the literature, but we could not get access to the original papers.

Table 6. ^{13}C-NMR data of oleanane-type pentacyclic triterpenoids isolated from Celastraceae species (2001–2021).

C	O1	O2	O3	O4	O5	O6	O7	O8	O9 [a]	O10 [a]
1	38.7	32.8	41.1	40.3	33.8	39.3	40.3	40.3	39.0	72.3
2	33.8	26.6	32.8	34.4	25.5	27.3	34.4	34.2	34.2	35.5
3	216.2	75.9	217.4	218.0	76.1	78.6	218.2	217.8	215.9	72.3
4	47.5	37.4	47.7	47.7	37.5	39.0	47.8	47.7	47.6	40.1
5	54.7	48.2	54.9	55.5	48.8	55.0	55.4	55.3	54.5	47.8
6	18.7	17.5	19.7	19.8	18.3	18.2	19.8	19.7	19.2	17.8
7	30.5	31.1	32.8	33.9	33.2	33.1	32.7	32.6	30.9	32.8
8	40.6	40.7	41.9	42.9	43.3	43.0	43.2	43.2	41.7	45.2
9	50.1	50.4	55.5	50.4	51.2	51.6	48.7	48.5	52.8	53.8
10	36.2	36.6	37.6	37.7	38.3	38.1	37.6	37.5	36.3	42.1
11	52.5	57.0	67.9	76.3	75.9	75.8	81.7	82.0	132.3	200.9
12	57.0	52.9	125.5	121.6	122.5	121.5	122.3	121.2	131.3	128.6
13	87.3	87.9	149.0	149.3	148.3	149.7	150.8	152.8	84.9	169.7
14	41.2	41.3	43.2	42.0	41.8	41.7	42.0	42.3	44.2	44.0
15	26.7	26.7	26.4	26.2	26.1	26.1	26.3	27.9	25.7	26.7
16	21.2	21.6	26.0	26.8	28.1	26.6	28.1	26.6	26.0	26.5
17	43.8	43.7	32.3	32.4	34.8	32.3	34.9	33.1	41.9	32.5
18	49.6	49.1	46.7	47.2	46.3	46.8	46.5	47.0	51.1	47.6
19	37.8	32.5	46.3	46.4	46.7	46.3	46.9	46.7	32.4	45.2
20	31.5	36.8	31.0	31.1	36.3	31.1	36.5	31.1	36.7	31.0
21	34.3	29.7	34.6	34.7	73.9	34.5	74.0	34.6	30.9	34.5
22	27.0	27.7	36.8	36.9	45.2	36.8	45.2	36.9	30.6	36.7
23	25.9	28.0	26.7	26.7	28.6	28.0	26.7	26.9	26.2	28.8
24	21.1	21.8	21.5	21.5	22.4	15.4	21.6	21.5	21.0	16.2
25	16.4	17.0	16.2	16.4	16.7	18.0	16.4	18.0	17.3	18.0
26	18.7	20.1	17.9	18.1	18.2	16.8	18.2	16.2	19.4	19.1
27	19.8	18.1	26.1	25.2	25.2	25.1	24.7	24.7	19.6	23.6
28	179.2	179.1	28.4	28.5	28.4	28.3	28.5	28.5	77.1	28.8
29	33.2	26.6	33.0	33.2	28.9	33.1	29.1	33.2	28.9	33.0
30	23.6	65.7	23.6	23.6	16.9	23.5	17.0	23.6	65.0	23.5
OCH$_3$				53.7	53.7	53.7				
Ref	[184]	[185]	[186]	[104]	[187]	[188]	[187]	[187]	[189]	[190]

C	O11	O12	O13	O14	O15	O16	O17	O18 [b]	O19 [a]	O20
1	39.6	39.4	39.3	40.0	38.1	39.8	48.6	47.6	47.7	46.6
2	33.8	34.2	34.2	34.5	24.0	34.0	69.1	69.6	68.7	69.0
3	218.1	217.7	217.8	217.5	80.3	218.3	83.1	85.9	85.7	84.0
4	47.0	47.5	47.5	48.1	37.5	47.2	39.3	44.3	43.9	39.3
5	54.6	55.4	55.3	55.7	55.0	54.8	55.5	57.3	56.5	55.4
6	19.4	19.7	19.6	19.1	18.8	19.6	17.7	19.8	19.2	18.4
7	33.7	32.2	32.1	32.3	32.2	33.7	34.0	34.3	33.5	33.1

Table 6. Cont.

8	40.6	39.9	39.7	45.2	39.6	40.6	43.5	40.9	40.0	39.4
9	50.1	46.9	46.8	61.3	47.2	50.4	56.8	49.2	48.3	47.6
10	36.7	36.7	36.7	37.0	38.0	36.8	40.1	39.2	38.3	38.3
11	21.3	23.8	23.7	199.3	23.3	21.5	72.3	25.1	24.2	23.6
12	25.7	123.6	122.4	128.2	123.1	26.0	211.1	124.6	122.4	124.5
13	37.6	142.5	143.8	168.9	142.6	38.9	82.4	144.4	144.4	140.5
14	43.6	41.9	41.8	43.8	41.8	43.2	45.3	42.5	42.1	42.6
15	27.3	25.6	26.0	26.1	25.2	27.3	22.6	30.3	29.1	24.4
16	36.3	23.6	28.2	30.9	18.0	36.8	30.5	27.9	23.1	25.3
17	36.9	39.6	35.0	37.3	31.7	34.5	33.6	47.0	47.0	35.3
18	142.9	48.0	46.7	43.0	46.1	147.5	49.0	45.0	41.8	43.5
19	127.7	41.4	47.0	45.5	40.0	124.7	38.5	82.3	46.2	33.9
20	40.0	151.6	36.3	31.3	42.6	28.0	31.4	35.9	30.8	39.6
21	73.1	69.1	74.0	34.1	36.6	27.9	34.1	29.4	34.0	39.9
22	43.5	45.9	45.3	21.8	75.2	37.3	39.0	33.1	32.2	83.2
23	26.6	26.6	26.5	21.7	23.8	26.8	28.7	23.8	24.2	28.7
24	20.7	21.6	21.5	26.7	16.5	20.9	16.6	66.1	65.7	16.9
25	16.3	15.3	15.2	16.0	15.3	16.5	17.2	17.5	17.3	17.0
26	15.7	16.8	16.7	18.7	16.5	15.9	20.7	17.7	17.4	17.1
27	13.9	25.8	25.8	23.7	25.9	14.5	18.5	25.1	26.1	24.1
28	26.7	68.8	28.3	69.6	24.9	25.5	31.3	178.6	176.6	25.0
29	28.5		29.1	23.6	177.9	70.5	25.1	28.6	33.2	182.5
30	21.4	103.6	16.9	33.2	20.3	25.6	31.9	25.2	23.8	21.1
OCH$_3$					51.7					
1′					173.5			93.9	93.7	
2′					34.6			78.6	78.6	
3′					23.3			78.0	79.0	
4′–11′					28.9–29.4			70.8	70.6	
5′								78.8	79.2	
6′								62.2	62.0	
1″								103.4	104.6	
2″								75.7	76.0	
3″								77.8	78.4	
4″								72.5	72.7	
5″								77.9	78.4	
6″								63.7	63.8	
12′					31.7					
13′					NR					
14′					13.9					
Ref	[191]	[61]	[96]	[192]	[68]	[191]	[193]	[194]	[194]	[61]

C	O21	O22 [a]	O23	O24	O25	O26	O27	O28 [a]	O29	O30
1	79.6	40.0	39.5	34.7	33.4	34.1	39.1	212.9	40.3	38.7
2	44.5	34.4	34.3	29.4	25.4	22.7	27.3	45.4	23.9	23.8
3	172.4	215.9	218.1	98.0	75.8	78.0	78.7	79.0	80.2	81.2
4	56.0	47.8	47.6	40.0	37.5	36.7	39.2	40.1	38.3	37.1
5	85.0	55.1	55.3	50.2	48.4	49.7	55.0	54.9	55.2	56.0
6	33.2	19.0	19.9	19.4	17.4	17.7	17.5	18.1	17.7	18.2
7	27.6	32.1	32.8	30.8	32.6	32.5	32.9	34.4	34.0	35.1
8	43.2	45.3	40.0	38.5	43.5	43.5	46.3	40.7	44.1	40.3
9	55.2	61.2	46.9	41.6	61.6	61.6	60.4	42.8	54.2	50.5
10	44.5	37.0	36.9	35.0	37.1	37.0	37.4	53.5	39.3	37.1
11	67.3	198.9	23.9	23.6	200.3	200.1	195.3	24.4	75.3	24.8
12	133.5	128.4	124.7	122.3	128.3	128.2	142.8	26.7	202.2	124.2
13	139.8	170.6	140.5	143.4	169.4	169.3	138.3	39.6	83.2	144.1
14	42.6	43.7	42.9	41.6	45.6	45.6	41.5	43.7	44.8	42.1
15	19.8	26.7	25.4	25.8	25.7	25.8	26.5	27.8	22.9	27.2

Table 6. Cont.

C										
16	34.1	26.9	24.5	28.0	22.0	21.5	26.7	31.3	39.1	27.5
17	51.4	32.5	35.5	36.0	37.0	37.0	31.3	32.5	33.5	32.0
18	37.8	47.3	43.7	47.0	42.7	42.7	38.7	140.7	49.1	47.4
19	40.5	40.8	40.0	46.5	44.9	45.0	41.3	132.6	38.3	46.9
20	47.7	36.0	39.7	36.0	31.0	31.0	30.9	32.5	31.5	31.1
21	215.9	29.9	39.7	73.8	33.9	33.8	34.7	33.4	34.5	34.8
22	40.4	36.4	83.3	45.0	30.6	30.6	36.9	31.1	30.0	37.2
23	21.0	21.5	27.0	27.0	28.4	28.0	28.0	28.6	28.2	28.0
24	20.3	26.5	21.7	18.0	22.3	21.9	15.5	16.6	16.4	16.6
25	13.8	15.9	15.2	67.7	16.3	16.3	16.5	16.1	16.4	15.5
26	19.1	18.5	16.7	17.0	18.6	18.6	18.7	17.0	20.9	16.5
27	22.6	23.4	25.3	25.1	23.6	23.5	23.2	15.1	18.9	25.7
28	26.8	28.7	24.2	28.1	69.7	69.7	29.2	63.6	31.4	28.7
29	28.4	28.2	182.7	16.6	32.9	32.9	33.1	31.4	32.3	33.2
30	21.3	65.3	21.2	28.8	23.4	23.4	23.4	29.7	24.5	23.7
OCH$_3$						21.7			21.4	
C=O						170.7			171.2	
1'									129.7	127.4
2'									130.1	115.4
3'									128.6	144.1
4'									133.2	146.6
5'									128.6	114.3
6'									130.1	122.2
7'									165.6	144.8
8'										116.0
9'										167.8
Ref	[117]	[190]	[195]	[196]	[154]	[154]	[54]	[190]	[197]	[198]

C	O31	O32	O33	O34	O36	O37	O38	O39	O40	O41
1	38.1	38.6	38.4	53.5	39.2	38.6	38.7	38.5	42.2	42.0
2	24.4	23.7	23.7	211.1	26.4	27.2	27.3	22.5	34.4	34.1
3	80.7	80.6	80.6	83.1	78.3	78.9	79.0	80.8	216.7	216.5
4	37.6	80.6	37.8	45.9	39.3	38.7	38.8	37.8	49.0	48.7
5	51.3	55.6	55.4	54.8	55.0	55.2	55.3	55.5	56.4	56.2
6	18.2	18.2	18.3	18.8	17.5	18.3	18.5	18.5	69.7	69.4
7	32.1	34.9	32.7	33.0	32.8	33.1	32.8	32.8	42.2	42.0
8	38.0	40.8	39.9	40.1	43.4	39.3	38.8	39.7	39.9	39.6
9	153.9	51.1	47.7	47.7	61.8	47.5	47.7	47.8	51.3	51.1
10	40.7	37.2	36.9	43.8	37.1	37.0	37.6	37.1	36.8	36.6
11	115.9	21.1	23.6	23.7	200.2	23.5	23.6	23.9	21.4	21.2
12	120.7	26.2	121.8	124.1	128.3	124.6	121.8	121.9	26.1	26.0
13	147.2	38.4	145.2	140.7	170.4	140.2	145.1	145.4	38.1	37.4
14	42.8	43.3	41.8	42.9	45.4	39.5	41.8	41.9	43.5	43.3
15	25.6	27.5	26.2	24.5	26.4	25.2	26.2	26.4	27.5	27.4
16	27.2	37.7	27.0	25.4	27.3	24.3	27.0	26.0	36.9	37.3
17	32.0	34.3	32.5	35.5	32.3	35.2	32.5	32.7	34.5	34.2
18	45.6	142.7	47.4	43.6	47.6	43.4	47.4	47.4	147.6	142.4
19	46.8	129.8	46.9	40.0	45.2	39.8	46.9	47.0	124.9	129.8
20	31.1	32.3	31.1	39.7	31.0	42.5	31.1	31.3	37.4	32.1
21	34.6	33.4	34.8	34.0	34.4	33.8	34.8	23.9	28.0	33.0
22	37.1	37.4	37.2	83.2	36.5	83.1	37.2	121.9	37.4	37.1
23	28.7	28.0	28.1	29.6	28.7	28.1	28.2	145.4	25.1	23.4
24	16.9	16.6	16.8	16.8	16.4	15.6	15.5	41.9	23.6	24.8
25	20.0	16.7	15.6	16.7	15.7	15.6	15.6	26.4	17.3	16.9
26	21.0	16.1	16.9	16.9	18.7	17.0	16.9	26.0	17.7	17.6
27	25.3	14.5	26.0	24.3	23.4	25.0	26.0	32.7	14.9	14.5
28	28.2	25.3	28.4	25.2	28.2	24.0	28.4	47.4	25.5	25.0
29	33.2	31.3	33.3	182.5	33.0	182.4	33.3	47.0	70.5	31.2
30	23.7	29.2	23.7	21.2	23.5	21.0	23.7	31.3	25.7	29.0
1'	127.8	173.7	173.5					173.9		

Table 6. *Cont.*

C	O42	O43 [c]	O44	O45	O47 [d]	O50	O51 [a]	O52 [c]	O53	O54
2′	115.5	34.5	34.9					35.0		
3′	143.8	25.2	25.2					25.4		
4′	146.2	29.2	29.2					29.4–29.9		
5′	114.3	29.3	29.3					29.4–29.9		
6′	122.4	29.3	29.4					29.4–29.9		
7′	144.4	29.7	29.6					29.4–29.9		
8′	116.4	29.7	29.7					29.4–29.9		
9′–13′	167.4	29.7	29.7					29.4–29.9		
14′		29.7	29.7					31.2		
15′		29.7	29.5					23.0		
16′		31.9	31.9					14.4		
17′		22.7	22.7							
18′		14.1	14.1							
Ref	[198]	[161]	[199]	[166]	[200]	[111]	[201]	[202]	[191]	[191]
C	**O42**	**O43** [c]	**O44**	**O45**	**O47** [d]	**O50**	**O51** [a]	**O52** [c]	**O53**	**O54**
1	38.9	42.5	40.6	128.4	40.1	39.3	47.8	34.6	41.5	40.1
2	34.0	35.3	34.8	143.9	35.0	34.2	68.6	42.2	175.2	37.4
3	216.9	217.0	218.6	201.2	219.0	217.9	83.8	76.6	180.3	213.6
4	47.1	48.6	48.4	44.0	48.0	47.5	39.8	37.2	46.0	44.8
5	52.4	56.8	55.9	53.9	56.0	55.3	55.9	48.5	48.9	53.6
6	30.3	20.9	20.1	18.8	20.3	19.7	18.9	37.9	20.7	23.9
7	72.9	32.2	31.5	32.7	26.7	32.1	33.2	18.1	32.1	31.6
8	43.3	44.3	44.4	40.5	40.5	39.8	48.2	32.8	39.3	39.1
9	46.9	55.4	50.7	43.1	47.5	46.8	48.2	43.2	39.4	45.3
10	36.7	38.9	38.1	38.4	37.8	36.7	38.6	55.6	41.6	36.7
11	23.5	68.3	76.6	23.6	24.0	23.7	23.5	67.1	23.9	22.0
12	122.5	128.7	122.8	123.0	123.4	123.3	122.5	126.1	124.6	122.5
13	143.8	147.2	148.9	143.4	143.5	143.2	144.9	147.7	140.1	143.8
14	45.3	43.0	42.2	41.9	43.4	41.9	42.2	41.4	43.3	41.8
15	29.2	27.3	26.5	25.7	26.3	25.4	28.3	25.1	24.4	27.6
16	22.3	28.0	27.2	28.1	29.0	22.2	23.7	26.5	25.3	23.0
17	36.7	33.1	32.3	37.4	39.5	36.9	46.7	31.5	35.4	46.5
18	43.0	49.0	48.1	43.9	49.8	41.1	41.7	47.0	43.6	41.3
19	46.4	43.3	42.6	39.9	44.2	39.9	42.0	30.9	39.8	45.8
20	31.0	44.8	43.3	42.5	46.0	42.4	31.0	43.5	39.6	30.7
21	34.1	34.2	33.3	36.2	40.6	28.5	34.3	30.9	33.9	33.8
22	30.8	39.5	38.6	75.8	78.7	29.8	33.3	38.0	83.2	32.4
23	26.4	27.4	27.0	27.3	26.6	21.6	29.4	28.4	28.5	116.0
24	21.5	22.2	21.9	21.9	21.7	26.6	16.9	22.1	23.4	
25	15.0	17.0	16.7	20.0	15.5	15.3	17.5	16.4	19.2	13.1
26	9.7	18.9	18.4	17.5	17.2	16.7	17.7	17.8	17.2	17.0
27	25.9	26.3	25.5	25.4	26.2	25.4	26.2	26.5	23.7	25.8
28	69.6	29.3	29.0	19.9	25.8	69.0	180.2	28.4	25.2	180.4
29	33.1	29.3	29.3	184.0	34.3	184.5	33.3	28.0	182.4	33.0
30	23.8	178.9	183.3	23.8	181.0	19.2	23.8	179.2	20.7	23.5
OCH$_3$			54.1						52.2	
Ref	[154]	[203]	[203]	[61]	[204]	[61]	[157]	[205]	[61]	[166]

C	O55	O56	O58 [a]	O60 [a]	O61	O62 [a]	O63 [b]	O65 [b]	O67 [a]	O68
1	36.8	39.6	33.6	38.9	33.6	37.9	38.2	38.9	38.9	38.5
2	32.4	34.4	26.2	28.1	26.2	25.1	26.6	28.2	27.6	26.7
3	217.7	217.8	75.2	78.0	70.8	81.9	79.7	79.6	73.7	78.7
4	39.1	47.7	37.8	39.7 *	47.4	55.4	38.0	40.1	42.9	38.6

Table 6. *Cont.*

5	55.3	55.6	49.1	55.8	49.1	46.9	48.0	56.5	48.8	55.1
6	19.5	20.0	18.6	18.8	19.7	20.3	19.5	19.0	18.7	18.2
7	33.8	32.5	32.9	33.2	32.7	32.6	33.7	33.4	33.6	32.5
8	39.3	40.1	40.3	37.3 *	39.8	39.9	39.6	41.2	39.8	39.7
9	47.4	47.1	47.7	48.1	46.7	47.1	49.6	48.9	48.2	47.5
10	46.9	37.0	37.3	37.9	37.6	37.9	39.9	37.8	37.3	36.8
11	23.6	24.0	23.8	23.8	23.5	23.5	20.2	24.4	23.8	23.4
12	122.4	123.0	123.2	122.7	121.7	120.9	124.3	123.5	122.7	123.1
13	143.6	144.3	144.4	144.7	145.1	143.0	144.7	144.5	145.0	143.2
14	41.7	42.1	42.0	42.2	41.9	42.0	43.2	43.5	42.2	41.5
15	32.2	27.3	25.9	28.3	26.0	35.9	24.7	26.5	28.4	25.2
16	21.4	26.3	23.0	24.0	26.9	73.1	27.9	29.0	23.8	21.8
17	46.6	32.7	37.6	46.6	32.5	48.6	41.2	39.5	46.7	36.7
18	41.0	46.4	41.7	41.6	47.3	41.2	56.6	49.9	42.0	41.1
19	45.8	40.5	41.4	42.0	46.7	47.5	40.0	44.3	46.5	40.1
20	30.7	42.8	42.8	35.9	31.1	30.9	43.9	46.2	31.0	42.1
21	32.4	29.2	29.5	29.6	34.7	35.8	76.0	40.7	34.3	28.4
22	26.4	36.1	30.8	32.9	37.1	32.8	41.5	78.8	33.3	29.7
23	27.7	26.8	29.2	28.4	183.2	206.4	28.7	28.3	68.2	27.9
24	15.0	21.7	22.7	16.5	24.2	10.2	17.4	16.1	13.1	15.4
25	15.0	15.5	15.7	15.3	13.1	15.4	16.1	15.7	16.0	15.3
26	17.0	17.0	17.0	17.4	16.7	17.2	16.3	17.3	17.5	16.5
27	25.8	26.1	26.0	26.2	25.9	27.1	26.7	26.3	26.2	25.7
28	183.9	28.4	68.3	180.2	28.4	174.8	21.0	25.8	180.4	68.6
29	33.0	185.1	181.5	28.2	33.3	33.2	182.3	34.1	33.3	181.7
30	23.5	19.4	20.1	65.5	23.7	24.5	25.2	181.1	23.8	19.1
Ref	[166]	[102]	[61]	[206]	[207]	[208]	[209]	[204]	[157]	[61]

C	O69 [b]	O70	O71 [a]	O72 [a]	O73	O75 [a]	O76	O77	O81	O82
1	42.0	38.1	213.0	38.5	38.5	49.6	38.3	78.7	34.9	32.5
2	28.0	23.4	45.4	28.0	27.4	170.6	23.5	172.2	25.2	26.0
3	80.0	81.0	78.4	78.1	78.7	182.0	80.8	110.9	76.0	76.1
4	39.7	37.7	40.0	39.5	38.7	45.4	37.9	36.4	37.4	37.3
5	57.1	55.4	54.7	55.3	55.2	55.2	55.2	48.8	48.7	47.3
6	68.7	18.2	18.2	18.8	18.3	20.7	18.3	17.6	18.3	18.3
7	41.6	32.6	33.2	32.9	32.6	33.4	32.6	34.5	33.0	32.5
8	40.7	39.4	39.7	41.1	39.3	41.1	39.8	41.1	43.6	37.0
9	49.4	47.6	39.7	54.8	47.6	45.9	47.5	41.0	56.4	48.9
10	37.6	37.0	52.7	37.1	37.0	38.5	36.8	40.1	38.2	33.0
11	24.5	23.4	25.7	126.0	23.1	74.5	23.7	22.1	67.6	23.4
12	124.0	122.6	123.2	127.0	122.1	121.4	121.6	24.2	126.1	122.6
13	144.4	143.6	144.1	136.6	143.4	149.8	145.2	133.5	148.0	143.7
14	43.3	41.6	42.5	42.5	41.6	43.1	41.7	45.2	41.8	41.7
15	28.8	27.7	28.3	33.3	27.7	25.8	26.1	26.4	26.0	26.0
16	24.1	22.9	23.7	25.6	23.4	27.5	26.9	39.3	28.0	32.5
17	47.9	46.6	46.8	48.7	46.6	32.9	32.5	34.6	34.7	25.2
18	41.9	41.0	42.3	133.8	41.3	47.2	47.2	133.9	45.8	35.0
19	41.4	45.7	46.2	41.0	45.8	41.6	46.7	38.7	46.7	46.6
20	36.8	30.7	31.0	32.8	30.6	36.1	31.1	33.3	36.3	50.0
21	29.3	33.6	34.3	37.5	33.8	30.0	34.7	35.4	73.9	74.0
22	33.1	32.5	33.2	36.3	32.3	36.6	37.1	36.4	45.1	36.3
23	28.4	28.1	29.0	28.5	28.1	28.2	28.1	19.3	28.3	45.4
24	17.6	17.2	16.8	16.0	15.6	23.3	16.8	24.3	22.3	28.2
25	17.3	15.2	15.0	18.4	15.3	17.7	15.6	15.0	16.7	22.3
26	18.8	16.7	18.1	17.1	16.8	17.1	16.9	18.1	18.1	16.8
27	26.5	25.9	26.0	20.1	26.0	25.5	26.0	21.3	26.3	15.2
28	181.8	184.0	180.1	178.9	181.0	28.4	28.4	23.8	28.5	26.0
29	74.4	33.1	33.3	32.4	33.1	28.2	33.3	32.3	28.9	28.3
30	19.5	23.6	23.8	24.4	23.6	65.6	23.7	24.0	16.9	29.0
1′								167.1		

Table 6. *Cont.*

C	O84	O85	O86	O88	O89	O90	O91	O92	O93	O94
2′							116.2			
3′							144.6			
4′							127.1			
5′							109.2			
6′							146.8			
7′							147.8			
8′							114.6			
9′							123.0			
OCH$_3$		21.3					55.9			
C=O		171.1								
Ref	[210]	[211]	[157]	[212]	[27]	[190]	[213]	[214]	[187]	[187]

C	O84	O85	O86	O88	O89	O90	O91	O92	O93	O94
1	39.5	39.2	38.6	38.6	38.2	39.6	39.9	41.3	41.4	38.6
2	27.4	28.2	27.2	19.0	23.7	33.8	34.0	27.6	27.7	27.1
3	78.7	78.1	79.0	79.0	80.9	218.0	217.8	78.5	78.5	78.7
4	39.0	39.4	38.8	38.8	37.8	47.0	47.2	39.5	39.1	38.6
5	55.1	55.9	55.2	55.2	55.2	54.6	55.0	55.7	55.9	55.2
6	18.4	18.9	18.4	18.4	18.3	19.4	19.7	17.9	18.0	18.0
7	32.9	33.3	32.6	32.6	32.6	33.4	34.0	35.5	35.6	34.5
8	43.3	40.1	39.8	39.8	39.8	40.3	40.7	42.8	42.9	40.7
9	49.7	48.1	47.6	47.6	47.7	50.2	50.6	56.3	56.5	50.8
10	37.9	37.3	36.9	36.2	36.9	33.7	37.0	39.4	39.5	36.9
11	81.7	23.9	23.6	23.5	23.5	21.3	21.7	70.9	71.1	20.7
12	121.2	122.5	122.3	122.2	121.7	25.9	26.3	38.3	38.6	25.7
13	153.2	144.9	144.2	144.8	145.2	38.7	38.6	37.7	37.3	37.5
14	41.8	42.5	41.7	41.7	41.7	42.5	43.4	42.8	43.0	43.5
15	26.4	26.5	25.6	26.1	28.3	27.2	27.6	27.3	27.5	27.1
16	27.4	28.7	22.0	27.2	26.2	33.9	37.7	31.5	37.6	36.4
17	32.3	38.0	36.9	32.9	32.5	40.6	34.4	39.0	34.3	37.0
18	46.9	45.4	42.3	46.3	47.3	143.4	142.6	137.5	141.6	143.1
19	46.9	46.9	46.5	29.0	46.8	127.8	130.0	134.5	129.9	127.5
20	31.2	30.9	31.0	41.0	31.2	45.1	32.4	32.2	32.4	37.0
21	34.7	42.3	34.1	36.2	34.7	215.2	33.4	33.2	33.4	73.1
22	37.0	75.6	31.0	36.9	37.1	52.3	37.4	31.1	37.3	43.7
23	28.2	28.8	15.5	15.6	28.0	24.5	26.9	28.2	28.2	27.7
24	15.5	15.9	28.1	28.1	16.7	20.7	21.0	15.5	15.6	15.8
25	18.3	16.6	15.5	15.5	15.6	13.3	16.0	16.8	17.4	16.4
26	16.8	17.3	16.7	16.8	16.8	15.6	16.0	17.4	16.9	15.8
27	24.7	25.8	25.9	26.0	25.9	14.7	14.5	14.4	14.3	13.9
28	28.5	28.8	69.7	16.8	26.9	25.6	25.3	65.4	25.3	26.7
29	33.3	33.3	33.2	26.9	33.3	28.8	31.3	29.7	31.3	28.5
30	23.7	21.2	23.6	74.8	23.7	26.3	29.2	30.5	29.2	21.4
OCH$_3$					21.3					
C=O					171.0					
Ref	[215]	[216]	[217]	[218]	[219]	[191]	[220]	[191]	[220]	[191]

C	O95	O96	O97	O98	O99	O100	O101	O102
1	38.9	40.8	38.5	38.8	31.6	38.8	34.8	38.3
2	27.4	27.5	27.4	27.9	25.9	27.9	33.0	22.7
3	78.9	79.1	79.0	78.6	75.7	78.6	217.0	80.9
4	38.9	39.7	39.0	38.9	37.7	38.9	47.2	37.9
5	55.4	55.6	55.7	51.2	44.9	51.2	48.1	55.3
6	18.2	69.0	18.3	18.4	18.2	18.4	19.7	18.3
7	34.6	42.2	34.7	32.2	31.9	32.2	30.4	32.5
8	40.7	39.7	40.8	37.0	40.7	37.0	46.6	39.8
9	51.1	51.8	51.3	154.3	154.9	154.3	53.7	47.5
10	37.2	36.8	37.3	40.7	38.8	40.7	37.8	36.9
11	21.0	21.2	21.2	115.8	115.3	115.8	72.1	23.5
12	26.1	26.2	26.2	120.8	121.4	120.8	120.5	121.6
13	38.8	37.6	39.0	147.1	145.2	147.1	152.8	145.2

Table 6. *Cont.*

14	43.2	43.5	43.4	42.8	42.8	42.8	41.7	41.7
15	27.4	27.6	27.6	25.7	25.5	25.7	26.7	26.1
16	36.9	37.3	37.7	27.3	28.5	27.3	26.6	26.9
17	34.5	34.3	34.4	32.2	34.5	32.2	32.8	32.6
18	147.8	142.8	142.8	45.6	44.9	45.6	48.0	47.2
19	124.4	129.9	129.8	46.9	47.1	46.9	46.1	46.8
20	27.1	32.3	32.3	31.1	36.3	31.1	31.1	31.1
21	28.0	33.3	33.4	34.7	73.9	34.7	34.6	34.7
22	37.5	40.8	37.4	37.2	45.3	37.2	36.9	37.1
23	27.9	27.6	28.0	28.0	28.3	28.8	27.9	28.1
24	15.4	16.8	15.4	16.6	22.4	15.1	19.4	16.8
25	16.7	18.1	16.1	15.5	25.1	20.1	19.1	15.5
26	16.1	17.5	16.7	16.5	20.9	21.0	99.1	16.8
27	14.7	14.8	14.6	25.7	20.1	25.3	23.6	25.9
28	25.5	25.2	25.3	28.7	28.6	28.3	28.7	28.4
29	70.6	31.3	31.3	33.2	28.9	23.7	33.3	33.3
30	25.7	29.1	29.2	23.7	16.9	33.2	23.7	23.7
1′								127.3
2′								109.3
3′								147.0
4′								146.6
5′								114.8
6′								120.8
7′								76.4
8′								76.0
9′								62.9
OCH$_3$								56.0
1′								128.5
2′								116.6
3′								143.8
4′								144.8
5′								117.5
6′								122.2
7′								143.7
8′								117.3
9′								167.0
CH$_3$								20.7
C=O								170.4
Ref	[191]	[191]	[220]	[198]	[187]	[68]	[221]	[126]

Ref: References; * Values bearing the same superscript are interchangeable; NR: Not reported; Solvent: CDCl$_3$; [a] C$_5$D$_5$N; [b] CD$_3$OD; [c] CD$_3$COCD$_3$; [d] CD$_3$OD; ^{13}C-NMR data of some compounds were not found. In these cases, the reported identification was performed by comparison of other physical data: O59 [222], O74 (m.p., IR, MS, ^{13}C-NMR of acetylated compound) [102], O78 (m.p., MS, ^{1}H-NMR, UV, [α]$_D$) [223], O79 (m.p., MS, IR, UV, [α]$_D$) [224], O80, O83 and 087. ^{13}C-NMR data of O35 [225], O46 [226], O48 [226], O49 [166], O57 [226], O64 [227], O66 [226] were reported, but the chemical shifts were not attributed to each carbon atom.

Table 7. ^{13}C-NMR data of ursane-type pentacyclic triterpenoids isolated from Celastraceae species (2001–2021).

C	U1	U2	U3 [a]	U4	U5	U6	U7	U8	U9	U10
1	41.4	41.1	42.5	40.8	40.8	41.4	39.2	39.0	39.0	47.2
2	34.2	34.5	35.1	27.4	27.7	34.3	34.5	28.3	28.3	69.1
3	217.6	220.9	217.2	78.4	78.6	217.6	220.2	76.9	77.0	85.0
4	47.6	51.1	48.3	39.0	39.1	47.7	51.3	52.9	52.9	43.4
5	55.0	55.5	56.1	54.8	54.9	55.6	55.9	56.3	56.4	56.1
6	19.9	19.3	20.5	18.4	18.4	19.7	19.4	18.5	18.5	18.3
7	36.4	36.6	34.4	37.2	37.8	33.3	36.8	37.5	37.6	34.4
8	43.7	43.6	43.5	43.7	43.8	37.6	43.8	43.9	43.9	42.9
9	52.8	52.4	53.8	51.6	50.5	54.4	44.9	45.3	45.1	46.4

Table 7. *Cont.*

C										
10	37.7	37.3	38.5	38.4	38.3	43.1	37.6	38.5	38.5	39.2
11	69.5	70.0	70.2	68.0	75.1	68.7	76.5	76.5	76.6	76.7
12	146.3	145.5	148.4	146.2	145.0	129.0	142.9	143.7	143.9	141.9
13	115.1	116.3	113.2	116.5	123.1	142.8	119.1	118.0	119.1	118.3
14	46.8	46.9	44.2	46.6	47.2	42.5	46.6	46.5	46.5	40.6
15	68.0	68.1	37.8	68.2	68.1	28.0	68.0	67.9	68.0	27.1
16	38.1	38.9	65.9	38.7	38.8	26.6	39.0	38.2	38.9	27.5
17	33.7	34.2	39.5	34.1	34.1	33.8	34.2	33.7	34.2	33.3
18	42.0	47.5	49.9	46.1	46.7	58.5	47.5	42.2	47.5	47.7
19	41.2	40.5	41.9	40.3	38.9	39.4	40.5	41.3	40.5	40.8
20	71.6	39.3	40.7	39.6	39.5	39.5	39.3	71.4	39.3	39.5
21	35.6	31.0	31.8	31.0	31.0	31.1	31.1	35.5	31.0	31.2
22	35.7	41.0	36.8	41.1	41.1	41.3	41.1	35.9	41.2	41.6
23	26.8	22.4	27.5	28.3	28.4	26.9	22.5	19.4	19.4	23.2
24	21.4	65.9	22.1	15.8	15.8	21.2	65.9	207.8	207.7	65.6
25	16.3	17.3	16.9	16.7	16.4	16.2	16.6	15.5	15.4	18.1
26	18.5	18.3	18.9	18.7	19.2	17.5	18.3	18.7	18.7	18.1
27	17.6	17.6	26.0	18.0	17.6	23.0	17.3	17.5	17.5	23.9
28	29.0	29.3	23.6	29.3	29.2	28.7	29.1	28.8	29.0	28.5
29	11.5	16.8	17.8	17.6	18.3	18.0	17.0	12.0	17.0	17.0
30	29.8	21.0	22.0	21.0	21.1	21.5	21.1	29.9	21.1	21.2
Ref	[193]	[193]	[190]	[54]	[54]	[228]	[193]	[193]	[193]	[193]

C	U11	U12	U13	U14	U15	U16	U17	U18 [a]	U19	U20
1	40.8	39.0	38.7	39.0	38.7	39.0	39.9	40.6	39.9	40.3
2	34.4	27.5	27.8	27.5	27.9	27.5	27.5	28.5	26.9	34.2
3	218.3	78.6	80.4	78.6	80.5	78.7	78.7	77.9	78.9	217.9
4	47.6	39.1	43.1	39.1	43.1	39.2	30.9	38.5	39.1	47.6
5	55.3	55.1	55.7	55.1	56.1	55.5	54.9	55.8	55.3	55.2
6	19.7	18.4	18.5	18.4	18.3	18.2	18.6	18.8	18.4	19.6
7	33.2	37.3	37.6	37.4	34.4	34.2	36.5	33.6	33.5	33.0
8	42.7	44.2	44.1	44.2	40.6	40.6	44.2	43.2	42.9	43.2
9	50.9	46.9	46.3	46.3	46.5	46.4	52.7	53.1	52.6	47.2
10	37.6	38.4	38.1	38.5	38.1	38.4	38.1	39.7	38.2	37.3
11	77.0	76.4	76.6	76.4	76.8	76.7	76.5	76.7	76.8	81.7
12	124.2	144.0	143.0	143.1	142.1	142.1	125.2	124.9	124.1	125.4
13	143.3	117.9	119.1	119.2	118.2	118.2	144.1	143.1	143.6	144.2
14	42.4	46.4	46.4	46.4	42.9	42.9	47.7	42.2	41.5	44.2
15	26.6	68.0	68.1	68.1	27.1	27.2	68.2	26.7	27.6	35.8
16	27.9	38.0	38.8	38.7	27.6	27.6	38.8	23.7	28.1	66.7
17	33.8	33.6	34.1	34.1	33.3	33.3	33.9	38.6	33.7	38.5
18	58.8	42.2	47.5	47.5	47.7	47.7	58.7	53.8	58.6	60.3
19	39.3	41.3	40.5	40.5	40.8	40.8	39.3	39.8	39.4	39.4
20	39.5	71.5	39.3	39.3	39.5	39.5	39.1	39.4	39.6	39.2
21	31.1	35.5	31.0	31.0	31.2	31.3	30.9	31.1	31.2	30.4
22	41.3	35.9	41.2	41.2	41.6	41.6	41.0	36.1	42.1	35.0
23	26.9	28.3	22.7	28.4	22.7	28.4	28.2	28.8	28.3	26.0
24	21.4	15.8	64.4	15.8	64.5	15.8	15.6	16.6	15.7	21.4
25	16.6	16.3	16.8	16.2	16.8	16.2	17.1	17.4	17.1	18.0
26	18.2	18.7	18.6	18.7	18.0	18.1	18.9	18.3	18.3	16.2
27	22.5	17.7	17.8	17.8	23.8	23.9	16.6	22.7	22.6	23.2
28	28.7	28.7	29.0	29.0	28.5	28.5	29.3	69.1	28.8	21.9
29	17.5	11.9	17.0	17.0	17.0	17.0	17.3	17.6	17.5	17.7
30	21.3	29.9	21.2	21.2	21.2	21.2	21.3	21.6	21.5	21.2
OCH_3	54.4	52.0	51.6	51.5	51.6	51.4	54.7		53.9	
Ref	[104]	[193]	[193]	[54]	[193]	[54]	[54]	[189]	[74]	[229]

C	U21	U22	U23	U24	U25	U26 [b]	U27	U28	U29 [a]	U31
1	40.1	39.3	39.9	39.4	39.4	40.2	38.3	39.5	38.6	39.2
2	34.2	34.3	34.1	34.3	34.4	26.6	27.1	27.7	27.6	33.9

Table 7. *Cont.*

C										
3	218.0	219.3	216.7	219.1	219.4	83.1	78.9	79.6	73.1	217.1
4	47.6	51.44	47.7	51.5	51.5	44.3	38.8	40.0	43.1	47.5
5	55.3	55.6	55.1	55.7	55.7	48.4	54.5	56.2	48.2	54.4
6	19.7	18.7	19.0	18.8	18.7	18.8	17.9	18.8	18.0	19.1
7	32.4	35.6	35.5	35.7	35.6	34.6	35.1	32.7	31.6	34.4
8	43.0	46.2	46.6	46.3	46.3	44.2	43.0	43.1	42.1	42.9
9	47.8	58.7	59.1	58.7	58.7	49.6	52.7	53.7	53.5	52.1
10	37.4	36.5	36.9	36.6	36.5	39.1	36.5	37.5	36.6	36.2
11	81.8	194.2	194.4	194.2	194.2	78.4	129.4	132.9	133.8	129.0
12	124.8	145.1	145.1	145.1	145.0	145.8	132.5	131.5	129.3	132.8
13	146.1	132.9	133.9	133.9	134.7	116.9	85.8	85.8	84.9	85.6
14	42.0	47.3	47.4	47.4	47.4	44.3	48.9	46.9	44.6	49.1
15	27.9	68.1	68.3	68.3	68.3	36.4	68.4	36.0	25.8	68.3
16	26.3	38.4	38.0	38.6	38.6	78.3	37.8	66.4	27.3	37.2
17	33.8	33.9	33.7	34.1	34.2	39.7	42.7	48.4	42.5	42.3
18	58.7	47.5	43.6	48.6	48.7	50.9	61.0	63.3	61.6	55.3
19	39.4	35.5	41.4	34.8	40.6	42.1	37.6	39.3	37.9	38.8
20	39.3	51.4	70.9	46.4	39.1	41.0	40.6	42.1	41.0	71.7
21	31.1	25.3	35.4	24.9	30.9	32.5	31.2	30.5	31.7	35.9
22	41.3	39.7	35.6	40.4	40.7	36.7	34.0	31.8	35.2	29.0
23	26.5	22.1	26.4	22.1	22.0	67.4	27.7	28.4	67.4	26.0
24	21.4	65.6	21.4	65.6	65.6	13.6	14.9	15.7	12.5	20.8
25	18.1	16.7	15.8	16.6	16.7	17.6	17.7	18.4	18.5	17.1
26	16.2	18.5	18.9	18.6	18.6	18.9	19.9	20.1	19.8	19.6
27	22.0	15.4	15.1	15.1	15.1	25.2	12.8	19.0	17.4	12.5
28	28.5	29.1	29.2	29.3	29.5	23.7	76.8	73.1	76.8	76.5
29	17.5	17.4	11.6	16.2	16.6	17.6	18.1	18.9	18.4	12.9
30	21.3	176.4	29.7	65.7	20.9	21.7	19.4	19.8	19.5	28.7
OCH$_3$						53.5				
1′						105.8				
2′						75.7				
3′						78.4				
4′						71.8				
5′						77.6				
6′						62.8				
1″						106.1				
2″						75.7				
3″						78.3				
4″						71.6				
5″						77.8				
6″						62.9				
Ref	[229]	[193]	[54]	[193]	[54]	[230]	[54]	[189]	[189]	[193]

C	U32	U33	U34 [a]	U35	U36 [a]	U37 [a]	U38	U39	U40	U41
1	40.9	40.9	39.5	47.1	48.0	47.9	38.6	39.8	39.8	33.5
2	28.2	27.4	34.2	68.6	68.7	68.8	23.8	25.9	34.2	25.3
3	76.9	78.7	217.8	84.9	85.7	85.8	81.1	NR	217.1	75.8
4	52.8	39.0	47.4	43.4	43.9	43.9	38.1	38.7	47.8	37.5
5	55.8	54.7	55.2	55.5	56.5	56.6	55.4	56.7	55.4	48.3
6	18.6	18.5	19.6	17.6	19.1	19.3	18.3	18.8	18.8	17.4
7	35.6	35.0	32.3	33.1	33.8	33.9	32.9	34.3	32.2	32.7
8	43.8	44.7	40.0	45.5	40.2	40.7	40.2	41.3	43.7	45.3
9	52.7	55.7	46.8	59.5	48.2	48.0	47.7	49.4	60.8	61.4
10	38.3	38.2	36.6	37.4	38.2	38.3	36.9	38.0	36.6	37.0
11	70.0	67.8	23.3	194.8	24.0	24.4	23.5	24.2	199.0	200.0
12	146.1	130.2	125.3	144.4	125.7	128.1	125.1	125.1	130.6	130.7
13	116.1	142.7	138.4	134.8	138.7	139.5	138.9	140.8	164.8	164.3
14	45.9	47.0	42.2	41.7	42.4	42.1	42.2	43.3	45.5	43.7
15	72.2	72.1	25.9	27.5	29.3	29.9	26.1	26.7	27.2	27.1
16	34.2	34.5	23.3	27.3	24.4	25.9	23.7	24.1	27.4	27.5

Table 7. *Cont.*

17	34.0	33.9	37.9	33.4	48.3	48.6	38.1	37.7	33.8	33.8
18	47.4	58.3	54.0	48.9	53.3	54.5	54.2	55.3	58.8	58.8
19	40.6	39.1	33.7	40.8	39.4	72.7	39.5	41.4	33.5	33.5
20	39.3	39.2	46.7	39.3	39.2	42.2	39.5	40.7	46.5	46.5
21	30.9	30.8	24.5	31.1	30.8	26.8	30.7	32.2	24.8	24.6
22	40.7	40.7	34.7	41.1	36.4	37.5	35.3	36.5	40.5	40.5
23	18.7	28.2	26.4	23.0	24.2	24.2	28.2	28.3	28.8	28.4
24	207.6	15.6	21.5	65.3	65.7	65.7	17.0	17.5	22.1	22.3
25	15.7	16.8	15.5	18.3	17.5	17.4	15.8	16.3	16.0	16.4
26	19.2	18.6	16.7	18.5	17.4	17.2	16.8	17.3	18.4	18.5
27	19.5	18.7	23.3	20.9	23.9	24.6	23.4	23.9	20.5	20.7
28	28.9	28.9	69.6	28.8	176.3	177.1	69.9	70.5	28.7	28.7
29	16.8	17.5	16.9	16.6	17.5	27.1	17.5	17.8	17.0	17.0
30	21.0	21.2	66.3	21.0	21.4	16.7	21.4	21.6	65.9	65.9
OCOPh	165.5	165.6								
iso	131.0	131.0								
orto	129.5	129.5								
meta	128.8	128.4								
para	132.9	132.8								
1′					93.7	93.8	127.2	127.3		
2′					79.2	79.1	114.0	129.5		
3′					78.9	79.1	144.8	115.8		
4′					70.7	70.8	147.2	157.4		
5′					79.2	79.2	115.3	115.8		
6′					62.1	62.2	122.1	129.5		
7′							150.0	143.8		
8′							115.7	116.4		
9′							167.9	167.2		
1″					104.8	104.7				
2″					76.0	75.9				
3″					78.4	78.4				
4″					72.7	72.9				
5″					78.3	78.3				
6″					63.7	63.9				
Ref	[54]	[54]	[185]	[193]	[194]	[194]	[231]	[204]	[185]	[185]
C	U42	U43	U44	U45	U46	U47	U48	U49	U50	U51 [a]
1	40.5	40.7	41.7	38.7	38.5	38.7	39.0	39.3	38.9	39.8
2	28.2	28.2	28.3	27.0	27.8	27.0	27.6	27.2	27.6	28.1
3	76.9	76.9	77.9	78.6	76.9	78.6	80.5	78.7	80.6	77.9
4	52.8	52.8	49.3	38.8	52.7	38.8	43.0	39.0	43.1	39.8
5	56.2	56.2	56.3	55.1	56.2	55.1	55.2	54.6	55.5	55.3
6	18.6	18.5	20.2	18.6	18.7	18.6	17.8	17.7	17.6	18.0
7	37.1	37.2	37.3	39.1	39.0	39.2	36.4	36.2	33.2	33.2
8	43.6	43.7	43.8	38.2	38.3	38.2	46.7	46.7	45.5	45.7
9	52.6	52.9	53.3	58.0	56.7	57.9	59.7	59.8	59.7	61.3
10	38.2	38.3	38.4	38.1	38.3	38.1	37.0	37.3	36.9	37.4
11	69.7	70.1	70.2	72.8	73.1	73.0	194.9	195.1	195.1	199.3
12	146.5	145.7	145.7	207.1	206.7	207.1	144.9	144.9	144.5	130.7
13	115.1	116.7	116.6	42.7	42.9	43.0	134.2	134.0	134.4	163.6
14	46.6	46.6	46.7	49.9	49.3	49.5	47.2	47.2	41.7	45.8
15	68.0	68.1	68.1	66.0	66.1	66.2	68.3	68.4	27.5	37.1
16	38.1	38.9	38.8	37.2	38.0	37.8	38.6	38.5	27.3	64.7
17	33.6	34.2	34.1	32.6	33.2	33.1	34.2	34.2	33.4	39.2
18	42.0	47.4	47.3	35.3	39.6	39.5	48.7	48.7	48.9	60.7
19	41.3	40.5	40.5	41.4	40.5	40.5	40.4	40.4	40.8	39.1
20	71.6	39.3	39.3	71.3	37.9	38.0	39.1	39.1	39.3	39.4
21	35.6	31.0	31.0	35.2	30.8	30.8	30.9	30.9	31.2	30.8
22	35.7	41.0	41.0	36.0	41.5	41.5	40.7	40.7	41.1	35.4
23	19.3	19.3	23.8	27.8	19.2	27.9	22.4	28.0	22.4	28.7
24	207.8	207.9	178.4	15.3	207.1	15.2	64.3	15.6	64.3	16.6

Table 7. *Cont.*

C										
25	15.8	15.8	14.4	17.3	16.2	17.3	17.1	16.6	17.1	17.0
26	18.7	18.6	18.5	21.2	20.9	21.1	19.0	19.1	18.4	18.7
27	17.7	18.0	17.8	18.1	17.7	17.9	15.2	15.3	20.9	21.9
28	29.0	29.3	29.3	27.9	28.2	28.2	29.4	29.4	28.8	23.0
29	11.4	16.8	16.7	11.7	16.6	16.6	16.6	16.5	16.5	17.6
30	29.9	21.1	21.0	29.1	20.5	20.5	20.9	20.9	21.0	21.2
OCH$_3$			51.3							
Ref	[193]	[54]	[54]	[193]	[193]	[193]	[193]	[54]	[54]	[190]

C	U52 [a]	U54	U55	U56	U57 [c]	U58	U59	U60	U62	U63 [a]
1	38.5	39.2	38.4	38.1	37.9	38.6	38.5	38.1	38.5	40.5
2	27.6	27.3	27.5	23.4	28.7	23.8	23.7	26.8	23.6	36.1
3	72.8	78.8	78.0	80.7	80.0	81.2	80.6	78.5	80.6	216.0
4	43.1	39.1	42.3	37.9	42.8	37.9	37.8	38.7	37.7	54.7
5	48.1	54.8	54.6	55.0	52.5	55.5	55.3	54.6	55.3	49.2
6	17.9	17.5	19.3	17.7	18.5	18.4	18.3	17.5	18.2	67.8
7	33.0	32.8	33.3	31.4	33.3	33.1	32.9	31.0	32.9	41.6
8	42.1	45.1	40.9	41.8	39.9	40.2	40.1	41.5	40.0	42.6
9	48.9	61.6	51.1	53.0	48.7	47.8	47.7	52.8	47.6	46.6
10	37.0	36.9	31.0	36.4	37.1	37.0	36.8	36.1	36.8	37.8
11	37.4	199.8	73.3	133.4	24.4	23.6	23.4	128.6	23.4	77.3
12	209.8	130.7	126.0	129.0	128.0	124.5	124.4	133.2	124.3	145.0
13	89.4	164.3	140.4	89.7	140.2	139.8	139.7	89.5	139.6	116.5
14	45.9	43.6	42.5	42.0	41.7	42.4	42.1	41.7	42.1	41.4
15	26.3	27.1	28.4	25.6	30.6	28.3	26.6	25.3	26.6	27.6
16	26.4	27.5	26.0	30.9	27.6	26.8	28.1	22.6	28.1	28.0
17	42.7	33.8	47.8	45.2	47.0	33.9	33.8	44.9	33.7	33.7
18	55.0	58.8	53.5	40.4	54.6	59.3	59.1	60.4	59.1	47.5
19	38.0	33.5	72.1	38.2	72.4	39.8	39.6	37.9	39.6	41.3
20	40.7	46.5	42.5	60.7	39.7	39.8	39.7	40.1	39.7	40.0
21	31.6	24.9	27.4	22.9	26.8	31.5	31.3	30.6	31.2	31.6
22	34.9	40.5	38.3	31.3	38.5	41.7	41.6	31.1	41.5	42.2
23	67.2	28.0	180.6	27.8	68.4	28.3	28.1	27.6	28.1	66.8
24	12.9	22.1	13.7	16.1	12.8	16.9	16.8	14.7	16.8	20.5
25	16.1	16.5	16.7	19.2	17.0	16.0	15.7	17.7	15.7	17.6
26	18.7	18.5	16.9	19.0	16.8	17.7	16.9	18.7	16.9	20.3
27	17.7	20.6	23.9	16.2	25.0	23.4	23.3	15.9	23.2	24.0
28	77.2	28.7	68.7	18.1	64.5	29.1	28.8	179.6	28.7	28.9
29	19.1	17.0	17.4	17.9	27.1	17.0	17.5	17.6	17.5	17.4
30	19.8	65.9	28.6	179.9	16.2	21.6	21.4	18.9	21.4	21.5
OCH$_3$			55.4							51.4
OCH$_3$			21.7	21.4	21.2	21.5				
C=O			170.4	171.1	169.9	171.5	173.5		173.7	
2'							34.9		34.9	
3'							25.2		25.2	
4'–13'							29.2–29.7		29.2–29.5	
14'							29.7		31.9	
15'							29.5		22.7	
16'							31.9		14.1	
17'							22.7			
18'							14.1			
Ref	[189]	[185]	[232]	[233]	[232]	[234]	[235]	[236]	[6]	[190]

C	U64 [c]	U65	U66 [a]	U67	U68	U69 [a]	U71	U72	U73 [c]	U74
1	38.9	43.0	39.4	39.5	39.5	48.0	39.7	33.6	34.1	38.4
2	29.3	175.5	34.8	34.2	34.2	68.6	34.1	25.8	25.7	26.8
3	208.1	207.7	216.2	217.9	217.8	83.7	217.3	74.2	76.3	77.4
4	47.2	50.7	47.4	47.5	47.5	39.9	47.7	37.4	42.6	38.4
5	55.6	47.7	55.2	55.3	55.3	55.9	55.2	48.1	54.1	54.8

Table 7. *Cont.*

6	19.8	20.4	19.8	19.7	19.7	18.9	18.6	17.1	19.0	18.2
7	34.0	32.1	32.5	32.3	32.4	33.5	32.4	32.9	33.2	32.3
8	40.2	40.0	40.3	40.2	40.1	40.1	44.5	44.6	41.3	39.6
9	46.6	40.4	47.1	46.9	46.9	47.0	60.7	61.2	48.1	47.2
10	37.5	42.3	36.7	36.7	36.7	38.5	36.1	37.2	37.4	38.4
11	24.5	23.8	23.8	23.7	23.7	23.8	199.5	199.3	73.1	23.0
12	127.9	125.1	125.2	125.5	125.8	125.5	130.7	130.3	127.3	124.8
13	138.8	139.1	139.3	138.0	137.9	139.3	163.5	163.3	139.4	138.0
14	42.0	42.7	43.1	42.6	42.2	42.6	43.8	43.6	43.1	42.2
15	28.7	26.5	26.5	26.3	25.9	28.7	28.5	28.3	27.7	25.5
16	26.6	29.2	21.4	27.5	23.3	24.9	23.7	23.9	25.9	20.3
17	48.4	35.2	39.7	38.0	37.9	48.1	47.7	47.0	48.9	36.4
18	54.3	58.6	58.3	52.8	53.0	53.6	52.8	52.8	54.5	57.2
19	73.2	38.4	34.8	34.0	34.1	39.4	38.6	38.1	71.9	33.6
20	41.8	47.1	51.2	45.1	51.8	39.5	38.6	38.4	44.0	49.7
21	27.0	71.8	34.4	32.4	25.1	31.0	30.3	30.1	27.2	33.1
22	39.1	50.3	77.6	74.8	34.3	37.5	36.7	36.0	38.6	76.7
23	23.5	23.9	26.7	26.7	26.7	29.4	26.6	28.8	177.4	28.0
24	181.1	19.5 *	21.6	21.6	21.6	17.7	21.3	22.1	13.1	15.3
25	15.0	19.4 *	15.4	15.5	15.5	17.0	15.5	16.0	16.4	15.7
26	17.0	16.9	16.9	16.8	16.8	17.5	19.0	18.7	17.7	16.4
27	25.3	22.9	23.8	23.8	23.5	23.9	20.9	20.4	25.2	23.1
28	65.4	28.4	25.2	21.7	69.3	179.9	180.0	180.0	179.7	24.3
29	26.9	17.4	19.0	18.5	18.0	17.5	17.0	16.5	27.7	17.9
30	16.8	15.8	178.2	182.1	181.6	21.4	21.0	20.4	17.9	177.0
Ref	[232]	[54]	[129]	[61]	[61]	[157]	[185]	[185]	[232]	[237]

C	U75	U76	U77 [d]	U78 [a]	U79	U80	U81	U82	U83 [a]	U84
1	42.7	38.1	39.0	48.0	40.0	40.9	40.4	40.8	33.9	38.8
2	34.5	23.8	28.3	68.6	34.4	27.3	27.8	27.4	28.4	27.3
3	74.6	80.8	79.1	83.9	NR	78.5	80.6	78.7	78.8	79.0
4	45.4	37.5	39.2	39.8	47.9	39.0	43.0	39.0	38.9	38.8
5	49.9	55.1	55.5	56.1	55.6	54.8	55.8	54.9	55.4	55.4
6	19.3	18.0	18.6	19.0	19.0	18.5	18.4	18.5	18.1	18.4
7	66.9	32.7	33.3	33.6	32.4	37.0	33.9	36.7	33.2	32.9
8	40.9	39.3	38.9	40.7	43.9	44.1	42.7	44.6	40.3	39.4
9	48.9	47.3	47.8	47.9	60.9	54.4	54.4	55.7	48.6	47.8
10	39.1	36.9	37.1	38.5	36.8	38.2	37.8	38.2	41.4	37.2
11	24.6	23.4	23.6	24.2	199.2	69.9	70.1	68.1	25.2	23.4
12	126.7	125.5	125.8	128.2	130.5	145.7	144.7	129.8	125.7	125.0
13	139.7	137.8	138.5	139.5	165.5	117.0	115.8	143.5	138.9	138.0
14	43.3	41.6	42.3	42.1	45.1	46.5	40.9	47.7	42.3	42.8
15	29.2	27.9	27.0	29.8	29.8	68.2	26.9	68.0	26.7	29.2
16	25.3	23.9	24.5	26.8	27.4	38.7	27.6	38.8	28.2	22.6
17	48.6	47.8	48.1	48.6	34.3	34.1	33.2	34.0	33.7	36.8
18	54.4	52.3	53.1	54.5	59.2	47.3	47.7	58.3	59.0	54.1
19	40.4	38.8	39.4	72.7	39.4	40.5	40.8	39.1	34.1	38.9
20	40.4	38.7	39.7	42.2	39.4	39.3	39.5	39.2	47.3	39.4
21	31.8	30.4	30.9	25.9	31.0	31.0	31.2	30.9	25.3	30.7
22	38.1	36.5	37.2	37.5	41.0	41.1	41.5	40.9	41.2	30.6
23	23.1	27.9	28.3	29.4	26.5	28.2	22.6	28.2	29.0	28.1
24	65.9	16.9	17.1	17.6	21.6	15.6	64.4	15.6	16.1	15.4
25	17.4	15.5	15.7	17.4	15.9	16.7	17.3	16.8	61.0	15.6
26	21.6	16.5	15.9	16.7	18.5	18.6	17.9	18.6	17.4	16.9
27	24.1	23.9	23.8	24.7	20.6	18.2	24.1	17.2	23.7	23.4
28	181.6	184.0	181.1	177.0	29.0	29.3	28.7	29.3	28.7	69.7
29	17.6	16.8	17.3	27.0	17.6	16.8	16.7	17.5	17.1	16.2
30	17.7	21.0	21.5	17.0	21.3	21.1	21.1	21.3	65.9	21.3

Table 7. *Cont.*

	[238]	[239]	[240]	[241]	[242]	[54]	[193]	[54]	[243]	[244]
1′				93.7						
2′				79.3						
3′				79.0						
4′				70.9						
5′				79.1						
6′				62.4						
1″				104.8						
2″				75.9						
3″				78.3						
4″				72.9						
5″				78.1						
6″				63.9						
Ref	[238]	[239]	[240]	[241]	[242]	[54]	[193]	[54]	[243]	[244]

C	U85	U86	U88
1	38.8	37.7	36.8
2	27.4	27.1	25.2
3	79.0	79.1	78.2
4	38.8	38.7	38.4
5	55.3	55.4	50.7
6	18.4	18.8	17.9
7	34.6	41.8	31.6
8	41.0	38.9	40.2
9	50.5	49.2	154.0
10	37.2	37.9	38.2
11	22.2	17.4	114.9
12	24.7	32.2	122.5
13	135.4	40.0	140.7
14	45.0	159.2	42.6
15	26.8	116.3	27.8
16	38.9	40.6	25.7
17	34.0	33.9	33.2
18	136.4	60.4	56.8
19	36.7	35.4	38.5
20	35.0	36.5	38.9
21	23.7	28.5	30.7
22	36.3	38.5	40.9
23	28.1	28.0	28.2
24	15.5	15.5	16.9
25	16.3	15.2	17.1
26	17.8	26.3	21.6
27	21.9	19.4	24.9
28	28.3	37.0	29.2
29	23.1	27.5	17.1
30	20.5	22.5	21.0
Ref	[245]	[245]	[246]

Ref: References; * Values bearing the same superscript are interchangeable; NR: Not reported; Solvent $CDCl_3$; [a] Pyridine-d_5; [b] CD_3OD; [c] $CDCl_3$ + DMSO-d_6; [d] $CDCl_3$+CD_3OD; ^{13}C-NMR data of some compounds were not found. In these cases, the reported identification were performed by comparison of other physical data: U30, U53 [247], U70 (^1H-NMR, IR, MS) [248] e U87 (m.p., $[\alpha]_D$, IR, ^1H-NMR) [249].

Table 8. ^{13}C-NMR data of pentacyclic triterpenoids classified as others isolated from Celastraceae species (2001–2021).

C	OT3	OT4	OT5	OT6	OT7	OT8	OT9 [a]	OT10 [a]	OT11 [a]	OT12
1	39.6	38.9	104.6	18.2	37.8	37.2	104.5	106.2	121.3	36.6
2	34.2	27.5	144.7	27.8	27.1	194.7	147.5	146.9	182.0	34.8
3	218.2	78.4	144.2	76.3	79.0	200.1	144.2	144.4	182.2	217.1
4	47.4	39.0	118.0	40.8	38.7	123.8	118.1	128.7	131.4	47.6
5	54.9	55.3	129.8	141.6	55.5	154.8	129.8	117.6	142.1	53.2
6	19.7	18.5	122.5	122.0	18.8	75.0	122.6	121.6	121.3	26.3
7	33.7	33.4	127.9	23.5	41.3	122.9	127.9	130.0	153.2	22.6
8	41.6	41.7	131.3	47.8	38.9	154.0	131.3	141.7	47.6	41.0
9	49.6	50.4	135.6	34.8	49.2	51.6	135.3	132.2	144.8	147.4
10	36.8	37.2	129.5	49.6	38.0	71.6	129.5	131.8	158.2	39.3
11	21.6	21.1	25.5	34.6	17.5	28.2	29.3	129.0	128.3	115.6
12	23.9	24.0	32.8	30.5	37.7	29.5	35.9	138.1	38.9	36.1
13	49.6	49.5	102.5	37.7	36.0	38.2	54.5	44.1	41.6	36.7
14	42.1	42.1	43.6	39.5	158.0	40.3	156.8	49.0	43.2	38.2
15	32.6	33.7	38.8	32.5	116.9	28.5	34.1	31.3	24.8	29.6
16	21.6	21.7	40.2	35.7	33.6	35.6	41.4	38.8	38.6	35.8
17	54.9	54.9	42.6	30.5	37.6	30.7	41.6	32.4	33.2	42.8
18	44.7	44.8	46.0	41.9	48.0	43.4	50.4	43.2	48.2	52.0
19	41.9	41.9	32.8	39.3	31.1	30.6	36.8	32.7	32.6	20.1
20	27.3	27.4	40.2	33.1	33.8	41.7	41.9	41.9	41.7	28.2
21	46.4	46.5	36.8	29.5	28.1	29.9	30.8	31.0	31.1	59.6
22	148.6	148.6	36.9	27.9	35.3	35.9	35.5	36.2	34.5	30.7
23	26.6	28.2	11.6	28.9	28.0	9.5	11.6	11.6	11.2	22.0
24	21.1	15.7		25.4	15.5					25.5
25	15.7	16.7	20.5	16.1	15.4	27.0	20.7	19.5	27.7	21.6
26	16.4	15.9	23.2	18.1	25.9	21.5	28.0	20.1	20.5	16.9
27	16.6	16.7	23.2	20.4	21.1	28.7	108.3	23.4	20.5	15.3
28	16.1	16.1	25.1	32.0	29.9	31.7	31.3	31.7	31.8	13.9
29	110.1	110.2	180.5	74.4	73.9	178.8	184.7	183.6	183.2	22.1
30	25.0	25.0	26.3	26.0	24.6	32.5	25.9	32.5	34.0	23.0
Ref	[94]	[94]	[139]	[213]	[213]	[250]	[139]	[139]	[139]	[251]

C	OT13	OT14	OT15	OT17	OT18	OT19	OT20	OT21
1	110.7	110.6	98.9	38.5	18.2	23.6	36.1	38.1
2	163.7	163.7	143.2	27.4	27.8	18.1	27.8	27.3
3			178.1	79.0	76.3	76.2	79.0	79.2
4	24.7	28.2	40.1	39.0	40.8	39.2	39.6	39.1
5	103.7	105.6	43.3	55.7	141.7	141.6	52.3	55.7
6	126.4	125.8	28.2	18.3	122.0	121.9	21.4	19.0
7	115.9	115.9	17.3	34.7	23.8	27.7	26.7	35.3
8	161.1	160.4	47.7	40.8	45.7	43.0	41.0	38.9
9	39.6	39.5	38.8	51.3	34.8	34.8	148.9	48.9
10	165.2	166.1	47.8	37.3	49.7	46.6	39.1	37.9
11	33.4	33.3	33.9	21.2	29.9	34.6	114.3	17.7
12	29.7	29.7	28.7	26.2	29.7	30.3	36.0	35.9
13	40.5	40.3	38.9	39.0	38.4	37.7	36.8	37.9
14	44.1	44.1	37.6	43.4	38.6	40.7	38.2	158.1
15	28.4	28.4	28.8	27.6	34.3	32.0	29.7	117.0
16	35.5	35.5	35.9	37.7	36.3	35.9	35.9	36.9
17	38.2	38.2	30.1	34.4	32.9	30.0	43.0	38.1
18	43.4	43.4	44.2	142.8	44.7	47.4	52.1	49.4
19	32.0	32.1	30.2	129.8	35.8	35.1	20.2	41.4
20	41.9	41.9	40.3	32.3	34.5	28.2	28.2	29.0
21	213.8	213.7	29.9	33.4	74.7	33.0	59.6	33.9
22	52.5	52.5	36.2	37.4	46.3	38.9	30.8	33.2
23			7.9	28.0	25.4	28.9	28.2	28.1
24			99.4	15.4	28.9	25.4	15.6	15.6
25	36.8	36.5	17.5	16.1	16.6	16.2	22.1	15.6

Table 8. *Cont.*

26	22.5	22.5	15.9	16.7	17.0	18.4	17.0	30.1
27	19.7	19.8	17.2	14.6	19.1	19.6	15.3	26.0
28	32.5	32.6	31.6	25.3	33.1	32.4	14.0	30.1
29			179.0	31.3	32.3	34.6	23.0	33.5
30	15.1	15.1	31.8	29.2	24.6	32.0	22.2	21.5
OCH$_3$	50.5	51.0	51.3					
Ref	[252]	[252]	[196]	[220]	[196]	[157]	[253]	[254]

Ref: References; NR: Not reported; Solvent CDCl$_3$; [a] CD$_3$OD; ^{13}C-NMR data of some compounds were not found. In these cases, the reported identification was performed by comparison of other physical data: OT1 (m.p., $[\alpha]_D$, IR, ^{1}H-NMR) [255], OT2 e OT16 (m.p., $[\alpha]_D$, IR, ^{1}H-NMR, MS) [256].

4. Conclusions

This review describes 504 pentacyclic triterpenoids isolated from Celastraceae species, classified as aromatics (29), dimers (50), friedelanes (103), lupanes (89), oleananes (102), quinonemethides (22), ursanes (88) and others (21). The data reported highlights the abundance and structural diversity of pentacyclic triterpenes isolated from plants of this family. The chemical complexity of these compounds helps to rationalize the various biological properties associated with these plant species, as well as these pure metabolites. The compilation of PCTTs ^{13}C-NMR data presented in this review represents a contribution to the structural elucidation of new compounds of this class of terpenes.

Supplementary Materials: The following are available online, Table S1: Pentacyclic triterpenoids isolated from Celastraceae species (2001–2021) [257–320].

Author Contributions: Conceptualization and investigation, K.C.C., L.P.D., G.F.S. and D.M.V.; methodology and data curation, K.C.C., R.G.d.C., M.G.d.A., A.R.A.M., E.L.M.M. and L.R.O.; validation, M.G.d.A., D.S. and A.R.A.M.; writing—original draft preparation, K.C.C.; writing—review and editing and formal analysis, L.P.D., G.F.S., D.M.V. and D.S.; supervision, L.P.D., G.F.S. and D.M.V. All authors have read and agreed to the published version of the manuscript.

Funding: This research was funded by Conselho Nacional de Desenvolvimento Científico e Tecnologico (CNPq), grant number 140434/2018-6.

Institutional Review Board Statement: Not applicable.

Informed Consent Statement: Not applicable.

Data Availability Statement: The information about the PCTTs was obtained from SciFinder, Scopus, and Web of Science, using as key search terms: "Celastraceae and triterpenes", "Celastraceae and com-pounds", "Celastraceae and phytochemistry" and "Celastraceae and metabolites".

Acknowledgments: The authors are thankful to Conselho Nacional de Desenvolvimento Científico e Tecnológico (CNPq), Coordenação de Aperfeiçoamento de Pessoal de Nível Superior (CAPES) and Fundação de Amparo à Pesquisa do Estado de Minas Gerais (FAPEMIG) for financial support.

Conflicts of Interest: The authors declare no conflict of interest.

References

1. Bukhari, S.; Jantan, I.; Seyed, M. Effects of plants and isolates of Celastraceae family on cancer pathways. *Anticancer. Agents Med. Chem.* **2015**, *15*, 681–693. [CrossRef] [PubMed]
2. Christenhusz, M.J.M.; Byng, J.W. The number of known plants species in the world and its annual increase. *Phytotaxa* **2016**, *216*, 201–217. [CrossRef]
3. Caneschi, C.; Muniyappa, M.; Duarte, L.; Silva, G.; Santos, O.; Spillane, C.; Filho, S. Effect of constituents from samaras of *Austroplenckia populnea* (Celastraceae) on human cancer cells. *J. Intercult. Ethnopharmacol.* **2015**, *4*, 6–11. [CrossRef] [PubMed]
4. Rodrigues, A.C.B.C.; Oliveira, F.P.; Dias, R.B.; Sales, C.B.S.; Rocha, C.A.G.; Soares, M.B.P.; Costa, E.V.; Silva, F.M.A.; Rocha, W.C.; Koolen, H.H.F.; et al. In vitro and in vivo anti-leukemia activity of the stem bark of *Salacia impressifolia* (Miers) A. C. Smith (Celastraceae). *J. Ethnopharmacol.* **2019**, *231*, 516–524. [CrossRef]

5. Menezes, L.D.; Gomes, G.O.; Antônio, J.; Da, P.; Neto, S.; Laundry, M.M.; Ferreira, V.M.; Valero, S.M. Evaluation of anti-inflammatory and antinociceptive activities of the *Austroplenckia populnea* extract in topical formulations. *Afr. J. Pharm. Pharmacol.* **2014**, *8*, 1180–1185. [CrossRef]

6. Rodrigues, V.G.; Duarte, L.P.; Silva, R.R.; Silva, G.D.F.; Mercadante-Simões, M.O.; Takahashi, J.A.; Matildes, B.L.G.; Fonseca, T.H.S.; Gomes, M.A.; Vieira Filho, S.A. *Salacia crassifolia* (Celastraceae): Chemical constituents and antimicrobial activity. *Quim. Nova* **2015**, *38*, 237–242. [CrossRef]

7. Cruz, W.; Ferraz, A.; Lima, W.; Silva, M.T.; Ferreira, F.; Siqueira, F.J.; De Brito, M.C.; Duarte, L.; Vieira Filho, S.; De Magalhães, J. Evaluation of the activity of *Tontelea micrantha* extracts against bacteria, candida and Mayaro virus. *J. Pharm. Negat. Results* **2018**, *9*, 21–26. [CrossRef]

8. Magalhães, C.G.; De Fátima, S.G.D.; Duarte, L.P.; Takahashi, J.A.; Santos, V.R.; Figueiredo, R.C.; Filho, S.A.V. *Maytenus salicifolia* Reissek (Celastraceae): Evaluation of the activity of extracts and constituents against *Helicobacter pylori* and oral pathogenic microorganisms. *Rev. Virtual Quim.* **2016**, *8*, 1524–1536. [CrossRef]

9. Magalhães, C.G.; Ferrari, F.C.; Guimarães, D.A.S.; Silva, G.D.F.; Duarte, L.P.; Figueiredo, R.C.; Filho, S.A.V. *Maytenus salicifolia*: Triterpenes isolated from stems and antioxidant property of extracts from aerial parts. *Rev. Bras. Farmacogn.* **2011**, *21*, 415–419. [CrossRef]

10. Ferreira, P.G.; Ferraz, A.C.; Figueiredo, J.E.; Lima, C.F.; Rodrigues, V.G.; Taranto, A.G.; Ferreira, J.M.S.; Brandão, G.C.; Vieira-Filho, S.A.; Duarte, L.P.; et al. Detection of the antiviral activity of epicatechin isolated from *Salacia crassifolia* (Celastraceae) against Mayaro virus based on protein C homology modelling and virtual screening. *Arch. Virol.* **2018**, *163*, 1567–1576. [CrossRef]

11. Veloso, C.C.; Rodrigues, V.G.; Azevendo, A.O.; Oliveira, C.C.; Gomides, L.F.; Duarte, L.P.; Duarte, I.D.; Klein, A.; Perez, A.C. Antinociceptive effects of *Maytenus imbricata* Mart. Ex. Reissek (Celastraceae) root extract and its tingenone constituent. *J. Med. Plants Res.* **2014**, *8*, 68–76. [CrossRef]

12. Zhang, L.; Ji, M.Y.; Qiu, B.; Li, Q.Y.; Zhang, K.Y.; Liu, J.C.; Dang, L.S.; Li, M.H. Phytochemicals and biological activities of species from the genus *Maytenus*. *Med. Chem. Res.* **2020**, *29*, 575–606. [CrossRef]

13. Alvarenga, N.; Ferro, E.A. Bioactive triterpenes and related compounds from Celastraceae. *Stud. Nat. Prod. Chem.* **2006**, *33*, 239–307. [CrossRef]

14. Truc, D.T.T.; Vy, C.T.H.; Phu, D.H.; Hoang, N.M.; Nhan, N.T. Lupan-Type Triterpenoids from the stems of *Salacia chinensis* L. (Celastraceae) and their α-glucosidase inhibitory activities. *Vietnam J. Chem.* **2019**, *57*, 433–437. [CrossRef]

15. Renda, G.; Gökkaya, İ.; Şöhretoğlu, D. Immunomodulatory properties of triterpenes. *Phytochem. Rev.* **2021**, 1–27. [CrossRef] [PubMed]

16. Zhou, M.; Zhang, R.H.; Wang, M.; Xu, G.B.; Liao, S.G. Prodrugs of triterpenoids and their derivatives. *Eur. J. Med. Chem.* **2017**, *131*, 222–236. [CrossRef] [PubMed]

17. Ríos, J.L.; Máñez, S. New pharmacological opportunities for betulinic acid. *Planta Med.* **2018**, *84*, 8–19. [CrossRef] [PubMed]

18. Xu, F.; Huang, X.; Wu, H.; Wang, X. Beneficial health effects of lupenone triterpene: A review. *Biomed. Pharmacother.* **2018**, *103*, 198–203. [CrossRef] [PubMed]

19. Shen, Y.; Chen, B.L.; Zhang, Q.X.; Zheng, Y.Z.; Fu, Q. Traditional uses, secondary metabolites, and pharmacology of Celastrus species—A Review. *J. Ethnopharmacol.* **2019**, *241*, 111934. [CrossRef]

20. Silva, F.C.O.; Ferreira, M.K.A.; Da Silva, A.W.; Matos, M.G.C.; Magalhães, F.E.A.; Da Silva, P.T.; Bandeira, P.N.; De Menezes, J.E.S.A.; Santos, H.S. Bioativities of plant-isolated triterpenes: A brief review. *Rev. Virtual Quim.* **2020**, *12*, 234–247. [CrossRef]

21. Ghiulai, R.; Roşca, O.J.; Antal, D.S.; Mioc, M.; Mioc, A.; Racoviceanu, R.; Macaşoi, I.; Olariu, T.; Dehelean, C.; Creţu, O.M.; et al. Tetracyclic and pentacyclic triterpenes with high therapeutic efficiency in wound healing approaches. *Molecules* **2020**, *25*, 5557. [CrossRef] [PubMed]

22. Huang, Y.Y.; Chen, L.; Ma, G.X.; Xu, X.D.; Jia, X.G.; Deng, F.S.; Li, X.J.; Yuan, J.Q. A review on phytochemicals of the genus *Maytenus* and their bioactive studies. *Molecules* **2021**, *26*, 4563. [CrossRef]

23. Duarte, L.P. Núcleo de Estudo de Plantas Medicinais—Departamento de Química ICEX. Available online: http://zeus.qui.ufmg.br/~{}neplam/principal.htm. (accessed on 22 December 2021).

24. Chung, P.Y. Novel targets of pentacyclic triterpenoids in *Staphylococcus aureus*: A Systematic Review. *Phytomedicine* **2020**, *73*, 152933. [CrossRef] [PubMed]

25. Silva, F.C.; Duarte, L.P.; Vieira Filho, S.A. Celastraceae family: Source of pentacyclic triterpenes with potential biological activity. *Rev. Virtual Quim.* **2014**, *6*, 1205–1220. [CrossRef]

26. Thimmappa, R.; Geisler, K.; Louveau, T.; O'Maille, P.; Osbourn, A. Triterpene biosynthesis in plants. *Annu. Rev. Plant Biol.* **2014**, *65*, 225–257. [CrossRef]

27. Mahato, S.B.; Kundu, A.P. 13C NMR spectra of pentacyclic triterpenoids-A compilation and some salient features. *Phytochemistry* **1994**, *37*, 1517–1575. [CrossRef]

28. Dewick, P.M. *Medicinal Natural Products: A Biosynthetic Approach*, 3rd ed.; John Wiley & Sons: Hoboken, NJ, USA, 2009; ISBN 9780470741689.

29. Shan, W.G.; Zhang, L.W.; Xiang, J.G.; Zhan, Z.J. Natural friedelanes. *Chem. Biodivers.* **2013**, *10*, 1392–1434. [CrossRef]

30. Pavia, D.L.; Lampman, G.M.; Kriz, G.S.; Vyvyan, J.R. *Introdução á Espectroscopia*; Cengage Learning: Bellingham, WA, USA, 2010; ISBN 9788522123384.

31. Bazzocchi, I.L.; Núñez, M.J.; Reyes, C.P. Diels–alder adducts from Celastraceae species. *Phytochem. Rev.* **2018**, *17*, 669–690. [CrossRef]

32. Corsino, J.; De Carvalho, P.R.F.; Kato, M.J.; Latorre, L.R.; Oliveira, O.M.M.F.; Araújo, A.R.; Bolzani, V.D.S.; França, S.C.; Pereira, A.M.S.; Furlan, M. Biosynthesis of friedelane and quinonemethide triterpenoids is compartmentalized in *Maytenus aquifolium* and *Salacia campestris*. *Phytochemistry* **2000**, *55*, 741–748. [CrossRef]

33. González, A.G.; Bazzocchi, I.L.; Moujir, L.; Jiménez, I.A. Ethnobotanical uses of Celastraceae. bioactive metabolites. *Stud. Nat. Prod. Chem.* **2000**, *23*, 649–738. [CrossRef]

34. Oliveira, F.F. Estudo Químico de *Vismia parviflora* e Síntese e Derivados Antroquinônicos. Ph.D. Thesis, Universidade Federal de Minas Gerais, Belo Horizonte, Brazil, 1997.

35. Chen, M.X.; Wang, D.Y.; Guo, J. 3-Oxo-11β-hydroxyfriedelane from the roots of *Celastrus monospermus*. *J. Chem. Res.* **2010**, *34*, 114–117. [CrossRef]

36. Silva, F.C.; Duarte, L.P.; Silva, G.D.F.; Filho, S.A.V.; Ivana, S.L.; Takahashi, J.A.; Tm, W.S. Chemical constituents from branches of *Maytenus gonoclada* (Celastraceae) and evaluation of antimicrobial activity. *J. Braz. Chem. Soc.* **2011**, *22*, 943–949. [CrossRef]

37. Silva, F.C.; Rodrigues, V.G.; Duarte, L.P.; Silva, G.D.F.; Miranda, R.R.S.; Filho, S.A.V. A new friedelane triterpenoid from the branches of *Maytenus gonoclada* (Celastraceae). *J. Chem. Res.* **2011**, *35*, 555–557. [CrossRef]

38. Oliveira, M.L.G.; Duarte, L.P.; Silva, G.D.F.; Filho, S.A.V.; Knupp, V.F.; Alves, F.G.P. 3-Oxo-12α-hydroxyfriedelane from *Maytenus gonoclada*: Structure elucidation by ^{1}H and ^{13}C chemical shift assignments and 2D-NMR spectroscopy. *Magn. Reson. Chem.* **2007**, *45*, 895–898. [CrossRef] [PubMed]

39. Kishi, A.; Morikawa, T.; Matsuda, H.; Yoshikawa, M. Structures of new friedelane and norfriedelane-type triterpenes and polyacylated eudesmane-type sesquiterpene from *Salacia chinensis* Linn. (S. Prinoides DC., Hippocrateaceae) and radical scavenging activities of principal constituents. *Chem. Pharm. Bull.* **2003**, *51*, 1051–1055. [CrossRef]

40. Hisham, A.; Kumar, G.J.; Fujimoto, Y.; Hara, N. Salacianone and salacianol, two triterpenes from *Salacia beddomei*. *Phytochemistry* **1995**, *40*, 1227–1231. [CrossRef]

41. Matsuda, H.; Murakami, T.; Yashiro, K.; Yamahara, J.; Yoshikawa, M. Antidiabetic principles of natural medicines. IV. Aldose reductase and α-glucosidase inhibitors from the Roots of *Salacia oblonga* WALL. (Celastraceae): Structure of a new friedelane-type triterpene, kotalagenin 16-acetate. *Chem. Pharm. Bull.* **1999**, *47*, 1725–1729. [CrossRef]

42. Kuo, Y.H.; Kuo, L.M.Y. Antitumour and anti-aids triterpenes from *Celastrus hindsii*. *Phytochemistry* **1997**, *44*, 1275–1281. [CrossRef] [PubMed]

43. Duarte, L.P.; Figueiredo, R.C.; Sousa, G.F.; Soares, D.B.D.S.; Rodrigues, S.B.V.; Silva, F.C.; De Fátima Silva, G.D.; Vieira Filho, S.A. Chemical constituents of *Salacia elliptica* (Celastraceae). *Quim. Nova* **2010**, *33*, 900–903. [CrossRef]

44. Duarte, L.P.; Silva De Miranda, R.R.; Rodrigues, S.B.V.; De Fátima Silva, G.D.; Filho, S.A.V.; Knupp, V.F. Stereochemistry of 16α-hydroxyfriedelin and 3-oxo-16-methylfriedel- 16-ene established by 2D NMR spectroscopy. *Molecules* **2009**, *14*, 598–607. [CrossRef]

45. Sousa, G.F.; Aguilar, M.G.; Dias, D.F.; Takahashi, J.A.; Moreira, M.E.C.; Vieira Filho, S.A.; Silva, G.D.F.; Rodrigues, S.B.V.; Messias, M.C.T.B.; Duarte, L.P. Anti-Inflammatory, antimicrobial and acetylcholinesterase inhibitory activities of friedelanes from *Maytenus robusta* branches and isolation of further triterpenoids. *Phytochem. Lett.* **2017**, *21*, 61–65. [CrossRef]

46. Sousa, G.F.; Soares, D.C.F.; Mussel, W.D.N.; Pompeu, N.F.E.; Silva, G.D.; de, F.; Filho, S.A.V.; Duarte, L.P. Pentacyclic triterpenes from branches of *Maytenus robusta* and *in vitro* cytotoxic property against 4t1 cancer cells. *J. Braz. Chem. Soc.* **2014**, *25*, 1338–1345. [CrossRef]

47. Rodríguez, F.M.; Perestelo, N.R.; Jiménez, I.A.; Bazzocchi, I.L. Friedelanes from *Crossopetalum lobatum*. A new example of a triterpene anhydride. *Helv. Chim. Acta* **2009**, *92*, 188–194. [CrossRef]

48. Ardiles, A.E.; González-Rodríguez, Á.; Núñez, M.J.; Perestelo, N.R.; Pardo, V.; Jiménez, I.A.; Valverde, Á.M.; Bazzocchi, I.L. Studies of naturally occurring friedelane triterpenoids as insulin sensitizers in the treatment type 2 diabetes mellitus. *Phytochemistry* **2012**, *84*, 116–124. [CrossRef]

49. Somwong, P.; Suttisri, R.; Buakeaw, A. A new 1,3-diketofriedelane triterpene from *Salacia verrucosa*. *Fitoterapia* **2011**, *82*, 1047–1051. [CrossRef]

50. Setzer, W.N.; Setzer, M.C.; Lynton Peppers, R.; McFerrin, M.B.; Meehan, E.J.; Chen, L.; Bates, R.B.; Nakkiew, P.; Jackes, B.R. Triterpenoid constituents in the bark of *Balanops australiana*. *Aust. J. Chem.* **2000**, *53*, 809–812. [CrossRef]

51. Niero, R.; Mafra, A.P.; Lenzi, A.C.; Cechinel-Filho, V.; Tischer, C.; Malheiros, A.; De Souza, M.M.; Yunes, R.A.; Delle Monache, F. A new triterpene with antinociceptive activity from *Maytenus robusta*. *Nat. Prod. Res.* **2006**, *20*, 1315–1320. [CrossRef]

52. Ferreira, F.L.; Hauck, M.S.; Duarte, L.P.; de Magalhães, J.C.; da Silva, L.S.M.; Pimenta, L.P.S.; Lopes, J.C.D.; Mercadante-Simões, M.O.; Vieira Filho, S.A. Zika virus activity of the leaf and branch extracts of *Tontelea micrantha* and its hexane extracts phytochemical study. *J. Braz. Chem. Soc.* **2019**, *30*, 793–803. [CrossRef]

53. Tamboli, A.R.; Namdeo, A.G. Isolation and characterization of *Salacia chinensis* and its evaluation of antioxidant activity. *Int. J. Pharmacogn.* **2020**, *7*, 126–132. [CrossRef]

54. Kaweetripob, W.; Mahidol, C.; Prawat, H.; Ruchirawat, S. Lupane, friedelane, oleanane, and ursane triterpenes from the stem of *Siphonodon celastrineus* Griff. *Phytochemistry* **2013**, *96*, 404–417. [CrossRef]

55. Giner, R.M.; Gray, A.I.; Gibbons, S.; Waterman, P.G. Friedelane triterpenes from the stem bark of *Caloncoba glauca*. *Phytochemistry* **1993**, *33*, 237–239. [CrossRef]

56. Nozaki, H.; Matsuura, Y.; Hirono, S.; Kasai, R.; Chang, J.J.; Lee, K.H. Antitumor agents, 116.1 cytotoxic triterpenes from *Maytenus diversifolia*. *J. Nat. Prod.* **1990**, *53*, 1039–1041. [CrossRef] [PubMed]

57. Patra, A.; Chaudhuri, S.K. Assignment of Carbon-13 nuclear magnetic resonance spectra of some friedelanes. *Magn. Reson. Chem.* **1987**, *25*, 95–100. [CrossRef]

58. Chavez, H.; Estevez-Braun, A.; Ravelo, A.G.; Gonzalez, A.G. Friedelane triterpenoids from *Maytenus macrocarpa*. *J. Nat. Prod.* **1998**, *61*, 82–85. [CrossRef] [PubMed]

59. Itokawa, H.; Shirota, O.; Ikuta, H.; Morita, H.; Takeya, K.; Iitaka, Y. Triterpenes from *Maytenus ilicifolia*. *Phytochemistry* **1991**, *30*, 3713–3716. [CrossRef]

60. Moiteiro, C.; Justino, F.; Tavares, R.; Marcelo-Curto, M.J.; Florêncio, M.H.; Nascimento, M.S.J.; Pedro, M.; Cerqueira, F.; Pinto, M.M.M. Synthetic secofriedelane and friedelane derivatives as inhibitors of human lymphocyte proliferation and growth of human cancer cell lines *in vitro*. *J. Nat. Prod.* **2001**, *64*, 1273–1277. [CrossRef]

61. Duan, H.; Takaishi, Y.; Momota, H.; Ohmoto, Y.; Taki, T.; Tori, M.; Takaoka, S.; Jia, Y.; Li, D. Immunosuppressive terpenoids from extracts of *Tripterygium wilfordii*. *Tetrahedron* **2001**, *57*, 8413–8424. [CrossRef]

62. Wang, K.W.; Zhang, H.; Pan, Y.J. Novel triterpenoids from *Microtropis triflora* with antitumor activities. *Helv. Chim. Acta* **2007**, *90*, 277–281. [CrossRef]

63. Maregesi, S.M.; Hermans, N.; Dhooghe, L.; Cimanga, K.; Ferreira, D.; Pannecouque, C.; Berghe, D.A.V.; Cos, P.; Maes, L.; Vlietinck, A.J.; et al. Phytochemical and biological investigations of *Elaeodendron schlechteranum*. *J. Ethnopharmacol.* **2010**, *129*, 319–326. [CrossRef]

64. Setzer, W.N.; Holland, M.T.; Bozeman, C.A.; Rozmus, G.F.; Setzer, M.C.; Moriarity, D.M.; Reeb, S.; Vogler, B.; Bates, R.B.; Haber, W.A. Isolation and frontier molecular orbital investigation of bioactive quinone-methide triterpenoids from the bark of *Salacia petenensis*. *Planta Med.* **2001**, *67*, 65–69. [CrossRef]

65. Li, Y.Z.; Li, Z.L.; Yin, S.L.; Shi, G.; Liu, M.S.; Jing, Y.K.; Hua, H.M. Triterpenoids from *Calophyllum inophyllum* and their growth inhibitory effects on human leukemia HL-60 cells. *Fitoterapia* **2010**, *81*, 586–589. [CrossRef]

66. Ramaiah, P.A.; Devi, P.U.; Frolow, F.; Lavie, D. 3-Oxo-friedelan-20α-oic acid from *Gymnosporia emarginata*. *Phytochemistry* **1984**, *23*, 2251–2255. [CrossRef]

67. Gottlieb, H.E.; Ramaiah, P.A.; Lavie, D. ^{13}C NMR signal assignment of friedelin and 3a-hydroxyfriedelan-2-one. *Magn. Reson. Chem.* **1985**, *23*, 616–620. [CrossRef]

68. Reyes, C.P.; Jiménez, I.A.; Bazzocchi, I.L. Pentacyclic triterpenoids from *Maytenus cuzcoina*. *Nat. Prod. Commun.* **2017**, *12*, 675–678. [CrossRef] [PubMed]

69. Pereira, R.C.G.; Soares, D.C.F.; Oliveira, D.C.P.; de Sousa, G.F.; Vieira-Filho, S.A.; Mercadante-Simões, M.O.; Lula, I.; Silva-Cunha, A.; Duarte, L.P. Triterpenes from leaves of *Cheiloclinium cognatum* and their *in vivo* antiangiogenic activity. *Magn. Reson. Chem.* **2018**, *56*, 360–366. [CrossRef]

70. Chang, C.W.; Wu, T.S.; Hsieh, Y.S.; Kuo, S.C.; Lee Chao, P.D. Terpenoids of *Syzygium formosanum*. *J. Nat. Prod.* **1999**, *62*, 327–328. [CrossRef]

71. Yang, G.Z.; Li, Y.C. Cyclopeptide and Terpenoids from *Tripterygium wilfordii* HOOK. F. *Helv. Chim. Acta* **2002**, *85*, 168–174. [CrossRef]

72. Duarte, M.C.; Tavares, J.F.; Madeiro, S.A.L.; Costa, V.C.O.; Filho, J.M.B.; De Fátima Agra, M.; Filho, R.B.; Da Silva, M.S. Maytensifolone, a new triterpene from *Maytenus distichophylla* Mart. Ex Reissek. *J. Braz. Chem. Soc.* **2013**, *24*, 1697–1700. [CrossRef]

73. Gao, H.Y.; Guo, Z.H.; Cheng, P.; Xu, X.M.; Wu, L.J. New Triterpenes from *Salacia hainanensis* Chun et How with α-glucosidase inhibitory activity. *J. Asian Nat. Prod. Res.* **2010**, *12*, 834–842. [CrossRef]

74. Fujita, R.; Duan, H.; Takaishi, Y. Terpenoids from *Tripterigyum hypoglaucum*. *Phytochemistry* **2000**, *53*, 715–722. [CrossRef]

75. Santos, J.P.; Rodrigues, B.L.; Oliveira, W.X.C.; Silva, F.C.; De Sousa, G.F.; Vieira Filho, S.A.; Duarte, L.P.; Silva, R.R. Caryopristimerin, the first example of a sesquiterpene-triterpene homo diels-alder adduct, and a new 29-*nor*-friedelane from roots of *Salacia crassifolia*. *J. Braz. Chem. Soc.* **2019**, *30*, 1558–1565. [CrossRef]

76. Li, K.; Duan, H.; Kawazoe, K.; Takaishi, Y. Terpenoids from *Tripterygium wilfordii*. *Phytochemistry* **1997**, *45*, 791–796. [CrossRef]

77. Gonzalez, A.G.; Alvarenga, N.L.; Ravelo, A.G.; Jimenez, I.A.; Bazzocchi, I.L. Two triterpenes from *Maytenus canariensis*. *J. Nat. Prod.* **1995**, *58*, 570–573. [CrossRef]

78. Addae-Mensah, I.; Adu-Kumi, S.; Waibel, R.; Oppong, I.V. A Novel D:A-friedooleanane triterpenoid and other constituents of the stem bark of *Dichapetalum barteri* Engl. *Arkivoc* **2007**, *2007*, 71–79. [CrossRef]

79. Wu, X.Y.; Qin, G.W.; Fan, D.J.; Xu, R.S. 1-Hydroxy-2,5,8-trimethyl-9-fluorenone from *Tripterygium wilfordii*. *Phytochemistry* **1994**, *36*, 477–479. [CrossRef]

80. Yeboah, E.M.O.; Majinda, R.R.T.; Kadziola, A.; Muller, A. Dihydro-β-agarofuran sesquiterpenes and pentacyclic triterpenoids from the root bark of *Osyris lanceolata*. *J. Nat. Prod.* **2010**, *73*, 1151–1155. [CrossRef]

81. Lu, C.H.; Zhang, J.X.; Gan, F.Y.; Shen, Y.M. Chemical constituents of the suspension cell cultures of *Maytenus hookeri*. *Acta Bot. Sin.* **2002**, *44*, 603–610.

82. Orabi, K.Y.; Al-Qasoumi, S.I.; El-Olemy, M.M.; Mossa, J.S.; Muhammad, I. Dihydroagarofuran alkaloid and triterpenes from *Maytenus heterophylla* and Maytenus *arbutifolia*. *Phytochemistry* **2001**, *58*, 475–480. [CrossRef]

83. Ngassapa, O.; Soejarto, D.D.; Pezzuto, J.M.; Farnsworth, N.R. Quinone-methide triterpenes and salaspermic acid from *Kokoona ochracea*. *J. Nat. Prod.* **1994**, *57*, 1–8. [CrossRef]

84. Zhang, X.W.; Wang, K.W.; Zhou, M.Q. Cytotoxic triterpenoids from the stalks of *Microtropis triflora*. *Chem. Biodivers.* **2017**, *14*, e1700066. [CrossRef]

85. Klass, J.; Tinto, W.F.; Mclean, S.; Reynolds, W.F. Friedelane triterpenoids from *Peritassa compta*: Complete ^{1}H and ^{13}C assignments by 2D NMR spectroscopy. *J. Nat. Prod.* **1992**, *55*, 1626–1630. [CrossRef]

86. Wandji, J.; Wansi, J.D.; Fuendjiep, V.; Dagne, E.; Mulholland, D.A.; Tillequin, F.; Fomum, Z.T.; Sondengam, B.L.; Nkeh, B.C.; Njamen, D. Sesquiterpene lactone and friedelane derivative from *Drypetes molunduana*. *Phytochemistry* **2000**, *54*, 811–815. [CrossRef]

87. Sousa, G.F.; Ferreira, F.L.; Duarte, L.P.; Silva, G.D.F.; Messias, M.C.T.B.; Vieira Filho, S.A. Structural determination of 3β,11β-dihydroxyfriedelane from *Maytenus robusta* (Celastraceae) by 1D and 2D NMR. *J. Chem. Res.* **2012**, *36*, 203–205. [CrossRef]

88. Costa, P.M.; Carvalho, M.G. New triterpene isolated from *Eschweilera longipes* (Lecythidaceae). *An. Acad. Bras. Cienc.* **2003**, *75*, 21–25. [CrossRef] [PubMed]

89. Salazar, G.C.M.; Silva, G.D.F.; Duarte, L.P.; Vieira Filho, S.A.; Lula, I.S. Two epimeric friedelane triterpenes isolated from *Maytenus truncata* Reiss: [1]H and ^{13}C chemical shift assignments. *Magn. Reson. Chem.* **2000**, *38*, 977–980. [CrossRef]

90. Gunatilaka, A.A.L.; Nanayakkara, N.P.D.; Sultanbawa, M.U.S.; Wazeer, M.I.M. [13]C nuclear magnetic resonance spectra of some naturally occurring friedelanones. *Org. Magn. Reson.* **1980**, *14*, 415–417. [CrossRef]

91. Dhanabalasingham, B.; Karunaratne, V.; Tezuka, Y.; Kikuchi, T.; Gunatilaka, A.A.L. Biogenetically important quinonemethides and other triterpenoid constituents of *Salacia reticulata*. *Phytochemistry* **1996**, *42*, 1377–1385. [CrossRef]

92. Yang, J.H.; Luo, S.D.; Wang, Y.S.; Zhao, J.F.; Zhang, H.B.; Li, L. Triterpenes from *Tripterygium wilfordii* Hook. *J. Asian Nat. Prod. Res.* **2006**, *8*, 425–429. [CrossRef]

93. Nakano, K.; Oose, Y.; Takaishi, Y. A novel epoxy-triterpene and nortriterpene from callus cultures of *Tripterygium wilfordii*. *Phytochemistry* **1997**, *46*, 1179–1182. [CrossRef]

94. Sousa, G.F.; Duarte, L.P.; Alcântara, A.F.C.; Silva, G.D.F.; Vieira-Filho, S.A.; Silva, R.R.; Oliveira, D.M.; Takahashi, J.A. New Triterpenes from *Maytenus robusta*: Structural elucidation based on NMR experimental data and theoretical Calculations. *Molecules* **2012**, *17*, 13439–13456. [CrossRef]

95. Vieira Filho, S.A.; Duarte, L.P.; Suva, G.D.F.; Lula, I.S.; Dos Santos, M.H. Total assignment of ^{1}H And ^{13}C NMR spectra of two 3,4-secofriedelanes from *Austroplenckia populnea*. *Magn. Reson. Chem.* **2001**, *39*, 746–748. [CrossRef]

96. Mena-Rejón, G.J.; Pérez-Espadas, A.R.; Moo-Puc, R.E.; Cedillo-Rivera, R.; Bazzocchi, I.L.; Jiménez-Diaz, I.A.; Quijano, L. Antigiardial activity of triterpenoids from Root Bark of *Hippocratea excelsa*. *J. Nat. Prod.* **2007**, *70*, 863–865. [CrossRef] [PubMed]

97. Zhang, K.; Liu, J.; Wang, Y.; Huang, H.; Chen, Y. Constituents of triterpenes from *Celastrus monospermus* Roxb. *Zhongshan Da Xue Xue Bao. Zi Ran Ke Xue Ban = Acta Sci. Nat. Univ. Sunyatseni* **1998**, *37*, 85–88.

98. Nogueiras, C.; Spengler, I.; Ravelo, A.G.; Jiménez, I.A. Triterpenes from the roots of Maytenus buxifolia. *Rev. Latinoam. Química* **2001**, *29*, 32–39.

99. Martínez, V.M.; Corona, M.M.; Vélez, C.S.; Rodríguez-Hahn, L.; Joseph-Nathan, P. Terpenoids from *Mortonia diffusa*. *J. Nat. Prod.* **1988**, *51*, 790–796. [CrossRef]

100. Betancor, C.; Freire, R.; Gonzalez, A.G.; Salazar, J.A.; Pascard, C.; Prange, T. Three Triterpenes And Other Terpenoids from *Catha cassinoides*. *Phytochemistry* **1980**, *19*, 1989–1993. [CrossRef]

101. Joshi, B.S.; Kamat, V.N.; Viswanathan, N. Triterpenes of *Salacia prinoides* DC. *Tetrahedron* **1973**, *29*, 1365–1374. [CrossRef]

102. Sousa, J.R.; Silva, G.D.F.; Pedersoli, J.L.; Alves, R.J. Friedelane and oleanane triterpenoids from bark wood of *Austroplenckia populnea*. *Phytochemistry* **1990**, *29*, 3259–3261. [CrossRef]

103. Espindola, L.S.; Dusi, R.G.; Demarque, D.P.; Braz-Filho, R.; Yan, P.; Bokesch, H.R.; Gustafson, K.R.; Beutler, J.A. Cytotoxic Triterpenes from *Salacia crassifolia* and metabolite profiling of Celastraceae species. *Molecules* **2018**, *23*, 1494. [CrossRef]

104. Almeida, M.T.R.; Ríos-Luci, C.; Padrón, J.M.; Palermo, J.A. Antiproliferative terpenoids and alkaloids from the roots of *Maytenus vitis-idaea* and *Maytenus spinosa*. *Phytochemistry* **2010**, *71*, 1741–1748. [CrossRef]

105. Magalhães, C.G.; De Fátima Silva, G.D.; Duarte, L.P.; Bazzocchi, I.L.; Diaz, A.J.; Moujir, L.; López, M.R.; Figueiredo, R.C.; Vieira Filho, S.A. Salicassin, an unprecedented chalcone-diterpene adduct and a quinone methide triterpenoid from *Maytenus salicifolia*. *Helv. Chim. Acta* **2013**, *96*, 1046–1054. [CrossRef]

106. Thiem, D.A.; Sneden, A.T.; Khan, S.I.; Tekwani, B.L. Bisnortriterpenes from *Salacia madagascariensis*. *J. Nat. Prod.* **2005**, *68*, 251–254. [CrossRef] [PubMed]

107. Likhitwitayawuid, K.; Bavovada, R.; Lin, L.Z.; Cordell, G.A. Revised structure of 20-hydroxytingenone and ^{13}C NMR assignments of 22β-hydroxytingenone. *Phytochemistry* **1993**, *34*, 759–763. [CrossRef]

108. Gunatilaka, A.A.L.; Fernando, H.C.; Kikuchi, T.; Tezuka, Y. ^{1}H and ^{13}C NMR Analysis of three quinone-methide triterpenoids. *Magn. Reson. Chem.* **1989**, *27*, 803–807. [CrossRef]

109. Jeller, A.H.; Silva, D.H.S.; Lião, L.M.; Bolzani, V.D.S.; Furlan, M. Antioxidant phenolic and quinonemethide triterpenes from *Cheiloclinium cognatum*. *Phytochemistry* **2004**, *65*, 1977–1982. [CrossRef]

110. Sneden, A.T. Isoiguesterin, a new antileukemic bisnortriterpene from *Salacia madagascariensis*. *J. Nat. Prod.* **1981**, *44*, 503–507. [CrossRef]

111. Santos, J.P. Estudo fitoquímico e da atividade biológica de constituintes das raízes e galhos de *Salacia crassifolia* (Celastraceae). Ph.D. Thesis, Universidade Federal de Minas Gerais, Belo Horizonte, Brazil, 2019.

112. González, A.G.; Alvarenga, N.L.; Ravelo, A.G.; Bazzocchi, I.L.; Ferro, E.A.; Navarro, A.G.; Moujir, L.M. Scutione, a New bioactive norquinonemethide triterpene from *Maytenus scutioides* (Celastraceae). *Bioorganic Med. Chem.* **1996**, *4*, 815–820. [CrossRef]

113. Tezuka, Y.; Kikuchi, T.; Dhanabalasingham, B.; Karunaratne, V.; Gunatilaka, A.A.L. Studies on terpenoids and steroids, 25. complete ¹H- and ¹³C-NMR spectral assignments of salaciquinone, a new 7-oxo-quinonemethide dinortriterpenoid. *J. Nat. Prod.* **1994**, *57*, 270–276. [CrossRef]

114. Khalid, S.A.; Friedrichsen, G.M.; Christensen, S.B.; El Tahir, A.; Satti, G.M. Isolation and characterization of pristimerin as the antiplasmodial and antileishmanial agent of *Maytenus senegalensis* (Lam.) Exell. *Arkivoc* **2007**, *2007*, 129–134. [CrossRef]

115. Lião, L.M.; Silva, G.A.; Monteiro, M.R.; Albuquerque, S. Trypanocidal activity of quinonemethide triterpenoids from *Cheiloclinium cognatum* (Hippocrateaceae). *Z. Nat. C* **2008**, *63*, 207–210. [CrossRef]

116. Sotanaphun, U.; Suttisri, R.; Lipipun, V.; Bavovada, R. Quinone-methide triterpenoids from *Glyptopetalum sclerocarpum*. *Phytochemistry* **1998**, *49*, 1749–1755. [CrossRef]

117. Ochieng, C.O.; Opiyo, S.A.; Mureka, E.W.; Ishola, I.O. Cyclooxygenase inhibitory compounds from *Gymnosporia heterophylla* Aerial Parts. *Fitoterapia* **2017**, *119*, 168–174. [CrossRef] [PubMed]

118. Kennedy, M.L.; Llanos, G.G.; Castanys, S.; Gamarro, F.; Bazzocchi, I.L.; Jiménez, I.A. Terpenoids from *Maytenus* species and assessment of their reversal activity against a multidrug-resistant *Leishmania tropica* Line. *Chem. Biodivers.* **2011**, *8*, 2291–2298. [CrossRef] [PubMed]

119. Rodríguez, F.M.; López, M.R.; Jiménez, I.A.; Moujir, L.; Ravelo, A.G.; Bazzocchi, I.L. New phenolic triterpenes from *Maytenus blepharodes*. Semisynthesis of 6-deoxoblepharodol from pristimerin. *Tetrahedron* **2005**, *61*, 2513–2519. [CrossRef]

120. Sotanaphun, U.; Lipipun, V.; Bavovada, R. Constituents of the pericarp of *Glyptopetalum sclerocarpum*. *Fitoterapia* **2004**, *75*, 606–608. [CrossRef]

121. González, A.G.; Alvarenga, N.L.; Ravelo, A.G.; Jiménez, I.A.; Bazzocchi, I.L.; Canela, N.J.; Moujir, L.M. Antibiotic phenol nor-triterpenes from *Maytenus canariensis*. *Phytochemistry* **1996**, *43*, 129–132. [CrossRef]

122. Shirota, O.; Morita, H.; Takeya, K.; Itokawa, H.; Iitaka, Y. Cytotoxic aromatic triterpenes from *Maytenus ilicifolia* and *Maytenus chuchuhuasca*. *J. Nat. Prod.* **1994**, *57*, 1675–1681. [CrossRef]

123. Rodrigues, V.G.; Duarte, L.P.; Silva, G.D.F.; Silva, F.C.; Góes, J.V.; Takahashi, J.A.; Pimenta, L.P.S.; Filho, S.A.V. Evaluation of antimicrobial activity and toxic potential of extracts and triterpenes isolated from *Maytenus imbricata*. *Quim. Nova* **2012**, *35*, 1375–1380. [CrossRef]

124. Gupta, M.; Alvarenga, N.; Rodriguez, F.; Ravelo, A.; Ignacio, J.; Bazzocchi, I. New phenolic and quinone-methide triterpenes from *Maytenus* species (Celastraceae). *Nat. Prod. Lett.* **1995**, *7*, 209–218. [CrossRef]

125. Ankli, A.; Heilmann, J.; Heinrich, M.; Sticher, O. Cytotoxic cardenolides and antibacterial terpenoids from *Crossopetalum gaumeri*. *Phytochemistry* **2000**, *54*, 531–537. [CrossRef]

126. Núñez, M.J.; Kennedy, M.L.; Jiménez, I.A.; Bazzocchi, I.L. Uragogin and blepharodin, unprecedented hetero-diels-alder adducts from Celastraceae species. *Tetrahedron* **2011**, *67*, 3030–3033. [CrossRef]

127. Morota, T.; Yang, C.X.; Ogino, T.; Qin, W.Z.; Katsuhara, T.; Xu, L.H.; Komatsu, Y.; Miao, K.L.; Maruno, M.; Yang, B.H. D:A-friedo-24-noroleanane triterpenoids from *Tripterigium wilfordii*. *Phytochemistry* **1995**, *39*, 1159–1163. [CrossRef]

128. Gutiérrez, F.; Estévez-Braun, A.; Ravelo, Á.G.; Astudillo, L.; Zárate, R. Terpenoids from the medicinal plant *Maytenus ilicifolia*. *J. Nat. Prod.* **2007**, *70*, 1049–1052. [CrossRef] [PubMed]

129. Takaishi, Y.; Wariishi, N.; Tateishi, H.; Kawazoe, K.; Nakano, K.; Ono, Y.; Tokuda, H.; Nishino, H.; Iwashima, A. Triterpenoid inhibitors of interleukin-1 secretion and tumour-promotion from *Tripterygium wilfordii* Var. Regelii. *Phytochemistry* **1997**, *45*, 969–974. [CrossRef]

130. Yang, G.Z.; Xi, M.L.; Li, Y.C. Two novel phenolic triterpenes from *Tripterygium wilfordii*. *J. Asian Nat. Prod. Res.* **2001**, *3*, 83–88. [CrossRef]

131. Gamlath, C.B.; Gunaherath, K.B.; Leslie Gunatilaka, A.A. Studies on terpenoids and steroids. Part 10.1 Structures of four new natural phenolic D:A-friedo-24-noroleanane triterpenoids. *J. Chem. Soc. Perkin Trans. 1* **1987**, 2849–2854. [CrossRef]

132. Sotanaphun, U.; Lipipun, V.; Suttisri, R.; Bavovada, R. Antimicrobial activity and stability of tingenone derivatives. *Planta Med.* **1999**, *65*, 450–452. [CrossRef]

133. González, A.G.; Kennedy, M.L.; Rodríguez, F.M.; Bazzocchi, I.L.; Jiménez, I.A.; Ravelo, A.G.; Moujir, L. Absolute configuration of triterpene dimers from *Maytenus* species (Celastraceae). *Tetrahedron* **2001**, *57*, 1283–1287. [CrossRef]

134. Shirota, O.; Morita, H.; Takeya, K.; Itokawa, H. Revised structures of cangorosins, triterpene dimers from *Maytenus ilicifolia*. *J. Nat. Prod.* **1997**, *60*, 111–115. [CrossRef]

135. Shirota, O.; Morita, H.; Takeya, K.; Itokawa, H. New geometric and stereoisomeric triterpene dimers from *Maytenus chuchuhuasca*. *Chem. Pharm. Bull.* **1998**, *46*, 102–106. [CrossRef]

136. Shirota, O.; Sekita, S.; Satake, M.; Morita, H.; Takeya, K.; Itokawa, H. Nine regioisomeric and stereoisomeric triterpene dimers from *Maytenus chuchuhuasca*. *Chem. Pharm. Bull.* **2004**, *52*, 739–746. [CrossRef] [PubMed]

137. Shirota, O.; Sekita, S.; Satake, M.; Morita, H.; Takeya, K.; Itokawa, H. Nine triterpene dimers from *Maytenus chuchuhuasca*. *Helv. Chim. Acta* **2004**, *87*, 1536–1544. [CrossRef]

138. Mesa-Siverio, D.; Chávez, H.; Estévez-Braun, A.; Ravelo, Á.G. Cheiloclines A-I. First examples of octacyclic sesquiterpene-triterpene hetero-diels-alder adducts. *Tetrahedron* **2005**, *61*, 429–436. [CrossRef]

139. Wu, J.; Zhou, Y.; Wang, L.; Zuo, J.; Zhao, W. Terpenoids from root bark of *Celastrus orbiculatus*. *Phytochemistry* **2012**, *75*, 159–168. [CrossRef] [PubMed]

140. Shirota, O.; Morita, H.; Takeya, K.; Itokawa, H. Five new triterpene dimers from *Maytenus chuchuhuasca*. *J. Nat. Prod.* **1997**, *60*, 1100–1104. [CrossRef]

141. Shirota, O.; Sekita, S.; Satake, M.; Morita, H.; Takeya, K.; Itokawa, H. Two cangorosin a type triterpene dimers from *Maytenus chuchuhuasca*. *Chem. Pharm. Bull.* **2004**, *52*, 1148–1150. [CrossRef]

142. Salazar, G.D.C.M.; de Sousa, G.F.; Duarte, L.P.; Silva, R.R.; Vieira-Filho, S.A.; de Fátima Silva, G.D. Truncatin: A triterpene dimer isolated from *Maytenus truncata* Reissek (Celastraceae). *WORLD J. Pharm. Pharm. Sci.* **2021**, *10*, 60–71.

143. González, A.G.; Alvarenga, N.L.; Estévez-Braun, A.; Ravelo, A.G.; Bazzocchi, I.L.; Mouijir, L. Structure and absolute configuration of triterpene dimers from *Maytenus scutioides*. *Tetrahedron* **1996**, *52*, 9597–9608. [CrossRef]

144. Corbett, R.E.; Cong, A.N.T.; Wilkins, A.L.; Thomson, R.A. Lichens and fungi. Part 17. The synthesis and absolute configuration at C-20 of the (*R*)- and (*S*)-Epimers of some 29-substituted lupane derivatives and of some 30-norlupan-20-ol derivatives and the crystal tructure of (20*R*)-3*β*-Acetoxylupan-29-ol. *J. Chem. Soc. Perkin Trans.* **1985**, 2051–2056. [CrossRef]

145. Yamashita, H.; Matsuzaki, M.; Kurokawa, Y.; Nakane, T.; Goto, M.; Lee, K.H.; Shibata, T.; Bando, H.; Wada, K. Four new triterpenoids from the bark of *Euonymus alatus* Forma Ciliato-Dentatus. *Phytochem. Lett.* **2019**, *31*, 140–146. [CrossRef]

146. Callies, O.; Bedoya, L.M.; Beltrán, M.; Muñoz, A.; Calderón, P.O.; Osorio, A.A.; Jiménez, I.A.; Alcamí, J.; Bazzocchi, I.L. Isolation, structural modification, and HIV inhibition of pentacyclic lupane-type triterpenoids from *Cassine xylocarpa* and *Maytenus cuzcoina*. *J. Nat. Prod.* **2015**, *78*, 1045–1055. [CrossRef] [PubMed]

147. Reyes, C.P.; Núñez, M.J.; Jiménez, I.A.; Busserolles, J.; Alcaraz, M.J.; Bazzocchi, I.L. Activity of lupane triterpenoids from *Maytenus* species as inhibitors of nitric oxide and prostaglandin E2. *Bioorganic Med. Chem.* **2006**, *14*, 1573–1579. [CrossRef] [PubMed]

148. Silva, S.R.D.; Silva, G.D.S.; Barbosa, L.C.A.; Duarte, L.P.; Vieira Filho, S.A. Lupane pentacyclic triterpenes isolated from stems and branches of *Maytenus imbricata* (Celastraceae). *Helv. Chim. Acta* **2005**, *88*, 1102–1109. [CrossRef]

149. Fern, A.; Hern, A.; Garc, T.H.; Spengler-salabarr, I.; Naturales, P.; Habana, L. Triterpenes with anti-inflammatory activity isolated from the bark of the endemic species *Maytenus elaeodendroides*, Griseb. *Rev. Cuba. Química* **2020**, *32*, 61–73.

150. Carpenter, R.C.; Sotheeswaran, S.; Sultanbawa, M.U.S.; Ternai, B. ^{13}C NMR studies of some lupane and taraxerane triterpenes. *Org. Magn. Reson.* **1980**, *14*, 462–465. [CrossRef]

151. Núñez, M.J.; Reyes, C.P.; Jiménez, I.A.; Moujir, L.; Bazzocchi, I.L. Lupane triterpenoids from *Maytenus* species. *J. Nat. Prod.* **2005**, *68*, 1018–1021. [CrossRef]

152. Satiraphan, M.; Pamonsinlapatham, P.; Sotanaphun, U.; Sittisombut, C.; Raynaud, F.; Garbay, C.; Michel, S.; Cachet, X. Lupane triterpenes from the leaves of the tropical rain forest tree *Hopea odorata* Roxb. and their cytotoxic activities. *Biochem. Syst. Ecol.* **2012**, *44*, 407–412. [CrossRef]

153. Chen, I.H.; Du, Y.C.; Lu, M.C.; Lin, A.S.; Hsieh, P.W.; Wu, C.C.; Chen, S.L.; Yen, H.F.; Chang, F.R.; Wu, Y.C. Lupane-type triterpenoids from *Microtropis fokienensis* and *Perrottetia arisanensis* and the apoptotic effect of 28-hydroxy-3-oxo-lup-20(29)-en-30-al. *J. Nat. Prod.* **2008**, *71*, 1352–1357. [CrossRef]

154. Zhou, J.; Wei, X.H.; Chen, F.Y.; Li, C.J.; Yang, J.Z.; Ma, J.; Bao, X.Q.; Zhang, D.; Zhang, D.M. Anti-inflammatory pentacyclic triterpenes from the stems of *Euonymus carnosus*. *Fitoterapia* **2017**, *118*, 21–26. [CrossRef]

155. Monaco, P.; Previtera, L. Isoprenoids from the leaves of *Quercus suber*. *J. Nat. Prod.* **1984**, *47*, 673–676. [CrossRef]

156. Rashid, M.A.; Gray, A.I.; Waterman, P.G.; Armstrong, J.A. Coumarins from *Phebalium tuberculosum* Ssp. *Megaphyllum and Phebalium filifolium*. *J. Nat. Prod.* **1992**, *55*, 851–858. [CrossRef] [PubMed]

157. Kang, H.R.; Eom, H.J.; Lee, S.R.; Choi, S.U.; Sung Kang, K.; Lee, K.R.; Kim, K.H. Bioassay-guided isolation of antiproliferative triterpenoids from *Euonymus alatus* Twigs. *Nat. Prod. Commun.* **2015**, *10*, 1929–1932. [CrossRef] [PubMed]

158. Yu, M.H.; Shi, Z.F.; Yu, B.W.; Pi, E.H.; Wang, H.Y.; Hou, A.J.; Lei, C. Triterpenoids and α-Glucosidase Inhibitory Constituents from *Salacia hainanensis*. *Fitoterapia* **2014**, *98*, 143–148. [CrossRef] [PubMed]

159. Fuchino, H.; Satoh, T.; Tanaka, N. Chemical evaluation of betula species in japan. I. Constituents of *Betula ermanii*. *Chem. Pharm. Bull.* **1995**, *43*, 1937–1942. [CrossRef]

160. Delgado-Méndez, P.; Herrera, N.; Chávez, H.; Estévez-Braun, A.; Ravelo, Á.G.; Cortes, F.; Castanys, S.; Gamarro, F. New terpenoids from *Maytenus apurimacensis* as MDR reversal agents in the parasite leishmania. *Bioorganic Med. Chem.* **2008**, *16*, 1425–1430. [CrossRef] [PubMed]

161. Miranda, R.R.S.; Silva, G.D.D.F.; Duarte, L.P.; Vieira Filho, S.A. Triterpene esters isolated from leaves of *Maytenus salicifolia* REISSEK. *Helv. Chim. Acta* **2007**, *90*, 652–658. [CrossRef]

162. Lai, Y.C.; Chen, C.K.; Tsai, S.F.; Lee, S.S. Triterpenes as α-glucosidase inhibitors from *Fagus hayatae*. *Phytochemistry* **2012**, *74*, 206–211. [CrossRef]

163. Burns, D.; Reynolds, W.F.; Buchanan, G.; Reese, P.B.; Enriquez, R.G. Assignment of ^{1}H and ^{13}C spectra and investigation of hindered side-chain rotation in lupeol derivatives. *Magn. Reson. Chem.* **2000**, *38*, 488–493. [CrossRef]

164. Dantanarayana, A.P.; Kumar, N.S.; Muthukuda, P.M.; I, M.; Wazeer, M. A lupane derivative and the ^{13}C NMR chemical shifts of some lupanols from *Pleurostylia opposita*. *Phytochemistry* **1982**, *21*, 2065–2068. [CrossRef]

165. Menezes-De-Oliveira, D.; Aguilar, M.I.; King-Díaz, B.; Vieira-Filho, S.A.; Pains-Duarte, L.; De Fátima Silva, G.D.; Lotina-Hennsen, B. The triterpenes 3*β*-lup-20(29)-en-3-ol and 3*β*-lup-20(29)-en-3-yl acetate and the carbohydrate 1,2,3,4,5,6-hexa-o-acetyl-dulcitol as photosynthesis light reactions inhibitors. *Molecules* **2011**, *16*, 9939–9956. [CrossRef]

166. Li, J.J.; Yang, J.; Lü, F.; Qi, Y.T.; Liu, Y.Q.; Sun, Y.; Wang, Q. Chemical constituents from the stems of *Celastrus orbiculatus*. *Chin. J. Nat. Med.* **2012**, *10*, 279–283. [CrossRef]

167. Kuroyanagi, M.; Shiotsu, M.; Ebihara, T.; Kawai, H.; Ueno, A.; Fukushima, S. Chemical studies on *Viburnum awabuki* K. KOCH. *Chem. Pharm. Bull.* **1986**, *34*, 4012–4017. [CrossRef]

168. Pieroni, L.G.; De Rezende, F.M.; Ximenes, V.F.; Dokkedal, A.L. Antioxidant activity and total phenols from the methanolic extract of *Miconia albicans* (Sw.) Triana Leaves. *Molecules* **2011**, *16*, 9439–9450. [CrossRef]

169. Choi, C.I.; Lee, S.R.; Kim, K.H. Antioxidant and α-glucosidase inhibitory activities of constituents from *Euonymus alatus* Twigs. *Ind. Crops Prod.* **2015**, *76*, 1055–1060. [CrossRef]

170. Aguiar, R.M.; David, J.P.; David, J.M. Unusual naphthoquinones, catechin and triterpene from *Byrsonima microphylla*. *Phytochemistry* **2005**, *66*, 2388–2392. [CrossRef] [PubMed]

171. Savona, G.; Bruno, M.; Rodríguez, B.L.; Marco, J. Triterpenoids from *Salvia deserta*. *Phytochemistry* **1987**, *26*, 3305–3308. [CrossRef]

172. Huang, J.; Guo, Z.H.; Cheng, P.; Sun, B.H.; Gao, H.Y. Three new triterpenoids from *Salacia hainanensis* Chun et How showed effective anti-α-glucosidase activity. *Phytochem. Lett.* **2012**, *5*, 432–437. [CrossRef]

173. Pistelli, L.; Noccioli, C.; Giachi, I.; Dimitrova, B.; Gevrenova, R.; Morelli, I.; Potenza, D. Lupane-triterpenes from *Bupleurum flavum*. *Nat. Prod. Res.* **2005**, *19*, 783–788. [CrossRef]

174. Souza, A.D.L.; Rocha, A.F.I.; Pinheiro, M.L.B.; Andrade, C.H.D.S.; Galotta, A.L.D.A.Q.; Santos, M.D.P.S.S.D. Constituintes químicos de *Gustavia augusta* L. (Lecythidaceae). *Quim. Nova* **2001**, *24*, 439–442. [CrossRef]

175. Ahmad, V.U.; Bano, S.; Voelter, W.; Fuchs, W. Chemical Examination of *Nepeta hindostana* (Roth) Haines the structure of nepeticin. *Tetrahedron Lett.* **1981**, *22*, 1714–1718. [CrossRef]

176. Tanaka, R.; Masuda, K.; Matsunaga, S. Lup-20(29)-en-3β,15α-diol and ocotillol-II from the stem bark of *Phyllanthus flexuosus*. *Phytochemistry* **1993**, *32*, 472–474. [CrossRef]

177. Ulubelen, A.; Topcu, G.; Lotter, H.; Wagner, H.; Eriş, C. Triterpenoids from the aerial parts of *Salvia montbretii*. *Phytochemistry* **1994**, *36*, 413–415. [CrossRef]

178. Chaturvedula, V.S.P.; Schilling, J.K.; Johnson, R.K.; Kingston, D.G.I. New cytotoxic lupane triterpenoids from the twigs of *Coussarea paniculata*. *J. Nat. Prod.* **2003**, *66*, 419–422. [CrossRef]

179. Hisham, A.; Kumar, G.J.; Fujimoto, Y.; Hara, N. 20,29-Epoxysalacianone and 6β-hydroxysalacianone, two lupane triterepenes from *Salacia beddomei*. *Phytochemistry* **1996**, *42*, 789–794. [CrossRef]

180. Razdan, T.K.; Harkar, S.; Qadri, B.; Qurishi, M.A.; Khuroo, M.A. Lupene derivatives from *Skimmia laureola*. *Phytochemistry* **1988**, *27*, 1890–1892. [CrossRef]

181. Ngassapa, O.D.; Soejarto, D.D.; Che, C.T.; Pezzuto, J.M.; Farnsworth, N.R. New cytotoxic lupane lactones from *Kokoona ochracea*. *J. Nat. Prod.* **1991**, *54*, 1353–1359. [CrossRef] [PubMed]

182. Djerassi, C.; Farkas, E.; Liu, H.L.; Thomas, G.H. Terpenoids. XVII. The cactus triterpenes thurberogenin and stellatogenin. *J. Am. Chem. Soc.* **1955**, *77*, 5330–5336. [CrossRef]

183. Djerassi, C.; Hodges, R. Terpenoids. XXIII. Interconversion of thurberogenin and betulinic acid. *J. Am. Chem. Soc.* **1956**, *78*, 3534–3538. [CrossRef]

184. Castellanos, L.; De Correa, R.S.; Martínez, E.; Calderon, J.S. Oleanane triterpenoids from *Cedrela montana* (Meliaceae). *Z. Naturforsch.—Sect. C J. Biosci.* **2002**, *57*, 575–578. [CrossRef]

185. Anoda, N.; Matsunaga, M.; Kubo, M.; Harada, K.; Fukuyama, Y. Six new triterpenoids from the aerial parts of *Maytenus diversifolia*. *Nat. Prod. Commun.* **2016**, *11*, 1085–1088. [CrossRef]

186. Xiao, Y.; Gou-liang, W.; Fu-Jun, G. Studies on triterpenoid constituents isolated from the roots of *Sabia schumanniana*. *J. Integr. Plant Biol.* **1994**, *36*, 154–158.

187. Cáceres-Castillo, D.; Mena-Rejón, G.J.; Cedillo-Rivera, R.; Quijano, L. 21β-Hydroxy-oleanane-type triterpenes from *Hippocratea excelsa*. *Phytochemistry* **2008**, *69*, 1057–1064. [CrossRef] [PubMed]

188. Mathias, L.; Vieira, I.J.C.; Braz-Filho, R.; Rodrigues Filho, E. A New pentacyclic triterpene isolated from *Myroxylon balsamum* (Syn. *Myroxylon peruiferum*). *J. Braz. Chem. Soc.* **2000**, *11*, 195–198. [CrossRef]

189. Chen, I.H.; Chang, F.R.; Wu, C.C.; Chen, S.L.; Hsieh, P.W.; Yen, H.F.; Du, Y.C.; Wu, Y.C. Cytotoxic triterpenoids from the leaves of *Microtropis fokienensis*. *J. Nat. Prod.* **2006**, *69*, 1543–1546. [CrossRef]

190. Chen, I.H.; Du, Y.C.; Hwang, T.L.; Chen, I.F.; Lan, Y.H.; Yen, H.F.; Chang, F.R.; Wu, Y.C. Anti-inflammatory triterpenoids from the stems of *Microtropis fokienensis*. *Molecules* **2014**, *19*, 4608–4623. [CrossRef]

191. Osorio, A.A.; Muñóz, A.; Torres-Romero, D.; Bedoya, L.M.; Perestelo, N.R.; Jiménez, I.A.; Alcamí, J.; Bazzocchi, I.L. Olean-18-ene triterpenoids from Celastraceae species inhibit HIV replication targeting NF-KB and Sp1 dependent transcription. *Eur. J. Med. Chem.* **2012**, *52*, 295–303. [CrossRef]

192. Shirota, O.; Tamemura, T.; Morita, H.; Takeya, K.; Itokawa, H. Triterpenes from brazilian medicinal plant "Chuchuhuasi" (*Maytenus krukovii*). *J. Nat. Prod.* **1996**, *59*, 1072–1075. [CrossRef]

193. Kaweetripob, W.; Mahidol, C.; Thongnest, S.; Prawat, H.; Ruchirawat, S. Polyoxygenated ursane and oleanane triterpenes from *Siphonodon celastrineus*. *Phytochemistry* **2016**, *129*, 58–67. [CrossRef]

194. Gao, L.; Duan, L.K.; Feng, J.E.; Jiang, Y.T.; Gao, J.; Fan, J.T.; Dai, R.; Jiang, Z.Y. Four new triterpene glucosides from *Salacia cochinchinensis* Lour. *Nat. Prod. Res.* **2020**, *12*, 1–8. [CrossRef]

195. Wang, K.W.; Wang, S.W. Chemical constituents of *Euonymus bockii*. *Chem. Nat. Compd.* **2014**, *50*, 948–949. [CrossRef]

196. Núñez, M.J.; Ardiles, A.E.; Martínez, M.L.; Torres-Romero, D.; Jiménez, I.A.; Bazzocchi, I.L. Triterpenoids from *Cassine xylocarpa* and *Celastrus vulcanicola* (Celastraceae). *Phytochem. Lett.* **2013**, *6*, 148–151. [CrossRef]
197. Niampoka, C.; Suttisri, R.; Bavovada, R.; Takayama, H.; Aimi, N. Potentially cytotoxic triterpenoids from the root bark of *Siphonodon celastrineus* Griff. *Arch. Pharm. Res.* **2005**, *28*, 546–549. [CrossRef]
198. Wang, H.; Tian, X.; Chen, Y.Z. Chemical constituents of the aerial part of *Celastrus hypoleucus*. *J. Chin. Chem. Soc.* **2002**, *49*, 433–436. [CrossRef]
199. Vieira Filho, S.A.; Duarte, L.P.; Silva, G.D.F.; Howarth, O.W.; Lula, L.S. 3β-(Stearyloxy)olean-12-ene from *Austroplenckia populnea*: Structure elucidation by 2D-NMR and quantitative ^{13}C-NMR spectroscopy. *Helv. Chim. Acta* **2003**, *86*, 3445–3449. [CrossRef]
200. Martins, L.R.; Takahashi, J.A. Rearrangement and oxidation of β-amyrin promoted by growing cells of *Lecanicillium muscarinium*. *Nat. Prod. Res.* **2010**, *24*, 767–774. [CrossRef] [PubMed]
201. Knight, S.A. Carbon-13 NMR spectra of some tetra- and pentacyclic triterpenoids. *Org. Magn. Reson.* **1974**, *6*, 603–611. [CrossRef]
202. Wang, K.W. A new fatty acid ester of triterpenoid from *Celastrus rosthornianus* with anti-tumor activitives. *Nat. Prod. Res.* **2007**, *21*, 669–674. [CrossRef]
203. Muhammad, I.; El Sayed, K.A.; Mossa, J.S.; Al-Said, M.S.; El-Feraly, F.S.; Clark, A.M.; Hufford, C.D.; Oh, S.; Mayer, A.M.S. Bioactive 12-oleanene triterpene and secotriterpene acids from *Maytenus undata*. *J. Nat. Prod.* **2000**, *63*, 605–610. [CrossRef]
204. Piacente, S.; Dos Santos, L.C.; Mahmood, N.; Pizza, C. Triterpenes from *Maytenus macrocarpa* and evaluation of their anti-HIV activity. *Nat. Prod. Commun.* **2006**, *1*, 1073–1078. [CrossRef]
205. Mokoka, T.A. Isolation and Characterization of Compounds Active against *Cryptococcus neoformans* from Maytenus *undata* (Thunb.) Blakelock (Celastraceae) Leaves. Master's Thesis, Universiteit van Pretoria, Pretoria, South Africa, 2007.
206. Ye, Y.; Kinoshita, K.; Koyama, K.; Takahashi, K.; Kondo, N.; Yuasa, H. New triterpenes from *Machaerocereus eruca*. *J. Nat. Prod.* **1998**, *61*, 456–460. [CrossRef]
207. Culioli, G.; Mathe, C.; Archier, P.; Vieillescazes, C. A lupane triterpene from *Frankincense* (*Boswellia* Sp., *Burseraceae*). *Phytochemistry* **2003**, *62*, 537–541. [CrossRef]
208. Sati, S.C.; Sati, D.; Sharma, A.; Pandey, L.P. Chemical constituents from bark of Euonymus tingen (Celastraceae). *J. Emerg. Technol. Innov. Res.* **2012**, *7*, 1494–1497.
209. Din, A.U.; Uddin, G.; Hussain, N.; Choudary, M.I. Ficusonic Acid: A new cytotoxic triterpene isolated from *Maytenus royleanus* (Wall. Ex M. A. Lawson) Cufodontis. *J. Braz. Chem. Soc.* **2013**, *24*, 663–668. [CrossRef]
210. Luo, Y.; Xu, Q.L.; Dong, L.M.; Zhou, Z.Y.; Chen, Y.C.; Zhang, W.M.; Tan, J.W. A new ursane and a new oleanane triterpene acids from the whole plant of *Spermacoce latifolia*. *Phytochem. Lett.* **2015**, *11*, 127–131. [CrossRef]
211. Kolak, U.; Topçu, G.; Birteksöz, S.; Ötük, G.; Ulubelen, A. Terpenoids and steroids from the roots of *Salvia blepharochlaena*. *Turkish J. Chem.* **2005**, *29*, 177–186.
212. Ikuta, A.; Kamiya, K.; Satake, T.; Saiki, Y. Triterpenoids from callus tissue cultures of *Paeonia* species. *Phytochemistry* **1995**, *38*, 1203–1207. [CrossRef]
213. Aguilar-Gonzalez, A.R.; Mena-Rejón, G.J.; Padilla-Montano, N.; Toscano, A.; Quijano, L. Triterpenoids from *Hippocratea excelsa*. The crystal structure of 29-hydroxytaraxerol. *Z. Nat. B* **2005**, *60*, 577–584. [CrossRef]
214. Liu, S.J.; Liao, Z.X.; Liu, C.; Yao, G.Y.; Wang, H.S. A new triterpenoid and eremophilanolide from *Ligularia przewalskii*. *Phytochem. Lett.* **2014**, *9*, 11–16. [CrossRef]
215. Ikuta, A.; Morikawa, A. Triterpenes from *Stauntonia hexaphylla* Callus tissues. *J. Nat. Prod.* **1992**, *55*, 1230–1233. [CrossRef]
216. Kinjc, J.E.; Miyamotc, I.; Nohara, T.; Murakamj, K.; Kids, K.; Tomimatsu, T.; Yamasaki, M. Oleanene-sapogenols from *Puerariae radix*. *Chem. Pharm. Bull.* **1985**, *33*, 1293–1296. [CrossRef]
217. Hui-Zheng, X.; Zhi-Zhen, L.; Chohachi, K.; Soejarto, D.D.; Cordell, G.A.; Fong, H.H.S.; Hodgson, W. 3β-(3,4-Dihydroxycinnamoyl)-erythrodiol and 3β-(4-hydroxycinnamoyl)-erythrodiol from *Larrea tridentata*. *Phytochemistry* **1988**, *27*, 233–235. [CrossRef]
218. Ragasa, C.Y.; De Luna, R.D.; Hofilena, J.G. Antimicrobial terpenoids from *Pterocarpus indicus*. *Nat. Prod. Res.* **2005**, *19*, 305–309. [CrossRef] [PubMed]
219. Choi, S.Z.; Sang, U.C.; Kang, R.L. Pytochemical constituents of the aerial parts from solidago *Virga-aurea* Var. Gigantea. *Arch. Pharm. Res.* **2004**, *27*, 164–168. [CrossRef] [PubMed]
220. Gonzalez, A.G.; Fraga, B.M.; Gonzalez, P.; Hernandez, M.G.; Ravelo, A.G. ^{13}C NMR spectra of olean-18-ene derivatives. *Phytochemistry* **1981**, *20*, 1919–1921. [CrossRef]
221. Mba'ning, B.M.; Lenta, B.N.; Ngouela, S.; Noungoué, D.T.; Tantangmo, F.; Talontsi, F.M.; Tsamo, E.; Laatsch, H. Salacetal, an Oleanane-type triterpene from *Salacia longipes* Var. Camerunensis. *Z. Naturforsch.—Sect. B J. Chem. Sci.* **2011**, *66*, 1270–1274. [CrossRef]
222. Djerassi, C.; Henry, J.A.; Lemin, A.J.; Rios, T.; Thomas, G.H. Terpenoids. XXIV.1 The Structure Of The Cactus Triterpene Queretaroic Acid. *J. Am. Chem. Soc.* **1956**, *78*, 3783–3787. [CrossRef]
223. Shibata, S.; Takahashi, K.; Yano, S.; Harada, M.; Saito, H.; Tamura, Y.; Kumagai, A.; Hirabayashi, K.; Yamamoto, M.; Nagata, N. Chemical modification of glycyrrhetinic acid in relation to the biological activities. *Chem. Pharm. Bull.* **1987**, *35*, 1910–1918. [CrossRef]
224. Bandaranayake, W.M. Terpenoids of *Canarium zeylanicum*. *Phytochemistry* **1980**, *19*, 255–257. [CrossRef]
225. El-Seedia, H.R.; Hazella, A.C.; Torssell, K.B.G. Triterpenes, lichexanthone and an acetylenic acid from *Minquartia guianensis*. *Phytochemistry* **1994**, *35*, 1297–1299. [CrossRef]

226. Kutney, J.P.; Hewitt, G.M.; Lee, G.; Piotrowska, K.; Roberts, M.; Rettig, S.J. Studies with tissue cultures of the chinese herbal plant, *Tripterygium wilfordii*. isolation of metabolites of interest in rheumatoid arthritis, immunosuppression, and male contraceptive activity. *Can. J. Chem.* **1992**, *70*, 1455–1480. [CrossRef]
227. Nozaki, H.; Suzuki, H.; Hirayama, T.; Kasai, R.; Wu, R.W.; Lee, K.H. Antitumour triterpenes of *Maytenus diversifolia*. *Phytochemistry* **1986**, *25*, 479–485. [CrossRef]
228. Luis, J.G.; Andrés, L.S. New ursane type triterpenes from *Salvia mellifera* Greene. *Nat. Prod. Lett.* **1999**, *13*, 187–194. [CrossRef]
229. Lima, F.V.; Malheiros, A.; Otuki, M.F.; Calixto, J.B.; Yunes, R.A.; Cechinel Filho, V.; Delle Monache, F. Three new triterpenes from the resinous bark of *Protium kleinii* and their antinociceptive activity. *J. Braz. Chem. Soc.* **2005**, *16*, 578–582. [CrossRef]
230. Terazawa, S.; Uemura, Y.; Koyama, Y.; Kawakami, S.; Sugimoto, S.; Matsunami, K.; Otsuka, H.; Shinzato, T.; Kawahata, M.; Yamaguchi, K. Microtropins Q–W, Ent-Labdane Glucosides: Microtropiosides G–I, ursane-type triterpene diglucoside and flavonol glycoside from the leaves of *Microtropis japonica*. *Chem. Pharm. Bull.* **2017**, *65*, 930–939. [CrossRef]
231. Nakagawa, H.; Takaishi, Y.; Fujimoto, Y.; Duque, C.; Garzon, C.; Sato, M.; Okamoto, M.; Oshikawa, T.; Ahmed, S.U. Chemical constituents from the colombian medicinal plant *Maytenus laevis*. *J. Nat. Prod.* **2004**, *67*, 1919–1924. [CrossRef]
232. Odak, J.A.; Manguro, L.O.A.; Wong, K.C. New compounds with antimicrobial activities from *Elaeodendron buchananii* Stem Bark. *J. Asian Nat. Prod. Res.* **2018**, *20*, 510–524. [CrossRef]
233. Tanaka, N.; Duan, H.; Takaishi, Y.; Kawazoe, K.; Goto, S. Terpenoids from *Tripterygium doianum* (Celastraceae). *Phytochemistry* **2002**, *61*, 93–98. [CrossRef]
234. Okoye, N.N.; Ajaghaku, D.L.; Okeke, H.N.; Ilodigwe, E.E.; Nworu, C.S.; Okoye, F.B.C. *b*-Amyrin and *a*-amyrin acetate isolated from the stem bark of *Alstonia boonei* Display Profound Anti-Inflammatory Activity. *Pharm. Biol.* **2014**, *2014*. *52*, 1478–1486. [CrossRef]
235. Miranda, R.R.S.; Silva, G.D.F.; Duarte, L.P.; Fortes, I.C.P.; Vieira Filho, S.A. Structural determination of 3β-stearyloxy-urs-12-ene from *Maytenus salicifolia* by 1D and 2D NMR and quantitative ^{13}C NMR spectroscopy. *Magn. Reson. Chem.* **2006**, *44*, 127–131. [CrossRef]
236. Tkachev, A.V.; Denisov, A.Y.; Gatilov, Y.V.; Bagryanskaya, I.Y.; Shevtsov, S.A.; Rybalova, T.V. Stereochemistry of hydrogen peroxide - acetic acid oxidation of ursolic acid and related compounds. *Tetrahedron* **1994**, *50*, 11459–11488. [CrossRef]
237. Zhang, D.M.; Yu, D.Q. Structure of tripterygic acid A: A New triterpene of *Tripterygium wilfordii*. *Planta Med.* **1990**, *56*, 98–100. [CrossRef] [PubMed]
238. Mazumder, K.; Siwu, E.R.O.; Nozaki, S.; Watanabe, Y.; Tanaka, K.; Fukase, K. Ursolic acid derivatives from bangladeshi medicinal plant, *Saurauja roxburghii*: Isolation and cytotoxic activity against A431 and C6 glioma cell lines. *Phytochem. Lett.* **2011**, *4*, 287–291. [CrossRef]
239. Talapatra, S.K.; Sarkar, A.C.; Talapatra, B. Two pentacyclic triterpenes from *Rubia cordifolia*. *Phytochemistry* **1981**, *20*, 1923–1927. [CrossRef]
240. Gnoatto, S.C.B.; Dassonville-Klimpt, A.; Da Nascimento, S.; Galéra, P.; Boumediene, K.; Gosmann, G.; Sonnet, P.; Moslemi, S. Evaluation of ursolic acid isolated from *Ilex paraguariensis* and derivatives on aromatase inhibition. *Eur. J. Med. Chem.* **2008**, *43*, 1865–1877. [CrossRef] [PubMed]
241. Zhang, Y.; Adelakun, T.A.; Qu, L.; Li, X.; Li, J.; Han, L.; Wang, T. New terpenoid glycosides obtained from *Rosmarinus officinalis* L. aerial parts. *Fitoterapia* **2014**, *99*, 78–85. [CrossRef]
242. González, A.G.; Andres, L.S.; Ravelo, A.G.; Luis, J.G.; Bazzocchi, I.L.; West, J. Terpenoids from *Salvia mellifera*. *Phytochemistry* **1990**, *29*, 1691–1693. [CrossRef]
243. Santos, J.P.; Oliveira, W.X.C.; Vieira-Filho, S.A.; Pereira, R.C.G.; Souza, G.F.; Gouveia, V.A.; de Paula Sabino, A.; Evangelista, F.C.G.; Takahashi, J.A.; Moura, M.A.F.; et al. Phytochemical and biological studies of constituents from roots of *Salacia crassifolia* (CELASTRACEAE). *Quim. Nova* **2020**, *43*, 558–567. [CrossRef]
244. Siddiqui, S.; Hafeez, F.; Begum, S.; Siddiqui, B.S. Kaneric Acid, A new triterpene from the leaves of *Nerium oleander*. *J. Nat. Prod.* **1986**, *49*, 1086–1090. [CrossRef]
245. Shan, H.; Wilson, W.K.; Castillo, D.A.; Matsuda, S.P.T. Are isoursenol and γ-amyrin rare triterpenes in nature or simply overlooked by usual analytical methods? *Org. Lett.* **2015**, *17*, 3986–3989. [CrossRef]
246. Maia, R.M.; Barbosa, P.R.; Cruz, F.G.; Roque, N.F.; Fascio, M. Triterpenes from the resin of protium *Heptaphyllum march* (Burseraceae): Characterization in binary mixtures. *Quim. Nova* **2000**, *23*, 623–626. [CrossRef]
247. Ding, P.; Zhou, M.Q.; Wang, K.W. A new cytotoxic triterpenoid from *Microtropis triflora* Merr. et Freem. *Chin. Pharm. J.* **2016**, *24*, 1557–1561.
248. Tsichritzis, F.; Jakupovic, J.F. Diterpenes and other constituents from *Relhania* Species. *Phytochemistry* **1990**, *29*, 3173–3187. [CrossRef]
249. Lai, J.; Ito, K. Studies on the constituents of *Marsdenia formosana* Masamune. V. Isolation and structure of a new triterpenoid, marsformosanone. *Chem. Pharm. Bull.* **1979**, *27*, 2248–2251. [CrossRef]
250. Rodriguez Perez, F.M. Estudio fitoquímico de especies de la familia Celastraceae (flora Panameña): *Maytenus blepharodes* y *Crossopetalum lobatuum*. Ph.D. Thesis, Univesity of La Laguna, San Cristobal de La Laguna, Spain, 2000.
251. Akihisa, T.; Yamamoto, K.; Tamura, T.; Nambara, T.; Iida, T.; Kimura, Y.; Chang, F.C. Triterpenoid ketones from *Lingnania chungii* Mcclure: Arborinone, friedelin and glutinone. *Chem. Pharm. Bull.* **1992**, *40*, 789–791. [CrossRef]

252. Lião, L.M.; Vieira, P.C.; Rodrigues-Filho, E.; Fernandes, J.B.; Silva, M.F.G.F. Isomeric triterpenoids from *Peritassa campestris*. *Z. Naturforsch.—Sect. C J. Biosci.* **2002**, *57*, 403–406. [CrossRef] [PubMed]
253. Sun, X.B.; Zhao, P.H.; Xu, Y.J.; Sun, L.M.; Cao, M.A.; Yuan, C.S. Chemical constituents from the roots of *Polygonum bistorta*. *Chem. Nat. Compd.* **2007**, *43*, 563–566. [CrossRef]
254. Sakurai, N.; Yaguchi, Y.; Inoue, T. Triterpenoids from *Myrica rubra*. *Phytochemistry* **1986**, *26*, 217–219. [CrossRef]
255. Hussein Ayoub, S.M.; Babiker, A.I. Monechmol, A. New pentacyclic triterpene from *Monechma debile*. *Planta Med.* **1984**, *50*, 520–521. [CrossRef] [PubMed]
256. Weeratunga, G.; Kumar, V. D:B-Friedoolean-5-ene-3β,29-diol, an angular methyl oxygenated d: B-friedooleanene from *Elaeodendron balae*. *Phytochemistry* **1985**, *24*, 2369–2372. [CrossRef]
257. Oliveira, M.L.G.; Assenco, R.A.G.; Silva, G.D.F.; Lopes, J.C.D.; Silva, F.C.; Lanna, M.C.S.; de Magalhães, J.C.; Duarte, L.P.; Vieira Filho, S.A. Cytotoxicity, anti-poliovirus activity and *in silico* biological evaluation of constituents from *Maytenus gonoclada* (Celastraceae). *Int. J. Pharm. Pharm. Sci.* **2014**, *6*, 130–137.
258. Zhang, Y.; Nakamura, S.; Wang, T.; Matsuda, H.; Yoshikawa, M. The absolute stereostructures of three rare D:B-friedobaccharane skeleton triterpenes from the leaves of *Salacia chinensis*. *Tetrahedron* **2008**, *64*, 7347–7352. [CrossRef]
259. Wang, Y.; Chen, W.S.; Wu, Z.J.; Xi, Z.X.; Chen, W.; Zhao, G.J.; Li, X.; Sun, L.N. Chemical constituents from *Salacia amplifolia*. *Biochem. Syst. Ecol.* **2011**, *39*, 205–208. [CrossRef]
260. Yoshikawa, M.; Shimoda, H.; Nishida, N.; Takada, M.; Matsuda, H. *Salacia reticulata* and its polyphenolic constituents with lipase inhibitory and lipolytic activities have mild antiobesity effects in rats. *J. Nutr.* **2002**, *132*, 1819–1824. [CrossRef] [PubMed]
261. Facundo, V.A.; Oliveira Meneguetti, D.U.; Militão, J.S.L.T.; Lima, R.A.; Hurtado, F.B.; Casseb, A.A.; Teixeira, L.F.; da Silva, I.; do, C.; da Silva, G.V.J.; et al. Chemical constituents from *Maytenus guianensis* Klotzsch Ex Reissek (Celastraceae) amazon rainforest. *Biochem. Syst. Ecol.* **2015**, *58*, 270–273. [CrossRef]
262. Lima, R.A.; Bay-Hurtado, F.; Meneguetti, D.U.O.; Facundo, J.B.; Militão, J.S.L.T.; Matos, N.B.; Facundo, V.A. Microbiological evaluation of isolated compounds from the bark of *Maytenus guianensis* Klotzsch Ex Reissek (Celastraceae). *Rev. Eletrônica Gestão Educ. Tecnol. Ambient.* **2016**, *20*, 592. [CrossRef]
263. Mba'ning, B.M.; Ateba, J.E.T.; Awantu, A.F.; Amaral, L.S.; Happi, G.M.; Neumann, B.; Stammler, G.; Lenta, B.N.; Ngouela, S.A.; Malavazi, I.; et al. Chemical constituents from the leaves and liana of *Salacia nitida* (Benth.) N.E.Br. (Celastraceae) and their antimicrobial activities. *Trends Phytochem. Res. Trends Phytochem. Res* **2019**, *3*, 2019–2083.
264. Andrade, S.F.; Comunello, E.; Noldin, V.F.; Delle Monache, F.; Filho, V.C.; Niero, R. Antiulcerogenic activity of fractions and 3,15-dioxo-21α-hydroxy friedelane isolated from *Maytenus robusta* (Celastraceae). *Arch. Pharm. Res.* **2008**, *31*, 41–46. [CrossRef]
265. Ferreira, F.L.; Rodrigues, V.G.; Silva, F.C.; Matildes, B.L.G.; Takahashi, J.A.; Silva, G.D.F.; Duarte, L.P.; Oliveira, D.M.; Filho, S.A.V. *Maytenus distichophylla* and *Salacia crassifolia*: Source of products with potential acetylcholinesterase inhibition. *Rev. Bras. Farmacogn.* **2017**, *27*, 471–474. [CrossRef]
266. Vieira Filho, S.A.; Duarte, L.P.; Silva, G.D.; de, F.; Mazaro, R.; Di Stasi, L.C. Constituintes químicos e atividade antiespermatogênica em folhas de *Austroplenckia populnea* (Celastraceae). *Rev. Bras. Farmacogn.* **2002**, *12*, 123–124. [CrossRef]
267. Torres-Romero, D.; King-Díaz, B.; Strasser, R.J.; Jiménez, I.A.; Lotina-Hennsen, B.; Bazzocchi, I.L. Friedelane triterpenes from *Celastrus vulcanicola* as photosynthetic inhibitors. *J. Agric. Food Chem.* **2010**, *58*, 10847–10854. [CrossRef]
268. Pereira, R.C.G.; Evangelista, F.C.G.; dos Santos Júnior, V.S.; de Paula Sabino, A.; Maltarollo, V.G.; de Freitas, R.P.; Duarte, L.P. Cytotoxic activity of triterpenoids from *Cheiloclinium cognatum* branches against chronic and acute leukemia cell Lines. *Chem. Biodivers.* **2020**, *17*, e2000773. [CrossRef]
269. Tu, G.H.; Shi, X.W.; Zhao, Y.; Zheng, C.D.; Gao, J.M. Triterpenes of *Euonymus alatus* and their cytotoxic activity; *Chem. Nat. Compound.* **2011**, *47*, 656–657. [CrossRef]
270. Wang, K.W.; Ju, X.Y.; Zhang, C.C.; Zhang, J.Y. Phytochemical and chemotaxonomic study on *Microtropis triflora*. *Biochem. Syst. Ecol.* **2014**, *52*, 1–3. [CrossRef]
271. Arciniegas, A.; Ramírez Apan, M.T.; Pérez-Castorena, A.L.; De Vivar, A.R. Anti-inflammatory constituents of *Mortonia greggii* Gray. *Zeitschrift fur Naturforsch. - Sect. C J. Biosci.* **2004**, *59*, 237–243. [CrossRef] [PubMed]
272. He, M.F.; Liu, L.; Ge, W.; Shaw, P.C.; Jiang, R.; Wu, L.W.; But, P.P.H. Antiangiogenic activity of *Tripterygium wilfordii* and its terpenoids. *J. Ethnopharmacol.* **2009**, *121*, 61–68. [CrossRef]
273. de Figueiredo, P.T.R.; Silva, E.W.R.; Cordeiro, L.V.; Barros, R.P.C.; Lima, E.; Scotti, M.T.; da Silva, M.S.; Tavares, J.F.; de, O.; Costa, V.C. Lupanes and friedelanes, the first chemical constituents of the aerial parts of *Maytenus erythroxylon* Reissek. *Phytochem. Lett.* **2021**, *45*, 19–24. [CrossRef]
274. Andrade, S.F.; Silva Filho, A.A.; Resende, D.D.O.; Silva, M.L.A.; Cunha, W.R.; Nanayakkara, N.P.D.; Bastos, J.K. Antileishmanial, antimalarial and antimicrobial activities of the extract and isolated compounds from *Austroplenckia populnea* (Celastraceae). *Z. Nat. C* **2008**, *63*, 497–502. [CrossRef]
275. Duarte, L.P.; Vieira Filho, S.A.; Silva, G.D.D.F.; De Sousa, J.R.; Pinto, A.D.S. Anti-trypanosomal activity of pentacyclic triterpenes isolated from *Austroplenckia populnea* (Celastraceae). *Rev. Inst. Med. Trop. Sao Paulo* **2002**, *44*, 109–112. [CrossRef]
276. Sousa, J.R.; Silva, G.D.F.; Miyakoshi, T.; Chen, C.L. Constituents of the root wood of *Austroplenckia populnea* Var. Ovata. *J. Nat. Prod.* **2006**, *69*, 1225–1227. [CrossRef]
277. Lindsey, K.L.; Budesinsky, M.; Kohout, L.; van Staden, J. Antibacterial activity of maytenonic acid isolated from the root-bark of *Maytenus senegalensis*. *S. Afr. J. Bot.* **2006**, *2006*. *72*, 473–477. [CrossRef]

278. Kamtcha, D.W.; Tene, M.; Bedane, K.G.; Knauer, L.; Strohmann, C.; Tane, P.; Kusari, S.; Spiteller, M. Cardenolides from the stem bark of *Salacia staudtiana*. *Fitoterapia* **2018**, *127*, 402–409. [CrossRef] [PubMed]

279. Aguilar, M.G.; Sousa, G.F.; Evangelista, F.C.G.; Sabino, A.P.; Vieira Filho, S.A.; Duarte, L.P. Imines and lactones derived from friedelanes and their cytotoxic activity. *Nat. Prod. Res.* **2020**, *34*, 810–815. [CrossRef] [PubMed]

280. Morales, S.A.T.; Aguilar, M.G.; Pereira, R.C.G.; Duarte, L.P.; Sousa, G.F.; de Oliveira, D.M.; Evangelista, F.C.G.; Sabino, A.P.; Viana, R.O.; Alves, V.S.; et al. Constituents from roots of *Maytenus distichophylla*, antimicrobial activity and toxicity for cells and *Caenorhabditis elegans*. *Quim. Nova* **2020**, *43*, 1066–1073. [CrossRef]

281. El Deeb, K.S.; Al-Haidari, R.A.; Mossa, J.S.; Ateya, A.M. Phytochemical and pharmacological studies of *Maytenus forsskaoliana*. *Saudi Pharm. J.* **2003**, *11*, 184–191.

282. Martucciello, S.; Balestrieri, M.L.; Felice, F.; Estevam, C.; dos, S.; Sant'Ana, A.E.G.; Pizza, C.; Piacente, S. Effects of triterpene derivatives from *Maytenus rigida* on VEGF-induced kaposi's sarcoma cell proliferation. *Chem. Biol. Interact.* **2010**, *183*, 450–454. [CrossRef] [PubMed]

283. Valladao, F.N.; De Miranda, R.R.S.; De Oliveira, G.S.; Silva, G.D.F.; Duarte, L.P.; Filho, S.A.V. Constituents of fruit pulp of *Maytenus salicifolia* and complete 1D/2D NMR data of 3β-hydroxy-D:B-friedo-olean-5-ene. *Chem. Nat. Compd.* **2010**, *46*, 686–691. [CrossRef]

284. Mokoka, T.A.; McGaw, L.J.; Mdee, L.K.; Bagla, V.P.; Iwalewa, E.O.; Eloff, J.N. Antimicrobial activity and cytotoxicity oftriterpenes isolated from leaves of *Maytenus undata* (Celastraceae). *BMC Complement. Altern. Med.* **2013**, *13*, 1–9. [CrossRef] [PubMed]

285. Silva, F.M.A.; Paz, W.H.P.; Vasconcelos, L.S.F.; da Silva, A.L.B.; da Silva-Filho, F.A.; de Almeida, R.A.; de Souza, A.D.L.; Pinheiro, M.L.B.; Koolen, H.H.F. Chemical constituents from *Salacia impressifolia* (Miers) A. C. Smith collected at the amazon rainforest. *Biochem. Syst. Ecol.* **2016**, *68*, 77–80. [CrossRef]

286. de Souza e Silva, S.R.; de Fátima Silva, G.D.; de Almeida Barbosa, L.C.; Duarte, L.P.; King-Diaz, B.; Archundia-Camacho, F.; Lotina-Hennsen, B. Uncoupling and inhibition properties of 3,4-*seco*-friedelan-3-oic acid isolated from *Maytenus imbricata*. *Pestic. Biochem. Physiol.* **2007**, *87*, 109–114. [CrossRef]

287. Yelani, T.; Hussein, A.A.; Meyer, J.J.M. Isolation and identification of poisonous triterpenoids from *Elaeodendron croceum*. *Nat. Prod. Res.* **2010**, *24*, 1418–1425. [CrossRef]

288. Oliveira, D.M.; Silva, G.D.F.; Duarte, L.P.; Vieira Filho, S.A. Chemical constituents isolated from roots of *Maytenus acanthophylla* Reissek (Celastraceae). *Biochem. Syst. Ecol.* **2006**, *34*, 661–665. [CrossRef]

289. Silva, T.M.; Carvalho, C.M.; Lima, R.A.; Facundo, V.A.; da Cunha, R.M.; Meneguetti, D.U.; de, O. Antibacterial activity of fractions and isolates of *Maytenus guianensis* Klotzsch Ex Reissek (Celastraceae) chichuá amazon. *Rev. Soc. Bras. Med. Trop.* **2018**, *51*, 533–536. [CrossRef] [PubMed]

290. Pina, E.S.; Silva, D.B.; Teixeira, S.P.; Coppede, J.S.; Furlan, M.; França, S.C.; Lopes, N.P.; Pereira, A.M.S.; Lopes, A.A. Mevalonate-derived quinonemethide triterpenoid from in vitro roots of *Peritassa laevigata* and their localization in root tissue by MALDI imaging. *Sci. Rep.* **2016**, *6*, 22627. [CrossRef] [PubMed]

291. Mohamad, T.A.S.T.; Naz, H.; Jalal, R.S.; Hussin, K.; Rahman, M.R.A.; Adam, A.; Weber, J.F.F. Chemical and pharmacognostical characterization of two malaysian plants both known as ajisamat. *Rev. Bras. Farmacogn.* **2013**, *23*, 724–730. [CrossRef]

292. Furukawa, M.; Furukawa, M.; Makino, M.; Uchiyama, T.; Fujimoto, Y.; Matsuzaki, K. New sesquiterpene pyridine alkaloids from *Hippocratea excelsa*. *Nat. Prod. Commun.* **2018**, *13*, 957–960. [CrossRef]

293. Taddeo, V.; Castillo, U.; Martínez, M.; Menjivar, J.; Jiménez, I.; Núñez, M.; Bazzocchi, I. Development and validation of an HPLC-PDA method for biologically active quinonemethide triterpenoids isolated from *Maytenus chiapensis*. *Medicines* **2019**, *6*, 36. [CrossRef]

294. Santos, V.A.F.F.M.; Leite, K.M.; Da Costa Siqueira, M.; Regasini, L.O.; Martinez, I.; Nogueira, C.T.; Galuppo, M.K.; Stolf, B.S.; Pereira, A.M.S.; Cicarelli, R.M.B.; et al. Antiprotozoal activity of quinonemethide triterpenes from *Maytenus ilicifolia* (Celastraceae). *Molecules* **2013**, *18*, 1053–1062. [CrossRef]

295. Mejía-Manzano, L.A.; Barba-Dávila, B.A.; Gutierrez-Uribe, J.A.; Escalante-Vázquez, E.J.; Serna-Saldívar, S.O. Extraction and isolation of antineoplastic pristimerin from *Mortonia greggii* (Celastraceae). *Nat. Prod. Commun.* **2015**, *10*, 1923–1928. [CrossRef]

296. Nizer, W.S.C.; Ferraz, A.C.; Moraes, T.; de, F.S.; Lima, W.G.; dos Santos, J.P.; Duarte, L.P.; Ferreira, J.M.S.; de Brito Magalhães, C.L.; Vieira-Filho, S.A.; et al. Pristimerin isolated from *Salacia crassifolia* (Mart. Ex Schult.) G. Don. (Celastraceae) roots as a potential antibacterial agent against *Staphylococcus aureus*. *J. Ethnopharmacol.* [CrossRef]

297. Ruphin, F.P.; Baholy, R.; Emmanue, A.; Amelie, R.; Martin, M.T.; Koto-te-Nyiwa, N. Antiplasmodial, cytotoxic activities and characterization of a new naturally occurring quinone methide pentacyclic triterpenoid derivative isolated from *Salacia leptoclada* Tul. (Celastraceae) originated from madagascar. *Asian Pac. J. Trop. Biomed.* **2013**, *3*, 780–784. [CrossRef]

298. Rodrigues-Filho, E.; Barros, F.A.P.; Fernandes, J.B.; Braz-Filho, R. Detection and identification of quinonemethide triterpenes in *Peritassa campestris* by Mass Spectrometry. *Rapid Commun. Mass Spectrom.* **2002**, *16*, 627–633. [CrossRef] [PubMed]

299. Silva, F.C.; Guedes, F.A.F.; Franco, M.W.; Barbosa, F.A.R.; Marra, C.A.; Duarte, L.P.; Silva, G.D.F.; Vieira-Filho, S.A. Algistatic effect of a quinonamethide triterpene on *Microcystis novacekii*. *J. Appl. Phycol.* **2013**, *25*, 1723–1728. [CrossRef]

300. Meneguetti, D.U.O.; Lima, R.A.; Hurtado, F.B.; Passarini, G.M.; Macedo, S.R.A.; de Barros, N.B.; Oliveira, F.A.D.S.; de Medeiros, P.S.D.M.; Militão, J.S.L.T.; Nicolete, R.; et al. Screening of the *in vitro* antileishmanial activities of compounds and secondary metabolites isolated from *Maytenus guianensis* Klotzsch Ex Reissek (Celastraceae) chichuá amazon. *Rev. Soc. Bras. Med. Trop.* **2016**, *49*, 579–585. [CrossRef]

301. Moujir, L.; López, M.R.; Reyes, C.P.; Jiménez, I.A.; Bazzocchi, I.L. Structural requirements for antimicrobial activity of phenolic *nor*-triterpenes from Celastraceae species. *Appl. Sci.* **2019**, *9*, 2957. [CrossRef]

302. León, L.; Beltrán, B.; Moujir, L. Antimicrobial activity of 6-oxophenolic triterpenoids. Mode of action against *Bacillus subtilis*. *Planta Med.* **2005**, *71*, 313–319. [CrossRef] [PubMed]

303. An, H.; Zhu, Y.; Xu, W.; Liu, Y.; Zhang, J.; Lin, Z. Evaluation of immunosuppressive activity of demethylzeylasteral in a beagle dog kidney transplantation model. *Cell Biochem. Biophys.* **2015**, *73*, 673–679. [CrossRef] [PubMed]

304. Mthethwa, N.S.; Oyedeji, B.A.O.; Obi, L.C.; Aiyegoro, O.A. *Anti-Staphylococcal*, anti-HIV and cytotoxicity studies of four south african medicinal plants and isolation of bioactive compounds from *Cassine transvaalensis* (Burtt. Davy) codd. *BMC Complement. Altern. Med.* **2014**, *14*, 1–9. [CrossRef]

305. Jeong, S.Y.; Zhao, B.T.; Kim, Y.H.; Min, B.S.; Woo, M.H. Cytotoxic and antioxidant compounds isolated from the cork Of *Euonymus alatus* Sieb. *Nat. Prod. Sci.* **2013**, *19*, 366–371.

306. Li, C.; Li, B.; Ye, J.; Zhang, W.; Shen, Y.; Yin, J. A new norditerpenoid from *Euonymus grandiflorus* Wall. *Nat. Prod. Res.* **2013**, *27*, 1716–1721. [CrossRef] [PubMed]

307. Pimenta, A.A.; De Souza, S.S.R.; De Fátima Silva, G.D.; De Almeida Barbosa, L.C.; Ellena, J.; Doriguetto, A.C. A Pentacyclic triterpene from *Maytenus imbricata*: Structure elucidation by X-ray crystallography. *Struct. Chem.* **2006**, *17*, 149–153. [CrossRef]

308. Chander, M.P.; Kumar, K.V.; Shriram, A.N.; Vijayachari, P. Anti-leptospiral activities of an endemic plant *Glyptopetalum calocarpum* (Kurz.) prain used as a medicinal plant by nicobarese of andaman and nicobar islands. *Nat. Prod. Res.* **2015**, *29*, 1575–1577. [CrossRef] [PubMed]

309. Bhavita, D.; Lakshmi, B.; Zaveri, M. Phytochemical investigation, isolation and characterization of betulin from leaf of *Gymnosporia montana*. *Int. J. Pharm. Sci. Res.* **2021**, *12*, 1752–1756. [CrossRef]

310. Alarcon, J.; Cespedes, C.L. Triterpenes and β-agarofurane sesquiterpenes from tissue culture of *Maytenus boaria*. *Bol. Latinoam. Caribe Plantas Med. Aromat.* **2016**, *15*, 206–214.

311. Moo-Puc, J.A.; Martín-Quintal, Z.; Mirón-López, G.; Moo-Puc, R.E.; Quijano, L.; Mena-Rejón, G.J. Isolation and antitrichomonal activity of the chemical constituents of the leaves of *Maytenus phyllanthoides* Benth. (Celastraceae). *Quim. Nova* **2014**, *37*, 85–88. [CrossRef]

312. Herrera-España, A.D.; Mena-Rejón, G.J.; Hernández-Ortega, S.; Quijano, L.; Mirón-Lópeza, G. Crystal structure of ochraceolide a isolated from *Elaeodendron trichotomum* (Turcz.) Lundell. *Acta Crystallogr. Sect. E Crystallogr. Commun.* **2017**, *73*, 1475–1478. [CrossRef] [PubMed]

313. Zhao, Q.; Li, H.M.; Chen, X.Q.; Li, R.T.; Liu, D. Terpenoids from *Tripterygium hypoglaucum* and their anti-inflammatory activity. *Chem. Nat. Compd.* **2018**, *54*, 471–474. [CrossRef]

314. Chen, X.Q.; Zan, K.; Wang, Q. Olean-type triterpenes of *Celastrus orbiculatus*. *Biochem. Syst. Ecol.* **2012**, *44*, 338–340. [CrossRef]

315. Doriguetto, A.C.; Duarte, L.P.; Ellena, J.A.; Silva, G.D.F.; Mascarenhas, Y.P.; Cota, A.B. 3-Oxoolean-12-en-20-yl α-methylcarboxylate. *Acta Crystallogr. Sect. E Struct. Rep. Online* **2003**, *59*, 164–166. [CrossRef]

316. Khan, H.; Khan, I.; Shah, M.; Shahidullah, A.; Khalil, N.; Aziz, A.; Wajid, A.; Jan, H.; Rahman, I. Antibacterial and antifungal activities of compounds isolated from *Gymnosporia royleana* (Celastraceae). *West Indian Med. J.* **2017**, *68*, 136–141. [CrossRef]

317. Singha, S.; Yotmanee, P.; Yahuafai, J.; Siripong, P.; Prabpai, S.; Sutthivaiyakit, S. Siphonagarofurans A-J: Poly-o-acylated β-dihydroagarofuran sesquiterpenoids from the fruits of *Siphonodon celastrineus*. *Phytochemistry* **2020**, *174*, 112345. [CrossRef]

318. Chang, X.; Wang, Z.Y.; Chen, X.; Ma, Y.N.; Zhang, H.Y.; Zhao, T.Z. Two new sesquiterpene pyridine alkaloids from root barks of *Celastrus angulatus*. *J. Asian Nat. Prod. Res.* **2019**, *21*, 1043–1051. [CrossRef] [PubMed]

319. Núñez, M.J.; Ardiles, A.E.; Martínez, M.L.; Torres-Romero, D.; Jiménez, I.A.; Bazzocchi, I.L. Unusual D:B-Friedobaccharane and oxygenated friedelane-type triterpenoids from *Salvadorean* Celastraceae Species. *Phytochem. Lett.* **2012**, *5*, 244–248. [CrossRef]

320. Yan, L.H.; Liu, X.Q.; Zhu, H.; Xu, Q.R.; Wang, W.M.; Zhang, S.M.; Zhang, Q.W.; Zhang, S.S.; Wang, Z.M. Chemical Constituents of *Euonymus fortunei*. *J. Asian Nat. Prod. Res.* **2015**, *17*, 952–958. [CrossRef] [PubMed]

 *molecules*

Review

Naturally Occurring Chromone Glycosides: Sources, Bioactivities, and Spectroscopic Features

Yhiya Amen [1,2,†], Marwa Elsbaey [2,†], Ahmed Othman [1,3,†], Mahmoud Sallam [3,†] and Kuniyoshi Shimizu [1,*]

[1] Department of Agro-Environmental Sciences, Graduate School of Bioresources and Bioenvironmental Sciences, Kyushu University, Fukuoka 819-0395, Japan; yhiaamen@mans.edu.eg (Y.A.); ah.othman@azhar.edu.eg (A.O.)

[2] Department of Pharmacognosy, Faculty of Pharmacy, Mansoura University, Mansoura 35516, Egypt; marwaelsebaey1611@mans.edu.eg

[3] Department of Pharmacognosy, Faculty of Pharmacy, Al-Azhar University, Cairo 11371, Egypt; m.sallam@azhar.edu.eg

* Correspondence: shimizu@agr.kyushu-u.ac.jp; Tel.: +81-92-802-4675

† Authors contributed equally to this work.

Abstract: Chromone glycosides comprise an important group of secondary metabolites. They are widely distributed in plants and, to a lesser extent, in fungi and bacteria. Significant biological activities, including antiviral, anti-inflammatory, antitumor, antimicrobial, etc., have been discovered for chromone glycosides, suggesting their potential as drug leads. This review compiles 192 naturally occurring chromone glycosides along with their sources, classification, biological activities, and spectroscopic features. Detailed biosynthetic pathways and chemotaxonomic studies are also described. Extensive spectroscopic features for this class of compounds have been thoroughly discussed, and detailed ^{13}C-NMR data of compounds **1–192**, have been added, except for those that have no reported ^{13}C-NMR data.

Keywords: chromone glycosides; chemical structure; activity; benzo-γ-pyrone; ^{13}C-NMR data

Citation: Amen, Y.; Elsbaey, M.; Othman, A.; Sallam, M.; Shimizu, K. Naturally Occurring Chromone Glycosides: Sources, Bioactivities, and Spectroscopic Features. *Molecules* **2021**, *26*, 7646. https://doi.org/10.3390/molecules26247646

Academic Editor: Jacqueline Aparecida Takahashi

Received: 16 November 2021
Accepted: 13 December 2021
Published: 16 December 2021

Publisher's Note: MDPI stays neutral with regard to jurisdictional claims in published maps and institutional affiliations.

1. Introduction

Chromone glycosides are a class of secondary metabolites with various medicinal properties. They are widely distributed in many plant genera and, to a lesser extent, in some fungal species and other sources [1]. Several biological activities have been reported for various chromone glycosides. For example, aloesin and its analogues, from Aloe, are used in cosmetic preparations to treat hyperpigmentation induced by UV radiation, owing to their role in inhibition of tyrosinase enzyme [2,3]. Additionally, 8-[C-β-D-[2-O-(E)-cinnamoyl]glucopyranosyl]-2-[(R)-2-hydroxypropyl]-7-methoxy-5-methylchromone), isolated from certain Aloe species, was reported to have potent topical anti-inflammatory activity comparable to the effect of hydrocortisone without affecting thymus weight [3]. Macrolobin, from *Macrolobium latifolium*, has a remarkable acetylcholinesterase inhibitory activity with an IC_{50} value of 0.8 μM. Uncinosides A and B, isolated from the Chinese herbal medicine *Selaginella uncinata*, showed potent anti-RSV (respiratory syncytial virus) activity with IC_{50} values of 6.9 and 1.3 μg/mL. Taking into consideration the broad biological activities of chromone glycosides, this review summarizes the naturally occurring chromone glycosides and categorizes these compounds on their structural basis, in addition to their sources, bioactivities and spectroscopic features. Importantly, this review will shed more light toward the NMR features of chromone glycosides to help natural product researchers in the identification of various chemical structures. Scientific databases as SciFinder, PubMed, and Google Scholar were used to collect the relevant literature data.

2. Biosynthesis

Chromones are biosynthesized through the acetic acid pathway by the condensation of five acetate molecules. These compounds, generally, have a methyl group at C-2 and are oxygenated at C-5 and C-7 [4]. Pentaketide Chromone Synthase (PCS) is a key enzyme in the biosynthesis process that catalyzes the formation of a pentaketide chromone (5,7-dihydroxy-2-methylchrome) from five-step decarboxylative condensations of malonyl-CoA, followed by the Claisen cyclization reaction to form an aromatic ring. However, it is unclear whether the heterocyclic ring closure of the pentaketide chromone is enzymatic or not, because the ring closure can take place due to spontaneous Michael-like ring closure, as in the case of flavanone formation from chalcone in vitro. PCS also accepts acetyl-CoA, resulting from decarboxylation of malonyl-CoA, as a starter substrate, but it is a poor substrate for PCS [5]. The pentaketide chromone has been isolated from several plants and is known to be the biosynthetic precursor of the chromone derivatives with additional heterocyclic rings (e.g., furano-, pyrano- and oxepino-chromone glycosides). Scheme 1 ([6] with modifications) shows the sequence of steps utilized in the biosynthesis of these compounds, fully consistent with the biosynthetic rationale developed above. The key intermediate is 5,7-dihydroxy-2-methylchromone [5,6]. For many years, the cyclization had been postulated to involve an intermediate epoxide, such that nucleophilic attack of the phenol onto the epoxide group might lead to formation of either five-membered furan, six-membered pyran or the seven-membered oxepin heterocycles, as commonly encountered in natural products [6].

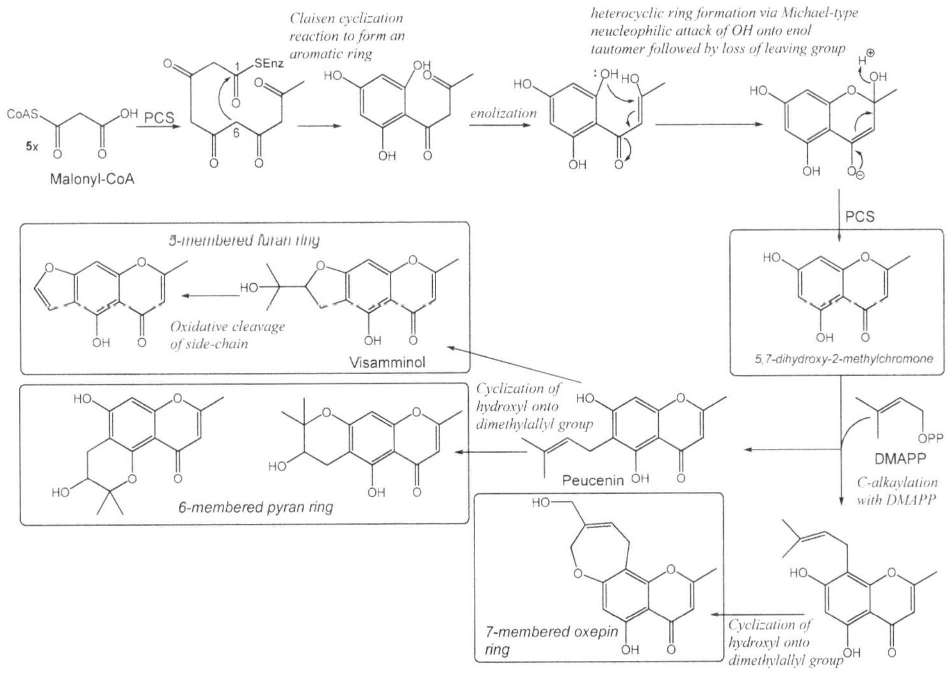

Scheme 1. Proposed mechanisms for the enzymatic formation of 5,7-dihydroxy-2-methylchromone and its derivatives.

Aloesone Synthase (ALS) (Scheme 2, [5] with modifications) is a key enzyme in the biosynthesis of heptaketide chromone aloesone derivatives, such as aloesone 7-*O*-β-D-glucopyranoside (**53**) in rhubarb and anti-inflammatory aloesone 8-*C*-β-D-glucopyranoside (aloesin, **98**) in Aloe (*A. arborescens*). ALS efficiently catalyzes the formation of a heptaketide aromatic pyrone 6-(2-(2,4-dihydroxy-6-methylphenyl)-2-oxoethyl)-4-hydroxy-2-

pyronechromone from acetyl-CoA and six molecules of malonyl-CoA through an aldol cyclization. The unstable heptaketide pyrone (or acid form) would then undergo subsequent spontaneous isomerization to the β-ketoacid chromone, which is followed by decarboxylation to produce the heptaketide aloesone [5].

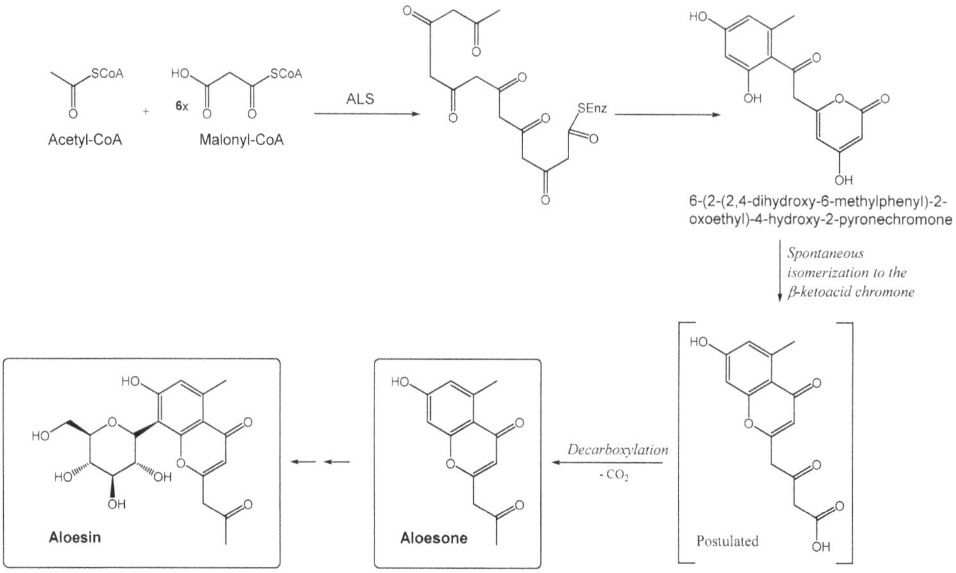

Scheme 2. Proposed mechanisms for the enzymatic formation of aloesone and its derivatives.

3. Taxonomy

We have reviewed the literature concerning the occurrence of chromone glycosides, and we have found that 192 different chromone glycosides have been isolated from different natural sources, including angiosperms, ferns, lichens, fungi and actinobacteria (Table 1). The occurrence of chromone glycosides is mostly confined to botanical families: Apiaceae, Fabaceae, Myrtaceae, Asphodelaceae, Ranunculaceae, Rubiaceae, Hypericaceae, Ericaceae, Amaryllidaceae, Polygonaceae and Araceae. However, few chromone glycosides are also present in Asteraceae, Eucryphiaceae, Saxifragaceae, Smilacaceae, Pentaphylaceae, Salicaceae, Meliaceae, Euphorbiaceae, Staphyleaceae, Amaranthaceae, Aquifoliaceae, Rosaceae, Bignoniaceae, Olacaceae, Pinaceae, Selaginellaceae, Gentianaceae, Cannabaceae, Euphorbiaceae, Cucurbitaceae, Thymelaeaceae and Poaceae. Many of the naturally occurring chromone-8-C-glycosides such as the well-known chromone glycoside aloesin (**98**) were reported from genus *Aloe*. Until now, chromone glycosides with additional heterocyclic moieties such as pyrano-, oxepino- and pyrido-chromone glycosides were only isolated from *Saposhnikovia divaricate*, *Eranthis* species and *Schumanniophyton magnificum*, respectively. Another interesting category comprises hybrids of furano-chromones with cycloartane triterpenes, which were reported from *Cimicifuga foetida*. Actinobacteria also constitute an important source of the chromone alkaloid aminoglycosides, which are isolated from *Streptomyces*, *Saccharothrix* and *Actinomycete* species.

Table 1. The distribution of chromone glycosides reported through this review.

		Family	Genus	Species	Compounds
Plants (Angiosperms)	1	Ericaceae	*Rhododendron*	*ovatum*	**2**
				spinuliferum	**3**
				collettianum	**55**
			Calluna	*vulgaris*	**13**
	2	Rubiaceae	*Schumanniophyton*	*magnificum*	**7, 25, 156**
			Knoxia	*corymbosa*	**20, 24, 35, 36**
			Adina	*rubescens*	**20**
			Neonauclea	*sessilifolia*	**26, 166**
	3	Amaryllidaceae	*Gethyllis*	*ciliaris*	**10**
			Pancratium	*biflorum*	**20, 66**
				maritimum	**20, 47**
	4	Polygonaceae	*Polygonum*	*capitatum*	**15**
			Rheum	*austral*	**50**
				sp.	**52, 53**
	5	Apiaceae	*Ammi*	*visnaga*	**20, 140, 146**
			Peucedanum	*austriacum*	**20**
				japonicum	**149**
			Cnidium	*monnieri*	**57, 58, 123–130**
				juponicum	**123, 124**
			Bupleurum	*chinense*	**58**
			Angelica	*archangelica*	**125**
				genuflexa	**146, 149**
				japonica	**146, 149**
			Archangelica	*litoralis*	**125**
			Saposhnikovia	*divaricata*	**143–146, 149, 150**
			Ledebouriella	*seseloides*	**144**
			Diplolophium	*buchananii*	**144, 146, 151**
			Sphallerocarpus	*gracilis*	**144**
			Glehnia	*littoralis*	**149**
	6	Hypericaceae	*Hypericum*	*henryi*	**22, 23**
				erectum	**22, 38**
				sikokumontanum	**38, 39, 60, 61**
				japonicum	**82, 83**
	7	Ranunculaceae	*Delphinium*	*hybridum*	**28**
			Cimicifuga	*heracleifolia*	**141**
				foetida	**146, 147, 157–165**
			Eranthis	*hyemalis*	**131, 132, 152, 153, 154**
				cilicica	**133, 134, 153, 155**

Table 1. *Cont.*

		Family	Genus	Species	Compounds
Plants (Angiosperms)	8	Myrtaceae	*Myrtus*	*communis*	**34**
			Syzygium	*aromaticum*	**66, 71, 80, 86**
			Baeckea	*frutescens*	**66, 67, 70, 72, 85, 87**
			Kunzea	*ambigua*	**67, 70, 72–74, 80, 85, 88, 89**
			Eucalyptus	*globulus*	**80**
				maidenii	**82, 83**
				grandis	**83**
				urograndi	**83**
	9	Fabaceae	*Cassia*	*multijuga*	**51, 77**
				siamea	**59**
				obtusifolia	**68, 69**
				spectablis	**78**
				obtusifolia	**84**
			Macrolobium	*latifolium*	**64**
			Aspalathus	*linearis*	**65**
			Abrus	*mollis*	**80**
			Ononis	*vaginalis*	**135**
	10	Asphodelaceae	*Aloe*	*vera*	**75, 76, 90–98, 103, 104–110, 112–117, 120–122**
				barbadensis	**97, 98, 103, 107, 108**
				rupestris	**99**
				cremnophila	**101**
				nobilis	**111, 118, 119**
	11	Araceae	*Scindapsus*	*officinalis*	**18–20, 29, 31, 32, 56, 57**
	12	Asteraceae	*Mutisia*	*acuminate*	**1**
	13	Eucryphiaceae	*Eucryphia*	*cordifolia*	**4**
	14	Saxifragaceae	*Astilbe*	*thunbergii*	**4**
	15	Smilacaceae	*Smilax*	*glabra*	**4**
	16	Pentaphylaceae	*Eurya*	*japonica*	**5**
	17	Salicaceae	*Salix*	*matsudana*	**6**
	18	Meliaceae	*Dysoxylum*	*binectariferum*	**7**
	19	Euphorbiaceae	*Acalypha*	*fruticose*	**7**
	20	Staphyleaceae	*Staphylea*	*bumalda*	**7**
	21	Amaranthaceae	*Salicornia*	*europaea*	**12**
	22	Aquifoliaceae	*Ilex*	*hainanensis*	**13**
	23	Rosaceae	*Dasiphora*	*parvifolia*	**16, 17**
	24	Bignoniaceae	*Tecomella*	*undulata*	**20, 27**
	25	Olacaceae	*Scorodocarpus*	*borneensis*	**26**
	26	Pinaceae	*Pseudotsuga*	*sinensis*	**37**
	27	Selaginellaceae	*Selaginella*	*uncinata*	**46, 48**
	28	Gentianaceae	*Swertia*	*punicea*	**54**

Table 1. *Cont.*

		Family	Genus	Species	Compounds
Plants (Angiosperms)	29	Cannabaceae	*Humulus*	*lupulus*	**62**
	30	Euphorbiaceae	*Chrozophora*	*prostrata*	**79**
	31	Cucurbitaceae	*Cucumis*	*melo*	**136**
	32	Thymelaeaceae	*Aquilaria*	*sinensis*	**137, 139**
	33	Poaceae	*Imperata*	*cylindrical*	**138**
Ferns	1	Polypodiaceae	*Drynaria*	*fortunei*	**7, 28, 30, 32, 33, 57**
	2	Dryopteridaceae	*Dryopteris*	*fragrans*	**20, 21**
	3	Onocleaceae	*Matteuccia*	*intermedia*	**49**
Lichens			*Roccellaria*	*mollis*	**41, 42, 45**
			Schismatomma	*accedens*	**41, 42, 45**
			Roccella	*galapagoensis*	**41, 42, 45**
			Lobodirina	*cerebriformis*	**44**
Fungi			*Armillaria*	*tabescens*	**11**
			Orbiocrella	sp.	**40, 43**
			Stemphylium	*botryosum*	**63**
Actinobacteria			*Streptomyces*	*phaeoverticillatus* var. *takatsukiensis*	**168**
				pluricolorescens	**169–171**
				sp.	**172–178, 180**
				griseoruber	**179**
			Saccharothrix	sp.	**181–183**
			Actinomycete		**184–192**

4. Chromone Glycosides

Chromone glycosides belong to a group of oxygen-containing heterocyclic compounds with a benzo-γ-pyrone skeleton. Naturally occurring chromone glycosides can be either *O*-glycosides or *C*-glycosides. For *O*-glycosides, the most frequently encountered group is the 7-*O*-glycosides; however, 2-, 3-, 5-, 8-, 11- and 13-*O*-glycosides also exist but to a lower extent. For an example, only one 6-*O*-glycoside **11** has been reported from nature, and from fungi, not higher plants [1]. Glycosylation can also be detected at side chains for chromones, at C-11 and C-12 as in compounds **56–59**, at the hydroxyprenyl and hydroxy-isoprenyl side chains as in **123** and **128**, respectively, or at the phenyl ethyl moiety as **139**. The most abundant among chromone glycosides is the glucoside from. However, other sugar moieties such as xylose, arabinose and rhamnose were also detected in 3-, 7- and 11-*O*-glycosides.

4.1. Chromone O-glycosides

4.1.1. 2-O-Glycosides

This category includes compound **1** (Figure 1), 2-hydroxy-5-methylchromone-β-D-glucopyranoside, isolated from the aerial parts of *Mutisia acuminata* var. *hirsuta*, a member of family Asteraceae [7]. The authors did not report biological activity for this compound.

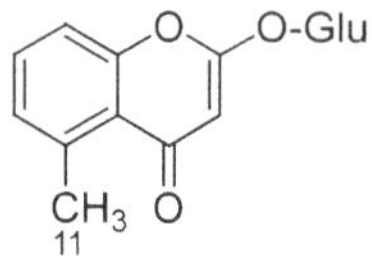

Figure 1. Structure of compound **1**.

4.1.2. 3-*O*-Glycosides

This category includes compounds **2–5**. They share the same aglycone nucleus but with different sugar moieties at C-3. Eucryphin **4** was reported as a new compound in 1979 [8]; however, it was reported again in 1996 as a new compound under the name smiglanin [9]. In addition, 3,5,7-trihydroxychromone 3-*O*-β-D-xylopyranoside **2** was first reported in 2005 from *Rhodadendron ovatum* [10], but it was reported again as a new compound in 2013 [11]. Compounds **2–5** are shown in Figure 2. The sources and the reported biological activities are summarized in Table 2.

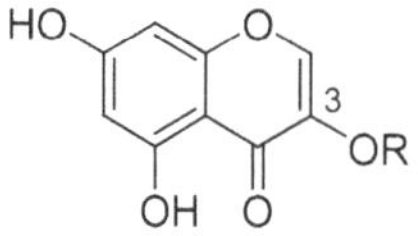

Figure 2. Structures of compounds **2–5**.

Table 2. 3-*O*-Chromone glycosides with their sources and biological activities.

No.	Compound	Source	Biological Activity
2	3,5,7-trihydroxychromone-3-*O*-β-D-xylopyranoside	*Rhododendron ovatum* roots [10] *Eurya japonica* stems [11]	Inhibitory effects on LPS (Lipopolysaccharide)-induced NO (Nitric Oxide) production with inhibition rate 36.24 ± 1.29% at 20 μg/mL [11]
3	3,5,7-trihydroxylchromone-3-*O*-α-L-arabinopyranoside	*Rhododendron spinuliferum* aerial parts [12]	Inhibition of NO production in LPS-stimulated RAW 264.7 cells with an IC$_{50}$ value more than 100 mM [12]
4	Eucryphin (5,7-dihydroxy-3-(α-*O*-L-rhamnopyranosyl)-4H-L-benzopyran-4-one)	*Eucryphia cordifolia* bark [8] *Astilbe thunbergii* rhizomes [13]	Norepinephrine-enhancing lipolytic effect 6.432 ± 0.014 FFA μmol/mL at 1000 μg [13] Enhancing effect on burn wound repair at 100 mg ointment per mouse [14]
	Smiglanin (3,5,7-trihydroxychromone-3-*O*-α-L-rhamnopyranoside)	*Smilax glabra* roots [9]	No reported biological activity
5	5,7-Dihydroxy-4H-chromen-4-one-3-*O*-β-D-glucopyranoside	*Eurya japonica* stems [11]	Inhibitory effects on LPS-induced NO production with inhibition rate 53.79 ± 1.78% at 20 μg/mL [11]

4.1.3. 5-*O*-Glycosides

Among the naturally occurring 5-*O*-glycosides, Staphylosides A and B (**8–9**), isolated from *Staphylea bumalda*, are characterized by a presence of a disaccharide moiety attached to C-5. The disaccharide chain in **8** is β-D-glucopyranosyl-(1→6)-β-D-glucopyranoside while in **9**, is α-D-glucopyranosyl-(1→6)-β-D-glucopyranoside. Compounds **6–10** are shown in Figure 3. The sources and the reported biological activities (if any) are summarized in Table 3.

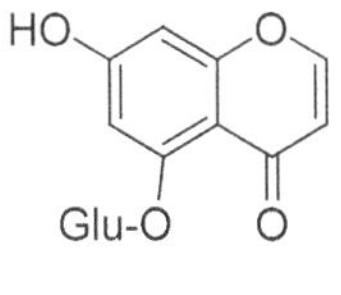

6

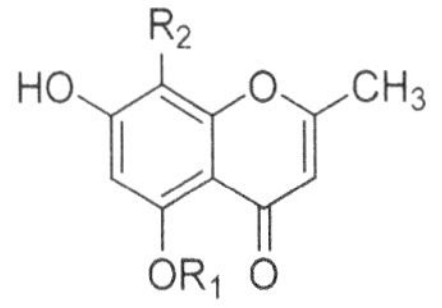

	R_1	R_2
7	Glu	H
8	Glu (6'→1'') β-Glu	H
9	Glu (6'→1'') α-Glu	H
10	Glu	CH_3

Figure 3. Structures of compounds **6**–**10**.

Table 3. 5-*O*-Chromone glycosides with their sources and biological activities.

No.	Compound	Source	Biological Activity
6	Matsudoside A (5-β-D-glucosyloxy-7-hydroxychromone)	*Salix matsudana* leaves [15]	No reported biological activity
7	Schumaniofioside A (2-methyl-5,7-dihydroxychromone 5-*O*-β-D-glucopyranoside)	*Schumanniophyton magnificum* root bark [16] *Dysoxylum binectariferum* fruits [17]. *Acalypha fruticose* aerial parts [18,19] *Drynaria fortune* rhizomes [13]	Inhibition of proinflammatory cytokines TNF-α (39.51 ± 1.21%) and IL-6 (22.21 ± 0.58%) at 5 μM [17] Inhibition of NF-kB transcriptional activity and iNOS with IC_{50} value of 29.5 ± 6.5 μg/mL [19]
8	Staphyloside A	*Staphylea bumalda* leaves [20]	No reported biological activity
9	Staphyloside B		
10	Isoeugenitol glucoside	*Gethyllis ciliaris* underground parts [21]	No reported biological activity

4.1.4. 6-*O*-Glycosides

Compound **11** (Figure 4) has a unique structure for bearing 4-*O*-methylglucopyranosyl unit. Chemically, it is 6-*O*-(4-*O*-methyl-β-D-glucopyranosyl)-8-hydroxy-2,7-dimethyl-4H-benzopyran-4-one, isolated from the rice culture of the fungus *Armillaria tabescens* [1]. Although such compounds are not common in higher plants, several of them have previously been isolated from fungi [1].

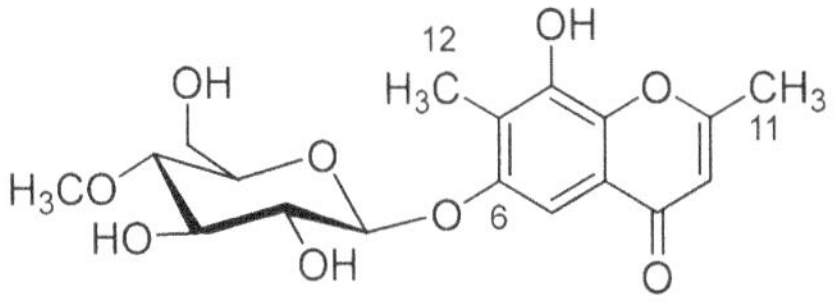

11

Figure 4. Structure of compound **11**.

4.1.5. 7-*O*-Glycosides

This subclass is characterized by the presence of sugar at C-7. Hyperimone A is the same as Urachromone A (**22**), reported at nearly the same time from different co-authors from the genus Hypericum. Takanechromone A (**38**) is the same as Hyperimone B, isolated from the same genus by different co-authors. They were reported each time as new compounds. We preferred to add only ^{13}C-NMR data of one set of these compounds (Table 23). Several biological activities have been reported to some members of this subclass. Compounds **12**–**53** are shown in Figure 5. The sources and the reported biological activities (if any) are summarized in Table 4.

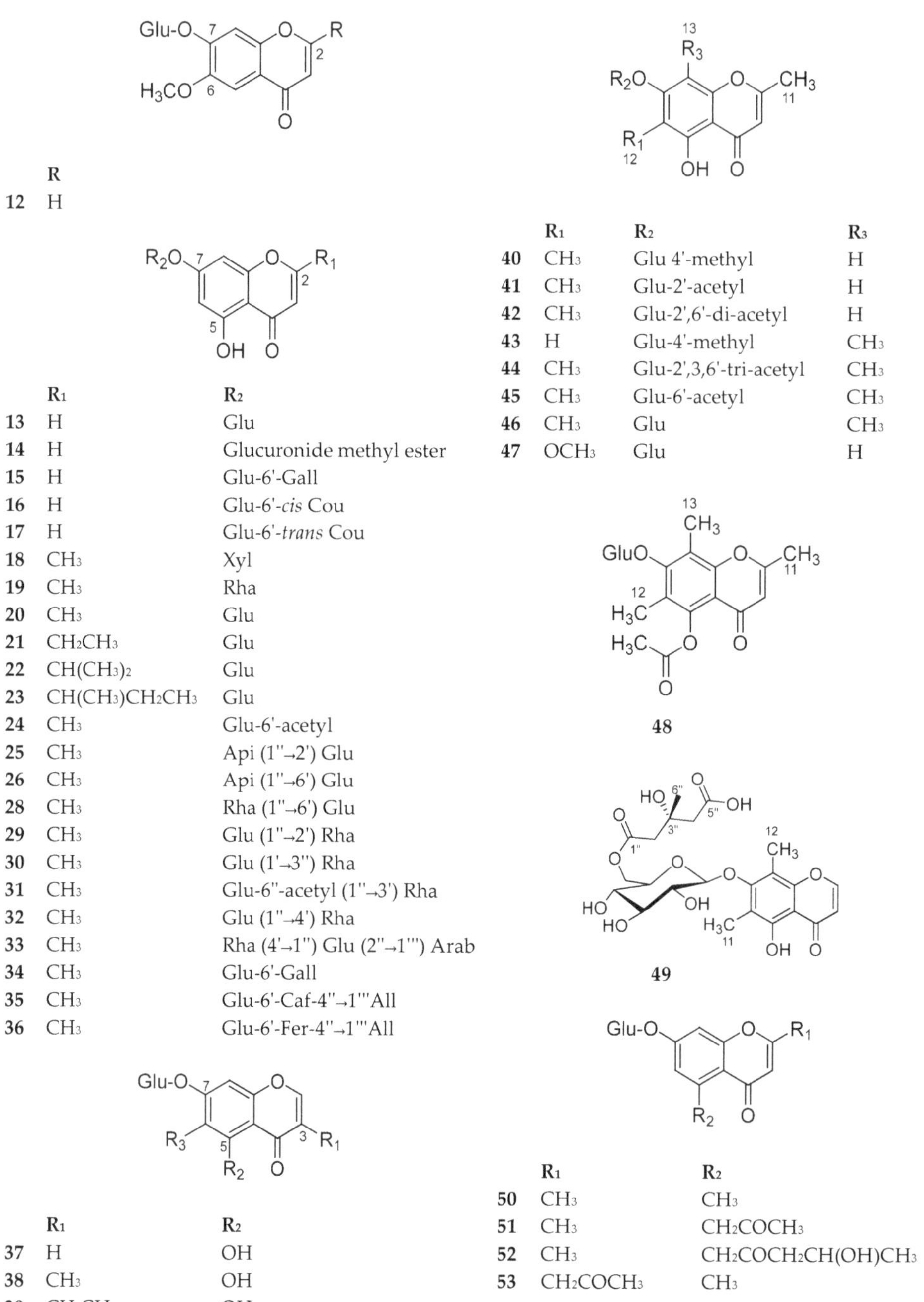

Figure 5. Structures of compounds **12–53**.

Table 4. 7-*O*-Chromone glycosides with their sources and biological activities.

No.	Compound	Source	Biological Activity
12	7-*O*-β-D-glucopyranosyl-6-methoxychromone	*Salicornia europaea* leaves and stems [25]	No reported biological activity
13	5,7-dihydroxychromone 7-*O*-β-D-glucopyranoside	*Ilex hainanensis* leaves [26] *Calluna vulgaris* flowers [27]	No reported biological activity
14	5,7-dihydroxychromone 7-*O*-β-D-glucuronide methyl ester	*Davallia mariesii* rhizomes [28]	No reported biological activity
15	7-*O*-(6′-galloyl)-β-D-glucopyranosyl-5-hydroxychromone	*Polygonum capitatum* aerial parts [29]	No reported biological activity
16	5-hydroxy-7-*O*-(6-*O*-*p*-*cis*-coumaroyl-β-D-glucopyranosyl)-chromone	*Dasiphora parvifolia* aerial parts [30]	No reported biological activity
17	5-hydroxy-7-*O*-(6-*O*-*p*-*trans*-coumaroyl-β-D-glucopyranosyl)-chromone	*Dasiphora parvifolia* aerial parts [30]	No reported biological activity
18	Officinaliside A	*Scindapsus officinalis* stems [31]	No reported biological activity
19	7-*O*-α-L-rhamnosyl-nereugenin	*Scindapsus officinalis* stems [31]	No reported biological activity
20	Undulatoside A (2-methyl-5,7-dihydroxychromone 7-*O*-β-D-glucopyranoside)	*Scindapsus officinalis* stems [31] *Ammi visnaga* fruits [32] *Knoxia corymbosa* [33] *Pancratium biflorum* roots [34] *Panacratium biflorum* flowering bulbs [35] *Pancratium maritimum* L. fresh bulbs [36] *Peucedanum austriacum* [37] *Tecomella undulata* bark [38] *Adina rubescens* leaves [39] *Dryopteris fragrans* [40] *Staphylea bumalda* leaves [20]	Immunomodulatory activity inhibited the proliferation of murine B lymphocytes in vitro at 10^{-5} M [33] Inhibition of nitric oxide production in lipopolysaccharide induced RAW 264.7 macrophages with an IC_{50} value of 49.8 μM [40] Weak antimigratory activity against human metastatic prostate cancer cells (PC-3M) at 50 μM [36]
21	Frachromone C (5-hydroxy-2-ethylchromone-7-*O*-β-D-glucopyranoside)	*Dryopteris fragrans* whole plant [40]	Inhibition of nitric oxide production in lipopolysaccharide induced RAW 264.7 macrophages with an IC_{50} value of 45.8 μM [40]
22	Urachromone A (5-hydroxy-2-isopropylchromone-7-*O*-β-D-glucopyranoside) Hyperimone A	*Hypericum henryi* aerial parts [41] *Hypericum erectum* [42]	No reported biological activity
23	Urachromone B	*Hypericum henryi* aerial parts [41]	No reported biological activity
24	Corymbosin K_2 (7-*O*-β-D-6-acetylglucopyranosyl-5-hydroxy-2-methylchromone)	*Knoxia corymbosa* [33]	Immunomodulatory activity inhibited the proliferation of murine B lymphocytes in vitro at 10^{-5} M [33]
25	Schumanniofioside B	*Schumanniophyton magnificum* root bark [16]	No reported biological activity
26	5-hydroxy-2-methylchromone-7-*O*-β-D-apiofuranosyl-(1→6)-β-D-glucopyranoside	*Neonauclea sessilifolia* roots [43] *Scorodocarpus borneensis* leaves [44] *Staphylea bumalda* leaves [20]	No reported biological activity

Table 4. *Cont.*

No.	Compound	Source	Biological Activity
27	Undulatoside B	*Tecomella undulata* [45]	No reported biological activity
28	2-methyl-chromone-5,7-diol 7-*O*-α-L-rhamnopyranosyl-(1-6)-β-D-glucopyranoside	*Delphinium hybridum* aerial parts [46] *Drynaria fortunei* rhizomes [22]	No reported biological activity
29	Officinaliside C (7-*O*-[β-D-glucopyranosyl-(1-2)-α-L-rhamnopyranosyl]-5-hydroxy-2-methyl-4H-1-benzopyran-4-one)	*Scindapsus officinalis* stems [31]	No reported biological activity
30	Drynachromoside C (5-hydroxy-2-methyl chromone-7-*O*-β-D-glucopyranosyl (1-3)-α-L-rhamnopyranoside)	*Drynaria fortunei* rhizomes [22]	Inhibitory activity on triglyceride accumulation at 10 μM [22]
31	Officinaliside B (7-*O*-[6-acetyl-β-D-glucopyranosyl-(1-3)-α-L-rhamnopyranosyl]-5-hydroxy-2-methyl-4H-1-benzopyran-4-one)	*Scindapsus officinalis* stems [31]	Inhibition of NO production in LPS-stimulated RAW 264.7 cells with an IC_{50} value of 16.1 μM [31]
32	Drynachromoside A (5-hydroxy-2-methyl-4H-benzopyran-4-one-7-*O*-β-D-glucopyranosyl-(1-4)-α-L-rhamnopyranoside)	*Scindapsus officinalis* stems [31] *Drynaria fortunei* rhizomes [47]	Proliferative activity 10.1% on MC3T3-E1 (Mouse osteoblast) cells at 25 μg/mL [47]
33	Drynachromoside D (5-hydroxy-2-methyl chromone-7-*O*-α-L-arabinopyranosyl(1-2)-β-D-glucopyranosyl(1-4)-α-L-rhamnopyranoside)	*Drynaria fortunei* rhizomes [22]	Inhibitory activity on triglyceride accumulation (inhibited PPARγ, C/EBPα and aP2 expression by 50%, 43% and 37% at 10 mM) [22]
34	Undulatoside A 6′-*O*-gallate	*Myrtus communis* leaves [48]	
35	Corymbosin K3 (7-*O*-[6-*O*-(4-*O*-*trans*-caffeoyl-β-D-allopyranosyl)]-β-D-glucopyranosyl-5-hydroxy-2-methylchromone)	*Knoxia corymbosa* [33]	Immunomodulatory activity inhibited the proliferation of murine B lymphocytes *in vitro* at 10^{-5} M [33]
36	7-*O*-[6-*O*-(4-*O*-*trans*-feruloyl-β-D-allopyranosyl)]-β-D-glucopyranosyl-5-hydroxy-2-methylchromone	*Knoxia corymbosa* [33]	No reported biological activity
37	5-hydroxy-6-methylchromone-7-*O*-β-D-glucopyranoside	*Pseudotsuga sinensis* [49]	No reported biological activity
38	Takanechromone A (5,7-dihydroxy-3-methylchromone-7-*O*-β-D-glucopyranoside) Hyperimone B	*Hypericum sikokumontanum* aerial parts [50] *Hypericum erectum* [42]	No reported biological activity
39	Takanechromone B (5,7-dihydroxy-3-ethylchromone-7-*O*-β-D-glucopyranoside)	*Hypericum sikokumontanum* aerial parts [50]	No reported biological activity

Table 4. *Cont.*

No.	Compound	Source	Biological Activity
40	7-*O*-(4-*O*-Methyl-β-D-glucopyranosyl)eugenitol	The scale-insect pathogenic fungus *Orbiocrella* sp. [23]	No reported biological activity
41	Mollin	Lichens (*Roccellaria mollis, Schismatomma accedens, Roccella galapagoensis*) [51]	No reported biological activity
42	Roccellin	Lichens (*Roccellaria mollis, Schismatomma accedens, Roccella galapagoensis*) [51]	No reported biological activity
43	7-*O*-(4-*O*-Methyl-β-D-glucopyranosyl)isoeugenitol	The scale-insect pathogenic fungus *Orbiocrella* sp. [23]	No reported biological activity
44	Lobodirin	*Lobodirina cerebriformis* lichen [51]	No reported biological activity
45	Galapagin	Lichens (*Roccellaria mollis, Schismatomma accedens, Roccella galapagoensis*) [51]	No reported biological activity
46	Uncinoside A (5-hydroxy-2,6,8-trimethylchromone 7-*O*-β-D-glucopyranoside)	*Selaginella uncinate* Herb [24]	Antiviral activity against respiratory syncytial virus (RSV) with an IC_{50} value of 6.9 μg/mL, against parainfluenza type 3 virus (PIV 3) with an IC_{50} value of 13.8 μg/mL [24]
47	Pancrichromone	*Pancratium maritimum* L. fresh bulbs [36]	No reported biological activity
48	Uncinoside B (5-acetyoxyl-2,6,8-trimethylchromone 7-*O*-β-D-glucopyranoside)	*Selaginella uncinate* herb [24]	Antiviral activity against respiratory syncytial virus (RSV) with an IC_{50} value of 1.3 μg/mL, against parainfluenza type 3 virus (PIV 3) with an IC_{50} value of 20.8 μg/mL [24]
49	Matteuinterin B	*Matteuccia intermedia* rhizomes [52]	
50	2,5-dimethylchromone-7-*O*-β-D-glucopyranoside	*Rheum austral* D. Don underground parts [53] *Rumex gmelini* Turcz. roots [54]	Anti-oxidant activity (DPPH radical scavenging capacity with an IC_{50} value of 66.9 ± 1.3 μM) [53]
51	5-acetonyl-7-β-D-glucopyranosyl-2-methylchromone	*Cassia multijuga* leaves [55,56]	No reported biological activity
52	2-methyl-5-(2′-oxo-4′-hydroxyphenyl)-7-hydroxychromone 7-*O*-β-D-glucopyranoside	Chinese rhubarb (Rhei Rhizoma) [57]	No reported biological activity
53	Aloesone 7-*O*-β-D-glucopyranoside	Chinese rhubarb (Rhei Rhizoma) [57]	No reported biological activity

Drynachromosides C (**30**) and D (**33**) exhibited inhibitory activity on triglyceride accumulation [22]. The effects of these compounds on mRNA expression of the three adipogenesis-related marker genes, PPARγ, C/EBPα and Ap2, in 3T3-L1 were investigated. The mRNA expression levels of PPARγ, C/EBPα and Ap2 were found to be dramatically downregulated. Compounds **40** and **43**, having a unique sugar unit of 4-*O*-methyl-β-D-glucopyranose, were isolated from the scale-insect pathogenic fungus *Orbiocrella* sp. BCC 33248 [23]. Uncinosides A (**46**) and B (**48**) [24], isolated from the Chinese herbal medicine *Selaginella uncinata*, showed potent anti-RSV (respiratory syncytial virus) activity with IC_{50} values of 6.9 and 1.3 μg/mL, respectively. Uncinoside B (**48**) was found to have a TI value of 64.0, a large therapeutic index comparable to that of ribavirin with a TI value of 24.0, which is an approved drug for the treatment of RSV infection in humans. They also showed moderate antiviral activities against PIV 3 (parainfluenza type 3 virus) with IC$_{50}$ values of 13.8 and 20.8 µg/mL and TI values of 6.0 and 4.0, respectively.

4.1.6. 8-*O*-Glycosides

Only two compounds **54–55** were reported in nature. They are shown in Figure 6. The sources and the reported biological activities (if any) are summarized in Table 5.

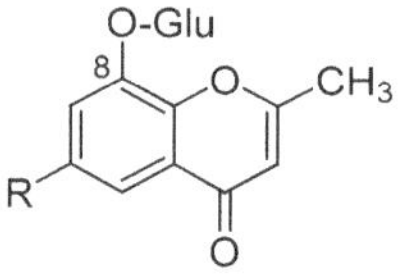

Figure 6. Structures of compounds **54–55**.

Table 5. 8-*O*-Chromone glycosides with their sources and biological activities.

No.	Compound	Source	Biological Activity
54	8-*O*-β-D-Glucopyranosyl-2-methylchromone	*Swertia punicea* whole herb [58]	No reported biological activity
55	8-*O*-β-D-Glucopyranosyl-6-hydroxy-2-methyl-4H-1-benzopyrane-4-one	*Rhododendron collettianum* aerial parts [59]	Inhibitory activity against tyrosinase enzyme with an IC$_{50}$ value of 256.97 µM [59]

4.1.7. 11- and 13-*O*-Glycosides

Compound **57** was reported in 2012 as Monnieriside A [60] and was then reported as Drynachromoside B [22,31,47]. Compounds **56–59** are shown in Figure 7. The sources and the reported biological activities (if any) are summarized in Table 6.

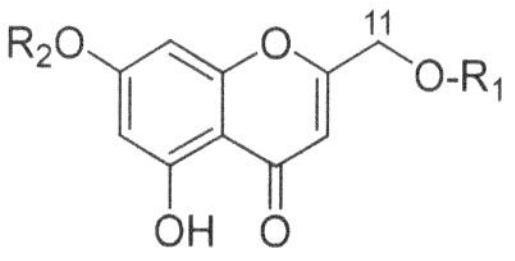

Figure 7. Structures of compounds **56–59**.

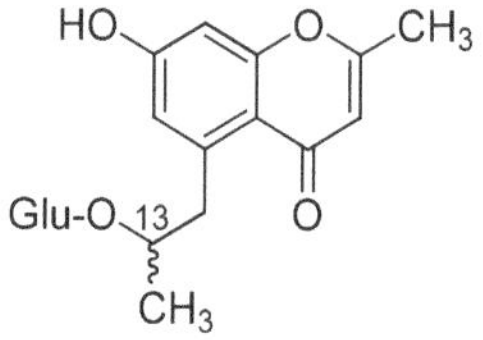

Table 6. 11, 13-*O*-chromone glycosides with their sources and biological activities.

No.	Compound	Source	Biological Activity
56	Officinaliside D (2-hydroxymethyl-5,7-dihydroxy-4H-benzopyran-4-one-1'-*O*-α-L-arabinopyranoside)	*Scindapsus officinalis* stems [31]	Inhibition of NO production in LPS-stimulated RAW 264.7 cells with an IC$_{50}$ value of 19.1 µM [31]
57	Drynachromoside B	*Drynaria fortune* rhizomes [22,47] *Scindapsus officinalis* stems [31]	Mild inhibitory activity against MC3T3-E1 (mouse osteoblast) cells at 3.125 to 100 µg/ml [47] Triglyceride accumulation inhibitory effect at 0.1 to 10 µM [22]
	Monnieriside A	*Cnidium monnieri* fruits [60]	No reported biological activity
58	Saikochromoside A	*Bupleurum chinense* [61] *Cnidium monnieri* fruits [60]	No reported biological activity
59	2-Methyl-5-propyl-7,12-dihydroxychromone-12-*O*-β-D-glucopyranoside	*Cassia siamea* stem [62]	No reported biological activity

4.1.8. Chromanone Glycosides

Chromanone glycosides or 2,3-dihydrochromone glycosides are not abundant in nature. Reviewing the literature, we encountered only four examples **60**, **61**, **62** and **63**. Their structures are shown in Figure 8. The sources and biological activities (if any) of these compounds are summarized in Table 7.

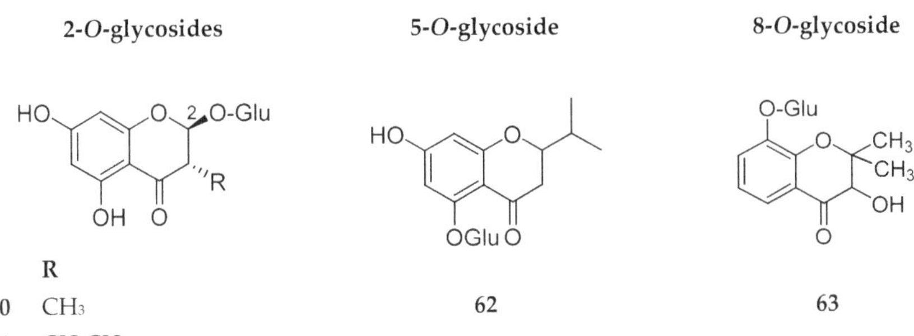

Figure 8. Structures of compounds **60–63**.

Table 7. Chromanone glycosides with their sources and biological activities.

No.	Compound	Source	Biological Activity
60	Takanechromanone A	*Hypericum sikokumontanum* aerial parts [50]	Anti-*Helicobacter pylori* at 100 µg/disc [50]
61	Takanechromanone B		
62	5-β-D-glucopyranosyloxy-7-hydroxy-2-isopropyl-chromanone	*Humulus lupulus* L. bracts [63]	No reported biological activity
63	Stemphylin (3-hydroxy-2, 2-dimethyl-5-α-D-glucopyranoside-2, 3-dihydrochromone)	The liquid culture of the fungus *Stemphylium botryosum* [64]	Phytotoxic activity [64]

4.2. Chromone C-Glycosides

In contrast to chromone *O*-glycosides, which are widely distributed and of common occurrence, C-glycoside derivatives are rarely found out.

4.2.1. 3-C-Glycosides

This subclass includes the unusual 5,7-dihydroxychromone-3α-D-C-glucoside, named macrolobin, isolated from the aerial parts of *Macrolobium latifolium* [65]. Its structure is shown in Figure 9. Its source and biological activities are summarized in Table 8.

Figure 9. Structure of compound **64**.

Table 8. 3-C-Chromone glycoside with its source and biological activities.

No.	Compound	Source	Biological Activity
64	Macrolobin (5,7-dihydroxychromone-3α-D-C-glucoside)	*Macrolobium latifolium* aerial parts [65]	Inhibition of acetylcholinesterase enzyme with an IC_{50} value of 0.8 μM Antimicrobial activity against *P. aeruginosa* and *Salmonella* at 0.73 and 0.44 μM, respectively [65]

4.2.2. 6-C-Glycosides

Compounds **65–79** are shown in Figure 10. The sources and the reported biological activities (if any) are summarized in Table 9.

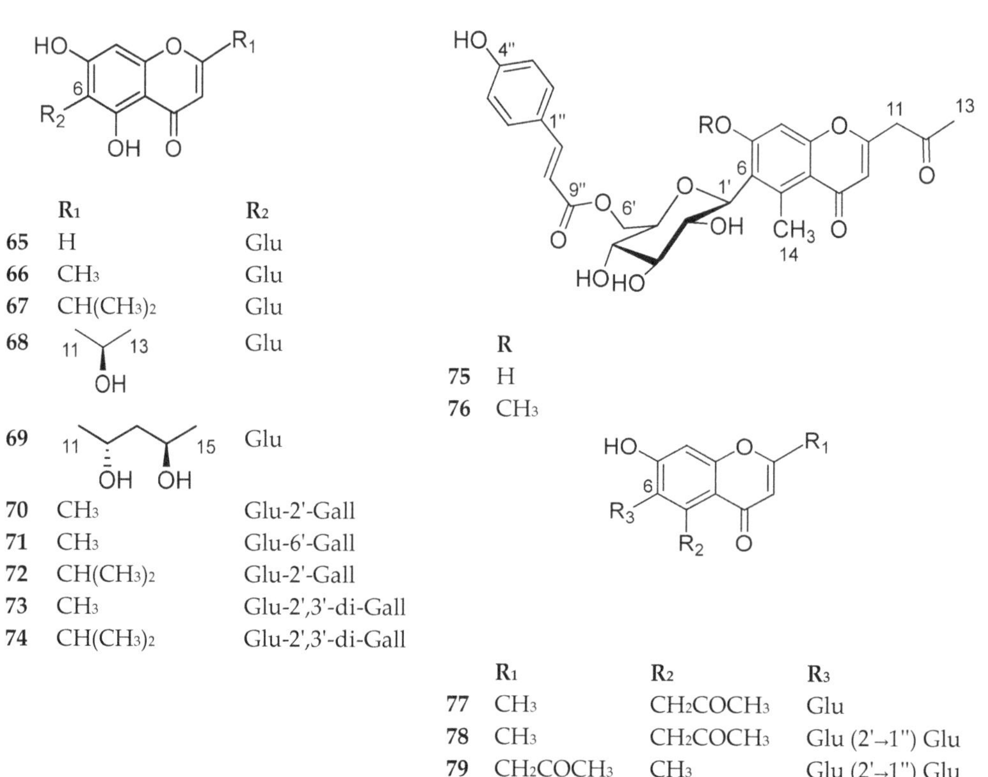

Figure 10 structure variable tables:

	R_1	R_2
65	H	Glu
66	CH₃	Glu
67	CH(CH₃)₂	Glu
68	(C11–C13 isopropyl-OH)	Glu
69	(C11–C15 diol)	Glu
70	CH₃	Glu-2'-Gall
71	CH₃	Glu-6'-Gall
72	CH(CH₃)₂	Glu-2'-Gall
73	CH₃	Glu-2',3'-di-Gall
74	CH(CH₃)₂	Glu-2',3'-di-Gall

	R
75	H
76	CH₃

	R_1	R_2	R_3
77	CH₃	CH₂COCH₃	Glu
78	CH₃	CH₂COCH₃	Glu (2'→1") Glu
79	CH₂COCH₃	CH₃	Glu (2'→1") Glu

Figure 10. Structures of compounds **65–79**.

Table 9. 6-C-Chromone glycosides with their sources and biological activities.

No.	Compound	Source	Biological Activity
65	5,7-dihydroxy-6-C-glucosyl-chromone	*Aspalathus linearis* fermented rooibos (red-brownish dry leaves) [66]	No reported biological activity
66	Biflorin (6-β-C-glucopyranosyl-5,7-dihydroxy-2-methylchromone)	*Pancratium Biflorum* roots [34] *Syzygium aromaticum* L. flower buds [67,68] *Baeckea frutescens* leaves [69,70]	Inhibitory activity to phosphodiesterase and spared cyclic nucleotides at 10^{-9} M [34] Inhibition of LPS-induced production of nitric oxide (NO) and prostaglandin E_2 (PGE2) in RAW 264.7 macrophages with IC_{50} values of 51.7 and 37.1 μM, respectively [67]
67	6-β-C-glucopyranosyl-5,7-dihydroxy-2-isopropylchromone	*Baeckea frutescens* leaves [69,70] *Kunzea ambigua* leaves [71]	Inhibitory activity 70.4% against EBV-EA (Epstein–Barr virus early antigen) activation induced by 12-O-tetradecanoylphorbol 13-acetate (TPA) at 500 mol ratio/TPA [71]
68	Obtusichromoneside C	*Cassia obtusifolia* seeds [72]	Weak inhibitory activity against human organic anion/cation transporters (OATs/OCTs) and organic anion transporting polypeptides (OATPs) at 50 μM [72]
69	Obtusichromoneside A		
70	Kunzeachromone C	*Kunzea ambigua* leaves [71] *Baeckea frutescens* leaves [70]	Inhibition of copper-induced LDL oxidation with an IC_{50} value of 3.35 ± 0.36 μM [70]
71	6-C-β-D-(6′-O-galloyl)glucosylnoreugenin	*Syzygium aromaticum* flower buds [68,73]	Cytotoxicity against human ovarian cancer cells (A2780) with an IC_{50} value of 66.78 ± 5.49 μM [68] Prolyl endopeptidase inhibitory effects with an IC_{50} value of 1.74 ± 0.03 μM [73]
72	6-β-C-(2′-O-galloylglucopyranosyl)-5,7-dihydroxy-2-isopropylchromone	*Baeckea frutescens* leaves [69,70] *Kunzea ambigua* leaves [71]	Inhibitory activity 68.4% against EBV-EA activation induced by TPA at 500 mol ratio/TPA [71] Inhibition of copper-induced LDL oxidation with an IC_{50} value of 3.90 ± 0.24 μM [70]
73	Kunzeachromone D	*Kunzea ambigua* leaves [71]	No reported biological activity
74	Kunzeachromone A		
75	Aloeveraside B	*Aloe vera* resin [74–76]	Inhibition of urease enzyme (55% and 62%, respectively) at 1 mg/mL concentration, significant growth inhibition (70.5 and 76.4%) of the breast cancer cell line MDA-MB-231 at 100 μM, and antioxidant (80% and 60%) at 1 mg/mL [74] Anti-lipid peroxidation activity with IC_{50} values of 432.1 ± 0.6 and 469.5 ± 0.4 μmol/L, respectively [75]
76	Aloeveraside A		
77	Acetonyl-6-glycosyl -7-hydroxy-2-methylchromone	*Cassia multijuga* leaves [55,56]	No reported biological activity
78	5-acetonyl-7-hydroxy-6-C-glucopyranosyl-2-methyl chromone 2″-O-glucopyranoside	*Cassia spectablis* seeds [77]	No reported biological activity
79	2-acetonyl-5-methyl-7-hydroxy-6-C-glucopyranosyl chromone 2″-O-glucopyranoside	*Chrozophora prostrata* roots [78]	No reported biological activity

4.2.3. 8-C-Glycosides

Many of the naturally occurring chromone-8-C-glycosieds can be found in genus *Aloe*. Approximately 26 chromone-8-C-glycosides were reported in the perennial plant *Aloe vera*, which is a well-known pharmaceutical herb used in traditional Chinese medicine [76]. Some significant bioactive chromone-8-C-glycosides were isolated and identified in *Aloe vera*, including Aloesin (**98**), aloeresin E (**109**), isoaloeresin D (**110**), aloeresin A (**114**) and other derivatives. For instance, aloeresin A (**114**) exhibited a promising therapeutic activity toward α-glucosidase enzyme [79], while the compound isobiflorin (**80**), isolated from the flower buds of *Syzygium aromaticum*, had the capacity to inhibit LPS-induced production of nitric oxide (NO) and prostaglandin E$_2$ (PGE2) in RAW 264.7 macrophages [67]. A chromone-8-C-glycoside, 5,7-dihydroxy-2-isopropylchromone-8-β-D-glucoside, reported in *Hypericum japonicum*, showed an activity against Epstein–Barr virus [71]. Additionally, BACE1 (β-secretase), which is a possible potential target in the treatment of Alzheimer's disease, was inhibited by some compounds as aloesin (**98**) [80], 7-*O*-methyl-aloeresin A (**115**) [81] and 2′-feruloyl-7-*O*-methylaloesin (**119**) [80]. Furthermore, tyrosinase, which is the key enzyme for controlling the production of melanin, was inhibited by aloeresin E (**109**) and isoaloeresin D (**110**) [82]. The compounds **80–122** are shown in Figures 11–13. The sources and the reported biological activities (if any) are summarized in Table 10.

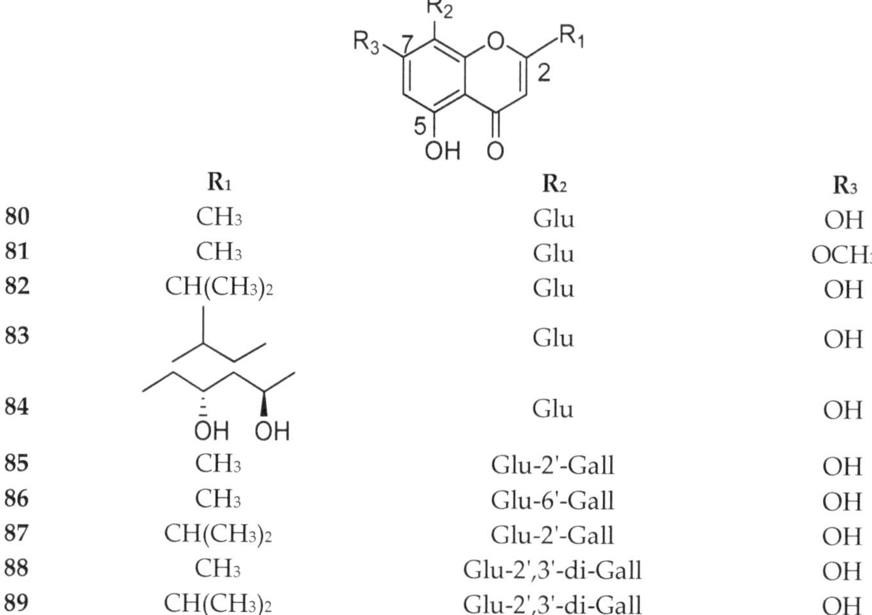

	R$_1$	R$_2$	R$_3$
80	CH$_3$	Glu	OH
81	CH$_3$	Glu	OCH$_3$
82	CH(CH$_3$)$_2$	Glu	OH
83		Glu	OH
84		Glu	OH
85	CH$_3$	Glu-2′-Gall	OH
86	CH$_3$	Glu-6′-Gall	OH
87	CH(CH$_3$)$_2$	Glu-2′-Gall	OH
88	CH$_3$	Glu-2′,3′-di-Gall	OH
89	CH(CH$_3$)$_2$	Glu-2′,3′-di-Gall	OH

Figure 11. Structures of compounds **80–89**.

	R₁	R₂	R₃
90	CH₃	CH₃	Glu
91	(CH=CH–CH₃)	CH₃	Glu
92		H	Glu
93		CH₃	Glu
94		H	Glu
95		CH₃	Glu
96		CH₃	Glu
97	CH₂COCH₃	H	Glucofuranose
98	CH₂COCH₃	H	Glu
99	CH₂COCH₃	CH₃	Glu
100	CH₂COCH₃	H	Glu-2',3',4',6'-tetra-acetate
101	CH₂COCH₃	H	Glu-2'-tigloyl
102	COOH	H	Glu
103		CH₃	Glu-2'-cin
104		CH₃	Glu-2'-*p*-Cou
105		CH₃	Glu-2'-Caf
106		CH₃	Glu-2'-(Z)-Cou
107		CH₃	Glu-2'-Cou-6'-acetyl

	R₁	R₂	R₃
108		CH₃	Glu-2'-*p*-methoxy-Cin
109		CH₃	Glu-2'-Cin
110		CH₃	Glu -2'-*p*-Cou
111		H	Glu-2'-*p*-methoxy-Cinn
112		CH₃	Glu-2'-Caf
113		CH₃	Glu-2'-Cin
114	CH₂COCH₃	H	Glu-2'-*p*-Cou
115	CH₂COCH₃	CH₃	Glu-2'-*p*-Cou

Figure 12. Structures of compounds **90–115**.

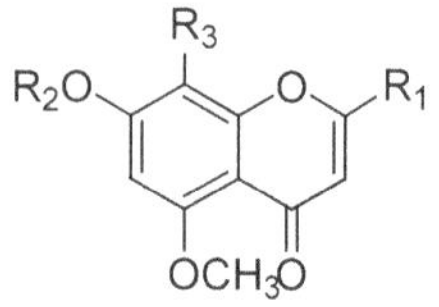

	R₁	**R₂**	**R₃**
116	CH₂COCH₃	H	Glu-6'-*p*-Cou
117	CH₂COCH₃	CH₃	Glu-6'-*p*-Cou
118	CH₂COCH₃	H	Glu-2'-Fer
119	CH₂COCH₃	CH₃	Glu-2'-Fer
120	C(OH)₂COCH₃	CH₃	Glu-2'-Cin
121	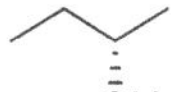	CH₃	Glu-2'-*p*-Cou-4'-Glu
122	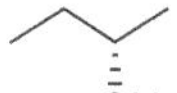	CH₃	Glu-2'-(*Z*)-Cou-4'-Glu

Figure 13. Structures of compounds **116–122**.

Table 10. 8-C-Chromone glycosides with their sources and biological activities.

No.	Compound	Source	Biological Activity
80	Isobiflorin	*Abrus mollis* Hance. aerial parts [83] *Syzygium aromaticum* L. flower buds [67] *Kunzea ambigua* (SM.) Druce. leaves [71] *Eucalyptus globulus* leaves [84,85] *Eugenia caryophyllata* flower buds [86]	Inhibition of LPS-induced production of nitric oxide (NO) with an IC_{50} > 60 μM and prostaglandin E2 (PGE2)) with an IC_{50} value of 46.0 μM [67]
81	7-methoxy-isobiflorin	Zhuyeqing Liquor; a famous traditional Chinese functional health liquor [87]	No reported biological activity
82	5,7-Dihydroxy-2-isopropylchromone-8-β-D-glucoside	*Hypericum japonicum* aerial parts [71,88,89] *Eucalyptus maidenii* bark [84]	Inhibit Epstein–Barr virus early antigen induced by 12-*O*-tetradecanoylphorbol 13-acetate (TPA) in Raji cells (70.4%) at 500 mol ratio/TPA [71]
83	5,7-Dihydroxy-2-(1-methylpropyl)chromone-8-β-D-glucoside	*Hypericum japonicum* aerial parts [71,88] *Eucalyptus grandis*, *Eucalyptus urograndi*, and *Eucalyptus maidenii* bark [90]	No reported biological activity
84	Obtusichromoneside B	*Cassia obtusifolia* seeds [72]	Inhibitory activity against human organic anion/cation transporters (OATs/OCTs) and organic anion transporting polypeptides (OATPs) at 50 μM [72]
85	Kunzeachromone E [8-β-C-(2'-galloylglucopyranosyl)-5,7-dihydroxy-2-methylchromone]	*Kunzea ambigua* leaves [71] *Baeckea frutescens* leaves [70]	Inhibition activity toward copper-induced LDL oxidation with IC_{50} value of 3.98 ± 0.24 μM [70]

Table 10. *Cont.*

No.	Compound	Source	Biological Activity
86	8-C-β-D-(6′-O-galloyl)glucosylnoreugenin [2-Methyl-5,7-dihydroxy-chromone-8-β-D-(6′-O-galloyl)-glucopyranoside]	*Syzygium aromaticum* L. leaves [91] *Syzygium aromaticum* L. flower buds [68,73]	Cytotoxicity against human ovarian cancer cells (A2780) with an IC_{50} value of 87.50 ± 1.56 µM [68] Significant inhibition capacity against Prolyl Endopeptidase with IC_{50} value of 1.48 ± 0.02 µM [73]
87	8-β-C-(2′-galloylglucopyranosyl)-5,7-dihydroxy-2-isopropylchromone	*Baeckea frutescens* leaves [70]	Active against copper-induced LDL oxidation with an IC_{50} value of 3.91 ± 0.18 µM [70]
88	Kunzeachromone F [2-Methyl-5,7-dihydroxy-chromone-8-β-D-(2′,3′-di-O-galloyl)-glucopyranoside]	*Kunzea ambigua* leaves [71]	No reported biological activity
89	Kunzeachromone B [2-Isopropyl-5,7-dihydroxy-chromone-8-β-D-(2′,3′-di-O-galloyl)-glucopyranoside]	*Kunzea ambigua* leaves [71]	No reported biological activity
90	2,5-dimethyl-8-C-β-D-glucopyranosyl-7-hydroxy-chromone	*Aloe vera* [92]	No reported biological activity
91	2-(E)-propenyl-7-methoxy-8-C-β-D-glucopyranosyl-5-methylchromone	*Aloe vera* [76,80]	BACE1 (β-secretase) inhibitory activity with an IC_{50} value of 20.5 µM [80]
92	8-C-β-D-glucosyl-(R)-aloesol	*Aloe vera* [76,80]	BACE1 (β-secretase) inhibitory activity (39.2%) at 100 µM [80]
93	8-C-β-D-glucosyl-7-O-methyl-(R)-aloesol	*Aloe vera* [76,80] and anerobic incubation of aloesin with bacterial mixture [93]	BACE1 (β-secretase) inhibitory activity (26.8%) at 100 µM [80]
94	8-C-β-D-glucosyl-(S)-aloesol	*Aloe vera* [76] and anerobic incubation of aloesin with bacterial mixture [93]	No reported biological activity
95	8-C-β-D-glucosyl-7-O-methyl-(S)-aloesol	*Aloe vera* [76] and anerobic incubation of aloesin with bacterial mixture [93]	No reported biological activity
96	8-C-β-D-glucosyl-7-O-methylaloediol	*Aloe vera* [76,80]	No reported biological activity
97	Neoaloesin A	*Aloe vera* [76] *Aloe barbadensis* leaves [94]	No reported biological activity
98	Aloesin	*Aloe vera* [76,80] *Aloe barbadensis* leaves [95]	Antioxidant activity (50 ± 1 µM trolox equivalent) at 100 mg of soluble solid/L solution [95] BACE1 inhibitory activity (37.5%) at 100 µM [80] Suppresses hyperpigmentation (40%) at 100 mg/g polyethylene glycol [2]
99	7-O-methylaloesin	*Aloe rupestris* leaves exudate [96]	No reported biological activity
100	Aloesin-2″,3″,4″,6″-tetra-O-acetate	Anerobic incubation of aloesin with bacterial mixture [93]	No reported biological activity

Table 10. *Cont.*

No.	Compound	Source	Biological Activity
101	2″-*O*-tigloylaloesin	*Aloe cremnophila* leaves exudate [97]	No reported biological activity
102	8-*C*-β-D-glucopyranosyl-7-hydroxy-5-methylchromone-2-carboxylic acid	Herbal tea *"muti"* [98]	No reported biological activity
103	8-[*C*-β-D-[2-*O*-(*E*)-cinnamoyl]glucopyranosyl]-2-[(*R*)-2-hydroxypropyl]-7-methoxy-5-methylchromone	*Aloe vera* [76] *Aloe barbadensis* leaves [99]	Topical anti-inflammatory activity at 200 µg/ear [99]
104	Aloeresin D	*Aloe vera* [76]	No reported biological activity
105	Rabaichromone	*Aloe vera* [76]	No reported biological activity
106	Allo-aloeresin D	*Aloe vera* [76]	No reported biological activity
107	Aloeresin K	*Aloe vera* [76] *Aloe barbadensis* leaf skin [100]	No reported biological activity
108	Aloeresin J	*Aloe vera* [76] *Aloe barbadensis* leaf skin [100]	No reported biological activity
109	Aloeresin E	*Aloe vera* leaves [76,82]	Inhibition of tyrosinase enzyme (40% and 80% at 50 and 100 ppm, respectively) [82]
110	Isoaloeresin D	*Aloe vera* leaves [76,82]	Inhibition of tyrosinase enzyme (20% and 40% at 50 and 100 ppm, respectively) [82] Antiviral activity against Pepper mild mottle virus; PMMoV (37.5 ± 6.5% at 1.5 mg/mL) [81]
111	2′-*O*-[*p*-methoxy-(*E*)-cinnamoyl]-(*S*)-aloesinol	*Aloe nobilis* leaves [12]	BACE1 inhibitory activity (34.1%) at 100 µM [80]
112	Iso-rabaichromone	*Aloe vera* [76,82]	No reported biological activity
113	8-*C*-glucosyl-(2′-*O*-cinnamoyl)-7-*O*-methylaloediol B	*Aloe vera* leaves [76]	No reported biological activity
114	Aloeresin A	*Aloe vera* [76]	Antioxidant activity [101] α-glucosidase inhibitory activities, with IC$_{50}$ values of 11.94 and 2.16 mM against rat intestinal sucrase and maltase [79]
115	7-*O*-methyl-aloeresin A	*Aloe vera* [76]	Tyrosinase inhibitory activity with an IC$_{50}$ value of 9.8 µM [81]
116	6′-*O*-coumaroyl-aloesin	*Aloe vera* [76]	Anti-lipid peroxidation activity with an IC$_{50}$ value of 476.4 ± 0.9 µM [75]
117	7-Methoxy-6′-*O*-coumaroyl-aloesin	*Aloe vera* [76]	Weak anticancer activity against breast cancer cell line, MDA-MB-231 (induce 30% decline in cell survival at 25 µM) [102]
118	2′-Feruloylaloesin	*Aloe nobilis* leaves [80]	Inhibition activity against β-secretase (36.4%) at 100 µM [80] Inhibition effect against mushroom tyrosinase (27 ± 0.57%) at 0.4 µM [103]
119	2′-Feruloyl-7-*O*-methylaloesin	*Aloe nobilis* leaves [80]	Inhibition activity against BACE1 (β-secretase) (48.7%) at 100 µM [80]

Table 10. *Cont.*

No.	Compound	Source	Biological Activity
120	9-Dihydroxyl-2′-*O*-(*Z*)-cinnamoyl-7-methoxy-aloesin	*Aloe vera* [76]	Inhibition of tyrosinase enzyme (9.5 ± 9.0%) at 100 µM Antiviral against Pepper mild mottle virus; PMMoV (31.5 ± 4.2% inhibition at 1.5 mg/mL) [81]
121	4′-*O*-β-D-glucosyl-isoaloeresin DI	*Aloe vera* [76]	No reported biological activity
122	4′-*O*-β-D-glucosyl-isoaloeresin DII	*Aloe vera* [76]	No reported biological activity

4.3. Phenyl and Isoprenyl Chromone Glycosides

This category is characterized by a hydroxyl prenyl moiety at C-6 or C-8, or a hydroxyl isoprenyl moiety at C-6 only. The sugar moiety can be either situated at C-7 hydroxyl of the chromone nucleus or C-4′ of the hydroxyl prenyl or C-2′ of the hydroxyl isoprenyl moiety. Most of the compounds in this category were reported from the genus Cnidium, belonging to family Apiaceae. The reported biological activity associated with several compounds in this category is their significant inhibition of fat accumulation in differentiated adipocytes employing 3T3-L1 preadipocyte cells as an assay system [60]. The compounds **123–134** are shown in Figure 14. The sources and the reported biological activities (if any) are summarized in Table 11.

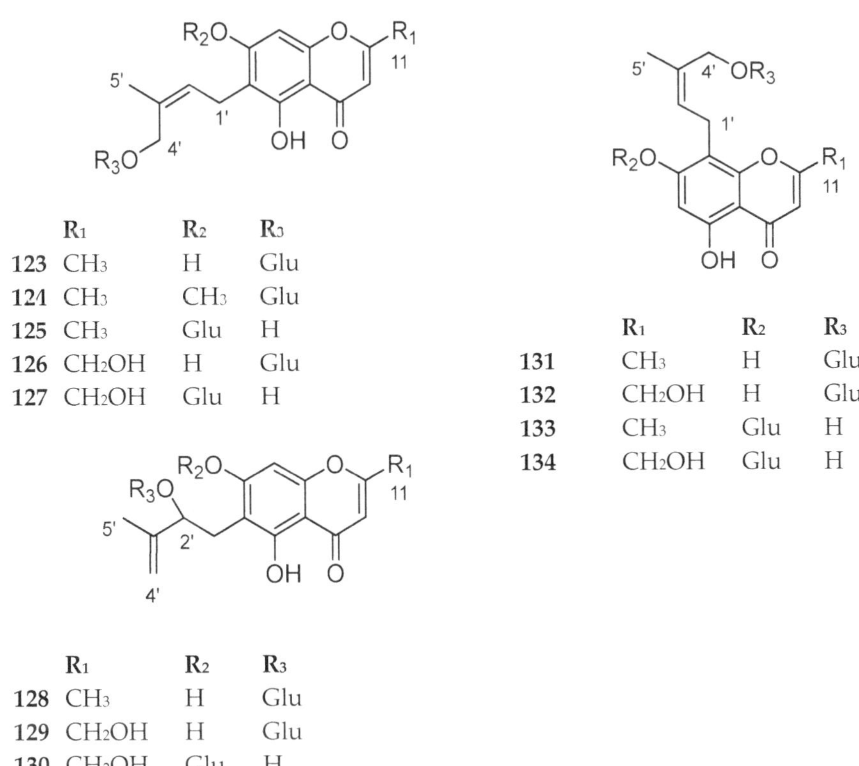

Figure 14. Structures of compounds **123–134**.

Table 11. Prenyl and isoprenyl chromone glycosides with their sources and biological activities.

No.	Compound	Source	Biological Activity
123	Cnidimoside A	*Cnidium Juponicum* whole plant [104,105] *Cnidium monnieri* fruits [60,106]	Significant inhibition of fat accumulation at 300 µM in differentiated adipocytes [60] Antitumor and antimetastatic actions at 0.1–100 µM (in vitro) and 20, 50 mg/kg, twice daily (in vivo) [105]
124	Cnidimoside B	*Cnidium Juponicum* whole plant [104] *Cnidium monnieri* fruits [60]	Significant ihibition of fat accumulation at 300 µM in differentiated adipocytes [60]
125	2-methyl-5-hydroxy-6-(2-butenyl-3-hydroxymethyl)-7-(β-D-glucopyranosyloxy)-4H-1-benzopyran-4-one	*Cnidium monnieri* fruits [60] *Angelica archangelica* [107] *Archangelica litoralis* [107]	No reported biological activity
126	Hydroxycnidimoside A	*Cnidium monnieri* fruits [60,106]	Significant inhibition of fat accumulation at 300 µM in differentiated adipocytes [60]
127	Monnieriside B	*Cnidium monnieri* fruits [60,106]	
128	Monnieriside C		
129	Monnieriside D	*Cnidium monnieri* fruits [60]	No reported biological activity
130	Monnieriside E		
131	7,8-Secoeranthin-β-D-glucopyranoside (8-{(2E)-4-[β-D-glucopyranosyl)oxy]-3-methylbut-2-enyl}-5,7-dihydroxy-2-methyl-4H-L-benzopyran-4-one)	*Eranthis hyemalis* tubers [108]	No reported biological activity
132	2-C-Hydroxy-7,8-seroeranthipn-β-D-glucopyranoside (8-{(2E)-4-[β-D-glucopyranosyl)oxy]-3-methylbut-2-enyl}-5,7-dihydroxy-2-(hydroxymethyl)-4H-1-benzopyran-4-on2)		
133	7-[(β-D-glucopyranosyl)oxy]-5-hydroxy-8-[(2E)-4-hydroxy-3-methylbut-2-enyl]-2-methyl-4H-1-benzopyran-4-one	*Eranthis cilicica* tubers [109]	No reported biological activity
134	7-[(β-D-glucopyranosyl)oxy]-5-hydroxy-2-hydroxymethyl-8-[(2E)-4-hydroxy-3-methylbut-2-enyl]-4H-1-benzopyran-4-one		

4.4. Phenyl Ethyl Chromone Glycosides

Reviewing the literature, we encountered five phenyl ethyl chromone glycosides. The phenyl ethyl moiety is usually located at C-2 of the chromone nucleus. The sugar moiety is attached to C-7 of the chromone skeleton in compounds **135–137**, while in compound **138**, the sugar is attached to C-8. In compound **139**, the sugar is not attached directly to the basic chromone skeleton. Compounds **135–139** are shown in Figure 15. Their sources are summarized in Table 12. There are no reported biological activities of these compounds.

7-*O*-glycosides

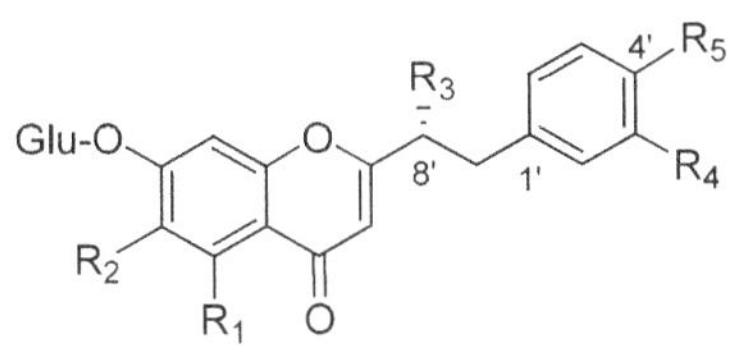

8-*O*-glycoside

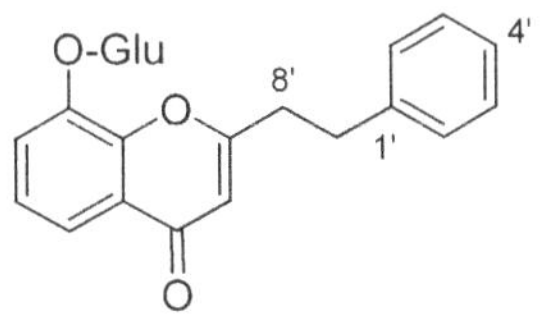

138

	R₁	R₂	R₃	R₄	R₅
135	H	H	OH	H	OH
136	HO	H	H	H	OH
137	OH	OCH₃	H	OH	OCH₃

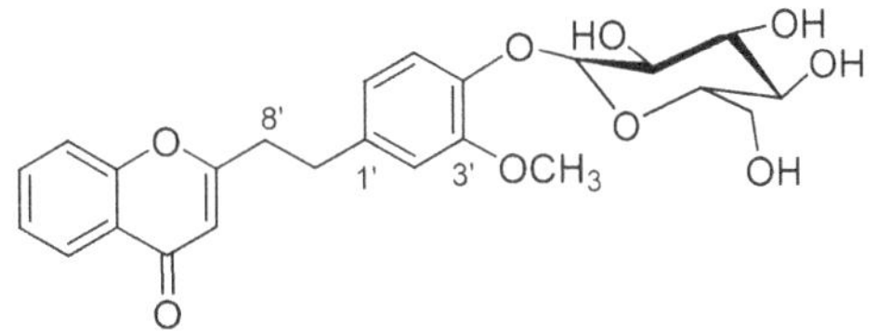

139

Figure 15. Structures of compounds **135–139**.

Table 12. Phenyl ethyl chromone glycosides with their sources.

No.	Compound	Source
135	Ononin glucoside	*Ononis vaginalis* whole plant [110]
136	7-Glucosyloxy-5-hydroxy-2-[2-(4-hydroxyphenyl)ethyl]chromone	*Cucumis melo* seeds [111]
137	Aquilarinoside C (6,4′-dimethoxy-3′-hydroxy-2-(2-phenylethyl)chromone 7-*O*-β-D-glucopyranoside)	*Aquilaria sinensis* stems [112]
138	2-(2-phenylethyl) chromone-8-*O*-β-D-glucopyranoside	*Imperata cylindrical* rhizomes [113]
139	2-[2-(4-glucosyloxy-3-methoxyphenyl)ethyl]chromone	*Aquilaria sinensis* resinous heartwood [114]

4.5. Chromone Glycosides with Additional Heterocyclic Moieties

This category of chromone glycosides is further classified based on the additional heterocyclic moiety into furano-chromone glycosides, pyrano-chromone glycosides, oxepino-chromone glycosides and pyrido-chromone glycosides.

4.5.1. Furano-Chromone Glycosides

This subclass of compounds is characterized by presence of an additional furan, or a tetrahydrofuran ring fused with the benzo-δ-pyrone. Khellol glucoside (**140**), isolated from *Ammi visnaga*, is one of the important members in this subclass. It possess potent coronary vasodilator and bronchodilator activities [115]. It was reported to have a significant hypocholesterolemic effect. It lowered low-density lipoprotein cholesterol (LDL-C) by 73%, high-density lipoprotein cholesterol (HDL-C) by 23%, and total-C by 44%, after a single oral dose of 20 mg/kg per day after two weeks [116]. Compounds **140–148** are shown in Figure 16. The sources and the reported biological activities (if any) are summarized in Table 13.

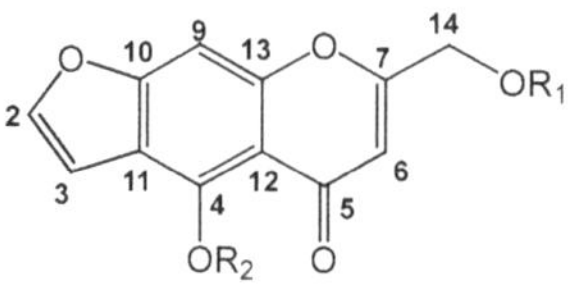

	R₁	R₂		R₁	R₂	R₃
140	Glu	CH₃	**143**	H	H	Glu
141	Api (1″→6′) Glu	H	**144**	H	CH₃	Glu
142	Glu (1″→6′) Glu	H	**145**	H	H	Api (1‴→6″) Glu
			146	Glu	CH₃	H
			147	Glu-6′-Fer	CH₃	H

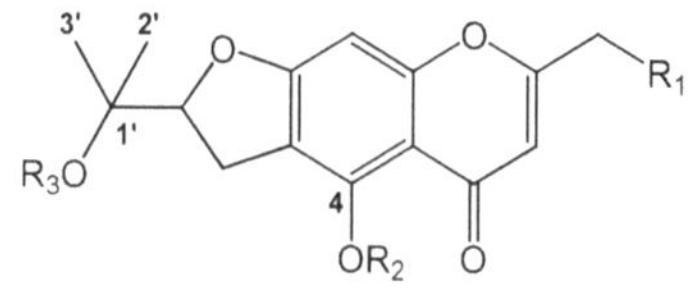

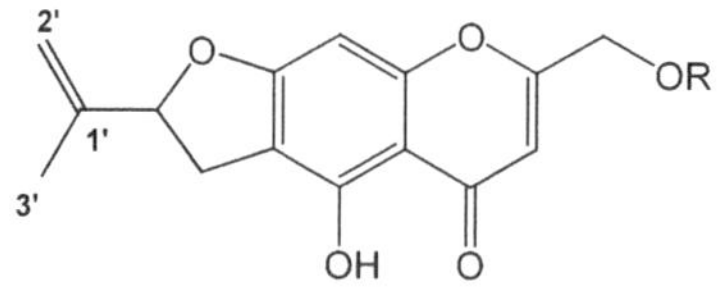

	R
148	Glu

Figure 16. Structures of compounds **140–148**.

Table 13. Furano-chromone glycosides with their sources and biological activities.

No.	Compound	Source	Biological Activity
140	Khellol glucoside (Khellinin; Khelloside)	*Ammi visnaga* fruits [117] *Eranthhis hyernalis* tubers [108]	Potent coronary vasodilator and bronchodilator [115] Hypocholesterolemic effect at 20 mg/kg per day [116]
141	Norkhelloside	*Cimicifuga heracleifolia* rhizomes [118]	No reported biological activity
142	7-[(*O*-β-D-glucopyranosyl-(1→6)-β-D-glucopyranosyl)oxy]methyl-4-hydroxy-5*H*-furo[3,2-g][1]benzopyran-5-one	*Eranthis cilicica* tubers [109]	No reported biological activity
143	4′-*O*-β-D-glucopyranosylvisamminol (Monnieriside G)	*Cnidium monnieri* fruits [60] *Saposhnikovia divaricata* roots [119]	Antitumor activity against SK-OV-3 with an IC₅₀ value of 93.91 μM [120]
144	4′-*O*-β-D-glucopyranosyl-5-*O*-methylvisamminol	*Ledebouriella seseloides* roots and rhizomes [121] *Saposhnikovia divaricata* roots [119,122] *Diplolophium buchananii* aerial parts [123] *Sphallerocarpus gracilis* roots [124]	Analgesic, antipyretic, anti-inflammatory, and anti-platelet aggregation activities [125,126] Antitumor activity with against H-460 cell line with an IC₅₀ value of 86.91 μM [120]
145	(2′*S*)-4′-*O*-β-D-apiofuranosyl-(1→6)-β-D-glucopyranosylvisamminol	*Saposhnikovia divaricata* roots [119]	Antitumor activities against PC-3 and SK-OV-3 cell lines with IC₅₀ values of 48.5, 81.91 μM, respectively [120]

Table 13. *Cont.*

No.	Compound	Source	Biological Activity
146	*prim-O*-glucosylcimifugin	*Ammi visnaga* fruits [127] *Angelica genuflexa* roots [128] *Eranthis hyemalis* tubers [108] *Angelica japonica* roots [129] *Cimicifuga foetida* rhizomes [130] *Diplolophium buchananii* aerial parts [123] *Saposhnikovia divaricata* roots [131]	Analgesic, antipyretic, anti-inflammatory, anti-platelet aggregation and antitumor activities [125,126,131]
147	Cimifugin-4'-*O*-[6''-feruloyl]-β-D-glucopyranoside	*Cimicifuga foetida* rhizomes [132]	No reported biological activity
148	Monnieriside F	*Cnidium monnieri* fruits [60]	No reported biological activity

4.5.2. Pyrano-Chromone Glycosides

This subclass of compounds is characterized by the presence of an additional pyran ring fused with the benzo-δ-pyrone. Only three compounds were reported from nature until now. Of them, 3'-*O*-glucopyranosylhamaudol (*Sec-O*-glucopyranosylhamaudol) (**149**) and (3'*S*)-3'-*O*-β-D-apiofuranosyl-(1→6)-β-D-glucopyranosylhamaudol (**150**), isolated from the *Saposhnikovia divaricata*, showed weak anti-cancer activity. Both compounds were screened against three cancer cell lines, namely human prostatic cancer cell (PC-3), human ovarian carcinoma cell (SK-OV-3), and human lung cancer cell (H460) using the conventional MTT assay. Compound **149** showed a weak activity against H460, with an IC_{50} value of 94.25 ± 1.45 µM while compound **150**, showed an activity against SK-OV-3 with an IC_{50} value of 86.21 ± 1.03 µM [119]. Compounds **149–151** are shown in Figure 17. The sources and the reported biological activities (if any) are summarized in Table 14.

	R
149	Glu
150	Api(1'''→6'')Glu

	R
151	Glu

Figure 17. Structures of compounds **149–151**.

Table 14. Pyrano-chromone glycosides with their sources and biological activities.

No.	Compound	Source	Biological Activity
149	3'-*O*-glucopyranosylhamaudol (*Sec-O*-glucopyranosylhamaudol)	*Angelica genuflexa* roots [128] *Angelica japonica* roots [129] *Glehnia littoralis* roots [133] *Peucedanum japonicum* roots [134] *Saposhnikovia divaricata* roots [119,122]	Antitumor activity against H-460 cell line with an IC_{50} value of 94.25 ± 1.45 µM [119]
150	(3'*S*)-3'-*O*-β-D-apiofuranosyl-(1→6)-β-D-glucopyranosylhamaudol	*Saposhnikovia divaricata* roots [119]	Antitumor activity against SK-OV-3 with an IC_{50} value of 86.21 ± 1.03 µM [119]
151	(2'*S*)-2'-hydroxy-7-*O*-methylallopeucenin 2'-*O*-β-D-glucopyranoside	*Diplolophium buchananii* aerial parts [123]	No reported biological activity

4.5.3. Oxepino-Chromone Glycosides

This subclass of compounds is characterized by the presence of an additional oxepin fused with the benzo-δ-pyrone. Only four compounds were reported from nature until now, and all of them were reported from *Eranthis* species. The compounds **152–155** are shown in Figure 18. The sources are summarized in Table 15. There are no reported biological activities for these compounds.

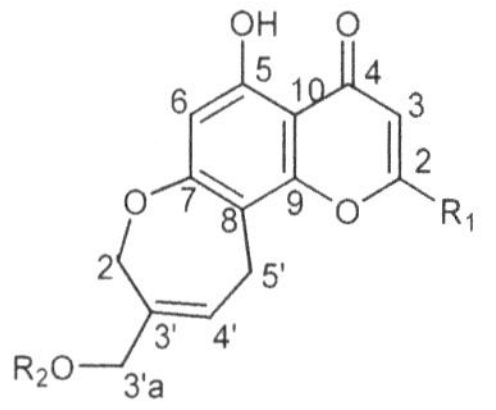

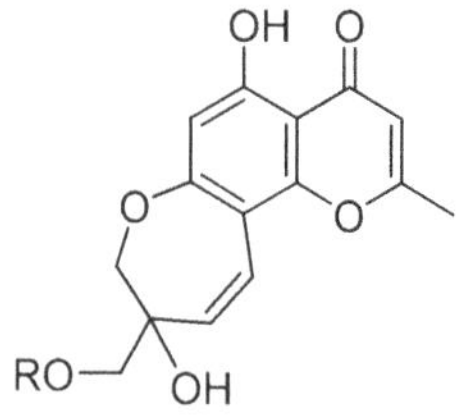

	R₁	R₂		R
152	CH₃	Glu	155	Glu(1‴→6″)Glu
153	CH₃	Glu(1‴→6″)Glu		
154	CH₂OH	Glu		

Figure 18. Structures of compounds **152–155**.

Table 15. Oxepino-chromone glycosides with their sources.

No.	Compound	Source
152	Eranthin *β*-D-glucopyranoside	*Eranthis hyemalis* tubers [108]
153	Eranthin *β*-D-gentiobioside	*Eranthis cilicica* tubers [109] *Eranthis hyemalis* tubers [108]
154	2-C-Hydroxyeranthin *β*-D-glucopyranoside	*Eranthis hyemalis* tubers [108]
155	9-[(O-*β*-D-glucopyranosyl-(1→6)-*β*-D-glucopyranosyl)oxy]methyl-8,11-dihydro-5,9-dihydroxy-2-methyl-4*H*pyrano[2,3-g][1]benzoxepin-4-one	*Eranthis cilicica* tubers [109]

4.5.4. Pyrido-Chromone Glycosides

This subclass includes only the chromone alkaloidal glycoside; Schumanniofoside. This compound was found to reduce the lethal effect of black cobra (*Naja melanoleuca*) venom in mice [135]. The authors proved that this effect is greatest when the venom is mixed and incubated with the extract or schumanniofoside. They concluded that the mode of action is by oxidative inactivation of the venom. *Schumanniophyton magnificum* is used extensively in African ethno-medicine for the treatment of various diseases and, most commonly, the treatment of snake bites [135]. Its structure is shown in Figure 19. Its source and biological activity are summarized in Table 16.

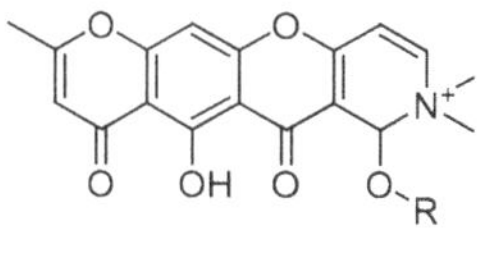

	R
156	Rha(1″→2')Glu

Figure 19. Structure of compound **156**.

Table 16. Pyrido-chromone glycosides with its source and biological activity.

No.	Compound	Source	Biological Activity
156	Schumanniofoside	*Schumanniophyton magnificum* stem bark [135]	Anti-snake venom activity at 0.01–0.16 g/kg [135]

4.6. Hybrids of Chromones with Other Classes of Secondary Metabolites

This is an interesting category, as the chromone skeleton is conjugated to another high molecular weight compound, as shown in the following subclasses.

4.6.1. Hybrids of Furano-Chromones with Cycloartane Triterpenes

This subclass of compounds is a hybrid of cycloartane triterpene and chromone. The reported compounds were isolated from the rhizomes of *Cimicifuga foetida*. The compounds **157–165** are shown in Figure 20. The sources and the reported biological activities (if any) are summarized in Table 17.

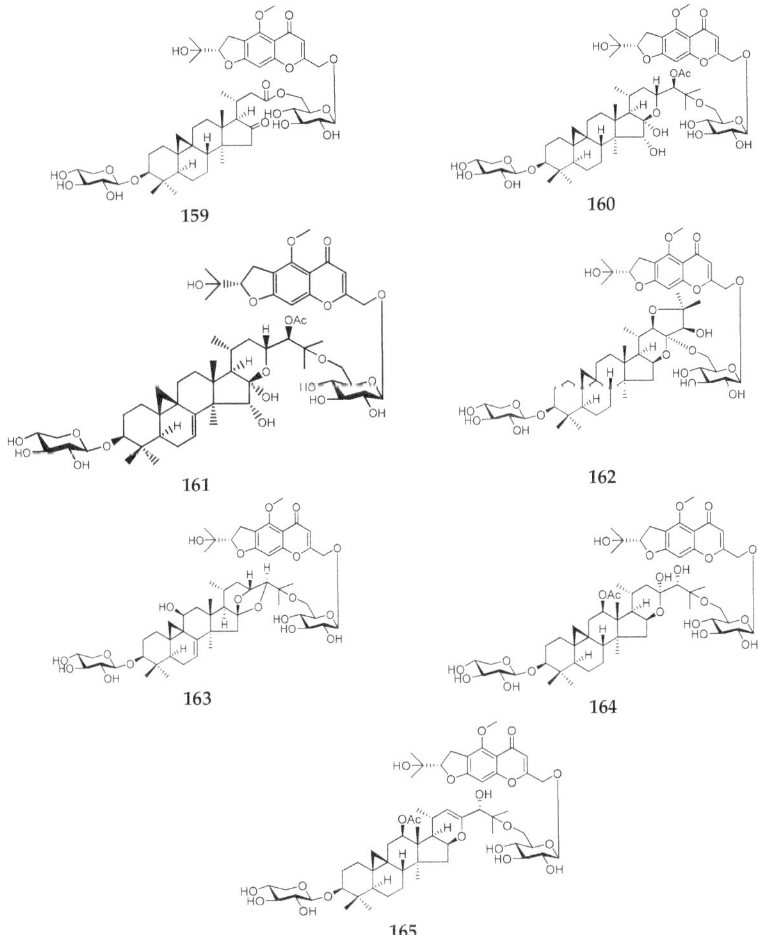

Figure 20. Structures of compounds **157–165**.

Table 17. Hybrids of furanochromones with cycloartane triterpenes with their sources and biological activities.

No.	Compound	Source	Biological Activity
157	Cimitriteromone A	*Cimicifuga foetida* rhizomes [130]	No reported biological activity
158	Cimitriteromone B	*Cimicifuga foetida* rhizomes [130]	Anti-proliferative activity with an IC_{50} value of 15.73 ± 0.59 µM [130]
159	Cimitriteromone C	*Cimicifuga foetida* rhizomes [130]	No reported biological activity
160	Cimitriteromone D	*Cimicifuga foetida* rhizomes [130]	Anti-proliferative activity with an IC_{50} value of 24.21 ± 0.61 µM [130]
161	Cimitriteromone E	*Cimicifuga foetida* rhizomes [130]	No reported biological activity
162	Cimitriteromone F	*Cimicifuga foetida* rhizomes [130]	No reported biological activity
163	Cimitriteromone G	*Cimicifuga foetida* rhizomes [130]	No reported biological activity
164	Cimitriteromone H	*Cimicifuga foetida* rhizomes [136]	No reported biological activity
165	Cimitriteromone I	*Cimicifuga foetida* rhizomes [136]	Anti-proliferative activity with an IC_{50} value of 27.14 ± 1.38 µM [136]

4.6.2. Hybrids of Chromones with Secoiridoids

There are only two compounds (Figure 21) belonging to this class, sessilifoside (**166**) and 7″-*O*-β-D-glucopyranosylsessilifoside (**167**). Both compounds were isolated from the roots of *Neonauclea sessilifolia* roots [41]. The authors did not report biological activities for these compounds.

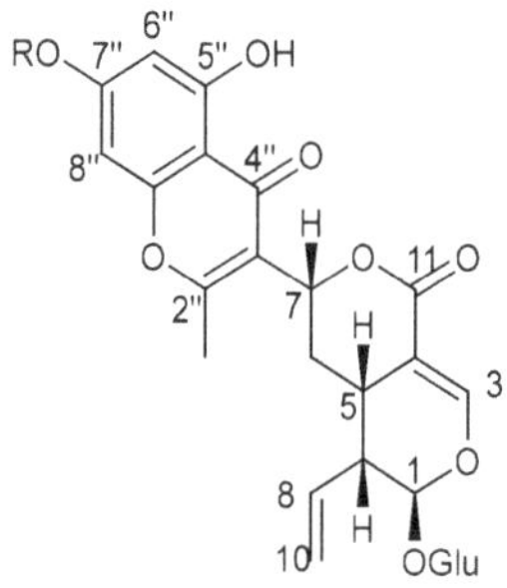

	R
166	H
167	Glu

Figure 21. Structures of compounds **166–167**.

4.6.3. Chromone Alkaloids Aminoglycosides

This category includes compounds **168–192**. Compounds **168–180** were reported from a strain of Streptomyces, isolated from a soil sample. These compounds showed antimicrobial activity against Gram-positive bacteria, as well as a potent antitumor activity. Conversely, compounds **181–183** were isolated from *Saccharothrix* species, while compounds **184–192** were reported from *Actinomycete* and exhibited antitumor and antimicrobial activities [137]. Compounds **168–192** are shown in Figures 22 and 23. The sources and the reported biological activities (if any) are summarized in Table 18.

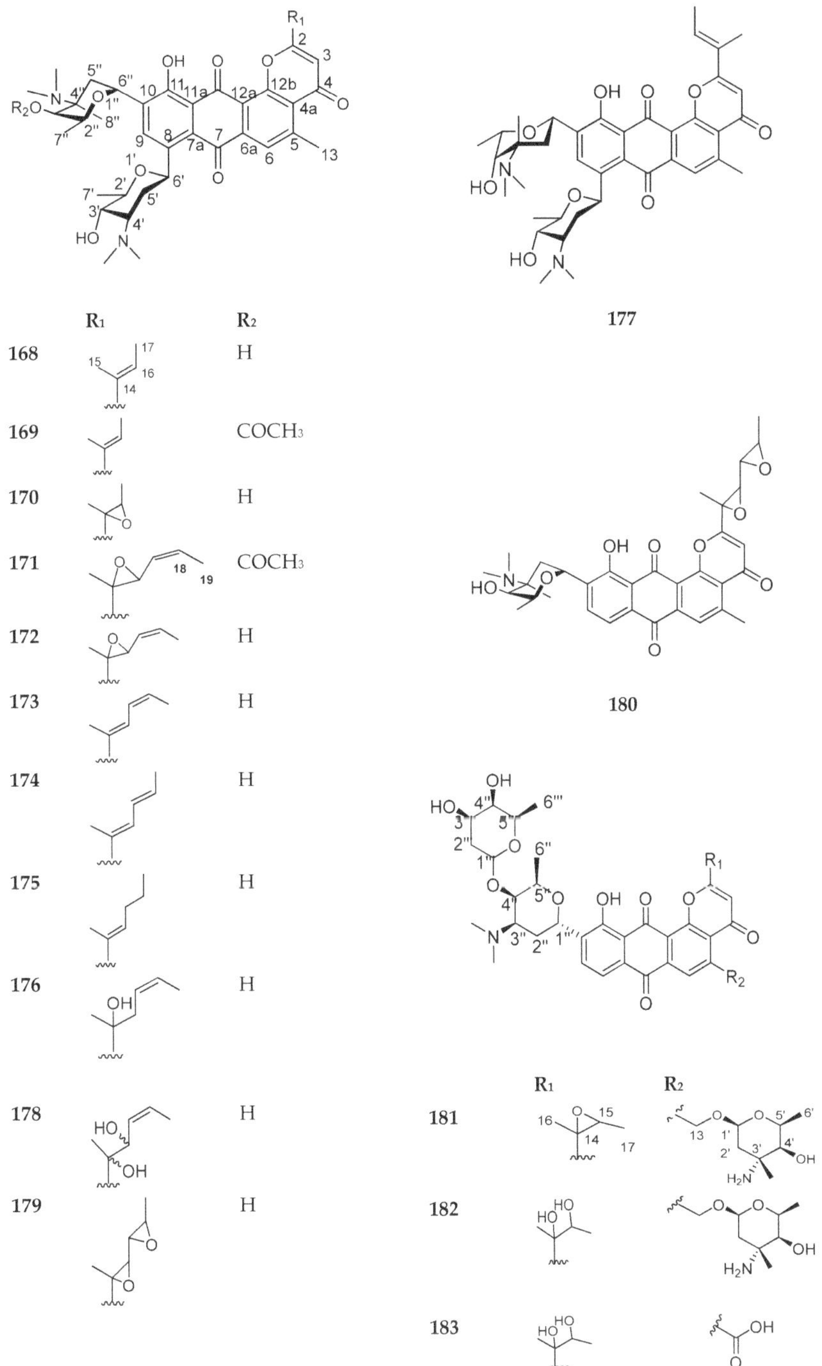

Figure 22. Structures of compounds **168–183**.

	R₁	R₂	R₃
184	NHCH₃	OH	OH
185	NH(CH₃)₂	OH	OH
186	NHCH₃	H	OH
187	NH(CH₃)₂	H	OH
188	NHCH₃	OH	H
189	NH(CH₃)₂	OH	H
190	NH₂	OH	OH

	R₁
191	NHCH₃
192	NH(CH₃)₂

Figure 23. Structures of compounds **184–192**.

Table 18. Chromone alkaloids aminoglycosides with their sources and biological activities.

No.	Compound	Source	Biological Activity
168	Kidamycin (rubiflavin B)	*Streptomyces phaeoverticillatus* var. *takatsukiensis* [138]	Antibiotic with MIC (minimum inhibitory concentration) ranges from 0.19–1.56 µg/mL and potent antitumor activity [138]
169	Neopluramycin	*Streptomyces pluricolorescens* [139,140]	Antibiotics and potent antitumor activity [140]
170	14, 16-Epoxykidamycin	*Streptomyces pluricolorescens* [141]	Antibiotic against Gram-positive bacteria with MIC ranges from 0.5 to 10 µg/mL and antitumor activity [141]
171	Pluramycin A	*Streptomyces pluricolorescens* [139,140]	Antibiotic and antitumor activity [140]
172	Rubiflavin A		
173	Rubiflavin C-1		
174	Rubiflavin C-2		
175	Rubiflavin D	*Streptomyces* species [142]	Antibiotic and antitumor activity [142]
176	Rubiflavin E		
177	Rubiflavin F (isokidamycin)		
178	PD 121,222	*Streptomyces* species [143]	Antibiotic and potent antitumor activity [143]
179	Hedamycin	*Streptomyces griseoruber* [144,145]	Antibiotic and potent antitumor activity against HeLa cells [144]
180	Ankinomycin (dean-golosaminylhedamycin)	*Streptomyces* species [146]	Antibiotic against Gram-positive bacteria with MICs ranges from 0.39–1.56 µg/mL and potent antitumor activity [146]
181	Pluraflavin A		
182	Pluraflavin B	*Saccharothrix* species [147]	Antibiotic and potent antitumor activity [147]
183	Pluraflavin E		

Table 18. *Cont.*

No.	Compound	Source	Biological Activity
184	Altromycin A		
185	Altromycin B		
186	Altromycin C	*Actinomycete*, AB 1246E-26 [148,149]	
187	Altromycin D		
188	Altromycin E		Antibiotic against Gram-positive bacteria and potent antitumor activity [137]
189	Altromycin F		
190	Altromycin G	*Actinomycete*, AB 1246E-26 [137]	
191	Altromycin H		
192	Altromycin I		

5. Spectroscopic Features

5.1. UV Features

Most of the published work on chromones show several strong bands in the range of 200–320 nm [150,151]. In contrast to chromone, the pyrone ring of 4-chromanone contains no double bond. The ultraviolet absorption spectra of chromones and chromanones are summarized in Table 19 [150].

Table 19. UV Band maxima of chromones and chromanone in 3-methylpentane at 25 °C.

	Chromones		Chromanones
Band system	λ_{max}		λ_{max}
A	360, 352, 345, 337, 324		363, 347
B	301, 290, 283		322(sh), 312 252, 246
C	225, 246, 239, 227, 223, 216, 202		219(sh), 213

The UV spectrum of chromones in alcohol shows two strong bands at λ_{max} 245 and 299 nm [152–154]. Some data reported three bands at λ_{max} 245, 303 and 297 nm [150]. 2-methyl-5,7-dihydroxy chromone shows bands at λ_{max} 250, 255, 295 and 325 nm, meanwhile 2-methyl-5-hydroxy-7-*O*-glycosyl chromone shows bands at λ_{max} 248, 255 and 290 nm [154]. The presence of an electron attracting group at C-2 resulted in a bathochromic shift in all bands [151]. The information gained from applying spectral shift reagents with flavonoids can be also applied to chromones. In the case of $AlCl_3$, a bathochromic shift of 20–70 nm, which is non-reversible with acids, indicates a free hydroxyl group at position 5. Meanwhile, a bathochromic shift with NaOAc can be diagnostic for the presence of a free 7-hydroxyl group [154,155].

5.2. IR Features

Carbonyl region: The IR carbonyl stretching frequency for a chromone is observed at 1640~1660 cm^{-1}, which is slightly higher than that of δ-pyrone (1650 cm^{-1}) but is much lower than that of coumarins (1720–1740 cm^{-1}) [25,153]. Despite that the OH group attached to C-5 of the chromone nucleus chelates strongly with the CO group, this intramolecular H-bonding has only a slight bathochromic effect on the CO stretching frequency [156]. All 5-hydroxychromones possess three significant maxima in the 1580–1700 cm^{-1} region. The two higher frequencies are intense at 1660 and 1630 cm^{-1}, with a constant wavenumber separation of 34 ($\pm$ 5) cm^{-1} in both carbon tetrachloride and chloroform.

Hydroxyl region: The IR hydroxyl stretching vibration for a chromone was observed at 2500–3650 cm^{-1}. A strong chelation in 5-hydroxychromones does not produce a considerable bathochromic shifts in both the OH and CO stretching bands [156].

Chelated 5-hydroxychromones produce no absorption maxima in the 3300–3600 cm^{-1} region, but a weak absorption envelope extends from 2400 to 3300 cm^{-1}. The entire envelope is associated with various stretching modes of the chelated 5-OH group [156]. For 7-hydroxychromones, a steric buttressing effect is observed when the 7-OH group is flanked by a bulky substituent in the ortho-position (6 or 8). The free OH band appears as a doublet centered at 3615 cm^{-1}, the separation of the components being ~26 cm^{-1}. When a prenyl moiety is located in the ortho-position to the 7-OH group, an intramolecular OH interaction occurs, resulting in two OH stretching frequencies. When a 7-OH group is flanked by an OMe group, intense intramolecularly bonded OH stretching frequencies are found at ~3513 and 3517 cm^{-1}, respectively [156]. The 2-hydroxymethyl group exhibits a free stretching frequency at $\approx$3615 cm^{-1}. At concentrations higher than 0.15 M, a broad-bonded OH frequency at 3400 cm^{-1} occurs due to intermolecular H-bonding, and it consequently disappears on dilution [156].

5.3. ^{1}H-NMR Features

In the following text, we try to give insight about the most characteristic ^{1}H-NMR features of the benzo-δ-pyrone skeleton (Figure 24) and its glycosides. The δ-pyrone ring has two olefinic protons assigned as H-2 and H-3. In 2, 3 unsubstituted chromones, for example compound **12**, the ^{1}H-NMR spectrum shows two ortho-coupled doublets (J = 6.0 Hz), located downfield at δ_H 8.19 (H-2) and upfield at δ_H 6.26 (H-3) [25]. For 2-alkyl and 2-*O*-glycosyl chromones (compounds **21** and **1**, respectively), they are characterized by an upfield singlet proton (H-3) at δ_H 6.11 and 5.98, respectively [7,40]. Meanwhile, 3-alkyl and 3-*O*-glycosyl chromones (compounds **39** and **2**, respectively) are characterized by a downfield singlet proton (H-2) at δ_H 7.93 and 8.07, respectively [10,50].

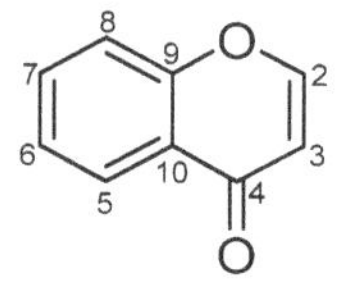

Figure 24. Basic skeleton of Chromone.

Chromanone glycosides or 2,3-dihydrochromone glycosides are characterized by an oxygenated proton (H-2) at δ_H 4.12 and 5.44 as in compounds **62** [63] and **60** [50], respectively. The splitting pattern of H-2 can be either d, dd or ddd, depending on the number of neighboring protons. A small coupling constant between H-2 and H-3 (J= 2.8 Hz) can determine that they are located in the equatorial–equatorial position [50]. Further information on the detailed configuration can be clarified from observing NOESY correlations. Unsubstituted chromanones at C-3, as in **62,** show two geminal protons at δ_H 2.50 and 2.70. Their splitting pattern shows geminal ($J_{3a\text{-}3b}$ = 16.2 Hz) and vicinal ($J_{ax\text{-}eq}$ = 2.7 Hz or $J_{ax\text{-}ax}$ = 12.6 Hz) couplings [63]. In 3-alkyl substituted chromanones, as in **60** and **61**, H-3 is also detected at δ_H 2.79 [50].

Naturally occurring chromones often bear a hydroxyl or methoxy group at C-5 and/or C-7 and a methyl group at C-2 and/or C-5 [153]. The C-5 methyl is usually observed in 6-C and 8-C glycosides. In aprotic solvents such as DMSO-d$_6$, the chelated 5-OH is detected as a singlet at δ_H 12.57; meanwhile, the 7-OH is detected at δ_H 10.00, as in compound **56 [31]**. The C-2 methyl in Schumaniofioside A **7** can be detected at δ_H 2.33 (3H, s) [17]. Meanwhile, those located at C-5, can be detected more downfield at δ_H 2.64 (3H, s) as in **79 [78]**.

For the phenyl part of the benzo-δ-pyrone skeleton, the protons show chemical shift and coupling constant values similar to those observed for protons in substituted benzenes.

Sugar moiety: Xylosyl, arabinosyl and glucosyl chromones show an anomeric proton signal at δ_H ~4.73 (d, J = 7–7.6 Hz) [10–12]. The former moieties can be differentiated by the number of the oxygenated protons at the δ_H 3-5 region, in addition to the difference in ^{13}C-NMR values. Rhamnosyl chromones show a distinct signal at δ_H ~1.25 (3H, d,

J = 6.0 Hz) corresponding to CH_3-6' of α-L-rhamnose [8]. The most abundant chromone of C-glycosides is the 8-C-glycoside form, followed by 6-C-glycosides. However, we encountered a unique 3-C-glycoside named macrolobin **64** [65]. The anomeric proton in macrolobin is detected at δ_H 5.32 (d, J = 1.5 Hz), the small coupling constant being indicative of the α-anomer [65]. Biflorin **66** and isobiflorin **80**, as representative for 6-C and 8-C-glycosides, respectively, show the anomeric proton signal at δ_H 4.55 (d, J = 9.8 Hz), and 4.63 (d, J = 9.8 Hz), respectively [69]. The former coupling constant value is higher than that observed in case of O-glycosides (J = 7–7.6 Hz) [11,12]. In 5-O, 7-O, 6-C, furano-, pyrano-, oxepino-chromone glycosides, the sugar moiety can be further substituted with another sugar, as in **8–9**, **25–32**, **78–79**, **141–142**, **150** and **153,** respectively. In 6-C, 8-C and, to a lesser extent, 7-O-glycosides, the sugar moiety can be mono- or disubstituted with a phenolic acid moiety, commonly at C-2' or C-6' or C-2' and C-3'. The most commonly occurring phenolic moiety is gallic acid, but other phenyl propanoids such as cinnamoyl, coumaroyl, feruloyl and coniferoyl moieties also exist. The galloyl moiety is characterized by a singlet aromatic signal integrated for two protons at δ_H 6.75 [27]. The cinnamoyl moiety is confirmed by two trans-coupled olefinic protons at δ_H 7.43 (d, J = 15.8 Hz) and 6.25 (d, J = 15.8 Hz), in addition to the aromatic signals of the benzene ring [30]. The presence of coumaroyl substitution is characterized by AA' BB' system for two pairs of ortho-coupled aromatic protons at δ_H 7.31 (2H, d, J = 8.6 Hz) and δ_H 6.74 (2H, d, J = 8.6 Hz), a trans-olefinic proton signals at δ_H 7.35 (1H, d, J = 16.1 Hz) and δ_H 6.03 (1H, d, J = 15.9 Hz) [29].

Prenyl and Isoprenyl chromone glycosides: In the case of 7-O-glycoside **127**, hydroxyl prenyl moiety can be easily characterized by the allylic methylene protons at δ_H 3.53 (2H, m, H-1'), an olefinic proton at δ_H 5.39 (t, J = 6.4, H-2'), oxygenated methylene protons δ_H 4.20 and 4.46 (1H each, d, J = 12.0 Hz, H-4'), and an olefinic methyl at δ_H 1.75 (3H, d, J = 1.2, H-5'') [60]. Its isomeric 4'-O-glycoside **126** showed similar signals; however, the hydroxyl methylene protons (H-4') were slightly downfield at δ_H 4.34 and 4.65 due to O-glycosidation [106]. Meanwhile, the hydroxyl isoprenyl group in the 7-O-glycoside **130** is characterized by the methylene protons at δ_H 2.96 (2H, m, H-1'), an oxygenated methine proton at δ_H 4.35 (H-2', m), exomethylene protons at δ_H 4.74 and 4.91 (H-4') and an olefinic methyl at δ_H 1.87 (3H, s, H-5'). Its isomeric 2'-O-glycoside **129** showed similar signals, but the oxygenated methine proton (H 2') is shifted downfield at δ_H 4.74 due to O-glycosidation [60].

Phenyl ethyl chromone glycosides: The presence of the phenyl ethyl moiety in compound **136** can be detected by the methylene proton signals at δ_H 2.75 (2H each, dd, J = 14.8, 6.4 Hz, H-7') and 3.19 (2H each, J = 14.8, 3.1 Hz, H-8'), in addition to the aromatic protons of the phenyl ring. The 8'-hydroxy phenyl ethyl moiety in **135** shows an oxygenated proton signal δ_H 5.85 (dd, J = 3.5, 5.9 Hz, H-8') [110].

5.4. ^{13}C-NMR Features

For better understanding of the differences in chemical shifts related to the substituents on the chromone moiety, we preferred to add the ^{13}C-NMR data in Tables 20–40. For the numbering of the skeleton, the following figure (Figure 25) gives few examples for the numbering system of the skeleton with multiple substituents. Briefly, the basic chromone nucleus was assigned numbers 1–10. In the case of a substitution at C-2, numbers 11, 12 ... etc. were given to the substituents, followed by substitution at C-3 and so on. Sugar moiety, and substituents attached to it, were assigned numbers 1', 2', ... and then 1'', 2'', ... etc. For better understanding, the following figure shows representative examples for the numbering system. Some complicated structures have their own numbering system, shown on them within the review.

Table 20. The ^{13}C-NMR spectral data of compounds **1–14** except those which have no reported ^{13}C-NMR data.

C	1	2	3	4	5	7	11	13	14
2	161.2	147.4	147.7	147.8	147.4	167.2	167.8	158.6	159.5
3	93.1	141.2	141.0	138.1	141.5	111.7	109.1	112.0	112.8
4	166.6	178.4	178.5	177.0	178.4	180.3	183.2	183.6	184.4
5	137.1	163.4	163.3	161.5	163.4	160.2	93.9	163.4	163.9
6	132.0	100.1	100.2	98.8	100.2	104.6	162.0	101.3	101.9
7	127.8	166.4	166.3	164.3	166.5	164.7	110.2	164.9	165.2
8	114.9	95.0	95.0	93.8	95.0	99.1	159.6	96.2	96.9
9	154.3	159.3	159.3	157.2	159.4	161.1	156.6	159.5	160.2
10	114.5	106.2	106.2	104.8	106.1	109.2	106.1	108.4	109.3
11	23.2	-	-	-	-	19.9	20.3	-	-
12	-	-	-	-	-	-	8.2	-	-
1′	100.1	104.6	104.4	100.6	104.2	105.0	101.7	101.6	102.1
2′	73.2	74.6	72.0	69.9	74.8	74.6	75.1	74.7	75.1
3′	77.5	77.1	73.6	70.2	77.4	77.3	78.4	77.9	77.8
4′	69.6	70.8	69.2	71.5	71.3	71.3	80.5	71.2	73.9
5′	76.7	67.1	67.1	69.7	78.5	78.7	78.1	78.4	77.5
6′	60.7	-	-	17.7	62.6	62.5	60.9	62.4	171.5
OCH$_3$	-	-	-	-	-	-	-	-	53.8
Solvent	DMSO-d_6	CD$_3$OD	CD$_3$OD	DMSO-d_6	CD$_3$OD	CD$_3$OD	C$_5$D$_5$N	CD$_3$OD	CD$_3$OD
References	[7]	[10]	[12]	[9]	[11]	[17]	[1]	[27]	[28]

Table 21. The ^{13}C-NMR spectral data of compounds **15–23** except which has no reported ^{13}C-NMR data.

C	15	16	17	18	20	21	22	23
2	158.1	158.4	158.5	168.9	161.6	174.3	175.5	174.6
3	110.8	111.9	112.0	108.8	108.8	107.8	106.2	106.6
4	181.8	183.6	183.6	182.5	182.5	184.3	183.3	183.1
5	161.3	163.1	163.2	161.2	157.9	163.0	162.2	162.7
6	99.8	101.2	101.2	99.8	100.3	101.0	100.6	100.7
7	162.9	164.6	164.8	163.1	168.9	164.8	164.2	164.2
8	94.6	96.2	96.2	94.9	95.0	95.9	95.3	95.3
9	157.7	159.4	159.4	157.9	163.4	159.5	158.4	158.4
10	106.8	108.5	108.5	105.5	105.5	107.0	106.5	107.4

Table 21. Cont.

C	15	16	17	18	20	21	22	23
11	-	-	-	20.5	20.5	28.3	33.3	40.4
12	-	-	-	-	-	11.2	19.9	27.6
13	-	-	-	-	-	-	19.9	17.7
14	-	-	-	-	-	-	-	11.7
1′	99.7	101.5	101.5	100.6	99.9	101.6	101.7	101.7
2′	73.1	74.7	74.7	73.3	73.5	74.7	74.8	74.8
3′	76.2	77.9	77.9	76.6	77.6	77.8	78.5	78.5
4′	69.7	71.9	72.0	69.6	70.7	71.2	71.2	71.1
5′	74.1	75.8	76.0	66.2	76.8	78.4	79.3	79.2
6′	63.4	64.2	64.7	-	61.1	62.4	62.4	62.3
1″	119.5	127.4	127.3	-	-	-	-	-
2″	108.8	133.7	131.2	-	-	-	-	-
3″	145.6	115.8	117.0	-	-	-	-	-
4″	138.6	160.0	161.4	-	-	-	-	-
5″	145.6	115.8	117.0	-	-	-	-	-
6″	108.8	133.7	131.2	-	-	-	-	-
7″	165.9	145.1	146.9	-	-	-	-	-
8″	-	116.0	115.1	-	-	-	-	-
9″	-	168.1	168.9	-	-	-	-	-
Solvent	DMSO-d_6	CD$_3$OD	CD$_3$OD	DMSO-d_6	DMSO-d_6	CD$_3$OD	C$_5$D$_5$N	C$_5$D$_5$N
References	[29]	[30]	[30]	[31]	[157]	[40]	[41]	[41]

Table 22. The ^{13}C-NMR spectral data of compounds **24–32** except which has no reported ^{13}C-NMR data.

C	24	25	26	28	29	30	31	32
2	168.7	157.3	167.9	168.2	168.9	169.2	168.9	168.5
3	108.7	108.3	108.8	108.8	108.8	109.3	108.8	108.4
4	182.4	181.9	182.7	182.8	182.5	184.2	182.5	182.1
5	161.5	161.1	162.4	162.4	161.7	163.1	161.7	161.3
6	99.9 *	108.7	100.6	100.7	100.2	100.8	100.0	99.6
7	162.9	162.6	163.9	163.9	161.9	163.4	161.9	161.6
8	94.9	108.2	95.3	95.1	95.2	95.7	95.1	94.7
9	157.7	168.2	158.3	158.3	157.9	159.5	157.9	157.5

Table 22. *Cont.*

C	24	25	26	28	29	30	31	32
10	105.5	105.0	106.3	106.3	105.7	106.8	105.6	105.2
11	20.3	19.9	19.9	20.1	20.5	20.4	20.5	20.1
1′	99.8 *	99.3	101.9	102.0	98.8	99.6	98.5	98.1
2′	73.3	77.0	74.6	74.6	81.2	71.0	69.8	69.5
3′	76.5	75.8	78.5	78.5	70.4	82.5	82.1	70.3
4′	70.2	68.9	71.5	71.6	70.9	72.4	70.7	81.4
5′	74.1	76.0	77.4	77.5	69.4	70.9	68.7	68.5
6′	63.7	60.5	69.0	68.0	18.3	18.1	18.0	17.9
1″	-	108.7	111.2	102.7	105.0	105.9	104.8	104.4
2″	-	76.7	77.9	72.0	74.5	75.4	74.1	74.5
3″	-	79.2	80.3	72.8	77.2	77.8	76.8	77.1
4″	-	73.9	75.1	74.1	70.0	71.1	70.6	70.1
5″	-	64.1	65.8	69.8	76.7	77.7	74.8	76.7
6″	-	-	-	18.6	61.5	62.2	64.1	61.2
CH$_3$CO	20.9, 170.5	-	-	-	-	-	21.1, 170.6	-
Solvent	DMSO-d_6	CDCl$_3$, DMSO-d_6	C$_5$D$_5$N	C$_5$D$_5$N	DMSO-d_6	DMSO-d_6	DMSO-d_6	DMSO-d_6
References	[33]	[16]	[44]	[46]	[31]	[22]	[31]	[47]

* Data interchangeable.

Table 23. The ^{13}C-NMR spectral data of compounds **33–41**.

C	33	34	35	36	37	38	39	40	41
2	168.3	168.2	168.3	168.3	157.8	154.8	154.7	168.6	168.4
3	108.2	108.7	108.3	108.2	110.6	120.3	125.8	108.7	108.4
4	181.4	182.7	181.9	181.9	181.5	183.8	183.5	182.5	182.3
5	161.1	162.3	161.1	161.2	157.7	163.0	163.1	158.4	185.1
6	99.5	100.6	99.8	99.5	108.9	100.9	100.8	109.1	108.5
7	161.4	163.7	162.6	162.6	160.9	164.6	164.6	161.1	155.56
8	94.5	94.6	94.3	94.5	93.3	95.8	95.8	93.3	97.7 *
9	157.3	158.3	157.4	157.4	155.4	159.4	159.5	155.9	160.3
10	105.0	106.3	105.1	105.1	106.0	107.4	107.6	105.1	105.0
11	19.9	19.9	19.8	19.8	7.4	10.2	19.3	20.5	19.9
12	-	-	-	-	-	-	13.4	7.9	6.9

Table 23. *Cont.*

C	33	34	35	36	37	38	39	40	41
1′	98.0	101.6	99.5	98.1	100.1	101.6	101.6	100.2	93.1 *
2′	69.5	74.4	72.9	73.0	73.1	74.7	74.7	73.8	77.3
3′	69.8	78.1	76.1	76.2	76.4	77.8	77.8	76.5	74.5
4′	80.9	71.3	69.9	69.9	69.6	71.2	71.2	79.4	69.7
5′	68.4	75.8	73.9	73.9	77.1	78.4	78.4	76.1	73.3
6′	17.9	64.5	63.4	63.4	60.6	62.4	62.4	60.7	60.5
1″	102.3	121.0	128.5	127.9	-	-	-	-	-
2″	81.4	110.4	114.8	114.9	-	-	-	-	-
3″	76.6	147.4	146.9	148.9	-	-	-	-	-
4″	69.8	140.8	147.7	149.3	-	-	-	-	-
5″	77.6	147.4	120.7	122.5	-	-	-	-	-
6″	61.0	110.4	115.9	115.8	-	-	-	-	-
7″	-	167.1	144.6	144.7	-	-	-	-	-
8″	-	-	115.7	115.8	-	-	-	-	-
9″	-	-	166.1	166.2	-	-	-	-	-
1‴	103.5	-	99.5	99.5	-	-	-	-	-
2‴	71.2	-	70.3	71.2	-	-	-	-	-
3‴	71.7	-	71.0	71.6	-	-	-	-	-
4‴	66.4	-	67.1	67.1	-	-	-	-	-
5‴	64.2	-	75.0	74.8	-	-	-	-	-
6‴	-	-	61.0	61.0	-	-	-	-	-
OCH$_3$	-	-	-	55.8	-	-	-	60.1	-
CH$_3$CO	-	-	-	-	-	-	-	-	20.7, 169.6
Solvent	DMSO-d_6	C$_5$D$_5$N	DMSO-d_6	DMSO-d_6	DMSO-d_6	CD$_3$OD	CD$_3$OD	DMSO-d_6	DMSO-d_6
References	[47]	[48]	[33]	[33]	[49]	[50]	[50]	[23]	[51]

* Data interchangeable.

Table 24. The ^{13}C-NMR spectral data of compounds **43–52** except those which have no reported ^{13}C-NMR data.

C	43	45	46	47	48	49	50	52
2	168.7	168.4	168.4	169.5	170.0	160.0	164.9	164.9
3	108.4	108.0	108.0	108.4	109.1	112.5	101.9	118.6
4	182.9	182.7	182.7	182.4	185.0	185.0	178.8	177.6
5	159.5	152.6	155.9	152.5	157.8	158.8	141.8	137.9
6	98.3	109.7	114.3	135.2	116.3	117.3	116.6	110.5
7	160.9	158.5	158.8	157.8	160.2	161.2	160.4	159.8
8	104.6	114.3	109.8	94.7	111.6	112.9	111.5	99.6
9	155.0	155.9	152.6	155.6	154.6	155.5	159.3	158.5
10	105.2	106.5	106.5	106.2	108.2	110.6	117.1	115.7
11	20.5	20.0	20.1	20.2	20.5	10.3	22.8	19.4
12	8.1	8.8	9.0	-	8.9	10.5	19.9	48.7
13	-	8.9	9.1	-	9.5	-	-	205.3
14	-	-	-	-	-	-	-	52.0
15	-	-	-	-	-	-	-	62.7
16	-	-	-	-	-	-	-	23.6
1′	100.3	104.3	104.4	100.8	105.7	106.5	100.3	102.1
2′	73.9	76.0	74.1	74.1	75.3	76.5	77.6	73.0
3′	76.5	73.9	76.3	78.0	75.6	78.5	73.6	76.3
4′	79.4	70.0	69.9	69.9	71.7	72.5	70.1	69.4
5′	76.1	73.4	77.0	77.0	77.7	76.4	76.9	77.0
6′	60.7	63.0	61.0	61.2	64.3	65.3	61.0	60.5
1″	-	-	-	-	-	173.1	-	-
2″	-	-	-	-	-	47.3	-	-
3″	-	-	-	-	-	71.4	-	-
4″	-	-	-	-	-	46.9	-	-
5″	-	-	-	-	-	176.6	-	-
6″	-	-	-	-	-	28.4	-	-
OCH$_3$	60.1	-	-	56.7	-	-	-	-
CH$_3$CO	-	20.4, 170.1	-	-	20.4, 172.5	-	-	-
Solvent	DMSO-d_6	DMSO-d_6	*	DMSO-d_6	*	CD$_3$OD	CDCl$_3$	DMSO-d_6
References	[23]	[51]	[24]	[36]	[24]	[52]	[53]	[57]

* The authors did not report the NMR solvent.

Table 25. The ^{13}C-NMR spectral data of compounds **53–64** except those which have no reported ^{13}C-NMR data.

C	53	54	55	56	57	60	61	62	64
2	161.0	166.3	169.9	166.6	166.5	105.9	104.7	83.5	148.0
3	117.2	110.7	109.3	107.6	106.9	46.2	52.9	42.2	140.5
4	178.1	177.4	184.0	182.2	182.5	199.2	198.9	193.7	178.9
5	141.3	117.8	96.0	162.0	162.0	165.3	165.2	161.5	163.6
6	113.0	125.0	162.9	99.5	98.8	97.5	97.5	100.0	100.2
7	159.9	119.2	101.0	165.1	164.8	168.4	168.4	166.6	166.2
8	99.7	147.4	164.7	94.5	93.6	97.0	97.0	99.1	95.3
9	158.8	147.3	159.3	158.1	158.2	160.4	160.2	166.2	159.3
10	116.1	125.4	119.0	104.3	104.2	102.0	102.4	106.7	106.6
11	47.3	20.1	20.3	66.1	66.0	13.8	23.7	33.2	-
12	202.5	-	-	-	-	-	11.8	18.2 *	-
13	29.8	-	-	-	-	-	-	-	-
14	22.3	-	-	-	-	-	-	-	-
1′	101.3	102.4	101.6	102.8	102.7	104.1	104.1	104.0	102.1
2′	73.1	74.8	74.7	73.9	73.6	75.0	75.0	74.6	72.0
3′	76.3	78.6	78.3	76.4	76.6	77.9	77.9	77.4	74.0
4′	69.5	71.2	71.4	67.7	70.1	71.0	71.0	71.2	71.5
5′	77.0	79.2	77.8	63.6	76.8	78.0	78.0	78.5	72.4
6′	60.9	62.4	62.4	-	61.3	62.5	62.5	62.5	62.5
Solvent	DMSO-d_6	C$_5$D$_5$N	CD$_3$OD	DMSO-d_6	CD$_3$OD	CD$_3$OD	CD$_3$OD	CD$_3$OD	CD$_3$OD
References	[57]	[58]	[59]	[31]	[60]	[54]	[50]	[63]	[65]

* Data interchangeable.

Table 26. The ^{13}C-NMR spectral data of compounds **66–72**.

C	66	67	68	69	70	71	72
2	167.3	174.7	168.3	168.4	167.8	169.4	174.4
3	107.8	105.1	108.6	108.8	107.0	109.2	105.3
4	181.8	182.2	181.8	181.8	182.0	184.3	182.2
5	160.6	160.6	160.6	160.6	160.4	162.3	160.4
6	108.7	108.7	108.6	108.8	108.1	108.9	107.0
7	163.2	163.3	163.1	163.1	162.4	165.0	163.4
8	93.3	93.4	93.3	93.3	93.9	95.2	93.7
9	156.6	156.7	156.6	156.6	157.0	159.4	156.9
10	103.0	103.3	103.2	103.2	102.9	105.1	103.1

Table 26. *Cont.*

C	66	67	68	69	70	71	72
11	19.8	32.3	43.1	66.2	20.0	20.4	32.4
12	-	19.8	64.1	46.0	-	-	19.8
13	-	19.8	23.3	63.8	-	-	19.8
14	-	-	-	43.8	-	-	-
15	-	-	-	23.5	-	-	-
1′	73.0	73.0	72.9	72.9	70.8	75.6	70.7
2′	70.1	70.2	70.0	70.0	72.0	72.6	72.0
3′	78.9	78.9	78.8	78.8	76.7	80.1	76.6
4′	70.6	70.6	70.5	70.5	70.8	71.9	70.7
5′	81.4	81.6	81.4	81.4	82.0	80.2	81.8
6′	61.4	61.5	61.4	61.4	61.6	65.2	61.5
1″	-	-	-	-	119.9	121.6	119.9
2″	-	-	-	-	108.9	110.4	108.7
3″	-	-	-	-	145.4	146.6	145.4
4″	-	-	-	-	138.2	140.0	138.1
5″	-	-	-	-	145.4	146.6	145.4
6″	-	-	-	-	108.9	110.4	108.9
7″	-	-	-	-	164.8	168.6	164.7
Solvent	DMSO-d_6	DMSO-d_6	DMSO-d_6	DMSO-d_6	DMSO-d_6	CD$_3$OD	DMSO-d_6
References	[69]	[69]	[72]	[72]	[71]	[73]	[69]

Table 27. The ^{13}C-NMR spectral data of compounds **73–79** except those which have no reported ^{13}C-NMR data.

C	73	74	75	76	79
2	167.7	174.9	162.3	163.1	160.6
3	106.3	105.3	113.2	111.0	112.4
4	182.0	182.2	181.9	181.8	178.5
5	161.1	160.9	143.1	146.1	*
6	108.1	106.4	127.0	128.3	126.5
7	163.1	163.2	160.1	161.1	160.8
8	93.4	93.4	119.4	121.1	100.6
9	157.1	157.0	161.3	157.9	159.0
10	103.0	103.1	115.8	114.3	114.6
11	19.9	32.4	48.7	48.5	47.8

Table 27. *Cont.*

C	73	74	75	76	79
12	-	19.78	204.4	204.6	202.4
13	-	19.75	29.8	30.7	29.8
14	-	-	23.3	23.3	22.5
CH$_3$O	-	-	-	55.8	-
1′	70.7	70.7	75.5	75.6	71.4
2′	69.7	69.7	71.9	72.3	81.7
3′	77.6	77.6	80.0	79.9	78.2
4′	68.7	68.7	72.9	72.1	70.1
5′	81.8	81.7	79.8	78.5	81.1
6′	61.2	61.2	65.2	65.4	60.9
1″	119.7	119.7	133.6	128.3	105.1
2″	108.9	108.9	131.2	131.0	74.3
3″	145.5	145.4	116.8	115.3	76.1
4″	138.5	138.4	161.3	160.4	69.3
5″	145.5	145.4	116.8	115.3	76.1
6″	108.9	108.9	131.2	131.0	60.3
7″	165.3	165.4	146.2	146.2	-
8″	-	-	114.9	116.0	-
9″	-	-	169.2	169.1	-
1‴	119.1	119.0	-	-	-
2‴	108.7	108.7	-	-	-
3‴	145.3	145.3	-	-	-
4‴	138.4	138.3	-	-	-
5‴	145.3	145.3	-	-	-
6‴	108.7	108.7	-	-	-
7‴	164.6	164.4	-	-	-
Solvent	DMSO-d_6	DMSO-d_6	CD$_3$OD	CD$_3$OD	DMSO-d_6
References	[71]	[71]	[74]	[74]	[78]

* The authors missed assigning this position.

Table 28. The ^{13}C-NMR spectral data of compounds **80–89**.

C	80	81	82	83	84	85	86	87	88	89
2	167.0	169.3	174.4	173.0	168.2	167.6	167.2	174.8	167.7	174.5
3	107.4	108.7	105.0	106.2	108.3	107.8	107.5	105.4	108.0	105.6
4	181.9	184.3	182.2	181.9	181.9	182.2	181.9	182.4	182.2	182.4
5	160.4	149.4	160.4	160.4	160.3	160.8	160.5	160.7	161.0	161.0
6	98.4	94.5	98.5	98.7	98.3	97.6	98.3	97.8	98.0	97.9
7	162.5	162.8	162.8	164.2	162.7	162.2	162.6	162.3	162.4	162.6
8	104.3	105.2	104.1	104.7	108.4	103.9	104.0	102.8	103.9	102.1
9	156.1	164.6	156.5	156.6	158.9	157.1	156.3	157.2	157.1	157.2
10	103.5	106.3	102.0	103.3	103.8	102.8	103.5	104.0	102.0	104.0
11	19.6	20.3	32.7	39.1	42.1	11.0	19.7	33.1	20.1	33.1
12	-	-	20.0	27.0	66.5	-	-	19.8	-	20.3
13	-	-	19.5	11.4	45.8	-	-	20.3	-	20.1
14	-	-	-	17.4	64.1	-	-	-	-	-
15	-	-	-	-	23.6	-	-	-	-	-
1′	73.1	74.9	73.1	73.3	73.2	70.6	73.3	70.7	70.7	70.8
2′	71.0	72.8	71.2	71.2	70.3	72.5	70.0	72.5	70.1	70.2
3′	78.5	80.0	78.5	78.7	78.6	76.1	78.1	76.0	77.0	77.0
4′	70.3	71.7	70.8	70.9	71.0	70.8	70.7	71.2	68.5	68.9
5′	81.1	82.4	81.5	81.5	81.4	81.7	78.3	82.0	81.5	81.9
6′	61.3	62.9	61.8	61.7	61.5	61.5	63.8	61.8	61.0	61.3
1″	-	-	-	-	-	119.7	119.4	119.6	119.6	119.6
2″, 6″	-	-	-	-	-	108.8	108.5	108.7	108.9	108.9
3″, 5″	-	-	-	-	-	145.5	145.5	145.4	145.6	145.5
4″	-	-	-	-	-	138.8	138.3	138.3	138.7	138.7
7″	-	-	-	-	-	165.0	165.8	165.0	165.5	165.4
1‴	-	-	-	-	-	-	-	-	118.7	118.7
2‴, 6‴	-	-	-	-	-	-	-	-	108.7	108.7
3‴, 5‴	-	-	-	-	-	-	-	-	145.5	145.4
4‴	-	-	-	-	-	-	-	-	138.5	138.6
7‴	-	-	-	-	-	-	-	-	164.8	164.8
OCH$_3$	-	56.8	-	-	-	-	-	-	-	-
Solvent	DMSO-d_6	CD$_3$OD	DMSO-d_6	DMSO-d_6	DMSO-d_6	DMSO-d_6	DMSO-d_6	DMSO-d_6	DMSO-d_6	DMSO-d_6
References	[86]	[87]	[158]	[158]	[72]	[71]	[91]	[69]	[71]	[71]

Table 29. The ^{13}C-NMR spectral data of compounds **91–100** except which has no reported ^{13}C-NMR data.

C	91	93	94	95	96	97	98	99	100
2	162.7	164.9	167.1	167.0	169.2	160.2	160.2	160.8	159.6
3	109.8	111.2	112.2	111.9	110.8	112.6	112.4	113.0	113.7
4	182.7	178.7	182.3	181.9	182.4	178.3	178.5	178.9	179.1
5	144.0	141.0	143.1	143.7	144.1	139.5	140.1	141.3	144.1
6	112.8	111.5	112.2	112.6	113.1	116.9	116.3	111.9	118.5
7	162.5	160.0	162.3	162.1	162.7	160.2	159.5	160.3	159.0
8	113.3	112.8	116.1	113.1	113.7	107.1	110.0	113.1	106.2
9	158.8	157.2	160.0	158.9	159.1	155.9	157.8	157.3	159.6
10	117.4	115.9	118.5	117.0	117.6	114.3	114.7	115.9	113.9
11	124.7	43.2	44.3	44.2	76.2	47.2	47.7	47.8	48.2
12	139.0	63.8	66.7	66.3	69.5	202.5	202.4	202.5	200.7
13	18.4	23.8	23.6	23.6	19.7	22.2	22.6	30.0	30.2
14	23.6	22.8	23.3	23.6	23.7	29.9	29.9	23.0	23.1
15	56.9	56.4	-	56.7	56.9	-	-	56.6	-
1′	75.1	73.0	76.0	74.6	74.9	80.2	73.5	72.9	68.1
2′	72.4	70.9	73.2	72.7	72.9	77.5	71.0	71.1	73.7
3′	80.3	78.8	80.1	80.0	80.3	74.8	78.7	79.1	73.7
4′	72.3	70.8	71.8	71.9	72.2	80.5	70.4	70.7	70.2
5′	82.9	81.8	82.7	82.4	82.6	68.3	81.5	81.8	76.6
6′	63.0	61.7	62.8	63.0	63.3	63.8	61.4	61.8	61.6
2′-OCOCH$_3$	-	-	-	-	-	-	-	-	168.6, 20.7
3′-OCOCH$_3$	-	-	-	-	-	-	-	-	169.4, 20.7
4′-OCOCH$_3$	-	-	-	-	-	-	-	-	170.3, 20.7
6′-OCOCH$_3$	-	-	-	-	-	-	-	-	170.6, 20.7
Solvent	CD$_3$OD	DMSO-d_6	CD$_3$OD	CD$_3$OD	CD$_3$OD	DMSO-d_6	DMSO-d_6	CD$_3$OD	CDCl$_3$
References	[80]	[159]	[160]	[82]	[160]	[94]	[94]	[96]	[93]

Table 30. The ^{13}C-NMR spectral data of compounds **103–110** except which has no reported ^{13}C-NMR data.

C	103	104	106	107	108	109	110
2	165.0	165.0	167.4	167.2	167.4	167.1	167.2
3	111.3	111.3	112.5	112.6	112.6	111.9	111.9
4	178.6	178.8	182.3	182.2	182.2	181.9	182.0
5	141.7	141.8	144.6	144.8	144.6	144.3	144.3
6	111.0	111.3	112.7	112.6	112.5	112.3	112.4
7	159.5	159.8	162.1	162.0	162.0	161.6	161.6
8	110.6	110.7	111.8	111.4	111.8	111.6	111.3
9	157.4	157.5	159.8	159.6	159.6	159.3	159.1
10	115.8	115.8	117.2	117.2	117.2	116.9	116.9
11	43.1	43.3	44.6	44.9	44.6	44.5	44.3
12	64.4	63.9	65.9	66.3	66.0	66.6	66.5
13	23.3	23.7	23.6	23.7	23.6	23.5	23.6
14	22.8	22.9	23.7	23.7	23.6	23.5	23.3
15	56.5	56.5	57.0	57.1	57.0	56.9	57.0
1′	70.6	70.6	72.8	72.8	72.8	71.9	71.8
2′	72.6	72.3	73.4	73.7	73.9	74.0	73.7
3′	75.7	75.9	77.7	77.7	77.9	77.6	77.4
4′	70.4	70.9	72.4	72.4	72.3	72.4	72.4
5′	81.8	82.0	83.0	80.1	82.9	82.7	82.4
6′	61.5	61.6	63.1	64.4	63.1	62.9	62.7
1″	133.9	125.0	127.0	127.0	128.2	135.3	126.7
2″	128.9	130.3	133.2	131.1	130.9	128.9	130.9
3″	128.2	115.8	115.7	116.9	115.4	129.7	116.6
4″	130.4	159.6	160.0	161.3	163.2	131.3	160.7
5″	128.2	115.8	115.7	116.9	115.4	129.7	116.6
6″	128.9	130.3	133.2	131.1	130.9	128.9	130.9
7″	144.1	144.4	145.0	146.6	146.1	146.1	146.4
8″	117.8	114.0	115.7	114.6	115.6	118.1	114.2
9″	165.0	165.4	167.6	168.0	167.8	167.2	167.8
OCH$_3$	-	-	-	-	55.9	-	-
COCH$_3$	-	-	-	173.0, 20.9	-	-	-
Solvent	DMSO-d_6	DMSO-d_6	CD$_3$OD	CD$_3$OD	CD$_3$OD	CD$_3$OD	CD$_3$OD
References	[99]	[159]	[80]	[100]	[100]	[82]	[82]

Table 31. The ^{13}C-NMR spectral data of compounds **111–122** except which has no reported ^{13}C-NMR data.

C	111	112	113	114	115	116	117	119	120	121	122
2	167.2	167.6	169.7	160.3	163.0	160.1	163.1	160.6	164.0	167.6	166.8
3	112.1	112.2	110.5	1125	113.8	112.5	111.0	112.5	112.4	112.2	112.2
4	182.2	182.3	182.4	178.6	182.1	178.4	181.8	178.6	182.0	182.3	182.3
5	143.5	144.6	144.8	141.0	144.7	140.4	146.1	141.8	144.9	144.6	144.8
6	116.3	112.6	112.7	115.8	112.2	115.7	128.3	111.5	113.1	112.6	112.7
7	161.3	162.0	162.1	159.1	162.1	159.4	161.1	159.7	162.3	161.0	162.2
8	110.0	111.9	111.8	110.2	112.0	110.7	121.1	110.8	111.6	111.9	111.9
9	160.4	159.6	159.4	158.3	159.6	157.9	157.9	157.4	159.5	159.6	159.6
10	117.0	117.3	117.5	114.8	117.1	114.8	114.3	115.6	117.6	117.2	117.3
11	44.3	44.6	75.7	48.1	49.1	47.8	48.5	47.9	98.6	44.6	44.6
12	66.7	66.8	69.2	202.4	204.7	202.3	204.6	202.1	203.7	66.8	66.8
13	23.5	23.6	19.7	30.4	29.9	29.6	30.7	29.6	23.7	23.7	24.0
14	23.5	23.6	23.7	22.7	23.7	22.6	23.3	22.7	25.0	23.6	23.5
15	-	57.1	57.1	-	57.1	-	55.8	55.5	57.2	57.1	57.1
1′	73.0	72.1	72.8	70.2	72.1	73.3	75.6	70.4	72.6	72.6	72.6
2′	74.0	74.0	74.1	72.3	73.9	70.8	72.3	72.2	74.3	74.1	73.6
3′	77.9	77.9	77.8	76.0	77.9	78.5	79.9	75.8	78.0	77.8	77.6
4′	72.1	72.7	72.3	70.2	72.6	70.4	72.1	70.6	71.9	78.3	78.1
5′	82.8	82.9	83.0	81.8	82.8	78.4	78.5	81.9	83.3	82.9	82.9
6′	63.1	63.1	63.1	61.8	63.3	64.8	65.4	61.5	63.5	63.0	63.0
1″	128.3	127.5	135.7	125 I	127.0	125.0	128.3	125.5	135.7	129.8	129.7
2″	131.0	115.1	129.2	1301	130.8	130.3	131.0	111.1	129.3	130.8	132.6
3″	116.7	149.9	130.1	115.8	116.7	115.7	115.3	147.9	130.1	118.0	116.7
4″	161.2	146.6	131.6	159.6	161.3	159.7	160.4	149.2	131.6	163.8	159.7
5″	116.7	116.6	130.1	115.8	116.7	115.7	115.3	115.5	130.1	-	-
6″	131.0	123.0	129.2	1301	130.8	130.3	131.0	122.9	129.3	-	-
7″	146.2	147.1	146.4	144.4	145.3	144.9	146.2	144.6	146.3	146.0	144.7
8″	115.8	114.5	118.4	114.1	115.1	114.0	116.0	114.2	118.5	116.5	117.6
9″	168.0	168.2	167.4	165.4	168.1	166.7	169.1	165.4	167.4	167.8	167.7
OCH$_3$	56.1	-	-	-	-	-	-	-	56.3	-	-	-
1‴	-	-	-	-	-	-	-	-	-	101.9	101.7
2‴	-	-	-	-	-	-	-	-	-	74.8	74.9
3‴	-	-	-	-	-	-	-	-	-	72.1	72.6

Table 31. *Cont.*

C	111	112	113	114	115	116	117	119	120	121	122
4‴	-	-	-	-	-	-	-	-	-	71.3	71.2
5‴	-	-	-	-	-	-	-	-	-	78.0	77.9
6‴	-	-	-	-	-	-	-	-	-	62.5	62.4
Solvent	CD$_3$OD	CD$_3$OD	CD$_3$OD	DMSO-d_6	DMSO-d_6	DMSO-d_6	CD$_3$OD	DMSO-d_6	CD$_3$OD	CD$_3$OD	CD$_3$OD
References	[80]	[82]	[160]	[161]	[162]	[163]	[74]	[164]	[81]	[160]	[160]

Table 32. The ^{13}C-NMR spectral data of compounds **123–129** except which has no reported ^{13}C-NMR data.

C	123	124	126	127	128	129
2	167.7	168.3	171.3	170.7	167.7	169.9
3	108.0	108.5	105.6	105.5	108.3	108.4
4	182.2	182.4	184.3	183.0	182.7	182.4
5	158.8	157.8	160.2	158.1	159.5	159.6
6	110.4	111.1	112.5	112.8	107.4	105.2
7	162.3	163.1	163.8	161.0	162.9	163.1
8	93.2	90.5	94.4	93.0	92.8	92.9
9	156.1	156.6	157.8	156.1	156.7	156.4
10	103.3	104.4	106.8	105.6	103.4	103.8
11	19.9	19.9	61.6	60.0	18.8	60.0
OCH$_3$	-	56.5	-	-	-	-
1′	20.5	20.5	22.1	20.5	26.4	26.4
2′	126.9	126.4	128.7	124.6	79.9	79.8
3′	131.8	132.3	132.9	134.1	143.7	143.7
4′	66.1	66.1	68.4	61.0	114.2	114.2
5′	21.2	21.3	21.9	20.0	15.2	15.2
1″	101.5	101.6	102.7	100.1	99.4	99.4
2″	73.6	73.6	75.3	73.2	73.6	73.6
3″	76.9	77.0	78.3	76.7	76.3	76.3
4″	70.2	70.2	71.8	67.7	70.2	70.2
5″	77.0	77.0	77.9	77.0	76.7	76.7
6″	61.1	61.1	62.9	60.1	61.2	61.2
Solvent	DMSO-d_6	DMSO-d_6	CD$_3$OD	CD$_3$OD	CD$_3$OD	CD$_3$OD
References	[104]	[104]	[106]	[60]	[60]	[60]

Table 33. The ^{13}C-NMR spectral data of compounds **130–139**.

C	130	131	132	133	134	135	136	137	138	139
2	170.7	167.6	170.7	168.3	171.4	168.9	165.0	171.0	171.5	171.8
3	105.5	107.4	104.9	108.1	105.5	113.9	*	110.4	110.7	110.7
4	183.0	181.8	181.9	182.5	182.5	176.6	*	179.3	180.4	180.6
5	158.8	158.8	158.9	159.5	159.5	132.6	*	106.6	119.2	126.2
6	111.4	97.6	97.8	98.3	98.4	111.4	99.3	150.7	126.2	126.6
7	162.1	159.9	160.1	160.6	160.8	158.2	162.9	153.1	122.1	135.6
8	93.3	107.5	107.4	108.1	108.2	103.6	94.8	106.9	147.8	119.3
9	156.3	153.8	153.5	154.4	154.1	163.4	*	149.8	149.0	158.0
10	105.7	104.3	104.8	105.0	105.4	112.7	*	116.2	125.0	124.3
11	60.0	19.5	59.3	20.2	59.5	127.9	121.6	102.2	141.4	136.2
1′	28.2	20.3	20.3	21.0	20.9	131.9	128.5	74.3	129.5	114.0
2′	74.6	120.5	120.4	121.1	121.0	115.7	115.7	78.1	129.6	150.7
3′	147.9	134.9	134.9	135.6	135.7	157.1	161.5	71.5	127.4	146.4
4′	109.5	65.9	65.8	66.5	66.5	115.7	115.7	78.0	129.6	118.1
5′	16.6	13.1	13.1	13.8	13.7	131.9	128.5	62.8	129.5	122.0
6′	-	-	-	-	-	39.3	39.1	-	33.8	33.6
7′	-	-	-	-	-	86.3	37.5	-	37.0	37.1
8′	-	-	-	-	-	127.9	121.6	102.2	141.4	136.2
1″	101.2	100.1	100.1	100.7	100.7	101.8	100.3	131.5	103.0	102.9
2″	73.6	72.9	72.9	73.5	73.5	74.6	*	133.0	75.0	74.9
3″	76.5	76.7	76.7	76.7	76.7	78.4	*	116.3	78.2	77.8
4″	69.8	69.2	69.2	69.8	69.8	71.2	*	164.1	71.3	71.3
5″	77.1	76.1	76.1	77.3	77.3	78.2	*	116.3	78.2	78.1
6″	61.1	60.2	60.2	60.8	60.8	62.5	*	120.9	62.5	62.4
7″	-	-	-	-	-	-	-	33.9	-	-
8″	-	-	-	-	-	-	-	37.1	-	-
OCH$_3$	-	-	-	-	-	-	-	6-OCH$_3$ 56.3 4″-OCH$_3$ 55.1	-	56.6
Solvent	CD$_3$OD	DMSO-d_6	DMSO-d_6	DMSO-d_6	DMSO-d_6	CD$_3$OD	DMSO-d_6	DMSO-d_6	CD$_3$OD	CD$_3$OD
References	[60]	[108]	[108]	[109]	[109]	[110]	[111]	[112]	[113]	[114]

* The authors missed assigning these positions.

Table 34. The ^{13}C-NMR spectral data of compounds **140–148**.

C	140	141	142	143	144	145	146	147	148
2	147.2	146.9	147.3	89.9	89.0	92.9	92.1	92.3	88.2
3	106.1	105.0	104.9	24.3	28.3	28.1	27.9	28.0	30.1
4	153.7	160.5	159.2	165.5	164.5	160.3	165.3	156.3	158.5
5	177.2	185.9	184.7	176.9	176.4	184.7	176.6	176.6	182.9
6	109.7	107.8	107.2	111.9	109.8	109.5	110.9	111.4	107.2
7	163.7	169.0	168.7	163.3	164.0	169.7	162.7	162.6	166.7
9	95.8	92.0	92.0	94.8	95.2	90.4	94.0	94.2	88.8
10	157.8	156.1	154.5	159.9	160.1	168.3	159.7	165.4	166.6
11	117.0	114.3	113.4	118.7	117.2	111.3	118.1	118.5	108.7
12	152.8	106.9	106.3	112.3	111.4	106.6	112.5	112.8	102.9
13	155.6	155.7	155.2	156.4	157.7	157.9	156.1	159.9	158.6
14	65.7	67.8	66.3	21.4	20.7	20.8	66.4	66.8	66.1
4-OCH$_3$	61.9	-	-	-	*	-	60.8	61.0	-
1′	103.0	104.4	104.4	77.3	74.3	79.7	70.8	70.8	143.7
2′	74.0	75.0	74.2	23.4	24.1	24.3	26.0	26.2	115.5
3′	77.2	77.9	77.6	22.3	23.5	23.0	25.6	25.8	15.7
4′	70.6	71.6	70.9	-	-	-	-	-	-
5′	77.1	78.0	76.7	-	-	-	-	-	-
6′	61.7	68.7	69.5	-	-	-	-	-	-
1″	-	111.1	103.3	98.9	100.1	99.5	104.0	104.4	102.8
2″	-	77.2	74.4	75.1	74.6	75.6	74.8	75.0	73.7
3″	-	80.6	77.4	78.8	78.6	78.5	78.4	78.3	76.7
4″	-	74.9	70.9	71.3	71.4	72.0	71.4	71.3	70.3
5″	-	65.5	77.7	77.1	78.2	76.8	78.1	75.7	76.9
6″	-	-	61.9	62.4	63.1	68.5	62.6	64.5	61.4
1‴	-	-	-	-	-	111.2	-	126.5	-
2‴	-	-	-	-	-	78.6	-	111.4	-
3‴	-	-	-	-	-	81.0	-	149.0	-
4‴	-	-	-	-	-	75.6	-	151.2	-
5‴	-	-	-	-	-	66.2	-	116.8	-
6‴	-	-	-	-	-	-	-	123.9	-
7‴	-	-	-	-	-	-	-	145.9	-
8‴	-	-	-	-	-	-	-	115.1	-
9‴	-	-	-	-	-	-	-	167.8	-
3‴-OCH$_3$	-	-	-	-	-	-	-	55.9	-
Solvent	DMSO-d_6, C$_6$D$_6$	CD$_3$OD	DMSO-d_6	CDCl$_3$, CD$_3$OD	CDCl$_3$, CD$_3$OD	CD$_3$OD	C$_5$D$_5$N	C$_5$D$_5$N	CD$_3$OD
References	[117]	[118]	[109]	[122]	[122]	[119]	[121]	[132]	[60]

* The authors missed assigning this position.

Table 35. The ^{13}C-NMR spectral data of compounds **149–155**.

C	149	150	151	152	153	154	155
2	167.4	169.5	162.7	167.6	167.6	170.9	168.7
3	108.7	108.0	111.1	107.5	107.4	105.6	108.6
4	182.7	184.2	175.3	181.9	181.8	182.5	182.6
5	160.0	160.9	160.7 *	158.6	158.5	159.2	160.3
6	104.1	105.0	104.9	103.1	103.1	103.8	102.5
7	159.6	160.4	152.7 *	163.5	163.3	164.1	164.3
8	94.9	95.8	91.1	109.8	109.9	110.5	105.8
9	156.3	157.7	157.8	152.9	152.7	153.1	155.0
10	104.4	105.0	107.7	105.5	105.4	106.5	106.1
2′	78.4	79.3	76.6	69.6	69.6	70.2	74.2
3′	74.3	75.0	72.4	135.2	135.0	135.8	73.5
4′	22.3	22.7	22.4	124.7	124.7	125.2	132.7
5′	-	-	-	20.6	20.3	20.9	117.3
1″	102.4	102.0	100.4	101.2	100.9	101.9	103.9
2″	74.9	74.9	**	72.8	72.6	73.3	73.7
3″	78.4	78.1	76.9	76.8	76.1	76.8	76.5
4″	71.8	71.9	70.2	70.1	69.4	70.1	70.9
5″	78.4	75.0	76.9	76.6	75.2	76.6	76.0
6″	63.0	68.8	61.3	60.6	67.7	61.6	68.5
1‴	-	111.0	-	-	102.7	-	103.4
2‴	-	77.9	-	-	72.8	-	73.7
3‴	-	80.5	-	-	76.1	-	76.9
4‴	-	75.0	-	-	69.4	-	70.0
5‴	-	65.7	-	-	75.9	-	77.0
6‴	-	-	-	-	60.4	-	61.1
2a	20.1	20.4	19.0	19.4	19.2	59.7	20.1
2′a	22.3	22.4	21.2	-	-	-	-
2′b	25.7	26.1	25.2	-	-	-	-
3′a	-	-	-	69.6	69.6	70.2	73.0
7-OCH$_3$	-	-	56.0	-	-	-	-
Solvent	CDCl$_3$	CD$_3$OD	DMSO-d_6	DMSO-d_6	DMSO-d_6	DMSO-d_6	DMSO-d_6
References	[121]	[119]	[123]	[108]	[108]	[108]	[109]

* Interchangeable data. ** The authors missed the assignment of this position.

Table 36. The ^{13}C-NMR spectral data of compounds **157–165**.

C	157	158	159	160	161	162	163	164	165
Chromone moiety									
2′	92.4	92.2	92.2	92.2	92.2	92.2	92.2	92.2	92.2
3′	27.9	27.8	27.8	27.8	27.8	27.8	27.8	27.8	27.8
4′	156.3	156.2	156.2	156.2	156.2	156.2	156.2	156.2	156.1
5′	176.2	176.2	176.2	176.2	176.2	176.2	176.3	176.3	176.3
6′	111.5	110.9	111.1	110.9	110.9	110.8	111.1	111.2	111.1
7′	162.1	162.5	162.4	162.4	162.7	162.5	162.6	162.6	162.5
9′	94.1	94.0	94.0	94.0	94.1	94.0	94.0	94.1	94.0
10′	165.4	165.2	165.2	165.2	165.2	165.2	165.2	165.2	165.2
11′	118.5	118.4	118.4	118.4	118.4	118.4	118.4	118.4	118.4
12′	112.9	112.8	112.8	112.5	112.8	112.8	112.8	112.8	112.7
13′	159.9	159.8	159.8	159.7	159.8	159.7	159.8	159.9	159.8
14′	66.3	66.3	66.5	66.2	66.4	66.2	66.5	66.3	66.5
4-OCH$_3$	60.9	60.9	60.9	60.8	60.9	60.9	60.9	60.9	60.8
1″	70.6	70.6	70.6	70.6	70.6	70.6	70.6	70.7	70.6
2″	26.0	26.1	26.1	26.1	26.1	26.1	26.1	26.2	26.1
3″	25.6	25.7	25.7	25.6	25.7	25.7	25.6	25.6	25.7
Glu-1	101.7	104.1	104.2	104.1	104.1	104.0	104.2	103.9	104.3
Glu-2	74.5	74.9	74.8	74.8	74.9	74.9	74.9	74.8	74.9
Glu-3	79.1	78.3	78.1	78.4	78.4	78.4	78.4	78.2	78.1
Glu-4	69.3	71.2	71.4	71.4	71.5	71.3	71.5	72.0	71.9
Glu-5	78.4	76.9	75.4	77.1	77.1	76.7	77.0	76.8	76.7
Glu-6	62.4	63.5	64.2	62.7	62.7	62.8	62.6	62.9	62.7
Triterpene moiety									
1	31.8	31.9	31.9	32.3	30.3	32.0	27.4	31.9	31.9
2	29.8	29.8	30.0	30.0	29.5	30.0	29.8	29.6	29.8
3	87.8	88.0	88.3	88.4	88.1	88.3	88.3	88.0	88.0
4	41.1	40.3	41.2	41.2	41.2	41.2	40.7	41.1	41.1
5	47.1	47.0	47.4	47.4	42.6	47.4	43.8	47.0	46.9
6	20.5	20.4	20.8	20.9	21.7	20.8	22.0	20.4	20.3
7	25.6	25.8	25.9	26.4	113.2	26.3	114.0	25.8	25.7

Table 36. *Cont.*

C	157	158	159	160	161	162	163	164	165
8	46.0	45.9	47.1	49.0	149.1	47.3	148.6	45.8	46.4
9	19.8	19.9	19.2	19.9	21.1	19.5	27.6	20.0	20.5
10	26.5	26.6	26.6	26.5	28.2	26.4	29.1	26.6	27.0
11	36.7	36.6	26.3	26.4	25.4	26.1	63.2	36.7	36.4
12	77.2	76.8	31.4	34.0	33.9	33.3	48.3	77.2	76.9
13	48.8	48.7	42.2	41.8	41.2	46.6	45.4	48.7	48.8
14	47.2	47.4	45.2	46.5	49.8	45.1	48.1	47.8	47.8
15	43.4	43.4	50.8	82.5	80.4	42.8	45.2	43.8	45.6
16	70.7	72.8	218.6	103.0	103.3	72.5	114.5	71.1	74.6
17	55.7	55.6	60.4	60.5	60.3	51.4	61.0	56.8	52.4
18	13.5	13.5	18.8	20.3	22.5	20.5	20.7	13.5	12.9
19	29.9	30.7	30.0	30.7	28.2	30.1	18.7	29.8	30.0
20	26.9	25.4	29.1	27.6	27.5	34.2	23.7	26.0	24.6
21	21.1	20.8	19.9	21.4	21.6	17.3	19.6	21.1	25.7
22	41.5	38.6	40.3	33.6	33.6	86.2	37.8	41.9	106.0
23	104.2	102.9	173.4	74.2	74.4	109.6	71.7	101.9	152.9
24	83.1	212.4	-	80.5	80.4	77.3	88.5	77.7	75.9
25	79.1	35.0	-	76.4	76.5	83.5	76.4	81.3	78.2
26	19.4	19.2	-	21.6	21.6	27.3	23.2	23.7	22.3
27	29.8	19.7	-	24.1	24.1	24.5	20.2	24.9	23.1
28	19.4	19.4	19.6	11.8	18.2	19.4	27.4	19.5	20.7
29	25.6	25.6	25.6	25.5	25.6	25.6	25.8	25.6	25.6
30	15.2	15.2	15.2	15.5	14.2	15.3	14.5	15.2	15.2
Xyl-1	107.4	107.5	107.5	107.4	107.4	107.4	107.4	107.5	107.5
Xyl-2	75.5	75.5	75.5	75.5	75.5	75.5	75.5	75.5	75.5
Xyl-3	78.5	78.6	78.5	78.5	78.5	78.5	78.5	78.5	78.5
Xyl-4	71.1	71.0	71.1	71.1	71.2	71.1	71.1	71.2	71.1
Xyl-5	67.0	67.0	67.1	67.0	67.0	67.0	67.0	67.0	67.0
$COCH_3$	170.4 21.5	170.4 21.5	-	171.0 21.1	171.0 21.1	-	-	170.5 21.6	170.6 21.1
Solvent	C_5D_5N	C_5D_5N	C_5D_5N	C_5D_5N	C_5D_5N	C_5D_5N	C_5D_5N	C_5D_5N	C_5D_5N
References	[130]	[130]	[130]	[130]	[130]	[130]	[130]	[136]	[136]

Table 37. The ^{13}C-NMR spectral data of compounds **166** and **167**.

C	166	167
1	98.1	98.1
3	154.5	154.5
4	105.6	105.6
5	28.7	28.6
6	30.0	29.9
7	75.1	75.0
8	133.3	133.3
9	43.9	43.8
10	121.1	121.1
11	168.6	168.5 [a]
1′	99.8	99.7
2′	74.8	74.7 [b]
3′	77.9	77.8 [c]
4′	71.6	71.2 [d]
5′	78.5	78.4 [e]
6′	62.7	62.4 [f]
2″	167.5	168.1 [a]
3″	118.2	118.6
4″	181.6	181.8
5″	163.5	163.1
6″	100.2	101.1
7″	166.2	164.9
8″	94.7	95.7
9″	159.2	158.7
10″	104.9	106.6
-CH$_3$	18.9	19.0
1‴	-	101.6
2‴	-	74.8 [b]
3‴	-	77.9 [c]
4‴	-	71.6 [d]
5‴	-	78.5 [e]
6‴	-	62.7 [f]
NMR	CD$_3$OD	CD$_3$OD
References	[43]	[43]

[a–f] Values with the same superscript are interchangeable.

Table 38. The ^{13}C-NMR spectral data of compounds **168–180** except those which have no reported ^{13}C-NMR data.

C	168	169	170	171	172	178	179	180
2	163.7	159.3	168.0	159.6	167.5	170.2	166.3	166.3
3	108.7	108.8	109.7	109.9	110.0	110.8	110.0	110.0
4	179.2	179.5	178.9	178.8	179.0	179.0	178.7	178.8
4a	125.8	126.2	125.9 *	126.4	125.9 *	125.8 *	125.8 *	126.5
5	149.6	149.6	149.9	150.0	149.8	150.1	149.7	149.9
6	125.4	125.4	125.8	125.9	125.9 *	125.8 *	125.9	126.0
6a	137.0	137.2	137.0	137.0	137.4	137.2	137.3	136.3
7	183.0	183.3	183.2	183.5	183.3	183.1	183.1	181.5
7a	125.8	127.3	126.3 *	126.5	126.4 *	126.1 *	126.2 *	130.7
8	140.0	140.6	139.7	140.0	140.1	139.9	140.2	119.3
9	133.0	132.6	132.9	132.2	133.1	133.3	133.1	133.2
10	138.4	138.5	137.8	137.0	138.5	138.6	138.6	140.7
11	159.7	163.9	159.7	167.6	159.9	159.9	159.8	159.5
11a	116.0	115.8	116.1	116.0	116.2	116.0	116.1	116.1
12	188.1	188.0	187.9	187.7	188.1	188.3	188.0	187.7
12a	118.9	118.9	119.1	119.1	119.2	118.9	119.2	119.8
12b	155.7	155.8	156.0	156.0	156.2	155.8	156.1	156.2
13	24.0	24.1	24.3	24.2	24.1	24.2	24.1	24.1
14	127.2	125.9	57.6	60.3	59.1	75.6	57.7	57.7
15	12.1	15.0	13.8	14.9	14.5	23.5	14.5	14.4
16	134.2	134.2	62.0	61.7	61.6	71.7 *	63.9	63.9
17	14.9	12.1	14.1	123.3	123.3	127.4	55.4	55.4
18	-	-	-	134.1	134.0	130.2	51.8	51.8
19	-	-	-	14.4	13.8	13.5	17.2	17.2
2′	77.3	77.7	77.2	77.6	77.3	77.2	77.3	67.9
3′	71.9	71.8	71.3	70.6	71.9	71.7 *	71.9	71.0
4′	67.4	67.5	67.8	68.1	67.4	67.3 *	67.4	57.7
5′	28.3	41.0	28.8	38.8	28.4	28.4	28.3	35.0
6′	75.2	75.4	74.8	74.0	75.2	75.0	75.2	68.2
7′	18.9	18.9	18.8	18.5	18.9	18.9	18.9	17.5
8′	-	-	-	-	-	-	-	13.2

Table 38. *Cont.*

C	168	169	170	171	172	178	179	180
4′-N(CH$_3$)$_2$	40.4	40.5	40.3	40.4	40.4	40.3	40.4	37.1
2″	67.2	69.8	67.5	69.8	67.2	67.4 *	67.3	-
3″	70.8	76.4	70.3	74.5	70.9	70.7	70.9	-
4″	57.4	57.6	57.6	59.1	57.3	57.9	57.3	-
5″	33.6	28.6	33.3	30.4	33.6	33.3	33.7	-
6″	69.5	64.9	69.4	65.7	69.7	69.6	69.6	-
7″	17.6	15.0	17.4	15.1	17.7	17.6	17.6	-
8″	12.3	13.7	13.1	13.9	12.3	12.6	12.3	-
3″-COCH$_3$	-	170.6 21.3	-	171.0 21.5	-	-	-	-
4″-N(CH$_3$)$_2$	36.8	39.3	36.9	38.8	36.8	36.8	36.8	-
Solvent	CDCl$_3$	CDCl$_3$	CDCl$_3$	CDCl$_3$	CDCl$_3$	CDCl$_3$	CDCl$_3$	CDCl$_3$
References	[165]	[140]	[141]	[140]	[142]	[143]	[165]	[146]

* Interchangeable values.

Table 39. The ^{13}C-NMR spectral data of compounds **181–183**.

C	181	182	183
2	168.4	176.4	176.5
3	112.0	111.1	111.3
4	180.3	181.3	178.4
4a	126.1	126.0	124.7
5	150.4	150.4	149.6
6	121.3	121.0	121.2
6a	138.3	138.1	139.0
7	182.7	182.7	182.4
7a	133.2	133.2	133.2
8	120.1	120.0	119.8
9	135.4	135.5	135.6
10	137.4	137.4	137.0
11	161.1	161.0	161.3
11a	118.1	118.0	118.0

Table 39. *Cont.*

C	181	182	183
12	189.1	189.2	189.2
12a	122.0	121.9	121.6
12b	157.7	157.4	156.9
13	70.8	70.8	175.3
14	61.2	77.7	77.7
15	63.7	72.6	72.6
16	20.2	23.9	23.9
17	13.7	17.0	17.1
1′	98.9	98.9	-
2′	37.2	37.2	-
3′	58.3	58.4	-
4′	71.2	71.1	-
5′	70.4	70.4	-
6′	17.2	17.2	-
7′	24.6	24.5	-
1″	70.2	70.1	70.3
2″	28.0	27.7	27.4
3″	65.2	65.1	64.9
4″	75.1	75.0	75.2
5″	72.6	72.6	72.3
6″	18.1	18.2	18.3
3″-N(CH$_3$)$_2$	43.2 41.7	43.3 41.4	42.5 42.5
1‴	101.4	101.4	101.4
2‴	33.4	33.4	33.4
3‴	66.6	66.6	66.6
4‴	72.0	72.0	72.0
5‴	69.5	69.5	69.5
6‴	17.5	17.5	17.5
Solvent	CD$_3$OD	CD$_3$OD	CD$_3$OD
References	[147]	[147]	[147]

Table 40. The ^{13}C-NMR spectral data of compounds **184–192**.

C	184	185	186	187	188	189	190	191	192
2	167.5	167.4	167.0	167.0	165.7	165.7	167.5	169.3	169.3
3	111.1	110.9	110.9	110.8	111.3	111.3	111.1	109.1	109.1
4	180.2	180.0	179.4	179.4	178.5	178.6	180.2	182.4	182.5
4a	126.6	126.4	126.5	126.4	126.3	126.3	126.6	113.3	113.3
5	149.2	149.1	149.4	149.2	148.2	148.2	149.3	166.7	166.7
6	122.5	122.4	122.9	122.8	124.1	124.0	122.5	110.7	110.7
6a	137.2	137.1	137.0	136.9	136.9	136.9	137.2	139.8	139.9
7	181.2	181.0	181.2	181.2	181.5	181.4	181.2	180.8	180.8
7a	130.3	130.2	130.4	130.3	130.5	130.5	130.5	130.6	130.4
8	119.8	119.7	119.7	119.7	119.5	119.5	119.8	119.4	119.5
9	133.6	133.7	133.4	133.6	133.3	133.5	133.7	132.4	132.8
10	141.3	141.1	141.0	140.9	141.0	140.9	140.7	140.2	140.9
11	159.1	159.2	159.3	158.9	159.3	159.3	159.4	159.1	159.1
11a	115.8	115.6	115.9	115.7	115.9	115.9	115.9	115.6	115.4
12	186.9	186.9	187.1	186.9	187.5	187.4	186.9	186.2	186.3
12a	121.8	121.6	121.6	121.6	120.8	120.8	121.8	112.3	112.4
12b	156.8	156.7	156.6	156.5	156.1	156.1	156.8	156.6	156.6
13	80.9	80.0	79.0	78.9	48.3	48.2	80.9	-	-
14	59.8	59.7	59.8	59.7	59.8	59.7	59.8	60.0	60.0
15	19.7	19.6	19.8	19.7	20.0	19.9	19.6	19.6	19.7
16	62.7	62.5	62.5	62.5	62.5	62.5	62.7	62.7	62.7
17	13.4	13.3	13.4	13.3	13.5	13.4	13.3	13.2	13.2
18	170.5	170.4	170.9	170.9	170.4	170.4	170.5	-	-
19	52.6	52.5	52.4	52.3	52.3	52.3	52.6	-	-
2′	73.8	73.8	74.9	74.9	74.3	74.2	73.8	-	-
3′	68.9	68.9	68.5	68.2	68.8	68.8	69.0	-	-
4′	80.2	80.1	74.8	74.9	81.5	81.4	80.2	-	-
5′	67.9	67.9	26.1	26.0	68.0	67.9	68.0	-	-
6′	73.7	73.6	70.3	70.2	73.9	74.0	73.7	-	-
7′	14.1	14.0	14.7	14.7	14.7	14.6	14.0	-	-
4′-OCH$_3$	57.9	57.8	55.6	55.5	57.1	57.1	58.0	-	-
2″	70.3	70.7	70.2	70.8	70.3	70.8	70.0	70.0	70.7
3″	77.7	82.6	77.8	82.7	77.8	82.8	76.0	77.3	82.8
4″	54.9	58.1	55.1	58.1	54.9	58.1	51.7	56.1	58.1

Table 40. *Cont.*

C	184	185	186	187	188	189	190	191	192
5″	40.3	44.7	40.4	44.7	40.4	44.8	44.9	38.5	44.8
6″	62.1	62.2	62.3	62.2	62.2	62.3	62.3	63.6	62.3
7″	14.7	13.5	14.8	13.5	14.7	13.6	14.1	15.4	13.6
8″	24.1	14.0	23.8	13.9	24.2	14.1	32.6	22.6	14.0
4″-N(CH$_3$)$_n$	27.9	40.3	27.8	40.3	28.1	40.4	-	27.2	40.3
1‴	93.4	94.4	93.7	94.5	93.4	94.5	93.3	94.5	94.5
2‴	30.8	31.1	30.9	31.1	30.9	31.1	30.8	31.1	31.1
3‴	74.8	74.9	74.8	74.9	74.9	75.0	74.8	75.0	75.0
4‴	72.1	72.1	72.1	72.0	72.2	72.2	72.2	71.5	72.2
5‴	65.4	65.0	65.7	65.0	65.4	65.1	65.4	66.6	65.0
6‴	17.7	17.6	17.7	17.6	17.8	17.7	17.8	17.5	17.7
3‴-OCH$_3$	55.9	56.1	56.1	56.1	55.9	56.2	56.0	56.4	56.1
Solvent	CDCl$_3$	CDCl$_3$	CDCl$_3$	CDCl$_3$	CDCl$_3$	CDCl$_3$	CDCl$_3$	CDCl$_3$	CDCl$_3$
References	[149]	[149]	[149]	[149]	[137]	[137]	[137]	[137]	[137]

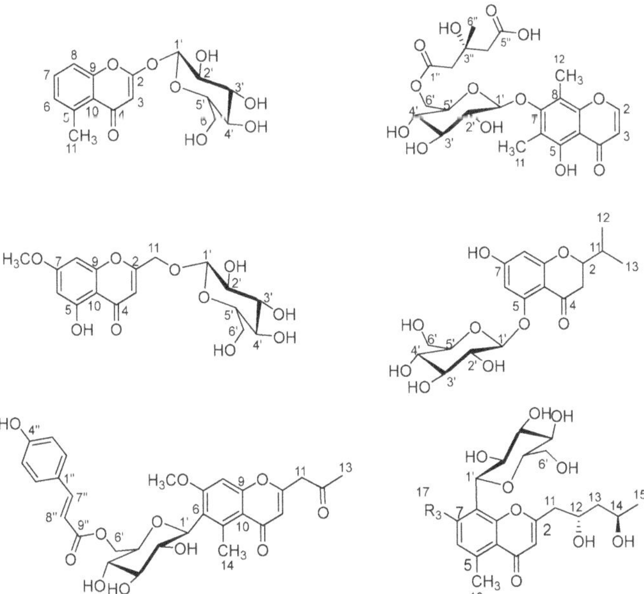

Figure 25. Representative guide figure for numbering of chromone glycosides attached to different substituents.

6. Conclusions

Chromone glycosides are one of the most important classes of secondary metabolites. In this review, we summarized 192 naturally occurring chromone glycosides with their sources, reported activities, and spectroscopic features. Basically, they were categorized into several classes: chromone-*O*-glycosides including compounds **1–59**, among them, four chromanone glycosides (**60–63**), chromone-*C*-glycosides including compounds **64–122**, prenyl and isoprenyl chromone glycosides including compounds **123–134**, phenyl ethyl chromone glycosides including compounds **135–139**, furano-chromone glycosides including compounds **140–148**, pyrano-chromone glycosides including compounds **149–151**, oxepino-chromone glycosides including compounds **152–155**, Pyrido-chromone glycoside including compound **156**, furanochromones with cycloartane triterpenes including compounds **157– 165**, glycoside derivatives of chromones with secoiridoids including compounds **166** and **167**, and chromone alkaloids aminoglycosides including compounds **168–192**. Diverse bioactivities were discovered for most of the reported chromone glycosides. Several chromone glycosides show potent biological activities as anti-viral, acetylcholinesterase inhibition, anti-tumor, anti-inflammatory, etc. This review directs the attention for further deep investigation of chromone glycosides for drug discovery.

Author Contributions: Y.A.: conceptualization, data collection, writing, reviewing. M.E., A.O. and M.S.: data collection, writing, reviewing. K.S.: supervision, reviewing. All authors have read and agreed to the published version of the manuscript.

Funding: This research received no external funding.

Institutional Review Board Statement: Not applicable.

Informed Consent Statement: Not applicable.

Data Availability Statement: Not applicable.

Conflicts of Interest: The authors declare no conflict of interest.

Abbreviations

All	Allose
Api	Apiose
Ara	Arabinose
Caf	Caffeic acid
Cin	Cinnamic acid
Cou	Coumaric acid
Fer	Ferulic acid
Gall	Gallic acid
Glu	Glucose
Rha	Rhamnose
Xyl	Xylose

References

1. Herath, H.M.T.B.; Jacob, M.; Wilson, A.D.; Abbas, H.K.; Nanayakkara, N.P.D. New secondary metabolites from bioactive extracts of the fungus *Armillaria tabescens*. *Nat. Prod. Res.* **2013**, *27*, 1562–1568. [CrossRef] [PubMed]
2. Choi, S.; Park, Y.-I.; Lee, S.-K.; Kim, J.-E.; Chung, M.-H. Aloesin inhibits hyperpigmentation induced by UV radiation. *Clin. Exp. Dermatol.* **2002**, *27*, 513–515. [CrossRef] [PubMed]
3. Machado, N.F.L.; Marques, M.P.M. Bioactive Chromone Derivatives-Structural Diversity. *Curr. Bioact. Compd.* **2010**, *6*, 76–89. [CrossRef]
4. Daniel, M. *Medicinal Plants: Chemistry and Properties*; Science Publishers: Hauppauge, NY, USA, 2006; ISBN 1578083958.
5. Mander, L.; Liu, H.-W. *Comprehensive Natural Products II: Chemistry and Biology*; Elsevier: Amsterdam, The Netherlands, 2010; Volume 1, ISBN 0080453821.
6. Dewick, P.M. *Medicinal Natural Products: A Biosynthetic Approach*; John Wiley & Sons: Hoboken, NJ, USA, 2002; ISBN 0471496413.
7. Daily, A.; Seligmann, O.; Nonnenmacher, G.; Fessler, B.; Wong, S.; Wagner, H. New chromone, coumarin, and coumestan derivatives from *Mutisia acuminata* var. *hirsuta*. *Planta Med.* **1988**, *54*, 50–52. [CrossRef]
8. Tschesche, R.; Delhvi, S.; Sepulveda, S.; Breitmaier, E. Eucryphin, a new chromone rhamnoside from the bark of *Eucryphia cordifolia*. *Phytochemistry* **1979**, *18*, 867–869. [CrossRef]
9. Li, Z.; Li, D.; Owen, N.L.; Len, Z.; Cao, Z.; Yi, Y. Structure determination of a new chromone glycoside by 2D INADEQUATE NMR and molecular modeling. *Magn. Reson. Chem.* **1996**, *34*, 512–517. [CrossRef]

10. Feng, Z.; Wang, Y.; Zhang, P. Chemical constituents of *Rhododendron ovatum* Planch. *Yaoxue Xuebao* **2005**, *40*, 150–152.
11. Kuo, L.-M.Y.; Zhang, L.-J.; Huang, H.-T.; Lin, Z.-H.; Liaw, C.-C.; Cheng, H.-L.; Lee, K.-H.; Morris-Natschke, S.L.; Kuo, Y.-H.; Ho, H.-O. Antioxidant Lignans and Chromone Glycosides from *Eurya japonica*. *J. Nat. Prod.* **2013**, *76*, 580–587.
12. Chen, G.; Jin, H.Z.; Li, X.F.; Zhang, Q.; Shen, Y.H.; Yan, S.K.; Zhang, W.D. A new chromone glycoside from *Rhododendron spinuliferum*. *Arch. Pharm. Res.* **2008**, *31*, 970–972. [CrossRef]
13. Han, L.K.; Ninomiya, H.; Taniguchi, M.; Baba, K.; Kimura, Y.; Okuda, H. Norepinephrine-augmenting lipolytic effectors from *Astilbe thunbergii* rhizomes. *J. Nat. Prod.* **1998**, *61*, 1006–1011. [CrossRef]
14. Sumiyoshi, M.; Kimura, Y. Enhancing effects of a chromone glycoside, eucryphin, isolated from Astilbe rhizomes on burn wound repair and its mechanism. *Phytomedicine* **2010**, *17*, 820–829. [CrossRef]
15. Liao, Y.Y. A new chromone glycoside from leaves of *Salix matsudana*. *Chinese Tradit. Herb. Drugs* **2014**, *24*, 2887–2889.
16. Tane, P.; Ayafor, J.F.; Sondengam, B.L.; Connolly, J.D. Chromone glycosides from *Schumanniophyton magnificum*. *Phytochemistry* **1990**, *29*, 1004–1007. [CrossRef]
17. Kumar, V.; Gupta, M.; Gandhi, S.G.; Bharate, S.S.; Kumar, A.; Vishwakarma, R.A.; Bharate, S.B. Anti-inflammatory chromone alkaloids and glycoside from *Dysoxylum binectariferum*. *Tetrahedron Lett.* **2017**, *58*, 3974–3978. [CrossRef]
18. Al-Taweel, A.; Alqasoumi, S.; Alam, P.; Abdel-Kader, M. Densitometric-high-performance thin-layer chromatographic estimation of diosmin, hesperidin, and ascorbic acid in pharmaceutical formulations. *J. Planar Chromatogr. Mod. TLC* **2013**, *26*, 336–342. [CrossRef]
19. Fawzy, G.A.; Al-Taweel, A.M.; Perveen, S.; Khan, S.I.; Al-Omary, F.A. Bioactivity and chemical characterization of *Acalypha fruticosa* Forssk. growing in Saudi Arabia. *Saudi Pharm. J.* **2017**, *25*, 104–109. [CrossRef]
20. Sueyoshi, E.; Yu, Q.; Matsunami, K.; Otsuka, H. Staphylosides A and B: Two new chromone diglucosides from leaves of *Staphylea bumalda* DC. *Heterocycles* **2008**, *76*, 845–849.
21. Elgorashi, E.E.; Coombes, P.H.; Mulholland, D.A.; Van Staden, J. Isoeugenitol, a cyclooxygenase-1 inhibitor from *Gethyllis ciliaris*. *South African J. Bot.* **2007**, *73*, 156–158. [CrossRef]
22. Han, L.; Zheng, F.; Zhang, Y.; Liu, E.; Li, W.; Xia, M.; Wang, T.; Gao, X. Triglyceride accumulation inhibitory effects of new chromone glycosides from *Drynaria fortunei*. *Nat. Prod. Res.* **2015**, *29*, 1703–1710. [CrossRef]
23. Isaka, M.; Haritakun, R.; Supothina, S.; Choowong, W.; Mongkolsamrit, S. N-Hydroxypyridone alkaloids, chromone derivatives, and tetrahydroxanthones from the scale-insect pathogenic fungus *Orbiocrella* sp. BCC 33248. *Tetrahedron* **2014**, *70*, 9198–9203. [CrossRef]
24. Ma, L.; Ma, S.; Wei, F.; Lin, R.; But, P.P.; Lee, S.H.; Lee, S.F. Uncinoside A and B, two new antiviral chromone glycosides from *Selaginella uncinata*. *Chem. Pharm. Bull. (Tokyo)* **2003**, *51*, 1264–1267. [CrossRef]
25. Arakawa, Y.; Chiji, H.; Izawa, M. Structural elucidation of two new chromones isolated from glasswort (*Salicornia europaea* L.). *Agric. Biol. Chem.* **1983**, *47*, 2029–2033.
26. Chen, X.-Q.; Zan, K.; Liu, H.; Yang, J.; Lai, M.-X.; Wang, Q. Triterpenes and flavonoids from *Ilex hainanensis* Merr. (Aquifoliaceae). *Biochem. Syst. Ecol.* **2009**, *37*, 678–682. [CrossRef]
27. Simon, A.; Chulia, A.J.; Kaouadji, M.; Delage, C. Quercetin 3-[triacetylarabinosyl(1→6)galactoside] and chromones from *Calluna vulgaris*. *Phytochemistry* **1994**, *36*, 1043–1045. [CrossRef]
28. Cui, C.B.; Tezuka, Y.; Kikuchi, T.; Nakano, H.; Tamaoki, T.; Park, J.H. Constituents of a fern, *Davallia mariessi* Moore. IV. Isolation and structures of a novel norcarotane sesquiterpene glycoside, a chromone glucuronide, and two epicatechin glycosides. *Chem. Pharm. Bull. (Tokyo)* **1992**, *40*, 2035–2040. [CrossRef]
29. Li, X.; Yu, M.; Meng, D.; Li, Z.; Zhang, L. A new chromone glycoside from *Polygonum capitatum*. *Fitoterapia* **2007**, *78*, 506–509. [CrossRef]
30. Murata, T.; Selenge, E.; Suganuma, K.; Asai, Y.; Batkhuu, J.; Yoshizaki, F. Chromone acyl glucosides and an ayanin glucoside from *Dasiphora parvifolia*. *Phytochem. Lett.* **2013**, *6*, 552–555. [CrossRef]
31. Yu, J.; Song, X.; Wang, D.; Wang, X.; Wang, X. Five new chromone glycosides from *Scindapsus officinalis* (Roxb.) Schott. *Fitoterapia* **2017**, *122*, 101–106. [CrossRef]
32. Cisowski, W. 2-Methyl-5,7-dihydroxychromone glucoside from *Ammi visnaga* Lamarck. *Pol. J. Chem.* **1986**, *60*, 837–840.
33. Wang, Y.-B.; Huang, R.; Zhang, H.-B.; Li, L. Chromone glycosides from *Knoxia corymbosa*. *J. Asian Nat. Prod. Res.* **2006**, *8*, 663–670. [CrossRef]
34. Ghosal, S.; Kumar, Y.; Singh, S.; Ahad, K. Biflorin, a chromone-C-glucoside from *Pancratium biflorum*. *Phytochemistry* **1983**, *22*, 2591–2593. [CrossRef]
35. Ghosal, S.; Singh, S.; Bhagat, M.P.; Kumar, Y. Three chromones from bulbs of *Pancratium biflorum*. *Phytochemistry* **1980**, *21*, 2943–2946. [CrossRef]
36. Ibrahim, S.; Mohamed, G.; Shaala, L.; Youssef, D. Non-alkaloidal compounds from the bulbs of the Egyptian plant *Pancratium maritimum*. *Z. Naturforsch. C.* **2014**, *69*, 92–98. [CrossRef] [PubMed]
37. Kozowska, V.; Zheleva, A.; Pangarova, T. 7-(β-D-Glucopyranosyloxy)-5-hydroxy-2-methyl chromone from *Peucedanum austriacum*. In *Proceedings of the International Conference of Chemistry and Biotechnology of Biologically Active Natural Products, Budapest, Hungary, 15–19 August 1983*; Bulgarian Academy of Sciences: Sofia, Bulgaria, 1981; Volume 3, pp. 114–116.
38. Gujral, V.K.; Gupta, S.R.; Verma, K.S. A new chromone glucoside from *Tecomella undulata*. *Phytochemistry* **1979**, *18*, 181–182. [CrossRef]

39. Brown, R.T.; Blackstock, W.P.; Chapple, C.L. Isolation of 5,7-dihydroxy-2-methylchromone and its 7-*O*-glycosides from *Adina rubescens*. *J. Chem. Soc. Perkin Trans.* **1975**, 1776–1778. [CrossRef]
40. Peng, B.; Bai, R.-F.; Li, P.; Han, X.-Y.; Wang, H.; Zhu, C.-C.; Zeng, Z.-P.; Chai, X.-Y. Two new glycosides from *Dryopteris fragrans* with anti-inflammatory activities. *J. Asian Nat. Prod. Res.* **2016**, *18*, 59–64. [CrossRef]
41. Chen, X.-Q.; Li, Y.; Cheng, X.; Wang, K.; He, J.; Pan, Z.-H.; Li, M.-M.; Peng, L.-Y.; Xu, G.; Zhao, Q.-S. Polycyclic Polyprenylated Acylphloroglucinols and Chromone O-Glucosides from *Hypericum henryi* subsp. uraloides. *Chem. Biodivers.* **2010**, *7*, 196–204. [CrossRef]
42. An, R.B.; Jeong, G.S.; Beom, J.-S.; Sohn, D.H.; Kim, Y.C. Chromone glycosides and hepatoprotective constituents of *Hypericum erectum*. *Arch. Pharm. Res.* **2009**, *32*, 1393–1397. [CrossRef]
43. Itoh, A.; Tanahashi, T.; Nagakura, N.; Nishi, T. Two chromone-secoiridoid glycosides and three indole alkaloid glycosides from *Neonauclea sessilifolia*. *Phytochemistry* **2003**, *62*, 359–369. [CrossRef]
44. Abe, F.; Yamauchi, T. Megastigmanes and flavonoids from the leaves of *Scorodocarpus borneensis*. *Phytochemistry* **1993**, *33*, 1499–1501. [CrossRef]
45. Gujral, V.K.; Gupta, S.R.; Verma, K.S. Structure of undulatoside-B, a new chromone glycoside from *Tecomella undulata*. *Indian J. Chem. Sect. B Org. Chem. Incl. Med. Chem.* **1979**, *17B*, 40–41.
46. Yoshimitsu, H.; Nishida, M.; Hashimoto, F.; Tanaka, M.; Sakata, Y.; Okawa, M.; Nohara, T. Chromone and flavonol glycosides from *Delphinium hybridum* cv. "Belladonna Casablanca". *J. Nat. Med.* **2007**, *61*, 334–338. [CrossRef]
47. Shang, Z.-P.; Meng, J.-J.; Zhao, Q.-C.; Su, M.-Z.; Luo, Z.; Yang, L.; Tan, J.-J. Two new chromone glycosides from *Drynaria fortunei*. *Fitoterapia* **2013**, *84*, 130–134. [CrossRef]
48. Tanaka, N.; Jia, Y.; Niwa, K.; Imabayashi, K.; Tatano, Y.; Yagi, H.; Kashiwada, Y. Phloroglucinol derivatives and a chromone glucoside from the leaves of *Myrtus communis*. *Tetrahedron* **2018**, *74*, 117–123. [CrossRef]
49. Yi, J.; Zhang, G.; Li, B. Studies on the chemical constituents of *Pseudotsuga sinensis*. *Yao Xue Xue Bao* **2002**, *37*, 352–354.
50. Tanaka, N.; Kashiwada, Y.; Nakano, T.; Shibata, H.; Higuchi, T.; Sekiya, M.; Ikeshiro, Y.; Takaishi, Y. Chromone and chromanone glucosides from *Hypericum sikokumontanum* and their anti-Helicobacter pylori activities. *Phytochemistry* **2009**, *70*, 141–146. [CrossRef]
51. Huneck, S.; Jakupovic, J.; Follmann, G. The final structures of the lichen chromones galapagin, lobodirin, mollin, and roccellin. *Zeitschrift Naturforsch. B* **1992**, *47*, 449–451. [CrossRef]
52. Li, X.; Niu, L.-T.; Meng, W.-T.; Zhu, L.-J.; Zhang, X.; Yao, X.-S. Matteuinterins A-C, three new glycosides from the rhizomes of *Matteuccia intermedia*. *J. Asian Nat. Prod. Res.* **2020**, *22*, 225–232. [CrossRef]
53. Hu, L.; Chen, N.-N.; Hu, Q.; Yang, C.; Yang, Q.-S.; Wang, F.-F. An Unusual Piceatannol Dimer from *Rheum austral* D. Don with Antioxidant Activity. *Molecules* **2014**, *19*, 11453–11464. [CrossRef]
54. Zhao, H.; Wang, Z.; Cheng, J.; Li, R.; Wang, Z. New chromone glucoside from roots of rumex gmelini. *Tianran Chanwu Yanjiu Yu Kaifa* **2009**, *21*, 189–191.
55. Singh, J. Photochemical investigations on *Cassia multijuga* leaves. Isolation and characterization of new chromone glycosides. *Pol. J. Chem.* **1981**, *55*, 1181–1183.
56. Singh, J. Two chromone glycosides from *Cassia multijuga*. *Phytochemistry* **1982**, *21*, 1177–1179. [CrossRef]
57. Kashiwada, Y.; Nonaka, G.I.; Nishioka, I. Chromone glucosides from rhubarb. *Phytochemistry* **1990**, *29*, 1007–1009. [CrossRef]
58. Mou, L.-Y.; Wei, M.; Wu, H.-Y.; Hu, L.-J.; Li, J.-L.; Li, G.-P. O-beta-D-Glucopyranosyl-2-methylchromone, a new chromone glycoside from the Tibetan medicine plant of *Swertia punicea* Hemsl. *Nat. Prod. Res.* **2020**, 1–9. [CrossRef]
59. Ahmad, V.; Ullah, F.; Hussain, J.; Farooq, U.; Zubair, M.; Khan, M.; Choudhary, M. Tyrosinase inhibitors from Rhododendron collettianum and their structure-activity relationship (SAR) studies. *Chem. Pharm. Bull. (Tokyo)* **2004**, *52*, 1458–1461. [CrossRef]
60. Kim, S.B.; Ahn, J.H.; Han, S.B.; Hwang, B.Y.; Kim, S.Y.; Lee, M.K. Anti-adipogenic chromone glycosides from *Cnidium monnieri* fruits in 3T3-L1 cells. *Bioorganic Med. Chem. Lett.* **2012**, *22*, 6267–6271. [CrossRef]
61. Liang, H.; Zhao, Y.Y.; Zhang, R.Y. A new chromone glycoside from *Bupleurum chinense*. *Chinese Chem. Lett.* **1998**, *9*, 69–70.
62. Lu, T.S.; Yi, Y.H.; Mao, S.L.; Zhou, D.Z.; Xu, Q.Z.; Tang, H.F.; Zhang, S.Y. A new chromone glycoside from *Cassia siamea* Lam. *Chinese Chem. Lett.* **2001**, *12*, 703–704.
63. Tanaka, Y.; Honma, D.; Tamura, M.; Yanagida, A.; Zhao, P.; Shoji, T.; Tagashira, M.; Shibusawa, Y.; Kanda, T. New chromanone and acylphloroglucinol glycosides from the bracts of hops. *Phytochem. Lett.* **2012**, *5*, 514–518. [CrossRef]
64. Barash, I.; Karr, A.L., Jr.; Strobel, G.A. Isolation and characterization of stemphylin, a chromone glucoside from *Stemphylium botrysum*. *Plant Physiol.* **1975**, *55*, 646–651. [CrossRef]
65. Do Nascimento, B.O.; da Silva Neto, O.C.; Teodoro, M.T.F.; de Oliveira Silva, E.; Guedes, M.L.S.; David, J.M. Macrolobin: A new unusual C-glycoside chromone from *Macrolobium latifolium* and its anticholinesterase and antimicrobial activities. *Phytochem. Lett.* **2020**, *39*, 124–127. [CrossRef]
66. Iswaldi, I.; Arráez-Román, D.; Rodríguez-Medina, I.; Beltrán-Debón, R.; Joven, J.; Segura-Carretero, A.; Fernández-Gutiérrez, A. Identification of phenolic compounds in aqueous and ethanolic rooibos extracts (*Aspalathus linearis*) by HPLC-ESI-MS (TOF/IT). *Anal. Bioanal. Chem.* **2011**, *400*, 3643–3654. [CrossRef] [PubMed]
67. Lee, H.-H.; Shin, J.-S.; Lee, W.-S.; Ryu, B.; Sik Jang, D.; Lee, K.-T. Biflorin, Isolated from the Flower Buds of *Syzygium aromaticum* L., Suppresses LPS-Induced Inflammatory Mediators via STAT1 Inactivation in Macrophages and Protects Mice from Endotoxin Shock. *J. Nat. Product.* **2016**, *79*, 711–720. [CrossRef] [PubMed]

68. Ryu, B.; Woo, J.-H.; Kim, H.M.; Choi, J.-H.; Jang, D.S. A new acetophenone glycoside from the flower buds of *Syzygium aromaticum* (cloves). *Fitoterapia* **2016**, *115*, 46–51. [CrossRef] [PubMed]

69. Satake, T.; Kamiya, K.; Saiki, Y.; Hama, T.; Fujimoto, Y.; Endang, H.; Umar, M. Chromone C-glycosides from *Baeckea frutescens*. *Phytochemistry* **1998**, *50*, 303–306. [CrossRef]

70. Kamiya, K.; Satake, T. Chemical constituents of *Baeckea frutescens* leaves inhibit copper-induced low-density lipoprotein oxidation. *Fitoterapia* **2010**, *81*, 185–189. [CrossRef]

71. Ito, H.; Kasajima, N.; Tokuda, H.; Nishino, H.; Yoshida, T. Dimeric Flavonol Glycoside and Galloylated C-Glucosylchromones from *Kunzea ambigua*. *J. Nat. Prod.* **2004**, *67*, 411–415. [CrossRef]

72. Wang, S.; Feng, K.; Han, L.; Fang, X.; Zhang, Y.; Yu, H.; Pang, X. Glycosidic compounds from *Cassia obtusifolia* seeds and their inhibitory effects on OATs, OCTs and OATPs. *Phytochem. Lett.* **2019**, *32*, 105–109. [CrossRef]

73. Han, A.-R.; Paik, Y.-S. Antioxidant and prolyl endopeptidase inhibitory capacities of chromone C-glucosides from the clove buds (*Syzygium aromaticum*). *J. Appl. Biol. Chem.* **2012**, *55*, 195–198. [CrossRef]

74. Rehman, N.U.; Hussain, H.; Khiat, M.; Al-Riyami, S.A.; Csuk, R.; Khan, H.Y.; Abbas, G.; Al-Thani, G.S.; Green, I.R.; Al-Harrasi, A. ChemInform Abstract: Aloeverasides A and B: Two Bioactive C-Glucosyl Chromones from Aloe vera Resin. *Helv. Chim.* **2016**, *47*. [CrossRef]

75. Rehman, N.U.; Al-Riyami, S.A.; Hussain, H.; Ali, A.; Khan, A.L.; Al-Harrasi, A. Secondary metabolites from the resins of *Aloe vera* and Commiphora mukul mitigate lipid peroxidation. *Acta Pharm.* **2019**, *69*, 433–441. [CrossRef]

76. Kahramanoglu, I.; Chen, C.; Chen, J.; Wan, C. Chemical constituents, antimicrobial activity, and food preservative characteristics of *Aloe vera* gel. *Agronomy* **2019**, *9*, 831. [CrossRef]

77. Singh, M.; Singh, J. Chemical examination of the seeds of *Cassia spectabilis*. *Zeitschrift Naturforsch. B* **1984**, *39B*, 1425–1426. [CrossRef]

78. Agrawal, A.; Singh, J. Glycosides of two xanthones and a chromone from roots of *Chrozophora prostrata*. *Phytochemistry* **1988**, *27*, 3692–3694. [CrossRef]

79. Jong-Anurakkun, N.; Bhandari, M.R.; Hong, G.; Kawabata, J. α-Glucosidase inhibitor from Chinese aloes. *Fitoterapia* **2008**, *79*, 456–457. [CrossRef]

80. Lv, L.; Yang, Q.-Y.; Zhao, Y.; Yao, C.-S.; Sun, Y.; Yang, E.-J.; Song, K.-S.; Mook-Jung, I.; Fang, W.-S. BACE1 (beta-secretase) inhibitory chromone glycosides from *Aloe vera* and *Aloe nobilis*. *Planta Med.* **2008**, *74*, 540–545. [CrossRef]

81. Kim, J.H.; Yoon, J.Y.; Yang, S.Y.; Choi, S.K.; Kwon, S.J.; Cho, I.S.; Jeong, M.H.; Ho Kim, Y.; Choi, G.S. Tyrosinase inhibitory components from *Aloe vera* and their antiviral activity. *J. Enzyme Inhib. Med. Chem.* **2017**, *32*, 78–83. [CrossRef]

82. Okamura, N.; Hine, N.; Harada, S.; Fujioka, T.; Mihashi, K.; Yagi, A. Three chromone components from *Aloe vera* leaves. *Phytochemistry* **1996**, *43*, 495–498. [CrossRef]

83. Wen, J.; Shi, H.-M.; Tu, P.-F. Chemical constituents of *Abrus mollis* Hance. *Biochem. Syst. Ecol.* **2006**, *34*, 177–179. [CrossRef]

84. Goodger, J.Q.D.; Seneratne, S.L.; Nicolle, D.; Woodrow, I.E. Foliar Essential Oil Glands of Eucalyptus Subgenus Eucalyptus (Myrtaceae) Are a Rich Source of Flavonoids and Related Non-Volatile Constituents. *PLoS ONE* **2016**, *11*, e0155568. [CrossRef]

85. Al-Sayed, E.; Hamid, H.A.; Abu El Einin, H.M. Molluscicidal and antischistosomal activities of methanol extracts and isolated compounds from *Eucalyptus globulus* and *Melaleuca styphelioides*. *Pharmaceut. Biol.* **2014**, *52*, 698–705. [CrossRef] [PubMed]

86. Zhang, Y.; Chen, Y. Isobiflorin, a chromone C-glucoside from cloves (*Eugenia caryophyllata*). *Phytochemistry* **1997**, *45*, 401–403. [CrossRef]

87. Gao, H.Y.; Wang, H.Y.; Li, G.Y.; Du, X.W.; Zhang, X.T.; Han, Y.; Huang, J.; Li, X.X.; Wang, J.H. Constituents from Zhuyeqing Liquor and their inhibitory effects on nitric oxide production. *Phytochem. Lett.* **2014**, *7*, 150–155. [CrossRef]

88. Gao, W.-N.; Luo, J.-G.; Kong, L.-Y. Quality evaluation of *Hypericum japonicum* by using high-performance liquid chromatography coupled with photodiode array detector and electrospray ionization tandem mass spectrometry. *Biomed. Chromatogr.* **2009**, *23*, 1022–1030. [CrossRef]

89. Luo, G.; Zhou, M.; Ye, Q.; Mi, J.; Fang, D.; Zhang, G.; Luo, Y. Phenolic derivatives from *Hypericum japonicum*. *Nat. Prod. Commun.* **2015**, *10*, 1934578X1501001224. [CrossRef]

90. Santos, S.A.O.; Villaverde, J.J.; Freire, C.S.R.; Domingues, R.M.M.; Neto, C.P.; Silvestre, A.J.D. Phenolic composition and antioxidant activity of *Eucalyptus grandis*, *E. urograndis* (*E. grandis* × *E. urophylla*) and *E. maidenii* bark extracts. *Ind. Crops Prod.* **2012**, *39*, 120–127. [CrossRef]

91. Tanaka, T.; Orii, Y.; Nonaka, G.; Nishioka, I. Tannins and related compounds. CXXIII. Chromone, acetophenone and phenylpropanoid glycosides and their galloyl and/or hexahydroxydiphenoyl esters from the leaves of *Syzygium aromaticum* Merr. et Perry. *Chem. Pharm. Bull. (Tokyo)* **1993**, *41*, 1232–1237. [CrossRef]

92. Yuan, A.; Kang, S.; Qin, L.; Ruan, B.; Fan, Y. Chemical constituents of the leaves of Chinese aloe (*Aloe vera* var. *chinensis*). *Zhongcaoyao* **1994**, *25*, 339–341.

93. Che, Q.-M.; Akao, T.; Hattori, M.; Kobashi, K.; Namba, T. Metabolism of aloesin and related cmpounds by human intestinal bacteria: A bacterial cleavage of the C-glucosyl bond and the subsequent reduction of the acetonyl side chain. *Chem. Pharm. Bull.* **1990**, *39*, 704–708. [CrossRef]

94. Park, M.K.; Park, J.H.; Shin, Y.G.; Kim, W.Y.; Lee, J.H.; Kim, K.H. Neoaloesin A: A new C-glucofuranosyl chromosome from *Aloe barbadensis*. *Planta Med.* **1996**, *62*, 363–365. [CrossRef]

95. Lucini, L.; Pellizzoni, M.; Pellegrino, R.; Molinari, G.P.; Colla, G. Phytochemical constituents and in vitro radical scavenging activity of different Aloe species. *Food Chem.* **2015**, *170*, 501–507. [CrossRef]

96. Bisrat, D.; Dagne, E.; Van Wyk, B.-E.; Viljoen, A. Chromones and anthrones from *Aloe marlothii* and *Aloe rupestris*. *Phytochemistry* **2000**, *55*, 949–952. [CrossRef]

97. Conner, J.M.; Gray, A.I.; Reynolds, T.; Waterman, P.G. Anthrone and chromone components of *Aloe cremnophila* and *A. jacksonii* leaf exudates. *Phytochemistry* **1990**, *29*, 941–944. [CrossRef]

98. Wang, W.; Cuyckens, F.; Van den Heuvel, H.; Apers, S.; Pieters, L.; Steenkamp, V.; Stewart, M.J.; Luyckx, V.A.; Claeys, M. Structural characterization of chromone C-glucosides in a toxic herbal remedy. *Rapid Commun. Mass Spectrom.* **2003**, *17*, 49–55. [CrossRef]

99. Hutter, J.A.; Salman, M.; Stavinoha, W.B.; Satsangi, N.; Williams, R.F.; Streeper, R.T.; Weintraub, S.T. Antiinflammatory C-Glucosyl Chromone from *Aloe barbadensis*. *J. Nat. Prod.* **1996**, *59*, 541–543. [CrossRef]

100. Zhong, J.S.; Huang, Y.Y.; Zhang, T.H.; Liu, Y.P.; Ding, W.J.; Wu, X.F.; Xie, Z.Y.; Luo, H.B.; Wan, J.Z. Natural phosphodiesterase-4 inhibitors from the leaf skin of *Aloe barbadensis* Miller. *Fitoterapia* **2015**, *100*, 68–74. [CrossRef]

101. Lee, S.; Do, S.G.; Kim, S.Y.; Kim, J.; Jin, Y.; Lee, C.H. Mass spectrometry-based metabolite profiling and antioxidant activity of *Aloe vera* (*Aloe barbadensis* Miller) in different growth stages. *J. Agric. Food Chem.* **2012**, *60*, 11222–11228. [CrossRef]

102. Rehman, N.U.; Hussain, H.; Khiat, M.; Khan, H.Y.; Abbas, G.; Green, I.R.; Al-Harrasi, A. Bioactive chemical constituents from the resin of *Aloe vera*. *Zeitschrift für Naturforsch. B* **2017**, *72*, 955–958. [CrossRef]

103. Yagi, A.; Kanbara, T.; Morinobut, N. Inhibition of Mushroom-Tyrosinase by Aloe Extract. *Planta Med.* **1987**, *53*, 515–517. [CrossRef]

104. Baba, K.; Kawanishi, H.; Taniguchi, M.; Kozawa, M. Chromone glucosides from *Cnidium japonicum*. *Phytochemistry* **1993**, *35*, 221–225. [CrossRef]

105. Kimura, Y.; Sumiyoshi, M.; Taniguchi, M.; Baba, K. Antitumor actions of a chromone glucoside cnidimoside A isolated from *Cnidium japonicum*. *J. Nat. Med.* **2008**, *62*, 308–313. [CrossRef] [PubMed]

106. Zhao, J.; Zhou, M.; Liu, Y.; Zhang, G.; Luo, Y. Chromones and coumarins from the dried fructus of *Cnidium monnieri*. *Fitoterapia* **2011**, *82*, 767–771. [CrossRef] [PubMed]

107. Cisowski, W.; Grimshaw, J. Glucoside of chromone from *Angelica archangelica* L. and *Archangelica litoralis* Fries. herbs. *Pol. J. Chem.* **1988**, *62*, 135–141.

108. Kopp, B.; Kubelka, E.; Reich, C.; Robien, W.; Kubelka, W. 4H-Chromenone Glycosides from *Eranthhis hyernalis* (L.) SALISBU. *Helv. Chim. Acta* **1991**, *74*, 611–616. [CrossRef]

109. Kuroda, M.; Uchida, S.; Watanabe, K.; Mimaki, Y. Chromones from the tubers of *Eranthis cilicica* and their antioxidant activity. *Phytochemistry* **2009**, *70*, 288–293. [CrossRef]

110. Bernhardt, M.; Shaker, K.H.; Elgamal, M.H.A.; Seifert, K. The New Bishomoflavone Ononin and Its Glucoside from Ononis vaginalis. *Zeitschrift für Naturforsch. C* **2000**, *55*, 516–519. [CrossRef]

111. Ibrahim, S.R.M. New 2-(2-Phenylethyl)chromone Derivatives from the Seeds of *Cucumis melo* L var. reticulatus. *Nat. Prod. Commun.* **2010**, *5*, 403–406. [CrossRef]

112. Quan, L.; Zhang, J.; Wu, Y. One Novel 2-(2-Phenylethyl)Chromone Glycoside from *Aquilaria sinensis*. *Chem. Nat. Compd.* **2017**, *53*, 635–637. [CrossRef]

113. Liu, X.; Zhang, B.; Yang, L.; Yu, G.; Wang, Z. Two new chromones and a new flavone glycoside from *Imperata cylindrical*. *Zhongguo Tianran Yaowu* **2013**, *11*, 77–80. [CrossRef]

114. Shao, H.; Mei, W.-L.; Kong, F.-D.; Dong, W.-H.; Li, W.; Zhu, G.-P.; Dai, H.-F. A new 2-(2-phenylethyl)chromone glycoside in Chinese agarwood "Qi-Nan" from *Aquilaria sinensis*. *J. Asian Nat. Prod. Res.* **2017**, *19*, 42–46. [CrossRef]

115. Zaher, A.; Boufellous, M.; Ouhssine, M.; Bourkhiss, B. Phytochemical Screening Of An Umbelliferae: *Ammi visnaga* L. (Lam.) In The Region Of Sidi Slimane- North-West Of Morocco. *J. Mater. Environ. Sci.* **2019**, *10*, 995–1002.

116. Stevens, T.J.; Jones, B.W.; Vidmar, T.J.; Moran, J.J.; Manni, D.S.; Day, C.E. Hypocholesterolemic effect of khellin and khelloside in female cynomolgus monkeys. *Arzneimittelforschung* **1985**, *35*, 1257–1260.

117. Günaydin, K.; Beyazit, N. The chemical investigations on the ripe fruits of *Ammi visnaga* (Lam.) Lamarck growing in Turkey. *Nat. Prod. Res.* **2004**, *18*, 169–175. [CrossRef]

118. Liu, Y.R.; Wu, Z.J.; Li, C.T.; Xi, F.M.; Sun, L.N.; Chen, W.S. Heracleifolinosides A-F, new triterpene glycosides from *cimicifuga heracleifolia*, and their inhibitory activities against hypoxia and reoxygenation. *Planta Med.* **2013**, *79*, 301–307. [CrossRef]

119. Ma, S.Y.; Shi, L.G.; Gu, Z.B.; Wu, Y.L.; Wei, L.B.; Wei, Q.Q.; Gao, X.L.; Liao, N. Two New Chromone Glycosides from the Roots of *Saposhnikovia divaricata*. *Chem. Biodivers.* **2018**, *15*, 2–7. [CrossRef]

120. Jiang, Y.; Guo, F.; Chen, L.; Xu, L.; Zhang, W.; Liu, B. The antitumor activity of naturally occurring chromones: A review. *Fitoterapia* **2019**, *135*, 114–129.

121. Sasaki, H.; Taguchi, H.; Endo, T.; Yosioka, I. The constituents of *Ledebouriella seseloides* Wolff. I. Structures of three new chromones. *Chem. Pharm. Bull.* **1982**, *3*, 3555–3562. [CrossRef]

122. Urbagarova, B.M.; Shults, E.E.; Taraskin, V.V.; Radnaeva, L.D.; Petrova, T.N.; Rybalova, T.V.; Frolova, T.S.; Pokrovskii, A.G.; Ganbaatar, J. Chromones and coumarins from *Saposhnikovia divaricata* (Turcz.) Schischk. Growing in Buryatia and Mongolia and their cytotoxicity. *J. Ethnopharmacol.* **2020**, *261*, 112517. [CrossRef]

123. Lemmich, J. Monoterpene, chromone and coumarin glucosides of *Diplolophium buchananii*. *Phytochemistry* **1995**, *38*, 427–432. [CrossRef]

124. Shao, Z.; Zhang, Y.; Jiang, K.; CHEN, J.; Zuo, B. Chemical constituents from the roots of *Sphallerocarpus gracilis*. *Nat. Prod. Res. Dev.* **2003**, *15*, 196–198.
125. Yan, Z.; Yang, X.; Wu, J.; Su, H.; Chen, C.; Chen, Y. Qualitative and quantitative analysis of chemical constituents in traditional Chinese medicinal formula Tong-Xie-Yao-Fang by high-performance liquid chromatography/diode array detection/electrospray ionization tandem mass spectrometry. *Anal. Chim. Acta* **2011**, *691*, 110–118. [CrossRef]
126. Semwal, R.B.; Semwal, D.K.; Combrinck, S.; Viljoen, A. Health benefits of chromones: Common ingredients of our daily diet. *Phytochem. Rev.* **2020**, *19*, 761–785. [CrossRef]
127. Khalil, N.; Bishr, M.; Desouky, S.; Salama, O. *Ammi visnaga* L., a potential medicinal plant: A review. *Molecules* **2020**, *25*, 301. [CrossRef]
128. An, R.-B.; Park, B.-Y.; Kim, J.-H.; Kwon, O.-K.; Lee, J.-K.; Min, B.-S.; Ahn, K.-S.; Oh, S.-R.; Lee, H.-K. Coumarins and chromones from *Angelica genuflexa*. *Nat. Prod. Sci.* **2005**, *11*, 79–84.
129. Baba, K.; Hata, K.; Kimura, Y.; Matsuyama, Y.; Kozawa, M. Chemical studies of *Angelica japonica* A. Gray. I. On the constituents of the ethyl acetate extract of the root. *Chem. Pharm. Bull. (Tokyo)* **1981**, *29*, 2565–2570. [CrossRef]
130. Shi, Q.-Q.; Lu, J.; Peng, X.-R.; Li, D.-S.; Zhou, L.; Qiu, M.-H. Cimitriteromone A–G, Macromolecular Triterpenoid–Chromone Hybrids from the Rhizomes of *Cimicifuga foetida*. *J. Org. Chem.* **2018**, *83*, 10359–10369. [CrossRef]
131. Sun, J.; Su, X.; Zhang, Z.; Hu, D.; Hou, G.; Zhao, F.; Sun, J.; Cong, W.; Wang, C.; Li, H. Separation of three chromones from *Saposhnikovia divaricata* using macroporous resins followed by preparative high-performance liquid chromatography. *J. Sep. Sci.* **2021**, *44*, 3287–3294. [CrossRef]
132. Lu, L.; Chen, J.C.; Li, Y.; Qing, C.; Wang, Y.Y.; Nian, Y.; Qiu, M.H. Studies on the constituents of *Cimicifuga foetida* collected in guizhou province and their cytotoxic activities. *Chem. Pharm. Bull.* **2012**, *60*, 571–577. [CrossRef]
133. Wang, H.; Xu, Y.; Yuan, Z. Isolation and identification of chemical constituents of roots of *Glehnia littoralis*. *J. Shenyang Pharm. Univ.* **2011**, *28*, 530–534.
134. Zheng, M.; Jin, W.; Son, K.-H.; Chang, H.-W.; Kim, H.-P.; Bae, K.-H.; Kang, S.-S. The constituents isolated from *Peucedanum japonicum* Thunb. and their cyclooxygenase (COX) inhibitory activity. *Korean J. Med. Crop Sci.* **2005**, *13*, 75–79.
135. Akunyili, D.N.; Akubue, P.I. Schumanniofoside, the antisnake venom principle from the stem bark of *Schumanniophyton magnificum* Harms. *J. Ethnopharmacol.* **1986**, *18*, 167–172. [CrossRef]
136. Shi, Q.-Q.; Gao, Y.; Lu, J.; Zhou, L.; Qiu, M.-H. Two new triterpenoid-chromone hybrids from the rhizomes of *Actaea cimicifuga* L.(syn. *Cimicifuga foetida* L.) and their cytotoxic activities. *Nat. Prod. Res.* **2020**, 1–7. [CrossRef] [PubMed]
137. Brill, G.M.; Jackson, M.; Whittern, D.N.; Buko, A.M.; Hill, P.; Chen, R.H.; Mcalpine, J.B. Altromycins E, F, G, H and I; additional novel components of the altromycin complex. *J. Antibiot. (Tokyo)* **1994**, *47*, 1160–1164. [CrossRef] [PubMed]
138. Kanda, N. A New Antitumor Antibiotic, Kidamycin. I Isolation, Purification and Properties of Kidamycin. *J. Antibiot. (Tokyo)* **1971**, *24*, 599–606. [CrossRef]
139. Kondo, S.; Wakashiro, T.; Hamada, M.; Maeda, K.; Takeuchi, T.; Umezawa, H. Isolation and characterization of a new antibiotic, neopluramycin. *J. Antibiot. (Tokyo)* **1970**, *23*, 354–359. [CrossRef]
140. Kondo, S.; Miyamoto, M.; Naganawa, H.; Takeuchi, T.; Umezawa, H. Structures of pluramycin A and neopluramycin. *J. Antibiot. (Tokyo)* **1977**, *30*, 1143–1145. [CrossRef]
141. Byrne, K.M.; Gonda, S.K.; Hilton, B.D. Largomycin FII chromophore component 4, a new pluramycin antibiotic. *J. Antibiot. (Tokyo)* **1985**, *38*, 1040–1049. [CrossRef]
142. Nadig, H.; Sèquin, U. Isolation and Structure Elucidation of Some Components of the Antitumor Antibiotic Mixture 'Rubiflavin'. *Helv. Chim. Acta* **1987**, *70*, 1217–1228. [CrossRef]
143. Nadig, H.; Séquin, U.; Bunge, R.H.; Hurley, T.R.; Murphey, D.B.; French, J.C. Isolation and structure of a new antibiotic related to rubiflavin A. *Helv. Chim. Acta* **1985**, *68*, 953–957. [CrossRef]
144. Schmitz, H.; Crook, K.E., Jr.; Bush, J.A. Hedamycin, a new antitumor antibiotic. I. Production, isolation, and characterization. *Antimicrob. Agents Chemother.* **1966**, *6*, 606–612.
145. Séquin, U.; Bedford, C.T.; Chung, S.K.; Scott, A.I. The structure of the antibiotic hedamycin. I. Chemical, physical and spectral properties. *Helv. Chim. Acta* **1977**, *60*, 896–906. [CrossRef]
146. Sato, Y.; Watabe, H.-O.; Nakazawa, T.; Shomura, T.; Yamamoto, H.; Sezaki, M.; Kondo, S. Ankinomycin, a potent antitumor antibiotic. *J. Antibiot. (Tokyo)* **1989**, *42*, 149–152. [CrossRef]
147. Vertesy, L.; Barbone, F.P.; Cashmen, E.; Decker, H.; Ehrlich, K.; Jordan, B.; Knauf, M.; Schummer, D.; Segeth, M.P.; Wink, J. Pluraflavins, potent antitumor antibiotics from *Saccharothrix* sp. DSM 12931. *J. Antibiot. (Tokyo)* **2001**, *54*, 718–729. [CrossRef]
148. Jackson, M.; Karwowski, J.P.; Theriault, R.J.; Hardy, D.J.; Swanson, S.J.; Barlow, G.J.; Tillis, P.M.; McAlpine, J.B. Altromycins, novel pluramycin-like antibiotics I. Taxonomy of the producing organism, fermentation and antibacterial activity. *J. Antibiot. (Tokyo)* **1990**, *43*, 223–228. [CrossRef]
149. Brill, G.M.; Mcalpine, J.B.; Whittern, D.N.; Buko, A.M. Altromycins, novel pluramycin-like antibiotics II. Isolation and elucidation of structure. *J. Antibiot. (Tokyo)* **1990**, *43*, 229–237. [CrossRef]
150. Gallivan, J.B. Spectroscopic studies on some chromones. *Can. J. Chem.* **1970**, *48*, 3928–3936. [CrossRef]
151. Griffiths, P.J.F.; Ellis, G.P. Benzopyrones—VI: The ultraviolet absorption spectra of chromone and 2-substituted chromones. *Spectrochim. Acta Part A Mol. Spectrosc.* **1972**, *28*, 707–713. [CrossRef]
152. Staunton, J. In *Comprehensive Organic Chemistry*; Barton, D., Ollis, W.D., Eds.; Pegamon Press: Oxford, UK,, 1979.

153. Nchinda, A. *Chemical Studies of Selected Chromone Derivatives*; Rhodes University: Makhanda, South Africa, 2001.
154. Dey, P.M. *Methods in Plant Biochemistry*; Academic Press: Cambridge, MA, USA, 2012; ISBN 9780080984186.
155. Markham, K.R.; Mabry, T.J. Ultraviolet-Visible and Proton Magnetic Resonance Spectroscopy of Flavonoids. In *The Flavonoids*; Springer: Boston, MA, USA, 1975; pp. 45–77.
156. Murray, R.D.H.; McCabe, P.H. Infrared studies of chromones—I: Carbonyl and hydroxyl regions. *Tetrahedron* **1969**, *25*, 5819–5837. [CrossRef]
157. Liu, S.; Zhu, T.; Qiu, Y.; Qi, W.; Wu, H.; Cai, B.; Lin, J. Chromones and tannins from the fruit of *Euscaphis japonica* var. *wupingensis*. *BioResources* **2019**, *14*, 5355–5364. [CrossRef]
158. Wu, Q.-L.; Wang, S.-P.; Du, L.-J.; Zhang, S.-M.; Yang, J.-S.; Xiao, P.-G. Chromone glycosides and flavonoids from Hypericum japonicum. *Phytochemistry* **1998**, *49*, 1417–1420. [CrossRef]
159. Speranza, G.; Dadà, G.; Lunazzi, L.; Phytochemistry, P.G. A C-glucosylated 5-methylchromone from *Kenya aloe*. *Phytochemistry* **1986**, *25*, 2219–2222. [CrossRef]
160. Okamura, N.; Hine, N.; Tateyama, Y.; Nakazawa, M.; Fujioka, T.; Mihashi, K. Five chromones from *Aloe vera* leaves. *Phytochemistry* **1998**, *49*, 219–223. [CrossRef]
161. Speranza, G.; Gramatica, P.; Dadá, G.; Manitto, P. Aloeresin C, a bitter C, O-diglucoside from Cape Aloe. *Phytochemistry* **1985**, *24*, 1571–1573. [CrossRef]
162. Rauwald, H.W.; Maucher, R.; Dannhardt, G.; Kuchta, K. Dihydroisocoumarins, Naphthalenes, and Further Polyketides from Aloe vera and A. plicatilis: Isolation, Identification and Their 5-LOX/COX-1 Inhibiting Potency. *Molecules* **2021**, *26*, 4223. [CrossRef] [PubMed]
163. Van Heerden, F.R.; Viljoen, A.M.; Van Wyk, B.E. 6′-O-Coumaroylaloesin from Aloe castanea—A taxonomic marker for Aloe section Anguialoe. *Phytochemistry* **2000**, *55*, 117–120. [CrossRef]
164. Holzapfel, C.W.; Wessels, P.L.; Van Wyk, B.E.; Marais, W.; Portwig, M. Chromone and aloin derivatives from *Aloe broomii, A. Africana* and A. speciosa. *Phytochemistry* **1997**, *45*, 97–102. [CrossRef]
165. Séquin, U.; Furukawa, M. The structure of the antibiotic hedamycin—III: 13C NMR spectra of hedamycin and kidamycin. *Tetrahedron* **1978**, *34*, 3623–3629. [CrossRef]

MDPI

St. Alban-Anlage 66

4052 Basel

Switzerland

www.mdpi.com

Molecules Editorial Office

E-mail: molecules@mdpi.com

www.mdpi.com/journal/molecules

9 783036 589732